AF443029

Stiff Sedimentary Clays

Genesis and Engineering Behaviour

Géotechnique Symposium in Print 2007

Edited by

Robert May

Chairman, Géotechnique Symposium in Print 2007 sub-committee

Related titles from ICE Publishing:
Selected papers on geotechnical engineering by P R Vaughan, FREng.
ISBN 978-0-7277-3620-8.
The Essence of Geotechnical Engineering: 60 years of Géotechnique.
ISBN 978-0-7277-3536-2
Rock Engineering. A. Palmström and H. Stille. ISBN 978-0-7277-4083-0
Specification for Tunnelling, Third edition. British Tunnelling Society and Institution
of Civil Engineers.
ISBN 978-0-7277-3477-8
A Short Course in Geology for Civil Engineers. M. Matthews, N. Simons and B.
Menzies. ISBN 978-0-7277-3350-4

ISBN 978-0-7277-4108-0

Typeset by Keytec Typesetting Ltd, Bridport Dorset
Printed and bound in Great Britain by CPI Antony Rowe, Chippenham and Eastbourne

Preface

The *Géotechnique* Symposium in Print on stiff and sedimentary clays was held on 14 May 2007. The Papers and Technical Notes for the Symposium were published in the February and March issues of *Géotechnique* for that year. The subject of the symposium aroused considerable interest with a wealth of papers being submitted and a very large attendance at the symposium meeting. Looking back at the proceedings from the perspective of three years of engineering practice, a substantial proportion of which has involved engineering in stiff clays, the symposium appears as something of a landmark in our understanding of stiff clays. Much new material was presented and there were lively exchanges of view between workers in different disciplines and between academics and engineers from different countries. Some views were controversial and the debate continues. The papers and discussions also drew attention to some important areas which remain insufficiently understood and need further careful investigation.

The book contains all of the papers and technical notes which formed part of the Symposium. In addition to these, the two excellent keynote papers delivered by Professor Richard Chandler and Dr Brian Simpson and transcribed discussions from the Symposium meeting are included in this volume. We have also taken the opportunity to add a small number of important papers from other issues of *Géotechnique* which were not originally part of the symposium but which hopefully serve to round out important aspects of the theoretical framework and engineering understanding of stiff sedimentary clays. I take full responsibility for this selection and apologise if your favourite papers are missing.

The arrangement of the papers in the February and March 2007 issues of *Géotechnique* was according to the geological formations addressed. Papers on the London Clay were published in February and papers on other formations and on more general topics were published in March. However, the arrangement of this book is in accordance with the themes of the four discussion sessions of the Symposium meeting. Inevitably some papers addressed a range of topics.

Each person approaching the body of papers in this volume will draw up a unique list of highlights and issues of particular interest. My personal highlights include the high quality body of field and laboratory test data from a range of stiff clay formations, including but certainly not limited to the London Clay. Insights into state of the art experimental and field testing techniques have proved of great value as has the theoretical basis underlying the interpretation of the data obtained. Finally, several of the papers provide fascinating insights into engineering design issue of major importance, including cut slopes, embedded retaining walls, dams and tunnels.

As on the one hand the papers provide answers to key technical issues, they also serve to highlight issues that require further work. The list could include issues such as the in situ effective horizontal stress ratio (K_0) generated in stiff clays due to substantial unloading followed by recent reloading. We often seem to measure values of K_0 below those we might expect from knowledge of the geological stress history and current theory. Another topic of interest is the relationship between anisotropic stiffness parameters as strains increase beyond the small strain range. The measurement of stiffness parameters remains difficult with more work required on the interpretation of shear wave velocity measurements on laboratory samples for example. The application of the intrinsic model appears much easier in younger clays than in those of greater age. Does everything become more curmudgeonly as it gets older (present readership excepted of course)? And, possibly related to the last observation, our understanding of creep in stiff clays appears to require more work.

The SiP could not have occurred without the dedicated inputs of a large number of people. We are all indebted to the *Géotechnique* SiP Subcommittee who put in countless long hours reviewing papers and to those who lead the various discussion sessions of the Symposium meeting – Robert May (sub-committee chair), Matthew Coop, Federica Cotecchia, Didier De Bruyn, Michael De Freitas, Eileen De Moor, Rameus Gallois, Michael Kavvadas, David Nash, Duncan Nicholson, John Powell, Kenichi Soga, Sarah Stallebrass, Lidija Zdravkovic. Throughout the gestation of the symposium we were given most able secretariat support from Mary Henderson, Ben Ramster and Kathleen Hollow of Thomas Telford. Our thanks also go out to the many authors whose work form the corpus of the Symposium including our keynote lecturers and to Quentin Leiper, ICE President (2006–07), who provided a fascinating opening address.

We live in interesting times. Human society faces major challenges including those of climate change, financial instability, population growth, resource shortages and inequalities. Never has the need for innovative engineering solutions been more pressing. And interestingly the geotechnical engineer, working in close collaboration with his or her academic colleagues has an important role to play. Whether the focus is on more energy efficient mass transport in tunnels beneath our major cities, provision of new flood defences or additional reservoirs or offshore wind turbine and safe storage of waste from nuclear power plants, safe and efficient engineering in stiff clays is increasingly important. My hope is that you will enjoy reading the papers in this volume and that they will inspire you to new engineering achievements and breakthroughs in our understanding of stiff clays.

Robert May

Contents

Keynote Speeches

Chandler, R. J. (2010). *Géotechnique* **60**, No. 12, 891–902 [**doi**: 10.1680/geot.07.KP.001]

Stiff sedimentary clays: geological origins and engineering properties

R. J. CHANDLER*

This Keynote Paper† summarises and extends earlier work on the characterisation of clays, with particular reference to stiff sedimentary clays. Examples of clay characterisation are given that include the various main geological events, from sedimentation to erosion and weathering, and other phenomena such as diagenesis and brecciation. The examples demonstrate the value of plotting basic data, such as the current stress state and undrained strength, in terms of the void index. These three parameters permit an immediate, simple categorisation, in terms of geological history, of sedimentary clays. Further examples illustrate the engineering consequences of weathering of stiff clays, and the associated phenomenon of brecciation produced by the processes of cambering and valley bulging. It is emphasised that, at all stages in a clay's geological history, it is the bond strength between the soil particles, not the geological preconsolidation stress, that controls the mechanical behaviour of sedimentary clays. At all geological stages the bond strength contributes to sensitivity in terms of both stress state and strength. These sensitivities are measured against two intrinsic behaviour patterns, the intrinsic compression line and the intrinsic strength line, both of which are defined, and both of which have considerable value for the characterisation of clays. In particular, the relations between yield stress in oedometer compression and triaxial undrained strength enable the quality of laboratory test data to be assessed.

KEYWORDS: clays; erosion; geology; laboratory tests; sedimentation

La présente communication programme† résume et développe des travaux précédents sur la caractérisation des argiles en se penchant, en particulier, sur des argiles sédimentaires rigides. Des exemples de caractérisation d'argiles sont fournis : ils comprennent les principaux évènements géologiques, de la sédimentation jusqu'à l'érosion et au vieillissement, ainsi que d'autres phénomènes, comme la diagénèse et la formation de brèches. Les exemples démontrent la valeur de données de base pour les levés, comme l'état de contrainte actuel et la résistance du sol non drainé, sur le plan de l'indice de vide. Ces trois paramètres permettent d'effectuer immédiatement une simple catégorisation, sur le plan de l'histoire géologique, des argiles rigides. D'autres exemples illustrent les conséquences techniques d'érosion d'argiles rigides, et le phénomène connexe de la formation de brèches, produit par les processus du cambrage et de bombage. On souligne qu'à tous les stades de l'histoire géologique d'une argile, c'est l'intensité de liaison entre les particules de sol, et non pas les contraintes de préconsolidation géologique, qui déterminent le comportement mécanique des argiles sédimentaires. A toutes les phases géologiques, l'intensité de liaison contribue à la sensibilité sur le plan de l'état de la contrainte et de la résistance. On mesure ces sensibilités en fonction de deux configurations de comportement intrinsèques : la ligne de compression intrinsèque et la ligne de résistance intrinsèque, ces deux configurations étant définies, et revêtant une grande importance dans la caractérisation des argiles. En particulier, les rapports entre la limite élastique, dans la compression en oedomètre, et la résistance non drainée triaxiale, permettent d'évaluer la qualité des données des essais effectués en laboratoire.

INTRODUCTION

A basic understanding of the geological controls on the geotechnical properties of soil is fundamental to geotechnical engineering. The purpose of this paper is to indicate how an understanding of the geological origins of stiff clays can aid their characterisation, and in so doing can provide a check that the basic properties of the material being studied have properly been understood.

One of the most significant papers written by the late Professor Sir Alec Skempton, though perhaps not the most widely read, was entitled 'The consolidation of clays by gravitational compaction', which he presented to a large audience at a general meeting of the Geological Society of London, and which was later published in the society's *Quarterly Journal of Geology* (Skempton, 1970). In it,

Skempton presented a series of profiles of natural water content with depth for normally consolidated clays ('Sedimentation compression curves'; Terzaghi, 1941), the various clays ranging widely in plasticity and in the depths from which the samples were taken, from the seabed to in excess of 3000 m.

Skempton's dataset was later used by Burland (1990), who introduced the concept of void index as a means by which to normalise the properties of different clays for plasticity, enabling the in situ stress states of natural clays to be compared with the oedometer compression behaviour of the same clays when reconstituted in the laboratory. The present writer extended these ideas further (Chandler, 2000), enabling correlation of the undrained strength of natural clays with the reconstituted strength.

For normally consolidated sedimentary clays consistent relationships exist between the in situ stress, the yield stress when undisturbed samples are loaded in the oedometer, and the undrained strength. Approximate empirical relationships between the same parameters have been derived for overconsolidated clays. These relationships can be very helpful when characterising the basic properties of natural clays, and provide a framework within which the measured properties of natural clays may be assessed, thus providing a check on such factors as the extent of sampling disturbance.

Manuscript received 30 November 2007; revised manuscript accepted 1 December 2009. Published online ahead of print 7 June 2010.
Discussion on this paper closes on 1 May 2011, for further details see p. ii.
* Department of Civil and Environmental Engineering, Imperial College London, UK.
† This paper was originally prepared for the 2007 *Géotechnique* Symposium in Print on *Stiff Sedimentary Clays: Genesis and Engineering Behaviour*.

In order to attempt to demonstrate relationships between geological history and geotechnical behaviour, this earlier work is summarised in part, and then extended to indicate the effect of such factors as weathering, and the disturbance caused by phenomena such as brecciation, on the mechanical behaviour of natural stiff sedimentary clays.

THE GENESIS OF STIFF CLAYS: CLAY STATES DURING SEDIMENTATION

Basic ideas: void index

The gradual process of geological compaction will result in clays having an undrained strength that exceeds 100 kPa, the threshold for a classification as stiff, once the vertical stress exceeds about 200 kPa (see the later Fig. 8). The sedimentation compression curves that form the core of Skempton's (1970) data, which are plotted as water content against vertical effective stress in Fig. 1, thus include a number of stiff clays. Irrespective of consistency, however, it is apparent that the position of the various sedimentation compression curves (SCC) depends on the liquid limit: SCCs with higher average liquid limits lie above those with lower values. Noting this, Skempton normalised the water content for plasticity using the liquidity index, plotted against vertical stress. A similar normalisation can be obtained using the void index, I_v: this latter procedure will be followed here to enable the behaviour of clays of different plasticity to be compared directly (Burland, 1990). I_v is defined as

$$I_v = \frac{e_0 - e^*_{100}}{C^*_c} \tag{1}$$

where e_0 is the current or in situ void ratio; e^*_{100} is the void ratio in an oedometer test at a vertical effective stress of 100 kPa (the * indicates 'intrinsic' behaviour, i.e. the behaviour of the soil reconstituted in the laboratory at a water content of about $1·25w_L$); and C^*_c is the compression index of the reconstituted clay over the stress range 100–1000 kPa.

The use of the void index effectively normalises the void ratio in terms of plasticity in much the same manner that the liquidity index, I_L, normalises water content in terms of plasticity. An advantage of the use of the void index rather than the liquidity index is that it relates directly (through e) to the proportion of voids in the soil (not indirectly, as with w), and the oedometer test defines C^*_c directly, rather than relying on an implicit relationship between geotechnical properties and plasticity as does the liquidity index.

I_v is best determined using equation (1), with C^*_c determined experimentally on a sample of the clay reconstituted

at $w = 1·25w_L$. Alternatively, if the liquid limit is known, e^* and C^*_c can be estimated from

$$e^*_{100} = 0·114 + 0·581e_L \tag{2}$$

and

$$C^*_c = 0·256e_L - 0·04 \tag{3}$$

where e_L is the void ratio at the liquid limit (Burland, 1990; Chandler, 2000). It is understood in this paper that reconstituted clays are those prepared in the laboratory at an initial water content of about $1·25w_L$.

When the void ratio is normalised as the void index, all clays have the same reconstituted oedometer compression line, the intrinsic compression line (ICL). Oedometer tests on six reconstituted clays are plotted in Fig. 2 (Burland, 1990). In Fig. 2(a) the positions of the compression curves plotted against void ratio depend on plasticity, as with the natural clays in Fig. 1. Normalised as void index, all six curves pass through a void index of zero at a vertical stress of 100 kPa, and a void index of $-1·0$ at 1000 kPa. This curve is the ICL, and in terms of void index is a unique relation that holds for all reconstituted clays. Its particular

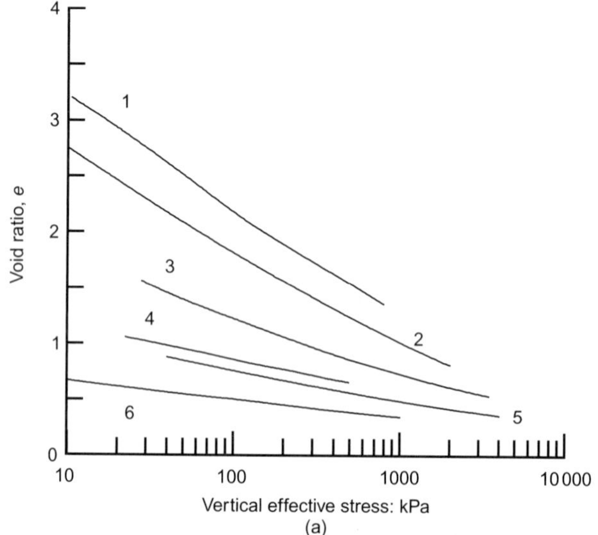

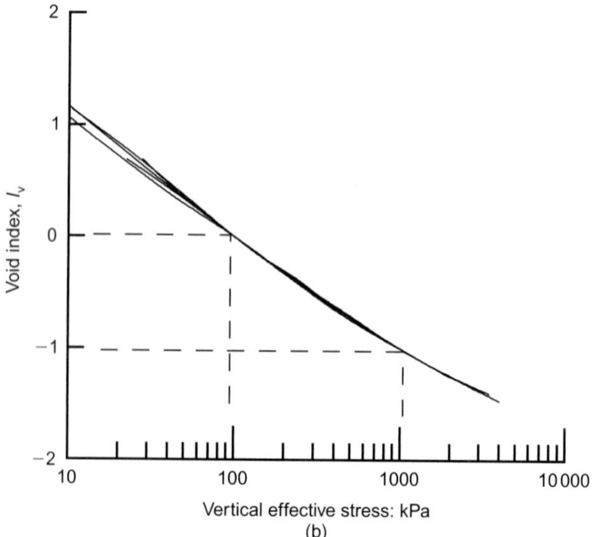

Fig. 2. Intrinsic compression: compression curves for six clays (1–6) in terms of (a) void ratio and (b) void index; samples consolidated from $w \approx 1·25w_L$ (from Burland, 1990)

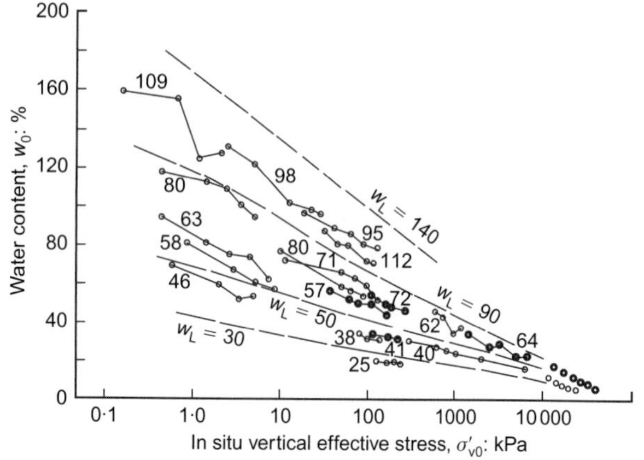

Fig. 1. Sedimentation compression curves used by Skempton (1970). Numbers indicate the average liquid limits (w_L) of the various SCCs

value is that it provides a datum with which the states of naturally occurring clays may be compared, and is therefore a valuable aid to soil characterisation.

Skempton's data from Fig. 1 are plotted as void index in Fig. 3: it can be seen that the natural clays lie to the right of the ICL. An average relationship for the natural clay data was referred to by Burland (1990) as the sedimentation compression line (SCL), which may be regarded as a typical relationship for marine sedimentary clays. Note that the corresponding term for an individual clay is the sedimentation compression curve (SCC), and that an SCC can only be determined using natural water content data from undisturbed samples, and thus cannot be reproduced in the laboratory.

The sensitivity, traditionally defined as the ratio of the undrained strength of the natural clay to the undrained strength remoulded at the same water content, is also shown in Fig. 3 where this is known from the original data. Note that, as Skempton (1970) observed (although he plotted the same data as liquidity index rather than void index), the various SCCs lie within a relatively narrow band, those with a higher sensitivity lying in the upper part of the range. The ICL provides a lower bound to the natural clay data, which exist at a higher void index than the ICL.

Although geological timescales cannot be reproduced in the laboratory, it appears that a more open structure than exhibited by reconstituted clays can be obtained by sedimenting clays in the laboratory, as shown by Stallebrass et al. (2007). Their data for London clay are replotted in Fig. 4, where it is seen that clay prepared from a suspension in a sedimentation column results in a sedimentation compression curve that lies above the ICL, though still below the SCL.

It must be appreciated that individual sedimentation compression curves are unlikely to be as regular as implied by those shown in Figs 1 or 3. Minor erosion cycles during deposition (e.g. de Freitas & Mannion, 2007; Hight et al., 2007) can be expected from time to time as natural sedi-

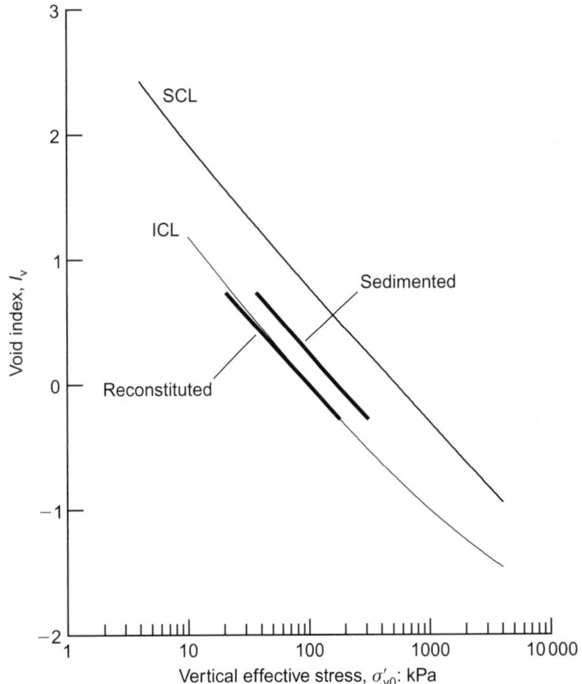

Fig. 4. Laboratory sedimentation, London clay (Stallebrass et al., 2007)

mentation proceeds, giving rise to minor unload–reload loops, as shown in Fig. 5.

Such events doubtless help to explain some of the apparent irregularities in the sedimentation curves of Figs 1 and 3.

Basic ideas: the sensitivity framework – stress sensitivity

The positions of the SCCs to the right of the ICL are the consequence of sedimentation over a far longer time period

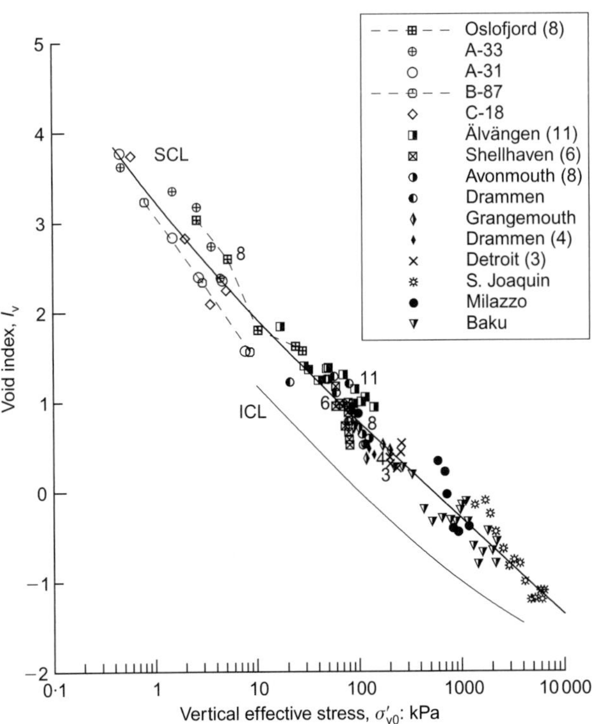

Fig. 3. Void index and effective stress: ICL and SCL (Burland, 1990), compared with sedimentation compression curve data from Skempton (1970), and from Fig. 1. The numbers indicate values of (strength) sensitivity from the original data, where known

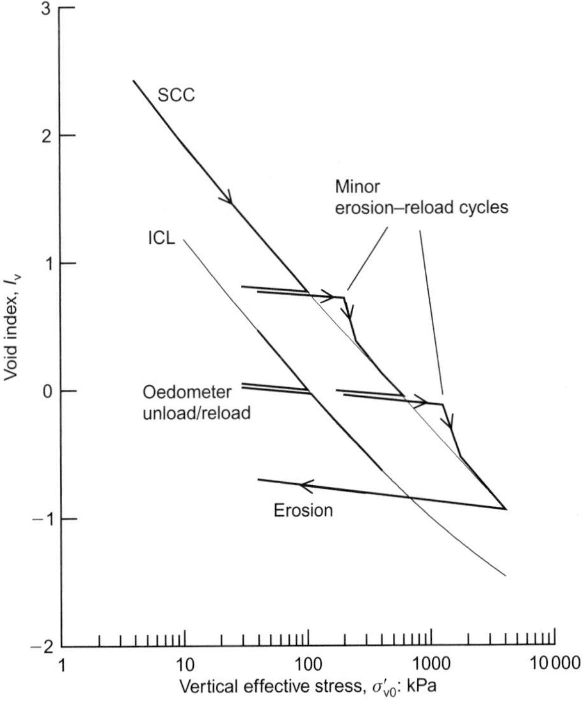

Fig. 5. Effect of minor erosion cycles during deposition (e.g. de Freitas & Mannion, 2007; Hight et al., 2007), compared with a laboratory unload–reload loop on reconstituted clay

than for the laboratory oedometer tests on reconstituted clays, resulting in the natural material developing a more open structure. The open structure implies that bonds have developed, capable of maintaining the open structure under the ever-increasing vertical loads. This open, bonded structure can apparently evolve relatively quickly at very shallow depths in the sediment, particularly with carbonate-rich clays. Such bonds are destroyed when the natural clay is reconstituted, so that the ratio of the in situ vertical stress to the stress on the ICL at the same void index provides a measure of the loss of bonding. This observation can be developed to provide a measure of the (stress) sensitivity of the natural clay to increases in the vertical stress.

Using the oedometer test, a measure of sensitivity in terms of vertical stresses, the stress sensitivity (S_σ; see Fig. 6) can be obtained. In the absence of bonding, as is the case with reconstituted clays, an unload–reload loop (see Fig. 5) can be expected to result in a very abrupt yield-point at the previous maximum vertical stress at or very close to the ICL; its stress sensitivity is close to 1·0. In contrast, an undisturbed sample of a natural clay tested in the oedometer, as shown in Fig. 6(a), will usually yield at a stress in excess of the in situ vertical stress on the clay's sedimentation compression curve, thus showing a stress sensitivity >1·0. Typically, however, the yield is not particularly abrupt, and the yield stress may be difficult to determine. The ratio of this yield stress (σ'_{vy}) to the vertical stress at the same void index on the ICL (intrinsic stress, σ^*_{ve}) is defined as the stress sensitivity (Cotecchia & Chandler, 2000). This, and other definitions, are shown in Fig. 6.

The term 'geological overconsolidation' is used here for clays where the process of erosion has followed deposition, and is quantified by the overconsolidation ratio (OCR), defined as the maximum vertical effective stress developed by the geological column of sediment as a ratio of the current, post-erosion vertical effective stress. It must be stressed that this concept is very simplistic: not only is it often very difficult to establish the maximum likely thickness of sediment from purely geological evidence, but with most stiff clays it is likely that there have been several cycles of loading and unloading prior to the current situation, so the stress history is almost certainly more complex

than as implied by the OCR. The foregoing discussion makes it clear, in general, that there is no unique relationship between the maximum geological stress and the yield stress, the latter being additionally influenced by the bond strength of the clay. Only an overconsolidated reconstituted clay will yield at its previous maximum consolidation stress.

Basic ideas: the sensitivity framework – strength sensitivity

The term 'sensitivity' applied to strength, as defined earlier, has been used for many years (e.g. Terzaghi, 1943). It is possible, however, to have a more rigorous definition of sensitivity, and moreover one that is consistent with the definition of stress sensitivity. A review of laboratory measurements of the undrained strength of reconstituted clays, consolidated under K_0 conditions, provides the strength equivalent of the ICL, the intrinsic strength line (IS$_u$L), as shown in Fig. 7 (Chandler, 2000). This assumes, of course, that there was no sample disturbance prior to measuring the strength. For comparison with the IS$_u$L it is the intact undrained strength of the natural clay that must be used, since we are concerned with the basic structure of the clay, unaffected by fissuring or similar phenomena. The traditional definition of sensitivity is thus replaced by a more formal definition: strength sensitivity, the ratio of the intact undrained strength of the natural undisturbed clay to the intrinsic strength at the same void index. Thus, provided the void index of a particular sample can be established, so can its strength sensitivity. Note that there is a constant ratio between the intrinsic strength and intrinsic compression lines, S^*_u/σ^*_{ve} (or c/p in Skempton's (1948) terminology), of about 0·33.

Soil test data from Skempton (1970) and Burland (1990), with additional information on the same clays given in Chandler (2000), where in situ stress state, yield stress in the oedometer test, and undrained strength are known, can be used to demonstrate the relationships between stress and strength sensitivity, shown in Fig. 8. It is seen that the various clays have in situ states to right of the ICL, and are clays usually referred to as normally or lightly overconsolidated. It is immediately apparent that the two diagrams in Fig. 8 are extremely similar, with the in situ stress data and

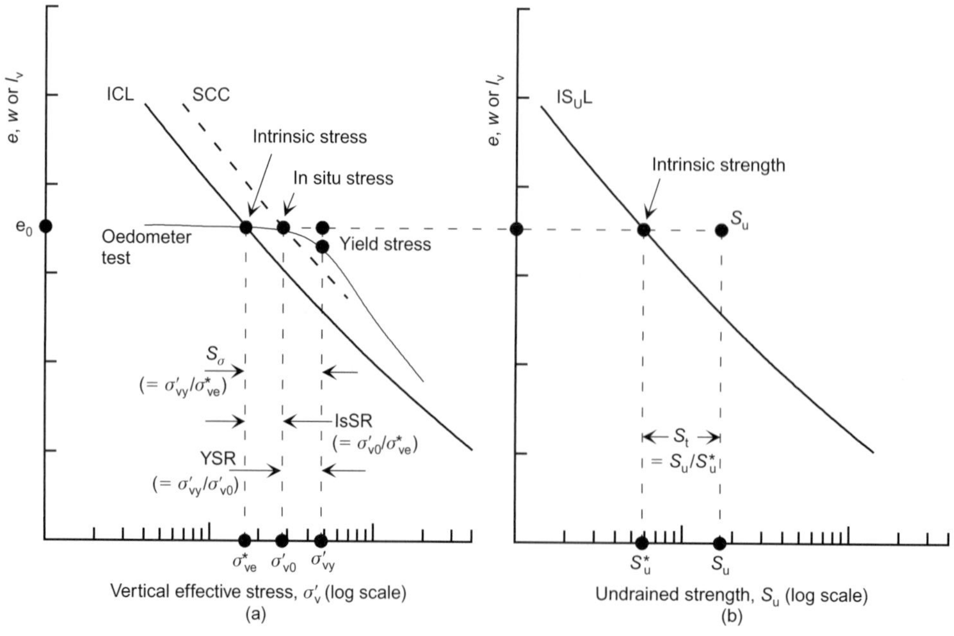

Fig. 6. Definitions of (a) yield stress and stress sensitivity (S_σ), and (b) strength sensitivity (S_t). The vertical scales are the same in (a) and (b)

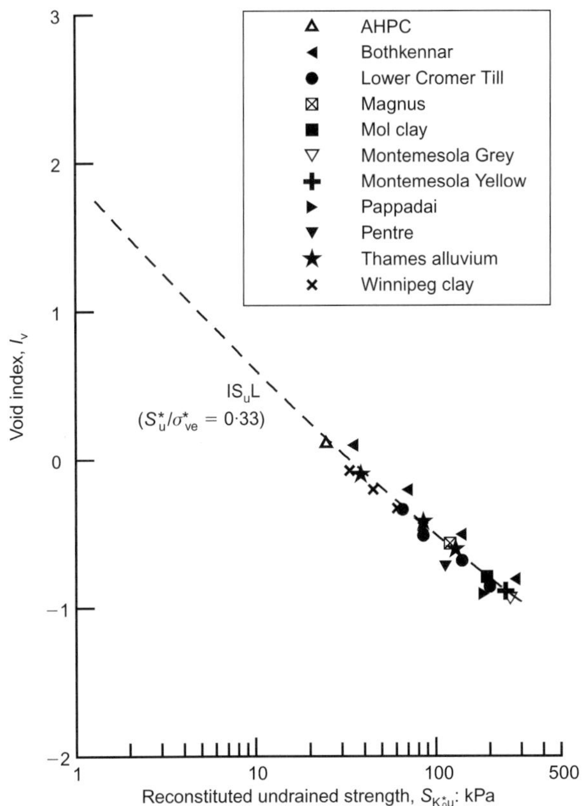

Fig. 7. Intrinsic strength line, IS_uL; dashed line fitted to $S_u/\sigma_v' = 0.33$ (Chandler, 2000)

the strength data of the various clays occurring in very similar positions in relation to the ICL and IS_uL respectively. Analysis and detailed discussion of the data in Fig. 8 by Chandler (2000) showed that, on average (scatter within $\pm25\%$), the simple relationship

$$\text{Stress sensitivity, } S_\sigma = \text{Strength sensitivity, } S_t \qquad (4)$$

holds for these clays whose stress states lie to the right of the ICL. Clays for which there is most uncertainty regarding this relationship are, not surprisingly, those with the highest sensitivities, which lie furthest to the right, and whose strength is often difficult to determine accurately.

Using equation (4), and with reference to Fig. 6, noting that $S_\sigma = \sigma_{vy}'/\sigma_{ve}^*$, $S_t = S_u/S_u^*$ and $S_u^*/\sigma_{ve}^* = 0.33$, it follows that

$$S_u = 0.33\sigma_{vy}' \qquad (5)$$

which is a useful rule of thumb for checking test data for clays whose stress states lie to the right of the ICL.

A comparable, but entirely empirical, expression for clays with stress states to the left of the IS_uL (i.e. overconsolidated or heavily overconsolidated clays), is

$$S_u = 0.33\sigma_{vy}'IsSR \qquad (6)$$

(Chandler, 2000), where IsSR is the in situ stress ratio (the ratio of the in situ vertical stress to the intrinsic vertical stress), defined in Fig. 6. Note that for stress states to the left of the ICL the IsSR is less than unity, as is S_t for values of S_u to the left of the IS_uL.

EVOLUTION: STIFF CLAY STATES POST-SEDIMENTATION

We now turn to consider the properties of stiff sedimentary clays with various different geological histories using the framework presented above.

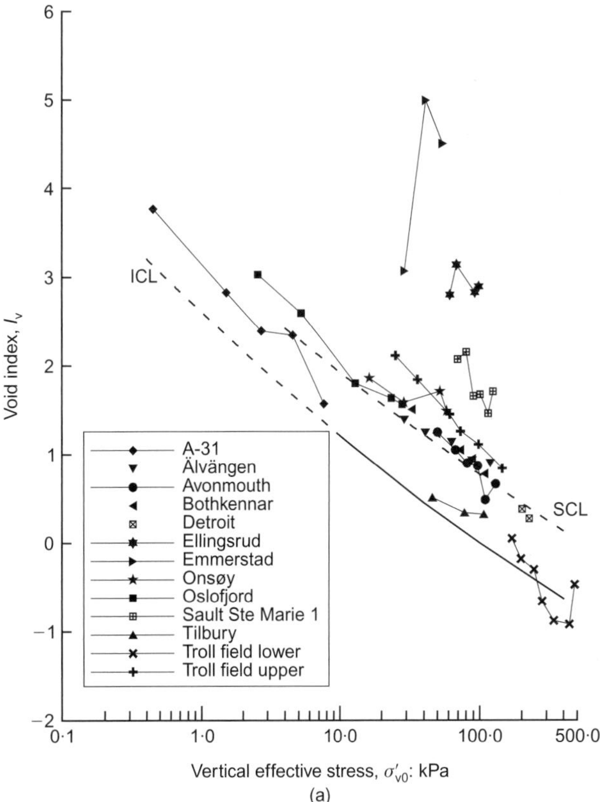

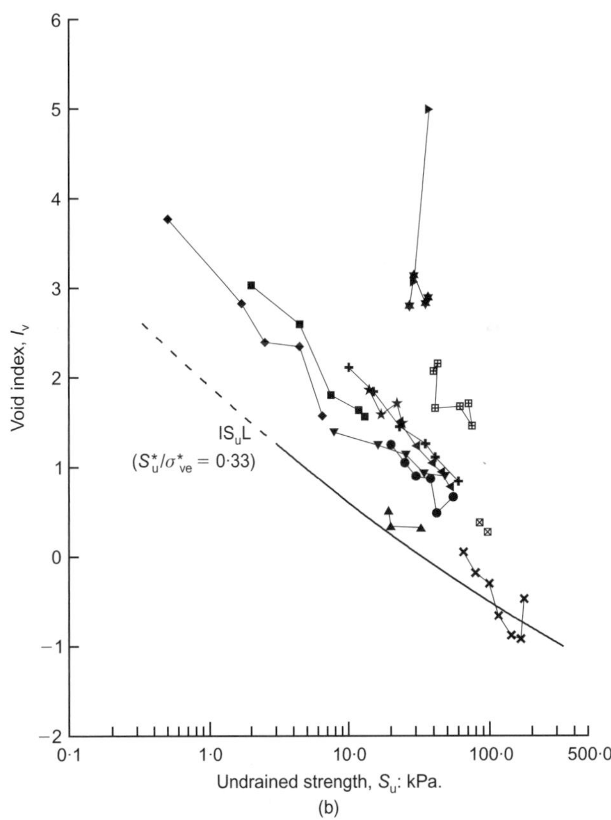

Fig. 8. (a) In situ states, and (b) undrained strength, for normally consolidated clays (Chandler, 2000; data in (a) from Burland, 1990)

Light overconsolidation

Perhaps the most important change of state of sedimentary clays occurs when erosion commences, and the clay swells in response to the unloading. A most interesting series of data relating to a deep, stiff clay that has been only slightly eroded, and thus is only lightly overconsolidated, is for the Belgian Boom clay, studied in detail as a possible material for a radioactive waste disposal facility at Mol. This clay, of middle Oligocene age ($\sim$30 Ma), has suffered only moderate erosion of about 40 m, so that at the depths from which samples were taken, of 223 m and 247 m, the geological overconsolidation ratios are about 1·18 and 1·16 respectively. Tests were carried out on block samples (by Baldi *et al.* (1991) Mair *et al.* (1992) and Coop *et al.* (1995) at 223 m depth; and by Horseman *et al.* (1987) at 247 m depth). At 223 m depth the block samples had average $w_0 = 28\%$, $G_s = 2·71$, $w_L = 72\%$ and $w_P = 26\%$. High-pressure oedometer tests were carried out, although not without some difficulty owing to the depth from which the samples were obtained and the consequent considerable stress relief, and undrained strengths were measured on triaxial samples reconsolidated to the estimated in situ stresses. At 247 m typical soil properties were $w_0 = 23·9 \pm 0·8\%$, $G_s = 2·70$, $w_L = 66 \pm 1\%$, and $w_P = 19 \pm 1\%$. High-pressure oedometer tests were carried out, and a single unconfined compression test, whose strength is used here, which was consistent with a series of effective stress triaxial tests. At both depths values of I_v were computed from the correlations with e^*_{100} and C^*_c (equations (2) and (3)), the consistency of the index test data leading to the expectation of reliable values of I_v when computed from equations (2) and (3). These test data are discussed in more detail in Chandler (2000).

Figure 9 shows the in situ stress states, oedometer tests and undrained strengths at the two depths. The stress states lie just to the left of the suggested SCC, as expected for the low geological OCRs. The oedometer tests show YSRs of about 2·0 at 223 m and 1·7 at 247 m, considerably greater than the corresponding geological OCRs (see Table 1), indicating considerable bonding, with stress sensitivities of 4·4 and 2·4 respectively. The undrained strengths, lying to the right of the IS_uL, are consistent with the in situ stress states, but may be lower than the true in situ values owing to the considerable stress relief. This suggestion is supported by the fact that higher values of S_u are computed from the oedometer yield stress using equation (5).

Overconsolidation

The classic study of London clay from the Ashford Common shaft, to the west of London, by Bishop *et al.* (1965), provides both stress state and undrained strength data for the overconsolidated unweathered London clay taken from a range of depths, levels A to F. London clay is of Eocene age ($\sim$55 Ma), and at Ashford Common has suffered erosion of about 200 m, and thus has a geological OCR of about 20 at 10 m depth (level A), falling to about 5 at 40m depth (level F). The geotechnical properties of the London clay at Ashford Common are very similar to those at Heathrow Terminal 5, discussed by Hight *et al.* (2007).

Figure 10 shows plots of I_v against vertical stress and undrained strength; the values of I_v are computed from equations (2) and (3). At each sampling level the liquid limit and natural water content were measured on a large number of samples, so that the computed values of I_v are regarded as reliable. The values of I_v plotted against in situ vertical

Table 1. Boom clay, Mol, Belgium: comparison of OCR and YSR

Depth: m	OCR	YSR
223	1·18	2·0
247	1·16	1·7

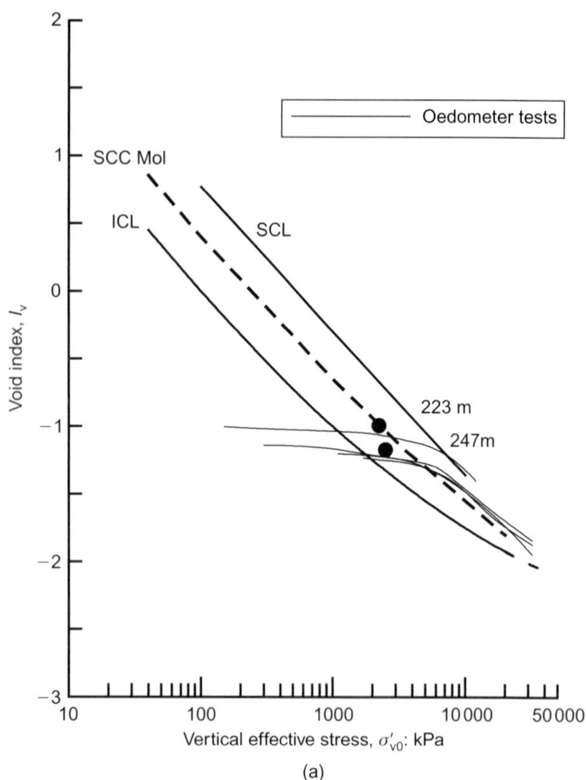

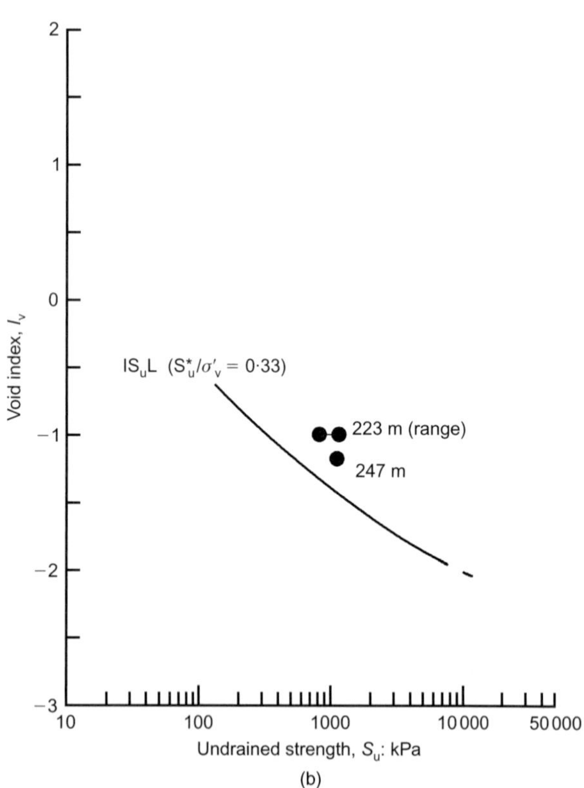

Fig. 9. (a) In situ states and (b) undrained strength, Mol clay (data from various sources: see Chandler, 2000)

stress for all six levels are shown in Fig. 10, together with the result of a triaxial K_0 consolidation test on a sample from level E at a depth of 34·8 m. The position of the six in situ stress states, well to the left of the ICL, clearly indicates their geologically overconsolidated nature.

The K_0 consolidation test from level E was on a sample allowed to swell from its in situ state before being recompressed, which must have resulted in some damage to its

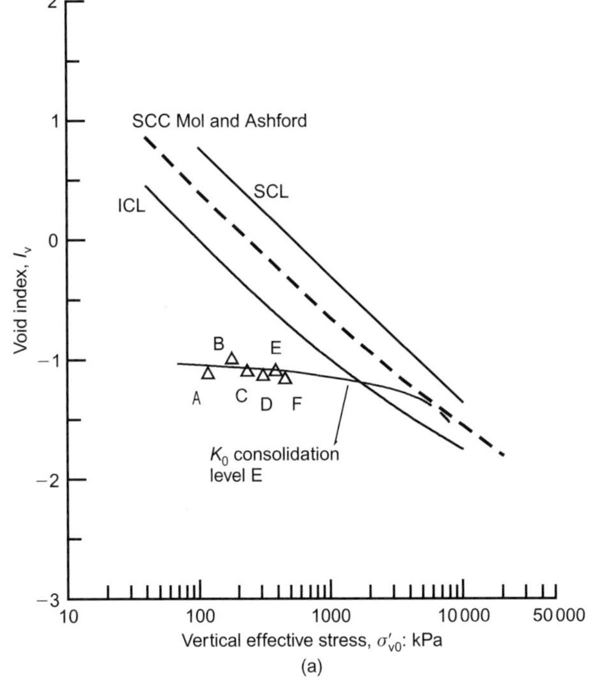

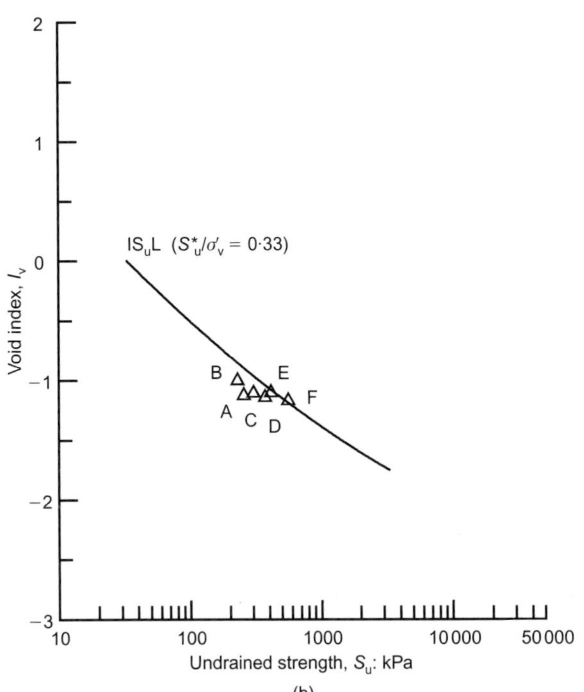

Fig. 10. London Clay, Ashford Common: (a) stress states and (b) undrained strength

IS_uL than the in situ states are to the ICL. Values of YSR can be estimated using equation (6), and are shown in Fig. 11, together with the estimated geological overconsolidation ratios. Again, the values of YSR are considerably greater than the values of OCR. A possible SCC for the Ashford Common clay would be one similar to that for Mol, noting that the K_0 consolidation yield stress is probably an underestimate, a suggestion supported by the higher value of yield stress implied by equation (6).

EVOLUTION: DIAGENESIS

Under conditions of increased temperature and stress, the latter resulting perhaps from depth of burial or tectonic forces, the nature of sediments can change. These changes result from mineral precipitation from the pore water, recrystallisation of component mineral grains, and similar phenomena, and are usually described as diagenesis or lithification. These processes result in irreversible changes to the nature of the sediment, resulting in a structure that has bonds whose strength bears no relationship to the sedimentary history of the clay, but which reflects the combined effects of changed chemistry and the imposed stresses at the time of diagenesis. Fig. 12 attempts to illustrate this, showing diagenesis or lithification occurring, perhaps resulting in a reduction in void index, although with the soil still remaining a particulate material. With increasing stress the modified soil is likely to be very stiff on reloading, and to yield at a much higher stress than that applied by the vertical stress. The behaviour post-yield is likely to be very different from that to be expected in the absence of diagenesis. Moreover, the approach to soil characterisation being suggested here can be unreliable, or even downright misleading, as the engineering properties will no longer match the soil's apparent plasticity.

An example of this problem is provided by the heavily

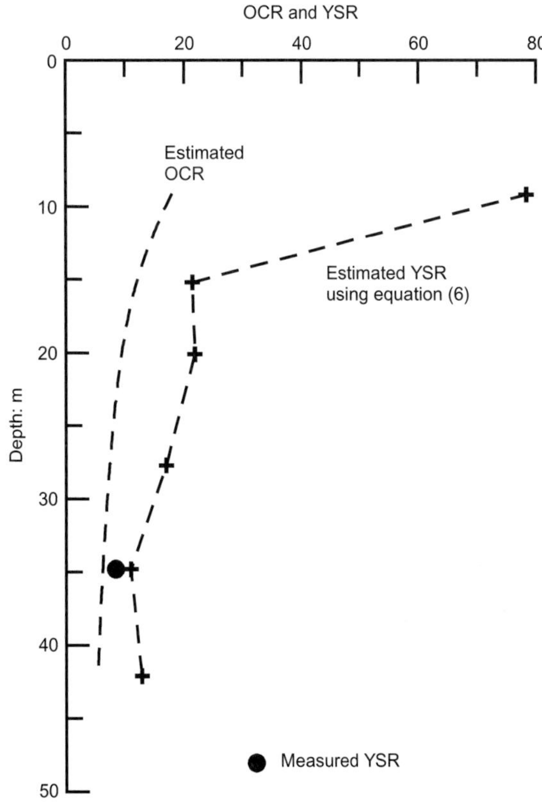

Fig. 11. London Clay, Ashford Common: comparison of geological OCR and YSR

structure, and the test was completed close to the yield stress, but unfortunately was not continued sufficiently to define yield unambiguously. However, following Burland (1990), the compression curve has been extrapolated, suggesting a yield stress of about 3·2 MPa. At level E this gives a yield stress ratio (YSR) of about 12, compared with the estimated OCR of about 6.

Strength tests, including unconsolidated undrained tests, were carried out at the six levels, the average results at each level being shown in Fig. 10. These strengths are assumed to give the intact strength of the London clay, since the samples were small (38 mm diameter), and were taken from depths where discontinuities will be widely spaced. They lie to the left of the IS_uL, although proportionately closer to the

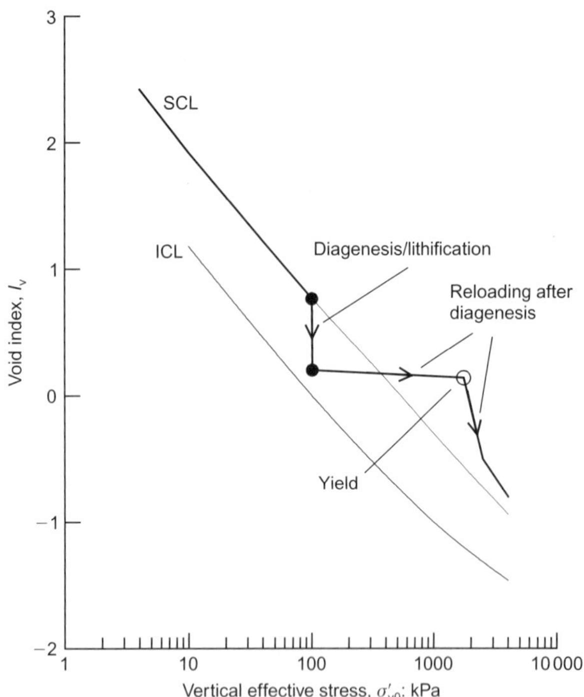

Fig. 12. Possible effects of diagenesis on the SCL

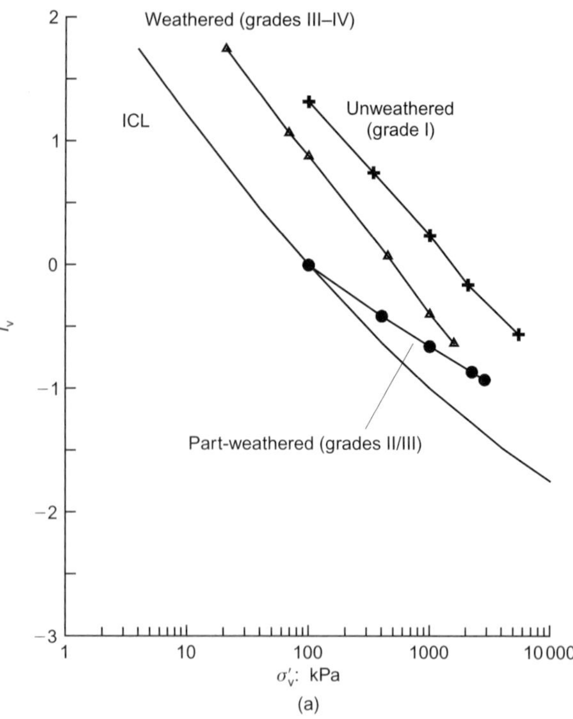

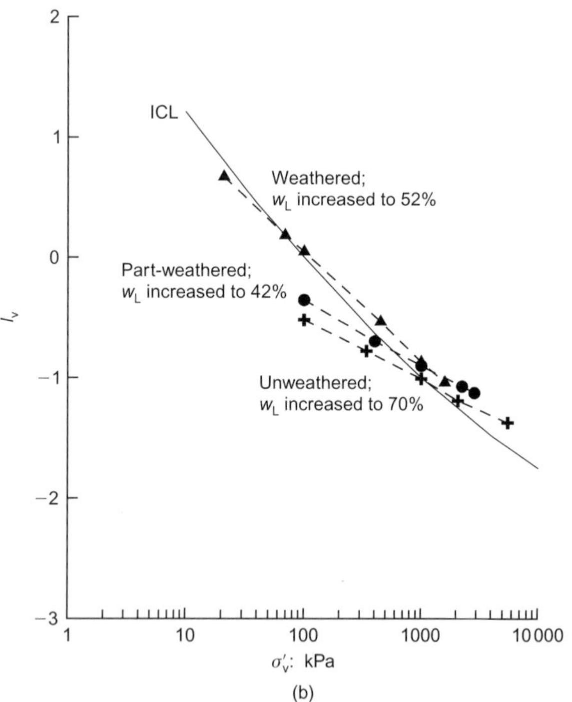

Fig. 13. ICLs for Mercia mudstone: (a) for measured w_L; (b) adjusted to 'true' w_L

overconsolidated and strongly bonded Mercia mudstone. This is of Triassic age, and has been subjected to an unknown but considerable degree of erosion, so the geological OCR is unknown. Probably as a result of flocculation immediately after deposition, this basically clay material now appears as silt-sized aggregates of strongly bonded clay, the bonding process being a form of diagenesis, the result of chemical changes. Thus much of its engineering behaviour (e.g. strength, permeability, consolidation) is controlled by its apparent silty nature. This results in engineering correlations with traditional plasticity index measurements that can be misleading, since the preparation of samples of Mercia mudstone for index testing involves some breakdown of the silt-sized aggregates. Consequently the index tests are no longer directly related to the soil structure, and hence have no particular relationship to the engineering behaviour of the in situ material (e.g. Chandler & Forster, 2001).

Some idea of the mismatch between index test data and engineering behaviour can be demonstrated by the relationships between void index and vertical effective stress for reconstituted Mercia mudstone. Fig. 13(a) shows that, if the ICL is computed using equations (2) and (3) together with the measured values of w_L, the ICLs of three differently weathered samples of Mercia mudstone all plot above the ICL, rather than on it, as should be the case. Note that the least weathered sample is the most remote from the ICL. The implication is that the standard plasticity index tests, all of which yielded liquid limits in the range 35–39%, are an 'incorrect' measure of the engineering behaviour of these strongly aggregated materials.

The mismatch can be examined further by using trial values of liquid limit in equations (2) and (3) to calculate the best fit to the ICL for each of the three mudstones. As can be seen in Fig. 13(b), relatively high values of liquid limit, in the range 42–70%, were needed to get values of void index appropriate to the ICL. The most weathered sample provided an approximate fit to the ICL, perhaps because this material would have broken down most easily during preparation for plasticity testing. A far worse fit was obtained for the part-weathered and unweathered samples. All that can be concluded is that the true clay content of the

Mercia mudstone is considerably higher than indicated by the measured plasticity of the unweathered mudstone, and that the extreme bonding exhibited by this material makes the use of plasticity tests for correlation purposes most unreliable.

EVOLUTION: WEATHERING

The effects of weathering in terms of the sensitivity framework have been considered in some detail by Chandler (2000), using London clay data from Chandler & Apted (1988). The clay tested was taken from a clay pit at South Ockendon, Essex, where the amount of erosion since the deposition of the London clay is probably similar to that at Ashford Common, that is, about 200 m. South Ockendon is about 30 km east of Ashford Common.

The clay pit had been excavated in gently sloping ground, so that it was possible to sample the same near-horizontal stratum at different depths and at different degrees of weathering; the range of sampling depths was 5–10 m, so that the corresponding geological OCRs were 30 and 20 respectively. Fig. 14 shows the relationships between void index, stress state and undrained strength S_u. It will be seen that there is a trend for the results from samples of the more weathered clay (grades III and IV) to lie above those for the less weathered clay (grades I and II), although there is some overlap. More details of the weathering scheme used are given in Chandler & Apted (1988).

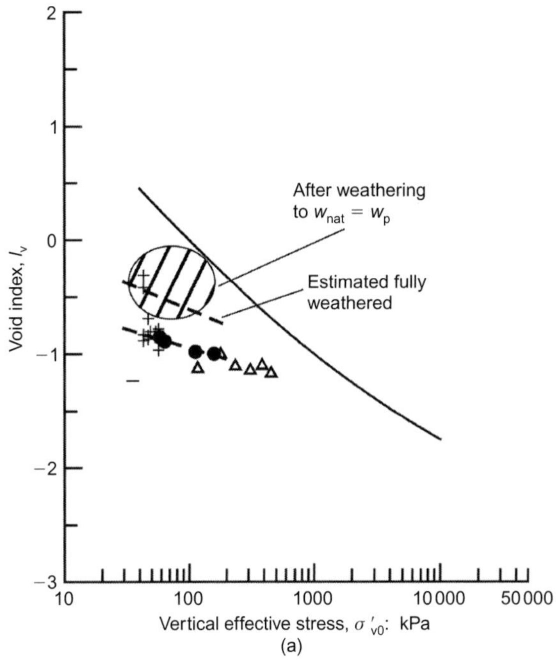

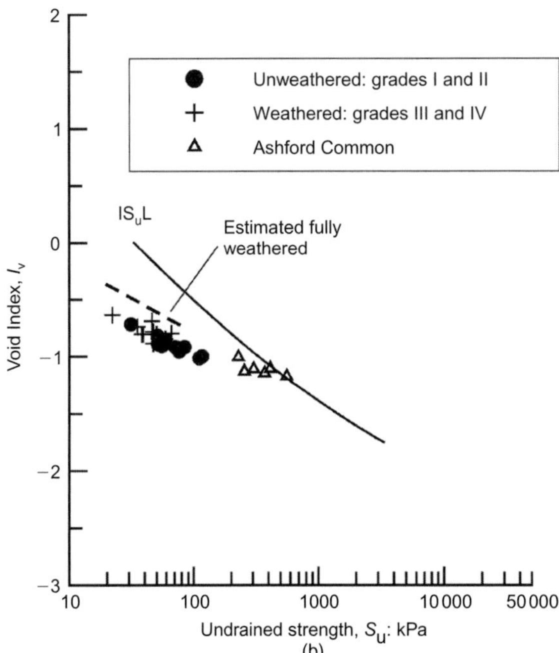

Fig. 14. Weathering: London Clay, South Ockendon; (a) in situ states, (b) undrained strength

Also shown in Fig. 14 are the corresponding data for Ashford Common, taken from Fig. 10. For all practical purposes the Ashford Common data continue the South Ockendon trend, which is consistent with the very similar geological history of the London clay at the two sites.

Weathering of the London clay, as with other stiff clays, produces a macro-fabric (grade III) that has stiffer, less weathered lumps of clay (lithorelicts) in a matrix of disturbed, softer clay. At South Ockendon completely weathered clay (grade IV), where the lithorelicts have totally degraded, has hardly developed. However, in many of the grade III test specimens it was possible to measure separately the matrix and lithorelict water contents, and, as shown in Fig. 14(b), to make an estimate of the likely water content (and hence void index) of the fully weathered clay, assuming it to be that of the matrix of the part-weathered clay. The liquid limit of all samples varied only slightly (±2%) from 80% (not surprising, since all samples came from the same geological horizon), and was used to calculate the intrinsic compression and strength lines.

It can be seen that weathering results in the stress state moving significantly towards the ICL (Fig. 14(a)). A suggested relationship between I_v and undrained strength for completely weathered (grade IV) clay is also shown (Fig. 14(b)), based on the assumption that the measured strengths were controlled by the matrix water contents. Again, the weathered clay strengths parallel the in situ weathered clay stresses shown in Fig. 14(a): the effect of weathering is to move the strength closer to the IS_uL. It can also be shown that weathering results in a considerable reduction in YSR; the unweathered clay can be inferred to have values of YSR of about 40 at shallow depths, falling to 20 at 15 m (Chandler, 2000). When the clay is fully weathered, the corresponding values of YSR could be expected to fall to 9 at 5 m, and 5 at 15 m.

A useful rule of thumb that appears to apply to many clay soils in temperate regions is that at shallow depths the weathered clay has a water content close to the plastic limit. Assuming that such clays will have index properties close to the A-line on the Casagrande plasticity chart, the stress states of weathered clays can be expected to lie within the ellipse shown in Fig. 14(a). The weathered South Ockendon clay is seen to be consistent with this observation.

EVOLUTION: BRECCIATION

The two associated phenomena of cambering and valley bulging were originally described in detail by Hollingworth *et al.* (1944), who encountered these features in the Jurassic strata of Northamptonshire, UK. In brief, camber involves limestones, usually horizontally bedded, that appear as anticlines on hill tops and valley sides, the anticlinal distortion resulting in a complicated series of relatively minor faults in the limestone ('dip-and-fault structure'), while in the valley floor the underlying strata, usually clays, are thrust up as a 'valley bulge'. The camber and bulge structures appear to be closely related, and have been explained as the consequence of the thawing of frozen ground at depth at the end of one or more of the cold phases during the Pleistocene. Thawing would have occurred both from below and from the ground surface. At depth, water released from melting ground ice in the clay probably escaped only slowly, and high pore water pressures controlled by the total overburden pressure (the geostatic pressure) are possible, leading to particularly low effective stresses. Under these conditions substantial deformations could have occurred under hillsides as the permafrost thawed, even though the inclination of the ground surface was quite low.

The most detailed investigation of cambering and bulging

made so far was carried out as part of the investigations for the Empingham dam (Horswill *et al.*, 1976), which retains the reservoir Rutland Water in the county of that name. At Empingham, Jurassic limestones at the crests of the valley slopes overlie the Upper Lias, a heavily overconsolidated stiff clay, which forms the valley slopes and also the floor of the valley, as shown in Fig. 15. The Lias clay was upthrust in an extensive bulge, and detailed investigations in the borrow pits that provided fill for the dam showed that there was a major shear surface of significant extent towards the base of the Lias that extended back into the hillside for an unknown distance. It was estimated that the total movement of the Lias clay into the valley exceeded 200 m. Such a mechanism must have resulted in the clay above the shear surface being subjected to a form of distortion akin to simple shear. Whatever the detail of the mechanism, samples of Lias clay involved in cambering show it to be considerably disturbed, or 'brecciated', consistent with extreme distortion, as seen in Figs 16 and 17.

Brecciated Lias has been shown to have a fabric that has lithorelicts of the original stiff clay set in a sparse matrix of disturbed clay of much higher water content (Chandler, 1972), comparable to a weathered macro fabric.

The construction of Empingham dam resulted in considerable movement within the dam foundations, which has been analysed by Kovacevic *et al.* (2007), who successfully modelled the movements of the dam, but to do so used strength parameters close to critical state values. The reason why critical state parameters should be appropriate seems to lie in the properties of the brecciated Lias, whose behaviour can be examined in the context of its stress state and strength. The Upper Lias data used for this illustration are taken from a cambered slope at Wothorpe, near Stamford, Lincolnshire, about 8 km east of Empingham dam (Chandler, 1972).

Again, void index is plotted against vertical effective stress and undrained strength (Fig. 18). The estimated SCC for the Lias at Wothorpe is shown in Fig. 18(a), which suggests a maximum vertical stress in the region of 20 000 kPa, equivalent to a maximum depth of burial of about 2000 m. Assuming the water content of the lithorelicts to represent the water content of the undisturbed but swelled clay, the corresponding void index ranges from $-1\cdot4$ to $-1\cdot7$, much lower than for London clay at similar depths ($-1\cdot0$ to $-1\cdot2$ at Ashford Common). This is to be expected, given the considerably greater apparent previous maximum stress to which the Lias has been subject. As with weathered

Fig. 16. Brecciated Upper Lias in a trial pit (width of section about 1 m)

Fig. 17. Brecciated Upper Lias clay, thin section (scale in mm)

clay, it seems likely that the matrix water content largely controls the undrained strength of the bulk material. Unfortunately, the water content of the very sparse matrix around the lithorelicts is not easily determined, owing to the restricted extent of softened clay. But assuming that the higher values that were obtained from water content measurements made on very small ($0\cdot3$ g) samples give an estimate of the matrix water content, the likely stress states for the matrix

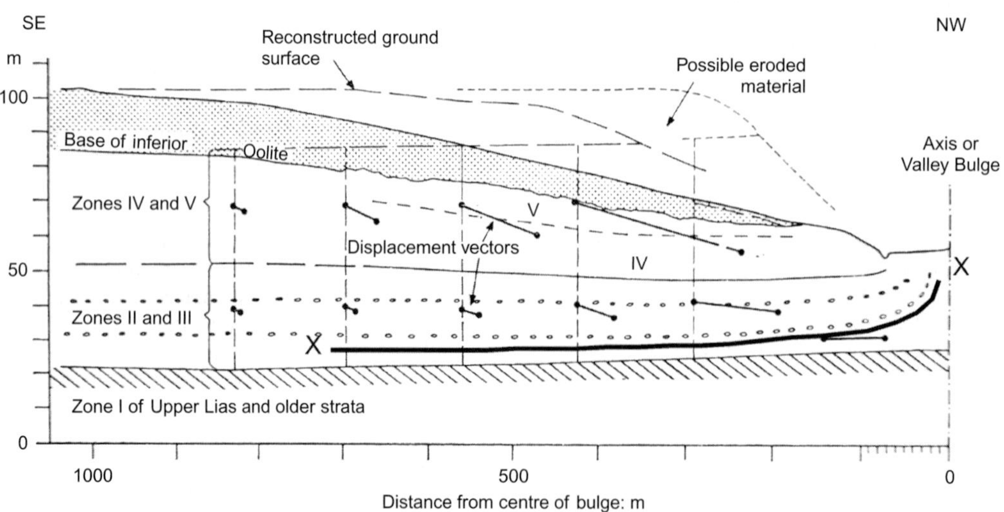

Fig. 15. Deformations in the camber slope at Empingham (modified from Horswill *et al.*, 1976). Zones I to V are fossil zones in the Upper Lias, X–X is the approximate position of the major shear surface (see text)

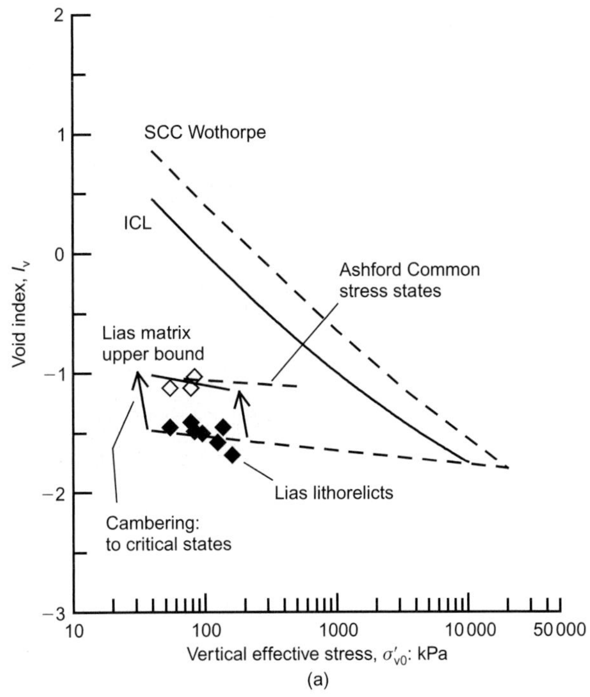

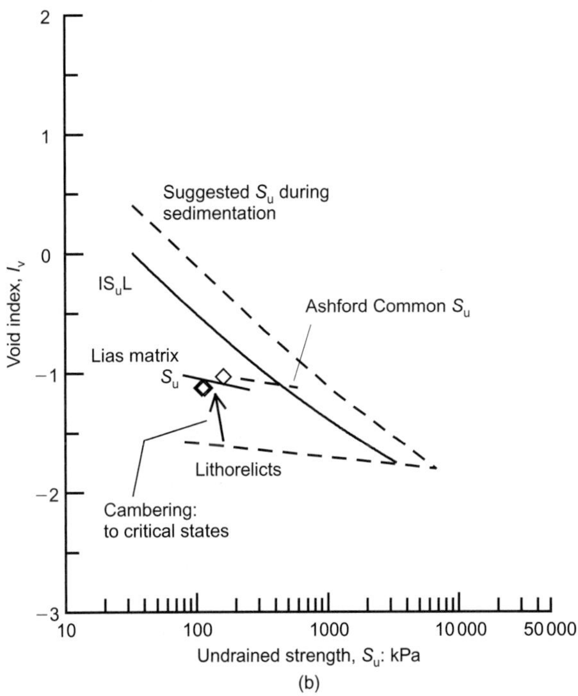

Fig. 18. Brecciated Upper Lias clay, Wothorpe: (a) stress states; (b) undrained strength

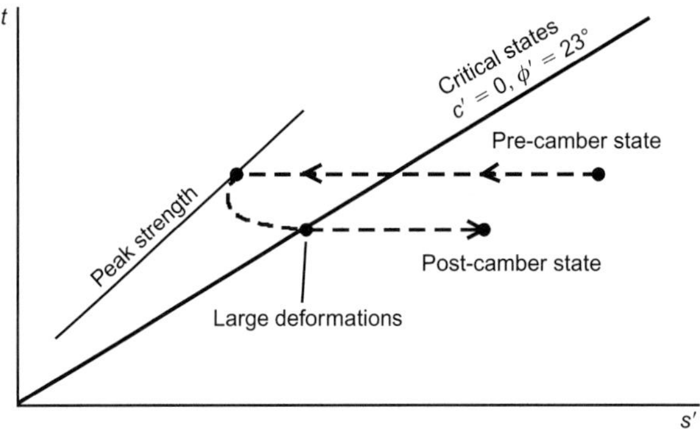

Fig. 19. Stress path for a typical soil element in a cambered slope

may be plotted in Fig. 18(a). It is seen that the disturbance resulting from cambering increases the void index substantially, to about −1·2. Undrained strengths measured at the same depths, assumed (as with weathered clay) to be controlled by the matrix water content, are shown in Fig. 18(b).

It is probably incorrect to assume that the present-day water contents reflect the increase of water content resulting from shearing during the cambering process, as the effective stresses when cambering was active were probably lower than at the present day, and no doubt there was some reconsolidation as the pore pressures subsequently dissipated. The postulated sequence of events is shown by a stress path for an element within the slope (Fig. 19). The slope in the pre-camber state was presumably frozen. As the thaw pro-

gressed, the effective stresses fell as positive pore pressures were generated with the release of water from the frozen ground. There would have been limited change in slope geometry initially, so the stress path moves horizontally; the slope material then failed at peak strength, the slope inclination fell, and the direction of the stress path is reversed to a condition at (or close to) critical state. The critical state parameters for the Upper Lias are probably about $c' = 0$, $\phi' = 23°$, the values that Kovacevic et al. (2007) found gave the best fit in their modelling of the Empingham dam foundation deformations. At this point there would have been large deformations within the slope as the camber process occurred, and the present-day slope was formed. Since the slopes at Empingham currently have an inclination around 4°, the effective stresses at the height of the camber process must have been very low indeed. With the dissipation of the camber pore pressures, and no further change in slope geometry, the stress path will have moved horizontally to the right, to the post-camber state.

The disturbance resulting from the cambering process has therefore raised the void index substantially, as indicated by the arrows in Fig. 18.

CONCLUSIONS

The mechanical behaviour of sedimentary clays is a consequence of the geological history, and the interparticle bonds that develop as a result of that history. The bond strength is not, as is sometimes assumed, controlled solely by the maximum vertical stress imposed by the overlying soil column. That is, the oedometer yield stress σ'_{vy} is a function of both the soil structure and the geological preconsolidation stress. As a result, clay samples from all depths, normally or overconsolidated, even those taken from just below the seabed, will usually have an oedometer yield stress σ'_{vy} greater than the geological preconsolidation stress, and thus YSR > geological OCR.

Knowledge of the stress states of sedimentary clays in terms of void index provides a simple and extremely useful starting point for the characterisation of soils in situ. Reference to Fig. 20 shows that the stress state (void index and σ'_{v0}) can give a useful confirmation of the clay type, and also its likely geological origin. The approximate range within which clays may be referred to as stiff (or stronger) is also shown in the diagram.

The two companion plots of in situ data, I_v against σ'_{v0} and I_v against S_u, are a valuable aid to the characterisation of most clays, particularly when comparison is made of the in situ state with the intrinsic compression line (ICL), and the undrained strength S_u with the intrinsic strength line (IS_uL). There are also some rules of thumb that are useful

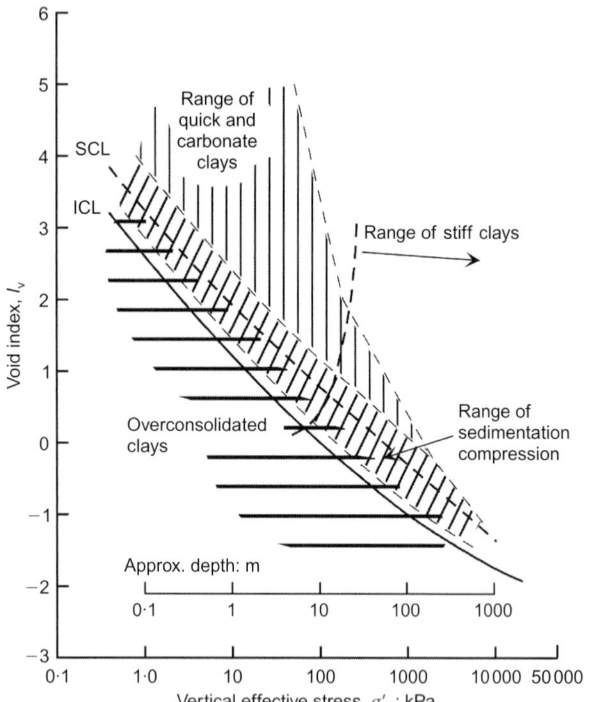

Fig. 20. Stress states related to geological history

when there is uncertainty regarding the undrained strength, perhaps resulting from sample disturbance. These include the approximate relationship between the intrinsic lines for one-dimensional compression (ICL) and triaxial compression undrained strength (IS_uL),

$$0{\cdot}33\,ICL = IS_uL$$

or

$$\frac{S_u}{\sigma^*_{ve}} = 0{\cdot}33$$

and the approximations for stress sensitivity S_σ, and strength sensitivity S_t

(*a*) for states to right (wet) of ICL

$$S_\sigma = S_t$$

(*b*) for states to left (dry) of ICL

$$S_\sigma = \frac{S_t}{IsSR}$$

These expressions are also given in slightly different form in equations (5) and (6).

The behaviour of stiff clays is modified by such factors as erosion/overconsolidation, diagenesis or lithification, brecciation and weathering. All these geological processes and their engineering consequences can be usefully illustrated in terms of the void index, in situ stress and undrained strength. Analysis in these simple terms adds significantly to the understanding of the engineering consequences of geological processes, and thus of the likely material behaviour when subject to engineering works.

ACKNOWLEDGEMENTS

The writer is grateful to the Symposium's Organising Committee for the invitation to give this Keynote Lecture, and to Dr Robert May for critically reviewing the manuscript.

REFERENCES

Baldi, G., Hueckel, A., Peano, A. & Pellegrini, R. (1991). *Developments in modelling of thermo-hydro-geomechanical behaviour of Boom clay and clay-based buffer materials*. Commission of the European Communities, Reference EUR 13964 EN.

Bishop, A. W., Webb, D. L. & Lewin, P. I. (1965). Undisturbed samples of London Clay from the Ashford Common shaft: strength–effective stress relationships. *Géotechnique* **15**, No. 1, 1–31, **doi**: 10.1680/geot.1965.15.1.1.

Burland, J. B. (1990). On the compressibility and shear strength of natural clays. *Géotechnique* **40**, No. 3, 329–378, **doi**: 10.1680/geot.1990.40.3.329.

Chandler, R. J. (1972). Lias clay: weathering processes and their effect on shear strength. *Géotechnique* **22**, No. 3, 403–431, **doi**: 10.1680/geot.1972.22.3.403.

Chandler, R. J. (2000). Clay sediments in depositional basins: the geotechnical cycle (The 3rd Glossop Lecture). *Q. J. Engng Geol. Hydrol.* **33**, No. 1, 5–39.

Chandler, R. J. & Apted, J. P. (1988). The effect of weathering on the strength of London Clay. *Q. J. Engng Geol.* **21**, No. 1, 59–68.

Chandler, R. J. & Forster, A. (2001). *Engineering in Mercia mudstone*, CIRIA Report C570. London: Construction Industry Research and Information Association.

Coop, M. R., Atkinson, J. H. & Taylor, R. N. (1995). Strength and stiffness of structured and unstructured soils. *Proc. 11th Eur. Conf. Soil Mech. Found. Engng, Copenhagen* **1**, 55–62.

Cotecchia, F. & Chandler, R. J. (2000). A general framework for the mechanical behaviour of clays. *Géotechnique* **50**, No. 4, 431–447, **doi**: 10.1680/geot.2000.50.4.431.

de Freitas, M. H. & Mannion, W. G. (2007). A biostratigraphy for the London Clay in London. *Géotechnique* **57**, No. 1, 91–112, **doi**: 10.1680/geot.2007.57.1.91, and this Symposium.

Hight, D. W., Gasparre, A., Nishimura, S., Minh, N. A., Jardine, R. J. & Coop, M. R. (2007). Characteristics of the London Clay from the Terminal 5 site at Heathrow Airport. *Géotechnique* **57**, No. 1, 3–18, **doi**: 10.1680/geot.2007.57.1.3, and this Symposium.

Hollingworth, S. E., Taylor, J. H. & Kellaway, G. A. (1944). Large scale superficial structures in the Northampton Ironstone field. *Q. J. Geol. Soc.* **100**, No. 1–4, 1–44.

Horseman, S. T., Winter, M. G. & Entwisle, D. C. (1987). *Geotechnical characterisation of Boom Clay in relation to disposal of radioactive waste*. Office for Official Publications of the European Communities.

Horswill, P., Horton, A. & Penman, A. D. M. (1976). Cambering and valley bulging in the Gwash valley at Empingham, Rutland [with an appendix by P. R. Vaughan; and discussion]. *Phil. Trans. R. Soc. Lond. A* **283**, No. 1315, 427–462.

Kovacevic, N., Higgins, K. G., Potts, D. M. & Vaughan, P. R. (2007). Undrained behaviour of brecciated Upper Lias Clay at Empingham Dam. *Géotechnique* **57**, No. 2, 181–196, **doi**: 10.1680/geot.2007.57.2.181, and this Symposium.

Mair, R. J., Taylor, R. N. & Clarke, B. G. (1992). *Repository tunnel constructions in deep clay formations*. Commission of the European Communities, Reference EUR 13964 EN.

Skempton, A. W. (1948). A study of the geotechnical properties of some post-glacial clays. *Géotechnique* **1**, No. 1, 7–22, **doi**: 10.1680/geot.1948.1.1.7.

Skempton, A. W. (1970). The consolidation of clays by gravitational compaction. *Q. J. Geol. Soc. London* **125**, No. 1–4, 373–412.

Stallebrass, S. E., Atkinson, J. H. & Mašín, D. (2007). Manufacture of samples of overconsolidated clay by laboratory sedimentation. *Géotechnique* **57**, No. 2, 249–253, **doi**: 10.1680/geot.2007.57.2.249, and this Symposium.

Terzaghi, K. (1941). Undisturbed clay samples and undisturbed clays. *J. Boston Soc. Civ. Engrs* **28**, No. 3, 211–231.

Terzaghi, K. (1943). *Theoretical soil mechanics*. New York: John Wiley.

Simpson, B. (2010). *Géotechnique* **60**, No. 12, 903–911 [**doi:** 10.1680/geot.07.KP.002]

Engineering in stiff sedimentary clays

B. SIMPSON*

The papers to the Symposium in Print are reviewed, looking in particular for insights and applications available to practising designers.† The papers display the complexity of the behaviour of stiff clays, and the very high level of expertise committed to researching it. The features of behaviour described, and to some extent quantified, are very important in understanding observed phenomena in conventional civil engineering situations. Studies of the stratigraphy of London Clay, as an example, show some helpful consistency across the deposit when results are plotted relative to its base, but also show that it is not a single, uniform material. Engineering at greater depth and with thermal effects, as required for nuclear repositories, provides some new challenges, involving both unfamiliar parameters and more familiar problems of characterising stiffness, strength and permeability, especially in bonded materials, set in a context of higher stresses. The combination of high-quality laboratory studies with field observations of full-scale behaviour remains essential to the development of geotechnical engineering, and both are well represented in the papers to the Symposium.

KEYWORDS: clays; constitutive relations; geology; laboratory tests; shear strength; stiffness

Les articles du Symposium in Print sont revus, se concentrant sur les applications utiles aux praticiens. Ces articles montrent la complexité du comportement des argiles dures et le haut niveau d'expertise mis en oeuvre pour les étudier. La description qualitative et parfois quantitative des argiles dures est essentielle à la compréhension de certains phénomènes observés en génie civil. L'étude de la stratigraphie de l'Argile de Londres partant de sa base montre des corrélations utiles et en même temps la non-uniformité du matériau. Construire à grande profondeur et avec des effets thermiques, comme pour le stockage des déchets nucléaires, amène de nouveaux défis impliquant des problèmes non familiers et d'autres plus courants tels la caractérisation de la raideur, résistance et perméabilité de matériaux cimentés soumis à des forces importantes. La combinaison de tests de laboratoire de haute qualité et d'observations de terrain grandeur nature reste essentielle pour le développement de la géotechnique, et sont bien représentés dans les articles de ce Symposium.

INTRODUCTION

The Symposium in Print consists of 17 papers and three technical notes, which depict many fascinating features of the behaviour of stiff sedimentary clays. Roughly half the papers relate to London Clay and half to other stiff clays in the United Kingdom and Europe. The contributions include geological insights, laboratory-based research work of very high quality, and large-scale field experiments and observations of actual constructions, together with numerical analysis. Examples can be found of both traditional civil engineering – deep excavations, retaining walls, slope stability, dams – and areas of practice likely to become more important in the future, particularly involving nuclear repositories characterised by very high stresses and important thermal effects. All the work is motivated by the needs of practical engineering, so it is pertinent to review how the scientific insights that have been produced will affect engineering design in the future.

Despite an enormous research effort over the last few decades, prediction of the deformation and strength of soils is still difficult and highly imprecise. We have gradually understood that the stress–strain behaviour is usually very complex, involving severe non-linearity, anisotropy of both stiffness and strength, time dependence, and effects of recent stress–strain paths. Many of these phenomena are difficult to describe in qualitative terms, let alone define quantitatively. Practising designers will need to find a way of acknowledging the complexities, so that they are not caught out by them, while proceeding with design on a much simplified basis.

In contrast, the physics of permeability is fairly well understood, but its evaluation is often a major uncertainty. The thermal properties of soil and water are additional parameters needed in the design of deep depositories: these new parameters probably have comparatively small uncertainty. All these features are explored in papers to the Symposium.

GEOLOGY OF THE LONDON CLAY
Stratigraphy

The London Clay has long been important, both for practical construction and as a major research material. It is possibly the most tested soil in the world, and is often treated by engineers as a uniform, reliable material. Those with field experience of pile or tunnel construction may well take a different view, knowing that its permeability, related to grain size, may be quite variable, and disturbed zones or weak joints may be encountered. Indeed, Kovacevic *et al.* (2007b) note the possible presence of tectonic shear zones in the area of Heathrow.

The lithological work of King (1981) provided a stratigraphy against which observations of fundamental properties of the deposit could be compared. Hight *et al.* (2007) and Pantelidou & Simpson (2007) show that the main stratigraphic changes can be identified by measurements of liquid limit. Since the consolidation history of the whole deposit is

Manuscript received 11 November 2009; manuscript accepted 12 November 2009. Published online ahead of print 24 August 2010.
Discussion on this paper closes on 1 May 2011, for further details see p. ii.
* Arup Geotechnics, London, UK.
† This paper was originally prepared for the 2007 *Géotechnique* Symposium in Print on *Stiff Sedimentary Clays: Genesis and Engineering Behaviour.*

essentially similar, this is also reflected in water contents, which are often more readily available than liquid limits. Pantelidou & Simpson note some promising results from conductivity testing, showing similarities between sites in central London several kilometres apart.

The London Clay Formation is variable in nature. Typically, it is described as silty clay, but it also contains bands of much more silty or sandy material, and it may be difficult to identify stratigraphic level on the basis of geotechnical tests alone. De Freitas & Mannion (2007) argue that the study of microfossils provides a quick and relatively cheap way of providing an unequivocal identification of stratigraphic level. This enables geomechanical data obtained from one site to be used with greater confidence at another.

The effects of stratigraphy on features of engineering behaviour are summarised as follows by Hight *et al.* (2007).

> There are important differences in the behaviour between the lithological units. Units B2(a), A3 and A2 are clearly more structured than the overlying units, having significant cohesive components of strength. Unit A2 is the most brittle, but is non-fissured. Anisotropy increases with the depth of unit and varies within units. Except for unit A2, there are no major differences between the modulus decay curves of the different units. ... As the units appear to be of consistent thickness, it is sufficient to establish the base of the London Clay at a site in order to anticipate the position of the lithological units within the profile.

Unit A2 is notably different from the overlying units: Kovacevic *et al.* (2007b) describe it as 'stiffer and much stronger'.

The stratigraphic units are important in terms of permeability, as demonstrated by Wongsaroj *et al.* (2007) in their study of long-term pore pressure changes around the Jubilee Line tunnels in St James's Park. Their findings by back-analysis are consistent with the hierarchies of permeability suggested by Hight *et al.* (2003) and Standing & Burland (2006): $k_{A2} > k_{A3ii(top)} > k_{A3i} \approx k_{A3ii(base)} > k_B$.

Horizontal effective stress

In situ horizontal effective stress in the ground is important for many design situations, including retaining walls and, as reported by Kovacevic *et al.* (2007b), cut slopes. Burland *et al.* (1979) postulated likely states of effective stress in London Clay, depending on overconsolidation, reloading by under-drainage, and reloading due to subsequent deposition of gravels, fill or other deposits. They assumed that the process of erosion and re-deposition would be sufficiently slow for the London Clay to be in a fully drained condition at all stages. Thus, at the point of maximum erosion, the effective stresses in clay near the eroded surface would be very low. Their conclusions were expressed in terms of the ratio $K_0 = \sigma'_h / \sigma'_v$, as shown in Fig. 1. More explanation of the derivation of this figure is given by Simpson *et al.* (1981).

Hight *et al.* (2007) reported values of K_0 for two situations: (a) London Clay overlain by gravel; and (b) London Clay previously overlain by 6 m of gravel, which was removed 25 years ago. Their data, shown in Fig. 2, are not consistent with the ideas of Burland *et al.* (1979); although gravel has been deposited back on top of the clay, the value of K_0 has not dropped to a low value, as the surface of the clay is approached.

Reacting to the question raised by this and similar data, De Freitas (personal communication, 2007) and others have suggested, on the basis of a geological interpretation, that the final stages of erosion of London Clay and deposition of several metres of gravel probably occurred very rapidly. This

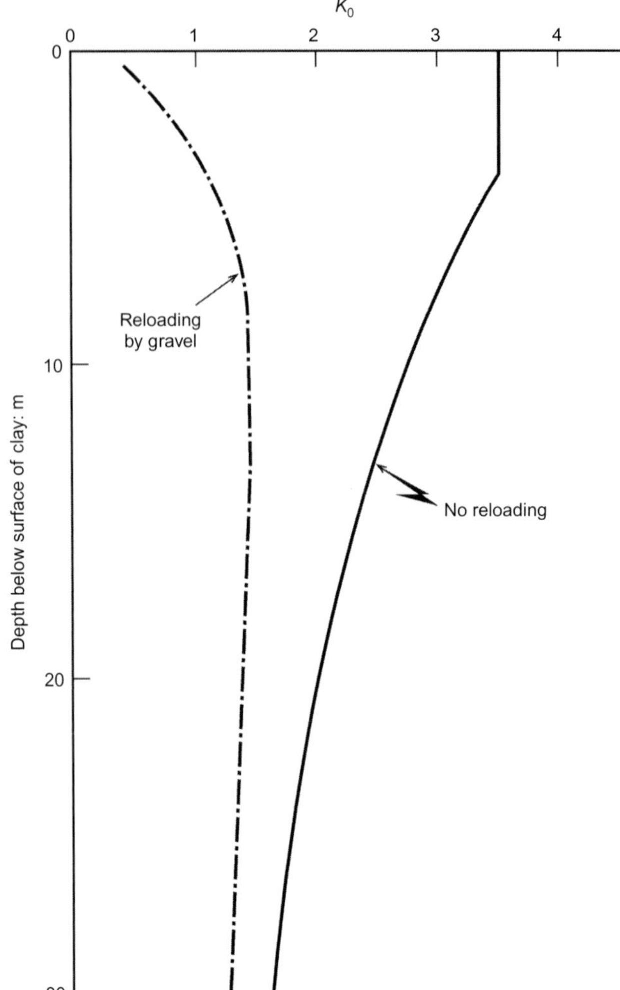

Fig. 1. K_0 **profiles postulated by Burland** *et al.* **(1979)**

would mean that the London Clay near the eroded surface was undrained through this process, and did not pass through a stage of effective stresses lower than its final stresses, under the new overburden of the gravel. In the author's opinion it remains an open question how high a horizontal effective stress can be sustained by London Clay when it is in a state of very low vertical effective stress, and over what timescale the clay swells so that the effective stresses stabilise. This will influence the K_0 value if it is later reloaded.

STRESS CHANGE AND STRAIN MAGNITUDES OF ENGINEERING SIGNIFICANCE

In order to set a context for some of the discussion to follow, it is relevant to consider the magnitudes of stress change or strain that are important to engineering design. These vary from one situation to another. In foundation design, bearing pressures on stiff clays will usually be in the range 100–400 kPa, so the stress changes in the ground that are relevant to deformation or failure will be tens or hundreds of kilopascals. A similar range could be relevant beneath a deep excavation or a high embankment, whereas the stress changes in the ground alongside a basement excavation are probably slightly smaller. It is extremely rare that stress changes of only a few kilopascals cause significant effects.

Similarly, in terms of strain, a direct strain of 0·001% is equivalent to 1 mm per 100 m – very rarely of engineering significance. Strain of 0·01% (1 mm per 10 m) might be

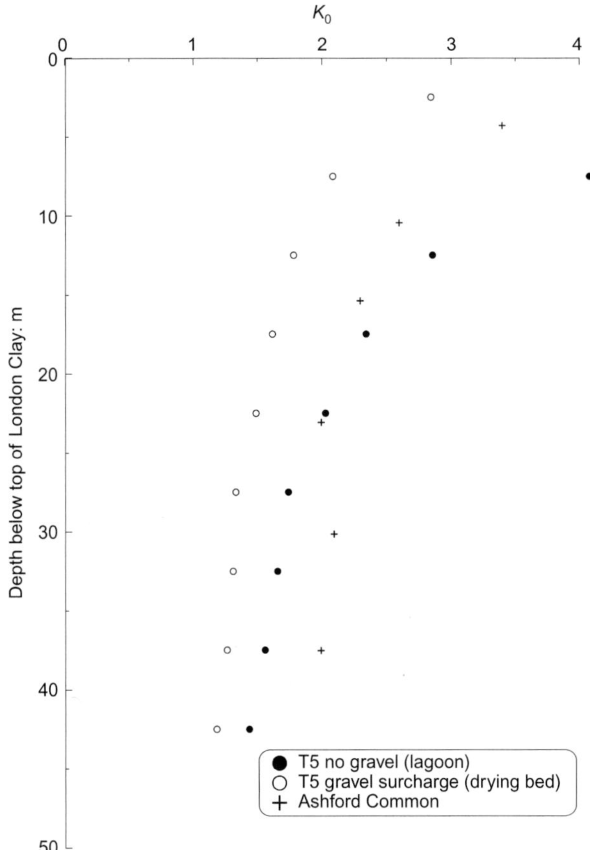

Fig. 2. K_0 **profile estimated on basis of suction measurements at Heathrow extension (Hight *et al.*, 2007, figure 11)**

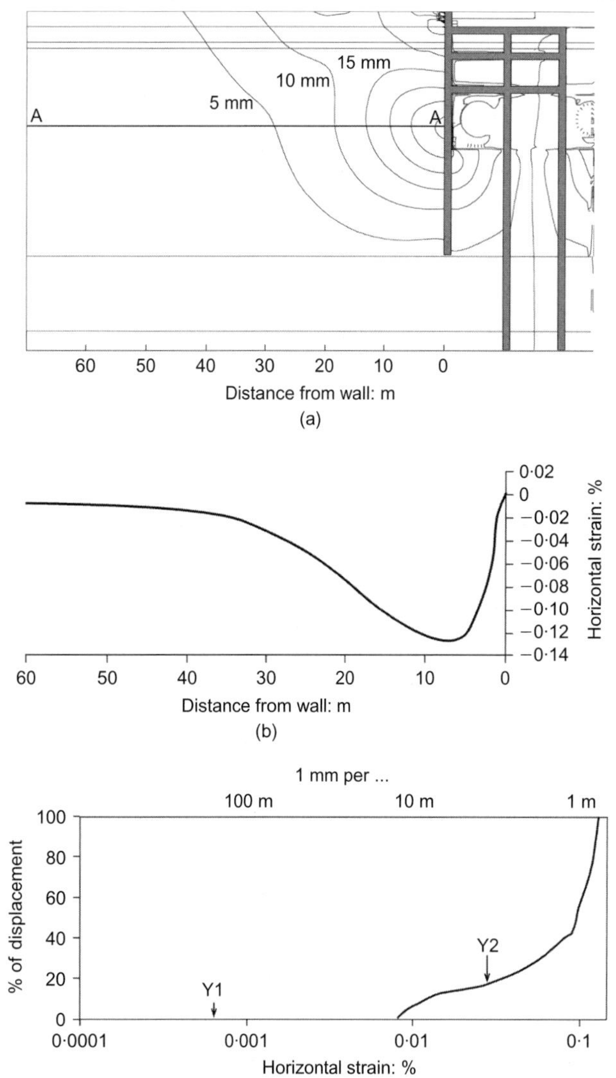

Fig. 4. **Strains that constitute displacement of a basement wall: (a) displacement contours; (b) horizontal strains on line AA; (c) percentage of the total displacement caused by strains of various magnitudes**

marginally significant, and 0·1% (1 mm per metre) would typically cause some damage. Thus engineering designers need to understand the behaviour of clays for strains in the range 0·01–0·1%, or more.

Figure 3 shows computed changes of horizontal effective stress in the ground adjacent to a three-level, 20 m deep basement in London Clay; the non-linear BRICK model was used for this purpose (Simpson, 1992, 1999). The deformation of the wall is probably governed by clay experiencing stress change in the range 20–80 kPa. Fig. 4(a) shows that the computed displacements are up to about 30 mm, a reasonably practical figure, and the horizontal strains that integrate to give this displacement are shown in Fig. 4(b). Fig. 4(c) shows the percentage of the total displacement caused by strains of various magnitudes. About 80% of the

displacement is caused by horizontal strains greater than 0·04%, and 50% of it is caused by strains greater than 0·1%.

The strains indicated for the Y1 and Y2 yield loci will be considered later. Since the clay is assumed to be undrained, and behaviour is two-dimensional, the magnitude of engineering shear strains would be double the horizontal strain.

A similar pattern holds for strains in other situations, such as around a tunnel. Fig. 5 shows the shear strains and resultant displacements computed for construction of a tunnel with 1·8% ground loss. Comparing the contours for 0·1% and 10 mm, it can be seen that at least two thirds of the displacement is generated by shear strains greater than 0·1%.

THE IMPORTANCE OF STRUCTURE

The term 'structure' is used in relation to the interaction of clay particles and aggregates of particles to indicate the combined effects of particle arrangement (fabric) and inter-particle bonding. Its origins may lie in the deposition of the particles, or it may develop with time, and it might be changed by the deformation of the clay; swelling may have

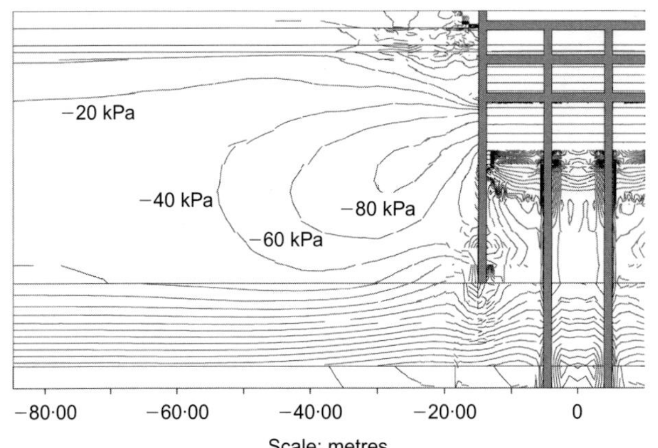

Fig. 3. **Change of horizontal effective stress (kPa) outside a 20 m deep basement**

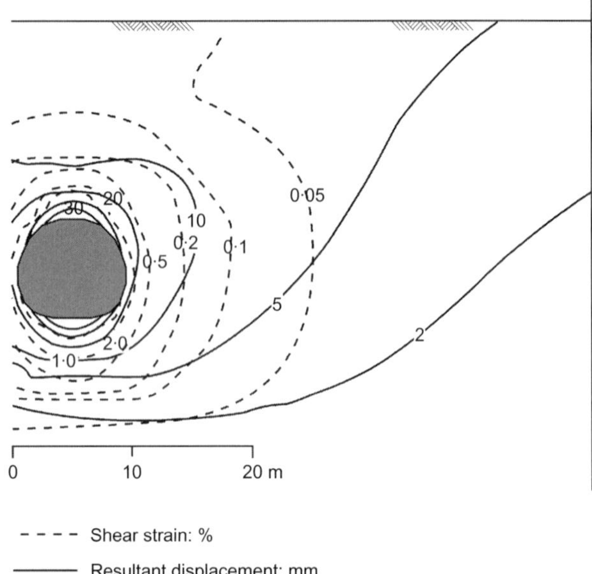

Fig. 5. Strains and displacement around a tunnel with 1·8% ground loss

a major effect in disrupting structure. Its effects on the strength, anisotropy, stiffness and permeability were subjects of many of the papers in the Symposium.

The effect of structure is a complex matter, which practising engineers may prefer to avoid. However, it has considerable practical significance, which has to be accommodated in design, even though it is not well understood. The effects of structure in terms of strength and stiffness may be beneficial in practice, so it is important also to know what processes might damage the structure and reduce its effects.

Significance of volume change

It is normally expected that soils may be taken to some form of yield or failure by shearing, after which they may suffer a loss of strength as the structure is damaged. In addition to this, it is found that structure can be damaged by both compression and, especially, swelling strains.

Amorosi & Rampello (2007) describe series of tests on Vallerica clay, comparing reconstituted clay with natural samples tested at stresses below or above their yield stresses. Even at very high stresses, the compression curves of the natural samples do not return to converge towards those of the reconstituted samples, indicating a significant effect of the fabric of the soil. However, differences between samples tested below or above the yield stresses also show that structure, probably the bonding component, can be damaged by plastic compression as well as by shearing.

Pinyol *et al.* (2007) discuss the diagenesis of bonds in 'clayey rocks', and the way the bonds can be damaged by swelling. A relatively small increase in water content, due to stress reduction and weathering, reduces the material to a residual strength similar to that found after large shear strains. Wetting–drying cycles, with release of suction at low total stress, can be particularly damaging. Although the paper concentrates on weak rocks, some of the features modelled may be present, although to a lesser extent, in many stiff clays, and the author suggests that they should be considered in the development of models for small-strain behaviour.

Cotecchia *et al.* (2007) note that the effect of swelling in

disrupting structure is greater than that of shearing, until shearing reaches 'gross yield'. On loading, the onset of gross yield is found to depend primarily on the magnitude of plastic volumetric strain, though the shear stiffness post gross yield is a function of the magnitude of plastic shear strain.

The effect of swelling on structure is discussed by Hight *et al.* (2007), who report that as swelling disrupts the structure it affects initial stiffness and yield in compression. However, it does not change the failure envelope in effective stress space, perhaps because the structure is in any case disrupted by shearing before the envelope is reached.

Hight *et al.* (2007) emphasise the very non-linear nature of swelling in London Clay samples, involving large strains as stresses drop below about 100 kPa. Soil models often assume log-linear behaviour for swelling (constant C_s or κ), although in the author's experience this is too frequently ignored in design analyses, being substituted by linear behaviour. In Fig. 6, log-linear swelling curves have been overlaid on the data of Hight *et al.* (2007) for London Clay, using $C_s = 0.02$ or 0.04, where C_s is the swelling index. Although the log-linear behaviour partly accounts for the non-linearity, the observed rate of swelling increases even more sharply as the stresses fall below 100 kPa. This could be seen as representing a breakdown of structure, which inhibits the swelling in its earlier stages.

This feature, even if understood only qualitatively, is important to the design of ground-bearing slabs beneath excavations and considerations of stand-up time of cuttings. The formation and opening of discontinuities during swelling add important changes to permeability, as noted by Bernier *et al.* (2007).

Sedimentation

To overcome the variabilities of natural ground, and of sampling processes, reconstituted samples are sometimes used to provide consistent sets of specimens for series of laboratory tests. Such samples lack important features of the structure of natural soils, gained from their sedimentation and stress history.

Provisional results from a small number of tests presented by Stallebrass *et al.* (2007) contain suggestions about the difference between samples derived from a sedimentary process, and reconstituted materials. Stallebrass *et al.* sedimented particles derived from London Clay in a large oedometer, and consolidated the clay one-dimensionally before testing. They compared test results with those of similar tests on reconstituted samples, remoulded at high water content, and found significant differences. Salt water was

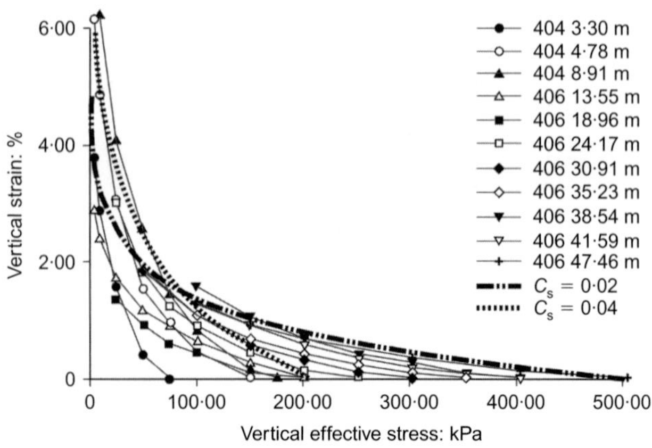

Fig. 6. One-dimensional swelling of intact London Clay (Hight *et al.*, 2007, figure 21)

used for the sedimentation process, so to check whether this was a critical factor, they tested samples reconstituted with either salt or distilled water. This showed that the type of water was not the cause of the differences.

The sedimented and reconstituted clays displayed different normal compression curves (their figure 5), with similar gradients but an offset between them. However, no difference of stiffness was detected by bender element tests made at intervals during a process of compression and swelling (their figure 6), and Fig. 7 shows the same result for stiffness at very small strain in constant-p' triaxial tests.

Figure 7 also shows the important observation that at slightly larger shear strains the results suggested that sedimented clay might be significantly stiffer than reconstituted, given the same stress state and stress history; tangent stiffness moduli are shown, which are often appropriate to theoretical developments. The stress–strain curves from this figure are shown in Fig. 8(a), and these might be thought to be fairly similar. The engineering significance of the differences is most easily understood in terms of secant moduli; the intersections of line AA with the curves in Fig. 8(b) show two secant moduli calculated for the same strain. These are compared in Fig. 8(c), which shows the ratio of moduli at the same strain (such as line AA in Fig. 8(b)). In the critical range of shear strain around 0·1% the sedimented clay is about 1·7 times stiffer than the reconstituted clay.

In practice, most analyses of ground behaviour are controlled by load, leading to proportionate stress change, rather than being strain controlled. The secant moduli then indicate directly how much strain will occur for a given, known, change of stress. Line BB in Fig. 8(b) shows two secant moduli calculated for the same stress change, and the secant moduli are plotted against stress change in Fig. 8(d). For an important range of stress change between 10 and 50 kPa (compare the discussion of Figs 3 and 4 above) the stiffness ratio exceeds 2, and it remains above 1·5 for stress change exceeding 100 kPa.

These results are preliminary, and require further investigation. Nevertheless, they might point the way towards a better understanding of some of the differences between laboratory test results and field behaviour.

STRENGTH OF STIFF SEDIMENTARY CLAYS

Papers in the Symposium showed the complexity of the strength exhibited by stiff clays, especially undrained strength. Although this is of obvious practical relevance, it will take some time for engineering designers to assimilate it sufficiently to apply it.

Besides the fact that clay in a practical situation may be drained, undrained, or somewhere in between, its strength may be characterised in many different ways, including

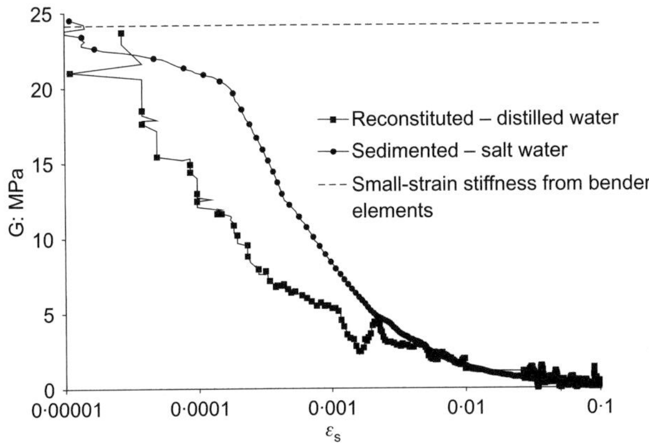

Fig. 7. Variation of shear stiffness with strain during constant-p' extension for reconstituted and sedimented samples (Stallebrass et al., 2007, figure 7)

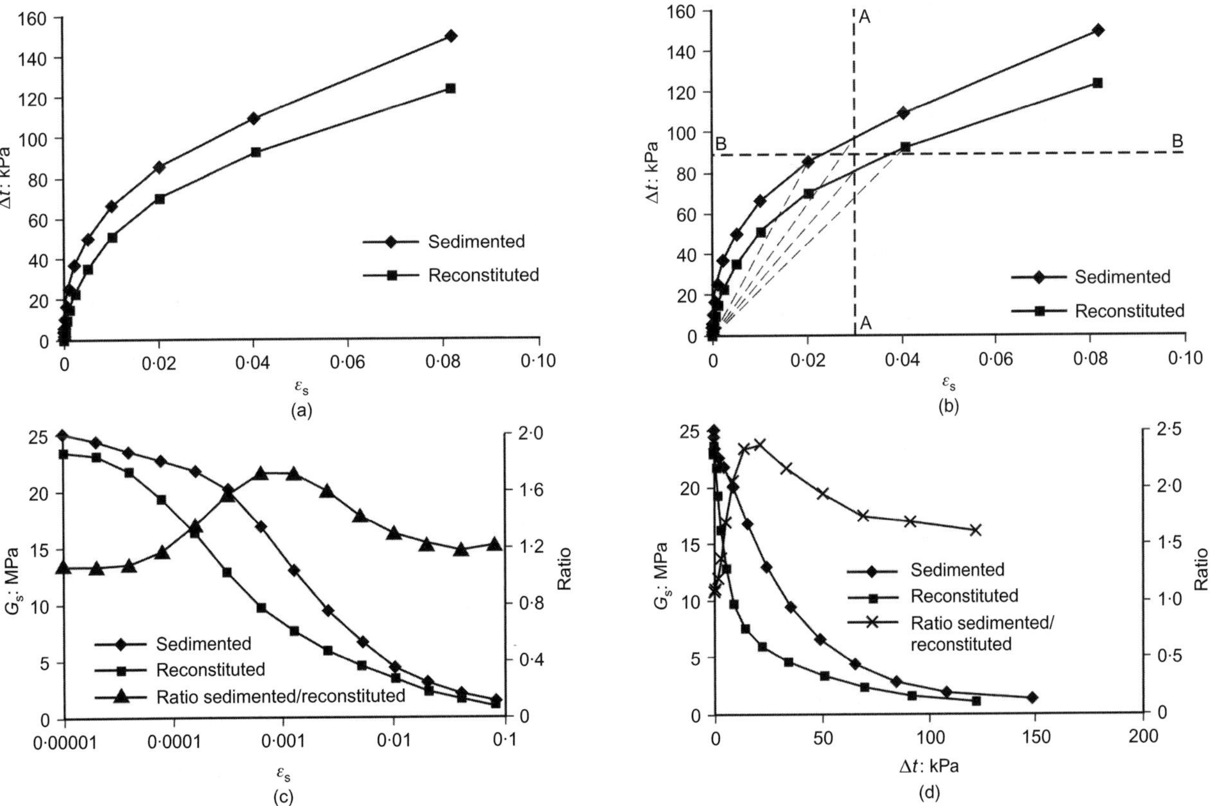

Fig. 8. Stiffness of sedimented and reconstituted clay: (a) stress–strain curves; (b) derivation of secant moduli; (c) secant moduli compared at the same strain; (d) secant moduli compared at the same stress change (Stallebrass et al., 2007, figure 7)

peak, post rupture, critical state, residual, anisotropic, plane strain, triaxial compression, triaxial extension, simple shear, ring shear, and various combinations of these. Codes of practice rely on designers assessing the relevant strength for a particular design situation: for example, Eurocode 7 (BSI, 2004) requires the designer to provide 'a cautious estimate of the value affecting the occurrence of the limit state', and safety factors may well not cover the differences between the various strengths. Strength may also be inferred from in situ tests such as penetrometers and pressuremeters, but these did not feature in the Symposium, and are not included here.

Using data for Gasparre (2005) and Nishimura *et al.* (2007), Hight *et al.* (2007) note the influence of pre-existing fissures on effective parameters for peak strength. Although a clear peak is observed in triaxial compression tests, this is much less so in extension tests and in plane strain, both of which seem to be more heavily affected by pre-existing fissures. In plane strain, however, the more plastic samples of the B2 unit exhibit a peak, defying the general trend that plastic samples are affected more by fissuring than those containing coarser grains.

The undrained strength of London Clay is found to be strongly anisotropic, owing to the combined effects of anisotropy of stiffness and the effective strength envelopes. Both drained and undrained strengths exhibit significant brittleness, particularly in the less fissured material of Unit A2, which occurs at greater depth.

In terms of effective stress, a simple approach to representing the strength of overconsolidated clay was suggested by Atkinson (2007), taking ideas from formulations used in rock mechanics and in critical state soil mechanics. This paper shows that a conventional c', ϕ' failure envelope can be replaced by a more realistic curve, as shown in Fig. 9, for which equations were given. This approach could be adopted immediately in practice, for example in slope stability programs or as a replacement for Mohr–Coulomb strength limits in a finite element program in drained analyses. If undrained strength is required, however, it is important to remember that computation of undrained strength from effective stress parameters requires a complete constitutive model, rather than simply an effective strength envelope, in order to avoid the same problems as 'Method A' which became notorious in the inquiry into the Nicoll Highway collapse in Singapore (Magnus *et al.*, 2005; Simpson *et al.*, 2008).

Critical state strength

Although the concept of a critical state strength belongs to a relatively simple view of soil behaviour, the critical state angle of shearing resistance, usually measured on reconstituted specimens, seems to offer a useful lower bound for the purpose of much engineering design. This includes both post-rupture behaviour and problems involving pre-existing fissures, provided they have not been polished by shear displacement. Gasparre *et al.* (2007a) note:

> The post-rupture angle of shearing resistance, ϕ'_{pr}, of intact samples and samples sheared along pre-existing fissures is also similar to that at the critical state for the reconstituted soil; it is possible that the post-rupture state involves a localised critical state on a shear plane. There was no discernible difference between the values of ϕ'_{pr} of the various units. ... The extension envelope of samples that failed on pre-existing fissures again shares the same angle of shearing resistance as compression tests in their post-rupture stages.

STRESS–STRAIN BEHAVIOUR
Elastic-plastic behaviour

Most constitutive models of soil behaviour assume a combination of elastic and plastic strains. The term 'elastic' is taken to refer to the component of strain for which strain energy is recovered on unloading. In the early development of the Cam-clay models, including modified Cam-clay, a single yield surface was assumed in stress space, as shown in Fig. 10. It was assumed that, within this yield surface, elastic volumetric strains occurred, but that there was no elastic shear strain: that is, the ratio of elastic shear to bulk modulus, G/K, was infinite, implying a Poisson's ratio of -1. With the advent of finite element programs this was an inconvenient assumption, which appeared unlikely, so values of Poisson's ratio generally between 0 and $+0.3$ were invoked, giving G/K in the range 1·5 to 0·46.

On the basis of an outstanding series of laboratory tests, Gasparre *et al.* (2007b) show that the Y1 yield locus, which defines the limits of strictly elastic strains, is extremely small, extending to stress changes of only about 0·4% of current stress and direct strains of less than 0·001% (1 mm per 100 m). This may be contrasted with the behaviour shown in Fig. 6, which shows very much larger recoverable volumetric strains. It appears that the original authors of Cam-clay, who had no means of measuring strains as small

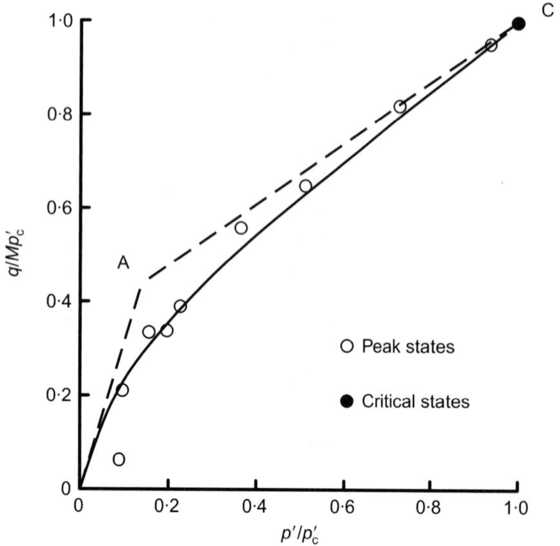

Fig. 9. Normalised peak and critical states of kaolin clay (Atkinson, 2007, figure 7)

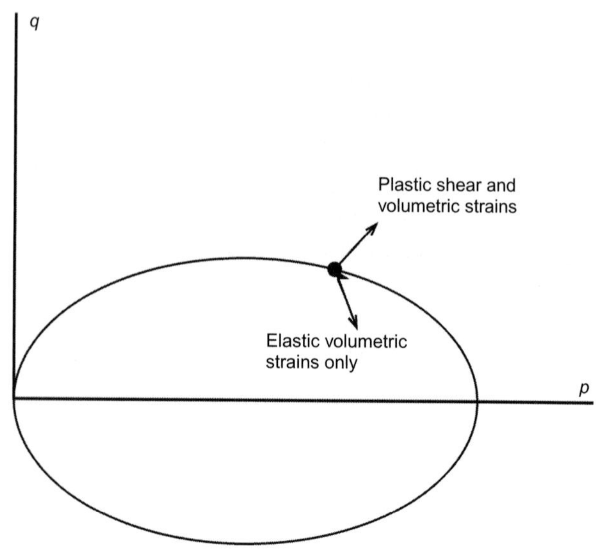

Fig. 10. Yield surface for modified Cam-clay

as 0·001%, may have been right to conclude that elastic shear strains were negligible compared with elastic (i.e. recoverable) volumetric strains. The location of the Y1 point in Fig. 4(c) confirms that the strain involved is too small to be of engineering relevance. The author suggests that it is important to separate the process of swelling, as shown in Fig. 6, from recovery of strain energy stored by the deformation of the solid particles; a similar principle is illustrated by the model for 'clayey rocks' presented by Pinyol *et al.* (2007), in which the swelling associated with micropores is considered separately from the macro behaviour.

Gasparre *et al.* (2007b) report that the Y2 yield surface for London Clay extends to stress changes of about 2·5% of current mean normal stress, with strains up to around 0·03% (3 mm per 10 m). These strains are of engineering significance, as illustrated by Fig. 4(c), but they are not fully elastic.

The significance of 'recent stress–strain history' – that is, the last few stages of straining, however long ago they occurred – has been controversial in recent years. It is assumed to be important in bubble models such as Stallebrass (1990) or the BRICK model of Simpson (1992), but this was challenged by Clayton & Heymann (2001), who concluded that many of the observed effects related to time dependence and creep. Gasparre *et al.* (2007b) report that the Y1 surface – the limit of true elastic behaviour – is affected by creep and 'forgets' recent stress–strain history, but the Y2 surface is unaffected by creep and remembers stress–strain history. It is noted here that the Y2 surface is of far greater engineering significance that the Y1 surface.

Using bender elements, it is relatively easy to measure the shear stiffness G_0 within the Y1 surface. From an engineering point of view, however, the author questions whether this can be regarded as any more than an index from which useful parameters might be derived by correlation, but which is not directly relevant in itself. In this respect, G_0 might be likened to the SPT. The author has often found that, unfortunately, 'the easier a quantity is to measure, the less use it is likely to be'.

Hight *et al.* (2007) note that the stiffnesses for London Clay at direct strains less than about 0·01%, which have been used successfully for some time, now appear to be too large, the data having been confused by creep effects. Although there could be error by a factor of up to 2 at 0·01% strain, this may not be important, as noted earlier, and the fact that the secant moduli were more nearly correct at 0·1% strain possibly avoided major problems in practical application.

Anisotropy
There is plenty of evidence that anisotropy of stiffness, strength and permeability may all be important in stiff clays.

Concentrating on very small stiffness of London Clay measured by bender elements and resonant column tests, Gasparre *et al.* (2007a) reported ratios of G_{hh} to G_{hv} of about 2 (their figure 17). For Boom Clay, Piriyakul & Haegeman (2007) noted that G_{hv} was not equal to G_{vh}, and commented:

> It is assumed that, because of anisotropic fissuring and layering of this overconsolidated Boom clay, the axes of testing in the triaxial cell are not equal to the axes of cross-anisotropy in situ, giving rise to differences between G_{vh} and G_{hv}.

In the author's understanding, inhomogeneity is the key reason why G_{vh} and G_{hv} could be unequal. They must always be equal in a homogeneous linear elastic material, regardless of anisotropy and its orientation. However, if fissures and layering effects cut across the paths of shear waves, the measured velocities V_{vh} and V_{hv} will be unequal, implying unequal G_{vh} and G_{hv} for the non-homogeneous material.

Although anisotropy is fairly easy to measure for very small strains, the degree of anisotropy is much less clear at the slightly larger strains that are of greater engineering relevance, as shown earlier (Figs 4 and 5).

Several papers in the past (e.g. Simpson *et al.*, 1996) have noted that anisotropy of stiffness might be part of the explanation for the relatively sharp settlement trough observed as a result of tunnel construction, which is not easily reproduced in finite element computations. Wongsaroj *et al.* (2007) analysed the long-term ground movements around the Jubilee Line tunnels in St James's Park, London, and found that anisotropy of both stiffness and permeability were important to the observed behaviour. Their model, like that of Simpson *et al.* (1996), assumes that the degree of stiffness anisotropy remains constant throughout the strain range. Wongsaroj *et al.* (2007) add the important observation that their anisotropic model predicts squatting of the tunnel during consolidation, as is generally observed, whereas their model with isotropic stiffness predicts decreasing horizontal diameter with time. They also note the importance of a good model of the permeabilities of both the ground and the tunnel lining.

Time dependence
The time dependence of stiffness of stiff clays has been reported by many authors over the years: examples include Mesri *et al.* (1978), Tatsuoka *et al.* (1999) and Pillai (Kanapathipillai) (1996), quoted by Simpson (1999). A fuller understanding of this behaviour is needed so that it can be included in constitutive models, and the laboratory work reported by Sorensen *et al.* (2007) goes some way towards this. Using a highly instrumented triaxial cell, Sorensen *et al.* (2007) tested both reconstituted and undisturbed London Clay, comparing also an artificial bonded material consisting of kaolin mixed with some cement. During the tests, they varied the strain rate in sharp steps. A distinction is drawn between 'isotach' behaviour, in which a separate stress–strain curve appears to exist for each strain rate, and TESRA behaviour (Temporary Effects of changes in Strain RAte), in which the specimen returns to the same unique stress–strain curve at any constant strain rate, but departs from it temporarily when the strain rate changes.

Sorensen *et al.* (2007) report that the materials tested showed mainly isotach behaviour, with a tendency to become TESRA as failure approached in all cases except the undisturbed London Clay, which remained isotach. The author considers that the set of data available from this work could readily form the basis of a numerical model.

SLOPE STABILITY ANALYSES
The use of finite element analysis to study slope stability was demonstrated in papers by Kovacevic *et al.* (2007a; 2007b). From the perspective of engineering design, possibly the most important lesson to draw from this work is the applicability and flexibility of the finite element method, replacing and exceeding the utility of conventional slope stability analysis. For the particular problems studied, the ability to model tectonic shear zones, time-dependent swelling and dissipation of suction, progressive failure, and the effect of the coefficient of earth pressure at rest (K_0) were all important, and could not readily have been reproduced in limit equilibrium calculations. As noted above, values of K_0 are uncertain. Reducing them removed the prediction of superficial slips but, since lower initial effective stresses implied that less suction developed during excavation, led to earlier predicted failure of deep seated slips.

The time-dependent computations are dependent on assumed values of the coefficient of permeability, and on the ability of the clay to sustain suctions. Both these parameters are difficult to evaluate for the large-scale body of clay on the basis of testing, so the comparisons presented here with field behaviour are particularly valuable.

FIELD MONITORING

In view of the emerging complexity of behaviour of stiff sedimentary clays, it is ever more essential that laboratory and theoretical developments are complemented and led by results from field observation.

Kovacevic *et al.* (2007a) show how field observations at Empingham Dam could be used to develop the parameters of a bubble model, enabling improved predictions in brecciated Upper Lias Clay. The author doubts whether the parameters could have been derived from soil testing, and certainly not without a much more extensive and expert test regime, so the process of deriving parameters by back-analysis is particularly important.

One paper in the Symposium reported a detailed monitoring exercise, of the station box for the Channel Tunnel Rail Link at Ashford. Richards *et al.* (2007) report results from the construction and long-term monitoring of the 10 m deep excavation in Hythe Beds, and Atherfield and Weald Clays, the Atherfield Clay having the dominant role at the section studied. The excavation was facilitated by installation of vertical drains, which are still operating successfully to reduce water pressures in the long term in the passive zone at the toe of the wall. The clay was sufficiently permeable that long-term pore pressures were set up within a few weeks, indicating that reliance could be placed on undrained behaviour only for a short period.

The case history is also very useful in recording no return of effective earth pressures to their initial values of K_0.

NUCLEAR REPOSITORIES

Research on possible repositories for nuclear waste is extending the scope of geotechnical engineering, first because it involves depths and stress levels greater than encountered in much conventional engineering, and second because it requires additional parameters, related particularly to temperature and loss of saturation due to release of total stress. Much valuable data is being recovered from monitoring both the civil engineering works and the thermal behaviour of experimental facilities, and analytical developments are following this, assisted by laboratory testing.

Gens *et al.* (2007) report an in situ heating test carried out on Opalinus Clay in the Mont Terri underground laboratory, investigating particularly the interaction of thermal phenomena with hydraulic and mechanical behaviour. The clay is subject to major reductions of total stress during formation of tunnels at 250–320 m depth. The clay is a bonded material, and the paper presents a model for the loss of bonding due to deformation. The finite element analysis is extended to include thermal properties of the clay and the variation of water viscosity with temperature. The strongest coupling is found from thermal to hydraulic and mechanical behaviour. Pore pressure generation is controlled primarily by temperature increase, and the largest contributor to deformation and displacements is thermal expansion.

In many conventional situations, stiffer materials, including clays, strain less than softer ones because the loading is stress controlled. However, when the loading is thermally controlled, strain may be much less affected by stiffness. Because of this, the author suggests that studies of damage to clay structure are potentially very relevant in cases of thermal loading.

The inclusion of thermal properties in finite element analysis adds complexity to the mathematics, but it involves well-understood physics, and parameter values that can be found to reasonable accuracy. Although breakage of bonding and anisotropy were modelled, the authors find that in this instance these features have little effect on their results. Thus the extension of the standard finite element model does not really add to the difficulties of ground prediction: the old enemies of stiffness, strength and permeability still remain the dominant uncertainties.

Under relief from very high stress, damage to the structure of the clay can change its properties significantly, as noted by Bernier *et al.* (2007). The paper describes improvements in excavation techniques aimed at reducing the damaged zone, and studies of the crack-sealing behaviour of the Boom Clay. As with other stiff clays, the stress–strain response of the Boom Clay is highly non-linear, and assessment of in situ stresses is uncertain. Cracking can lead to major changes in hydraulic properties, and this may account for the fact that tunnelling could be detected by changes of pore pressure at a distance of about 60 m (12·5 tunnel diameters) ahead of the excavation. Both the fracture process and the viscosity of the soil skeleton were found to be important features of the behaviour, required in a numerical model.

As total stresses are relieved, the clay may develop very large suctions. This is investigated for intact specimens of Boom Clay cut from block samples by Delage *et al.* (2007). The clay was found to hold suctions up to 5 MPa before starting to desaturate. The paper also reports the complex response of the specimens to sequences of loading and unloading.

Work of this type provides some very interesting new challenges for geotechnical engineers. Understanding of the physical behaviour of the ground has to be extended, additional parameters are required, and our grasp of basic geotechnical principles will be stretched as they are put to use in a new context.

CONCLUDING REMARKS

The Symposium in Print produced a set of outstanding papers displaying the complexity of the behaviour of stiff clays, and the very high level of expertise committed to researching it. The features of behaviour described – and to some extent quantified – are very important to understanding observed phenomena in conventional civil engineering situations. The complexity is such that applications to prediction can be developed only relatively slowly: full constitutive models are needed for this, and it is difficult to obtain reliable parameters for complex behaviour. Back-analysis of field observations provides some important confirmation, but as models involve more interacting parameters, derivation of values of each parameter from back-analysis becomes more difficult.

Simple extrapolations from 'index' parameters, such as G_0, will remain important. However, the stiffness that is really important to engineering design relates to strains much larger than those for which G_0 is measured, and researchers could do well to concentrate efforts in this region of larger strains. G_0 is relatively easy to measure using bender element tests in the laboratory or shear wave testing in the ground, but sadly the author suggests that 'the easier a quantity is to measure, the less use it is likely to be.'

Engineering at greater depth and with thermal effects, as required for nuclear repositories, provides some new challenges. Thermal behaviour is fairly well understood and the parameters are relatively easy to evaluate, so incorporation in finite element analysis should not be too difficult, in principle. However, the behaviour of bonded materials subject to relatively large thermal strains requires careful study, particularly if cracking grossly changes hydraulic conductivity.

Géotechnique is rarely an 'easy read', and this set of papers is no exception. The behaviour of stiff clays that is portrayed is very complex, the details are difficult to comprehend, and practical application will in some cases be some way off. The reader has to take time to reflect carefully on what is written, a luxury afforded to researchers more readily than to practising designers. Nevertheless, the author submits that these papers contain results and information that will be vital to the development of further research, understanding and practical prediction of behaviour. It may be that practising engineers will ultimately use the information only at second hand as it is included in software or design procedures, but it will be essential that they understand what they are assuming and using, and for that they will, when relevant, need to commit the time to some solid reading.

REFERENCES

Amorosi, A. & Rampello, S. (2007). An experimental investigation into the mechanical behaviour of a structured stiff clay. *Géotechnique* **57**, No. 2, 153–166, **doi:** 10.1680/geot.2007.57.2.153.

Atkinson, J. (2007). Peak strength of overconsolidated clays. *Géotechnique* **57**, No. 2, 127–135, **doi:** 10.1680/geot.2007.57.2.127.

Bernier, F., Li, X.-L. & Bastiaens, W. (2007). Twenty-five years' geotechnical observation and testing in the Tertiary Boom Clay formation. *Géotechnique* **57**, No. 2, 229–237, **doi:** 10.1680/geot.2007.57.2.229.

British Standards Institution (2004). *Eurocode 7: Geotechnical design – Part 1: General rules*. BS EN 1997-1. Milton Keynes: BSI.

Burland, J. B., Simpson, B. & St John, H. D. (1979). British state of the art report on movements around deep excavations. *Proc. 7th Eur. Conf. Soil Mech. Found. Engng, Brighton* **1**, 13–30.

Clayton, C. R. I. & Heymann, G. (2001). Stiffness of geomaterials at very small strains. *Géotechnique* **51**, No. 3, 245–255, **doi:** 10.1680/geot.2001.51.3.245.

Cotecchia, F., Cafaro, F. & Aresta, B. (2007). Structure and mechanical response of sub-Apennine Blue Clays in relation to their geological and recent loading history. *Géotechnique* **57**, No. 2, 167–180, **doi:** 10.1680/geot.2007.57.2.167.

De Freitas, M. H. & Mannion, W. G. (2007). A biostratigraphy for the London Clay in London. *Géotechnique* **57**, No. 1, 91–99, **doi:** 10.1680/geot.2007.57.1.91.

Delage, P., Le, T.-T., Tang, A.-M., Cui, Y.-J. & Li, X.-L. (2007). Suction effects in deep Boom Clay block samples. *Géotechnique* **57**, No. 2, 239–244, **doi:** 10.1680/geot.2007.57.2.239.

Gasparre, A. (2005). *Advanced laboratory characterisation of London Clay*. PhD thesis, Imperial College, London.

Gasparre, A., Nishimura, S., Coop, M. R. & Jardine, R. J. (2007a). The influence of structure on the behaviour of London Clay. *Géotechnique* **57**, No. 1, 19–31, **doi:** 10.1680/geot.2007.57.1.19.

Gasparre, A., Nishimura, S., Minh, N. A., Jardine, R. J. & Coop, M. R. (2007b). The stiffness of natural London Clay. *Géotechnique* **57**, No. 1, 33–47, **doi:** 10.1680/geot.2007.57.1.33.

Gens, A., Vaunat, J., Garitte, B. & Wileveau, Y. (2007). In situ behaviour of a stiff layered clay subject to thermal loading: observations and interpretation. *Géotechnique* **57**, No. 2, 207–228, **doi:** 10.1680/geot.2007.57.2.207.

Hight, D. W., Gasparre, A., Nishimura, S., Minh, N. A., Jardine, R. J. & Coop, M. R. (2007). Characteristics of the London Clay from the Terminal 5 site at Heathrow Airport. *Géotechnique* **57**, No. 1, 3–18, **doi:** 10.1680/geot.2007.57.1.3.

Hight, D. W., McMillan, F., Powell, J. J. M., Jardine, R. J. & Allenou, C. P. (2003). Some characteristics of London Clay. *Proceedings of the conference on characterisation and engineering properties of natural soils*, Singapore, Vol. 2, pp. 851–907.

King, C. (1981). The stratigraphy of the London Clay and associated deposits. *Tertiary Research Special Paper*, No. 6. Rotterdam: Backhuys.

Kovacevic, N., Higgins, K. G., Potts, D. M. & Vaughan, P. R. (2007a). Undrained behaviour of brecciated Upper Lias Clay at Empingham Dam. *Géotechnique* **57**, No. 2, 181–195, **doi:** 10.1680/geot.2007.57.2.181.

Kovacevic, N., Hight, D. W. & Potts, D. M. (2007b). Predicting the stand-up time of temporary London Clay slopes at Terminal 5, Heathrow Airport. *Géotechnique* **57**, No. 1, 63–74, **doi:** 10.1680/geot.2007.57.1.63.

Magnus, R., Teh, C. I. & Lau, J. M. (2005). *Report of the committee of inquiry into the incident at the MRT Circle Line worksite that led to the collapse of the Nicoll Highway on 20 April 2004*. Singapore: Ministry of Manpower.

Mesri, G., Ullrich, C. R. & Choi, Y. K. (1978). The rate of swelling of overconsolidated clays subjected to unloading. *Géotechnique* **28**, No. 3, 281–307, **doi:** 10.1680/geot.1978.28.3.281.

Nishimura, S., Minh, N. A. & Jardine, R. J. (2007). Shear strength anisotropy of natural London Clay. *Géotechnique* **57**, No. 1, 49–62, **doi:** 10.1680/geot.2007.57.1.49.

Pantelidou, H. & Simpson, B. (2007). Geotechnical variation of London Clay across central London. *Géotechnique* **57**, No. 1, 101–112, **doi:** 10.1680/geot.2007.57.1.101.

Pillai (Kanapathipillai), A. (1996). *Review of the BRICK model of soil behaviour*. MSc dissertation, Imperial College, London.

Pinyol, N., Vaunat, J. & Alonso, E. E. (2007). A constitutive model for soft clayey rocks that includes weathering effects. *Géotechnique* **57**, No. 2, 137–151, **doi:** 10.1680/geot.2007.57.2.137.

Piriyakul, K. & Haegeman, W. (2007). Void ratio function for elastic shear moduli for Boom Clay. *Géotechnique* **57**, No. 2, 245–248, **doi:** 10.1680/geot.2007.57.2.245.

Richards, D. J., Powrie, W., Roscoe, H. & Clark, J. (2007). Pore water pressure and horizontal stress changes measured during construction of a contiguous bored pile multi-propped retaining wall in Lower Cretaceous clays. *Géotechnique* **57**, No. 2, 197–205, **doi:** 10.1680/geot.2007.57.2.197.

Simpson, B. (1992). 32nd Rankine Lecture. Retaining structures: displacement and design. *Géotechnique* **42**, No. 4, 539–576, **doi:** 10.1680/geot.1992.42.4.539.

Simpson, B. (1999). Engineering needs. *Proc. 2nd Int. Symp. on Pre-failure Deformation Characteristics of Geomaterials, Turin* **2**, 1011–1026.

Simpson, B., Calabresi, G., Sommer, H. & Wallays, M. (1981). Design parameters for stiff clays. *Proc. 7th Eur. Conf. Soil Mech. Found. Engng, Brighton* **5**, 91–125.

Simpson, B., Atkinson, J. H. & Jovičić, V. (1996). The influence of anisotropy on calculations of ground settlements above tunnels. *Proceedings of the international symposium on geotechnical aspects of underground construction in soft ground*, London, pp. 591–594.

Simpson, B., Nicholson, D. P., Banfi, M., Grose, W. G. & Davies, R. V. (2008). Collapse of the Nicoll Highway excavation, Singapore. In *Forensic engineering: From failure to understanding* (ed. B. S. Neale), pp. 1–14. London: Thomas Telford.

Sorensen, K. K., Baudet, B. A. & Simpson, B. (2007). Influence of structure on the time-dependent behaviour of a stiff sedimentary clay. *Géotechnique* **57**, No. 1, 113–124, **doi:** 10.1680/geot.2007.57.1.113.

Stallebrass, S. E. (1990). *Modelling the effect of recent stress history on the deformation of overconsolidated soils*. PhD thesis, City University, London.

Stallebrass, S. E., Atkinson, J. H. & Masin, D. (2007). Manufacture of samples of overconsolidated clay by laboratory sedimentation. *Géotechnique* **57**, No. 2, 249–253, **doi:** 10.1680/geot.2007.57.2.249.

Standing, J. R. & Burland, J. B. (2006). Unexpected tunnelling volume losses in the Westminster area, London. *Géotechnique* **56**, No. 1, 11–26, **doi:** 10.1680/geot.2006.56.1.11.

Tatsuoka, F., Santucci de Magirtris, F., Momoya, M. & Maruyama, N. (1999). Isotach behaviour of geomaterials and its modelling. *Proc. 2nd Int. Symp. on Pre-failure Deformation Characteristics of Geomaterials, Turin* **1**, 491–500.

Wongsaroj, J., Soga, K. & Mair, R. J. (2007). Modelling of long-term ground response to tunnelling under St James's Park, London. *Géotechnique* **57**, No. 1, 75–90, **doi:** 10.1680/geot.2007.57.1.75.

Session 1

Formation and Engineering Geology of Stiff Clays

de Freitas, M. H. & Mannion, W. G. (2007). *Géotechnique* **57**, No. 1, 91–99

A biostratigraphy for the London Clay in London

M. H. DE FREITAS[*] and W. G. MANNION[*]

Biostratigraphy enables specific layers of strata to be given a unique stratigraphic address that can be used to identify the horizons in which in situ tests are conducted and from which samples for testing are collected. This is particularly valuable when working in fine-grained deposits, where stratigraphic position is not easily recognised by lithology alone, and permits any data bank of geotechnical values and performance to be used without fear of confusing one stratigraphic layer with another. Construction in the London Clay Formation, and its supporting research, has necessitated an accurate and speedy method for identifying, with a precision of centimetres, the stratigraphic layer from which geotechnical data have been obtained. The principles that enable biostratigraphy to be used for this purpose are explained and illustrated with reference to the London Clay Formation in London: this approach can be used in most sedimentary strata. An example of biostratigraphic data used in London is presented.

La biostratigraphie permet d'attribuer à des couches spécifiques de strate une localisation stratigraphique unique qui peut être utilisée pour identifier les emplacements dans lesquels les essais in situ sont réalisés et dans lesquels les échantillons à analyser sont prélevés. Ces informations sont particulièrement importantes pour travailler sur des dépôts granulaires fins, où les positions stratigraphiques sont difficilement reconnaissables par la seule méthode lithologique. Elles permettent d'utiliser toute base de données de valeurs et de performances géotechniques sans risquer de confondre une couche stratigraphique avec une autre. La construction dans la formation de London Clay (argile de Londres) et les travaux de recherche la soutenant ont nécessité une méthode rapide, avec une précision de l'ordre du centimètre, pour identifier la couche stratigraphique à partir de laquelle les données géotechniques ont été obtenues. Les principes qui permettent d'appliquer la stratigraphie à cet effet sont ici expliqués et illustrés en référence à la formation de London Clay (argile de Londres) à Londres. Il est possible d'utiliser cette approche pour la plupart des strates sédimentaires. Un exemple de données biostratigraphiques utilisé à Londres est ici présenté.

KEYWORDS: clays; geology; sedimentation

THE PROBLEM

Construction in London is intimately associated with the geotechnical properties of the London Clay Formation (LCF). Much is known about this material, but use of its geotechnical data is limited by the difficulty of knowing whether the values obtained at one site can be applied to another at a different location. This difficulty arises from three factors, as follows.

(*a*) The LCF is not exclusively a *clay* but a *formation*. That is, it is an assemblage of related sediments of similar age—in this case mainly silty clays but with more silty and sandier units at some levels. It is layered, and within any one layer the amounts of clay, silt and sand can vary laterally from place to place: this means that geotechnical values measured in any one layer may not represent values elsewhere in the same layer.

(*b*) Where lateral variation is thought to be small, as in two sites reasonably close to each other, some visual clues to similarity of layers can be attempted. Unfortunately, particle sizes in the LCF are small, are difficult to distinguish with the naked eye, and cannot easily be used for subdividing the formation. Attempts to compare the material from one location with that from another require the support of laboratory work to establish such characters as particle size distribution, void ratio, water content and plasticity. However, such data do not uniquely identify a specific layer. Thus despite such efforts it may not be possible to know whether the geotechnical properties measured at one location can be used elsewhere.

(*c*) The deposition of the LCF and its subsequent geological history have caused not only layers to vary from place to place beneath London but the vertical succession above them to have been subjected to differing amounts of loading and unloading across the region. Thus a layer of given particle sizes, plasticity and depth below ground level may be compared in the same way with another of similar character and depth elsewhere in London, without the data reflecting the different stress histories of the two locations; to reveal that might require the much more expensive in situ measurements of permeability and stiffness.

The summation of these three factors means that the considerable wealth of geotechnical data available for the LCF can be used in only a limited way because a unique comparison of one layer with another cannot easily be made. Biostratigraphy enables this comparison to be made and helps to unlock the geotechnical data bank for a deposit. This paper describes basic elements of the biostratigraphy of the LCF, in London, as an example of its usefulness to geotechnical engineering.

STRATIGRAPHY

Strata can be studied using many techniques: for example by using their particle characters (lithostratigraphy), their age (chronostratigraphy), the magnetism within them (magnetostratigraphy), and their fossil content (biostratigraphy). Applying many stratigraphic disciplines simultaneously, in an integrated way, vastly improves the interpretation of a sedimentary succession. A lithological subdivision of strata will be hampered, when used by itself, by the fact that such

Manuscript received 5 May 2006; revised manuscript accepted 29 August 2006.
Discussion on this paper closes on 1 July 2007, for further details see p. ii.
[*] Department of Civil and Environmental Engineering, Imperial College, London, UK.

units are commonly *diachronous*: that is, they are not the same age everywhere. In contrast, biostratigraphy uses the range of fossil species through time to correlate strata of similar age regardless of their lithology.

Consider a basin of deposition extending from a shoreline to deeper water offshore. Near-shore sediments are likely to be coarser than their deeper-water equivalents, so the lithology of any one horizon will change as the horizon is traced from deep water to shallow water (Fig. 1). However, the animals living in the sea and on, or in, the sediments at the sea floor will be of the same age across the basin at any given time, and the same can be said of their remains. Within a collection of animal remains (an *assemblage*) distinctive *marker* fossils may be present: such markers can be used to subdivide the LCF. Organisms that make good marker fossils are typically fast-evolving, widespread, abundant, readily preserved and easily recognisable. Microfossils are particularly suitable for this purpose, especially those whose abundance and diversity vary with depositional environment. These include foraminifera, ostracods, calcareous nannofossils, palynomorphs, radiolaria and diatoms: all these are widely used by biostratigraphers.

THE LONDON CLAY FORMATION

The LCF forms part of the Eocene Series of strata in southern England, and occurs approximately midway through the Palaeogene System, within the Cenozoic Era (formerly referred to as the Tertiary); it was deposited between 52 and 55 million years ago. At that time this area was close to the western shoreline of a sedimentary basin that extended into the North Sea and beyond, to Poland: the North Sea Basin (Fig. 2). The original continuity of the LCF has been broken by later tectonic events that occurred during the mid-Cenozoic Alpine orogeny, resulting in the LCF being preserved only in two separate synclinal structures known as the Hampshire and London Basins.

The stratigraphy of the LCF has been studied for more than 150 years (Wetherell, 1836; Prestwich, 1846, 1847, 1850; Whitaker, 1866; Sherborn & Chapman, 1886, 1889; Dewey, 1912; Stamp, 1921; Wrigley, 1924, 1940; Wooldridge, 1926; Hawkins, 1954; Davis & Elliott, 1957; Curry, 1965, 1966; Curry *et al.*, 1978; Wright, 1972; Murray & Wright, 1974; Stinton, 1975; Cooper, 1976; Bujak *et al.*, 1980; King, 1981, 1984, 1991; Plint, 1983, 1988; Aubry, 1985, 1986; Aubry *et al.*, 1986; Ellison *et al.*, 1994; Ellison & Zalasiewicz, 1996;

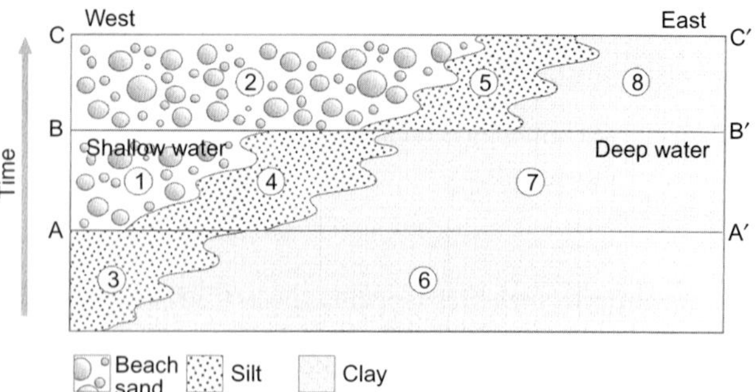

Fig. 1. Shoreline is seen to move from west to east with time. A lithostratigrapher would correlate sand units 1 and 2, silt units 3, 4 and 5, and clay units 6, 7 and 8. A biostratigrapher would correlate time units A–A', B–B' and C–C' (after Emery, 1996)

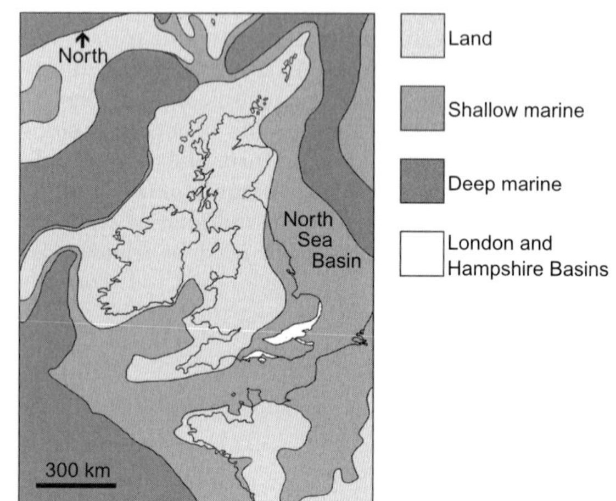

Fig. 2. Palaeogeographical reconstruction of the 'London Clay Sea' in the Early Eocene (after Ziegler, 1990)

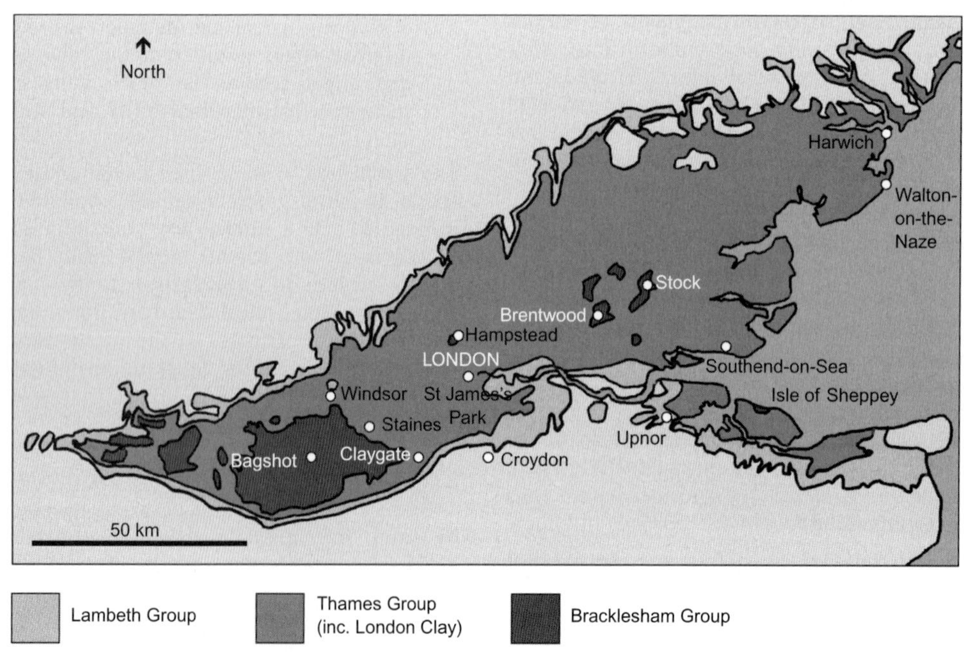

Fig. 3. Present outcrop of Palaeogene strata in the London Basin (after Ellison & Zalasiewicz, 1996)

Ellison & King, 2004; Knox, 1996a, 1996b; Neal, 1996). These studies have been hampered by the similarity of the silty clays that make up most of the LCF and by the lack of natural exposures, which are restricted to the coast (Fig. 3).

Lithostratigraphy

The maximum recorded vertical thickness of the LCF is now, after consolidation, approximately 150 m, and preserved in south Essex (see the Stock Borehole in King, 1981, pp. 45–47). In the Staines No. 5 Borehole, drilled for Thames Water Utilities just to the west of London, the LCF starts with about 10 m of silty sands interbedded with clay and passes up into approximately 60 m of predominantly silty clays (the material typically thought of as 'London Clay'): these silty clays are interbedded with silt and sand horizons towards the upper 30 m of the LCF at that site

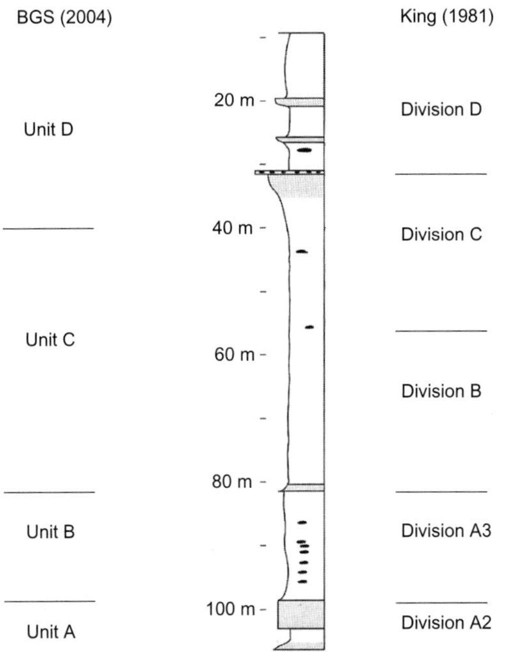

Fig. 4. British Geological Survey (Ellison & King, 2004) and King's (1981) divisions for the London Clay in the Staines No. 5 Borehole (after Ellison & King, 2004)

(Ellison & King, 2004) (Fig. 4). In the Hampstead Heath Borehole, drilled by the British Geological Survey in north London, these pass up into the fine sands and silts of the Claygate Member, which make up the rest of the formation. Attempts to subdivide the succession are hampered by the difficulties described earlier ('The problem'). However, the British Geological Survey have suggested five divisions based solely on lithology—units A–D and the Claygate Member (Fig. 4). It must be noted that this subdivision was published principally with mapping purposes in mind. King (1981) had already developed a higher-resolution scheme based on the recognition of cyclical changes in lithology that could be correlated using biostratigraphic data (Fig. 4). It is King's 1981 scheme that most workers refer to.

Biostratigraphy

The LCF contains the remains of animal communities that lived at the sea floor (benthonic fauna) mixed with those that floated in the sea above (planktonic fauna); their occurrences are very much influenced by physical and chemical conditions on, and in, the substrate, and in the water column. Generally, factors such as temperature and salinity limit distribution patterns while others, such as food and nutrient availability, influence abundances. Used together, the benthonic and planktonic microfauna can provide many insights into the changes that take place in depositional environments through time. The microfauna of the LCF has enabled repeated episodes of increasing and decreasing depths of seawater to be identified throughout its deposition (King, 1981, 1991), a sequence of events the identification and correlation of which are described as sequence stratigraphy.

Sequence stratigraphy

King (1981), following the work of Stamp (1921), recognised that faunal changes could be correlated with lithological changes, and on this basis subdivided the LCF succession into five *cycles*. Each cycle starts with a relatively thin sedimentary deposit containing coarse-grained material derived from underlying materials: a *transgressive lag* (Fig. 5). This records an episode of slow sedimentation and winnowing at the onset of a rise in sea level. Such a rise in sea level inundated the land mass to the west, driving the shoreline away from London where the depth of water

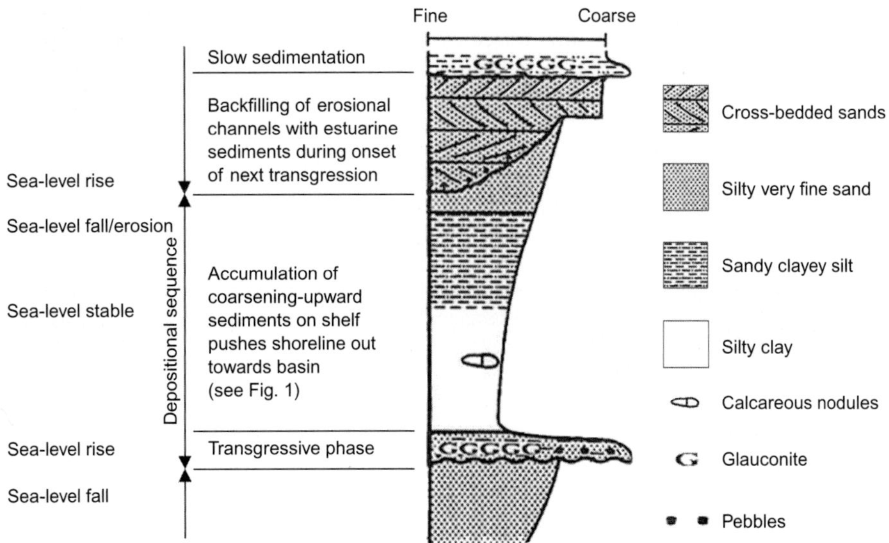

Fig. 5. An idealised depositional sequence in the London Clay of the Hampshire Basin (after King, 1991)

increased. The *lag* deposits are overlain by much finer-grained sediments that coarsen towards the top of each cycle: this coarsening-upwards phase records progression of the shoreline back to the east, towards London, during a subsequent period of sea-level stability. King informally named the divisions these cycles defined A, B, C, D and E.

These lithological changes are most clearly seen in sediments originally deposited near the ancient shoreline, where coarser sediments are most readily deposited, and are now preserved only in the Hampshire Basin, the westernmost remnant of the 'London Clay Sea'. Plint (1983, 1988) attempted to link the changes in sea level that such cycles recorded there to worldwide rises and falls in sea level, and this prompted King (1991) to reinterpret his divisions A to E as probably being *depositional sequences* in the sense invoked by Mitchum *et al.* (1977). A depositional sequence is a stratigraphic unit of conformable and related strata; its boundaries are defined by unconformities, which record falls in sea level relative to the land. Depositional sequences result from changes in the balance between the supply of sediment to the basin (basically controlled by climate over the land mass to its west) and the space available for its deposition, the latter being governed by relative changes in sea level consequent to regional subsidence. It is a collection of such sequences that gives the sediments of the LCF their vertical and lateral variability, and it is now appreciated that King's cycles (divisions A to E) are shorter-term cycles superimposed on the longer-term transgressive/regressive event that was responsible for the Thames Group as a whole. The Thames Group, of which the LCF constitutes much the greater part, is linked to subsidence on a regional scale following the onset of sea-floor spreading between Greenland and Scotland (Knox, 1996b).

Using a sequence stratigraphic approach, King (1991) was able to identify and successfully trace his divisions across the Hampshire Basin. He subsequently subdivided his earlier divisions A1 to E3 into A1a and b, A2, A3a, b and c, B1a and b, B2a to e, and so on, to reflect the smaller *parasequences* of which they themselves are composed. To date, only King's earlier subdivision of the LCF has been applied in London (King, 1981), but a higher-resolution scheme is being determined for use within this region based on King's (1991) work (Mannion, 2007).

Unconformities can be seen at the erosional bases of the Portsmouth, Whitecliff, Otterbourne and Hoe Members in upper parts of the LCF in the Hampshire Basin. These unconformities bound depositional sequences and pass laterally, basinward, into *correlative conformities*; such surfaces become progressively harder to detect as they are traced eastwards into the basin. Fortunately, accumulations of grains of glauconite and phosphate, and pebbles of flint (i.e. the lag deposits of a transgression), often occur at these levels and are used to detect such conformities. Influxes of new fossil species can also be used to detect these junctions as they typically first enter an area during a transgression. Within each depositional sequence, periods of slow sediment deposition can be marked by an increase in the percentage of fossils per unit volume of sediment, to form a *condensed section*. These form laterally extensive surfaces that can further aid subdivision of the succession and help improve regional correlations.

BIOSTRATIGRPAHY OF ST JAMES'S PARK BOREHOLE 5

The first significant studies of benthonic foraminifera in the LCF were conducted by Sherborn & Chapman (1886, 1889) on samples collected at Piccadilly. Their specimens, and those of Bowen (1954) and of Murray & Wright (1974), together with Thanetian specimens collected by Haynes (1956, 1958a, 1958b, 1958c), are held at the Natural History Museum, London, and have been examined by the second author. In addition, the published works that have been consulted to aid identification of the London faunas are Williams (1970), Murray *et al.* (1989), and King (1991), and those illustrating benthonic foraminiferal assemblages from the Eocene of the Netherlands and Belgium: Ten Dam (1944), Kaasschieter (1961), and Willems (1980). The classification scheme used follows that of Loeblich & Tappan (1964, revised 1988 and 1992).

The foraminiferal biostratigraphy of the LCF at St James's Park was studied to resolve the stratigraphic divisions at that location, following studies into unexpected volume losses associated with tunnelling of the Jubilee Line at that site (Standing & Burland 2006). St James's Park Borehole 5 was 40 m deep; twenty 100 g samples, taken at approximately 1·5 m intervals through the LCF section of the borehole, were studied. The samples were processed and their foraminiferal content examined following procedures similar to those outlined in Haynes (1981, pp. 12–15). The results are shown as a 'distribution chart' that records the number of each genera recovered from each sample (Table 1). The first appearances of many of the individual genera occur at similar levels within the LCF across the London and Hampshire Basins. The depths given in the following record are all measured in metres relative to Ordnance Datum (mOD) (Fig. 6).

Division A2 (King, 1981): Table 1

This is the lowest horizon penetrated by Borehole 5. Sample 1 (−33·10 mOD) was taken from interbedded silty clays and silty sandy clays, commonly containing particles of lignite. Calcareous foraminifera are absent from this sample, and this is characteristic of division A2, which has been decalcified throughout its extent (King, 1981, 1991). Standing & Burland (2006) noted that profiles of water content at St James's Park broadly reflected changes in lithology. Much scatter in water content was measured within division A2, reflecting the variable nature of the grain sizes at this stratigraphic level (Fig. 6). Here lithology, variation in water content, and biostratigraphy all agree with the known characters of division A2.

Division A3 (King, 1981): Table 1

The boundary separating division A2 from A3 is one of the easiest to identify in the succession using lithology alone: there is a marked contrast between the interbedded sediments below the boundary and the homogeneous silty clays above. The boundary is highlighted by a similarly striking change biostratigraphically: in the Hampshire Basin the base of division A3 is marked by an influx of calcareous benthonic foraminifera associated with the first occurrences of several new species of agglutinated foraminifera (King, 1991). This event is also seen in Borehole 5 (Table 1). Samples 2 to 5 (−31·60 mOD, −30·10 mOD, −28·60 mOD and −27·10 mOD) came from homogeneous silty clays containing occasional grey claystones (typical 'London Clay'). Fragments of *Ammodiscus* and *Miliammina* were found in Sample 2 (−31·60 mOD); King (1991) records the first appearances of these agglutinated foraminifera in the lower parts of division A3 in the Hampshire Basin. There follows an influx of calcareous benthonic foraminifera in Sample 3 (−30·10 mOD), which, together with the increase in abundance and diversity of the assemblage seen in Sample 4 (−28·60 mOD), confirms this level is the equivalent of division A3 in the Hampshire Basin (King, 1991). The presence of some planktonic foraminifera in Sample 5 (−27·10 mOD) may be explained by connection of the

Table 1. Distribution of foraminiferal genera from St James's Park Borehole 5 (based on 1535 specimens identified from 20 samples; not all genera shown here)

Stratigraphical assignation	Sample	Sample depth below ground level: m	Ammodiscus fragments	Miliammina	Cibicidoides	Anomalinoides	Alabamina	Spiroplectinella	Quinqueloculina	Lenticulina	Pulsiphonina	Planktonic foraminifera	Lagena	Pullenia	Spiroplectammina	Turrilina	Bathysiphon	? Haplophragmoides	Praeglobobulimina	Eponides	Gaudryinopsis	Clavulina fragments	Nodosaria	Laevidentalina & Svenia	Percultazonaria	Euuvigerina	Cibicides	? Nodogenerina	Gyroidinoides	Glomospirella
Div B	20	7·5	2		30	4		11		5	3	1	1	2			1			6					2		6			1
	19	9	6		16	2		20			6	1					1			2							1			
	18	11·4	7		14	5	7	5		3	7			1				1		16				2	2		2		5	
	17	12·9	2		35	3	1	7		23	13			3					1	12				1	3	1	3			
	16	14·4	5		24			3		7	6	1		1					1	12										
	15	15·9	1		15			11	2	5	7	2	1	1		1	2			4				2	2			1		
	14	18·9	2		16	6	7	9		1	5			1					25	8					1					
	13	20·4			45	20	22	79	1	19	161	8	1	13			1		11	11	1	4	1	8	5	3	4			
	12	21·9				1					1							1	2	1										
	11	23·4	6		1						1							3												
	10	24·9	5				1				1						1	1												
	9	26·4	3		3					1																				
Div A3	8	28	30		33	44		59		2		1	1	8	16	2														
	7	29·5			5	1		1			1	7		1																
	6	31	18		11	8		7	8	2	2	3	1																	
	5	32·5			17	24				3		2																		
	4	34			6	41		14	1	2	16																			
	3	35·5	3		17	13	1																							
	2	37	6	1																										
Div A2	1	38·5																												

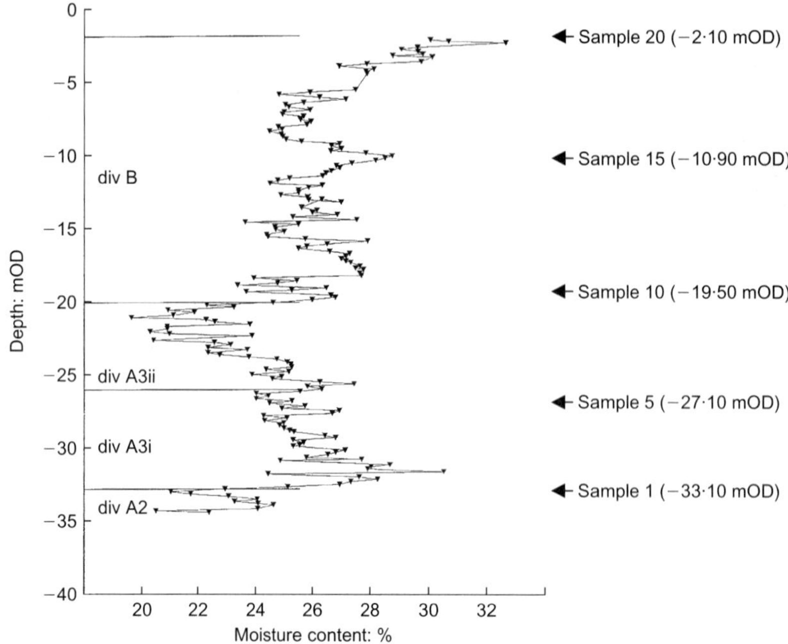

Fig. 6. Moisture content against depth for St James's Park Borehole 5 (after Standing & Burland, 2006). Depths given in metres below Ordnance Datum (mOD). Ground level = 5·40 mOD. Top of London Clay = −1·80 mOD

London Clay Sea to global oceanic circulation, hinting that water depths were increasing (see Wright, 1972). The lithological consequence of this is an increase in clay content of the sediment, which is reflected in the higher moisture content of the beds at this depth (Fig. 6). Samples 6, 7 and 8 (−25·60 mOD, −24·10 mOD and −22·60 mOD) were from silty clays with thin partings of very fine sand. There is a marked increase in fossil abundance and diversity in Sample 8 (−22·60 mOD); the assemblage from this sample includes specimens of *Spiroplectammina*, a foraminiferid that is restricted solely to the upper parts of division A3 in the Hampshire Basin (King, 1991).

Standing & Burland (2006) determined the boundary between divisions A3 and B to be at −20·00 mOD in Borehole 5, at the junction between underlying sandy clayey silts and overlying homogeneous silty clays. King (1991) found that a thin basal bed of glauconitic sandy clayey silt marks the boundary in the Hampshire Basin. Sample 9 (−21·00 mOD) contained abundant glauconite and quartz grains of sand size, indicating that the base of division B is probably 1 m or so below −20·00 mOD. Standing & Burland (2006) record an increase in water content above the division A3/B boundary (Fig. 6). Sample 9 (−21·00 mOD) plots on the borderline between the CH zone (clays of high plasticity) and the CI zone (clays of intermediate plasticity). The samples above 9 all plot within the CV zone (clays of very high plasticity).

Division B (King, 1981): Table 1

Sample 9 (−21·00 mOD) is from just above the base of division B. Samples 10, 11 and 12 (−19·50 mOD, −18·00 mOD and −16·50 mOD) contain sparse assemblages, but some of the foraminifera present are typical of division B in the Hampshire Basin, although none of these are exclusive markers. Planktonic foraminifera are absent from Samples 9 to 12 but reappear in Sample 13 (−15·00 mOD), where over 500 foraminiferal specimens were counted—over a third of the total number recorded in all 20 samples. This condensed section clearly records a significant flooding surface (perhaps the *maximum flooding surface* for the whole of the Thames Group), which is difficult to observe by

lithology alone as samples 10 to 15 all come from silty clays rendered homogeneous by bioturbation. The assemblage in Sample 13 is similar to that of middle and upper parts of division B in the Hampshire Basin. Sample 17 (−7·50 mOD) also contains a higher than average number of specimens, and probably represents another, rather less significant marine flooding surface (possibly a parasequence boundary) within the upper parts of division B.

Division C (King, 1981)

The boundary between divisions B and C lies in the passage of underlying silty clays containing weak silt and sand partings, at the top of division B, into overlying homogeneous silty clays at the bottom of division C (King, 1981). The junction is not easy to determine by lithological means alone in the London area: King (1981) describes the first appearance of the *Osangularia* as an excellent foraminiferid marker for this junction. It has not been observed in Borehole 5, so division C is probably absent at this location. Standing & Burland (1999, 2006), who considered lithologies only, also considered division C to be absent at St James's Park.

PROSPECTS

The biostratigraphic correlations described here have been established by the second author elsewhere within and around London, including sites at Staines, Heathrow, Stanmore Common, Portobello Dock, St James's Park, Crystal Palace and Whitechapel. Much geomechanical data exist for individual layers at different locations but have yet to be connected with confidence; biostratigraphic correlation will enable this to be achieved. The value of such a tool can be appreciated in Fig. 6: if water content alone had to be used to define position within the succession, a very large range of possible positions could be postulated for a given range of moisture contents taken from 1 m of borehole core. Biostratigraphy defines the position unambiguously. When this research for London is complete, it is likely that Borehole 5 could be divided into at least six units (Table 1).

To assist this work, an atlas of results is being compiled

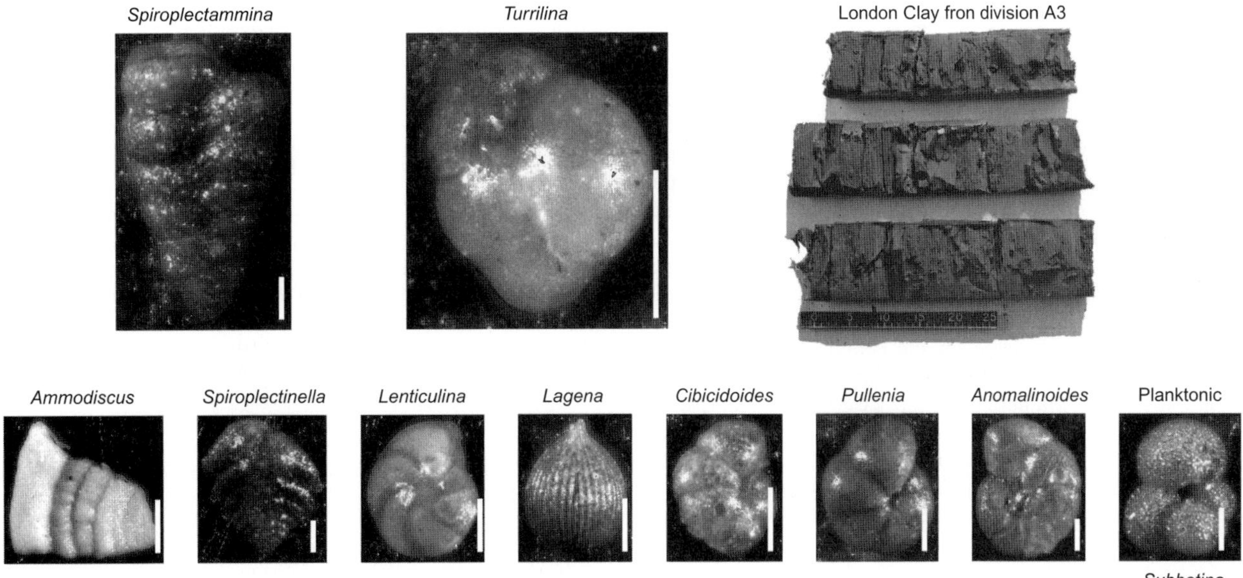

Fig. 7. Example of page contents from the stratigraphical atlas for the London Clay in London (Mannion, 2007). The highlighted genera, *Spiroplectammina* and *Turrilina*, first appear −22·60 mOD at St James's Park (division A3). Some genera ranging up from lower depths are also shown (see Table 1). Scale bars = 100 μm

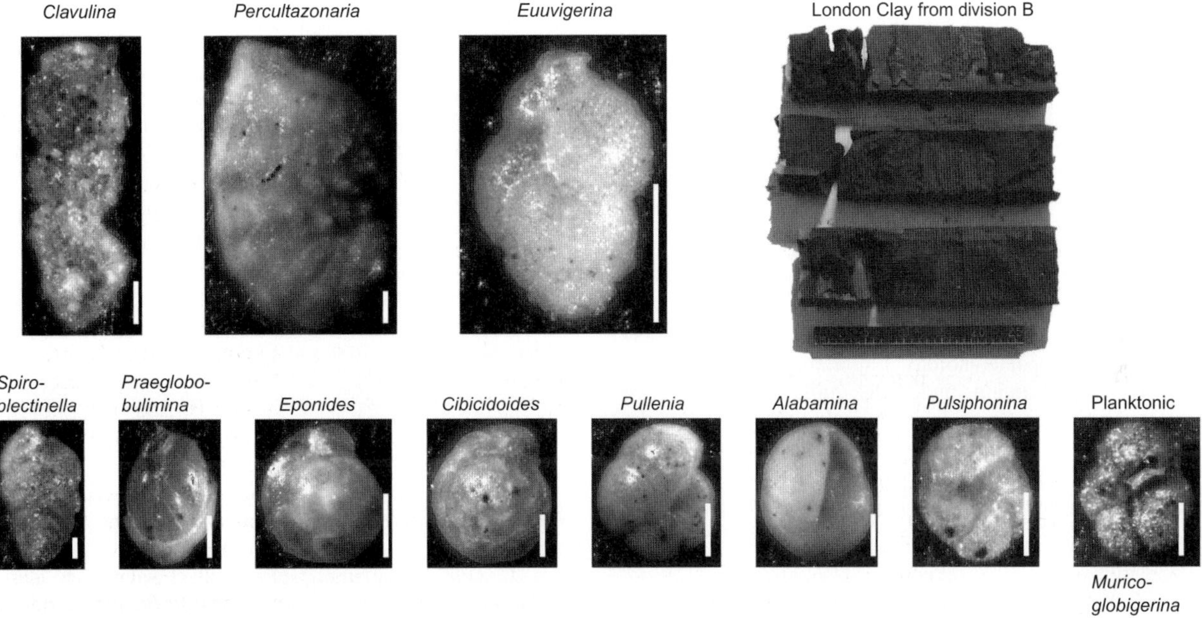

Fig. 8. The highlighted genera first appear −15·00 mOD at St James's Park (division B). Elements of the assemblage range up from division A3 but some, e.g. *Cibicidoides*, are represented by different species at this level. Scale bars = 100 μm

to provide a convenient, quick and inexpensive means of accurately determining stratigraphic position within the succession at any one site in London (Figs 7 and 8). The atlas records the stratigraphic occurrences of all the individual species at each of the sections studied in London. A collection of reference specimens available to assist future workers with their identifications is also being compiled.

Not every location within the LCF will be amenable to such studies. Coarse horizons tend to be poor in well-preserved microfaunal remains, and post-depositional dissolution of the calcareous foraminifera and ostracods, by acidic groundwaters, has occurred in the lower part of division A, and at levels where the LCF is coarser grained, and when it is above the water table. It remains uncertain as to how successful biostratigraphy will be for identifying LCF trans-

ported by solifluxion, but it holds the potential for resolving such issues where lithology and fabric may have been disturbed either by transport or by sampling.

Likewise it has the potential to reveal a history of loading and unloading that may accompany tectonics. Unexpected tectonic surfaces have been found in the London area, as recently described at Prospect Park, just to the north of London Heathrow Airport, by Chandler *et al.* (1998); biostratigraphy can identify zones that have been either subtracted or duplicated within the succession by tectonic activity.

CONCLUSIONS
(*a*) The division of monotonous sequences of fine-grained deposits, especially bioturbated clays, can be readily

achieved using biostratigraphy. It is quicker and more specific than mechanical alternatives such as the measurement of plasticity indices and water contents (as seen from Fig. 6), and provides an unequivocal identification of stratigraphic level.

(*b*) The principles applied for this exercise in the LCF can be used in any deposit where adequate faunal remains are available. The creation of data banks of faunal assemblages for formations of particular geotechnical concern is to be encouraged, as it is by these means that the wealth of case history evidence, geotechnical testing and in situ observation acquired for a formation are made fully available for the benefit of specific sites.

(*c*) These methods have permitted the wealth of biostratigraphic data obtained from the LCF in the Hampshire Basin (King, 1991) to be translated to London so that horizons of similar age can be identified across the capital and its suburbs. This enables geomechanical data obtained from one site to be used with greater confidence at another.

(*d*) Once established, the zones for the LCF can be used to reveal the absence of horizons, from either erosion or tectonics, and hence the presence of a history of loading and unloading that may be influential in defining the magnitude of in situ stresses but not be visible from lithology alone.

ACKNOWLEDGEMENTS

The research described in this paper is funded by the Engineering and Physical Sciences Research Council, and the authors are grateful for their support. The authors also acknowledge with gratitude the advice and help given by Dr Chris King, particularly with the lithological logging of the London Clay Formation and identification of its microfossil content. Samples of borehole cores have been provided by the British Geological Survey, and London Underground Limited, who have also kindly permitted the results obtained from them to be published. Processing of material was undertaken in the Micropalaeontology Laboratory at the Natural History Museum, London, under the guidance of Dr Clive Jones, Dr Andy Henderson and Dr Jackie Skipper. Finally, the authors thank their colleagues within the Civil and Environmental Engineering Department, Imperial College, London, for the supervision and assistance they have provided, especially Dr Mathew Coop, Dr Jamie Standing, Dr Liana Gasparre (now with GCG) and Dr Satoshi Nishimura.

REFERENCES

Aubry, M. P. (1985). Northwestern European Paleogene magnetostratigraphy, biostratigraphy and paleogeography: calcareous nannofossil evidence. *Geology* **13**, No. 3, 198–202.

Aubry, M. P. (1986). Palaeogene calcareous nannoplankton biostratigraphy of northwestern Europe. *Palaeogeogr. Palaeoclimatol. Palaeoecol.* **55**, Nos 2–4, 267–334.

Aubry, M. P., Hailwood, E. A. & Townsend, H. A. (1986). Magnetic and calcareous-nannofossil stratigraphy of the lower Palaeogene formations of the Hampshire and London Basins. *J. Geol. Soc. London* **143**, No. 5, 729–735.

Bowen, R. N. C. (1954). Foraminifera from the London Clay. *Proc. Geol. Assoc.* **65**, No. 2, 125–174.

Bujak, J. P., Downie, C., Eaton, G. L. & Williams, G. L. (1980). Dinoflagellate cysts and acritarchs from the Eocene of Southern England. *Spec. Pap. Palaeontol.* Special paper 24, 1–100.

Chandler, R. J., Willis, M. R., Hamilton, P. S. & Andreou, I. (1998). Tectonic shear zones in the London Clay Formation. *Géotechnique* **48**, No. 2, 257–270.

Cooper, J. (1976). British Tertiary stratigraphical and rock terms. *Tertiary Research Special Paper* **1**, 1–37.

Curry, D. (1965). The Palaeogene beds of south-east England. *Proc. Geol. Assoc.* **76**, No. 2, 151–174.

Curry, D. (1966). Problems of correlation in the Anglo-Paris-Belgian Basin. *Proc. Geol. Assoc.* **77**, No. 4, 437–467.

Curry, D., Adams, C. G., Boulter, M. C., Dilley, F. C., Eames, F. E., Funnell, B. M. & Wells, M. K. (1978). *A correlation of Tertiary rocks in the British Isles*, Special Report No. 12. London: Geological Society.

Dam, A., ten. (1944). Die stratigraphische Gliederdung des niederländischen Paläozäns und Eozäns nach Foraminiferen. *Mededelingen Geologische Stichting, Serie C* **3**, 1–142.

Davis, A. G. & Elliott, G. F. (1957). The palaeogeography of the London Clay Sea. *Proc. Geol. Assoc.* **68**, No. 4, 255–277.

Dewey, H. (1912). Report of an excursion to Claygate and Oxshott, Surrey. *Proc. Geol. Assoc.* **23**, No. 4, 237–242.

Ellison, R. A. & King, C. (2004). Palaeogene-Eocene. In *Geology of London. Memoir of the British Geological Survey, Sheets 256 (North London), 257 (Romford), 270 (South London) and 271 (Dartford)* (ed. R. A. Ellison), pp. 44–51. London: British Geological Survey.

Ellison, R. A. & Zalasiewicz, J. A. (1996). Palaeogene and Neogene. In *London and the Thames Valley* (ed. M. G. Sumbler), 4th edn, pp. 92–109. London: HMSO (British Geological Survey).

Ellison, R. A., Jolley, D. W., King, C. & Knox, R. W. O'B. (1994). A revision of the lithostratigraphic classification of the early Palaeogene strata in the London Basin and East Anglia. *Proc. Geol. Assoc.* **105**, No. 3, 187–197.

Emery, D. (1996). Historical perspective. In *Sequence stratigraphy* (eds D. Emery and K. Myers). London: Blackwell, 3–7.

Hawkins, H. L. (1954). The Eocene succession in the Eastern Part of the Enborne Valley, on the borders of Berkshire and Hampshire. *Q. J. Geol. Soc. London* **110**, No. 4, 409–430.

Haynes, J. (1956). Certain smaller British Paleocene Foraminifera Part 1. Nonionidae, Chilostomellidae, Epsitominidae, Discorbidae, Amphisteginidae, Globigerinidae, Globorotaliidae, and Gümbelinidae. *Contributions from the Cushman Foundation for Foraminiferal Research* **7**, No. 3, 79–101.

Haynes, J. (1958a). Certain smaller British Paleocene Foraminifera Part 3. Polymorphinidae. *Contributions from the Cushman Foundation for Foraminiferal Research* **9**, No. 1, 4–16.

Haynes, J. (1958b). Certain smaller British Paleocene Foraminifera Part IV. Arenacea, Lagenidea, Buliminidae, and Chilostemellidae. *Contributions from the Cushman Foundation for Foraminiferal Research* **9**, No. 3, 58–77.

Haynes, J. (1958c). Certain smaller British Paleocene Foraminifera Part V. Distribution. *Contributions from the Cushman Foundation for Foraminiferal Research* **9**, No. 4, 83–92.

Haynes, J. R. (1981). *Foraminifera*. London: Macmillan.

Kaasschieter, J. P. H. (1961). Foraminifera of the Eocene of Belgium. *Institut Royal des Sciences Naturelles de Belgique, Mémoire* **147**.

King, C. (1981). The stratigraphy of the London Clay and associated deposits. *Tertiary Research Special Paper* **6**, 1–158.

King, C. (1984). The stratigraphy of the London Clay Formation and Virginia Water Formation in the coastal sections of the Isle of Sheppey (Kent, England). *Tertiary Research* **5**, No. 3, 121–160.

King, C. (1991). *Stratigraphy of the London Clay (Early Eocene) in the Hampshire Basin*. PhD thesis, Kingston University.

Knox, R. W. O'B. (1996a). Correlation of the early Palaeogene in northwest Europe: an overview. In *Correlation of the Early Paleogene in Northwest Europe* (eds R. W. O'B. Knox, R. M. Corfield and R. E. Dunay), Special Publication No. 101, pp. 1–11. London: Geological Society.

Knox, R. W. O'B. (1996b). Tectonic controls on sequence development in the Palaeocene and earliest Eocene of southeast England: implications for North Sea stratigraphy. In *Sequence stratigraphy in British geology* (eds S. P. Hesselbo and D. N. Parkinson), Special Publication No. 103, pp. 209–230. London: Geological Society.

Loeblich, A. R. & Tappan, H. (1964). Sarcodina, chiefly 'Thecamoebians' and Foraminiferida. In *Treatise on Invertebrate Paleontology* (ed. R. C. Moore), Part C, Protista 2, vols. 1 and 2. Lawrence, Kansas: University of Kansas Press (Geological Society of America).

Mannion, W. G. (2007). *Stratigraphical applications relevant to geotechnical interpretations of the London Clay Formation*. PhD thesis, Imperial College, London.

Mitchum, R. M. Jr, Vail, P. R. & Thompson, S., III (1977). Seismic stratigraphy and global changes in sea level, part 2: The depositional sequence as a basic unit for stratigraphic analysis. In *Seismic stratigraphy: Applications to hydrocarbon exploration* (ed. C. E. Payton), pp. 53–62. Tulsa, Oklahoma: Memoir of the American Association of Petroleum Geologists.

Murray, J. W. & Wright, C. A. (1974). Palaeogene Foraminiferida and palaeoecology, Hampshire and Paris Basins and the English Channel. *Spec. Pap. Palaeontol.* **14**, 1–129.

Murray, J. W., Curry, D., Haynes, J. R. & King, C. (1989). Palaeogene, In *Stratigraphical atlas of fossil Foraminifera* (eds D. G. Jenkins and J. W. Murray), 2nd edn. Chichester: Ellis Horwood, 490–536.

Neal, J. E. (1996). A summary of Paleogene sequence stratigraphy in northwest Europe and the North Sea. In *Correlation of the Early Paleogene in Northwest Europe* (eds R. W. O'B. Knox, R. M. Corfield and R. E. Dunay), Special Publication No. 101, pp. 15–42. London: Geological Society.

Plint, A. G. (1983). Discussion on Eocene sedimentation and tectonics in the Hampshire Basin. *J. Geol. Soc. London* **139**, No. 3, 249–254.

Plint, A. G. (1988). Global eustacy and the Eocene sequence in the Hampshire Basin, England. *Basin Research* **1**, No. 1, 11–22.

Prestwich, J. (1846). On the Tertiary or Supracretaceous Formations of the Isle of Wight, as exhibited in the sections at Alum Bay and Whitecliff Bay. *Q. J. Geol. Soc. London* **2**, 223–259.

Prestwich, J. (1847). On the probable age of the London Clay, and its relations to the Hampshire and Paris Tertiary Systems. *Q. J. Geol. Soc. London* **3**, 354–377.

Prestwich, J. (1850). On the structure of the strata between the London Clay and the Chalk in the London and Hampshire Tertiary systems, Part I, The London Clay Basement Bed. *Q. J. Geol. Soc. London* **6**, 252–281.

Sherborn, C. D. & Chapman, F. (1886). On some microzoa from the London Clay exposed in the drainage works, Piccadilly, London, 1885. *J. R. Microsc. Soc.* **6**, 737–763.

Sherborn, C. D. & Chapman, F. (1889). Additional notes on the Foraminifera of the London Clay exposed in drainage works, Piccadilly, London, in 1885. *J. R. Microsc. Soc.*, 483–488.

Stamp, L. D. (1921). On cycles of sedimentation in the Eocene strata of the Anglo-Franco-Belgian Basin. *Geological Magazine* **58**, 108–114, 146–157, 194–200.

Standing, J. R. & Burland, J. B. (2006). Unexpected tunnelling volume losses in the Westminster area, London. *Géotechnique* **56**, No. 1, 11–26.

Stinton, F. C. (1975). *Fish otoliths from the English Eocene*. London: Palaeontographical Society, London.

Van Wagoner, J. C., Mitchum, R. M. Campion, K. M. & Rahmanian, V. D. (1990). *Siliciclastic sequence stratigraphy in well logs, cores and outcrops: concepts for high resolution correlation of time and facies*, Methods in Exploration Series No. 7. Tulsa, Oklahoma: American Association of Petroleum Geologists.

Wetherell, N. T. (1836). Observations on some of the fossils of the London Clay, and in particular those organic remains which have recently been discovered in the tunnel for the London and Birmingham Railroad. *London Edinburgh Phil. Mag.* **9**, 462–469.

Whitaker, W. (1866). On the 'Lower London Tertiaries' of Kent. *Q. J. Geol. Soc. London* **22**, 404–435.

Willems, W. (1980). *De foraminiferen van de Ieper-formatie (Onder-Eoceen) in het zuidelijk Nordseebecken*. PhD thesis, Rijksuniversiteit Gent.

Williams, G. M. (1970). *The stratigraphy and micropalaeontology of the London Clay*. PhD thesis, Imperial College, London.

Wooldridge, S. W. (1926). The structural evolution of the London Basin. *Proc. Geol. Assoc.* **37**, No. 2, 162–196.

Wrigley, A. (1924). Faunal divisions of the London Clay, illustrated by some exposures near London. *Proc. Geol. Assoc.* **35**, No. 3, 245–259.

Wrigley, A. (1940). The faunal succession in the London Clay, illustrated in some new exposures near London. *Proc. Geol. Assoc.* **51**, No. 3, 230–245.

Wright, C. A. (1972). The recognition of a planktonic foraminiferid datum in the London Clay of the Hampshire Basin. *Proc. Geol. Assoc.* **83**, No. 4, 413–420.

Ziegler, P. A. (1990). *Geological Atlas of Western and Central Europe*, 2nd edn. The Hague: Shell Internationale Petroleum Maatschappij B.V.

Pantelidou, H. & Simpson, B. (2007). *Géotechnique* **57**, No. 1, 101–112

Geotechnical variation of London Clay across central London

H. PANTELIDOU* and B. SIMPSON*

The paper presents classification and stress path test results from various sites across central London, and focuses on interpreting the observed behaviour within the context of the geology and geological history of the Formation. The large majority of the tests have been undertaken by commercial laboratories in the course of design projects. A good correlation is found between the classification tests and the stratification of the London Clay described by King in 1981. Based on the secant stiffness at a shear strain of 0·01%, the stiffness of the clay in triaxial stress path tests can reasonably be normalised in the way suggested by Viggiani & Atkinson or Simpson. However, variation with plasticity is not evident from these test results.

KEYWORDS: clays; geology; laboratory tests; plasticity; site investigation; stiffness

Cet article présente les résultats de tests de classification et de chemins de contrainte obtenus sur divers sites du centre de Londres. Il s'intéresse plus partiellement à l'interprétation du comportement observé, dans le contexte de la géologie et de l'histoire géologique de la formation. Des laboratoires commerciaux ont réalisé la grande majorité des essais dans le cadre de projets de conception industrielle. On a pu constater une bonne corrélation entre les tests de classification et la stratification de l'argile de Londres (London Clay) décrite par King en 1981. En se basant sur la rigidité sécante pour une déformation en cisaillement de 0,01 %, la rigidité de l'argile pour des essais à cheminements de contraintes triaxiales peut être raisonnablement normalisée suivant le moyen suggéré par Viggiani et Atkinson ou Simpson. Cependant il est difficile de mettre en évidence une variation avec la plasticité d'après ces résultats.

INTRODUCTION

London Clay is a stiff, overconsolidated marine clay, deposited across the London and Hampshire Basins of south-east England (strictly, the drowned margins of the North Sea basin). Although it is often treated as a uniform material, significant variations in strength, stiffness and consolidation characteristics within the Formation are directly linked to a variable depositional history.

The aim of this work was to collect available data on the geotechnical properties of the Formation, mainly from triaxial stress path testing from the Arup database, review them collectively, and examine whether they can be related to the geology and geological history. Most of the tests presented are good-quality commercial tests, which do not provide the same high resolution as would tests carried out in a research laboratory. All tests were carried out on rotary-cored samples of London Clay. In order to interpret the test results in terms of the variable nature of the Formation, other site investigation (SI) information, and in particular water content and Atterberg limits from the relevant sites, have also been used and are presented.

Similar London Clay data collection and interpretation have been attempted in the past, most recently by Hight *et al.* (2002). The present work aimed to complement previous attempts and take a step further towards a better understanding of the variability of the Formation.

BACKGROUND

The London Clay is a marine formation of Early Eocene age. The Hampshire and London Basins were connected to form a single depositional area at the time of the formation of the London Clay.

The present as well as the original (before erosion) thickness of the London Clay varies throughout both the London and Hampshire Basins. At some locations in the London Basin, the original thickness of the Formation is still present; King (1981) reports thicknesses of 150 m in Essex and the Isle of Sheppey, and 90 m at Reading.

The common reference point for this sedimentary deposit is its base: the material was deposited from the base upwards, with erosion, re-deposition and other geological events continuously changing the top of the Formation. In the following, all information is plotted with reference to the base of the Formation.

Depositional environment

According to King (1981), five sedimentary cycles (Divisions A to E) have been recognised within the London Clay, each recording an initial marine transgression followed by gradual shallowing of the sea. At the boundaries between subdivisions, there are abrupt changes in the coarse-grained content and mineralogy. A typical cycle starts with a bed containing scattered glauconitic grains and, in some cases, rounded flint pebbles. This is followed by a sequence of clays, which become progressively more silty and sandy upwards.

The most comprehensive work on the London Clay stratigraphy is that by King (1981). Based mainly on the study of the fossils found in the London Clay, he identified variations in the depositional environment. For silty clays in the eastern London Basin the fossil content indicates water depths of 200 m, whereas the content of some of the silty sands in the Hampshire Basin indicates water depths of less than 20 m. This variation in depositional environment resulted in a significant variation in the stratigraphy of the Formation.

King's stratigraphic diagram for central London (Fig. 1) shows the five subdivisions of the London Clay Formation, A–E. The general principle of a coarsening upward sequence in each sedimentary cycle is evident, a result of gradual shallowing of the water in each cycle.

Changes in overburden: overconsolidation

In most parts of the London Basin, substantial erosion has taken place, removing the upper parts of the London Clay,

Manuscript received 5 May 2006; revised manuscript accepted 30 November 2006.
Discussion on this paper closes on 1 July 2007, for further details see p. ii.
* Arup Geotechnics, London, UK.

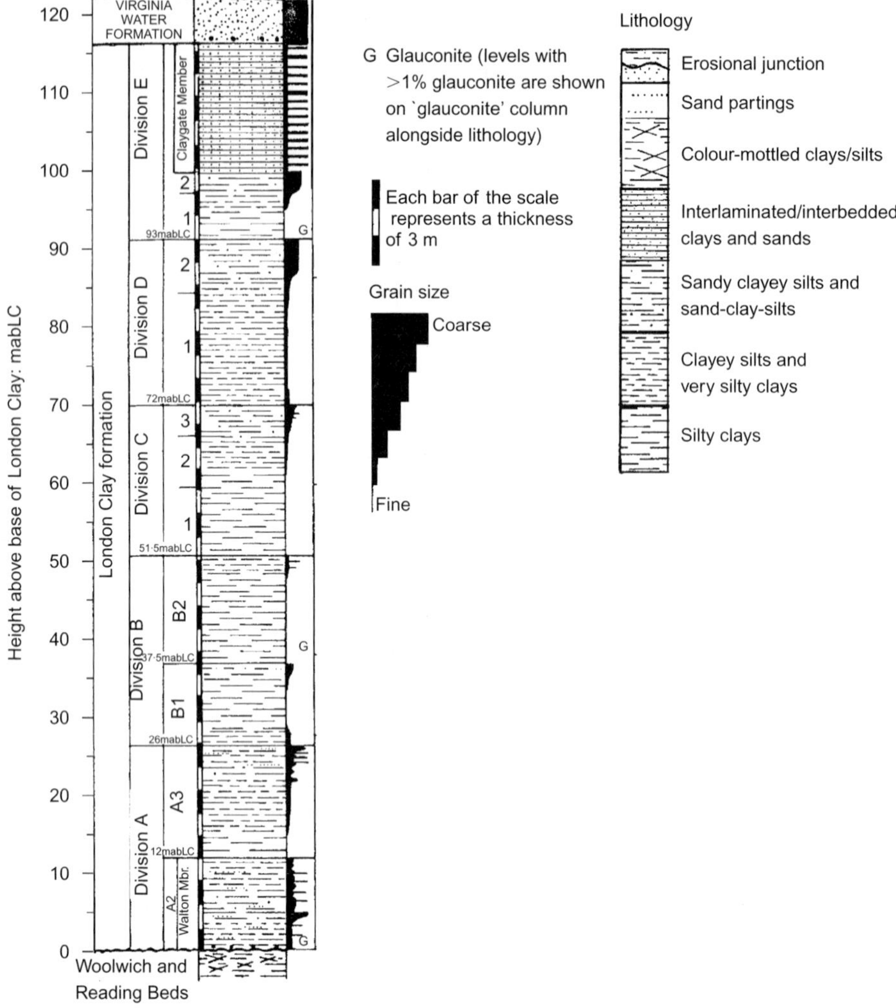

Fig. 1. King's (1981) stratigraphic sequence for central London

and any other Tertiary strata that had been present above it. Within the London area, the full thickness is present only at Hampstead Heath, where the overlying Bagshot Sand is found. A succession close to the maximum thickness is seen south of the Thames at Crystal Palace, where the Claygate Member caps the hill. Subdivisions C, D and E are most often absent in central London. According to King (1981), the total thickness of the London Clay in central London was about 130 m.

Estimating the original thickness of the strata above the London Clay, which principally constitute the Bagshot Formation, is more problematic. It is not possible to be accurate about the original thickness of these strata in central London. King (1981) records a 40 m thickness in the Bagshot and Chertsey area to the south-west of London, but at Ascot, just to the north, the thickness reduces markedly to only 7 m.

The thicknesses of Holocene alluvium and recent fill are not believed to have exceeded their present-day amounts.

Assessment of the thickness of the eroded strata is essential for the calculation of the overconsolidation ratio and hence the estimation of the horizontal stresses, strength and stiffness. Previous estimates based on laboratory testing by various authors are given in Table 1. This includes

Table 1. Previous estimates of erosion above London Clay in the London Basin

Reference	Location	Thickness of Tertiary strata removed by erosion: m
Burland *et al.* (1979)	Central London	170
Skempton & Henkel (1957)	Central London	152–213
Bishop *et al.* (1965)	Ashford Common: 20 km west of London	365–400
Smith (1978)	Regent's Park	190–396
Skempton (1961)	Bradwell, Essex: 80 km north-east of London	150
Henkel (1957)	North London	150–210

Note: The datum levels for the base of the London Clay are not quoted for any of these examples.

Bishop's estimate for Ashford Common, which does not accord well with the geological evidence. It is in an area, near Heathrow, where a large thickness of London Clay is still present, and the base of the London Clay is probably at -80 mOD.

Based on the above, the maximum total thickness of the Tertiary strata eroded has been assumed in this paper to be 170 m, with negligible deposition of Quaternary strata above them. The effect of the uncertainty in overburden on the geotechnical parameters is not examined here; it is expected to have only a minor effect on the geotechnical properties discussed in the paper.

DATA EXAMINED

Triaxial stress path tests and classification tests have been carried out as part of the ground investigation for various sites across London. These are collected here and briefly described in order to make their assessment easier. The site locations at which data have been collected are shown in Fig. 2.

Plasticity

Figure 3 presents the plasticity results for all London Clay specimens on the Casagrande chart. The clay varies from intermediate to very high plasticity. There is a large scatter in the data, without an obvious pattern relating to either stratigraphy or geographic location.

London Clay profile

Comparison with King's profile. In order to put the data within the geological context discussed in the previous section, the London Clay profile in terms of liquid limit (w_L) and natural water content (w) from all sites is plotted in Fig. 4 against height above the base of the Formation, the maximum thickness encountered being just under 60 m. Plasticity (I_p) and liquidity (I_L) indices are also included. Fig. 5 shows w and w_L taken from Fig. 4, but for only the bottom 35 m of the London Clay, from where the majority of the information has been obtained. It is proposed that the liquid limit (w_L) profile is most often adequate for the purposes of stratigraphic differentiation in the Formation. This is used in the following presentation of site information.

The parameters show a pattern generally consistent with King's suggested profile (shown here as a stratigraphic column for comparison); the differences are most clearly seen in w_L. There are marked changes in the profiles at 12 m and 26 m above the base of the London Clay (mabLC), which coincide with the boundaries between subdivisions A2 and A3 and A3 and B1 respectively. The wider scatter in the data from the lowest subdivision A2 (below 12 mabLC), probably indicates the random presence of sandier horizons in this unit across London. The data suggest a repeatable stratigraphic profile across all sites examined, albeit with differences in the magnitude of the coarse-grained content between sites. The actual coarse-grained content was measured in only a small number of specimens; in most cases such information is not available. It is unlikely, however, that the results presented are dominated by pockets of sand within the specimens.

The trend of an upward decrease of the liquid limit and water content also indicates an upward increase of the coarse-grain content within each unit. However, the trend in the A2 subdivision is obscured by the large scatter in the data. These observations are in good agreement with the depositional history of the Formation, described in the previous section: a deep marine environment is responsible for a laterally uniform material deposition, with a sequence of cycles of gradual shallowing of the waters, resulting in a coarsening upwards of the material within each cycle.

Of the above parameters, only the liquid limit and plasticity index are material properties; the natural water content and liquidity index are also state dependent and might be misleading, when comparing sites with different geological history. Hight *et al.* (2002) also presented profiles from different sites across London in a similar manner, although their preferred approach was to compare water contents rather than material properties.

Conductivity plots and alternative means of profiling. The stratigraphic variation of London Clay described here is directly relevant to the engineering characteristics of the Formation. Although the profiling is easily done using simple Atterberg limits, there is a lot of scatter, so a large amount of data is needed for a stratigraphic pattern to emerge; this is most often not available. Alternative means of profiling should therefore be considered. Hight *et al.* (2002) presented profiling using cone penetration tests. These follow well the stratigraphic boundaries. Similarly, profiling based on shear

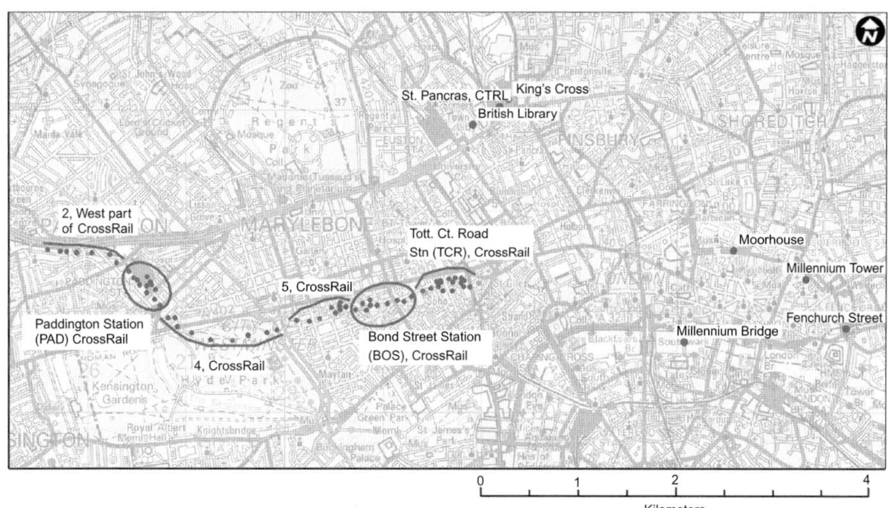

Fig. 2. Locations of data collection sites. The material contained on this map has been based on Ordnance Survey mapping with the permission of the Controller of Her Majesty's Stationery Office. © Crown Copyright reserved. Licence No. 100039628

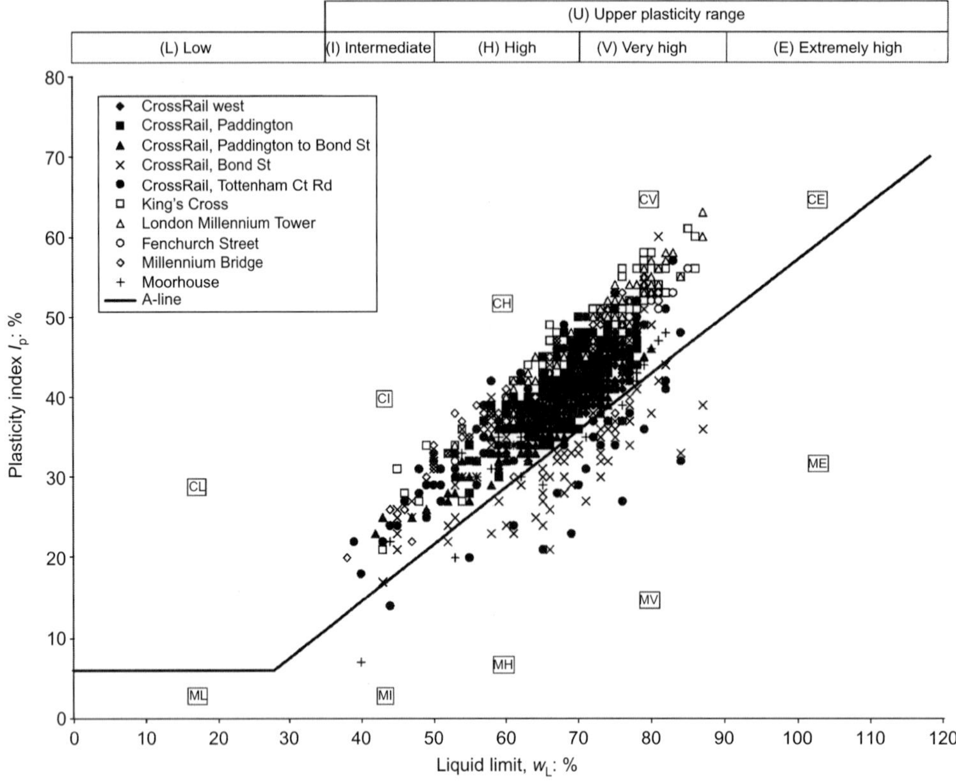

Fig. 3. Plasticity results for all specimens on the Casagrande chart

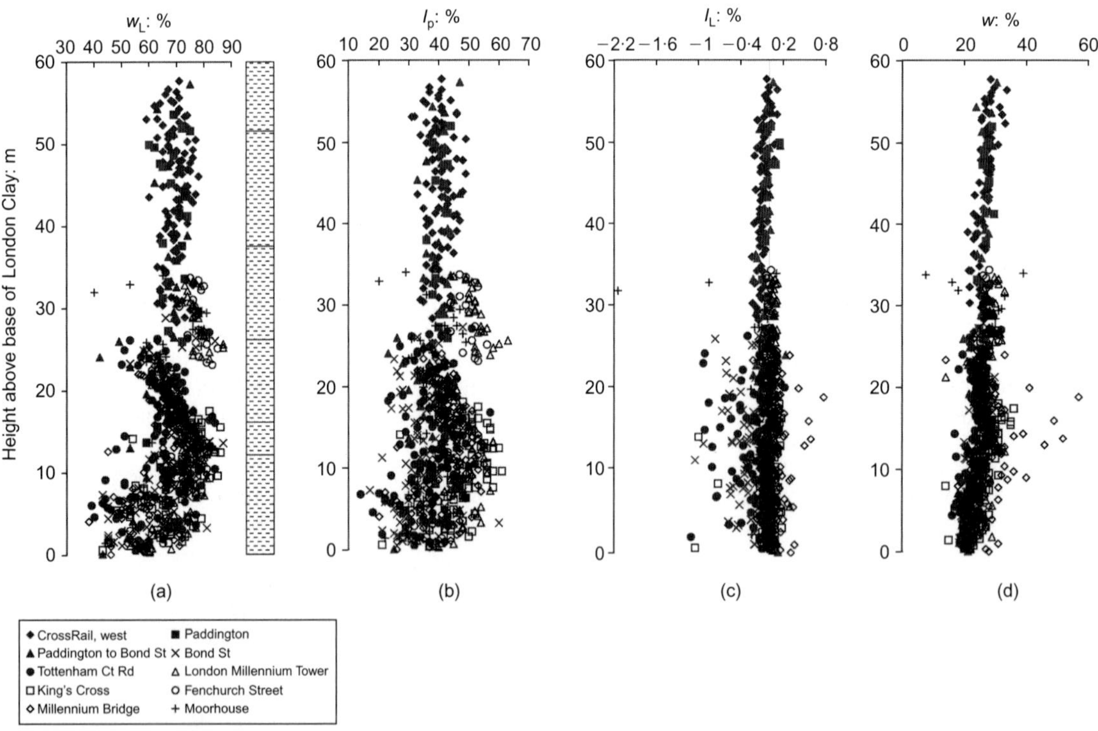

Fig. 4. London Clay profile: (a) liquid limit; (b) plasticity index; (c) liquidity index; (d) natural water content

wave velocity (also presented by Hight *et al.*, 2002) is also promising as a tool.

In Fig. 5, the profiles are also compared with the conductivity logs from geophysical investigations carried out at sites in King's Cross and Moorhouse, a distance of 3 km apart. Conductivity is dependent mainly on grain size and mineralogy, which makes it ideal for identifying variations in the material of the London Clay Formation. The main benefit of the conductivity testing is that it gives a contin-uous profile, so thin layers of differing materials are not missed. The magnitudes recorded are dependent on the particular installation conditions of each borehole, but the repeatability of the shape of the conductivity curves between boreholes appears to confirm that the units' successions and their thicknesses are fairly similar for the two sites. It is possible that the same pattern may be found across the London area, but the number of geophysical profiles of London Clay obtained across London is too limited to allow

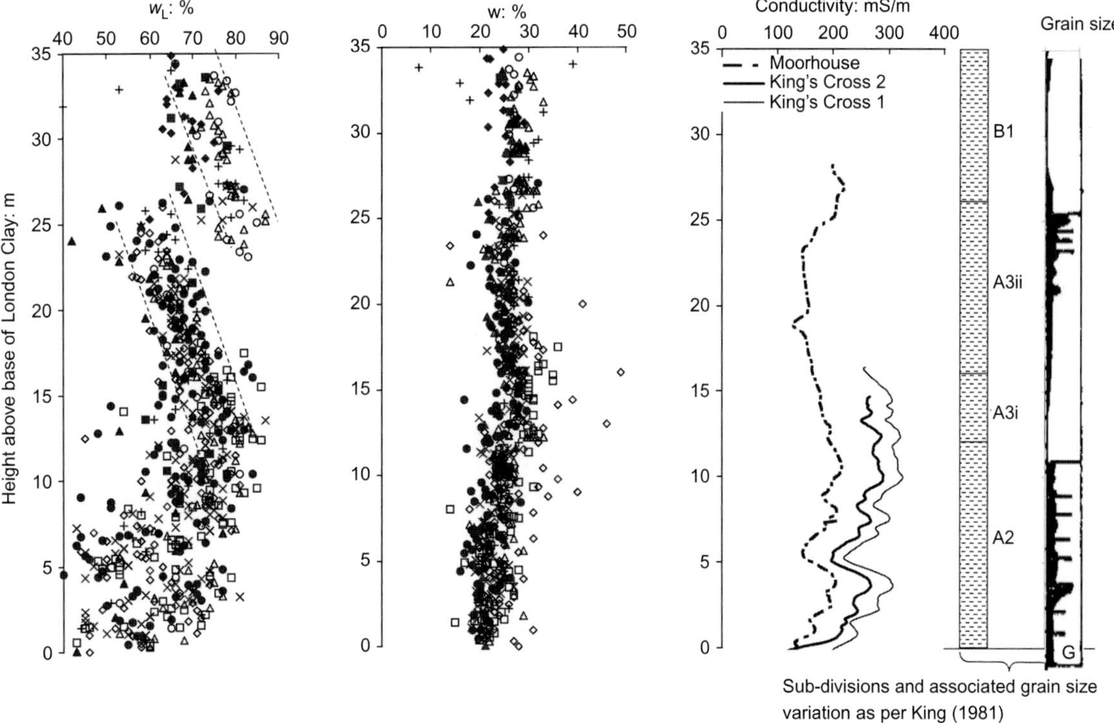

Fig. 5. Values for bottom 35 m of London Clay: (a) liquid limit; (b) natural water content; (c) conductivity

conclusions to be drawn with confidence. The conductivity reflects King's profile to some extent, picking up the distinct sandier band at 5 m above the base, as well as the gradual grain size increase (gradual drop in the conductivity) from the base of the A3 subdivision. It might show the sandier band at 12 to 13 mabLC, though it also seems to show a distinct feature at about 8 mabLC.

Triaxial stress path tests

Triaxial stress path tests carried out in three of the above sites are summarised here: the London Millennium Tower, section 1 of Crossrail, and King's Cross. All tests had axial local strain measurements using Hall-effect transducers. Local radial strain measurements were sometimes also used. All tests were sheared undrained to failure, starting from mean effective stresses similar to the in situ stresses. The minimal sampling disturbance was confirmed by checking the moisture content variation across the radius of each specimen for the King's Cross and Millennium Tower tests. However, for the Crossrail data the sample quality is not known.

For each site, the London Clay profiling in terms of w_L is shown, together with triaxial stress path tests carried out in the Formation. For each set of tests, the results are presented as

(*a*) effective stress paths (mean effective stress p' with deviator stress q)
(*b*) shear modulus normalised with mean effective stress at start of shearing, G/p'_0
(*c*) mobilised angle of shearing resistance ϕ_{mob} with shear (deviator) strain ε_q ($= \frac{2}{3}(\varepsilon_a - \varepsilon_r)$, where ε_a and ε_r are axial and radial strain respectively).

In all the effective stress path plots, the lines corresponding at the critical state lines for $\phi = 22°$ for both compression and extension are shown for comparison.

The main parameter discussed later in this paper is shear modulus, which is presented as secant values, not tangent values. Brief comments on other aspects of the test results are included below.

London Millennium Tower. London Clay is overlain by up to 5 m of Made Ground and Terrace Gravel on this site. The liquid limit profile for the site is shown in Fig. 6, indicating the relative positions of the triaxial stress path test specimens. The profile is compared with the stratigraphic sequence suggested by King.

All triaxial stress path test specimens were obtained from rotary coring using double tube core barrels fitted with a rigid plastic core liner to permit continuous 100 mm core recovery.

Figure 7 shows the results of the triaxial stress path tests from this site. Two distinct undrained shearing stages were specified, separated by a consolidation stage between them. The shearing stages were designed to simulate undrained excavation followed by recovery of pore water pressures before they were sheared to failure. The first shearing started from anisotropic stress conditions, after a series of drained stress path stages that aimed at reproducing the latter part of the assumed geological stress history at the site, with an assumed at-rest earth pressure coefficient K_0 between 1·1 and 1·3.

King's Cross. At the specimen location, London Clay is overlain by Made Ground about 5 m thick. The liquid limit profile for the site is also shown in Fig. 6.

All triaxial stress path test specimens were rotary core sampled and trimmed to 70 mm diameter. Fig. 8 shows the results of the triaxial stress path tests from this site. As with the Millennium Tower testing, two distinct undrained shearing stages were specified, separated by a consolidation stage between them. The first shearing started from isotropic conditions, with a mean effective stress equal to the estimated current in situ mean effective stress, with an estimated K_0 close to unity. In contrast to the Millennium

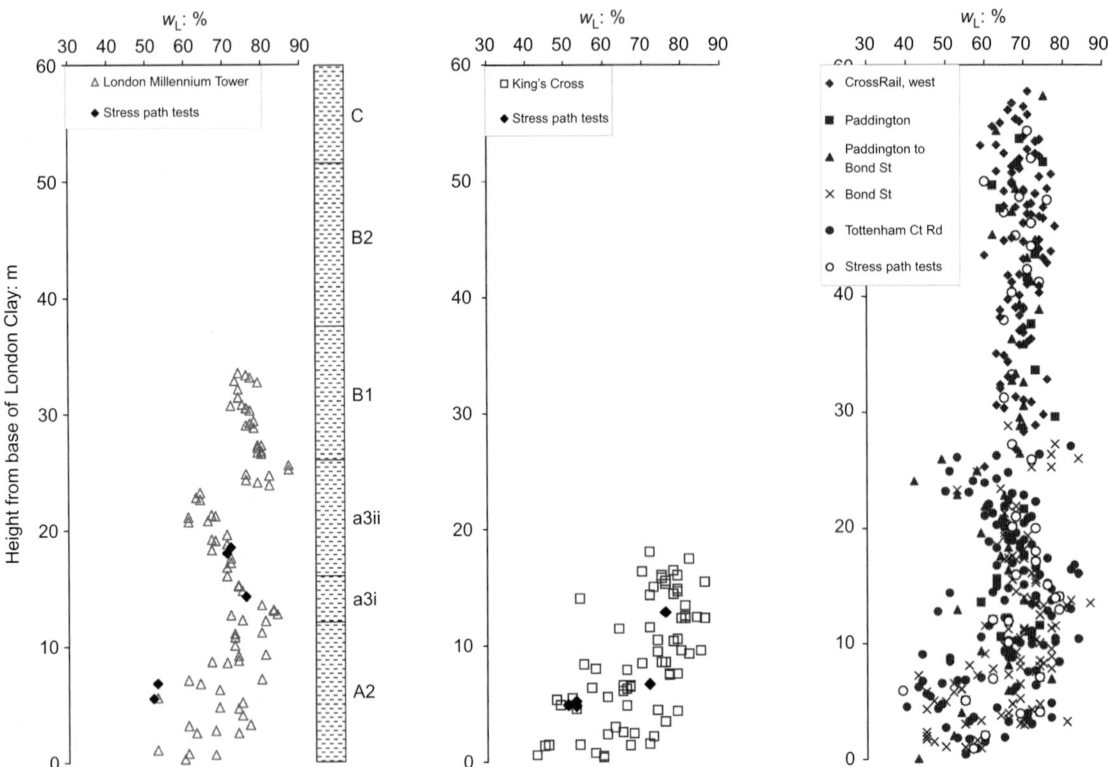

Fig. 6. Liquid limit profiles for the sites where stress path tests were carried out: (a) London Millenium Tower; (b) King's Cross; (c) Crossrail

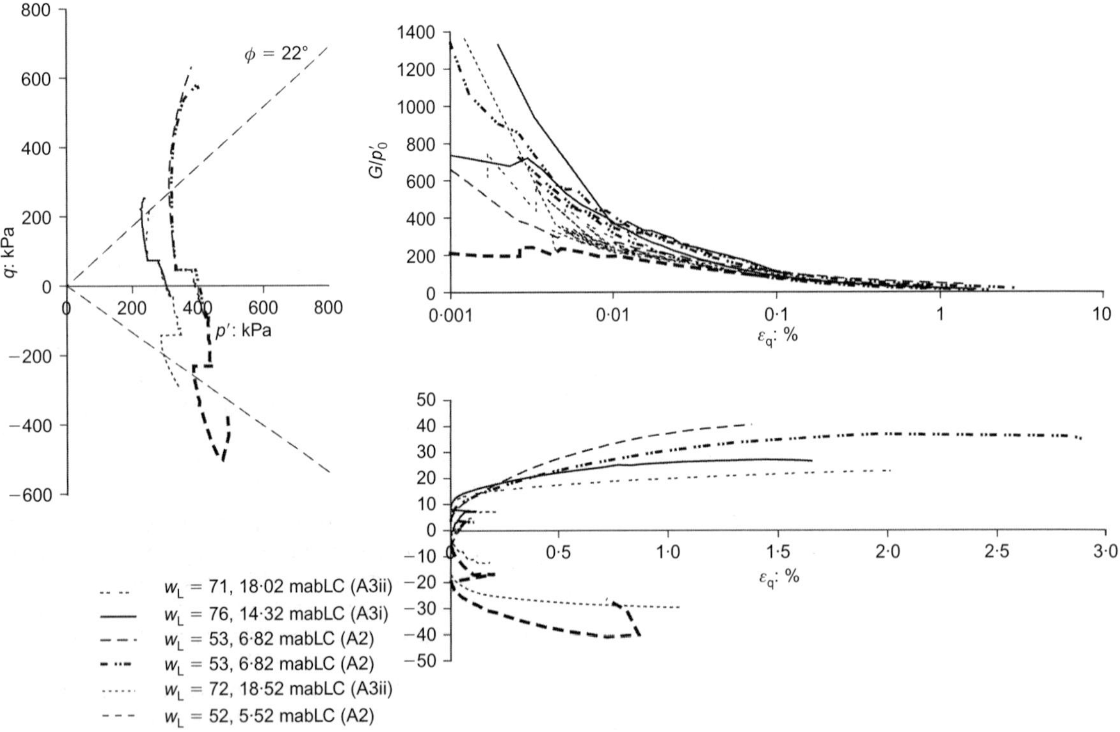

Fig. 7. London Millennium Tower: results of triaxial stress path tests

Tower testing, the latter part of the assumed geological history was not reproduced. Note that the triaxial stress path tests were carried out on specimens from the sandier horizons of the sequence.

CrossRail (west of Charing Cross Road only). A substantial number of tests were carried out as part of this site investigation along the Crossrail route, covering sites from west London (Torquay Street) to Charing Cross Road. The thickness of London Clay changes significantly from west to east, including subdivisions B and C in the west, but only up to A3i in the more central locations. Fig. 6 shows the liquid limit profiles from all CrossRail tests. Fig. 9 presents the triaxial stress path tests carried out as part of the CrossRail site investigation.

All triaxial stress path tests for Crossrail were carried out on high-quality thin-wall samples of London Clay. Un-

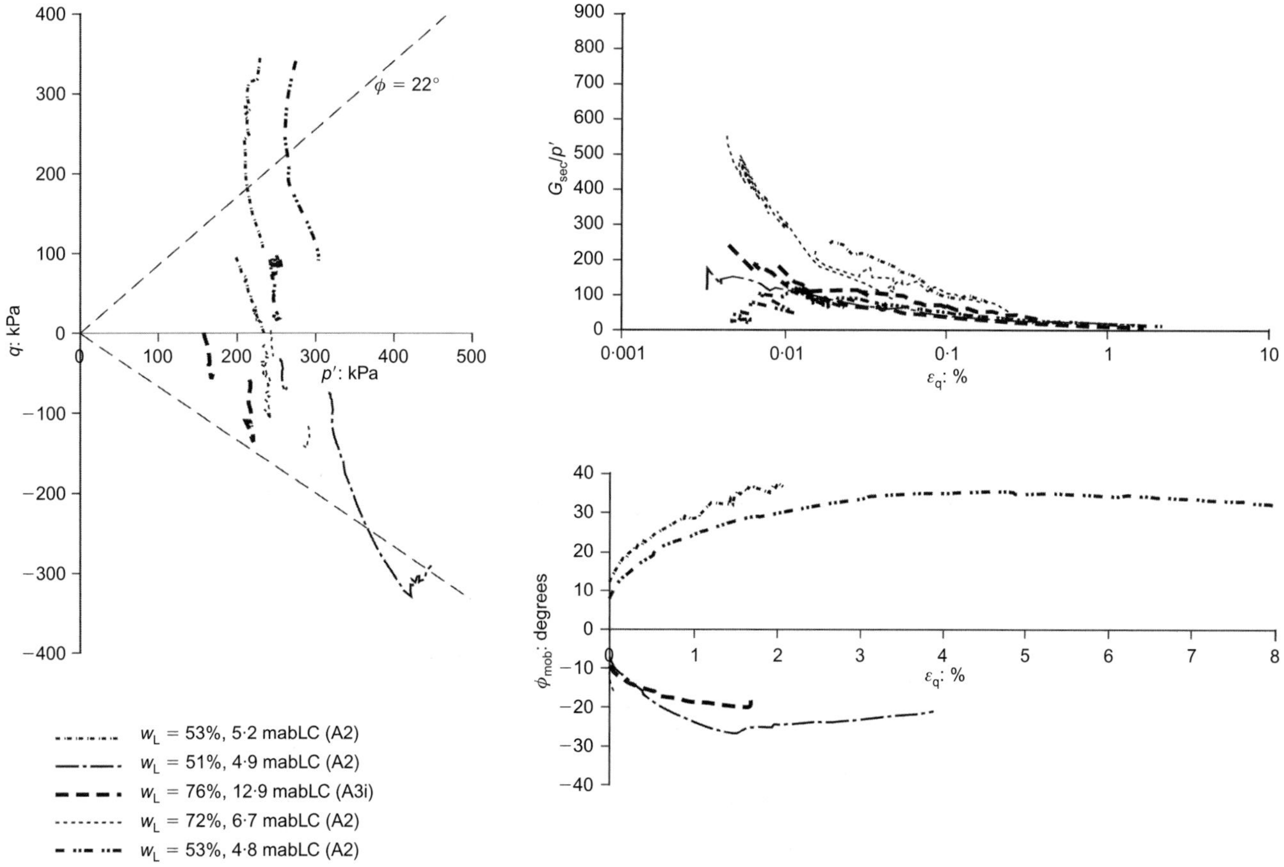

Fig. 8. King's Cross: results of triaxial stress path tests

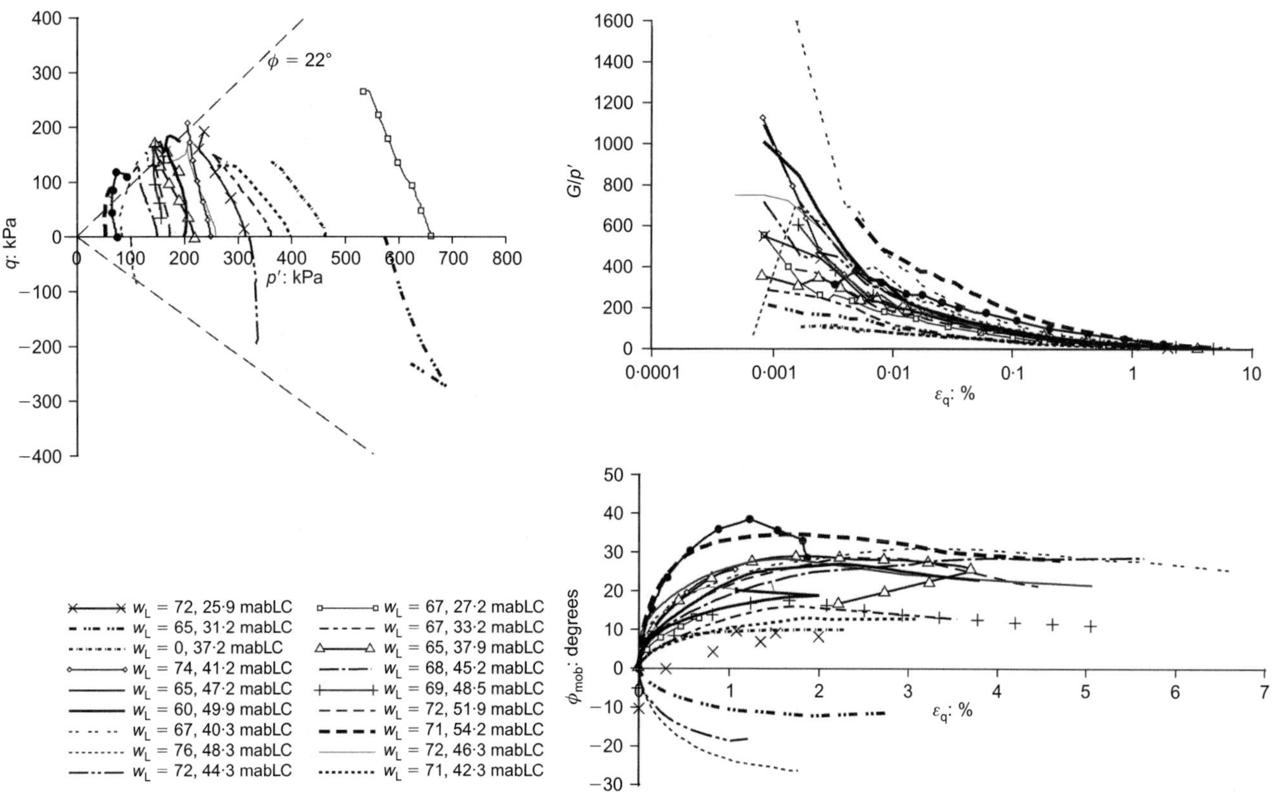

Fig. 9. CrossRail: results of triaxial stress path tests

drained shearing started from isotropic conditions, with the mean effective stress corresponding to the in situ stress, assessed using filter paper measurements and measurements of soil suction in the triaxial cell.

DISCUSSION

It can be expected that variations in the material will reasonably be expressed in terms of clay plasticity. However, its effect on the soil behaviour, as demonstrated in the stress path tests, might be masked by other factors. In the following, the factors affecting the soil behaviour during testing are identified and quantified in an effort to normalise the results and make comparison possible. These include stress state, geological and recent stress history and testing procedure. It is assumed here that all specimens were 'undisturbed', but in reality sample disturbance might also have affected the test results.

As noted earlier, the shear moduli discussed are secant values and not tangent.

Estimate of in situ stresses

Estimating the in situ stresses of an overconsolidated clay correctly is important, because the mean effective stress p' is needed in the normalisation of the test data. It involves good knowledge of the stress history as well as the present stress state of the soil. As previously stated, maximum overburden removal of 170 m has been assumed in this paper. In the following, the Mayne & Kulhawy (1982) approach has been used to estimate the in situ stresses for all specimens.

Dependence on stress and stress history

The dependence of small-strain stiffness on its current stress and stress history has been proposed by researchers such as Viggiani & Atkinson (1995) and Simpson (1992). They expressed this dependence of the very small-strain shear modulus G_0 in terms of the current mean effective stress p' and overconsolidation ratio $R_0 = p'/p'_{max}$.

BRICK approximation derived from Simpson (1992):

$$G_0 = A_B p' \left[1 + \beta(\lambda - \kappa)\ln(R_0) \right] \qquad (1)$$

where β, A_B are constants ($\beta = 4$, $A_B = 300$)

Viggiani & Atkinson (1995):

$$\frac{G_0}{p_r} = A\left(\frac{p'}{p_r}\right)^n R_0^m$$

or

$$G_0 = A_v p'^n R_0^m \qquad (2)$$

where p_r is reference pressure $= 1$ kPa, and $A_v = A/p_r^{n-1} =$ constant.

Although the two equations are mathematically different, they give similar results in the range of stresses and stress history relevant for London Clay (Wroth & Houlsby, 1985). A value of the material constant $A_B = 300$ was found to give a good fit to London Clay data. Viggiani & Atkinson proposed that the constants A, n and m are functions of plasticity index, approximated by the following expressions, which are plotted in Fig. 10.

$$A = 3750 \cdot 3 e^{-0 \cdot 0439 I_p} \qquad (3a)$$

$$n = 0 \cdot 0837 \ln(I_p) + 0 \cdot 4906 \qquad (3b)$$

$$m = 0 \cdot 0012 I_p + 0 \cdot 19 \qquad (3c)$$

It is noted that these were based on reconstituted samples only.

It has not yet been experimentally proven whether a similar dependence occurs at stiffnesses at higher strains. It seems unlikely that the same ratio would hold throughout the strain range to failure because, as noted by Simpson (1992), $\sin\phi'$ is directly proportional to G_0 for a given shape of stiffness degradation curve, yet it is known that $\sin\phi'$ does not increase in proportion to G_0. (This relationship was for plane strain. A similar, but more complex relationship can be derived for triaxial conditions.) Nevertheless, in the following, use of the parameters established for G_0 to normalise stiffness at all strains is investigated.

Stress path rotation

The various sets of tests had different approach stress paths to the start of shearing. Fig. 11 shows that tests carried out on CrossRail and King's Cross specimens were isotropically loaded or unloaded and then sheared in compression or extension; specimens from London Millennium Tower, on the other hand, went through quite complex drained stress paths, in an effort to reproduce the latter part of the previous stress history. The stress path rotation at the start of shearing is expressed with the angle θ, as specified by Atkinson *et al.* (1990).

Evaluating normalised stiffness

Material, current stress state and stress history. The effects of current stress and stress history can be taken into account by normalising the shear modulus using equations (1) and (2) (the BRICK approach and the Viggiani & Atkinson approach

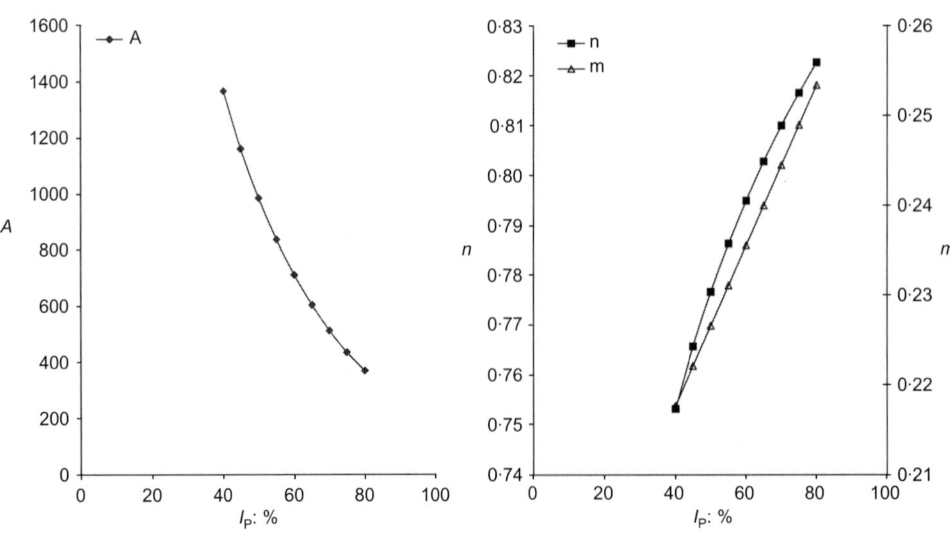

Fig. 10. Variation of stiffness parameters for G_0 with plasticity index (from Viggiani & Atkinson, 1995)

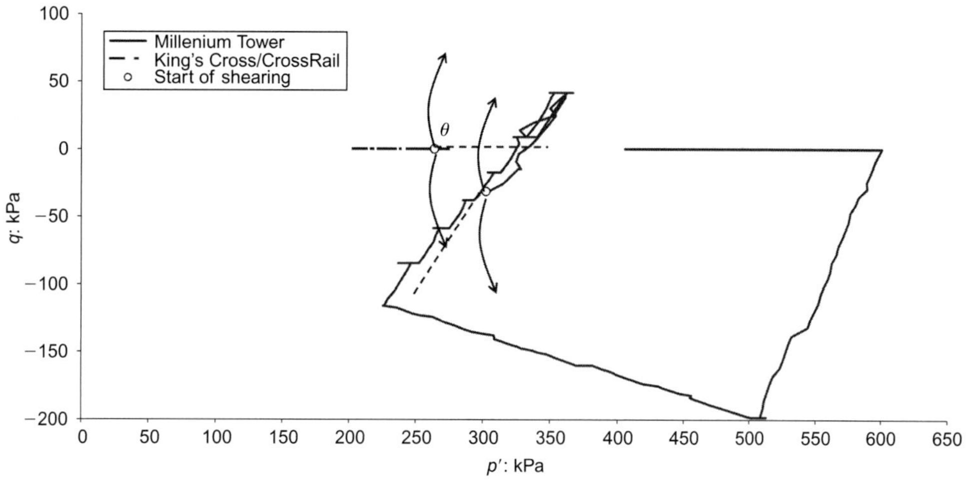

Fig. 11. Differences in the approach paths and stress path rotation at the beginning of shearing

respectively), assuming they apply to stiffness at higher strains. Of these, the Viggiani & Atkinson approach can also include the effect of material differences, because the constants in this equation have been related to clay plasticity (equations (3)). Fig. 12 presents a comparison between the different approaches to normalising shear modulus, using the shear modulus at 0·01%

(*a*) normalised with the mean effective stress at the start of shearing, $G_{0·01\%}/p'_0$

(*b*) normalised with the estimated in situ mean effective stress, $G_{0·01\%}/p'_{\text{in situ}}$ for the specimen

(*c*) normalised with mean effective stress and overconsolidation ratio, in line with the BRICK approach, $G_{0·01\%}/\{A_B p'_0[1 + \beta(\lambda - \kappa)\ln(R_0)/(1 + e)]\}$. The parameters β, λ and κ are material constants; as they have not been calibrated for the clay plasticity, the 'typical' parameters for London Clay have been used: $A_B = 300$, $\beta = 4$, $\lambda = 0·161$, $\kappa = 0·062$, $e = 0·6$.

(*d*) normalised with the mean effective stress and overconsolidation ratio, in line with the Viggiani & Atkinson approach, $G_{0·01\%}/[A(p^n/p_r^{n-1})R_0^m]$, where A, n and m are material constants expressed as a function of the I_P of each specimen.

The normalised stiffness is plotted against the difference between the mean effective stress at the start of shearing and that estimated to be the in situ value, $\Delta p' = p'_{\text{in situ}} - p'_0$. From the CrossRail project, a large number of tests were not tested at or near their in situ value.

Points on the (*a*) and (*b*) plots still include the effects of overconsolidation and material variation; in plot (*c*) material variation is still included, and in plot (*d*) all factors have been taken into account in the normalisation. In all plots, the effect of stress path rotation is still present. In plot (*b*), normalising with the in situ value could potentially reveal any effect of structure. It is notable that this plot provides perhaps the most consistent of the four correlations, giving relatively constant values, whereas correlations (*c*) and (*d*) show an increasing trend in values for positive $\Delta p'$.

The plots mainly show values less than unity, probably because the secant stiffness exhibited at 0·01% strain is less than that represented by G_0 at smaller strains. Unfortunately, the data are not good enough to allow evaluation at smaller strains. The normalisations presented here have not led to a reduction in bandwidth of the results, possibly because the range of overconsolidation ratio is relatively small.

Figure 13 plots the shear modulus at 0·01% shear strain for all tests against liquid limit, with normalisations carried out as per (*a*) BRICK (Simpson, 1992) and (*b*) Viggiani &

Atkinson (1995). Comparison of the two shows that it is difficult to distinguish any effect of plasticity on the stiffness.

Stress path rotation. The previous comparisons did not take into account the effect of different stress path rotations at the beginning of each shearing stage. This is now attempted in Fig. 14, where the normalised shear moduli at 0·01% strain (as per (*a*) BRICK and (*b*) Viggiani & Atkinson) are plotted against angle θ, as discussed above. Again, it is difficult to see any pattern emerging, perhaps because the data do not include values of angle θ close to 0 or to ±180°.

Summary. Figure 15 summarises the soil parameters discussed above from all tests relative to their stratigraphic sequence. More specifically, the secant shear modulus at 0·01% strain and the peak and critical state friction angles from all tests are plotted against height above base of London Clay. King's stratigraphic column is also shown for comparison. There is no apparent trend indicated in this figure.

In the above, no account has been taken of the effect of pre-existing fissuring on the samples on the behaviour during shearing, which may explain some of the apparent very low angles of shearing resistance. Even though it is anticipated that these should not have a major effect on the small-strain stiffness of the material, as they would for larger strain behaviour and strength, this may need to be explored further.

CONCLUSIONS
(*a*) Vertical variation in the material properties down the London Clay profile is evident and remarkably consistent across central London, closely following its geological subdivisions. This is a result of variations in the depositional environment, already identified by geologists.

(*b*) The London Clay material varies within each geological subdivision; for engineering purposes, a categorisation based on the Atterberg limits of the materials would be preferred.

(*c*) For the lower subdivision A2, consisting of the first 12 m of the Formation above its base, there appears to be a large scatter in the material properties, possibly manifesting a very variable coarse grain content.

(*d*) Atterberg limits, and in particular liquid limit, are a useful means of identifying the stratigraphic variation of the material, in the absence of any more accurate

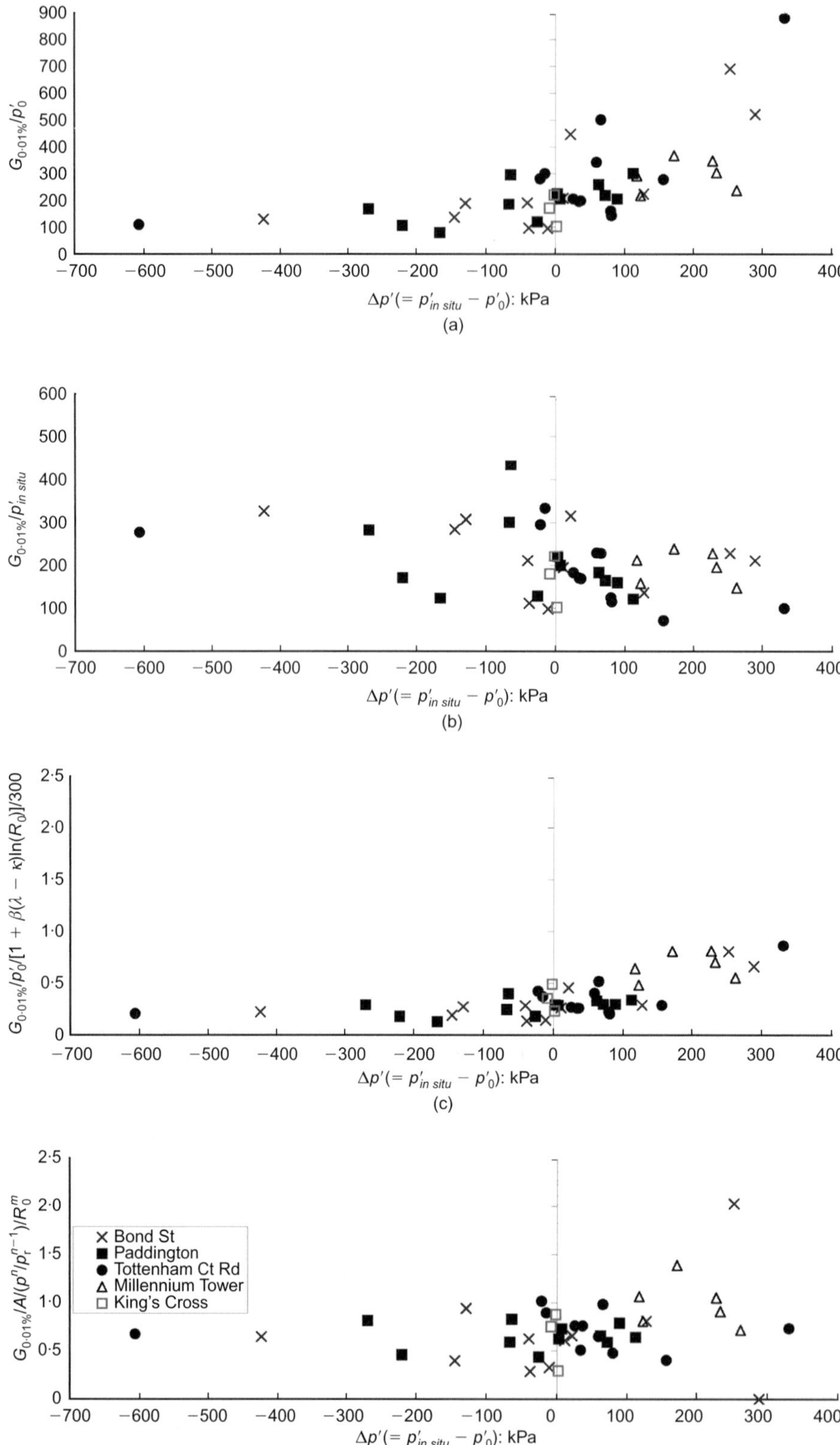

Fig. 12. Comparison of the various approaches to normalising shear modulus: (a) no correction; (b) normalisation with in situ mean effective stress, $p'_{\text{in situ}}$; (c) normalisation with p' and OCR (as in BRICK); (d) normalisation with p' and OCR (Viggiani & Atkinson, 1995)—constants A, n and m are dependent on I_p

information. The water content profile can also depict the stratigraphy, but this could be misleading, owing to its dependence on stress state, as well as material. Although liquid limit is an effective means of profiling, it usually requires a substantial number of tests to be clear. Conductivity logging has been tried in two sites in central London and is potentially a powerful tool for identifying stratigraphy within the London Clay, and presumably in other soils.

(e) The interpretation of triaxial stress path tests focused on the clay stiffness at 0·01% strain, (i) because the data become reasonably reliable at this strain, but most

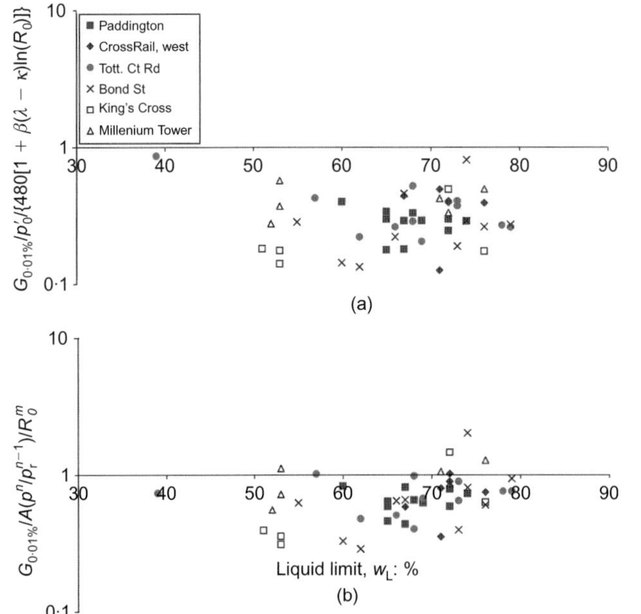

Fig. 13. Shear modulus at 0·01% shear strain plotted against liquid limit for all tests: (a) normalised as per BRICK (Simpson, 1992); (b) normalised as per Viggiani & Atkinson (1995)

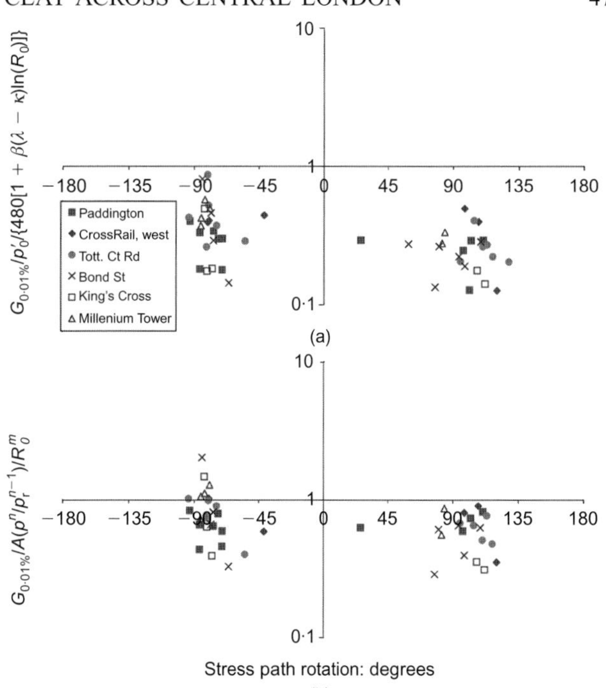

Fig. 14. Normalised shear moduli at 0·01% shear strain plotted against θ: (a) normalised as per BRICK (Simpson, 1992); (b) normalised as per Viggiani & Atkinson (1995)

importantly (ii) because it is the smallest scale of engineering interest: 0·01% strain is equivalent to 1 mm change in 10 m.

(f) Stiffness and strength were expected to follow the geological variation, and previous research suggests that the very small-strain shear modulus G_0 depends on plasticity. However, no correlation has been found in this paper between stiffnesses at 0·01% strain and the stratigraphic divisions or the index properties.

(g) Several methods of normalising secant stiffness at 0·01% shear strain have been compared. Normalisation with respect to the initial mean effective stress in the laboratory test was least successful. More successful were normalisation with respect to in situ mean

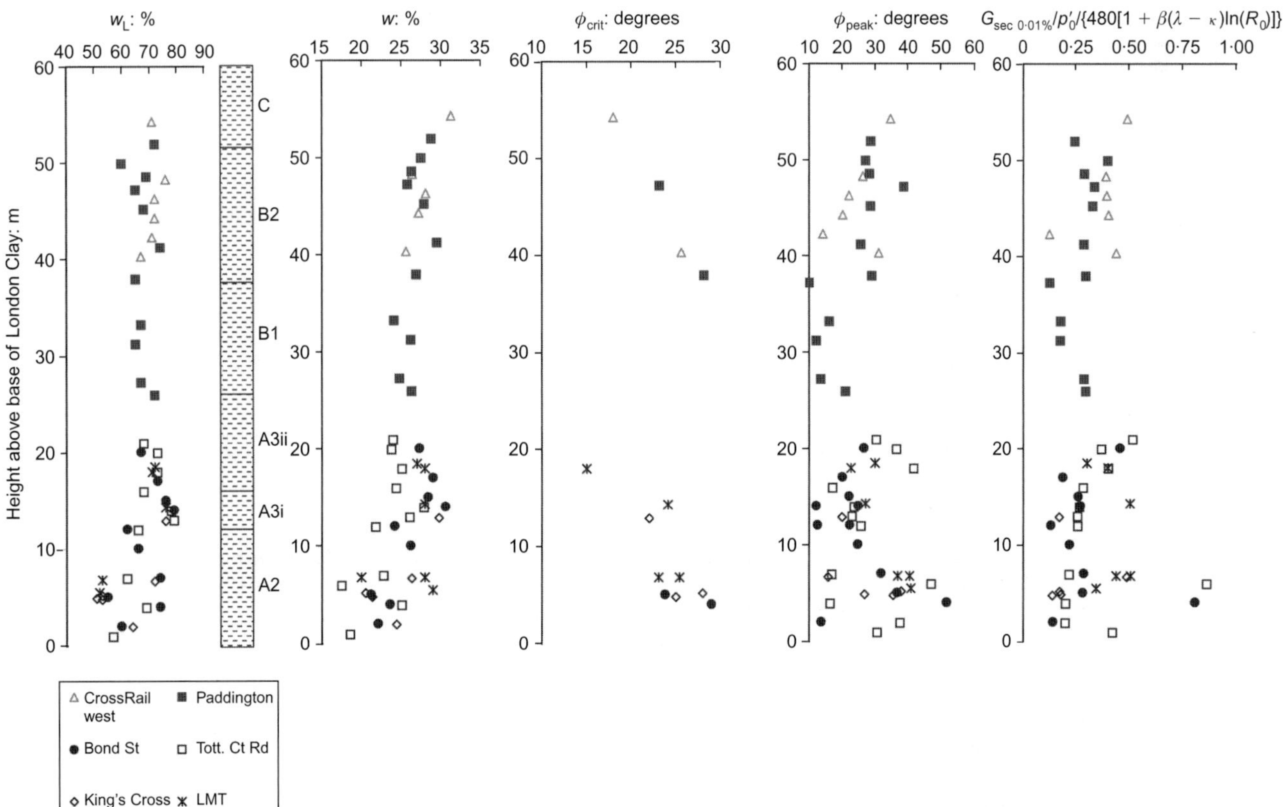

Fig. 15. Summary of soil parameters from all tests relative to their stratigraphic sequence

effective stress, or to functions combining laboratory initial stress and overconsolidation ratio, derived from the theories of Simpson (1992) or Viggiani & Atkinson (1995).

(*h*) From the same normalised stiffness plots, it appears that the dependence on in situ stress is stronger than the dependence on plasticity index.

(*i*) The data used in the present work have been drawn from site investigations, carried out as part of the design process of real projects, mainly in commercial laboratories. Although the tests were adequate to show that the samples lay within the general framework of understanding of London Clay, it has proved very difficult to find consistent correlations in a large body of data spread across six sites. Academic research, using higher-quality stress path testing, may be able to find clearer correlations between the soil properties, plasticity and detailed stratigraphy.

ACKNOWLEDGEMENTS

The authors would like to thank Cross London Rail Links and London Underground for allowing the use of site investigation data for the CrossRail and King's Cross sites respectively. The following colleagues are also gratefully acknowledged: Mr D. Pascal for his in-depth research in the stratigraphy of London Clay; Professor J. Atkinson for his invaluable advice on the interpretation of parts of the tests; and Mr S. Macklin and Drs P. Morrison and H. C. Yeow for their useful comments on the contents and presentation of the paper.

NOTATION

A, N, m materials constants
A_B, λ, κ material constants in the BRICK soil model
e voids ratio
G shear modulus
G_0 very small-strain shear modulus
$G_{0\cdot01\%}$ shear modulus at 0·01% strain
I_L liquidity index
I_p plasticity index
K_0 at-rest earth pressure coefficient
p' mean effective stress
p'_0 mean effective stress at start of shearing
p_r reference pressure
q deviator stress
R_0 overconsolidation ratio
w natural water content
w_L liquid limit
β, λ, κ material constants
ε_a axial strain
εB_q shear (deviator) strain
ε_r radial strain
θ stress path rotation at start of shearing
ϕB_{mob} mobilised angle of shearing resistance

REFERENCES

Atkinson, J. H., Richardson, D. & Stallebrass, S. E. (1990). Effect of recent stress history on the stiffness of overconsolidated soils. *Géotechnique* **40**, No. 4, 531–540.

Bishop, A. W., Webb, D. L. & Lewin, P. I. (1965). Undisturbed samples in London Clay from the Ashford Common Shaft: strength-effective stress relationships. *Géotechnique* **15**, No. 1, 1–31.

Burland, J. B., Simpson, B. & St John, H. D. (1979). Movements around excavations in London Clay. *Proc. 7th Eur. Conf. Soil Mech., Brighton* **1**, 13–30.

Henkel, D. J. (1957). Investigations of two long-term failures in London Clay slopes at Wood Green and Northolt. *Proc. 4th Int. Conf. Soil Mech. Found. Engng, London* **2**, 315–320.

Hight, D. W., McMillan, F., Powell, J. J. M., Jardine, R. J. & Allenou, C. P. (2002). Some characteristics of London Clay. In *Characterisation and engineering properties of natural soils* (eds T. S. Tan, K. K. Phoon, D. W. Hight and S. Leroueil), pp. 851–908. Rotterdam: Balkema.

King, C. (1981). The stratigraphy of the London Clay and associated deposits. *Tertiary Research Special Paper*, No. 6. Rotterdam: Backhuys.

Mayne, P. W. & Kulhawy, F. H. (1982). K_0–OCR relationships in clay. *J. Geotech. Engng Div. ASCE* **108**, No. GT6, 851–870.

Simpson, B. (1992). Retaining structures: displacement and design (32nd Rankine Lecture). *Géotechnique* **42**, No. 4, 541–576.

Skempton, A. W. (1961). Horizontal stresses in an overconsolidated Eocene clay. *Proc. 5th Int. Conf. Soil Mech. Found. Engng, Paris* **1**, 351–357.

Skempton, A. W. & Henkel, D. J. (1957). Tests on London Clay from deep borings at Paddington, Victoria and the South Bank. *Proc 4th Int. Conf. Soil Mech., London* **1**, 100–106.

Smith, T. J. (1978) *Consolidation and other geotechnical properties of shales with respect to ageing and consolidation*. PhD thesis, University of Durham.

Viggiani, G. & Atkinson, J. H. (1995). Stiffness of fine-grained soil at very small strains. *Géotechnique* **45**, No. 2, 249–265.

Wroth, C. P. & Houlsby, G. T. (1985). Soil mechanics: property characterisation, and analysis procedures. *Proc. 11th Int. Conf. Soil Mech., San Francisco* **1**, 1–55.

Cotecchia, F., Cafaro, F. & Aresta, B. (2007). *Géotechnique* **57**, No. 2, 167–180

Structure and mechanical response of sub-Apennine Blue Clays in relation to their geological and recent loading history

F. COTECCHIA*, F. CAFARO* and B. ARESTA*

The paper discusses the intrinsic properties, geological history, natural structure and mechanical behaviour of stiff Italian clays, part of the sub-Apennine Blue Clay formation, which are located at different sites within the marine basin where the clay formation was deposited in the early Pleistocene. For all the clay deposits being considered, diagenesis, unloading due to erosion, and weathering of the top strata represent the main stages of the geological history. However, the clays have undergone different levels of unloading, and have developed different levels of structure due to their different locations in the basin. These differences are accounted for in the analysis of their current states and behaviour. Based upon a large experimental database, the structure and mechanical behaviour of the clay samples from the different sites are compared and related to the clay history. In particular, the soil structure is characterised based upon SEM analysis results and compression–swelling test results. The comparison of the response to compression–swelling cycles of the natural clay with that of the same clay when reconstituted leads to the evaluation of parameters such as the stress sensitivity S_σ and the swell sensitivity C_s^*/C_s, which are used to assess the structural strength of the natural clay with respect to that of the reconstituted clay. The stress–strain behaviour of the different clays along different constant stress-ratio compression paths and shear paths (from low to high pressures) is discussed in detail. The variations in stress–strain response of samples subjected to different pre-shear consolidation paths are outlined. The corresponding trends of the plastic strain increment ratios are investigated. As a result of the analysis, the features of the gross yield curves of the different clay horizons are also assessed. Finally, the study provides elements for the characterisation of the hardening properties of stiff clays, and therefore of the evolution of the material anisotropy during plastic deformation.

KEYWORDS: anisotropy; consolidation; stiffness

Cet article discute des propriétés intrinsèques, de l'histoire géologique, de la structure naturelle et du comportement mécanique des argiles fermes italiennes qui font partie de la formation d'argile bleue subapennine et qui sont situées à divers emplacements du bassin marin où cette formation argileuse a été déposée, au début du Pléistocène. Pour tous les gisements d'argile considérés ici, les stades principaux de l'histoire géologique sont la diagénèse, la décompression due à l'érosion et la dégradation des couches supérieures. Cependant, les argiles ont subi différents niveaux de décompression et ont développé différents niveaux de structure du fait de leurs divers emplacements dans le bassin. Ces différences sont reflétées par l'analyse de leurs états et comportements actuels. À partir d'une grande base de données expérimentale, on a comparé la structure et le comportement mécanique des échantillons d'argile provenant de différents emplacements et on a relié ces caractéristiques à l'histoire de l'argile. En particulier, on a caractérisé la structure du sol grâce aux résultats de l'analyse SEM (Structural equation modeling ou modélisation d'équations structurelles) et de l'essai de compression-gonflement. En comparant la réponse aux cycles de compression-gonflement de l'argile naturelle avec celle de la même argile lorsque celle-ci a été reconstituée, on a pu évaluer des paramètres tels que la sensibilité à la contrainte S_σ et la sensibilité au gonflement C_s^*/C_s, ceux-ci étant utilisés pour évaluer la résistance structurelle de l'argile naturelle par rapport à celle de l'argile reconstituée. Cet article discute en détail du comportement de contrainte – déformation de différentes argiles selon des chemins de compression et des chemins de cisaillement différents à indice de contrainte constant. Il décrit les variations de la réponse contrainte – déformation d'échantillons soumis à divers chemins de consolidation pré-cisaillement. Il y a aussi des recherches sur les tendances correspondantes des indices d'incrément de contrainte plastique. Suite à cette analyse, cet article estime aussi les caractéristiques des courbes de rendement brut des différents horizons d'argiles. En dernier lieu, cette étude fournit des éléments pour la caractérisation des propriétés de durcissement des argiles fermes et par conséquent pour celle de l'évolution de l'anisotropie matérielle durant une déformation plastique.

INTRODUCTION

The mechanical response of clays depends on their consolidation state, their structure (the combination of fabric and bonding) and their loading history. For natural clays, the geological history may give rise to complex structures and may include peculiar loading processes, such as those developing in drying–wetting cycles due to weathering. One generally reported effect is that natural structure conveys extra strength to the intact soil with respect to the reconstituted soil (Burland, 1990), making its sensitivity higher than 1. However, loading may cause a loss of such structural strength, this being one of the basic differences between the hardening processes of natural and reconstituted clays.

The present paper discusses the behaviour of stiff marine clays, part of the sub-Apennine Blue Clays present in the Montemesola Basin (Apulia, Italy; Fig. 1), which have attained their strength from both consolidation and diagenesis. Therefore these clays are representative of that class of materials that have acquired additional bonding in their geological history with respect to that developed during sedimentation. The experimental study endeavours to make a

Manuscript received 16 May 2006; revised manuscript accepted 17 October 2006.
Discussion on this paper closes on 1 August 2007, for further details see p. ii.
* Department of Civil and Environmental Engineering, Politecnico di Bari, Italy.

contribution to the understanding of the effects on the mechanical behaviour of stiff clays of processes that are generally part of their geological history, such as overconsolidation due to erosion, diagenesis, preconsolidation due to drying, and degradation of structure due to weathering. The study is based on a comprehensive investigation of three different sites within the basin, site P in the centre (Pappadai) and two about the north-western edge, sites A and B in Fig. 1 (about 4–5 km distant from each other), and develops further previous analyses of the properties of the materials (Cotecchia, 1996; Cotecchia & Chandler, 1997; Cafaro, 1998; Cafaro & Cotecchia, 2001; Cotecchia, 2003). Comparison of the clays from the different sites gives an indication of the variability in composition, structure and mechanical properties within a stiff clay deposit. Furthermore, the study provides elements for the characterisation of the hardening properties of stiff clays and, in particular, of the effects of recent stress history, since the shear behaviour of specimens subjected to different consolidation paths is investigated.

In the Montemesola Basin the clays were deposited above carbonate bedrock, within a protected sea (Ciaranfi *et al.*, 1971). At present, in the entire basin the deep clays are grey and the upper ones are yellow, as a result of long-term weathering and oxidation. The clays were sampled down a borehole at site A and block-sampled from a quarry wall at site B (Fig. 1), as shown in the simplified section reported in Fig. 2(a). The borehole samples were taken at 5, 8, 11,

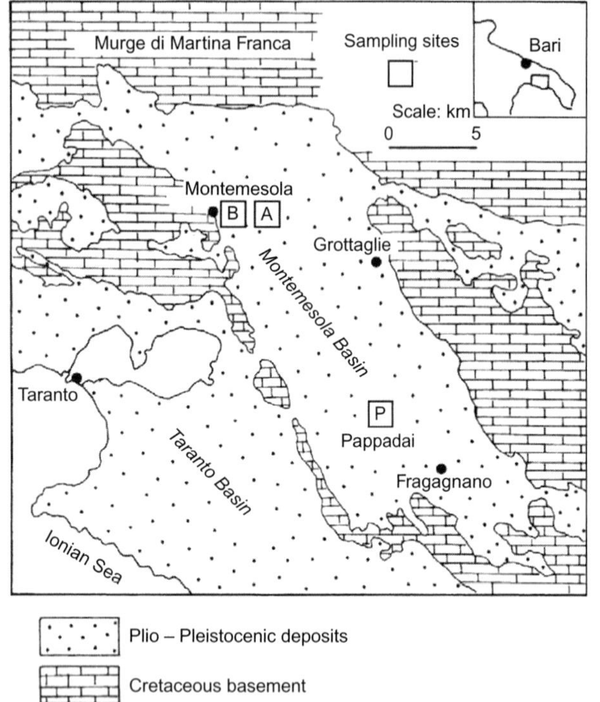

Fig. 1. Montemesola Basin and sampling sites (after De Marco *et al.*, 1981, modified)

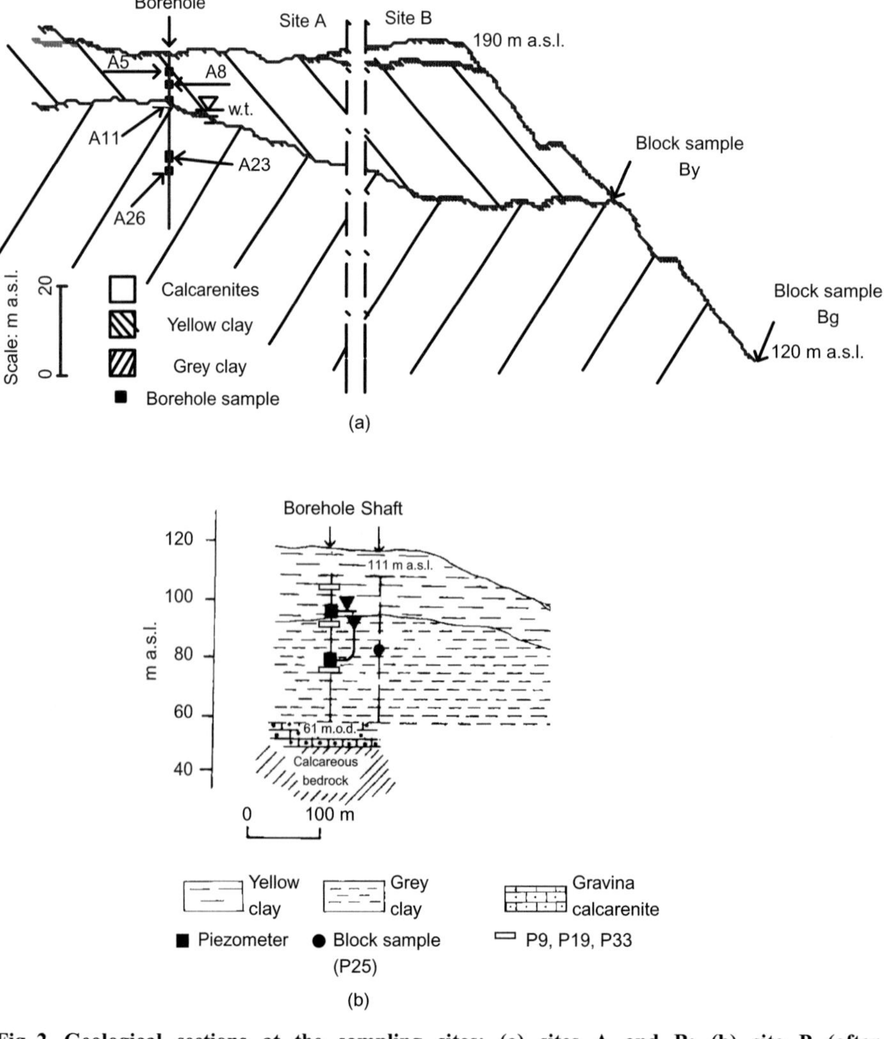

Fig. 2. Geological sections at the sampling sites: (a) sites A and B; (b) site P (after Cotecchia & Chandler, 1997)

23 and 26 m depth, from both the weathered (yellow) and unweathered (grey) clay horizons. The block samples were taken at the base of the quarry wall within the grey clays and at 35 m above the base within the yellow clays (Cafaro & Cotecchia, 2001). In the Pappadai valley (site P, Fig. 2(b)), block samples were taken at 25 m depth down a shaft (Cotecchia & Chandler, 1997) as well as at 9, 19 and 33 m depth down a borehole nearby. At all the sampling sites the weathered clay stratum is 20–30 m thick, and the transition from the grey to the yellow clays is gradual.

GEOLOGICAL HISTORY OF THE CLAYS

Cotecchia & Chandler (1995) report that clay sedimentation at Pappadai occurred in a quiet sea about 1·3 million years ago (early Pleistocene), with a reducing environment at the seabed. At both sites A and B the deposition conditions were more turbulent than at Pappadai, and it is likely that the sedimentation structures forming at these sites were different from those at site P. As discussed by Cotecchia & Chandler (1997), diagenesis took place in the clays, probably more in the centre of the basin than about the edges. Later on, unloading due to erosion caused overconsolidation in the whole basin, the thickness of the eroded stratum being up to 120 m in the central area (Ciaranfi *et al.*, 1988). Thereafter, in the Quaternary, in the whole basin the upper clays have been affected by important weathering processes, consisting mainly of cycles of deep drying and swelling (Cafaro & Cotecchia, 2001), which have caused changes in structure and mechanical properties for the upper clays, as well as oxidation of the iron minerals (Garavelli & Nuovo, 1974), with the consequent change in colour.

The oxidation of clays at depth is recognised as requiring thousands of years to develop (e.g. Lias Clay; Chandler, 1972). Nonetheless, vertical propagation of shrinkage–swelling cycles can be much faster and not necessarily associated with oxidation. Therefore, at some places in the basin, the grey clays immediately below the yellow ones may also have been affected by drying, and this is accounted for in the interpretation of their current state and mechanical properties, which is discussed later.

INTRINSIC PROPERTIES, IN SITU STATE AND STRUCTURE OF THE CLAYS

Table 1 reports the values of the index properties, carbonate content and clay fraction of the samples in Fig. 2, and Fig. 3 shows the profiles of some of the index properties at the three sites. The clay fraction varies between 42% and 56% and is mainly illitic, but with significant presence of both smectite and kaolinite (Garavelli & Nuovo, 1974; Cotecchia & Chandler, 1995). Often the high carbonate content of the clay is connected to the presence of a coarse calcareous fraction (Cotecchia & Chandler, 1995). There is not much difference in index properties between the grey and the yellow clays, because weathering has not changed the soil composition significantly.

The grey clays are currently saturated, whereas the yellow clays are partially saturated ($S_r = 85$–98%). Positive pore water pressures have been recorded in the grey clay (Fig. 2), which have been accounted for in the calculation of the in situ vertical effective stresses (σ'_{v0}) and overconsolidation ratios (OCR) reported in Fig. 4. By means of self-boring pressuremeter tests (Cotecchia & Chandler, 1997), earth pressure ratios K_0 in the range 1–1·5 have been measured in the grey clays and up to 2·5 in the yellow clays. The coefficients of permeability measured in the laboratory by means of oedometer testing have been about 10^{-11} m/s for all the samples from the three sites.

The current fabric of the clays has been investigated by means of scanning electron microscopy (SEM) (Cotecchia & Chandler, 1997; Cafaro & Cotecchia, 2001).

At Pappadai, the grey clay at 25 m depth (P25, Table 1) is laminated, and the coarse material in the laminations has

Table 1. Index properties

Site	Depth: m	Name	Group	G_s	CF: %	w_L: %	I_P: %	A	Carbonate content: %	w: %	I_C	S_r: %	e_0
A	5	A5 (yellow)	–	–	–	–	–	–	–	21·9	–	91·4	0·65
	8	A8 (yellow)	1	–	–	51·9	28·7	–	–	22·8	1·02	95–98	0·64
	11	A11 (yellow)	1	–	–	53·2	27·6	–	–	–	–	–	–
	23	A23 (grey)	1	–	–	42·5	24·0	–	–	22·6	0·83	98	0·66
	26	A26 (grey)	2	–	–	69·8	41·7	–	–	30·1	0·95	96	0·83
B	35	By (yellow)	1	2·72	42	49·0	26·4	0·63	27	17·1	1·19	85–90	0·55
	70	Bg (grey)	1	2·71	49	51·1	27·7	0·56	34	23·6	1·05	100	0·65
	50	BS (Simeone,1991)	1	2·77	–	58·4	32·1	0·80	–	24·1	1·07	97	0·69
P	9	P9 (yellow)	2	2·74	56	69·3	38·4	0·69	19	23·2	1·20	96	0·67
	19	P19 (grey)	1	2·74	52	51·8	28·8	0·55	26	22·7	1·01	100	0·63
	25	P25 (grey)	2	2·75	50	65·0	35·0	0·70	28	31·0	0·97	97	0·88
	33	P33 (grey)	1	2·74	48	53·4	27·5	0·57	42	27·0	0·96	98	0·75

G_s, specific gravity; CF, clay fraction; w_L, liquid limit; I_P, plasticity index; A, activity; w, water content; I_C, consistency index; S_r, saturation degree; e_0, current void ratio.

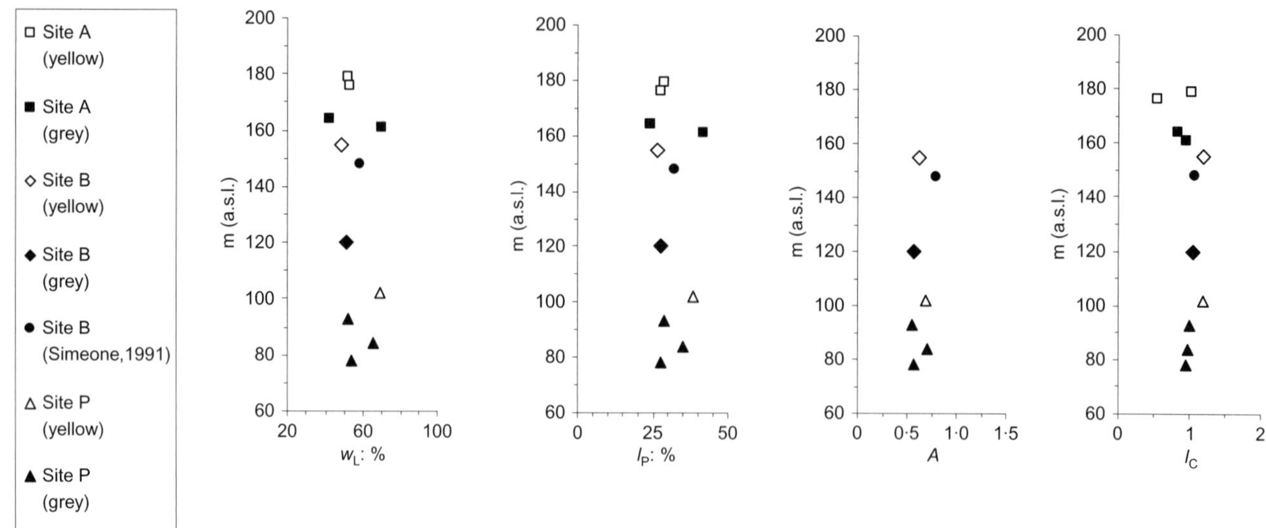

Fig. 3. Index property profiles (parameters as in Table 1)

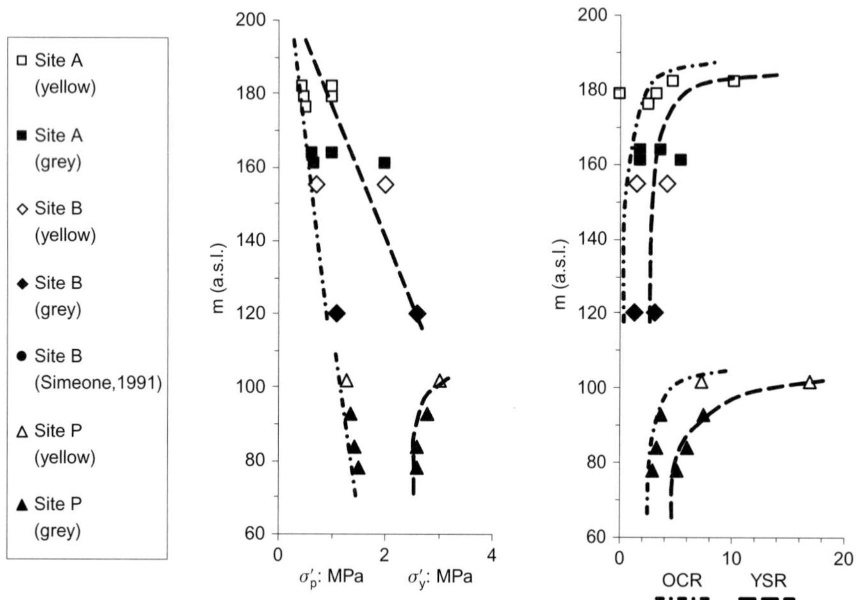

Fig. 4. Preconsolidation and gross yield pressures, overconsolidation ratios and yield stress ratios

been found to consist largely of nannofossils. As such, the clay is likely to have been sedimented in a very quiet environment, so that, despite being massive and stiff, it has a relatively high void ratio ($e_0 = 0.88$) and includes flocculated aggregates of high porosity together with closely packed stacks of domains (with particles in face-to-face contact). The presence of the flocculated fabric portions is indicative of a 'compacted bookhouse' fabric (Sides & Barden, 1970). The 'index of fabric orientation' measured for P25 by image processing (Martinez-Nistal *et al.*, 1999) is 0·277, which is indicative of a very oriented fabric. Using chemical microanalysis on the same sample, Cotecchia & Chandler (1997) detected a film of amorphous calcite coating the clay particles and their contacts, which represents a result of diagenesis giving rise to additional bonding. Therefore P25 has a very oriented bookhouse fabric with significant bonding.

Cafaro & Cotecchia (2001) show that at site B the fabric of the grey clay (Bg, Table 1) also conforms to a compacted bookhouse type, but it is more densely packed and less oriented than that of sample P25, probably as an effect of the more turbulent deposition conditions. In the whole basin the weathered clays generally have a higher consistency

index (I_c) than the grey ones (Fig. 3), owing to the further preconsolidation caused by drying. SEM analyses of the yellow clay at site B (By) have shown that its fabric is far more densely packed than that of the grey clay, and is formed of randomly oriented floccules with large interfloccular pores, which give rise to concentric layouts with distinct nuclei. This fabric is similar to that observed for clays subjected to shrinkage–swelling cycles in the laboratory (Basma *et al.*, 1996). Thus the intense drying that has occurred within the yellow clay horizon appears to have erased the original oriented clay fabric, giving rise to a quasi-isotropic fabric.

COMPRESSION BEHAVIOUR

The compression behaviour of the clays sampled in the basin has been studied based upon both oedometer and stress path test data. Part of these data has been reported by Cotecchia & Chandler (1997) and Cafaro & Cotecchia (2001), and part is the result of a new investigation.

As the clay compressibility is influenced by clay composition (Skempton 1970), samples of analogous composition have been grouped together, so that the differences in compressibility among samples of the same group are mainly an effect of differences in their structure. Group 1 includes samples of lower plasticity index (Table 1): P19 and P33 from Pappadai, A8, A11 and A23 from site A, and both the samples Bg and By from site B (Fig. 2). Group 2 includes the samples of higher plasticity: P9 and P25 from Pappadai and A26 from site A. According to the location of the samples shown in Fig. 2, both groups include both weathered and unweathered samples. The compression behaviour of the clay when reconstituted has been investigated for samples Bg, By, A11, P19 and P25. In the following, the asterisk * is used to denote reconstituted clay parameters.

Figure 5 shows the compression curves of samples from both groups, either natural or reconstituted. The symbols referring to samples of Group 1 are empty and those for group 2 are full. The corresponding compression parameters are shown in Table 2. The compression curves of the reconstituted samples represent the intrinsic compression lines (ICLs; Burland, 1990) of the clays. The ICLs of samples A11, Bg and P19 of Group 1 are very close and confirm the similarities in composition of these samples. However, the ICL of By, of the same group, is located on the left, probably due to its lower clay fraction (Table 1). At the same time, the ICL of P25, representative of the intrinsic compressibility for Group 2, is located on the right of the ICLs of Group 1, owing to the higher plasticity index of the clay. All the compression curves of the natural samples lie to the right of the corresponding ICL, as an effect of the higher strength of the natural structure with respect to that of the reconstituted clay.

Gross yield has been recognised by several authors (Graham et al., 1988; Hight et al., 1992; Gens & Nova, 1993; Kavvadas, 1994; Wood, 1995; Cotecchia & Chandler, 2000) to be a state of the soil, either natural or reconstituted, outside the elastic domain, at which the stiffness falls significantly owing to the onset of a rate of plastic straining and structure change that is far more important than pre-gross yield, so as to determine a change in the hardening rule. Cotecchia & Chandler (1998) show, by means of SEM analyses, the extent to which the Pappadai clay fabric changes upon gross yield in compression. For natural clays, gross yielding may be accompanied by a decay in structural strength, often called *destructuring* (Leroueil & Vaughan, 1990; Wood, 1995), which represents the main difference between the gross yield processes of natural and reconstituted clays.

The gross yield vertical stresses σ_y' of the natural samples in Fig. 5 have been identified along the compression curves according to the Casagrande construction and are reported in Table 3, together with the sample yield stress ratios, $\mathrm{YSR} = \sigma_y'/\sigma_{v0}'$; σ_y' and YSR are also plotted against elevation in Fig. 4. The difference between YSR and OCR (Fig. 4) is marked at all depths, both in the centre and at the margin of the basin. For the unweathered samples such a difference is mainly the effect of diagenesis (Cotecchia & Chandler, 2000), which increases the gross yield pressures. For the weathered clays, such a difference arises also from the additional preconsolidation due to drying, which has not been accounted for in calculating OCR (due only to erosion unloading).

Cotecchia & Chandler (2000) identify the stress sensitivity ratio, $S_\sigma = \sigma_y'/\sigma_e^*$ (where σ_e^* is the equivalent pressure on the ICL), as a parameter representative of the response that

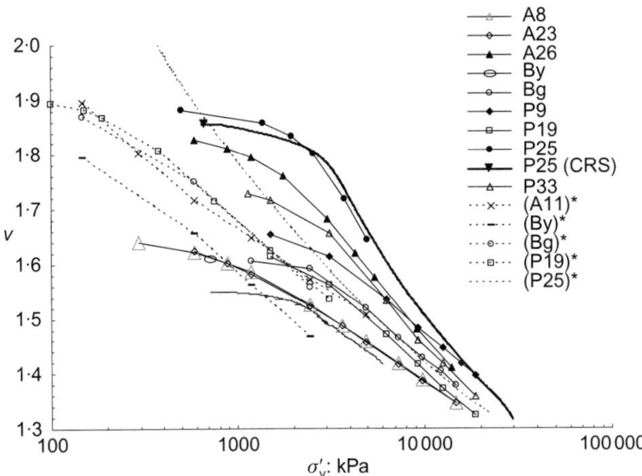

Fig. 5. One-dimensional compression behaviour of the clays sampled at the three sites; data from Cotecchia (1996), Cafaro (1998) and new data

Table 3. Gross yield pressures, yield stress ratios and stress sensitivity ratios

Sample	σ_y': kPa	YSR	S_σ	
			σ_y'/σ_e^* OED	p_{iy}'/p_{iy}^* ISO
A8	1000	6·3	–	–
A23	1000	3·0	–	–
A26	2000	5·5	–	–
By	2000	4·1	1·4	1·3
Bg	2600	3·1	2·2	1·7
P9	3000	17·0	–	–
P19	2800	7·5	1·3	–
P25	2600	6·1	3·5	3·2
P33	2600	5·1	–	–

Table 2. One-dimensional compression parameters

Sample	C_c	C_s pre-yield	C_s post-yield	C_c^*	C_s^*	Schmertmann ratio, C_s^*/C_s	
						Pre-yield	Post-yield
A5	0·214	0·038 (294–39 kPa)	0·067 (1176–78 kPa)	–	–	–	–
A8	0·236	–	–	–	–	–	–
A11	–	–	–	0·240	0·374	–	–
A23	0·230	0·051 (294–39 kPa)	0·097 (2451–147 kPa)	–	–	–	–
A26	0·396	–	–	–	–	–	–
By	0·242	0·047 (1176–78 kPa)	0·073 (1176–78 kPa)	0·296	0·367	1·0	0·7
Bg	0·293	0·011 (1176–147 kPa)	0·065 (1176–147 kPa)	0·300	0·379	3·0	0·5
P9	0·291	0·132 (600–80 kPa)	0·146 (600–80 kPa)	–	–	–	–
P19	0·308	0·064 (600–80 kPa)	0·107 (600–80 kPa)	0·305	0·377	–	–
P25	0·560	–	–	0·494	0·481	2·5	1·0
P33	0·387	0·053 (600–80 kPa)	0·096 (600–80 kPa)	–	–	–	–

the current clay structure is prone to exhibit up to gross yield. In this way it represents the strength of the current clay structure by comparison with that of the same clay when reconstituted, and it is constant along a swelling line (i.e. for states before gross yield), because only beyond gross yield may plastic straining trigger a significant degradation of structure. The S_σ values of the samples for which both natural and intrinsic compressibility have been investigated are shown in Table 3; for all of them $S_\sigma > 1$. Tables 2 and 3 show that, within each of the two groups of samples of similar composition, higher C_c values are associated with higher S_σ values, this being indicative of the non-stable nature of the structural elements adding strength to the clay, recently modelled as 'metastable' by Baudet & Stallebrass (2004).

If comparing unweathered samples of similar composition and depth, but different location in the basin, such as samples P25 and A26 of Group 2, it is found that the sample from the centre (P25) has its compression curve located at higher void ratios than that from the edge (A26). The same difference applies to samples P33 and Bg of Group 1 and is consistent with the more open fabric of the clay sedimented in the centre of the basin with respect to that deposited in the more turbulent environment at the edges.

The effects of weathering on compressibility can be evaluated by comparing the compression curves of samples A8 and By, which are the most weathered and oxidised within Group 1 (in particular By, on the quarry wall, has been affected by both regional and very recent weathering), and the curves of the other samples from the same group. Fig. 5 shows that the compression curves of both A8 and By are the most offset towards the left in the $e-\sigma'_v$ plane within the group. The location of the compression curves of both the unweathered samples P25 and A26 and that of the weathered sample P9 for Group 2 confirms that the effect of weathering is that of decreasing the void ratio and shifting the post-gross yield compression curve to the left in the $e-\sigma'_v$ plane. If comparing the compression curve of sample P33 with that of P19, both grey samples from the centre of the basin, and that of the grey samples Bg and A23, both from the margins, it is found that samples P19 and A23 have lower void ratios than the corresponding deeper samples and that, post-gross yield, they reach a compression curve located on the left of that of the deeper samples, P33 and Bg respectively. Such offsets might have resulted from further preconsolidation and disturbance of samples P19 and A23 due to the penetration at depth of drying processes, though these have not been such as to cause the oxidation of the clay.

Table 2 reports the values of the swelling indexes C_s measured in oedometer swelling stages carried out, both pre-

and post-gross yield, on some of the samples in Table 1, together with the corresponding values of swell sensitivity C_s^*/C_s (Schmertmann, 1969). The swell sensitivity of all samples is higher pre-gross yield, and this is indicative of a degradation of bonding occurring post-gross yield. The decay in C_s^*/C_s with gross yielding is more significant for the deepest unweathered samples (e.g. P25 and Bg), which have the highest C_s^*/C_s values pre-gross yield and are the most sensitive (Table 3). For the yellow clay, both the pre-gross yield swell sensitivity and its decay are smaller than for the grey clay, and this suggests that weathering has caused a degradation of the clay bonding.

The behaviour of the clays during different constant stress ratio η $(= q/p')$ compressions has been investigated by means of stress path testing. In particular, isotropic ($\eta = 0$), anisotropic ($\eta = $ const.) and one-dimensional (K_0) compressions have been carried out. The anisotropic compressions were carried out with ram pressure increment rates not higher than 7 N/h. In the K_0 compressions the radial strain was kept constantly about zero, with a tolerance not larger than $\varepsilon_r = \pm 0.08\%$; it was measured either by the use of a radial belt or by measuring volume changes and axial displacements (the latter were either measured locally or measured externally and corrected for apparatus compliances).

Figure 6 shows an example of comparison, for a given sample, between the oedometer and the K_0 compression curves (from stress path testing). The agreement between the test results confirms the validity of the radial strain control in the stress path test. The post-gross yield stress ratios, K_0 (NC) $= \sigma'_r/\sigma'_v$, measured when such agreement has been achieved, have been found to vary between 0·56 and 0·63

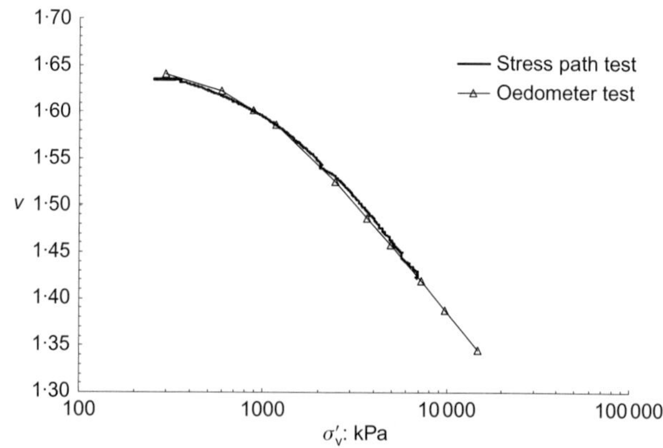

Fig. 6. Comparison between stress path and oedometer test results (sample A8)

Table 4. Samples subjected to stress-path testing: values of soil constants and gross yield pressures

Group		Sample	λ, λ^*	κ, κ^*	M_{nat}, M^*	K_0, K_0^* (NC)	$N - N_0$ or $N - N_\eta$ (measured)	$N - N_0$ (MCC)	p'_{iy}: kPa	$p'_{K_0\mathrm{y}}$: kPa
1	Weathered	A8	0·106	0·038	0·92	0·60	0·019	0·020	1800	1600
		(By)*	0·128	0·027	1·03	0·56	0·082	0·033	–	–
		By	0·107	0·015	1·07	0·60	0·020	0·029	1950	1700
	Unweathered	(Bg)*	0·127	0·02	1·14	–	0·070	0·029	–	–
		Bg	0·127	0·008	1·07	0·58	0·030	0·028	2300	2000
		P19	0·136–0·139	0·029–0·024	0·90	–	–	–	1800	–
		P33	0·193–0·185	0·023	0·95	–	–	–	1900	–
2	Weathered	P9	0·127	–	0·79	–	–	–	2300	–
	Unweathered	(P25)*	0·204	0·03	0·91	0·57	0·006	0·063	–	–
		P25	0·254	0·02	1·08	0·63	0·050	0·044	2260	2000
		A26	0·180	0·04	1·08	–	0·017	0·028	2250	2050

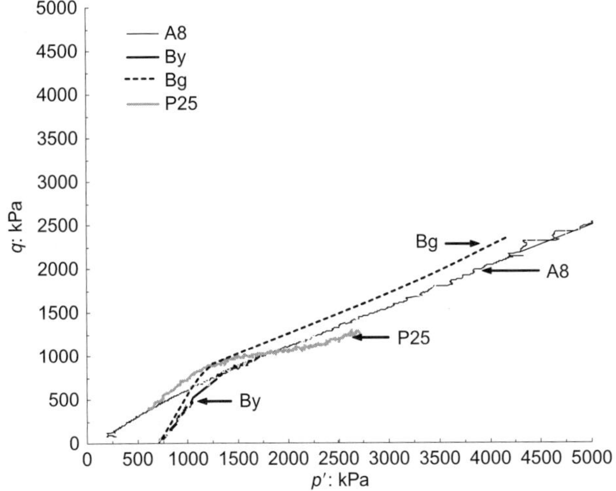

Fig. 7. Stress paths during K_0-compression of natural clay specimens

(Table 4). Fig. 7 shows the stress paths followed by the specimens in one-dimensional compression. The trend is that typically observed for structured clays (Leroueil & Vaughan, 1990; Coop *et al.*, 1995; Cotecchia & Chandler, 1997), which exhibit a drop in K_0 pre-gross yield, beyond which it increases to a constant value.

Figures 8, 9 and 10 report, in the v–p' plot, the different isotropic, K_0 and η ($\neq 0$) compression curves of both natural and reconstituted samples from the three sites. The corresponding compression parameters are reported in Table 4. In Figs 9 and 10 it can be seen that the compression curves of the natural samples lie to the right of the corresponding curves of the reconstituted samples, as already found for other sensitive clays (Smith *et al.*, 1992; Coop & Cotecchia, 1995; Coop *et al.*, 1995; Amorosi & Rampello, 1998; Callisto & Calabresi, 1998). Also, the INCL and K_0NCL (or ηNCL for $\varepsilon_r < 0$; Fig. 8) of the natural clay are parallel to each other, the same as for reconstituted clay. The INCL lies to the right of the K_0NCL, so that the mean effective gross yield pressure in anisotropic compression, p'_{K_0y}, is lower than that in isotropic compression, p'_{iy}. The vertical distances between the corresponding INCL and K_0NCL ($= N$

$- N_0$) in Figs 8–10 are reported in Table 4, together with the ($N - N_0$) values predicted by the Modified Cam Clay model (MCC; Roscoe & Burland, 1968). This distance is always larger for the reconstituted clay than for the natural clay. Among natural clays, it is largest for the unweathered clay of very oriented fabric and possessing amorphous calcite bonding, P25, being even higher than the ($N - N_0$) value of the weathered clays (e.g. A8 and By), of less oriented fabric.

The positive difference ($p'_{iy} - p'_{K_0y}$) observed for the clays being examined is not necessarily observed for soft sensitive clays, for which often $p'_{iy} < p'_{K_0y}$ (Leroueil *et al.*, 1979; Tavenas *et al.*, 1979; Tavenas, 1981; Leroueil & Vaughan, 1990). In much of the literature such a behavioural feature has been considered an effect of the inherent clay anisotropy connected to the fabric orientation resulting from the one-dimensional normal consolidation of the clay, whereas Cotecchia & Chandler (2000) suggest that it might be, at least in part, an effect of the high-sensitivity structures peculiar to soft clays. The data presented above confirm that such a feature of behaviour is not necessarily observed for clays possessing an oriented fabric. Rather, it may result from the response of various combinations of fabric and bonding, which have experienced mainly anisotropic loading and which respond to a change from anisotropic to isotropic loading with an immediate loss of stiffness.

SHEARING BEHAVIOUR

The shearing behaviour of isotropically consolidated specimens of Pappadai clay from 25 m depth (P25) has been discussed by Cotecchia & Chandler (1997) and Cotecchia (2003), and that of clay specimens from site B (By and Bg) has been presented by Cafaro & Cotecchia (2001). In the following, the framework of clay behaviour based upon these data is recalled, and compared with new test data. All the shear tests considered in the following were carried out on specimens of 38 mm diameter, except for two tests on 50 mm diameter specimens; this difference in diameter did not give any effect on the results. The axial strain rates used in undrained shear were generally below 0·006 mm/min, and those in drained shear were an order of magnitude lower than in undrained shear.

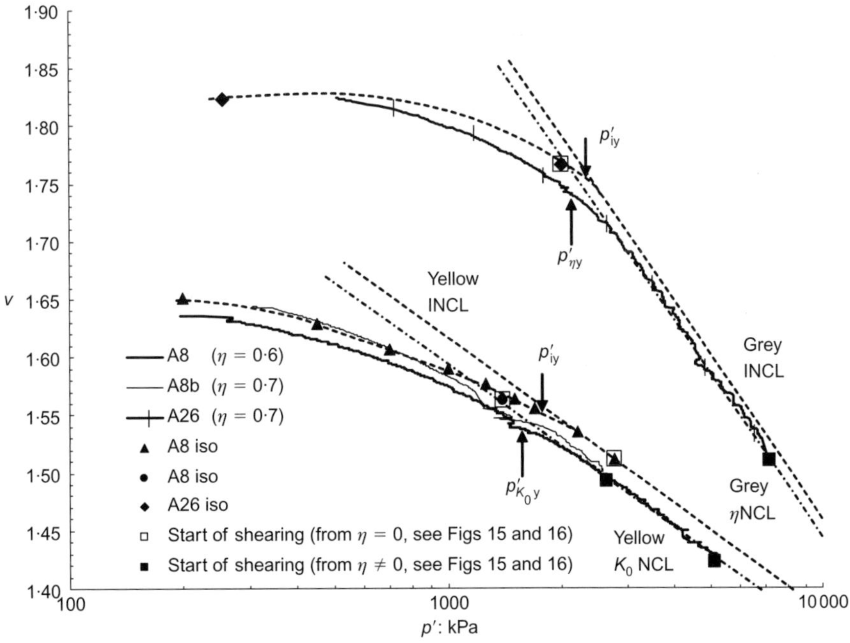

Fig. 8. Isotropic and anisotropic compression curves of natural clays (site A)

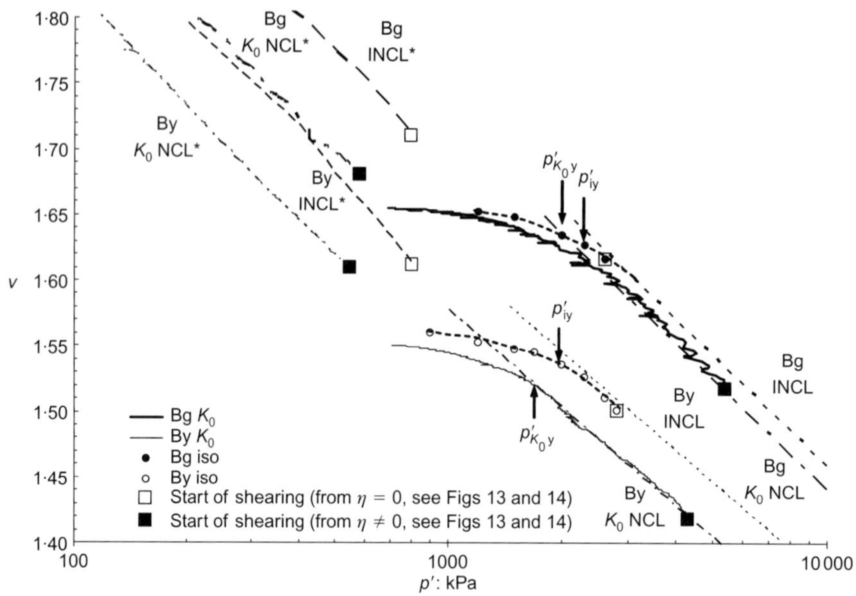

Fig. 9. Isotropic and anisotropic compression curves of natural and reconstituted specimens (site B; data from Cafaro, 1998)

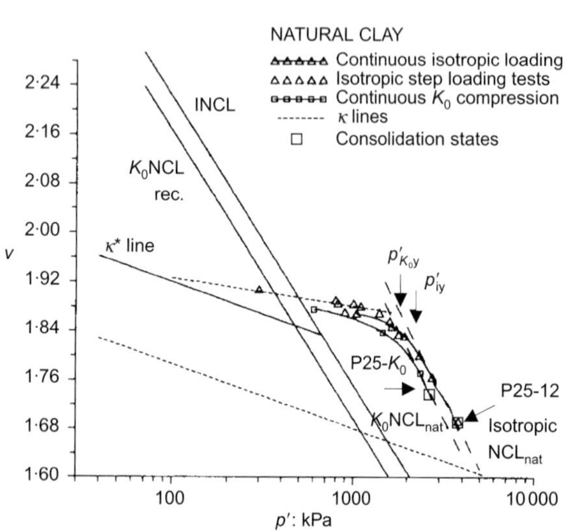

Fig. 10. Isotropic and anisotropic compression curves of natural and reconstituted specimens (site P; after Cotecchia & Chandler, 1997, modified)

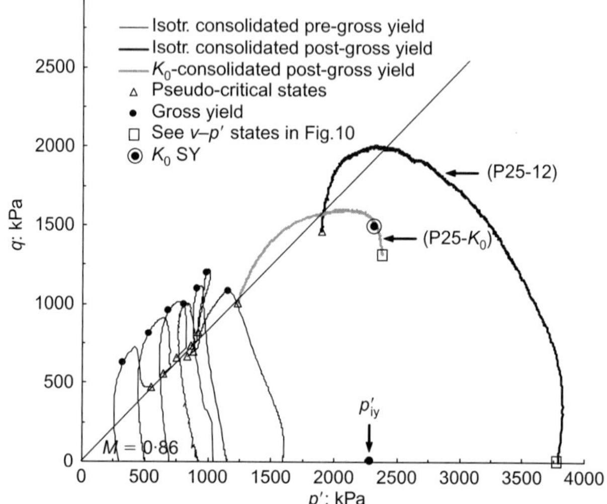

Fig. 11. Undrained stress paths of both isotropically and K_0-consolidated natural clay specimens (P25; after Cotecchia, 2003, modified)

Stress paths and stress–strain response

Cotecchia (2003) reports the stress–strain curves ($q/p'-\varepsilon_s$, $\varepsilon_v-\varepsilon_s$, $\Delta u-\varepsilon_s$, $v-p'$) of triaxial shear tests, either drained or undrained, performed on specimens from sample P25 isotropically consolidated over a large range of pressures ($p'_c = 250-3800\,\text{kPa}$; Fig. 10). Fig. 11 shows the effective stress paths followed by the clay in the undrained shear tests. As p'_{iy} is about 2260 kPa (Fig. 10) for sample P25, the shear data show that the clay exhibits a *dry* behaviour for yield stress ratios YSR_{is} ($= p'_{iy}/p'_c$) above 2 and a *wet* behaviour at lower YSR_{is}. Both on the *dry* and on the *wet* side, from small-medium to large shear strains the stress ratio q/p' is found to be related to the plastic dilation rate $\mathrm{d}\varepsilon_v^p/\mathrm{d}\varepsilon_s^p$ according to (Cotecchia & Chandler, 1997, 2000)

$$\frac{q}{p'} = Q - A\frac{\mathrm{d}\varepsilon_v^p}{\mathrm{d}\varepsilon_s^p} \qquad (1)$$

which is indicative of a basically frictional behaviour.

On the *dry* side, Cotecchia & Chandler (1997) locate the

first shear gross yield about the start of strain-softening (i.e. reducing q/p'; see Fig. 11), which is found to correspond to the maximum dilation rate, according to equation (1). Beyond this state equation (1) applies to the response of the clay (which goes on gross yielding with shear) until q/p' drops to post-rupture values. On the *wet* side, for $\text{YSR}_{is} > 1$ Cotecchia & Chandler (1997) locate the first shear gross yield about a drop in rate of the q/p' increase, whereas for $\text{YSR}_{is} = 1$ the clay is necessarily gross yielding since the start of shearing. During this gross yielding, the deviatoric stress increases according to equation (1) until, at very large strains, it reduces, with q/p' dropping to the same post-rupture value as that on the *dry* side (Fig. 11). Such a post-rupture ratio has been defined as a pseudo-critical stress ratio by Cotecchia & Chandler (2000). The drop in q/p' that does not comply with equation (1), on both the *dry* and the *wet* side, results both from a further increase in the rate of strain-induced degradation of structure (*destructuring*), which occurs at some stage post-shear gross yield, and from localisation of strain.

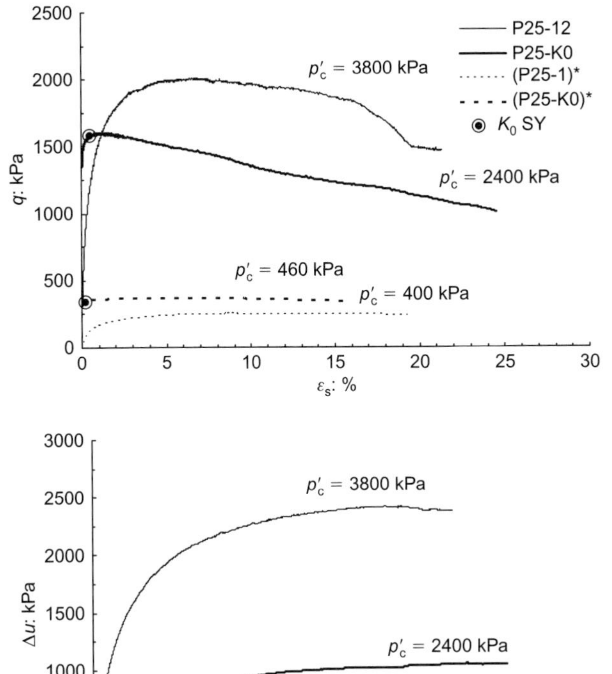

Fig. 12. Undrained stress–strain behaviour of natural and reconstituted clay (P25)

Figure 12 shows the stress–strain behaviour recorded in undrained shear for specimens P25-12 and P25-K_0, isotropically and K_0-consolidated post-gross yield respectively (Fig. 10); the corresponding stress paths are shown in Fig. 11. Fig. 12 also reports the stress–strain curves during shear of reconstituted specimens. These have been one-dimensionally consolidated from slurry up to $\sigma'_{vmax} = 200$ kPa and then either isotropically (P25-1)* or one-dimensionally (P25-K_0)* consolidated in the triaxial (Fig. 10). The data in Fig. 12 show that, as already reported in the literature, after anisotropic consolidation post-gross yield (YSR$_{K_0} = 1$) the clay, both if reconstituted (Henkel & Sowa, 1963; Amerasinghe, 1973; Parry & Nadarajah, 1973; Gens, 1982; Rossato *et al.*, 1992; Coop & Cotecchia, 1995) and if natural (Lefebvre *et al.*, 1983; Jamiolkowski *et al.*, 1985; Tavenas & Leroueil, 1985; Graham *et al.*, 1988; Smith *et al.*, 1992; Callisto & Calabresi, 1998; Amorosi & Rampello, 1998; Saihi *et al.*, 2002), exhibits a stiffer and more fragile response than when isotropically consolidated. The specimen isotropically consolidated post-gross yield (e.g. P25-12 and P25-1*) experiences a continuous positive strain-hardening up to large shear strains, with low dq/dε_s rates. The anisotropically consolidated specimen (e.g. P25K_0 and P25K_0^*) instead exhibits a response characterised by high dq/dε_s and low du/dε_s rates from very small to medium shear strains. In fact, in this stage the clay, despite experiencing shear gross yielding from the start of shearing, seems to stop deforming and destructuring at the rates applying to other post-gross yield loading paths. However, it reaches soon a new threshold of both stiffness decay and increase in Δu rate, which will be referred to as K_0-shear gross yield (K_0SY) in the following (see K_0SY in Fig. 12). K_0SY is a peculiar shear gross yield state, which corresponds to a new acceleration of structure change and plastic straining after a temporary stiffer response. The stress paths of both specimens P25K_0

(Fig. 11) and P25K_0* rise almost vertically to K_0SY, where they start bending to the left owing to the acceleration in the pore water pressure increase. Soon after K_0SY the deviatoric stress reaches a peak value and starts to decrease while the clay is still strain-hardening. However, the fragility of anisotropically consolidated Pappadai clay is less significant than that observed for other anisotropically consolidated clays (e.g. Lacasse *et al.*, 1985; Tavenas & Leroueil, 1985; Rossato *et al.*, 1992; Amorosi & Rampello, 1998).

Cafaro & Cotecchia (2001) report the stress–strain response (q–ε_s, Δu–ε_s) of both the weathered (By) and unweathered (Bg) clay samples from site B when subjected to undrained shear in the triaxial. The data show that also for both these clays the response changes from *dry* to *wet* with reducing YSR$_{is}$, confirming the framework of behaviour reported for Pappadai clay. Fig. 13 reports the undrained stress–strain curves resulting from new tests on both the clays from site B, consolidated either isotropically or one-dimensionally to states post-gross yield (see Fig. 9 for the consolidation states). Fig. 13 also reports the response in undrained shear of the same clays when reconstituted and one-dimensionally normally consolidated (Fig. 9). The data confirm the high initial stiffness and the fragility of the clay when consolidated anisotropically post-gross yield. Such a behaviour applies irrespective of the clay structure, as it is observed for both the unweathered and the weathered clay, in both the natural and the reconstituted state. The K_0SY states identified in the tests are shown in Fig. 13.

The effective stress paths corresponding to the tests in Fig. 13, as well as those resulting from undrained shear tests on specimens from site B isotropically consolidated on the *dry* side, are shown in Fig. 14. As for Pappadai clay, the stress paths of the clays from site B one-dimensionally consolidated post-gross yield are also almost vertical, or even inclined towards the right, up to K_0SY. Thereafter, the increase in rate of pore water pressure rise makes the stress path bend towards the left and renders it rather flat in the top part.

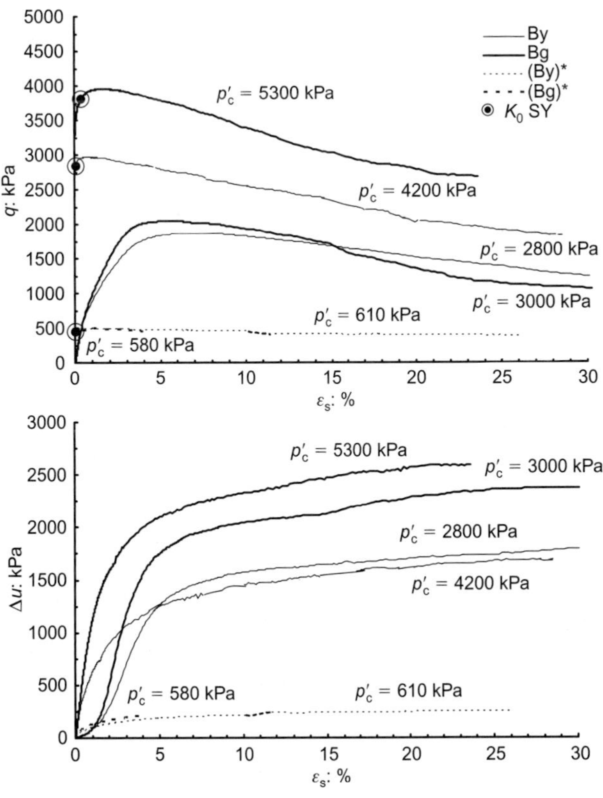

Fig. 13. Undrained stress–strain behaviour of natural and reconstituted clay (site B)

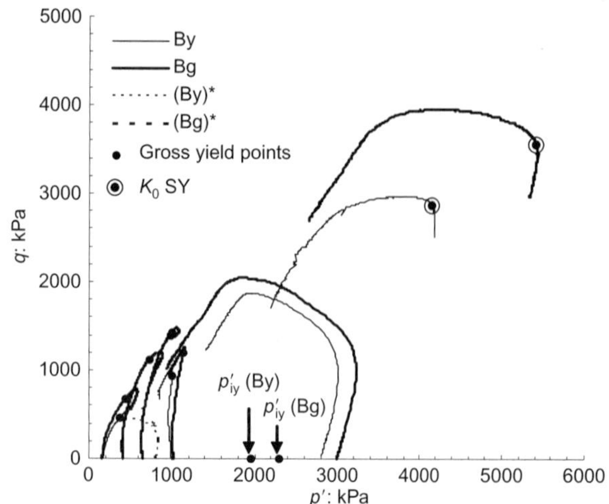

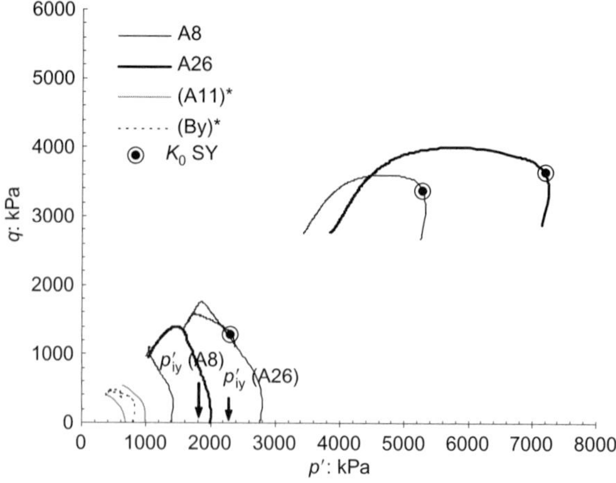

Fig. 16. Stress paths of yellow and grey clay specimens (sites A and B)

Fig. 14. Stress paths of natural and reconstituted clay specimens (site B; data from Cafaro & Cotecchia, 2001, and new data)

Figure 15 reports the stress–strain curves of undrained shear tests on specimens from samples A8 and A26 (Fig. 2a), which are of weathered and unweathered clay respectively. The stress paths of such tests are shown in Fig. 16, which also reports, for comparison, the results of shear tests on reconstituted specimens of the yellow clay samples A11 and BY. The natural specimens in the figures have been consolidated on the *wet* side both just before gross yield and post-gross yield, both isotropically and anisotropically (consolidation states in Fig. 8). Therefore they all exhibit a *wet* behaviour. The data confirm the higher stiffness and fragility of the anisotropically consolidated specimens and the change in clay response occurring at K_0SY.

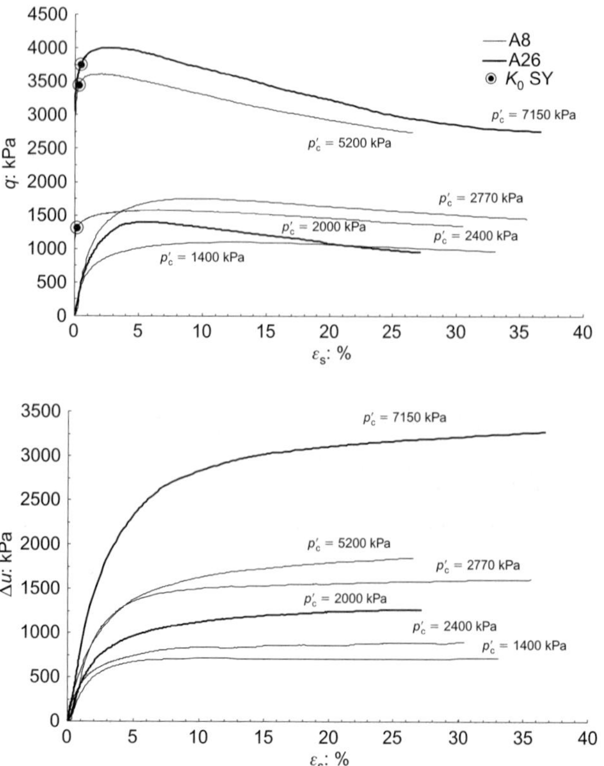

Fig. 15. Undrained stress–strain behaviour of natural clay specimens (site A)

Hardening post-gross yield

As recalled earlier, a substantial difference exists between the hardening laws pre- and post-gross yield, owing to the onset of major plastic strains and structure changes in the material post-gross yield, for both natural and reconstituted clays. Where the Cam Clay models (Roscoe & Schofield, 1963; Roscoe & Burland, 1968) simply make the first yield coincide with gross yield, in more advanced models for reconstituted soils gross yield occurs when the soil state reaches a bounding surface before which the soil is already elasto-plastic (e.g. Al-Tabbaa & Wood, 1989; Stallebrass & Taylor, 1997). Beyond such a state the hardening rule may also account for an evolution of the material anisotropy (e.g. Kavvadas & Amorosi, 2000). Advanced models for natural clays simulate the development of structure degradation with loading, making it particularly important when the soil state lies on the bounding surface (e.g. Kavvadas & Amorosi, 2000; Rouainia & Wood, 2000; Baudet & Stallebrass, 2004). In the following, the onset and evolution of gross yielding along different loading paths are discussed for the clays being studied. Such evolution will be referred to as *gross yield hardening* and will be compared between natural and reconstituted samples. It is worth pointing out that the reconstituted clays of reference in such a comparison have been K_0-consolidated from slurry up to stresses so high as to determine an anisotropic fabric ($\sigma_v' \geqslant 100$kPa; Sfondrini, 1975). Therefore both the natural and the reconstituted clay being compared are characterised by initial anisotropy.

The gross yield states in undrained shear of specimens of natural Pappadai clay (P25), isotropically consolidated pre-gross yield, are shown in Fig. 11. In Fig. 17 these gross yield states, together with those observed in drained shear tests on the same clay isotropically consolidated pre-gross yield, are normalised for volume by means of the equivalent pressure p_e^*, and for the effects of structure by means of the stress sensitivity, S_σ. p_e^* corresponds to the current specific volume on the INCL*. According to Cotecchia & Chandler (2000), S_σ can be calculated with reference to any gross yield pressure (i.e. $S_\sigma = \sigma_y'/\sigma_e^* = p_{K_0y}'/p_{K_0y}^* = p_{iy}'/p_{iy}^*$) if it is assumed that structure causes solely a change in size of the clay response with respect to that of the reconstituted clay. In Fig. 17 S_σ is calculated as p_{iy}'/p_{iy}^* (Table 3), and is constant (equal to 3·2) for specimens consolidated pre-gross yield. Fig. 17 also reports the normalised effective stress paths of specimens consolidated post-gross yield, either isotropically or K_0, and subjected to both undrained and drained shearing. In particular, those sheared undrained are specimens P25-12 ($S_\sigma = 2·6$) and P25-K_0 ($S_\sigma = 2·7$), also shown in Figs 11 and 12. The same figure also shows the

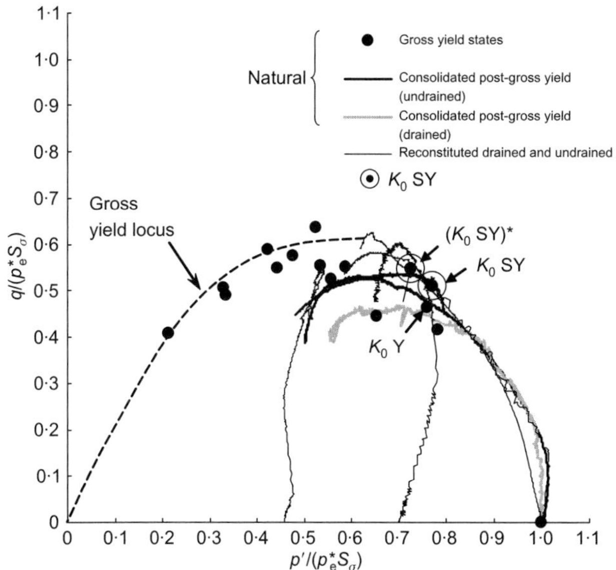

Fig. 17. Natural and reconstituted clay (P25): stress paths normalised for volume and structure (after Cotecchia, 2003, modified)

normalised effective stress paths of reconstituted specimens ($S_\sigma = 1$).

In Fig. 17, the gross yield states of all the specimens isotropically consolidated pre-gross yield and the stress paths of all the specimens isotropically consolidated post-gross yield, both natural and reconstituted, appear to plot about a single curve, if some scatter of the gross yield points is disregarded. In this case, this curve may be considered to represent the normalised gross yield curve of the isotropically consolidated clay, and the natural structure may be seen to give rise to an increase in size of the gross yield curve of the clay with respect to that of the reconstituted clay, not changing its shape significantly (Cotecchia & Chandler, 2000). In addition, the normalised gross yield curve is also seen to coincide with the normalised state boundary surface (SBS) of the isotropically consolidated specimens.

Figure 17 also reports the gross yield state of the natural clay in K_0 compression (K_0Y) and the threshold states of stiffness decay during shear of the K_0-consolidated specimens (both natural, P25-K_0, and reconstituted, P25-K_0*). It appears that, although the normalised stress paths of the K_0-consolidated specimens intersect the normalised stress paths of the isotropically consolidated ones, states K_0Y and K_0SY are close to the normalised gross yield curve shown in the figure. As these gross yield states lie within the same scatter as that of the gross yield states of the isotropically consolidated specimens, in a simplified model the same unique curve could be considered appropriate to represent the gross yield hardening of both K_0-consolidated and isotropically consolidated clay, without additional error. Therefore the curve in the figure could be considered the general normalised gross yield locus of the clay up to large shear strains. This implies that the clay gross yield hardening is controlled mainly by current volume, through p_e^*, and the structure at the end of consolidation, represented by S_σ (which is constant during shearing), until when, at large shear strains (post-q_{peak}), these parameters are not sufficient to normalise the soil behaviour any longer. At this stage additional factors influence gross yielding, for example ε_s^p. In addition, in order to provide an appropriate simulation of the clay behaviour, a constitutive model must account for the dependence of the shear stiffness of the clay on the consolidation stress ratio before shearing.

Figure 18 shows the stress paths and gross yield states of the clays from site B normalised for volume (p_e^*), structure ($S_\sigma = p'_{iy}/p^*_{iy}$) and composition (by means of the critical stress ratio of the reconstituted clay, M^*; Coop et al., 1995, Cotecchia & Chandler, 2000). It appears that for these clays also, the stress paths of the K_0-consolidated specimens intersect those of the specimens isotropically consolidated to $YSR_{is} = 1$, as observed for Pappadai clay. However, in Fig. 18 the gross yield states K_0Y and K_0SY also plot close to a unique normalised gross yield curve, with some approximation.

The unique normalised gross yield curve has been considered by Cotecchia & Chandler (2000) to be generally arch shaped. However, the new data show a bulging on the right side, characterising the normalised stress path of most of the specimens isotropically consolidated to $YSR_{is} = 1$ (both if natural and if reconstituted: Figs 17 and 18). According to the simplified scheme of behaviour being discussed, such bulging also characterises the shape of the normalised gross yield curve.

Figure 19 shows the undrained stress paths of sample A8 (Fig. 16), normalised for volume and structure. In this case the values of both p_e^* and S_σ have been calculated with

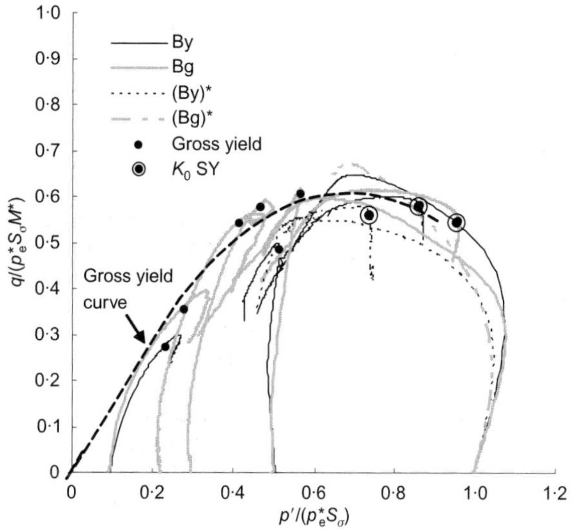

Fig. 18. Natural and reconstituted clay (site B): stress paths normalised for volume, structure and composition (data from Cafaro & Cotecchia, 2001 and new data)

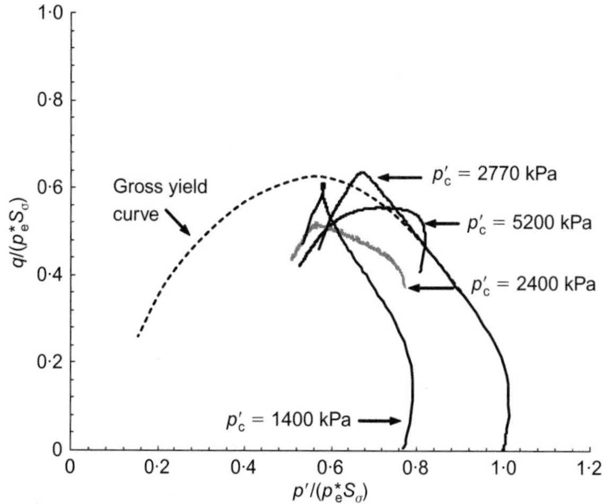

Fig. 19. Undrained stress paths of sample A8 normalised for volume and structure

respect to the INCL* of the weathered clay from site B (By, Fig. 9), as test data for the reconstituted sample A8 were not available. Such normalisation is not as successful as those in Figs 17 and 18 because of the inaccuracy of the values of the normalising parameters. However, it shows that the shape of the stress path of the specimen K_0-consolidated to high pressure ($p'_c = 5200$ kPa) diverges from that of the isotropically consolidated specimen more than for the specimen K_0-consolidated to medium pressure ($p'_c = 2400$ kPa). Therefore the figure emphasises that the material anisotropy evolves with consolidation, as the effects of the consolidation stress ratio on the clay response are seen to increase the further the clay is consolidated.

Figures 17, 18 and 19 give evidence of some of the gross yield hardening features of the clays being examined, as briefly summarised in the following.

If the scatter of the gross yield states about a single normalised gross yield curve is disregarded, the clay being loaded from a pre-gross yield consolidation state (YSR > 1) exhibits gross yield at a state depending solely on the S_σ value at the end of consolidation and on p'_e, that is, on current volume. This applies irrespective of the type of stress path being followed (e.g. drained shear, undrained shear, isotropic compression or anisotropic compression). Therefore the diversity in pre-gross yield plastic shear strains, ε_s^p, among different loading paths does not seem to affect the location of gross yield significantly. In addition, because a single value of S_σ normalises, for different paths, the effects on the gross yielding of both the clay structure at the end of consolidation and the pre-gross yield structure change, such a change does not vary significantly among different loading paths or, otherwise, is generally of a limited size. However, post-q_{peak}, $p_e^* S_\sigma$ alone does not normalise gross yielding any longer, and this gives evidence of the onset of a degradation of structure that is related to both shear and volumetric strains at this stage.

When comparing the stress paths of specimens either isotropically or anisotropically consolidated post-gross yield, Figs 17 and 18 show that, despite the differences in shape of such stress paths, their size is quite well normalised by $p_e^* S_\sigma$. This suggests that for the anisotropically consolidated clay also, a large part of the stress–strain response depends on volume and structure through $p_e^* S_\sigma$, the same as for the isotropically consolidated clay. Thus it can be concluded that, along any compression path post-gross yield, *destructuring* is sufficiently accounted for by S_σ and therefore is dependent mainly on ε_v^p (which S_σ depends on). However, the anisotropically consolidated clay exhibits, in between states $K_0 Y$ and $K_0 SY$, a shear response stiffer than that of the clay isotropically consolidated post-gross yield. Therefore ε_s^p during post-gross yield compression is found to influence significantly the stiffness of the clay upon shearing.

Dilation rates

As said earlier, Cotecchia & Chandler (1997) show that equation (1) applies to the stress–strain data of Pappadai clay subjected to shearing after consolidation pre-gross yield. The intercept Q is defined as M_{nat} for the natural clay and as M^* for the reconstituted clay ($M_{\text{nat}} = 1{\cdot}08$ and $M^* = 0{\cdot}91$). However, the A values are found to differ between drained ($A = 0{\cdot}72$, $A^* = 0{\cdot}53$) and undrained stress paths ($A = 1{\cdot}8$, $A^* = 1{\cdot}47$), and this shows that the shear behaviour is not purely frictional; rather, it depends on the differences in structure evolution along the different stress paths. Table 4 reports the values of M^* and M_{nat} measured for the clays examined in the paper during shearing of specimens consolidated pre-gross yield.

Figure 20 reports the $q/p' - \mathrm{d}\varepsilon_v^p/\mathrm{d}\varepsilon_s^p$ data for specimens

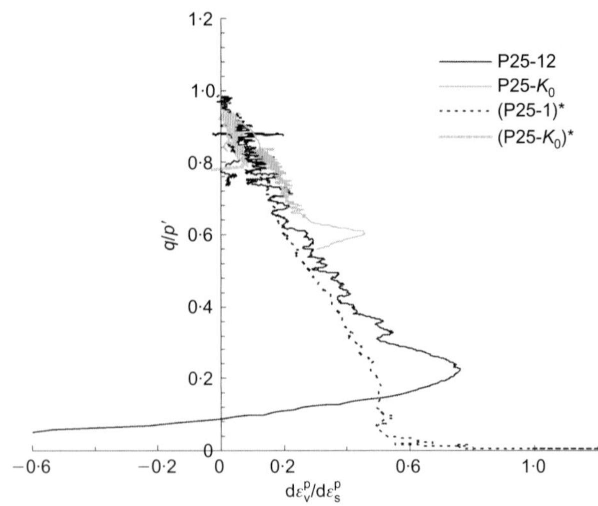

Fig. 20. Dilation rates in undrained shear for natural and reconstituted specimens, both isotropically and K_0-consolidated (site P)

P25-12 and P25-K_0, which were consolidated post-gross yield, and for the reconstituted clay specimens (P25-1)* and (P25-K_0)* (Fig. 12). The data of both the natural specimens fit equation (1) up to the peak q/p' with a single A value, but with a value of M_{nat} (= 1) lower than pre-gross yield and closer to M^*.

Figure 21 shows the $q/p' - \mathrm{d}\varepsilon_v^p/\mathrm{d}\varepsilon_s^p$ data for the tests in

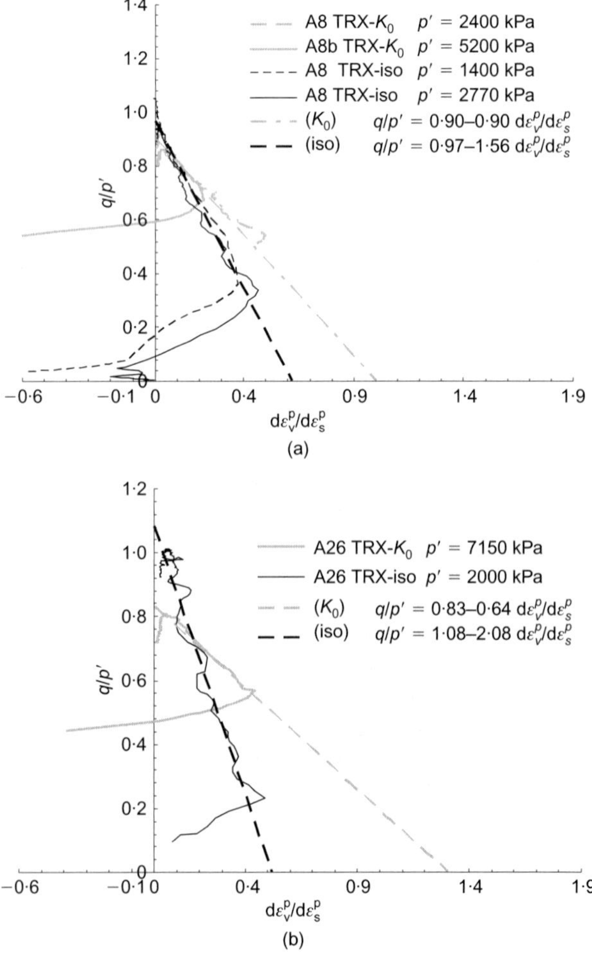

Fig. 21. Dilation rates in undrained shear for natural specimens from site A, both isotropically and K_0-consolidated: (a) yellow clay; (b) grey clay

Fig. 15. As already observed for Pappadai clay, the data fit equation (1) with intercepts M_{nat} that are very close for specimens consolidated, either anisotropically or isotropically, to comparable pressures, but in this case the A value is seen to vary slightly with the consolidation stress ratio. Post-gross yield, increasing p' is seen to cause a decrease in M_{nat}, as seen for Pappadai clay. Therefore equation (1) is seen to always apply to the clay shear response, but with changing values of the parameters. A varies mainly with the stress path and, to a lesser extent, with the consolidation stress ratio, whereas M_{nat} is seen to drop slowly with compression post-gross yield, owing to clay *destructuring*.

CONCLUSIONS

This study has succeeded in assessing similarities and differences between the nature, history, structure and mechanical behaviour of stiff illitic clays deposited in different areas of a marine basin. The index properties show the degree of variability that can characterise a clay deposit, despite being formed within a protected basin. Such variability is reflected in some differences of the intrinsic mechanical parameters of the clays at different locations (e.g. N^*, λ^*, M^*). Differences have been also identified between the structure of the clays in the centre and at the edges of the basin, the first of which is more oriented and sensitive, probably because of the quieter deposition conditions. In addition, weathering is seen to significantly affect the clay structure, causing its strength decay. Such an effect has been recorded in the same way in the centre and about the edges of the basin.

Despite these differences, the stiff clays being studied follow the same framework of behaviour and are characterised by similar gross yield properties. For such clays, gross yield appears not to be much affected by the shear strains taking place along the loading path pre-gross yield, the same as for the reconstituted clays. In addition, also in compression post-gross yield the shear strains do not seem to influence gross yield hardening, which depends mainly on ε_v^p. Therefore plastic deformations do not seem to cause an important evolution of the shape of the gross yield curve. However, the consolidation history and the corresponding ε_s^p have important effects on the clay stiffness decay with straining, and thus on the stress–strain curve and on the shape of the undrained stress path; much less significant are their effects on the size of the stress path. Finally, for stiff clays of the type being considered, the clay structure, despite being initially anisotropic, does not suffer from major loss of strength due to the change from anisotropic to isotropic loading conditions, so that $p'_{iy} > p'_{K_0 y}$.

NOTATION

C_c compression index
C_s swelling index
C_s^*/C_s swell sensitivity
e void ratio
G shear modulus
ICL intrinsic compression line
INCL isotropic normal compression curve
k coefficient of permeability
k gradient of swelling line in v–lnp'
K_0 coefficient of earth pressure at rest $= \sigma'_h/\sigma'_v$
LI liquidity index
M^* q/p' at critical state
M_{nat} intercept of the flow rule of the natural clay
N specific volume on isotropic normal compression line for $p' = 1$ kPa
N_0 specific volume on K_0 normal compression line for $p' = 1$ kPa

N_η specific volume on anisotropic normal compression line for $p' = 1$ kPa
OCR overconsolidation ratio $= \sigma'_p/\sigma'_{v0}$
p' mean normal effective stress
p'_c consolidation mean effective stress
p_e^* mean effective stress on either K_0 or isotropic reconstituted normal compression line at same specific volume as natural clay
$p_{iy}^*, p_{K_0 y}^*$ mean effective stress on reconstituted normal compression line at same specific volume as at gross yield for natural clay; subscripts i and K_0 refer to isotropic and K_0 states respectively
p'_{iy} mean effective stress at gross yield in isotropic compression
$p'_{K_0 y}$ mean effective stress at gross yield in K_0 compression
q deviatoric stress
R overconsolidation ratio in terms of p'
S_σ stress sensitivity $= \sigma'_y/\sigma_e^*$
u pore water pressure
v specific volume
YSR yield stress ratio $= \sigma'_y/\sigma'_{v0}$
ε_s^p plastic shear strain
ε_v^p plastic volumetric strain
η q/p'
λ gradient of normal consolidation lines with constant q/p' in v–lnp'
σ_e^* equivalent pressure, taken on ICL for specific volume of natural clay at gross yield
σ'_p vertical preconsolidation pressure
σ'_v vertical effective stress
σ'_y vertical effective stress at yield
$*$ denotes parameters applying to reconstituted clay

REFERENCES

Al-Tabbaa, A. & Wood, D. M. (1989). An experimentally based bubble model for clay. *Proc. 3rd Int. Symp. on Numerical models in geomechanics, Niagara Falls*, 91–99.

Amerasinghe, S. F. (1973). *The stress–strain behaviour of clay at low stress levels and high overconsolidation ratio*. PhD thesis, Cambridge University.

Amorosi, A. & Rampello, S. (1998). The influence of natural soil structure on the mechanical behaviour of a stiff clay. *Proc. HSSR'98, Napoli* 1, 395–402.

Basma, A. A., Al-Homoud, A. S., Husein Malkawi, A. I. & Al-Bashabsheh, M. A. (1996). Swelling-shrinkage behavior of natural expansive clays. *Appl. Clay Sci.* 11, Nos 2–4, 211–227.

Baudet, B. & Stallebrass, S. (2004). A constitutive model for structured soils. *Géotechnique* 54, No. 4, 269–278.

Burland, J. B. (1990). On the compressibility and shear strength of natural clays. *Géotechnique* 40, No. 3, 329–378.

Cafaro, F. (1998). *Influenza dell'alterazione di origine climatica sulle proprietà geotecniche di un'argilla grigio-azzurra pleistocenica*. PhD thesis, Technical University of Bari.

Cafaro, F. & Cotecchia, F. (2001). Structure degradation and changes in the mechanical behaviour of a stiff clay due to weathering. *Géotechnique* 51, No. 5, 441 453.

Callisto, L. & Calabresi, G. (1998). Mechanical behaviour of a natural soft clay. *Géotechnique* 48, No. 4, 495–513.

Chandler, R. J. (1972). Lias clay: weathering processes and their effect on shear strength. *Géotechnique* 22, No. 3, 403–431.

Ciaranfi, N., Nuovo, G. & Ricchetti, G. (1971). Le argille di Taranto e di Montemesola (studio geologico, geochimico e paleontologico). *Boll. Soc. Geol. Ital.* 90, No. 3, 293–314.

Ciaranfi, N., Pieri, P. & Ricchetti, G. (1988). Note alla carta geologica delle Murge e del Salento (Puglia Centromeridionale). *Mem. Soc. Geol. Ital.* 41, No. 1, 449–460.

Coop, M. R. & Cotecchia, F. (1995). The compression of sediments at the archeological site of Sibari. *Proc. 11th Eur. Conf. Soil Mech. Found. Engng, Copenhagen* 8, 19–26.

Coop, M. R., Atkinson, J. H. & Taylor, R. N. (1995). Strength, yielding and stiffness of structured and unstructured soils. *Proc. 11th Eur. Conf. Soil Mech. Found. Engng, Copenhagen* 8, 55–62.

Cotecchia, F. (1996). *The effects of structure on the properties of an Italian Pleistocene clay.* PhD thesis, University of London.

Cotecchia, F. (2003). Mechanical behaviour of the stiff clays from the Montemesola Basin in relation to their geological history and structure. *Proceedings of the international conference on characterisation and engineering properties of natural soils* (ed. T. S. Tan), Vol. 2, pp. 817–850. Lisse: Swets & Zeitlinger.

Cotecchia, F. & Chandler, R. J. (1995). The geotechnical properties of the Pleistocene clays of the Pappadai valley, Taranto, Italy. *Q. J. Engng Geol.* **28**, No. 1, 5–22.

Cotecchia, F. & Chandler, R. J. (1997). The influence of structure on the pre-failure behaviour of a natural clay. *Géotechnique* **47**, No. 3, 523–544.

Cotecchia, F. & Chandler, R. J. (1998). One-dimensional compression of a natural clay: structural changes and mechanical effects. *Proc. HSSR'98, Napoli* **1**, 103–114.

Cotecchia, F. & Chandler, R. J. (2000). A general framework for the mechanical behaviour of clays. *Géotechnique* **50**, No. 4, 431–447.

De Marco, A., Moresi, M. & Nuovo, G. (1981). Le argille dei bacini di Taranto e di Grottaglie-Montemesola: caratteri granulometrici, mineralogici e chimici. *Rend. Soc. Ital. Mineral. Petrol.* **37**, No. 1, 241–266.

Garavelli, C. L. & Nuovo, G. (1974). Le argille di Montemesola: dati mineralogici e chimici. *Rend. Soc. Ital. Mineral. Petrol.* **30**, 611–642.

Gens, A. (1982). *Stress–strain and strength characteristics of a low plasticity clay.* PhD thesis, University of London.

Gens, A. & Nova, R. (1993). Conceptual bases for a constitutive model for bonded soil and weak rocks. *Proc. Int. Conf. on Hard Soils – Soft Rocks, Athens*, 483–494.

Graham, J., Crooks, J. H. A. & Lau, S. L. K. (1988). Yield envelopes: identification and geometric properties. *Géotechnique* **38**, No. 1, 125–134.

Henkel, D. J. & Sowa, V. A. (1963). The influence of stress history on the stress paths in undrained triaxial tests on clays. In *Laboratory shear testing of soils,* ASTM Special Technical Publication No. 361, pp. 280–291. Philadelphia, PA: ASTM.

Hight, D. W., Bond, A. J. & Legge, J. D. (1992). Characterisation of the Bothkennar clay: an overview. *Géotechnique* **42**, No. 2, 303–347.

Kavvadas, M. (1994). On the mechanical behaviour of bonded soils. *COMETT Seminar on Large Excavations*, Brussels.

Kavvadas, M. & Amorosi, A. (2000). A constitutive model for structured soils. *Géotechnique* **50**, No. 3, 263–273.

Jamiolkowski, M., Ladd, C. C., Germaine, J. T. & Lancellotta, R. (1985). New developments in field and laboratory testing of soils. *Proc. 11th Int. Conf. Soil Mech. Found. Engng, San Francisco* **2**, 57–153.

Lacasse, S., Berre, T. & Lefebvre, G. (1985). Block sampling of sensitive clays. *Proc. 11th Int. Conf. Soil Mech. Found. Engng, San Francisco* **2**, 887–892.

Lefebvre, G., Ladd, C. C., Mesri, G., & Tavenas, F. (1983). *Report of the Testing Committee.* Committee of Specialists on Sensitive Clays on the NBR Complex, SEBJ, Montreal, Annexe I.

Leroueil, S. & Vaughan, P. R. (1990). The general and congruent effects of structure in natural soils and weak rocks. *Géotechnique* **40**, No. 3, 467–488.

Leroueil, S., Tavenas, F., Brucy, F., La Rochelle, P. & Roy, M. (1979). Behaviour of destructured natural clays. *Proc. ASCE* **105**, No. GT6, 759–778.

Martinez-Nistal, A., Veniale, F., Setti, M. & Cotecchia, F. (1999). A scanning electron microscopy image processing method for quantifying fabric orientation of clay geomaterials. *Appl. Clay Sci.* **14**, No. 4, 235–243.

Parry, R. H. & Nadarajah, V. (1973). A volumetric yield locus for lightly overconsolidated clay. *Géotechnique* **23**, No. 3, 450–454.

Roscoe, K. H. & Schofield, A. N. (1963). Mechanical behaviour of an idealised 'wet' clay. *Proc. Eur. Conf. Soil Mech. Found. Engng, Wiesbaden* **1**, 47–54.

Roscoe, K. H. & Burland, J. B. (1968). On the generalized stress-strain behaviour of wet clay. In *Engineering plasticity* (eds J. Heymann and F. A. Leckie), pp. 535–609. Cambridge: Cambridge University Press.

Rossato, G., Ninis, N. L. & Jardine, R. J. (1992). Properties of some kaolin-based model clay soils. *Geotech. Test. J.* **15**, No. 2, 166–179.

Rouainia, M. & Muir Wood, D. (2000). A kinematic hardening constitutive model for natural clays with loss of structure. *Géotechnique* **50**, No. 2, 153–164.

Saihi, F., Leroueil, S., La Rochelle, P. & French, I. (2002). Behaviour of the stiff and sensitive Saint-Jean-Vianney clay in intact, destructured and remoulded conditions. *Can. Geotech. J.* **39**, No. 2, 1075–1087.

Schmertmann, J. H. (1969). Swell sensitivity. *Géotechnique* **19**, No. 4, 530–533.

Sfondrini, G. (1975). Caratteristiche microtessiturali e microstrutturali di alcuni sedimenti argillosi connesse con la natura ed il tipo di sollecitazioni subìte. *Geol. Appl. Idrogeol.* **10**, No. 2, 300–320.

Sides, G. & Barden, L. (1970). The microstructure of dispersed and flocculated samples of kaolinite, illite and montmorillonite. *Can. Geotech. J.* **8**, No. 3, 391–400.

Simeone, V. (1991). *Fragilità e decadimento della resistenza di una argilla pleistocenica marina sovraconsolidata.* PhD thesis, University of Bari.

Skempton, A. W. (1970). The consolidation of clays by gravitational compaction. *Q. J. Geol. Soc. London* **125**, 373–412.

Smith, P. R., Jardine, R. J. & Hight, D. W. (1992). On the yielding of Bothkennar clay. *Géotechnique* **40**, No. 2, 257–274.

Stallebrass, S. E. & Taylor, R. N. (1997). The development and evaluation of a constitutive model for the prediction of ground movements in overconsolidated clay. *Géotechnique* **47**, No. 2, 235–253.

Tavenas, F. (1981). Some aspects of clay behaviour and their consequences on modelling technique. In Laboratory shear strength of soil, ASTM STP 740, pp. 667–677. West Conshohocken, PA: ASTM International.

Tavenas, F. & Leroueil, S. (1985). Discussion. *Proc. 11th Int. Conf. Soil Mech. Found. Engng, San Francisco* **5**, 2693–2694.

Tavenas, F., Des Rosiers, J. P., Leroueil, S., La Rochelle, P. & Roy, M. (1979). The use of strain energy as a yield and creep criterion for lightly overconsolidated clays. *Géotechnique* **29**, No. 3, 285–304.

Wood, D. M. (1995). Kinematic hardening model for structured soils. *Proc. 5th Int. Symp. on Numerical Models in Geomechanics, Davos*, 83–88.

INFORMAL DISCUSSION

Session 1
Formation and Engineering Geology of Stiff Clays

CHAIR: DR RAMUES GALLOIS AND DR MICHAEL DE FREITAS

Discussion:

The Chairman (Dr de Freitas) explained that questions referring to Prof. Chandler's Key Note Paper should be included in this Discussion period and further suggested four subjects for consideration:

1. What cannot be explained by lithology alone?
2. How is geological history converted in to engineering parameters?
3. What role will geology have to play?
4. How should stiff clay be described?

Mr Peter Hobbs, *British Geological Survey*

My feeling is that lithology description is essential but I think what we are missing is the detailed stratigraphy in a lot of important clay formations. There are great thicknesses of the material which in many cases have not been sub-divided as has been described for the London Clay here and I think if you look at the London Clay databases themselves you will find there is very little detailed stratigraphic data in many of them. My question is—is there some form of engineering surrogate for detailed stratigraphy that we can develop in the future that doesn't rely entirely on palaeontology or more specialised disciplines?

Dr Ramues Gallois *Chairman*

Yes, there is. The problem that has developed in recent years is that stratigraphical research carried out by universities and the Geological Survey over the past two hundred years has gone out of fashion. Britain now produces few palaeontologists and although there are plenty of stratigraphical data in the literature, much of it old but still valid, there are not many people who can take advantage of it and apply it to site investigations. One of the proxies for such studies is geophysical borehole logging. For example, natural-gamma-ray logs can be used to make detailed inter-borehole correlations over distances of tens to hundreds of metres in clays without the need for continuous cores or a knowledge of the fossils. All one has to do is match the gamma-ray traces in much the same way that one matches wallpaper patterns. It is easy to use, works through borehole casings, and is relatively inexpensive. A portable logger that produces digital and paper readouts currently costs about £7,000. One can log at the rate of about 1 metre a minute in most clays so standing-time costs are often small. There are other geophysical logging methods that can be used, but these all have practical and/or cost disadvantages for use in shallow site-investigation boreholes. Detailed stratigraphical studies are difficult and often costly to arrange because they require continuous cores and there are so few specialists available to carry out the work. I am therefore surprised natural gamma-ray logging, a simple technique that has been available for over 40 years, is so rarely used on site.

Dr Michael de Freitas *Chairman*

What about the way in which the materials change with time? Is there some way in which we can look at our materials in terms of the way they have changed or the rates at which they change. There is some very sophisticated work in these proceedings that has required an understanding of the way material changes, the way apparent consolidation pressures change with time, and the way permeability changes with time.

Dr Angus Skinner, *Engineering Consultant*

What has been described to us are materials which have undergone very substantial changes over very long periods of time. The stiff clays we have been discussing have been affected by pore water chemical concentration changes, temperature and time—three factors, each of which has a separate effect. For example, in terms of time you can strain the material at different rates. Geological time has been taken to produce a material which has different properties to that which can be produced in the time-scale available for laboratory testing. Everything that has been said this morning can actually be put into a strain rate and temperature framework. We are talking about strength, but what strength? Strength determined at what strain rates? Intrinsic compression has been mentioned, but the data for this is obtained using different techniques, different laboratory instrumentation, and different strain rates. If you produce the compression lines by incremental loading you get different results as compared with continuous loading at various loading rates. Loading over geological time has produced the properties of the stiff clay materials. The rates at which we test them affects the parameters we measure. Strain rate is a fundamental influence about which very little is spoken.

Dr Michael de Freitas *Chairman*

Rates of change can be very significant as Dr Skinner has pointed out. Taking account of the rate effects in the formation of stiff clays is a lot more sophisticated than taking samples and looking at material grading, water content and consistency limits. Perhaps we need better tools with which to examine our soils—tools which could be developed with both geological and engineering insights. We have heard how geophysics may have much more to offer in linking lithology to geotechnical properties. Possibly understanding strain rate responses could offer another tool.

Dr Robert May, *Atkins*

I'd like to ask a particular question if I may on that issue, it used to be common to run oedometer tests on stiff clays to determine the consolidation parameters including the preconsolidation pressure—that has become less fashionable and confidence has decreased in oedometer tests for stiff clays except perhaps for reconstituted materials, the intrinsic tests that Professor Chandler was talking to us about. Should we have confidence in oedometer tests and can we use such tests to develop a geological understanding of diagenesis in stiff clays and its links with their engineering parameters?

Professor David Potts, *Imperial College London*

I would like to raise some issues with regard to the discussion topic, 'What role will geology have to play?'

When analysing geotechnical structures, there are two main ingredients that are crucial to the analysis, namely the current soil stiffness and strength. Usually, if the soil is a clay, we have information on its undrained strength. However, undrained strength is not a fundamental property of the soil as it depends on the current state of the soil and the way it is subsequently stressed. Consequently, it is not an input parameter to many of the constitutive models that we currently use in our analyses. In this regard, I have a question for Professor Chandler. What is the point of presenting relationships between the state of the soil and a particular value of its undrained strength if this is not the strength we require for analysis? Surely it is the drained strength that is more useful. We have heard this morning that understanding the geology is both difficult and confusing. This leads me to raise the following provocative question (as devil's advocate). If this is so, why don't we forget about trying to understand the detailed geology and just test the soil to obtain information on its strength and stiffness, and use this with our past experience to obtain the input parameters for our analysis?

Professor Richard Chandler, *Imperial College*

Since Professor Potts mentioned my name, one of the things he said was that we use our experience. What is that experience? It's some knowledge of the background of the geology. What I was putting forward (in my keynote lecture) was a simple framework that tells you where you are in geological terms and how this relates to in situ properties. The in-situ can be related to intrinsic properties, which are indices. They are not going to give you absolute numbers that you can use for your numerical analyses. Those you have to get by way of more sophisticated testing. But what you do need is some overall framework within which to

characterise your material. When you are dealing with London Clay in London you know an enormous amount about it. You may know exactly where it fits in every calibration curve. Other materials are not quite so straightforward. You need to know what it is in terms of the material type and its history; where it has come from; and what processes it has been through, because these all affect the overall behaviour. I think we need to go back occasionally and ask ourselves what is our basic framework when we are looking at the materials. How and why are we reaching decisions on key aspects of material behaviour. For example, why at Empingham had no-one really taken on board the very complex geology of those cambered slopes? Understanding the geology and having a good framework of material behaviour helps you to ask the right questions to determine key parameters for engineering analyses.

Dr Michael de Freitas *Chairman*

Professor Potts raises a very important point and there are people who actually believe it—that it doesn't really matter how the soil develops its properties, all we need is the number for our analysis. The problem is we still do not know how to link the two and there is much work still to do. This research field is wide open for our young engineering geologists.

Mr David Beadman, *Tony Gee and Partners*

In my opinion, the most valuable test is a case history because that is testing the full size structures. The difficulty with using many case histories is deciding whether the particular case history is applicable to your current project. This is where the geology is important. Because many case histories are not properly set in their geological context, it is difficult to know whether or not the results are relevant to your current project. I suggest that good geological descriptions would greatly improve the usefulness of many case histories.

Dr Michael de Freitas *Chairman*

I completely agree; editors should not delete the geological details from case histories in the belief they are unnecessary: they are essential.

Session 2

Laboratory and In Situ Techniques and their Interpretation

Gasparre, A., Nishimura, S., Coop, M. R. & Jardine, R. J. (2007). *Géotechnique* **57**, No. 1, 19–31

The influence of structure on the behaviour of London Clay

A. GASPARRE*, S. NISHIMURA†, M. R. COOP† and R. J. JARDINE†

An intensive investigation is described into the London Clay units from Heathrow Terminal 5. Intrinsic properties and composition were established, and relating the behaviour of intact and reconstituted samples allowed the effects of the clay's natural structure to be identified at all depths. Structure varied between units, but some general features emerged that have not been seen in other stiff clays. In particular, intact samples follow paths under isotropic or K_0 compression that fail to provide well-defined gross yield points, or converge with the unit's intrinsic compression lines. Structure contributes to the enhanced shear strength of intact unfissured clay and affects the initial stiffness relationships. Some identifiable features of the intact units' fabric correlated directly with their behaviour. Natural fissures within the clay, the most important mesofabric feature, had an important impact on shear strength and led to an unusual pattern of directional dependence, and particle orientation trends identified by scanning electron microscopy governed the strong elastic stiffness anisotropy. The potential for destructuration through swelling to low effective stresses was also studied and found significant for subsequent volumetric compression behaviour, but not for shearing. Hight *et al.* synthesise the data from this and companion papers presented by the authors, and discuss the practical consequences of the results obtained.

KEYWORDS: compressibility; fabric/structure of soils; geology; laboratory tests; stress path

Cet article présente une étude approfondie d'unités de London clay (argile de Londres) issues du Terminal 5 de l'aéroport d'Heathrow. Les propriétés intrinsèques et la composition de cette argile y sont déterminées. La comparaison du comportement d'échantillons intacts et reconstitués a également permis d'identifier les effets de la structure naturelle de l'argile quelle que soit la profondeur. La structure variait en fonction des unités, mais certaines caractéristiques générales sont apparues qui n'avaient pas été observées pour d'autres argiles raides. En particulier, les spécimens intacts ont suivi des chemins en compression à K_0 ou isotrope qui n'ont pas permis d'obtenir des points de limites d'élasticité bien définis ou convergent avec les lignes de compression intrinsèque de l'unité. La structure contribue à la résistance au cisaillement accrue de l'argile non fracturée intacte et influe sur les relations relatives à la rigidité initiale. Certaines caractéristiques identifiables de l'organisation des unités intactes affichent une corrélation directe avec leur comportement. Les fissures naturelles dans l'argile, représentant la caractéristique de mésostructure la plus importante, ont montré un impact important sur la résistance au cisaillement et ont conduit à un profil inhabituel de dépendance directionnelle. Les profils d'orientation des particules identifiés par microscopie électronique à balayage gouvernent la forte anisotropie de la rigidité élastique. Cet article étudie également le potentiel de déstructuration par gonflement à des contraintes effectives faibles. Il a été reporté comme étant significatif pour un comportement en compression volumétrique subséquent, mais pas pour le cisaillement. Hight *et al.* ont synthétisé ces données et celles des articles associés présentés par les auteurs, et discutent des conséquences pratiques relevant des résultats obtenus.

INTRODUCTION

A key characteristic of geologically old, often overconsolidated, stiff clays is a post-sedimentation structure, as defined by Cotecchia & Chandler (2000), that adds to the sedimentation structure associated with deposition and virgin compression (Burland, 1990). Such clays, which tend to have isotropic or K_0 compression yield stresses significantly greater than any previously applied maxima, have been examined by Coop *et al.* (1995), Burland *et al.* (1996), Cotecchia & Chandler (1997, 2000), Amorosi & Rampello (1998) and others. However, their interest tended to focus on samples from a single depth, or within a weathering profile (e.g. Cafaro & Cotecchia, 2001). The availability of high-quality samples from two deep rotary boreholes and multiple blocks taken at Heathrow Terminal 5 (T5) enabled an investigation of how the post-sedimentation structure varies with depth, and between geological sub-units, in a single relatively thick sequence of unweathered London Clay.

King (1981, 1991) describes the detailed geological history of the London Clay, which was deposited in an Early Eocene sedimentary basin covering much of northern and north-eastern Europe. The soils deposited in the UK were eroded from the surrounding Jurassic shales, Greensand, Chalk and lateritic Eocene soils. Their mineralogy varies spatially, and is dominated in Central London by illite and smectite, becoming more kaolinitic westwards (Huggett, pers. comm., 2005). In all cases the soils were thoroughly bioturbated after deposition.

Sea level generally rose during the Early Eocene, but a series of transgressive-regressive cycles also took place, leading to water depth changes and sediment variability. One result is the relatively coarse nature of the Hampshire Basin clays, which were closer to the edge of the depositional basin. Another is the vertical system of units and sub-units associated with changes in depositional environment and age. Each unit was associated with reducing water depth and sediments that coarsened upwards until sea level rises took place and a new sequence began; any sea level reductions led to erosion and the deposition of coarser soils. Deposition was slower in the deeper water at the site of the London

Manuscript received 5 May 2006; revised manuscript accepted 14 November 2006.
Discussion on this paper closes on 1 July 2007, for further details see p. ii
* Geotechnical Consulting Group, London, UK; formerly Imperial College, London.
† Imperial College, London, UK.

Basin, leading to less pronounced changes than at the Hampshire edge of the depositional basin. The resulting stratigraphic changes correlate with index property and microfossil profiles (Hight *et al.*, 2002; de Freitas & Mannion, 2007).

The units present in Central London were buried by continuous deposition before experiencing tectonic activity during the Alpine orogeny. In addition to the London Basin synclinal folding, faulting with a relative slippage of several metres may have developed, leading to discontinuities in the units (Chandler *et al.*, 2004). Substantial erosion took place later during the Tertiary and Pleistocene, removing much of the clay originally deposited in the London Basin. Around 175 m is estimated to have been eroded at Heathrow, excising all of King's units D and E and part of C, as is typical of Central London. The remaining units (A, B and the base of C) provide a total thickness of 52 m. Thames terrace gravels were deposited during the late Quaternary, which protected the underlying clay from the effects of weathering. Six metres existed at Heathrow, restricting clay alteration to the uppermost metre (Hight *et al.*, 2002). These gravel beds were removed at some T5 locations when sewage treatment lagoons were constructed in the 1930s.

Bishop *et al.* (1965), Burland (1990) and others have reported variations with depth in London Clay Formation properties, but their studies were performed without the benefit of King's geological framework. Due account has been taken more recently by Hight *et al.* (2002), Standing & Burland (2006) and others, but these case studies could not include fully comprehensive research into the units' mechanical properties. The work described herein placed particular emphasis on investigating the influences of structure through a sophisticated laboratory testing programme that was designed and interpreted in the context of the new geological framework.

SAMPLING, APPARATUS AND PROCEDURES

Samples were obtained throughout the depth of the London Clay from two adjacent and continuously sampled rotary boreholes, both advanced using a triple-barrel arrangement and a natural polymer foam flush technique. The samples were retrieved from their liners as soon as they were recovered from the borehole, and an outer annulus of around 5 mm thickness was trimmed off before preserving the samples in layers of clingfilm (plastic wrap) and wax. Additional 300 mm cube block samples were taken from the benches of an excavation about 140 m away; the blocks were cut at depths below the top of the clay of 1·2, 5·2 and 10·5 m. Gravel was present at the borehole locations but had been removed at the block sampling sites. Further details about the site and sampling procedures are given by Hight *et al.* (2007).

The tests described in this paper were carried out in a variety of stress path apparatus, oedometers, a high-pressure triaxial apparatus with a 5 MPa capacity, and a hollow cylinder/resonant column apparatus (HCA). While the HCA had a fixed sample size of about 180 mm in length and 38 mm and 70 mm inner and outer diameters, sample sizes of 38 mm and 100 mm diameter were used for the triaxial tests on natural samples so that the influence of sample size could be investigated. Tests performed with a second HCA are described by Gasparre *et al.* (2007) and Nishimura *et al.* (2007). The high-pressure triaxial apparatus used samples of 50 mm diameter, and all the reconstituted samples were 38 mm in diameter. In all cases a 2:1 height to diameter ratio was used. The triaxial apparatuses were equipped with local axial strain sensors, either of an inclinometer type (Jardine *et al.*, 1984) or LVDTs (Cuccovillo & Coop, 1997).

Radial strain belts, mid-height pore pressure probes and bender elements were also deployed in some cells. The latter were either platen mounted, measuring G_{vh} or mounted across the mid-height of the sample, measuring G_{hh} and G_{hv}. All of the triaxial and HCA samples used radial drainage to minimise test times, which nevertheless frequently extended up to a month. The HCA was of a Drnevich type with a Hardin resonant column oscillator that measured G_{vh} on vertically oriented samples. Full details of the apparatus and testing programmes are given by Gasparre (2005) and Nishimura (2006).

The intact samples were trimmed by hand to the required dimensions. The inner cavities of the HCA samples were cut on a metal working lathe, inserting gradually increasing drill bit sizes and finishing with a reaming tool. The reconstituted triaxial samples were made from the trimmings of the intact samples, mixed at water contents about 1·25 times the liquid limit and compressed to around 30 kPa vertical stress in 38 mm diameter consolidometers before being trimmed to length and placed in the triaxial apparatus. The reconstituted HCA samples, made from slurry with water contents of 1·5 times the liquid limit, were compressed in a 90 mm diameter consolidometer to 730 kPa before being trimmed in a similar way to the intact samples. In all cases the void ratios were measured from the initial and final water contents, bulk unit weight and dry unit weight, with an average being taken.

The intact triaxial and HCA samples were consolidated either isotropically to their in situ p' levels, or to their estimated anisotropic in situ q and p' stress points following approach paths that reproduced the site's recent geological history of erosion and then terrace gravel deposition. In the HCA tests the in situ stresses were calculated and applied for each individual sample; in the triaxial tests representative stresses for each unit were adopted. Fig. 1 shows the typical approach paths followed when the samples were consolidated to their in situ states. Note that with shallower samples the estimated K_0 values could not be reached, either because of the proximity of the failure envelope or perhaps an overestimation of K_0 in situ, as described by Gasparre *et al.* (2007) and Nishimura *et al.* (2007). The path adopted was to approach the estimated K_0 as closely as possible without exceeding a volumetric strain of about 1%. Shearing of the triaxial samples was typically strain controlled and undrained, in either compression or extension, because of the long test durations required for drained shearing. The HCA samples were also sheared undrained, but with control of the σ_1' axis direction (angle α with respect to the vertical) and the intermediate stress ratio b $(= (\sigma_2' - \sigma_3')/(\sigma_1' - \sigma_3'))$. Some of the intact triaxial samples were compressed to stresses higher than in situ to examine the shape of the intact boundary surface, whereas others were swelled isotropically to low stresses (10 kPa) before being recompressed (typically to 100 kPa) before shearing, to investigate the effects of destructuration through swelling. In the analysis of

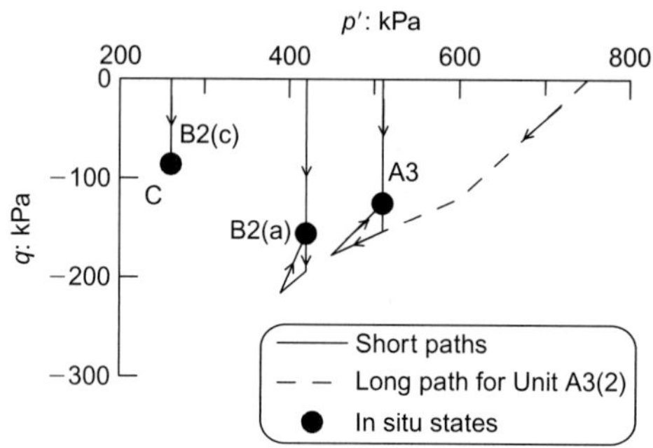

Fig. 1. Approach stress paths for different lithological units

the triaxial tests, standard area corrections have been used, assuming a right-cylinder unless a shear plane formed when the correction of Chandler (1966) was adopted. Corrections were also made for the restraint of the membrane (La Rochelle *et al.*, 1988).

The reconstituted triaxial samples were all isotropically compressed to a variety of stress levels and then sheared undrained, considering only behaviour in compression. The reconstituted HCA samples were K_0-consolidated to stresses that had the same stress ratio (q/p') and overconsolidation ratio as the intact samples tested, but with lower stress levels. This approach was adopted because of the practical difficulty of reproducing the actual stress history in the laboratory with large samples.

The intact oedometer samples were typically set up at their in situ σ_v' values, and were then either (a) loaded to as high a stress as possible or (b) swelled first to a lower σ_v' (25 or 10 kPa). To reach the highest stresses, the oedometer samples had diameters of either 50 mm or 38 mm. In addition to the tests performed by the authors, eight commercial tests carried out by Surrey Geotechnical Consultants are also included in the analysis of the data.

NATURE AND STRUCTURE OF THE CLAYS

Site profiles of the most relevant properties are summarised in Table 1, and Atterberg limits and natural water contents are shown in Fig. 2 along with the unit boundaries identified by King (1981) and applied to the Heathrow site by Hight *et al.* (2002). Unit A1 is now recognised separately as the Harwich Formation, and unit B1 is a thin sandier layer from which no samples were recovered. Unit B2 has been separated into three sub-units in the basis of their biostratigraphy (de Freitas & Mannion, 2007). From these data the differences between units are not particularly clear, although the highest plasticity seems to be in sub-units B2(a) and B2(b), with lower-plasticity soils above and below in units A, B2(c) and C. A clearer representation of the difference in the nature of the soils is given by the particle size distributions in Fig. 3. Units A3(2), B2(c) and C contain 3–6% fine sand, whereas sub-units B2(a) and B2(b) had fine sand fractions of 0·5–1·5%. X-ray diffraction analyses of three samples gave the results shown in Table 2. Units C and B2(a) gave similar results, in contradiction to the grading and index test data, whereas unit A3(2) has a greater proportion of illite compared with smectite. A3(2) also has the lowest clay to quartz ratio, which correlates with its lower plasticity and higher sand content. The chlorite content of the sample from unit C confirms that it has not been weathered.

Scanning electron micrographs (SEMs) of three samples from different units are shown in Fig. 4. The sample from unit A3(2) clearly shows that the clay structure is more packed and orientated at depth, whereas the shallower unit C has a more open and disturbed fabric. Anisotropy might therefore be expected to increase with depth. Units A3(2)

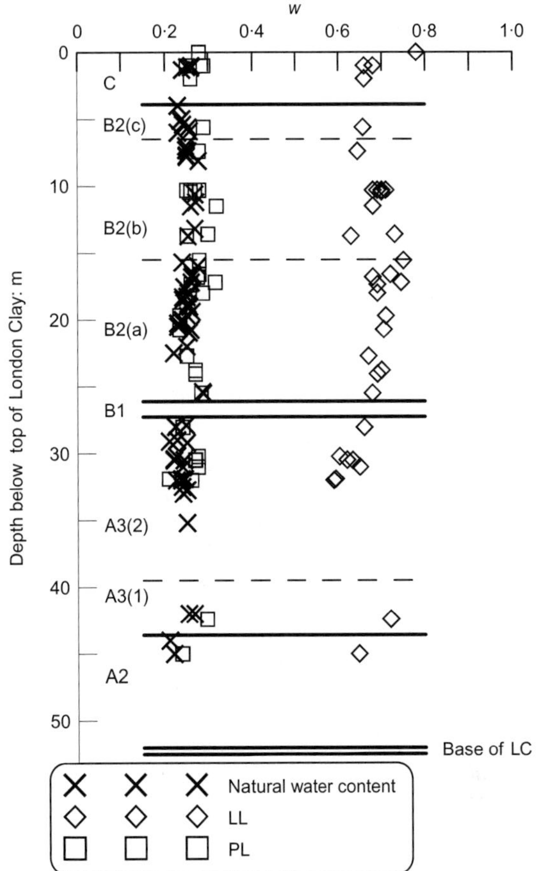

Fig. 2. Profiles of Atterberg limits and natural water contents

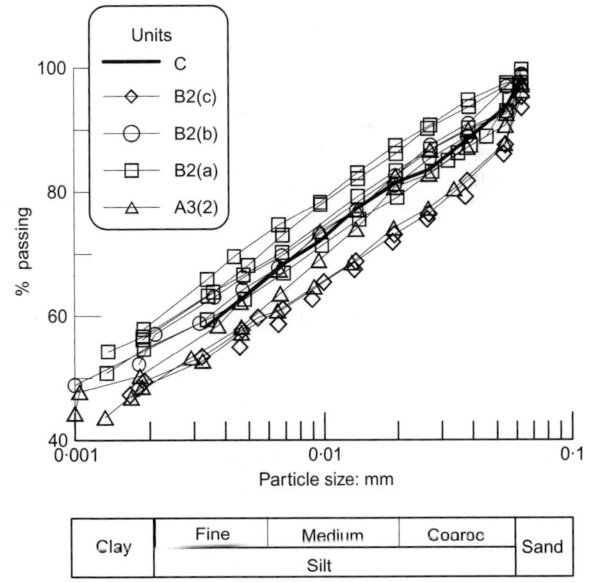

Fig. 3. Particle size distributions

Table 1. Site profile of index properties and in situ void ratios (single values indicate that only one test was conducted)

London Clay Unit	G_s	e_0	w_0: %	LL: %	PL: %	PI: %	CF: %	A
C	2·74	1·68–1·76	23–26	78–66	25–29	37–50	49–62	0·69–0·82
B2(c)	2·65	1·66–1·69	23–26	65–66	28–29	37	41–50	0·73
B2(b)	2·77	1·64–1·76	24–28	63–75	25–32	38–47	54–63	0·67–0·83
B2(a)	2·76	1·65–1·82	22–29	67–75	23–32	40–48	55–62	0·69–0·77
A3(2)	2·77	1·58–1·72	21–25	60–66	24–28	33–62	51–54	0·71–0·78
A3(1)		1·67–1·69	23–25	59–60	21–26	33–39	49	0·67
A2		1·61–1·69	21–27	65–72	24–30	41–42		

Table 2. X-ray diffraction analyses of three samples

London Clay Unit	Depth from top of London Clay: m	Illite: %	I-rich illite-smectite: %	Random S-rich illite-smectite: %	Chlorite: %	Kaolinite: %	Clay:quartz
C	1	22	2	58	4	15	34·0
B2(a)	16	21	3	63	3	11	36·8
A3(2)	27	38	2	40	6	14	22·4

(a)

(b)

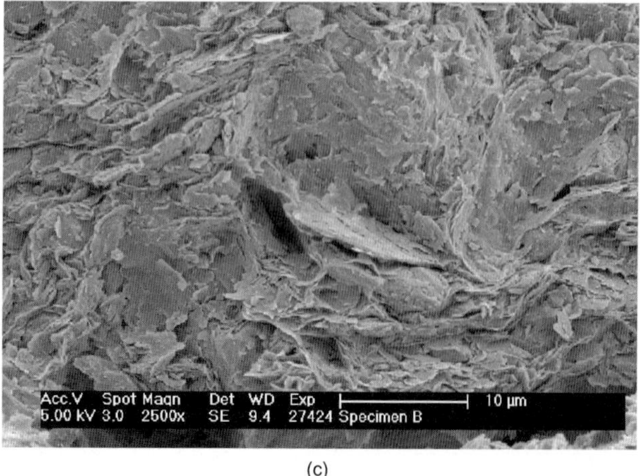

(c)

Fig. 4. Scanning electron micrographs of three samples from units: (a) C; (b) B2(a); (c) A3(2)

and C also have a greater proportion of coarser particles, as seen in Fig. 3, whereas unit B2(a) has a more homogeneous fabric, perhaps because of bioturbation. Although there is a calcite crystal in the image for unit C, there is no evidence of any amorphous calcite coating that might provide a cementing agent. The presence of calcite, together with the absence of decalcified sediments and cryptocrystalline iron, confirms that the unit C sample, taken 1 m below the clay surface, is unweathered.

The principal mesofabric (i.e. visible without magnification) features are the silt and sand partings and the fissures. Many of the larger silt and sand laminations observed in the past may well correspond to unit boundaries, while King (1981) observed that the smaller partings are more common in unit A than in B or C. Observations from block sampling and sample preparation in the laboratory revealed that the natural fissures tend to have a variable spacing from a few centimetres to a few tens of centimetres. The pattern of the natural fissures is distinguished from that formed if the samples were allowed to dry (which led to a predominantly horizontal set of additional cracks), or due to testing. Unlike tectonic shears, which are far less frequent (Chandler *et al.*, 1998), the natural fissures tended to be discontinuous and extend typically up to around 15 cm, with orientations that were typically ±30° from either the horizontal or vertical (Skempton *et al.*, 1969). Most previous authors have noted that fissure frequency reduces with depth (e.g. Skempton *et al.*, 1969; Chandler & Apted, 1988). A detailed analysis of frequency and orientation was not attempted here, and the fissures were often hard to identify in samples prior to testing. Analysis of the triaxial compression tests on 100 mm diameter samples indicated that the percentage of samples failing on pre-existing fissures varied, with about 50% for units B2(a) and B2(b) combined, dropping to 17% for A3(2) and down to zero for B2(c), C and A2. The tendency to fail on fissures correlated with the plasticity as strongly as with depth. Fewer samples were tested in extension than in compression, and extension tests were not successfully completed on samples from deeper units A3 and A2. However, the percentage of extension tests failing on pre-existing fissures increased to 70% for unit B2(a), 33% for B2(b), 50% for B2(c) and 80% for unit C. The greater tendency to fail on pre-existing fissures in extension than in compression reflects the flatter orientation of the former's planes of maximum stress obliquity (τ/σ'_n) and their closer alignment with the generally sub-horizontal pattern of natural fissuring.

INTRINSIC BEHAVIOUR

The oedometer test data for reconstituted samples are plotted in Fig. 5(a) and the isotropic compression curves from the triaxial tests in Fig. 5(b). Consistency was generally observed between the slopes of the one-dimensional and isotropic compression curves for the different lithological units. Reconstituted samples from unit A3(1) were tested

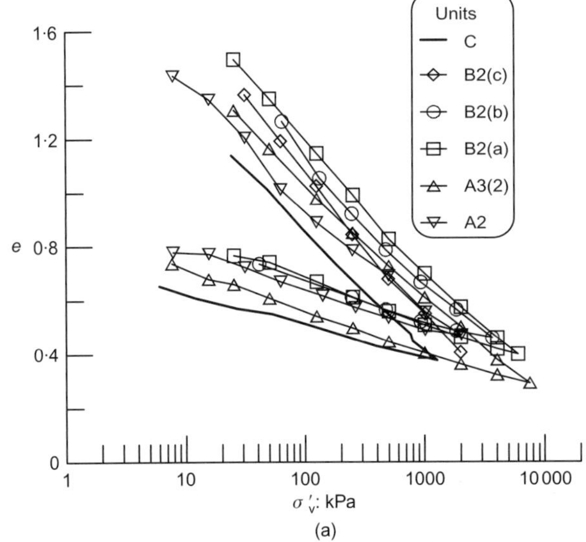

(a)

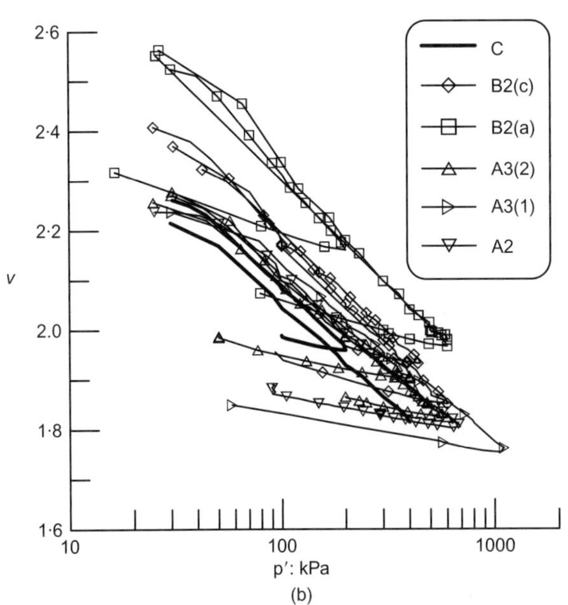

p': kPa

(b)

Fig. 5. Compression curves of reconstituted samples: (a) one-dimensional curves from oedometer tests; (b) isotropic compression curves from triaxial tests

only isotropically (in the triaxial apparatus), whereas those from unit B2(b) were tested only in the oedometer. Although there is some variation in the plasticity and grading of the soils, the gradients of the intrinsic compression and swelling lines were similar for all units, giving a C_c^* of 0·386 and C_s^* of 0·184. The effect of plasticity is revealed in the vertical location of the intrinsic compression lines (ICLs), which are significantly higher for the more plastic sub-units B2(a) and B2(b). Within each unit or sub-unit, however, there appears to be relatively little scatter, indicating that differences between the units are more important than differences within each one.

In Fig. 6 the shearing data from the undrained triaxial tests on reconstituted samples have been normalised, each with respect to an isotropic ICL for its unit, by means of an equivalent pressure, p_e^*,

$$p_e^* = \exp\left(\frac{N - v}{\lambda}\right) \quad (1)$$

where λ and N are the gradient of the isotropic ICL and its

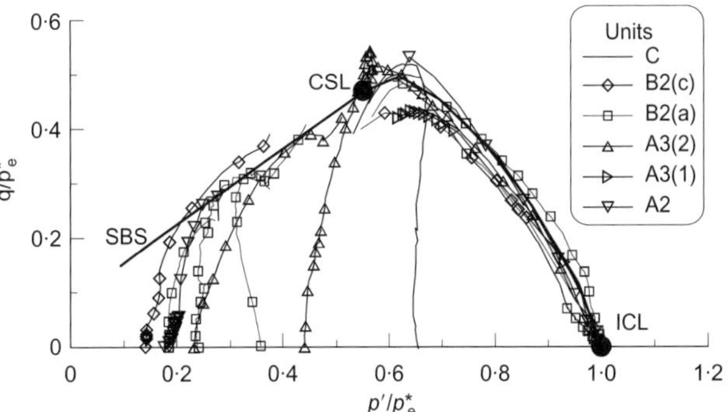

Fig. 6. Normalised shearing data from undrained triaxial tests on reconstituted samples

intercept at 1 kPa on a v–$\ln p'$ plot (v = specific volume). Once the effect of the different units on the ICL location has been accounted for, there is no remaining effect of unit on the intrinsic state boundary surface (SBS*) identified from isotropically consolidated samples, and the separation of the ICL and critical state line is similar for each unit, as is the critical state value of q/p' ($= M$) of about 0·85 (ϕ_{cs}' of 21·3°). (Samples consolidated under different regimes may develop a range of local boundary surfaces within their true overall state boundary surfaces; see Jardine et al., 2004.)

INFLUENCE OF STRUCTURE ON COMPRESSION BEHAVIOUR

The oedometric compression data for the intact samples of the various units are given in Fig. 7. The data have been normalised with respect to the location of the one-dimensional ICL of each unit by means of the void index I_v (Burland, 1990):

$$I_v = \frac{e - e_{100}^*}{e_{100}^* - e_{1000}^*} = \frac{e - e_{100}^*}{C_c^*} \quad (2)$$

where e_{100}^* and e_{1000}^* are the void ratios on the ICL corresponding to vertical effective stresses $\sigma_v' = 100$ kPa and 1000 kPa. At higher stresses, some of the ICLs diverge, showing slightly different curvatures.

The compression paths for the intact samples indicate little variability within each unit, but with significant differences between units. The I_v normalisation applied in Fig. 7 suggests, perhaps misleadingly, that the effects of structure reduce with

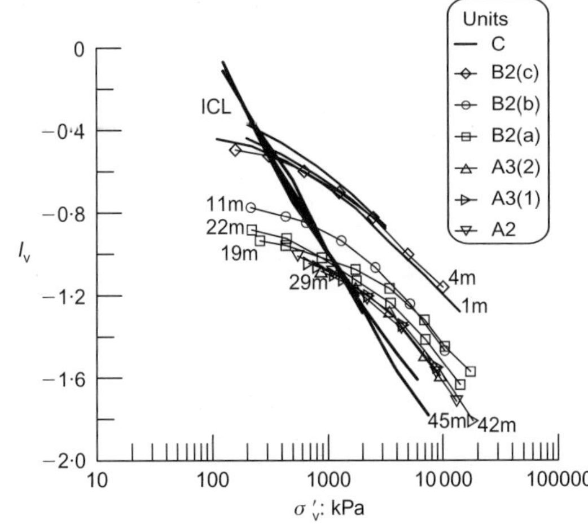

Fig. 7. Compression curves of natural samples normalised by the void ratio

depth, with the distances by which the intact samples' compression paths extend beyond the ICL decreasing with depth. The onset of gross yield is poorly defined in the semi-log plot given in Fig. 7, and does not become clearer when log–log axes are adopted (Gasparre, 2005). Unusually, the graph also shows a tendency for some of the compression paths to continue to diverge from the ICL, even at high stresses. It is possible that convergence might have been achieved if still higher stresses had been applied. The observed profile of swell sensitivity S_s (= C_s^*/C_s; Schmertmann, 1969) is plotted in Fig. 8 along with the values obtained by substituting the C_s values found after compression to the maximum stress for each sample. The initial 'in situ' values are typically in the range 2 to 3·2, indicating a marked influence of structure but (unlike I_v) no clear trend with depth. However, lower ratios of 1·1 to 1·6 were found after compression to high stresses (up to 18 MPa). Noting that completely destructured samples would be expected to give $S_s = 1$, it is clear that significant destructuration developed within the samples under the applied compressive pressures, despite the natural samples' lack of convergence with ICLs. As a result of the progressive destructuration in compression and the poorly defined gross yield points, the 'oedometer sensitivity' parameters that have been applied by Smith (1992) and Cotecchia & Chandler (1997) to quantify the effects of structure in lower OCR sediments were not found as effective with the London Clay. Gasparre & Coop (2007) discuss normalisation of compression data and the effects of structure in greater detail.

The lack of convergence of the compression curves of the intact samples with the ICL, and the relatively low initial values of swell sensitivity, are consistent with Gasparre's interpretation that the main structural differences between the intact and reconstituted London Clay lie in their microstructural interparticle arrangements, or fabric, rather than bonding. As noted earlier, no clear evidence was seen of cementing in the SEM images.

INFLUENCE OF INTACT STRUCTURE ON PEAK SHEAR STRENGTH

The stress–strain behaviour of all lithological units was essentially dilatant and strain-softening, but with strain localisation truncating the dilatant response at various stages and so promoting shear plane development. Full details of the stress–strain behaviour of the many tests undertaken are given by Gasparre (2005) and Nishimura (2006). Two examples showing the range of observed shearing behaviour are shown in Fig. 9, plotting data from (a) one sample that mobilised its full intact strength and (b) a second that failed by shearing along a pre-existing fissure. The arrow on each curve indicates the point at which strain localisation was first observed. Generally, samples that mobilised their intact strength developed their peak strengths at axial strains between 2% and 4% before manifesting sharp falls and reaching a post-rupture strength at larger strains. With the 100 mm diameter samples the shear plane usually formed before the peak strength was developed, and the subsequent failure process involved the progressive propagation of the shear surface through the sample. Samples from unit B2(c) also showed a tendency to develop multiple shear planes, which seemed to suppress the peak dilation. In contrast, the 38 mm samples that showed intact strength reached their peak strengths at the same time as the shear surface was first observed. Samples that sheared along pre-existing fissures showed a similar initial response to intact samples, but were unable to develop the same peak strengths. Their stress–strain curves were truncated by slipping on fissures, and showed earlier convergence towards their post-rupture strength.

Figure 10(a) presents the peak shear strength envelopes for each unit, as established from triaxial tests, excluding data from those samples that failed on pre-existing fissures (indicated by crosses). The data show good agreement with the peak envelopes defined by Bishop *et al.* (1965) at Ashford Common. In order to identify the effect of the samples' natural structure, each point is normalised with respect to the equivalent pressure p_e^*, defined above, using the isotropic ICL applying to the particular unit. No triaxial tests on

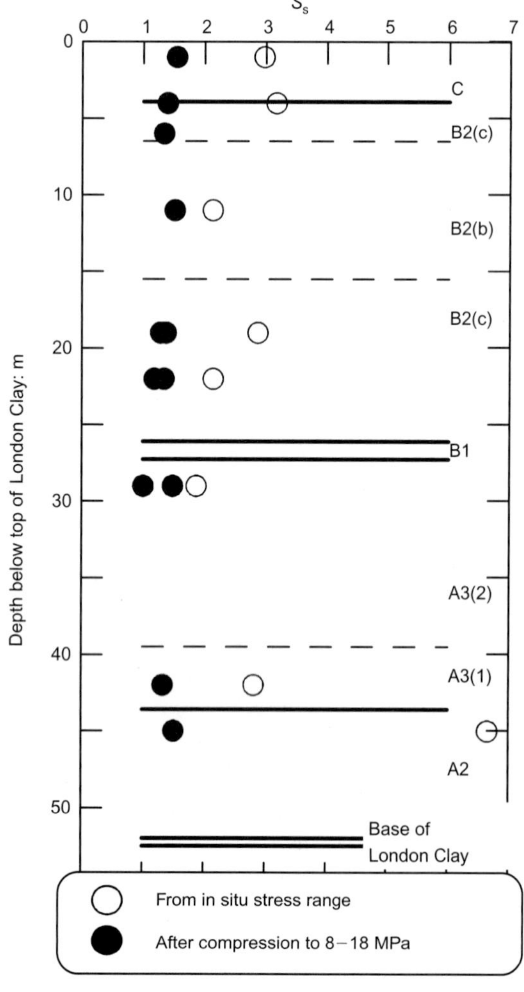

Fig. 8. Swell sensitivity

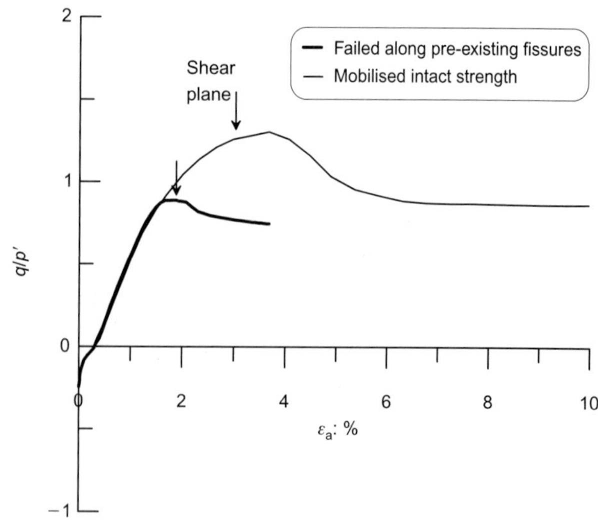

Fig. 9. Typical stress–strain curves for undrained shearing in compression (example from unit A3(2))

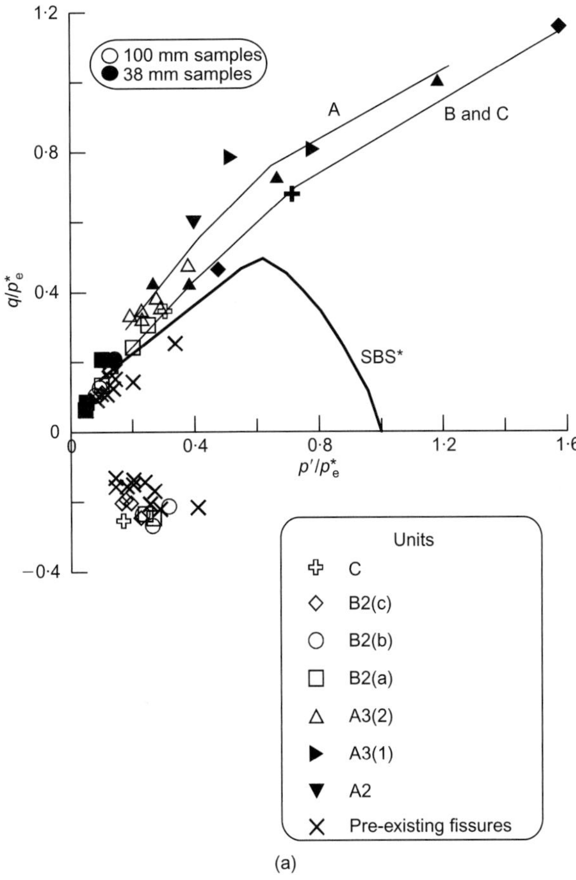

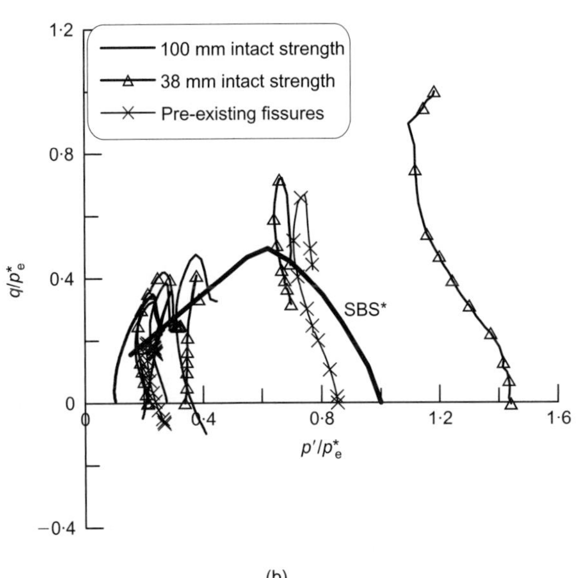

Fig. 10: Normalised stresses: (a) peak states for samples from different units (open symbols 38 mm samples and closed 100 mm; units A3(1) and A2 38 mm samples only); (b) normalised stress paths for unit A3(2)

reconstituted samples were carried out for unit B2(b), so the isotropic ICL for unit B2(a) was adopted for normalisation; oedometer tests on reconstituted samples had shown that these two sub-units had similar compression behaviours.

The normalised effective stress paths for unit A3(2) are given in Fig. 10(b), and are typical of the paths seen for all units. In compression, the peak states plot significantly higher than the intrinsic SBS* for isotropically consolidated samples, indicating one feature of the clay's natural structure. Again there are clear differences between units, but the variability within each unit or sub-unit is less significant. A

single envelope may be defined for all of unit A, with a lower envelope for the whole of units B and C. The higher normalised envelope for unit A suggests an increase of the effect of structure with depth, unlike the normalised oedometer data in Figs 6 and 7. The ratio of the normalised strength taken at the intrinsic critical state is about 1·2 for intact samples from unit B2 and 1·35 for unit A3.

Some researchers (e.g. Amorosi & Rampello, 1998; Cotecchia & Chandler, 2000) have used high-pressure triaxial testing to identify the wet side of the intact SBS of stiff clays. However, the clays they tested had better-defined gross yield points in compression than the London Clay. Only a limited number of high-pressure tests were conducted on the London Clay because the trends seen in the K_0 and isotropic compression tests indicated that very high pressures would be required to define the wet side of the intact SBS. As mentioned earlier, these showed that compression paths were mostly still diverging from the ICL at 18 MPa even though the associated changes of S_s indicated that significant destructuration had occurred. It follows that any SBS identified from high-pressure probing tests would not reflect the full structure of the intact soil. Bishop et al. (1965) had shown stress paths for the wet side of the NCL, and tried to define the shape of the intact state boundary surface for samples from three main depths at Ashford Common, possibly corresponding to units B2, A3 and A2. The effects of destructuration discussed above might, however, have affected the shape interpreted for the three depths.

The samples that failed on pre-existing fissures, which are indicated by crosses in Fig. 10(a), plot significantly below the peak intact envelopes. As discussed earlier, many of the extension samples failed on pre-existing fissures. Considering only the compression and extension tests that were not influenced by fissures, Fig. 10(a) indicates similar peak trends for 38 mm and 100 mm samples for p'/p_e^* values of less than 0·4, suggesting no clear effect of sample size. However, the 100 mm samples are more likely to contain favourably orientated pre-existing fissures than 38 mm samples, and this leads to fissures affecting many samples, and therefore a scale effect for the average strength. A key problem in the laboratory testing of London Clay is that the deposit's fissure spacings lead to dimensions for a representative element volume (REV) that would give repeatable behaviour, which are impractically large. Instead, even at the 100 mm sample size there is a bipolar dataset, with either a fissure present that has an orientation that influences the strength, or not. A true REV would typically give strengths that are intermediate between the boundaries formed by the intact peak strength and the single fissure behaviour (e.g. Vitone & Cotecchia, 2007), rather than the lower bound that is often used for London Clay (Marsland, 1971).

When a sample fails on a discrete shear plane the strength can no longer be analysed in terms of axisymmetric stress invariants, and the stresses acting on the shear plane should be determined by means of a Mohr's circle analysis using the angle of the shear plane measured from the sample after testing. Failures that develop on surfaces that are markedly steeper or shallower than the planes of maximum stress obliquity (τ/σ_n') and/or develop at unexpectedly low values of ϕ_p' are indicative of the influence of pre-existing fissures. Mohr's circle analysis was the principal means by which fissure-dominated failures were identified in the triaxial series, as fissures were generally not visible during sample set-up. However, fissure patterns were more evident prior to testing with the HCA specimens that involved a different preparation procedure.

Figure 11 shows the stresses mobilised on the eventual failure plane both at peak and in the post-peak or 'post-rupture' state (Burland, 1990), both for newly created failure

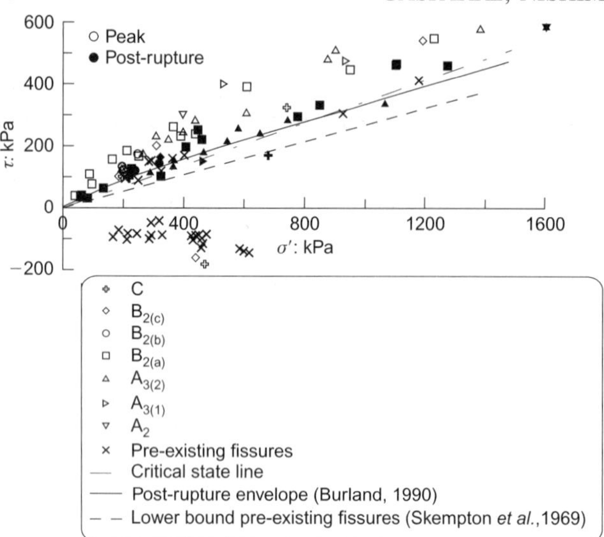

Fig. 11. Stresses mobilised both at peak and in the post-rupture state

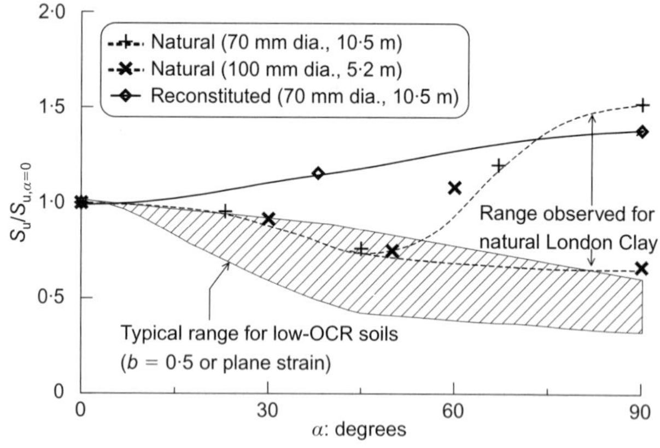

Fig. 12. Variation of S_u with α for London Clay and various low-OCR K_0-reconstituted soils

planes through intact soil and for pre-existing fissures. There is little difference between the post-rupture strengths of the initially intact samples and those with pre-existing fissures, indicating that the natural fissures have not been subjected to large movements. The post-rupture angle of shearing resistance, ϕ'_{pr}, of intact samples and samples sheared along pre-existing fissures is also similar to that at the critical state for the reconstituted soil (about $21\cdot6°$); it is possible that the post-rupture state involves a localised critical state on a shear plane. There was no discernible difference between the values of ϕ'_{pr} of the various units.

Only samples from the more plastic units were tested in extension, and many of these failed on pre-existing fissures. Nevertheless, the differences between the peak and post-rupture extension strengths plotted in Fig. 11 appear to be less pronounced than in compression. The strength envelopes for each unit have not been drawn in Fig. 11 to avoid obscuring the data. However, given the limited number of samples showing intact strengths, by inspection the effects of lithology on the peak extension envelopes are difficult to distinguish. Plotted in $\tau-\sigma'_n$ space this peak extension envelope appears to fall below the compression envelope of the same units. However, the extension envelope of samples that failed on pre-existing fissures again shares the same angle of shearing resistance as compression tests in their post-rupture stages.

Nishimura *et al.* (2007) discuss the anisotropic shear strength characteristics revealed under more general shearing conditions. The clay's natural micro- and mesoscale structure leads to the characteristics summarised in Fig. 12, where the effect on the undrained shear strength S_u of the angle of inclination of the major principal stress axis from the vertical, α, is investigated for experiments that were all conducted with the intermediate principal stress controlled to give $b = (\sigma'_2 - \sigma'_3)/(\sigma'_1 - \sigma'_3) = 0\cdot5$. Considering a dataset of block sample specimens taken from $5\cdot2$ m and $10\cdot5$ m below ground level, the peak values of S_u have been normalised with respect to that for $\alpha = 0$. The data range for various K_0-reconstituted soils, quoted from past studies, represents tests with b values between $0\cdot3$ and $0\cdot5$ and OCRs of $1-4$. It is clear that the heavily overconsolidated London Clay samples used here have a very different anisotropy, with shear strength increasing rather than decreasing with α for the reconstituted samples, reflecting the microstructure created by pre-compression. The influence of

α on the strength of the natural soil is significantly different, with a minimum developing with $\alpha \approx 45°$ as a result of the mesofabric and in particular the fissuring. Pre-existing fissures played a significant role in most of the tests conducted at higher α values, and the shear strength obtained at this α range is subject to a strong dependence on the fissures present in individual samples. These and other features are discussed further in the companion paper (Nishimura *et al.*, 2007).

An additional factor that contributes to undrained shear strength anisotropy is the non-vertical directions of the undrained effective stress paths developed in HCA and triaxial tests. The tendency for p' to fall as $(\sigma_2 - \sigma_\theta)$ increases for the intact samples can be seen from a comparison of Fig. 5 and Fig. 8(b), but is more clearly shown in Fig. 13, which compares HCA tests for undrained shearing with $b = 0\cdot5$ on natural and reconstituted samples with similar overconsolidation ratios (approximately 9). Here the data are normalised with respect to the initial mean normal effective stress p'_0. The inclined straight line plotted on this diagram is a theoretical prediction made by Nishimura (2006) assuming the strong elastic cross-anisotropy found experimentally, as discussed below, and by Gasparre *et al.*

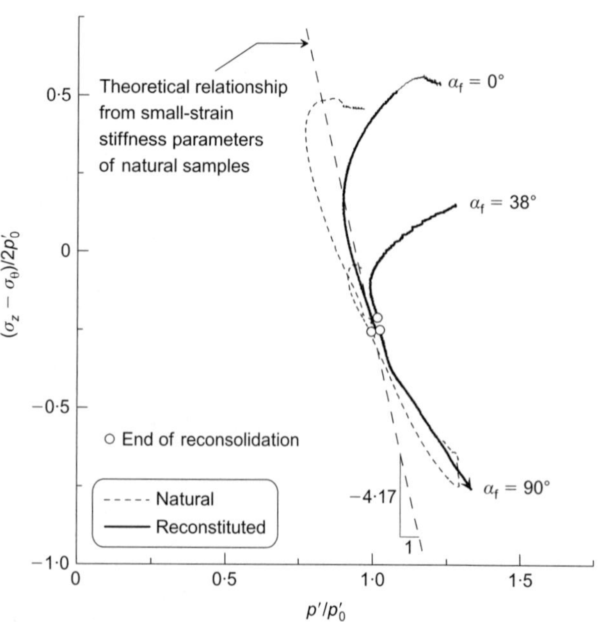

Fig. 13. Effective stress paths observed for natural and reconstituted London Clay from $10\cdot5$ m tested in HCA (α_f is α at failure)

(2007). The deviations of the intact samples' effective stress paths from the vertical are greater than for the reconstituted specimens, indicating a more highly anisotropic microfabric in the natural clay.

EFFECTS OF SWELLING TO LOW STRESS LEVELS

The compression paths for the oedometer samples that had been initially swelled to low stresses are given in Fig. 14. It is clear that the swelling had a significant effect on both initial C_s and the yield in compression. The major effect of swelling is an increase of compressibility of the samples, which is particularly evident for the more plastic units (B2(c) and B2(a)), which show C_s values around 75% higher, whereas those from the less plastic unit A3 grow by around a third. However, when the triaxial samples from unit B2(b) were swelled to low stresses before recompression to higher stresses and shearing, there was no discernible effect on the normalised strength envelope (Fig. 15). This is consistent with the results of shear tests performed by Takahashi et al. (2005) on London Clay samples from Units B2 and A3.

INFLUENCE OF STRUCTURE ON SHEAR STIFFNESS

Gasparre et al. (2007) describe how the anisotropic elastic stiffness properties of intact and reconstituted London Clay samples were measured, employing a range of dynamic and static triaxial and HCA testing techniques. Tests on reconstituted samples were confined to platen-mounted triaxial apparatus bender element tests and resonant column HCA experiments, both giving dynamic values of G_{vh}. Dual-axis bender element triaxial tests were conducted in the triaxial tests on intact specimens, along with resonant column tests on intact HCA samples and multi-axis static tests of both kind involving local strain sensors. The bender element measurements were interpreted with a first arrival technique with a sine input wave, but were checked using frequency domain methods, the differences being insignificant for both the intact and reconstituted soil. The reconstituted samples were all isotropically normally consolidated and then swelled to investigate the effects of overconsolidation. The normally

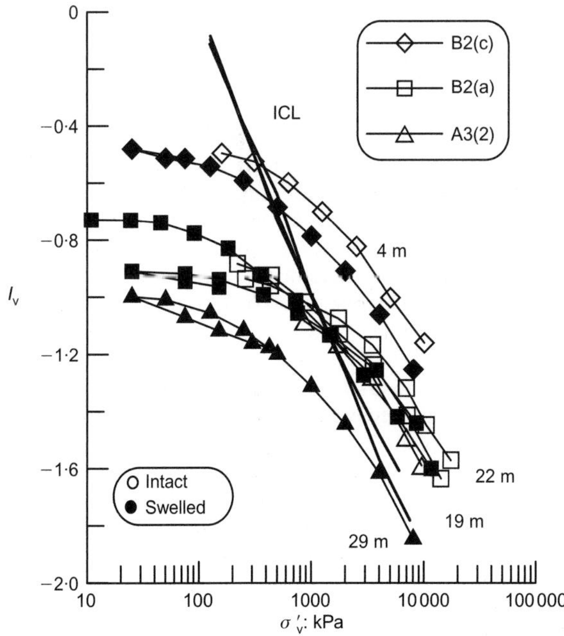

Fig. 14. Compression paths for the oedometer samples initially swelled to low stresses

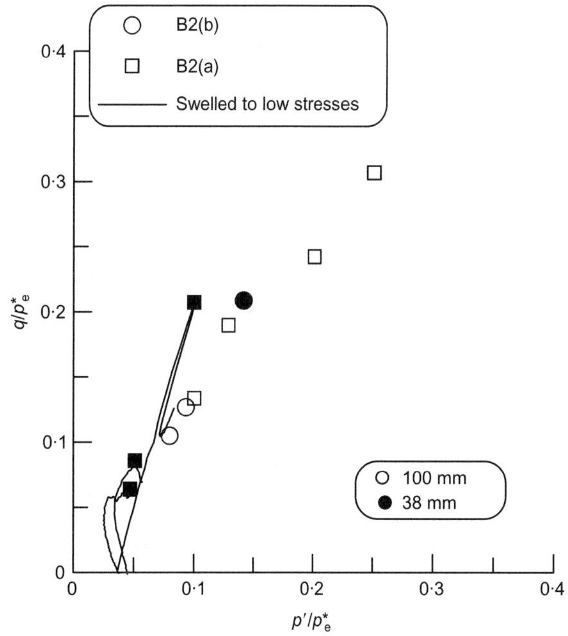

Fig. 15. Normalised stress paths of samples swelled to low stresses compared with peak states for samples that were not swelled

consolidated reconstituted samples may be used to define the shear stiffnesses G_0 developed on the isotropic normal compression line, which has the form $G_0 = A(p')^n$. Many authors have used a reference pressure p'_r to render the constants of this equation dimensionless, but here it was preferred to maintain the engineering units in Fig. 16, where the values of A and n for the units tested and the elastic stiffnesses for the natural and reconstituted samples in each unit are plotted. An n value around 0·74 appears to apply to the more plastic units C and B2(a), reducing slightly to 0·71 for the deeper unit A2. The A values increase with depth from 0·65 to 1·1, marking a significant difference in the location of these lines between units, depending again on the nature of the soil. The reconstituted HCA samples (labelled as 'reconstituted B2(a)-RC' in Fig. 16) had been normally consolidated to 730 kPa in a consolidometer and then swelled to an isotropic stress of 183 kPa in the HCA to OCR = 4. Despite the different consolidation procedures, the HCA resonant column data broadly confirm the triaxial bender element stiffnesses.

Figure 17 examines in more detail the initial shear stiffnesses of the intact samples, emphasising their anisotropy. The G_{hv} and G_{hh} profiles are from triaxial bender element tests, and the resonant column measurements provide G_{vh}. As Jovicic & Coop (1998) discussed, laboratory values of G_{vh} and G_{hv} are similar for the London Clay, and the anisotropy of the soil is reflected in the higher values of G_{hh}. Although the data for a particular unit are often scattered, by means of comparing data for the various units, a series of lines that are nearly parallel may be interpreted. Increases in p' lead to smaller changes in G_0 for the natural samples than for the reconstituted samples. To some extent this arises from their overconsolidation, but the gradients are lower even than those for the overconsolidated reconstituted samples, indicating an effect of structure. For tests conducted at the in situ stress level the stiffnesses tend to increase almost linearly with depth, as discussed by Gasparre et al. (2007), but Fig. 16(b) shows that this is not simply an effective stress level effect, because at a given stress level there are significant differences in the G–p' relationships applying to each unit. The anisotropy also

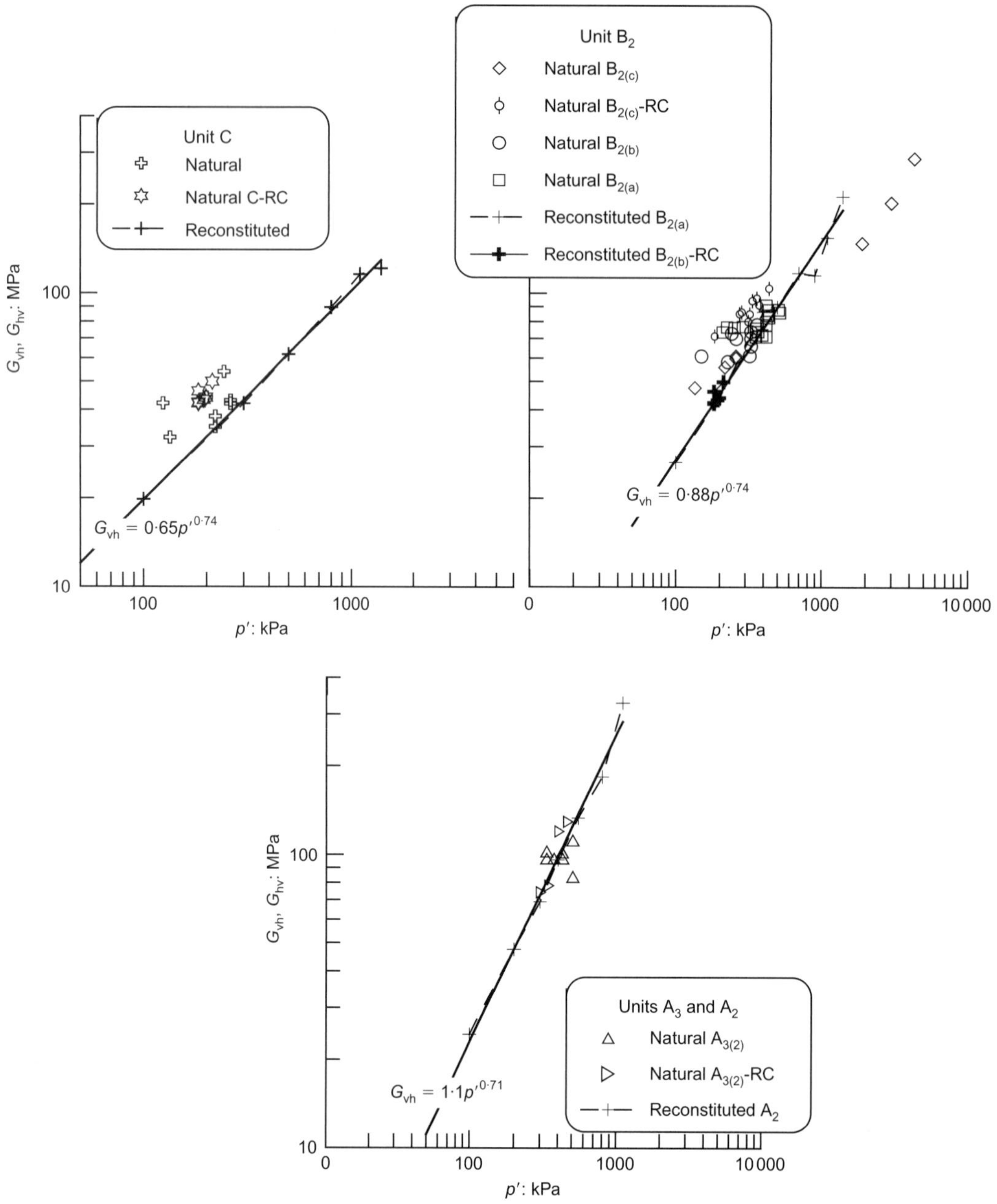

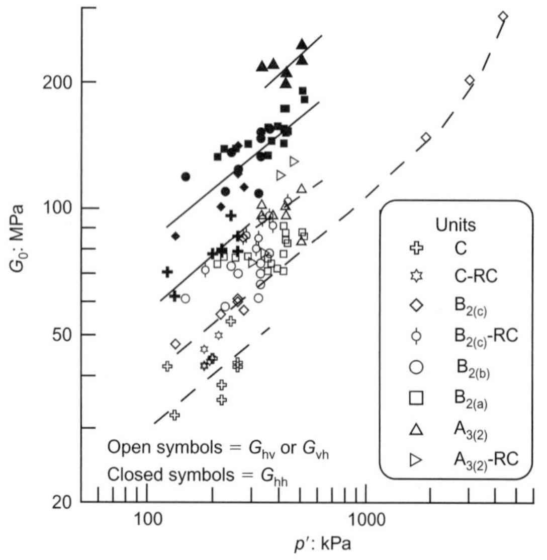

Fig. 16. Comparison between elastic stiffnesses of natural and reconstituted soil

Fig. 17. Elastic stiffnesses G_{hh}, G_{vh} and G_{hv} for intact samples

increases slightly with depth, with the ratio G_{hh}/G_{hv} rising from about 1·8 in units C and B2(c) to 2·1 in units B2(a) and A3(2), reflecting the increasing particle orientation seen with the SEM. These ratios did not change over the range of effective stresses applied to the samples. As the maximum pre-loading stresses experienced in situ by all units are similar, varying by perhaps 300 kPa compared with a mean of about 1500 kPa, the increased orientation of the particles for the lower units is not likely to be related to the stresses experienced. A perhaps surprising feature is that for all units the dynamic measurements of G_0 for intact samples are only slightly greater at their in situ stresses than would be expected with normally consolidated reconstituted samples, despite their overconsolidation (Fig. 16). These differences become still more marked if account is taken of the intact samples' void ratios e. Following Jamiolkowski *et al.* (1994), the intact results have been normalised by the function $f(e) = e^{-1\cdot3}$, and Fig. 18 shows how $G_0/e^{-1\cdot3}$ varies with p'. Although the differences between lithological units are muted by this normalisation, they remain evident. Comparing the normalised intact and reconstituted data in Fig. 19, it

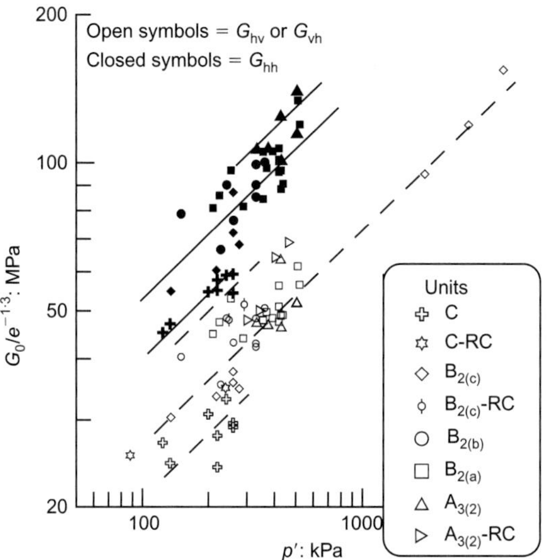

Fig. 18. Normalised elastic stiffnesses G_{hh}, G_{vh} and G_{hv} for intact samples

appears that natural structure has the effect of reducing the elastic shear stiffnesses. The latter feature is in complete contrast to the earlier described enhanced shear strength and much lower C_s values (or high normalised bulk stiffness) values shown by the intact samples.

It is useful to consider whether the relatively low shear stiffnesses of the intact London Clay results from fissuring. Fig. 20 presents undrained tangent shear stiffness ($G_{tan} = \delta 9/2\delta(\varepsilon_1 - \varepsilon_3)$) degradation curves for two 100 mm specimens from unit A3(2) that were subjected to similar tests. One sample ultimately failed on a pre-existing fissure without developing a significant peak, whereas the other failed through intact soil, with a pronounced peak strength. The shear stiffnesses at small to intermediate strains are very similar, indicating that local compression or slip along the fissure that eventually contributes to the failure surface does not commence until larger strains. It might be argued that other (perhaps sub-horizontal or vertical) fissures existed equally in both samples that affected stiffness more markedly, but that only in one sample did fissures affect shear strength. However, comparisons with the 38 mm diameter

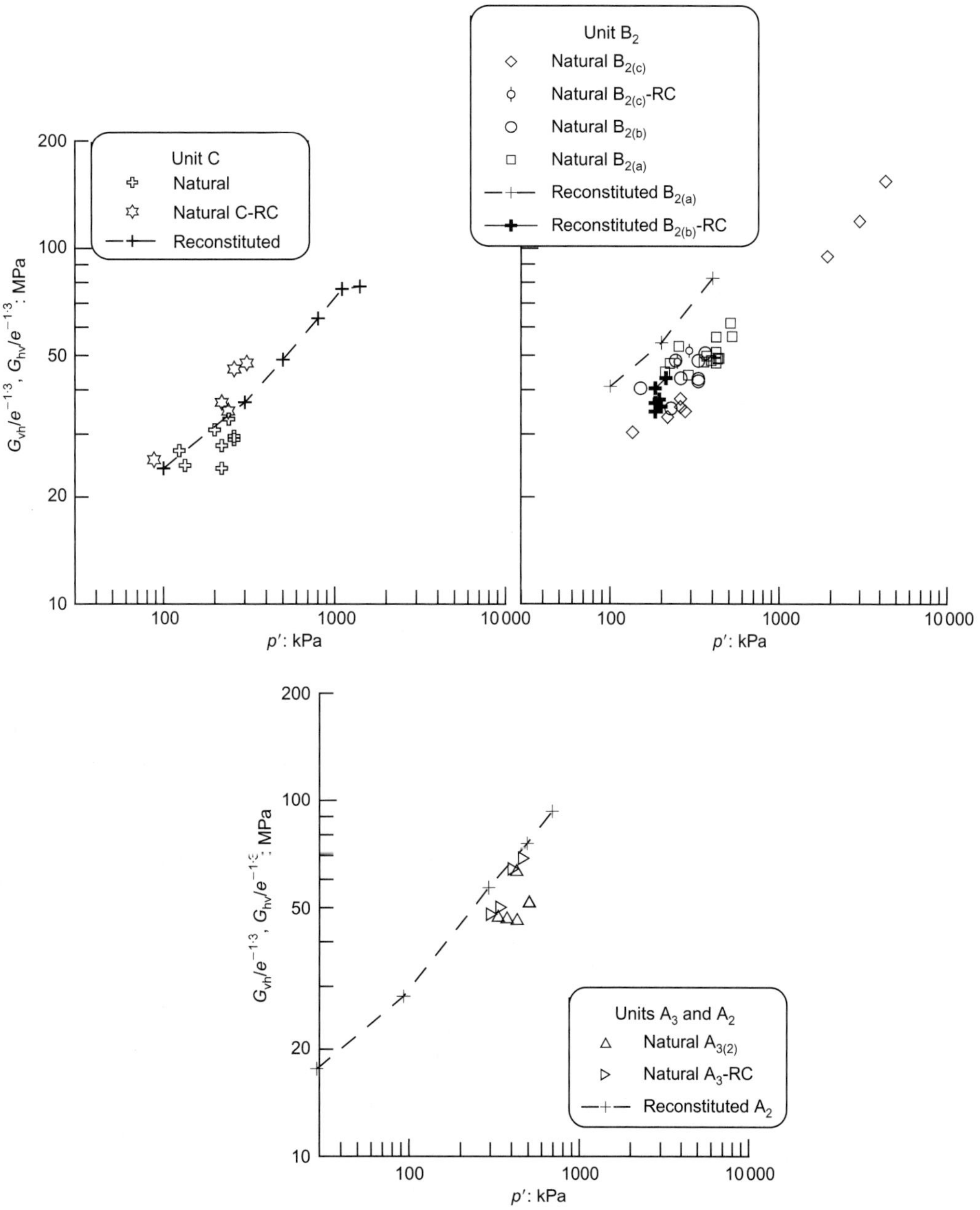

Fig. 19. Comparison between normalised elastic stiffnesses of natural and reconstituted soil

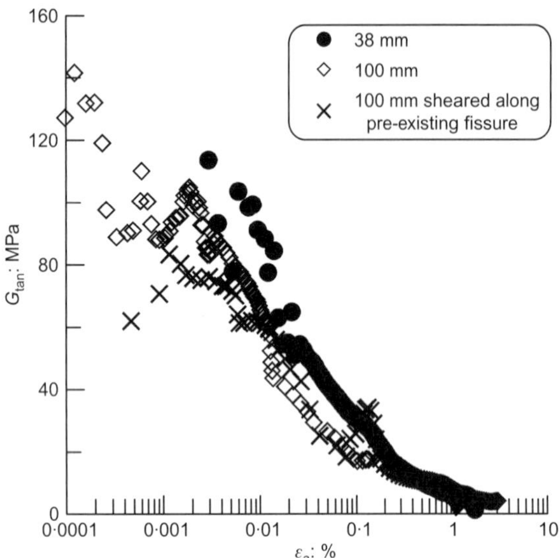

Fig. 20. Comparison of tangent stiffness backbone curves for a sample from unit A3(2) that failed on a pre-existing fissure with samples that failed on new surfaces

samples indicate that, unlike the average shear strength, there was no effect of sample size on stiffness. If the fissures affected the stiffness, then the generally less frequently fissured 38 mm samples should have had significantly higher stiffnesses. The generally good agreement of the resonant column G_{vh} and bender element G_{hv} data further reinforces the above view; the resonant column measures a mean stiffness throughout the vertical plane, whereas the bender elements transmitted shear waves through the mid-height of the sample on paths that would be far less affected by horizontal fissures at other levels. Gasparre *et al.* (2007) discuss the London Clay's stiffness behaviour in greater detail.

CONCLUSIONS

The significantly different behaviours observed in the London Clay units and sub-units sampled at the Heathrow T5 site arise from variations with depth in both their natural structure and their intrinsic properties. The individual units' microscale particle arrangements and mesoscale fissuring lead to mechanical behaviours that diverge from those of reconstituted specimens prepared from the same depths. Key features that are affected by structure include the brittle peak triaxial shear strength response, the strong shear strength anisotropy, and the compression–swelling behaviour. Certain anomalies have been identified that point to unresolved complexities in the effects of structure on material response.

(*a*) Structure augments bulk stiffness (reducing swelling index C_s), but its effect on shear stiffness is neutral or negative, with intact samples showing similar or lower shear stiffnesses to reconstituted samples under equivalent conditions.

(*b*) Natural fissuring of the type displayed has an important effect on shear strength, but does not appear to affect stiffnesses at smaller strains significantly.

(*c*) Structure enhances the peak intact shear strengths applying under in situ stress levels at all depths, but the extent to which it augments the gross yield ratio in compression tests reduces with depth.

(*d*) Structure can be damaged by compression or swelling, but the effects on stiffness and strength are neither uniform nor proportional.

SEM imaging and mineral analysis was useful in confirming the nature of the soil and in identifying the preferred particle orientations that affect stiffness anisotropy. However, it did not highlight features that could explain the anomalous effects of structure on strength and bulk stiffness.

ACKNOWLEDGEMENTS
The authors would like to thank Surrey Geotechnical Consultants for permission to use their test data, the EPSRC and London Underground for their financial support, BAA for the rotary cored samples and access to the site, and Dr J. Huggett of the Natural History Museum for her work and advice on the SEM and XRD analyses.

NOTATION

A	coefficient of stiffness model
b	intermediate principal stress ratio
C_s	swelling index of intact soil
C_c^*	intrinsic compression index
C_s^*	intrinsic swelling index
e	void ratio
G_0	generic elastic shear stiffness (G_{vh}, G_{hh}, G_{hv} or a third of undrained Young's modulus)
G_{tan}	tangent elastic shear stiffness
G_{vh}, G_{hh}, G_{hv}	elastic shear stiffnesses
I_v	void index
K_0	coefficient of earth pressure at rest
M	q/p' at critical state in compression
N	intercept of isotropic ICL at 1 kPa on $v - \ln p'$ plot
n	exponent of stiffness model
p'	mean normal effective stress
p'_0	initial mean normal effective stress
p_e^*	equivalent pressure
p'_r	reference pressure
q	deviatoric stress
S_s	swell sensitivity
S_u	undrained shear strength
v	specific volume
α	angle of inclination of major principal stress axis from vertical
α_f	value of α at failure
ε_a	axial strain
λ	gradient of isotropic ICL
σ'_1, σ'_2, σ'_3	principal effective stresses
σ'_n	normal effective stress
σ'_v	vertical effective stress
σ'_z, σ'_r, σ'_θ	normal stresses in cylindrical coordinate system
τ	shear stress
ϕ'_{cs}	critical state angle of shearing resistance
ϕ'_p	peak angle of shearing resistance taking $c' = 0$
ϕ'_{pr}	post-rupture angle of shearing resistance

REFERENCES
Amorosi, A. & Rampello, S. (1998). The influence of natural soil structure on the mechanical behaviour of a stiff clay. *Proc. 2nd Int. Symp. on Hard Soils—Soft Rocks, Naples* **1**, 395–402.

Bishop, A. W., Webb, D. L. & Lewin, P. I. (1965). Undisturbed samples of London Clay from Ashford Common shaft: strength–effective stress relationship. *Géotechnique* **15**, No. 1, 1–31.

Burland, J. B. (1990). On the compressibility and shear strength of natural soils. *Géotechnique* **40**, No. 3, 329–378.

Burland, J. B., Rampello, S., Georgiannou, V. N. & Calabresi, G. (1996). A laboratory study of the strength of four stiff clays. *Géotechnique* **46**, No. 3, 491–514.

Cafaro, F. & Cotecchia, F. (2001). Structural degradation and changes in the mechanical behaviour of a stiff clay due to weathering. *Géotechnique* **51**, No. 5, 441–453.

Chandler, R. J. (1966). The measurement of residual strength in triaxial compression. *Géotechnique* **16**, No. 3, 181–186.

Chandler, R. J. & Apted, J. P. (1988) The effects of weathering on the strength of London Clay. *Q. J. Engng Geol.* **21**, No. 1, 59–68.

Chandler, R. J., de Freitas, M. H. & Marinos, P. (2004). Geotechnical characterisation of soils and rocks: a geological perspective. *Advances in Geotechnical Engineering: The Skempton Conference*, Vol. 1, pp. 67–102. London: Thomas Telford.

Chandler, R. J., Willis, M. R., Hamilton, P. S. & Andreou, I. (1998). Tectonic shear zones in the London Clay formation. *Géotechnique* **48**, No. 2, 257–270.

Coop, M. R., Atkinson, J. H. & Taylor, R. N. (1995). Strength, yielding and stiffness of structured and unstructured soils. *Proc. 11th Eur. Conf. Soil Mech. Found. Engng, Copenhagen* **1**, 55–62.

Cotecchia, F. & Chandler, R. J. (1997). The influence of structure on the pre-failure behaviour of a natural clay. *Géotechnique* **47**, No. 3, 523–544.

Cotecchia, F. & Chandler, R. J. (2000). A general framework for the mechanical behaviour of clays. *Géotechnique* **50**, No. 4, 431–447.

Cuccovillo, T. & Coop, M. R. (1997). The measurement of local axial strains in triaxial tests using LVDTs. *Géotechnique* **47**, No. 1, 167–171.

de Freitas, M. H. & Mannion, W. (2007). A biostratigraphy for the London Clay in London. *Géotechnique* **57**, No. 1, 91–99.

Gasparre, A. (2005). *Advanced laboratory characterisation of London Clay.* PhD thesis, Imperial College, London. http://www.imperial.ac.uk/geotechnics/Publications/PhDs/phdsonline.htm

Gasparre, A. & Coop, M. R. (2007). The quantification of the effects of structure on the compression of a stiff clay. *Can. Geotech. J.* (submitted).

Gasparre, A., Nishimura, S., Minh, N. A., Coop, M. R. & Jardine, R. J. (2007). The stiffness of natural London Clay. *Géotechnique* **57**, No. 1, 33–47.

Hight, D. W., McMillan, F., Powell, J. J. M., Jardine, R. J. & Allenou, C. P. (2002). Some characteristics of London Clay. In *Characterisation and engineering properties of natural soils* (eds T. S. Tan, K. K. Phoon, D. W. Hight and S. Leroueil), pp. 851–908. Lisse: Swets & Zeitlinger.

Hight, D. W., Gasparre, A., Nishimura, S., Minh, N. A., Jardine, R. J. & Coop, M. R. (2007). Characteristics of the London Clay from the Terminal 5 site at Heathrow Airport. *Géotechnique* **57**, No. 1, 3–18.

Jamiolkowski, M., Lancellotta, R. & Lo Presti, D. C. F. (1994). Remarks on the stiffness of six Italian clays. *Proc. 1st Int. Conf. on Pre-failure Deformation Characteristics of Geomaterials, Hokkaido*, 855–885.

Jardine, R. J., Symes, M. J. & Burland, J. B. (1984). The measurement of soil stiffness in the triaxial apparatus. *Géotechnique* **34**, No. 3, 323–340.

Jardine, R. J, Gens, A., Hight, D. W. and Coop, M. R. (2004). Developments in understanding soil behaviour: Keynote paper. *Advances in Geotechnical Engineering: The Skempton Conference*, Vol. 1, pp. 103–207. London: Thomas Telford.

Jovicic, V. & Coop, M. R. (1998). The measurement of stiffness anisotropy in clays with bender element tests in the triaxial apparatus. *ASTM Geotech. Test. J.* **21**, No. 1, 3–10.

King, C. (1981). The stratigraphy of the London Basin and associated deposits. *Tertiary Research Special Paper*, **6**. Rotterdam: Backhuys.

King, C. (1991). *Stratigraphy of the London Clay (Early Eocene) in the Hampshire Basin.* PhD thesis, Kingston Polytechnic, London.

La Rochelle, P., Leroueil, S., Trak, B., Blais-Leroux, L. & Tavenas, F. (1988). Observational approach to membrane and area corrections in triaxial tests. in *Advanced triaxial testing of soil and rock* (eds R. T. Donaghe, R. C. Chaney and M. L. Silver). Philadelphia: ASTM, pp. 715–731.

Marsland, A. (1971) The shear strength of stiff fissured clays. In *Stress–strain behaviour of soils. The Roscoe Memorial Conference, Cambridge*, pp. 59–68.

Nishimura, S. (2006). *Laboratory study on anisotropy of natural London Clay.* PhD thesis, Imperial College, London. http://www.imperial.ac.uk/geotechnics/Publications/PhDs/phdsonline.htm

Nishimura, S., Minh, A. N. & Jardine, R. J. (2007) Shear strength anisotropy of natural London Clay. *Géotechnique* **57**, No. 1, 49–62.

Schmertmann, J. H. (1969). Swell sensitivity. *Géotechnique* **19**, No. 41, 530–533.

Skempton, A. W., Schuster, R. L. & Petley, D. J. (1969) Joints and fissures in the London Clay at Wraysbury and Edgware. *Géotechnique* **19**, No. 2, 205–217.

Smith, P. R. (1992). *The behaviour of natural high compressibility clay with special reference to construction on soft ground.* PhD thesis, Imperial College, London.

Standing, J. R. & Burland, J. B. (2006) Unexpected tunnelling volume losses in the Westminster area, London. *Géotechnique* **56**, No. 1, 11–26.

Takahashi, A., Jardine, R. J. & Fung, D. W. H. (2005). Swelling effects on mechanical behaviour of natural London Clay. *Proc. 16th Int. Conf. Soil Mech., Osaka*, 443–446.

Vitone, C. & Cotecchia, F. (2007) The mechanical behaviour of medium to intensely fissured clays. In preparation.

Gasparre, A., Nishimura, S., Minh, N. A., Coop, M. R. & Jardine, R. J. (2007). *Géotechnique* **57**, No. 1, 33–47

The stiffness of natural London Clay

A. GASPARRE*, S. NISHIMURA†, N. A. MINH‡, M. R. COOP† and R. J. JARDINE†

An investigation of natural London Clay is reported involving advanced triaxial, hollow cylinder apparatus (HCA) and dynamic testing techniques. Significant anisotropy was revealed at all scales of deformation, and the framework of cross-anisotropic elasticity was found to apply broadly to the initial elastic behaviour. The stiffness parameters obtained by independent techniques generally exhibited good agreement, with the greatest deviation being seen in the Poisson's ratios, which fell far from the values usually assumed in conventional foundation analysis. Probing tests established the limits to the elastic domain over a range of depths, showing that these scaled in proportion to the mean effective stress level, as did those of a second kinematic surface that surrounded the elastic domain. Once engaged, this second surface signified a new pattern of strain increment directions, faster elastic-plastic stiffness decay with strain, and also a greater dependence of behaviour on recent stress history. However, the two kinematic surfaces cover a relatively small proportion of the admissible stress space, and behaviour at larger strains is both anisotropic and strongly non-linear, features that affect profoundly the soil displacements induced by geotechnical construction in this deposit.

Cet article présente une étude réalisée sur des échantillons d'argile de Londres naturelle, qui utilise les techniques d'un appareil triaxial à cylindre creux (HCA) de technologie avancée et d'essais dynamiques. Ces tests ont révélé une anisotropie significative à toutes les échelles de déformation et ont montré que le cadre de l'élasticité en anisotropie croisée s'appliquait globalement au comportement élastique initial. Les paramètres de rigidité obtenus par des techniques indépendantes montrent dans l'ensemble un bon accord. Les rapports de Poisson affichent la déviation la plus importante, puisque leurs valeurs s'écartent très fortement de celles normalement supposées pour l'analyse de fondation conventionnelle. Des tests de sondage ont établi les limites du domaine élastique sur une plage de profondeurs, montrant que celles-ci s'adaptaient en proportion au niveau de contrainte effective moyen, comme le faisaient celles d'une seconde surface cinématique qui entourant le domaine élastique. Une fois engagée, cette seconde surface conduisait à une nouvelle tendance de directions d'incrément de résistance, un affaiblissement de la rigidité élastique-plastique plus rapide en fonction de la déformation et à une dépendance plus forte du comportement par rapport à l'historique de contrainte récent. Cependant, les deux surfaces cinématiques couvrent une proportion relativement faible de l'espace de contrainte admissible et le comportement pour des contraintes plus importantes est à la fois anisotrope et fortement non linéaire, caractéristiques qui affectent profondément les déplacements du sol induits par la construction géotechnique dans ce dépôt.

KEYWORDS: anisotropy; clays; constitutive relations; fabric/structure of soils; laboratory tests; stiffness

INTRODUCTION

It is well known that detailed information on the ground's highly non-linear stress–strain behaviour is essential if realistic predictions are to be made for the displacements induced by geotechnical construction: see for example Jardine *et al.* (1991). However, the influence on stiffness of several potentially important factors remains uncertain for almost all natural geomaterials. Several experimental and field studies have been undertaken on the London Clay to investigate the potential effects of sampling disturbance, anisotropy, loading rate, time and stress history (e.g. Jardine *et al.*, 1985; Butcher & Powell, 1996; Jovicic & Coop, 1998; Clayton & Heymann, 2001) and others have applied numerical methods to explore the potential significance of different constitutive frameworks and ranges of parameters (e.g. Simpson, 1992; Simpson *et al.*, 1996; Addenbrooke *et al.*, 1997; Potts & Zdravkovic, 2001; Grammatikopoulou, 2004). Significant efforts were made to investigate the London Clay Formation at Sizewell in Suffolk and at Heathrow Terminal

5 by Hight *et al.* (1997, 2002), synthesising data obtained with laboratory and field techniques. However, only limited attention could be given in these practical studies to exploring the nature of the stiffness response, the anisotropy developed over the full engineering strain range, any potential difference between dynamic and static measurements, possible variations of stiffness parameters with effective stress state, or the significance of geological variations between the London Clay stratigraphic units identified by King (1981).

Recent advances in soil testing have led to apparatus and techniques that offer new capabilities for investigating stiffness anisotropy, particularly at small strains. Probing tests with hybrid triaxial cells fitted with both local strain instrumentation and bender elements allow both static and dynamic test probes to be performed, and offer the possibility of measuring all the terms in an initial elastic stiffness matrix. Small-strain static probing tests can be combined with multiaxial body wave velocity measurements using either bender elements (Kuwano & Jardine, 1998; Lings *et al.*, 2000) or other P- and S-wave transducers (e.g. Bellotti *et al.*, 1996). However, it is necessary to assume that the soil is linearly elastic, cross-anisotropic and rate-independent over the probing stress increment range. Additional tests may be designed to move beyond the boundary of the kinematic elastic domain to explore inelastic behaviour, but only under conventional triaxial $(q–p')$ conditions. Locally instrumented hollow cylinder apparatus (HCA) are less restricted, allowing anisotropy to be studied up to and

Manuscript received 5 May 2006; revised manuscript accepted 14 November 2006.
Discussion on this paper closes on 1 July 2007, for further details see p. ii.
* Geotechnical Consulting Group, London, UK; formerly Imperial College, London, UK.
† Imperial College, London, UK.
‡ Atkins Ltd; formerly Imperial College, London, UK.

including failure, so avoiding the assumptions associated with hybrid wave velocity techniques (Zdravkovic & Jardine, 1997; HongNam & Koseki, 2005). However, high-resolution, stable stress and strain instrumentation is essential in both cases if the initial linear range is to be characterised successfully, and this is more difficult to achieve with HCA equipment. Hybrid HCA experiments combining static torsional shear and dynamic resonant column measurements provide a further useful source of stiffness data (Jardine, 1995; Hight *et al.*, 1997).

A comprehensive study has been completed recently by a team from Imperial College, London, into the natural London Clay encountered at the Heathrow Terminal 5 (T5) site. The research included multiple static and hybrid dynamic triaxial and HCA measurements on high-quality samples from a single location close to which field wave velocity measurements were made, which showed broad agreement with the laboratory data (Hight *et al.*, 2007). This paper reports the measurements and discusses the new insights offered into the London Clay's stiffness. The soil behaviour has been interpreted using the kinematic strain-hardening plasticity framework proposed by Jardine (1995), who identified two kinematic surfaces, named Y_1 and Y_2, that exist within the conventional main yield surface, termed Y_3. A scheme of the model is shown in Fig. 1. Within the zone limited by the Y_1 surface the soil response is linear elastic and the strains are fully recoverable. The Y_2 surface corresponds to the contour of a zone beyond which the strain increment vector may change direction and the rate of

plastic strain development accelerates. It has been speculated that this surface corresponds to the limit beyond which particle contacts fail and particle movements occur. The conventional yield surface Y_3 corresponds in normalised stress space to the local boundary surface (LBS), which cannot be crossed by undrained stress paths. The LBS exists within the more extensive state boundary surface (SBS), which provides the outermost boundary between admissible and non-admissible normalised effective stress states (Jardine *et al.*, 2004). The focus in this paper is on the London Clay's stiffness behaviour within its Y_3 surface. Other aspects of the work are reported in companion papers by Gasparre *et al.* (2007), Hight *et al.* (2007) and Nishimura *et al.* (2007); full details of the test procedures and data obtained are given in Gasparre (2005), Nishimura (2006) and Minh (2006).

Coordinate system

Cylindrical coordinates (a, r, θ) are the most appropriate to describe conditions in HCA or triaxial cylindrical tests. However, Cartesian coordinates (v, h_1, h_2) are often used to report anisotropic wave velocity or stiffness data. In this paper, the stiffness parameters are presented in the Cartesian v–h coordinate system, assuming the material remained cross-anisotropic throughout the small-strain shear tests, and the stresses and the strains are presented in the a–r–θ system, as shown in Fig. 2.

CROSS-ANISOTROPIC ELASTIC STIFFNESS MEASUREMENTS

For a structured soil to be truly cross-anisotropic, it should have horizontal bedding, have experienced orthogonal K_0 stress conditions, and not have been disturbed by prior tectonic or other directionally oriented actions. Although the London Clay may not meet these requirements fully, it is often considered that, at any particular depth, the deposit behaves as a cross-anisotropic elastic material at very small strains, and that its effective compliance equation can be written as

$$
\begin{Bmatrix}
\delta\varepsilon_x \\
\delta\varepsilon_y \\
\delta\varepsilon_z \\
\delta\gamma_{xy} \\
\delta\gamma_{yz} \\
\delta\gamma_{zx}
\end{Bmatrix}
=
\begin{bmatrix}
1/E'_h & -v'_{hh}/E'_h & -v'_{vh}/E'_v & 0 & 0 & 0 \\
-v'_{hh}/E'_h & 1/E'_h & -v'_{vh}/E'_v & 0 & 0 & 0 \\
-v'_{hv}/E'_h & -v'_{hv}/E'_h & 1/E'_v & 0 & 0 & 0 \\
0 & 0 & 0 & 1/G_{hv} & 0 & 0 \\
0 & 0 & 0 & 0 & 1/G_{vh} & 0 \\
0 & 0 & 0 & 0 & 0 & 1/G_{hh}
\end{bmatrix}
\times
\begin{Bmatrix}
\delta\sigma'_x \\
\delta\sigma'_y \\
\delta\sigma'_z \\
\delta\tau_{xy} \\
\delta\tau_{yz} \\
\delta\tau_{zx}
\end{Bmatrix}
\tag{1}
$$

where E'_v and E'_h are the drained Young's moduli in the vertical and horizontal directions respectively; v'_{hh} and v'_{hv} are the drained Poisson's ratios for horizontal strains due to horizontal and vertical strain respectively, and v'_{vh} is the drained Poisson's ratio for vertical strains due to horizontal strain; G_{vh} and G_{hv} are the shear moduli in the vertical

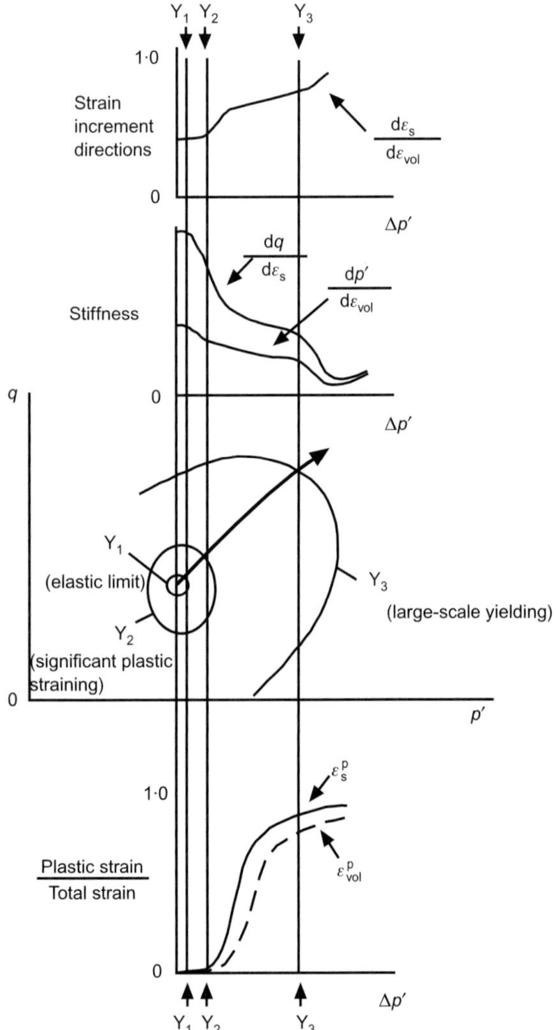

Fig. 1. Scheme of multiple yield surfaces (Jardine, 1995)

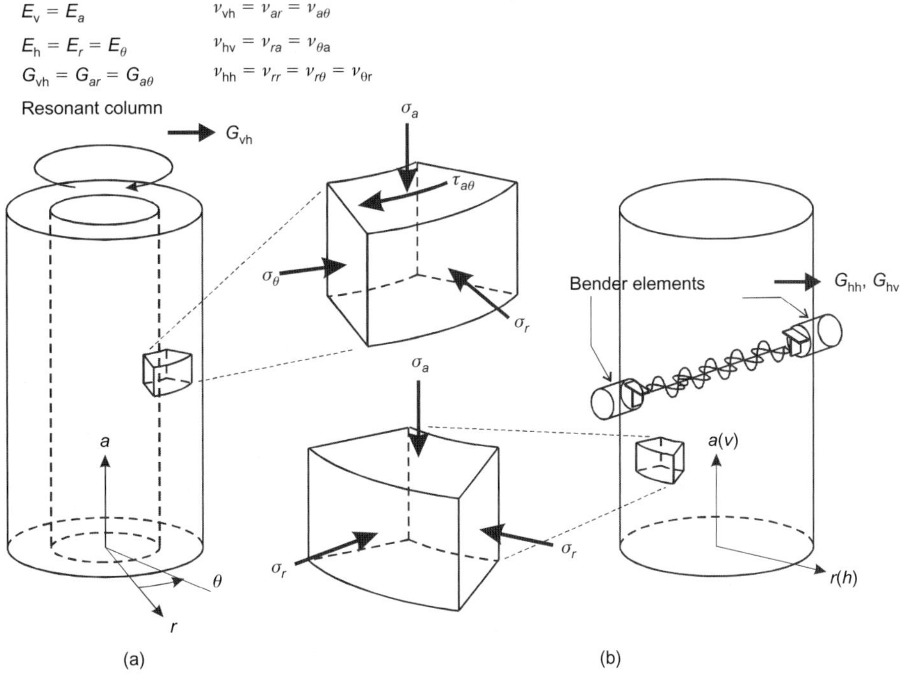

Fig. 2. Coordinate systems in triaxial and hollow cylinder tests: (a) HCA specimen; (b) (cylindrical) triaxial specimen

plane; and G_{hh} is the shear modulus in the horizontal plane. The z-axis is taken as the vertical here.

The following two constraints apply, in addition to $G_{vh} = G_{hv}$, leading to just five independent parameters to be identified.

$$\frac{v'_{hv}}{E'_h} = \frac{v'_{vh}}{E'_v} \quad (2)$$

$$G_{hh} = \frac{E'_h}{2(1 + v'_{hh})} \quad (3)$$

Kuwano & Jardine (1998) and Lings (2001) describe how G_{hh} and G_{vh} or G_{hv} may be obtained directly from bender element measurements and combined with static vertical and radial effective small-strain probes in hybrid triaxial tests to obtain all five independent parameters. Under conventional triaxial conditions (i. e. $\sigma'_v = \sigma'_z$ and $\sigma'_h = \sigma'_x = \sigma'_y$), equation (1) reduces to

$$\left\{ \begin{array}{c} \delta\varepsilon_v \\ \delta\varepsilon_h \end{array} \right\} = \left[\begin{array}{cc} \dfrac{1}{E'_v} & \dfrac{-2v'_{hv}}{E'_v} \\ \dfrac{-v'_{vh}}{E'_v} & \dfrac{1 - v'_{hh}}{E'_h} \end{array} \right] \left\{ \begin{array}{c} \delta\sigma'_v \\ \delta\sigma'_h \end{array} \right\} \quad (4)$$

Performing axial probes under constant radial stress probes ($\delta\sigma'_h = 0$), equation (4) reduces to

$$\delta\varepsilon_v = \frac{1}{E'_v}\delta\sigma'_v \quad (5)$$

$$\delta\varepsilon_h = -\frac{v'_{vh}}{E'_v}\delta\sigma'_v \quad (6)$$

allowing E'_v and v'_{vh} to be measured. With radial probes performed under constant axial stress ($\delta\sigma'_v = 0$), equation (4) reduces to

$$\delta\varepsilon_v = -\frac{2v'_{hv}}{E'_h}\delta\sigma'_h \quad (7)$$

$$\delta\varepsilon_h = \frac{1 - v'_{hh}}{E'_h}\delta\sigma'_h \quad (8)$$

Kuwano & Jardine (1998) show how the G_{hh} measurements are combined with equations (7) and (8) to derive E'_h, v'_{hh}, v'_{vh} and v'_{hv} and complete the analysis, assuming full compatibility between the dynamic and static measurements. Whereas in a triaxial test v'_{vh} may be measured directly from an axial loading probe, the other two Poisson's ratios, v'_{hv} and v'_{hh} have to be calculated indirectly from the derived E'_h values. The Poisson's ratio v'_{hh} can be obtained from equation (8), and v'_{hv} can be obtained from equation (2) or from equations (7) and (8), which can be rewritten as

$$v'_{hv} = -\frac{\delta\varepsilon'_v}{\delta\varepsilon_h}\frac{(1 - v'_{hh})}{2} \quad (9)$$

$$v'_{hv} = -\frac{E'_h}{2}\frac{\delta\varepsilon_v}{\delta\sigma'_h} \quad (10)$$

Equations (9) and (10) give very similar values, but these values are about three times those obtained from equation (2) and about 1·5 times those directly measured in the HCA (as will be discussed later). This discrepancy might be due to the indirectly measured quantities involved in equations (9) and (10) and in the consequent amplification of errors in the estimation of v'_{hv}.

HCAs that offer independent control of σ_θ, σ_r, σ_a and $\tau_{a\theta}$ (or σ_x, σ_y, σ_z and τ_{zx}) and accurate ε_θ, ε_r, ε_a or γ_θ (or ε_x, ε_y, ε_z and γ_{zx}) measurements allow direct determinations through suites of drained probes in which one component is varied at a time while all other effective stresses are held constant (Zdravkovic & Jardine, 1997):

$$E'_v = \frac{\delta\sigma'_z}{\delta\varepsilon_z} \text{ and } v'_{vh} = -\frac{\delta\varepsilon_x}{\delta\varepsilon_z}$$

(when $\delta\sigma'_x = 0$, $\delta\sigma'_y = 0$ and $\delta\tau_{zx} = 0$)

$$E'_h = \frac{\delta\sigma'_x}{\delta\varepsilon_x} \text{ and } v'_{hh} = -\frac{\delta\varepsilon_y}{\delta\varepsilon_x}$$

84 GASPARRE, NISHIMURA, MINH, COOP AND JARDINE

(when $\delta\sigma'_y = 0$, $\delta\sigma'_z = 0$ and $\delta\tau_{zx} = 0$)

$$G_{vh} = \frac{\delta\tau_{zx}}{\delta\gamma_{zx}}$$

(when $\delta\sigma'_x = 0$, $\delta\sigma'_y = 0$ and $\delta\sigma_z = 0$)   (11)

Hybrid HCA tests give independent dynamic measurements of G_{vh} that can be extended into the inelastic range by static torsional, or simple shear testing: $G_{vh} = \delta\tau_{zx}/\delta\gamma_{zx}$.

The above relationships hold for both drained and undrained conditions, although volume change requirements impose $\nu^u_{vh} = 0.5$. Knowing the full set of drained independent parameters the undrained set can be derived as described by Lings (2001). The probes discussed herein were aimed at determining the drained independent parameters, but the undrained Young's moduli in the vertical direction, E^u_v, at different depths were also measured with undrained axial compression and extension probes, allowing comparison with the values calculated from the drained parameters.

MATERIAL, APPARATUS AND TEST PROCEDURES
Material and sampling
Tables 1, 2 and 3 summarise the triaxial and HCA experiments performed for the present study on high-quality samples taken for the Imperial College project. The samples were retrieved from continuously sampled rotary boreholes and from blocks cut by hand in excavations at Heathrow T5. The stratigraphy of the site constitutes about 6 m of gravel overlying about 52 m of London Clay. About 175 m of clay were eroded (Skempton & Henkel, 1957; Chandler, 2000) before the deposition of the Quaternary gravel. At the location where the block samples were retrieved, the gravel had been removed during the 1930s. Hight *et al.* (2007) give further details of the sampling procedures and general site details. Lithological characterisation of the clay at the site was made through microfossil analysis (King, 1981; de Freitas and Mannion, 2007), which allowed the identification of three main units and several lithological sub-units. SEM and X-ray diffraction analyses showed that deeper units have a more compacted and orientated structure, while similarities in the nature and structure of samples within units were found (Gasparre *et al.*, 2007).

Triaxial apparatus
The hybrid triaxial cells employed to test 100 mm diameter, 200 mm high intact samples were fitted with the high-resolution axial and radial strain LVDT sensors described by Cuccovillo & Coop (1997) and laterally mounted

Table 1. Elastic probes performed in the triaxial apparatus

London Clay unit	Test name	Sample		Reconsolidation stress		Number of probes						Shear after probes
				p'_0: kPa	q_0: kPa	ac	ae	rc	re	ctq	ctp'	
C	7gUC	1·2 m (16·5 mOD)	Block	260	−86	2		3		1	1	Undrained compression
	7gUE					2	1	1			2	Undrained extension
B2(c)	11gUC	5·2 m (12·5 mOD)	Block	260	−86	2	1	2	1	2	1	Undrained compression
	12·5gUC	6·5 m (11 mOD)	Rotary core			2	1	1			1	Undrained compression
B2(a)	22·6gUC	16·6 m (0·9 mOD)	Rotary core	420	−156	1		1			1	Undrained compression
	23gUE	17 m (0·5 mOD)				2		1				Undrained extension
	24g37DC	18 m (−0·5 mOD)				2	1	2		1	1	
	24g37DC	18 m (−0·5 mOD)		510	−125	2	1	1		1	1	Drained compression
	31·4gUE	20 m (−13·9 mOD)				2		1		1	1	Undrained extension
A3(2)	36gUE	30 m (−12·5 mOD)	Rotary core	510	−125	1		2		1	1	Undrained extension
	36·5gDC	31 m (−13 mOD)				1						Drained compression

Note: ac = axial compression; ae = axial extension; rc = radial compression; re = radial extension; ctq = constant-q probe; ctp' = constant-p' probe.

Table 2. Conditions of tests performed in ICHCA II

London Clay unit	Test	Sample	Reconsolidation stress		Shear after probes
			p'_0: kPa	q_0: kPa	
B2(c)	IS0590	5·2 m (12·5 mOD) block	280	0	Undrained shear with $\alpha = 90°$ and $b = 0.5$
	HCDT	5·2 m (12·5 mOD) block	280	−140	Drained, $\Delta\tau_{a\theta} > 0$, $\Delta\sigma_a = 0$, $\Delta\sigma_r = 0$ and $\Delta\sigma_\theta = 0$
	HCDQ	5·2 m (12·5 mOD) block	280	−140	Drained, $\Delta\tau_{a\theta} = 0$, $\Delta\sigma_a = 0$, $\Delta\sigma_r = 0$ and $\Delta\sigma_\theta > 0$
	HCDZ	5·2 m (12·5 mOD) block	280	−110	Drained, $\Delta\tau_{a\theta} = 0$, $\Delta\sigma_a > 0$, $\Delta\sigma_r = 0$ and $\Delta\sigma_\theta = 0$

Note: α is the angle between σ_1 and the vertical, and $b = (\sigma_2 - \sigma_3)/(\sigma_1 - \sigma_3)$

Table 3. Samples and conditions for resonant column and simple shear tests

London Clay unit	Depth: m	Reconsolidation path	Sample	G_{vh}: MPa	p'_0: kPa	q_0: kPa	Water content: %
B2(c)	0·8	*	Rotary core	50·0	219	−134	29·0
	1·2	*	Block	72·7	260	−86	25·3
	3·0	*	Rotary core	60·1	242	−123	26·5
	7·9	*	Rotary core	89·2	294	−149	24·2
B2(b)	10·5	†	Block	86·4	323	−165	24·6
	10·6	†	Rotary core	84·4	323	−165	26·0
	10·8	†	Rotary core	84·2	323	−165	25·5
	11·5	†	Rotary core	78·7	315	−171	24·9
	14·6	†	Rotary core	76·9	349	−173	26·4
	15·3	*	Rotary core	92·2	362	−173	26·5
B2(a)	16·6	*	Rotary core	91·6	374	−171	26·5
	20·7	†	Rotary core	103·1	395	−168	25·0
	23·7	†	Rotary core	128·2	430	−160	23·8
	24·8	*	Rotary core	106·4	438	−151	26·4
	25·1	*	Rotary core	105·1	438	−151	25·2
	26·5	†	Rotary core	120·6	447	−141	25·2
A3(2)	29·0	*	Rotary core	116·4	470	−135	24·2
	29·9	†	Rotary core	127·9	470	−135	22·9

* As specified in Fig. 3.
† No reloading (i.e. as for unit B2(c)).

bender elements (see Fig. 2) to measure G_{hh} and G_{hv}, the shear moduli associated with horizontally propagating shear waves that are polarised in the horizontal and vertical planes respectively (Pennington *et al.*, 1997). No bender element measurements of G_{vh} were made (vertical propagation and horizontal polarisation), as Jovicic & Coop (1998) had found that $G_{vh} = G_{hv}$ in laboratory tests on London Clay. The latter feature, which is expected for a homogeneous elastic continuum, was found to be broadly true in larger-scale field measurements made at the Heathrow T5 site, but not in earlier work at Sizewell (Hight *et al.*, 1997, 2002). The LVDT devices allowed strain increments of around $\pm 3 \times 10^{-5}\%$ to be resolved, and the overall system (including the stress sensors) allowed the elastic stiffnesses of the samples to be measured with an accuracy of around $\pm 3\%$. Conventional pressure transducers and load cells were used for the cell pressure, pore pressure and deviatoric load, along with a miniature mid-height pore pressure probe to monitor local pore pressures and drainage conditions.

Gasparre & Coop (2006) describe the care needed to measure elastic stiffnesses in London Clay using local strain measurements. For example, the 0·7°C typical diurnal temperature range of the authors' laboratory was found to have an excessive influence, and the test cells were insulated to reduce the variations during probing tests to less than 0·1°C. Also important was the connection between the top platen and the internal load cell. Flat connections between the two (either bolted or involving a suction cap) often led to strain non-uniformity, no matter how accurately the sample was trimmed. The final arrangement consisted of a half ball (located in a top platen notch) combined with a suction cap to allow extension test paths

Triaxial reconsolidation procedures

Following sample setting-up, an undrained cell pressure was applied that exceeded the in situ mean stress in all cases, leading to measurable positive initial pore water pressures, which made initial effective stresses computable. Samples were then recompressed to a range of effective stress states prior to further testing. In cases where it was desired to match the estimated in situ stresses, a single final representative average stress point (q, p') was adopted for

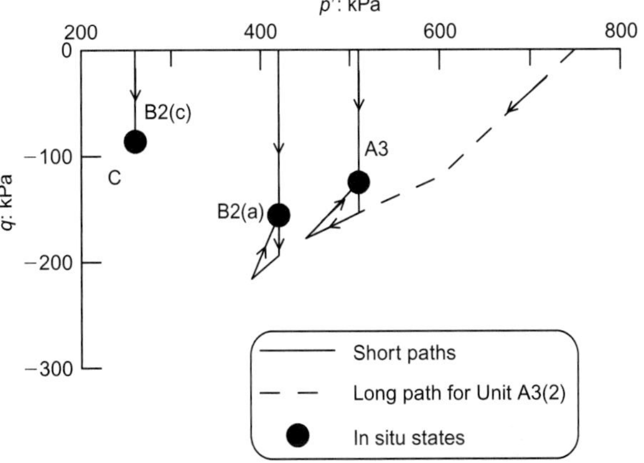

Fig. 3. Approach stress paths for different lithological units

each stratigraphic unit so as to aid comparisons between different samples. The in situ stresses were estimated from suction measurements made on site from the central portions of thin-walled tube samples that were extruded immediately after being taken (Hight *et al.*, 2002). Recompression involved isotropic stress changes prior to one of the four generic anisotropic final approach paths shown in Fig. 3, with the aim of reproducing the site's recent geological history of erosion and then terrace gravel deposition. Further points to note in connection with these paths are as follows.

(*a*) A problem was encountered in units B2(c) and C in applying the desired approach paths without failing the samples. This was due either to the vicinity of the effective stress state approaching the failure criterion at shallow depths or perhaps to overestimation of the in situ K_0 values. Such problems were not encountered in the deeper units. The approach path was modified as shown to involve moving (with constant p') to as high a K ratio as could be achieved reliably without developing tensile axial strains greater than 0·5%.

(*b*) Common in situ stresses were applied to units B2(c) and C, as the latter unit was believed initially to be absent from the site.

(*c*) The approach paths used for A3(2) were changed part way through the programme. A long approach stress path was initially chosen in order to retrace the geological history of the clay better; however, a shorter path was then adopted to minimise the strains induced in the samples. Although different volumetric strains were developed following the long and short paths (1·1% and 0·6% respectively), no significant difference was eventually found between the stiffnesses or yielding behaviour associated with the alternative reconsolidation routes.

Triaxial stress probe and bender element tests

The stress probe tests used to define the elastic parameters were performed from estimated in situ effective stresses, typically under drained conditions, and involved relatively small stress changes (around 2 kPa) that were designed to remain within the clay's initial elastic region. These small stress changes were applied over about 4–5 hours (0·3–0·5 kPa/h, corresponding to strain rates between 0·0003% and 0·0006% per hour within the elastic zone, depending on soil stiffness) to ensure pore pressure equalisation, with long resting periods (of about 1 week) being imposed between arriving at the 'in situ' stress state and the start of probing so that the creep would slow to rates (less than 0·0002%/day) that were insignificant in comparison with the probing tests (Gasparre & Coop, 2006). Slow monotonic load–unload cyclic tests were performed to determine the elastic parameters and define the Y_1 and Y_2 points that are discussed later; similar points were also identified from tests loaded at marginally faster rates (2–3 kPa/h), for example on samples being sheared undrained to failure, and it is recognised that the sizes of these yield loci are likely to depend on strain rate (Tatsuoka & Shibuya, 1992). A limited number of Y_2 points were also interpreted from tests on smaller samples equipped with local strain sensors that offered lower strain resolution (around 0·001%) in the small-strain region.

In addition to performing axial or radial small-strain probing tests, probes were also performed under constant-p' and constant-q conditions, from which the equivalent shear modulus, $G_{eq} = \delta q/3\delta\varepsilon_s$) and bulk modulus K' could be measured. Comparisons with predictions made from the uniaxial probing tests by applying the cross-anisotropic compliance matrix equation (1) allowed a check on the reliability of the measurements and underlying assumptions. The average difference between the measured and calculated values was 12%, indicating an encouraging but not perfect match.

The reported bender element stiffness parameters were found from tests involving a sinusoidal wave pulse and a first arrival timing technique; checks made with a frequency domain arrival time method using continuous steady wave input led to no significantly different values. Bender element determinations of G_{hh} and G_{hv} made while moving along the approach paths shown in Fig. 3 indicated little or no change in either parameter under constant-p' conditions, suggesting that G_{hh}/G_{hv} was hardly affected by the q/p' ratio for the paths applied.

Hollow cylinder apparatus (HCA)

Two different HCAs were employed in the present study: the Imperial College Mark II HCA (ICHCA II) and the hybrid Imperial College Resonant Column HCA (ICRCHCA), illustrations and further details of which are given by Nishimura *et al.* (2007). The nominal inner diameters, outer diameters and heights of specimens were 60 mm, 100 mm and 200 mm respectively in the ICHCA II, and 38 mm, 70 mm and 170–190 mm respectively in the ICRCHCA.

Local strain sensors were deployed in the reported ICHCA II tests. The axial and torsional shear strains were measured with an enhanced electrolevel system, and radial and circumferential strains were calculated from the outer and inner diameter changes monitored with a set of three proximity transducers and a laterally mounted LVDT respectively. Taking multiple readings and using an averaging routine allowed strains to be resolved down to around 0·0003%. The ICRCHCA was equipped with a Hardin oscillator and accelerometer assembly with which torsional resonant column tests were performed to obtain the dynamic shear modulus G_{vh} down to very small strains (less than 10^{-6}%). Static tests could also be performed in which the torsional shear strain was measured platen to platen with a system comprising proximity transducers and a cam; the other strains were measured globally and are not reported here. Minh (2006) and Nishimura (2006) give more detailed descriptions of the stress and strain calculations and transducer performance of the ICHCA II and ICRCHCA respectively.

HCA test procedures

Particular care was taken to minimise disturbance during the preparation of HCA specimens, and the procedures followed are described by Nishimura *et al.* (2007). After setting up in a similar way to the triaxial tests, specimens were reconsolidated following the scheme shown in Fig. 3 designed to match those in situ. The reloading paths were omitted in some of the tests, as indicated in Table 3.

Static tests performed in the ICHCA II are considered here, performed on block samples taken from 12·5 mOD (5·2 m below the top of the London Clay; the gravel was absent at the block sampling site). Over 30 small-strain drained probing experiments were conducted in which only one stress component was changed under drained conditions, while the others were held constant. Complete suites of such tests were performed on four specimens, at three effective stress states, in which individual samples were subjected to successive slow probing cycles involving changes in the σ_a, σ_θ and $\tau_{a\theta}$ components of around 2 kPa over a 1 h period (corresponding to principal strain rates of the order of 0·001–0·002%/h), one at a time, with a 2-day ageing period between each probing cycle. The five parameters E'_v, E'_h, G_{vh}, ν'_{vh} and ν'_{hh} were obtained by applying equation (9). Three of the samples were taken to failure after the final probing tests by increasing just one stress component, employing a strain rate of 2·4% per day under drained conditions. The sample for the fourth test, IS0590, was taken to failure undrained, and its final shearing data are not presented.

Also reported are ICRCHCA resonant column tests and undrained simple shear tests conducted on blocks from 1·2 m and 10·5 m depth (16·5 mOD and 7·2 mOD respectively) and rotary cores taken over a 29 m deep sequence below the top of the London Clay, as outlined in Table 3 (in this paper the depths quoted are from the top of the London Clay). The results of resonant column tests are fully reported in this paper, but static stress–strain data are presented from just three typical simple shear tests for reasons of space. Nishimura (2006) and Nishimura *et al.* (2007) give further information on the complete simple shear dataset. The 'in situ' stresses applied to resonant column specimens were those assessed for each particular sample's depth, rather than the 'representative stresses' applied to each unit in the triaxial testing. The HCA was configured to keep all axial, circumferential and radial strains constant in the simple shear tests while increasing $\gamma_{a\theta}$ under undrained conditions (see Nishimura, 2006). Hight *et al.* (2007) report indepen-

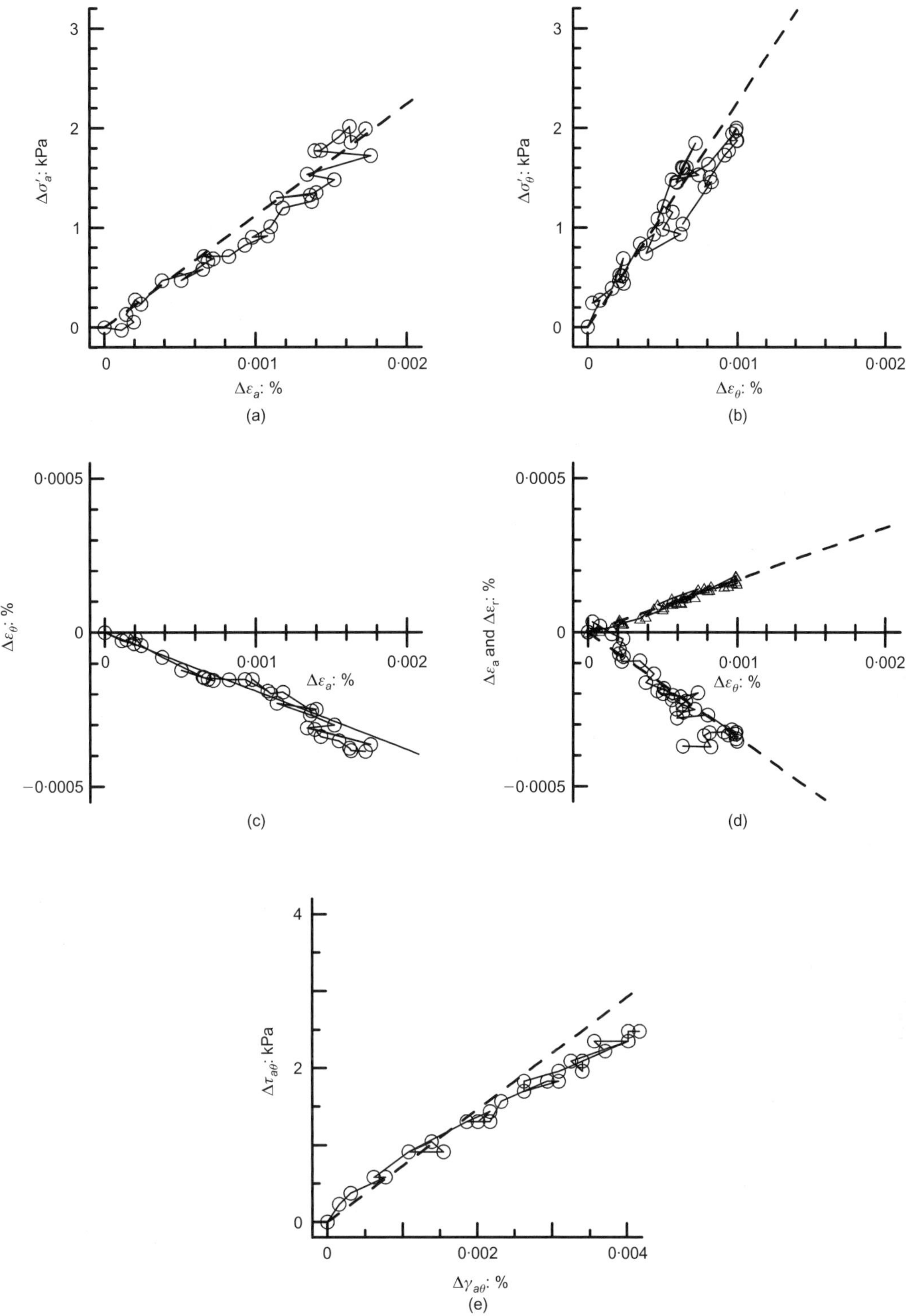

Fig. 4. Typical stress–strain and strain–strain relationships in HCA drained probes (HC-DT, $p' = 280$ kPa, $K = \sigma'_1/\sigma'_3 = 1\cdot7$): (a) $E'_v = 112$ MPa; (b) $E'_h = 226$ MPa; (c) $\nu'_{vh} = 0\cdot19$; (d) top, $\nu'_{hh} = -0\cdot17$, bottom, $\nu'_{hv} = 0\cdot35$; (e) $G_{vh} = 70$ MPa

dent field shear wave velocity measurements made at T5 and synthesise these with the laboratory measurements.

ELASTIC STIFFNESS AND ITS ANISOTROPY

Typical stress–strain data obtained from the HCA uniaxial drained probes conducted from in situ stress conditions are shown in Fig. 4. The stiffness could generally be resolved at a strain of about $10^{-3}\%$. The small-strain stiffness parameters calculated from these data are shown in Fig. 5 along with those obtained from the triaxial probing test series.

Note that the plotted bulk moduli were directly measured by constant-q triaxial probing tests, rather than deduced from the cross-anisotropic elastic parameters obtained by the hybrid procedure described above.

Strong stiffness anisotropy is evident in Figs 4 and 5, with $E'_h > E'_v$ and $G_{hh} > G_{vh}$, and Table 4 summarises the elastic parameters obtained under in situ stress conditions in B2c, the only unit in which all test types were performed, showing averages and ranges as the individual stiffness results show some scatter. The mean E'_v, E'_h and G_{vh} values generally agree well between test types, but the HCA

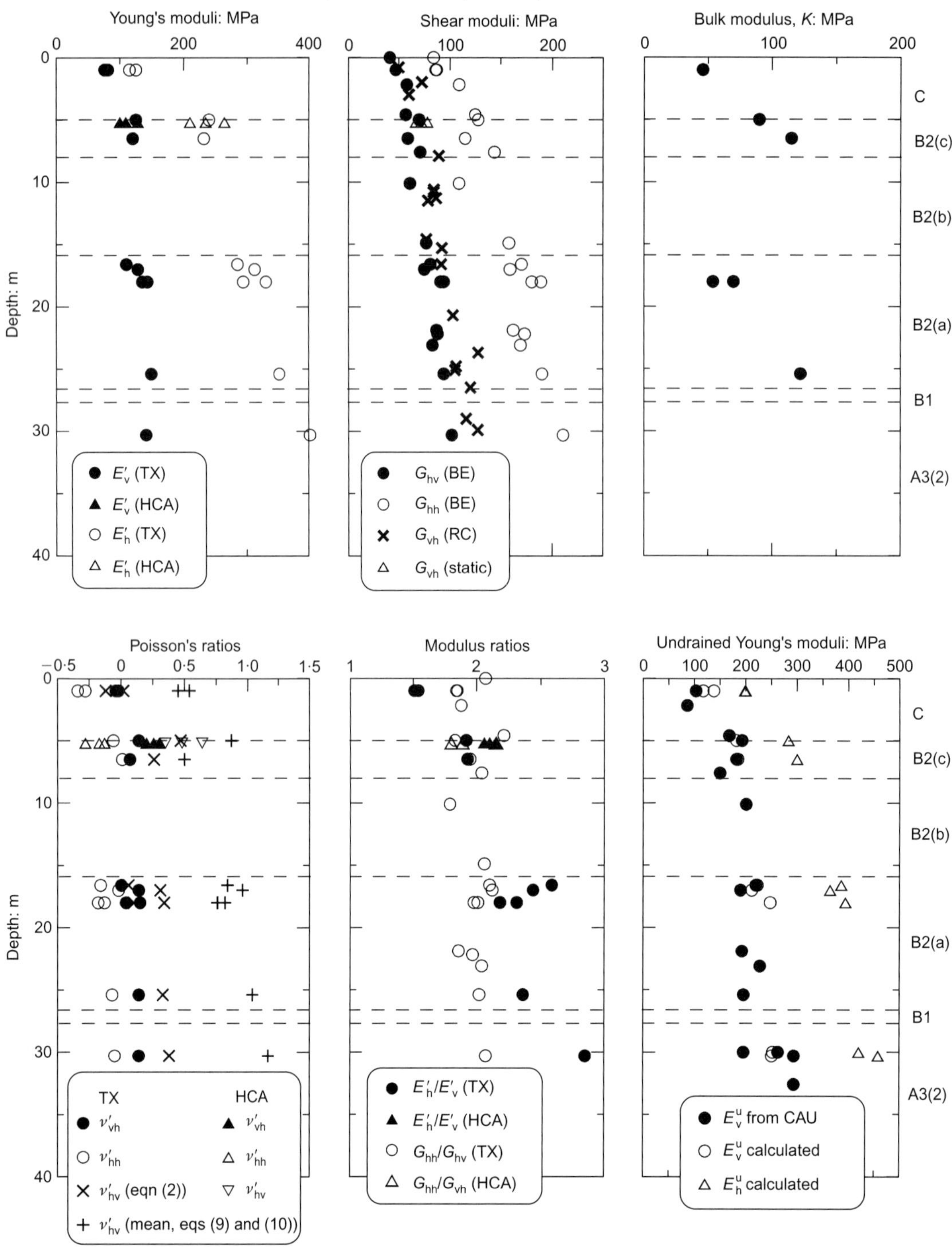

Fig. 5. Profiles of elastic stiffness parameters

resonant column G_{vh} data appear relatively high. It is interesting that Nishimura (2006) found a general trend at all depths (down to 29 m) for resonant column values to exceed the maxima seen in static simple shear tests on the same specimens by 10–30%. It was suggested in his work that the strain rate averaged over a cycle in the resonant column tests was about 1 000 000 times faster than those applied in static simple shear, and that the potential influence of a strain rate may well have played a role in these discrepancies. As can be seen in Table 4, however, the dynamic bender element measurements gave lower G_{hv} than the static measurements. Given the uncertainty entailed in interpretations of dynamic tests, the discrepancies encountered are insufficiently clear to test or quantify the hypothesised rate-dependence. The values of E_v^u measured from undrained axial compressions agreed well with the values calculated from the combination

of the elastic independent parameters, as shown in Fig. 5, with the differences between the calculated and measured values being generally in the range between 5% and 10%, rising to about 30% in only two cases.

The model of the elastic behaviour adopted in this paper to represent the behaviour of the London Clay is one that is cross-anisotropic and rate-independent. To some extent this is a pragmatic choice, because the complete set of parameters for full anisotropy cannot be derived, and to obtain the full set of cross-anisotropic parameters from triaxial tests requires data from both slow static and dynamic probes. Discrepancy between data may, to some extent, reflect some rate dependence or inaccuracy in the assumption of cross-anisotropy. Scatter within the data may reflect boundary conditions, strain non-uniformity and natural variability, but is believed to be mostly due to the accuracy with which the

Table 4. Comparison between stiffness parameters obtained in bender-element-aided triaxial tests and HCA tests for unit B2(c)

Stiffness parameters	Bender-element-aided triaxial tests	Static HCA tests	Resonant column
E_v': MPa	122 (± 3)	112 (± 14)	–
E_h': MPa	238 (± 4)	236 (± 27)	–
$G_{vh} = G_{hv}$: MPa	65 (± 1)	72 (± 6)	88 (± 3)
ν_{vh}'	0·10 ($\pm 0·14$)	0·25 ($\pm 0·05$)	–
ν_{hh}'	-0·02 ($\pm 0·07$)	-0·19 ($\pm 0·08$)	–
ν_{hv}'	0·71 ($\pm 0·15$)	0·49 ($\pm 0·15$)	–
E_v^u: MPa	184 (± 1)		

Note: Values given are mean and standard deviation.

various measurements could be made. It is estimated that each of the directly measured elastic stiffnesses, whether static or dynamic, is accurate to within ± 2–5%, whereas the Poisson's ratio terms are more susceptible to strain measurement errors. As many of the ratios are close to zero it is misleading to quote percentage errors, but a variation of up to $\pm 0·15$ is evident about the tabulated mean values of ν_{vh}', ν_{hh}' and ν_{hv}'. The HCA and triaxial tests give the same general hierarchy of values, and it is important to note that the ranges of values are far from those routinely assumed for London Clay. However, the HCA and triaxial results also differ significantly. The high-resolution triaxial tests should offer the clearest 'static' ν_{vh}' determinations and the HCA the more secure ν_{hh}' and ν_{hv}' information, because the triaxial values of the latter have to rely on rate independence and the synthesis of independent static and dynamic measurements. The main difference between the two datasets is the ratio ν_{vh}'/ν_{hv}', which should equal E_v'/E_h' to satisfy thermodynamic requirements; the purely static HCA measurements gave a closer match. Checks on the consistency between predictions made (from the drained elastic parameters) and independent direct measurements of parameters such as E_v^u, K and G indicated that the discrepancies discussed above are not unduly influential, and that rate-independent cross-anisotropic elasticity may offer an appropriate, if approximate, framework for describing the elastic stiffness of London Clay.

Figure 5 suggests that ν_{vh}' and ν_{hh}' increase only slightly with depth, whereas ν_{hv}' shows a more marked increase, in step with the ratio E_h'/E_v'. The absolute values of stiffness and the degree of anisotropy also increase, reflecting both higher effective stresses and changes in stratigraphy. Gasparre *et al.* (2007) present further tests to show that the stiffnesses within any particular sub-unit are less sensitive to applied changes in effective stresses than is implied by the profile with depth. Also, Fig. 5 shows the values E_h^u and E_v^u with depth. E_h^u was calculated from the drained independent parameters (Lings, 2001). The ratio E_h^u/E_v^u increases with depth from about 1·5 to about 1·8, although it is always lower than E_h'/E_v', which varies between 1·5 and 2·6.

YIELDING BEHAVIOUR

Examples of small-strain probes conducted in the HCA and triaxial apparatus are shown in Figs 4–6. The yield points at the end of the elastic region Y_1 were identified as the point where the stress–strain curves deviate from linear-

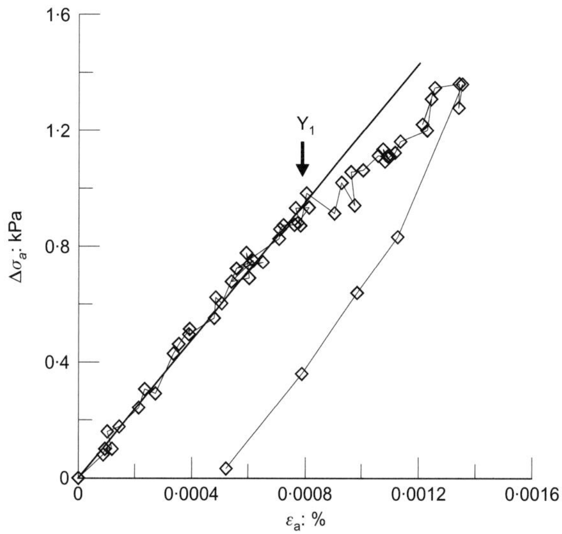

Fig. 6. Drained axial probe test on a sample from Unit B2(c), identifying Y_1 yielding

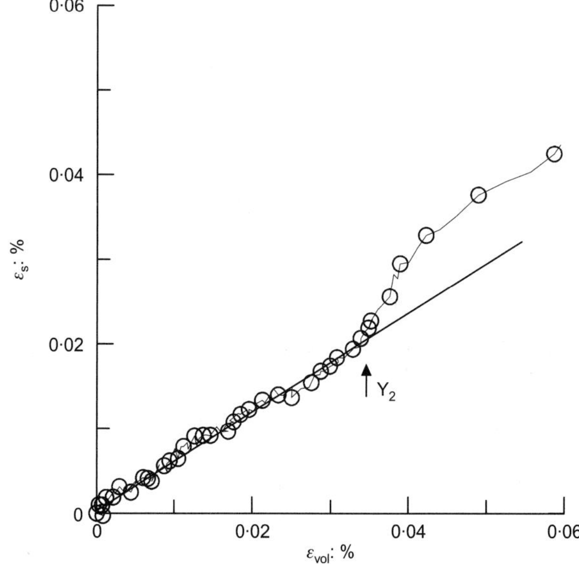

Fig. 7. Y_2 yield points in a drained test on a sample from Unit A3(2)

ity. Unloading prior to this point was reversible, within the scatter of the high-resolution triaxial data, whereas subsequent unloading led to plastic strains. The kinematic Y_1 surface can be dragged by the current effective stress point, and grows with mean effective stress. The Y_2 kinematic surface was identified from drained tests as points where the strain increment vectors rotate, as revealed by plotting volumetric against deviatoric strains (Jardine, 1995). An example is given in Fig. 7; Y_2 points can also be identified in undrained tests from changes in effective stress path direction (or change of pore pressure against strain), although this may be less clear. As shown later, the Y_2 points may also to correspond to points where stiffness degradation accelerates markedly with strain.

The Y_1 points identified by triaxial probing tests are shown in Fig. 8, plotting the increments $(\Delta q, \Delta p')$ required to reach Y_1 from the in situ stress conditions. The incremental presentation allows tests from different stress states (and hence units) to be compared more easily. The Y_1 surface is very small, far below the limits assumed in routine foundation engineering, and evidently increases in size with depth. When the $(\Delta q, \Delta p')$ values are divided by p'_0, the mean effective stress applying prior to probing, they tend towards a common surface (Fig. 9). Other tests in which samples from one unit were consolidated to the in situ stresses of another confirmed that the size of the yield surfaces is dependent only on the stresses applied and not on the structure of the soil. The Y_2 yield surface also expands with depth, increasing in diameter from about 10 to 25 kPa.

Most of the probes that were used to determine the yield points used slow drained loading to achieve a desired effective stress path direction. Typical loading rates were around 0·3–0·5 kPa/h. Several of the tests were sheared undrained, with loading rates around six times faster. The size of the Y_1 region can be expected to grow with strain rate, particularly when rates are changed by orders of magnitude (Tatsuoka & Shibuya, 1992). However, the faster undrained tests, which are identified in Fig. 9, fall within the broad scatter (around ±15%) of the drained yield points, which suggests that any rate effect was not strong enough to be evident at this scale. Those of the samples that developed ultimate failure mechanisms involving pre-existing fissures are also highlighted. As discussed further by Gasparre *et al.* (2007), comparisons between these and other samples indicate that fissuring has little effect on the Y_1 or Y_2 yielding behaviour.

When normalised by p'_0, the incremental Y_2 data points trend towards a second common shape of similar, slightly elliptical, geometry to the Y_1 surface (see Fig. 10). Both surfaces are slightly skewed in shape and are approximately centred on the in situ stress state, probably reflecting the extended creep and/or ageing. Their shapes, which are more rounded for shallower samples and more orientated for deeper samples, are likely to reflect the degree of anisotropy of the clay.

NON-LINEAR STIFFNESS BEHAVIOUR

The rate at which the elastic-plastic stiffness of the samples degrades with strain after reaching the Y_1 yield points has an important influence on most ground movement problems (Jardine *et al.*, 1991). Fig. 11 shows variations of the secant equivalent shear modulus, $G_{eq,sec} = \Delta q/(2\Delta$

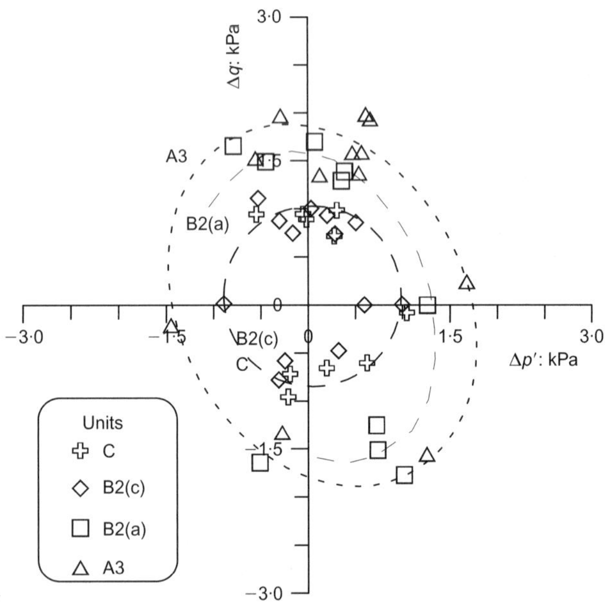

Fig. 8. Y_1 surfaces for different lithological units

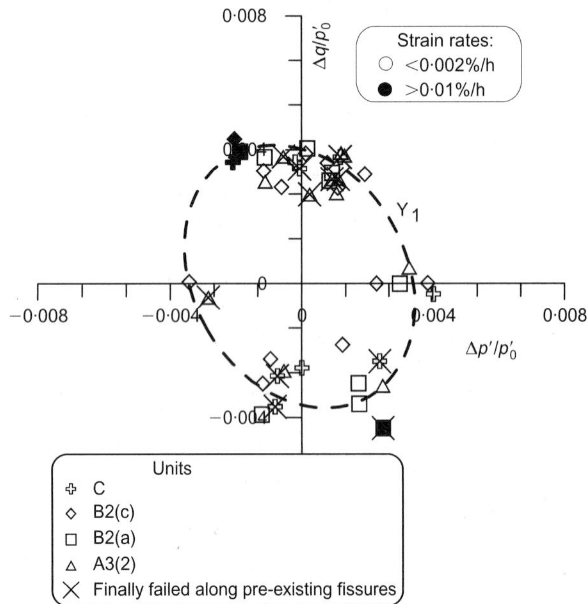

Fig. 9. Normalised contour for the Y_1 surface

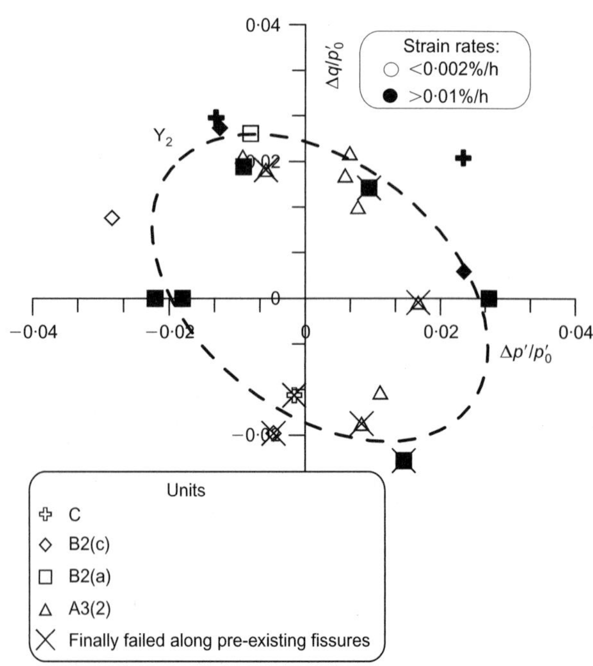

Fig. 10. Normalised contour for the Y_2 surface

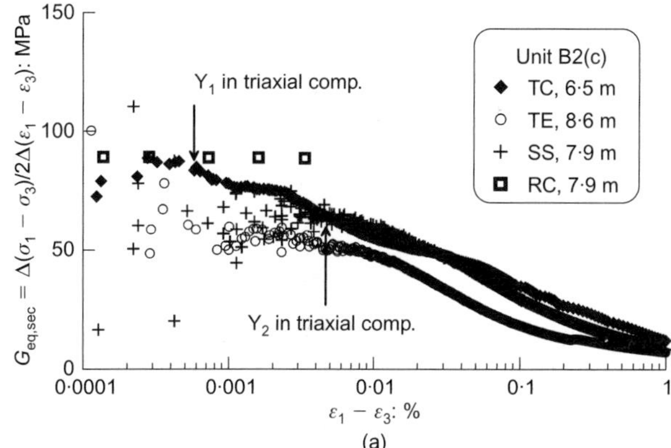

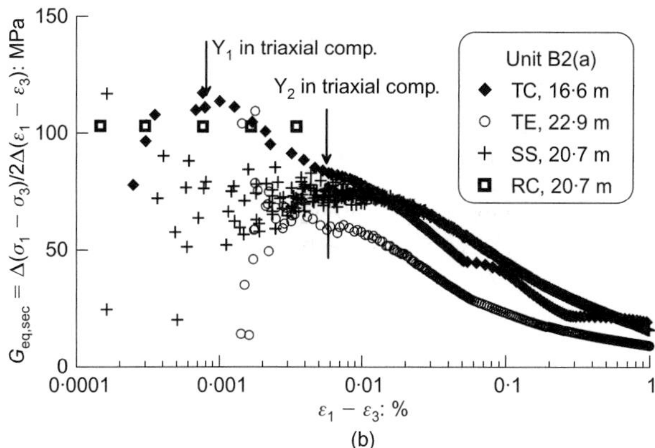

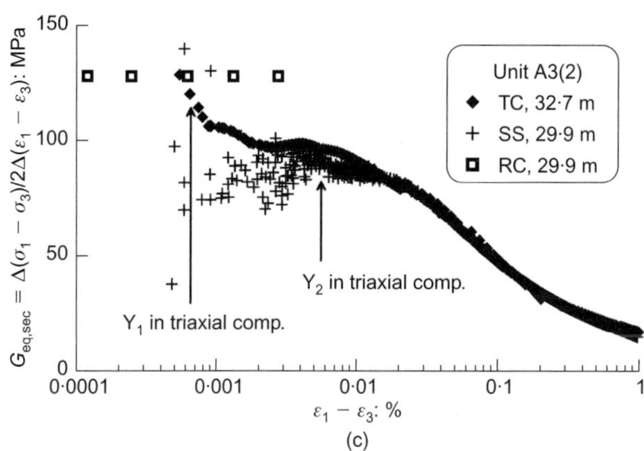

Fig. 11. Stiffness degradation observed during undrained triaxial compression, triaxial extension and simple shear tests: (a) Unit B2(c); (b) Unit B2(a); (c) Unit A3(2)

$(\varepsilon_1 - \varepsilon_3))$, against shear strain, $2/3(\varepsilon_1 - \varepsilon_3)$, observed for samples from three units during undrained triaxial compression, triaxial extension and simple shear tests (no triaxial extension test was conducted for unit A3(2)). In some cases the Y_2 yield point can be seen to be the start of more rapid stiffness degradation. This is more evident when tangent instead of secant values are plotted. The degradation curves are of similar shape for different units.

Provided identical samples are tested, undrained triaxial compression and extension degradation curves should start at the same elastic stiffness. Although this general trend was confirmed, the extension showed steeper degradation with strain because the probing effective stress point with $K_0 > 1$ was located relatively near to the triaxial extension failure envelope. For unit B2(c) the change of the direction of the shearing path compared with the constant p' unloading approach path may also have contributed to the differences seen, but for the other two units, whereas the triaxial com-

pression tests used the approach paths in Fig. 3, the triaxial extension and simple shear tests used a constant-p' unloading path to reach the in situ stresses. An example of the $G_{eq,sec}$ data from resonant column HCA measurements on rotary core samples is also shown for each unit, along with the corresponding undrained static simple shear test data. The latter had broadly similar stiffness degradation characteristics to those of the triaxial compression tests and, as noted earlier, gave maximum stiffnesses that fell significantly below the resonant column $G_{eq,sec}$ values. The interpreted values of $G_{eq,sec}$ declined very gently between the minimum and maximum γ_{vh} amplitudes applied, with the values recorded at around 0·005% shear strain falling less than 1% below those recorded at $\gamma_{vh} = 10^{-5}\%$. Referring to the results shown in Figs 4 and 6, the maximum dynamic strains were well beyond the static elastic limit. It should be noted, however, that γ_{vh} is not uniform in resonant column tests but changes sinusoidally with time and height, developing maxima at the fixed end and minima at the free (driven) end. The γ_{vh} value quoted is the average assessed across the sample (Nishimura, 2006), and each test involves many cycles of loading. Although it is useful for determining the elastic stiffness, the resonant column, or dynamic testing in general, may not be suitable for pinpointing the elastic limit applying to first-time monotonic loading, even if rate independence is assumed.

The variation of the secant moduli measured during the drained shearing to failure stages of the uniaxial HCA tests on block samples from 5·2 m depth are shown in Fig. 12. In all of these tests stiffness degradation began at early stages (from around 0·001% strain). Fig. 13 gives examples of how the equivalent bulk moduli varied with volume strain during the triaxial isotropic compression stages indicated in Fig. 3; the latter data are essential to ground movement predictions of the type described by Jardine *et al.* (1991).

INFLUENCE OF RECENT STRESS HISTORY

In order to investigate the effects of recent stress history on the intact clay's stiffness characteristics, sets of additional probing tests were performed, following a simplified version of the scheme described by Atkinson *et al.* (1990). The probing tests consisted of undrained compression or extension, all starting from the same near isotropic initial stress point, which had been approached by constant-p' drained paths of either 10 or 100 kPa in length, leading to plastic shear strains increments of around 0·005% and 0·05%

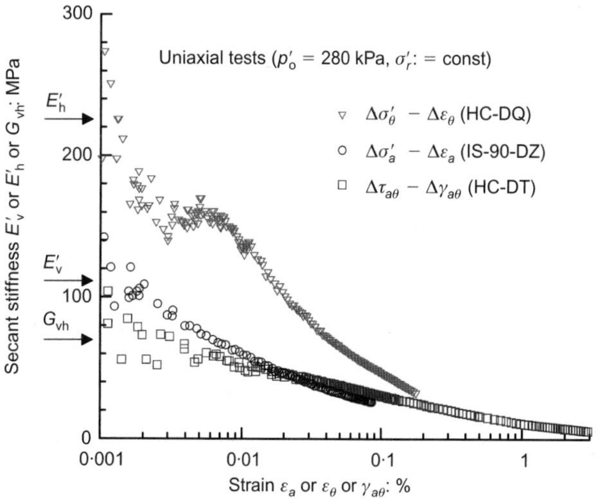

Fig. 12. Secant moduli–strain relationships in drained HCA uniaxial loading tests

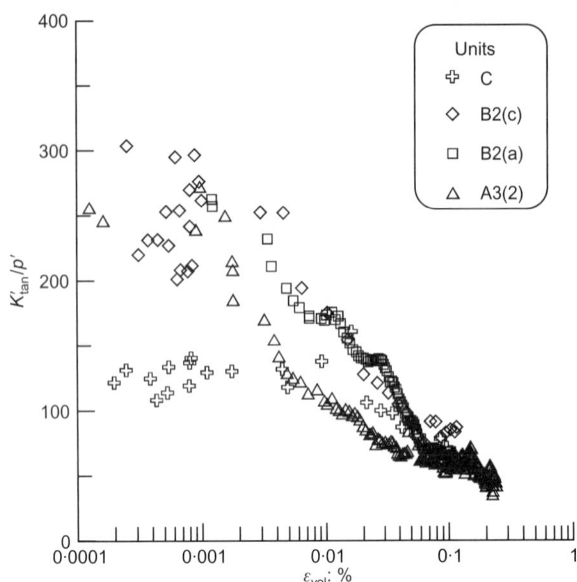

Fig. 13. Degradation curves for tangent bulk moduli

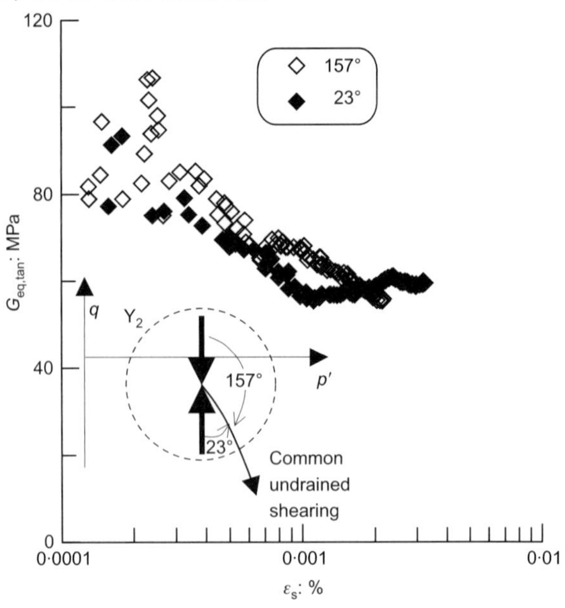

Fig. 14. Tangent stiffness degradation curves for probes with approach path within the Y_2 region and creep allowed

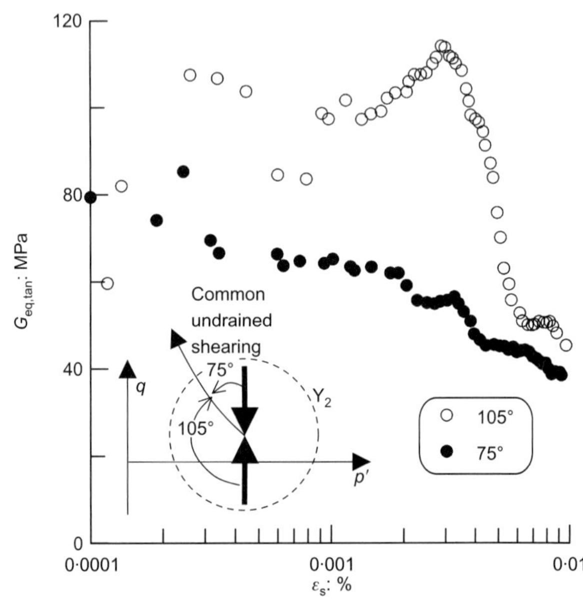

Fig. 15. Tangent stiffness degradation curves for probes with approach path within the Y_2 region and creep not allowed

respectively. (The initial stress point was chosen to avoid being close to either the compression or extension failure envelopes while avoiding a load cell compliance 'flat spot' found on the isotropic axis that affected the applied strain rates unduly.) Drained shear probes, although preferable, were not practical in this study, because of their durations. Just two samples were tested repeatedly to reduce the potential effects of the natural variability, but this introduced the potential for unwanted pre-shearing effects associated with the approach paths. To reduce the latter problem, the shorter approach paths were investigated first, followed by the longer paths. The samples tested were from 11·3 and 11·5 m below the top of the London Clay in Unit B2(b), and bender element check tests on one of the samples confirmed that the values of G_{hv} and G_{hh} remained unchanged between probing tests; as the two samples gave essentially the same results in equivalent tests, the results from only one specimen are discussed in this paper.

Three sets of probing tests are discussed. The first involved two tests in which the 'short' approach paths were applied, followed by an undrained extension common path. As discussed earlier, such paths should have remained within the current kinematic Y_2 yield surface. A creep period of about 7 days was then allowed, during which creep rates reduced to negligible values ($<0·0001\%$/h) before applying the common undrained extension at a rate of 5 kPa/h. With ideal isotropic elastic materials the angles (θ as defined by Atkinson et al., 1990) developed between the approach paths and probing paths would be either zero or 180°. However, the anisotropic London Clay samples developed different undrained effective stress path orientations, with $\theta = 23°$ and 157°. The tangent stiffness relationships developed in the two tests are plotted in Fig. 14. In this case the approach paths had no influence on the results, confirming the absence of stress history effects noted by Clayton & Heymann (2001) in tests on London Clay that involved comparably short approach paths and long creep periods. The second series of probes, on the same sample, reproduced the first up to the end of the approach path. However, only a 3 h pause period was allowed before commencing probing by undrained compression. The creep rates before testing were about 0·001%/h, and the compression was carried out at 5 kPa/h. As shown in Fig. 15, a clear stress history effect was found, with interactions between creep and renewed shearing; the larger stress path rotation gave a quite different stiffness degradation characteristic.

Finally, the first set of probing tests was repeated but with $\Delta q = 100$ kPa, so that the effective stress path would engage and relocate the Y_2 surface. A pause period of 10 days was required before creep rates declined to unresolvably low values ($<0·0001\%$/h) and undrained extension probing tests were performed. The data presented in Fig. 16 show that unlike the first series, where the paths remained within the initial Y_z surface, the stiffness decay relationships were strongly affected by the recent stress history, despite the extended creep/ageing period.

Additional insights were gained by plotting the Y_1 and Y_2 surfaces deduced from the pairs of probing tests and comparing these with the normalised surfaces proven for the clay under in situ stresses (given in Figs 9 and 10). As shown in Figs 17 and 18, in cases when the test paths did not engage and move the Y_2 surface, creep could erase the effects of the approach path on the outgoing stress paths, so that the Y_1 and Y_2 yield points for probes within Y_2 that allowed

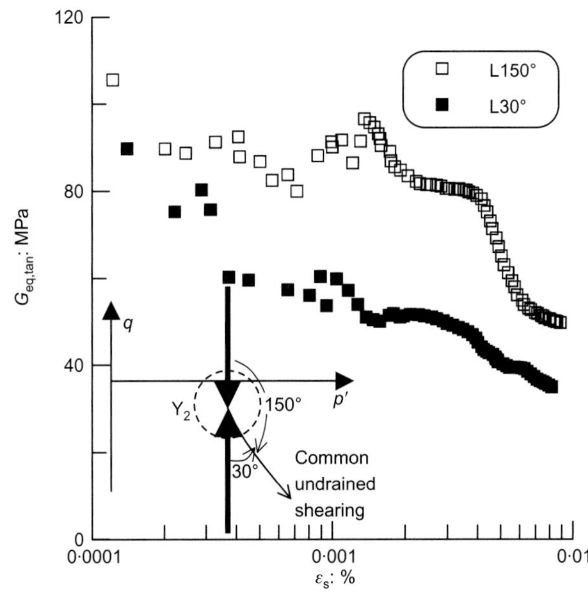

Fig. 16. Tangent stiffness degradation curves for probes with approach path above the Y_2 region and creep allowed

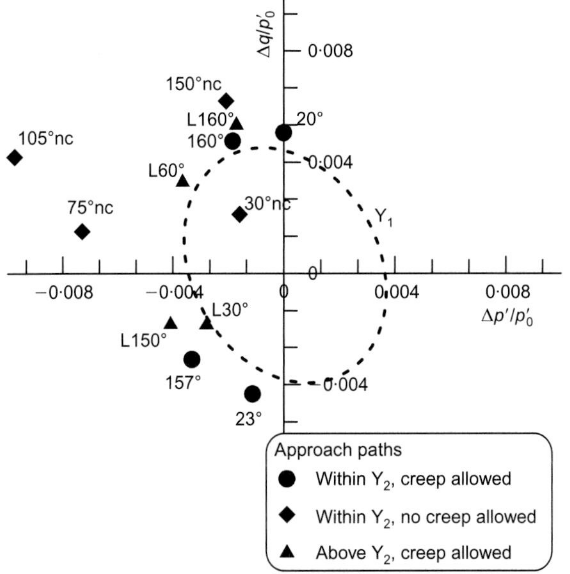

Fig. 17. Normalised Y_1 points for samples subjected to different stress path approaches

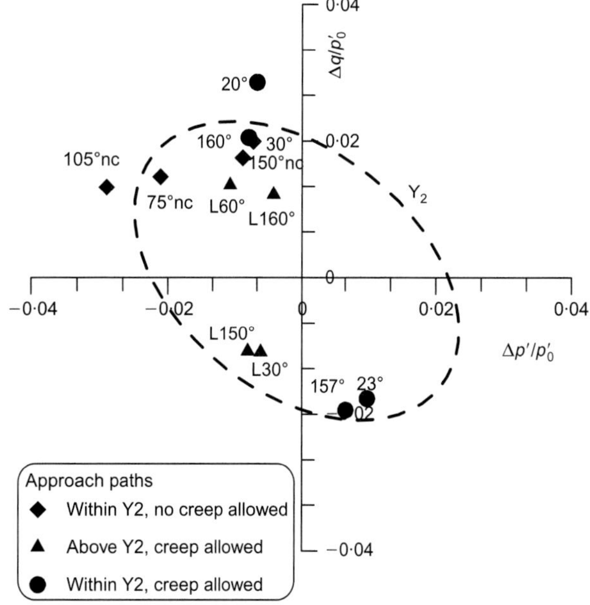

Fig. 18. Normalised Y_2 points for samples subjected to different stress path approaches

creep agree well with the previous envelopes. When creep was not allowed for probes within Y_2 the Y_2 points are unaffected, but Y_1 is dependent on the previous stress history. When the Y_2 surface had been engaged and relocated, with larger plastic strains developing, stress history effects were again evident despite extended creep periods. In this case the creep has re-centred Y_1 so that the Y_1 points agree with the previous envelopes, but the strains developed during the approach stress path affected the Y_2 points. These findings add a further potential significance to the Y_2 surface as a threshold above which hardening plastic strains occur. However, the effects of creep and ageing during pause periods that extend beyond those practical in the laboratory remain open to speculation.

CONCLUSIONS

The stiffness of natural London Clay has been explored through advanced static triaxial and HCA testing involving high-resolution transducers combined with dynamic bender element and resonant column techniques.

Significant anisotropy was revealed over a wide range of strain, with that applying at very small strains being quantified within the framework of cross-anisotropic elasticity. The stiffness parameters obtained by independent techniques generally exhibited good agreement, with the greatest deviation being seen in the Poisson's ratios, which fell far from the values usually assumed in conventional foundation analysis. A range of explanations exists for the modest discrepancies seen between different test types, including potential effects of strain rates (or cyclic frequencies) and test boundary conditions. However, the results point to clear trends between the various cross-anisotropic parameters and their variations with depth, in situ effective stresses and stratigraphical unit.

For any given depth, behaviour was approximately cross-anisotropic linear elastic within a relatively small Y_1 yield surface that surrounded the current effective stress point. Stiffness decayed with strain once this limit was reached, and data have been shown from a wide variety of test types that typify the anisotropic and steeply non-linear trends exhibited by tests continued to failure.

Probing tests established the sizes of the Y_1 surfaces over a range of depths, showing that these scaled in proportion to the mean effective stress level, as did those of the second Y_2 kinematic surface that surrounded Y_1. The significance of Y_2 was that, when engaged, the soil response gave a new pattern of strain increment directions. As shown in this paper, Y_2 in many cases appeared to correspond to faster elastic-plastic stiffness decay with strain, and it was also found that there was a greater dependence of behaviour on recent stress history for stress path approaches that exceeded Y_2.

Finally, experiments have been reported that explore the interaction between recent stress history, creep/ageing periods and probing paths. It has been shown that relatively short creep periods can erase the effects of recent stress perturbations that remained within the original Y_2 surface, while changes that engage and displace the latter impose a more enduring 'memory' of the recent stress history.

ACKNOWLEDGEMENTS

The authors thank: the EPSRC and London Underground for their financial support; BAA for the rotary-cored samples and access to the site; Dr A. Takahashi for his work on the sampling, HCA commissioning and control software; the Imperial College technicians for their assistance; and Dr H. C. Yeow of Arup Geotechnics for his advice. The third author was funded by the Vietnamese Government Overseas Scholarship Program Project 322.

NOTATION

b	intermediate principal stress ratio $(= (\sigma_2 - \sigma_3)/(\sigma_1 - \sigma_3))$
E'_h	drained Young's modulus in the horizontal direction
E^u_h	undrained Young's modulus in the horizontal direction
E'_v	drained Young's modulus in the vertical direction
E^u_v	undrained Young's modulus in the vertical direction
G_{eq}	equivalent shear modulus
$G_{eq,sec}$	secant equivalent shear modulus
$G_{eq,tan}$	tangent equivalent shear modulus
G_{hh}	shear modulus in the horizontal plane
$G_{vh} = G_{hv}$	shear moduli in the vertical plane
K_0	coefficient of earth pressure at rest
K'_{sec}	secant drained bulk modulus
p'	mean normal effective stress
p'_0	initial mean normal effective stress
q	deviatoric stress
q_0	initial deviatoric stress
α	angle between σ_1 axis and vertical
ε_s	shear strain $(= 2(\varepsilon_1 - \varepsilon_3)/3)$
ε^p_s	plastic shear strain
ε_{vol}	volumetric strain $(= \varepsilon_1 + \varepsilon_2 + \varepsilon_3)$
ε^p_{vol}	plastic volumetric strain
$\varepsilon_x, \varepsilon_y, \varepsilon_z$	normal strains in Cartesian coordinate system
$\varepsilon_a, \varepsilon_r, \varepsilon_\theta$	normal strains in polar coordinate system
$\varepsilon_1, \varepsilon_3$	principal strains
$\gamma_{xy}, \gamma_{yz}, \gamma_{zz}$	shear strains in Cartesian coordinate system
$\gamma_{a\theta}$	shear strain in polar coordinate system
ν'_{hh}	drained Poisson's ratio for horizontal strains due to horizontal strain
ν'_{hv}	drained Poisson's ratio for horizontal strains due to vertical strain
ν'_{vh}	drained Poisson's ratio for vertical strains due to horizontal strain
ν^u_{vh}	undrained Poisson's ratio for vertical strains due to horizontal strain
$\sigma_1, \sigma_2, \sigma_3$	principal stresses
$\sigma_a, \sigma_r, \sigma_\theta$	normal stresses in polar coordinate system
σ'_h	horizontal effective stress
σ'_v	vertical effective stress
$\sigma'_x, \sigma'_y, \sigma'_z$	normal effective stresses in Cartesian coordinate system
$\tau_{xy}, \tau_{yz}, \tau_{zx}$	shear stresses in Cartesian coordinate system
$\tau_{a\theta}$	shear stress in polar coordinate system

REFERENCES

Addenbrooke, T. I., Potts, D. M. & Puzrin, A. M. (1997). The influence of pre-failure soil stiffness on the numerical analysis of tunnel construction. *Géotechnique* 47, No. 3, 693–712.

Atkinson, J. H., Richardson, D. & Stallebrass, S. E. (1990). Effect of stress history on the stiffness of overconsolidated soil. *Géotechnique* 40, No. 4, 531–540.

Bellotti, R., Jamiolkowski, M., Lo Presti, D. C. F. & O'Neill, D. A. (1996). Anisotropy of small strain stiffness in Ticino sand. *Géotechnique* 46, No. 1, 115–131.

Butcher, A. P. & Powell, J. J. M. (1996). Practical considerations for field geophysical techniques used to assess ground stiffness. In *Advances in site investigation practice* (ed. C. Craig), pp. 701–714. London: Thomas Telford.

Chandler, R. J. (2000). Clay sediments in depositional basin: the geotechnical cycle. *Q. J. Engng Geol. Hydrogeol.* 33, No. 1, 7–39.

Clayton, C. R. I. & Heymann, G. (2001). Stiffness of geomaterials at very small strains. *Géotechnique* 51, No. 3, 245–255.

Cuccovillo, T. & Coop, M. R. (1997). The measurement of local axial strains in triaxial tests using LVDTs. *Géotechnique* 47, No. 1, 167–171.

de Freitas, M. H. & Mannion, W. H. (2007). A biostratigraphy for the London Clay in London. *Géotechnique* 57, No. 1, 91–99.

Gasparre, A. (2005). *Advanced laboratory characterisation of London Clay.* PhD thesis, Imperial College, London (http://www.imperial.ac.uk/geotechnics/Publications/PhDs/phdsonline.htm)

Gasparre, A. & Coop, M. R. (2006). Techniques for performing small strain probes in the triaxial apparatus. *Géotechnique* 56, No. 7, 491–495.

Gasparre, A., Nishimura, S., Coop, M. R., & Jardine, R. J. (2007). The influence of structure on the behaviour of London Clay. *Géotechnique* 57, No. 1, 19–31.

Grammatikopoulou, A. (2004). *Development, implementation and application of kinematic hardening models for overconsolidated clays.* PhD thesis, Imperial College, London.

Hight, D. W., Bennell, J. D., Chana, B., Davis, P. D., Jardine, R. J. & Porovic, E. (1997). Wave velocity and stiffness measurements of the Crag and Lower London Tertiaries at Sizewell. *Géotechnique* 47, No. 3, 451–474.

Hight, D. W., McMillan, F., Powell, J. J. M., Jardine, R. J. & Allenou, C. P. (2002). Some characteristics of London Clay. *Characterisation of engineering properties of natural soils* (eds D. W. Hight, S. Leroueil, K. K. Phoon and T. S. Tan), pp. 851–908. Rotterdam: Balkema.

Hight, D. W., Gasparre, A., Nishimura, S., Minh, N. A., Jardine, R. J. & Coop, M. R. (2007). Characteristics of the London Clay from the Terminal 5 site at Heathrow Airport. *Géotechnique* 57, No. 1, 3–18.

HongNam, N. & Koseki, J. (2005). Quasi-elastic deformation properties of Toyoura sand on cyclic and torsional loading. *Soils Found.* 45, No. 5, 19–38.

Jardine, R. J. (1995). One perspective on the pre-failure deformation characteristics of some geomaterials. Keynote lecture. *Proceedings of the international symposium on pre-failure deformation characteristics of geomaterials*, Hokkaido, Japan, Vol. 2, pp. 885–886.

Jardine, R. J., Fourie, A. B., Maswoswe, J. & Burland, J. B. (1985). Field and laboratory measurements of soil stiffness. *Proc. 11th Int. Conf. Soil Mech. Found. Engng, San Francisco* 2, 511–514.

Jardine, R. J., Gens, A., Hight, D. W. & Coop, M. R. (2004). Developments in understanding soil behaviour. In *Advances in geotechnical engineering, Proceedings of the Skempton Conference*, Vol. 1, pp. 103–206. London: Thomas Telford.

Jardine, R. J., St John H. D., Hight, D. W. & Potts, D. M. (1991). Some applications of a non-linear ground model. *Proc. 10th Eur. Conf. Soil Mech. Found. Engng, Florence* 1, 223–228.

Jovicic, V. & Coop, M. R. (1998). The measurement of stiffness anisotropy in clays with bender element tests in the triaxial apparatus. *Geotech. Test. J.* 21, No. 1, 3–10.

King, C. (1981). The stratigraphy of the London Basin and associated deposits. *Tertiary Research Special Paper* 6, Rotterdam: Backhuys.

Kuwano, R. & Jardine, R. J. (1998). Stiffness measurements in a stress-path cell. In *Pre-failure deformation behaviour of geomaterials*, pp. 391–394. London: Thomas Telford.

Lings, M. L. (2001). Drained and undrained elastic stiffness parameters. *Géotechnique* 51, No. 6, 555–565.

Lings, M. L., Pennington, D. S. & Nash, D. F. T. (2000). Anisotropic stiffness parameters and their measurement in a stiff natural clay. *Géotechnique* 50, No. 2, 109–125.

Minh, N. A. (2006). *An investigation of the anisotropic stress–strain–strength characteristics of an Eocene clay.* PhD thesis, Imperial College London (http://www.imperial.ac.uk/geotechnics/Publications/PhDs/phdsonline.htm)

Nishimura, S. (2006). *Laboratory study on anisotropy of natural London Clay.* PhD thesis, Imperial College, London. (http://www.imperial.ac.uk/geotechnics/Publications/PhDs/phdsonline.htm)

Nishimura, S., Minh, N. A. & Jardine, R. J. (2007). Shear strength anisotropy of natural London Clay. *Géotechnique* 57, No. 1, 49–62.

Pennington, D. S., Nash, D. F. T. & Lings, M. L. (1997). Anisotropy of G_0 shear stiffness of Gault clay. *Géotechnique* **47**, No. 3, 391–398.

Potts, D. M. & Zdravkovic, L. (2001). *Finite element analysis in geotechnical engineering: Application.* London: Thomas Telford.

Simpson, B. (1992). Retaining structures: displacement and design. 32nd Rankine Lecture. *Géotechnique* **42**, No. 4, 539–576.

Simpson, B., Atkinson, J. H. & Jovicic, V. (1996). The influence of anisotropy on calculation of ground settlements above tunnels. In *Geotechnical aspects of underground construction in soft ground* (eds R. J. Mair and R. N. Taylor), pp. 591–595. Rotterdam: Balkema.

Skempton, A. W. & Henkel, D. J. (1957). Tests on London Clay from deep borings at Paddington, Victoria and the South Bank. *Proc. 4th Int. Conf. Soil Mech. Found. Engng, London* **1**, 100–106.

Tatsuoka, F. & Shibuya, S. (1992). Deformation characteristics of soils and rocks from field and laboratory tests. Keynote lecture. *Proc. 9th Asian Conf. Soil Mech. Found. Engng, Bangkok* **2**, 101–170.

Zdravkovic, L. & Jardine, R. J. (1997). Some anisotropic stiffness characteristics of a silt under general stress conditions. *Géotechnique* **47**, No. 3, 407–437.

Nishimura, S., Minh, N. A. & Jardine, R. J. (2007). *Géotechnique* **57**, No. 1, 49–62

Shear strength anisotropy of natural London Clay

S. NISHIMURA[*], N. A. MINH[†] and R. J. JARDINE[*]

The paper reports the shear strength anisotropy of the natural, highly overconsolidated, London Clay from Heathrow Terminal 5 as established by comprehensive hollow cylinder apparatus (HCA) testing. Multiple high-quality block samples from 5·2 m and 10·5 m below ground level provided samples for suites of undrained stress-path shear tests performed after consolidation to effective stress states similar to those estimated in situ. The direction of the major principal stress axis, α (or of the principal stress increment, $\alpha_{d\sigma}$), and relative magnitude of the intermediate principal stress, $b = (\sigma_2 - \sigma_3)/(\sigma_1 - \sigma_3)$, were chosen as the controlled stress-related variables and their influence on the peak shear strength was investigated. Strong shear strength anisotropy was proven, and a potential effect of the parameter b was also detected. With $b = 0·5$, for example, a minimum was noted in peak q/p' at around $\alpha = 45$–67°, and the maximum value, which was larger by some 40%, developed at $\alpha = 0°$. A limited set of data obtained with larger test specimens suggested a possibility that loading with $\alpha = 90°$ could also lead to relatively low shear strength. The pre-failure pore water pressure development reflected anisotropy of the deformation characteristics at smaller strains and contributed to the total-stress undrained shear strength anisotropy. Regarding the influence of b, the Mohr–Coulomb failure line passing through the data for $b = 0$ or 1 provided a lower bound of all the data for a given value of α. The investigation of anisotropy was extended to deeper horizons of the stratum through profiling tests involving triaxial extension and simple shear conditions. Although the results confirmed the general patterns of anisotropy observed at 5·2 and 10·5 m below ground level, the degree of anisotropy appeared to become stronger with depth.

KEYWORDS: anisotropy; clays; fabric; shear strength; stress path; torsion

Cet article rapporte les résultats d'anisotropie de résistance au cisaillement de la London Clay (argile de Londres) du Terminal 5 de l'aéroport d'Heathrow, une argile naturelle hautement surconsolidée, tels qu'estimés par des essais complets réalisés avec un appareil à cylindre creux. Plusieurs échantillons de bloc de haute qualité, de 5,2 m et 10,5 m de profondeur, ont permis de prélever des spécimens pour une série d'essais à chemin de contrainte en cisaillement non drainés réalisés après consolidation à des états de contrainte effectifs similaires à ceux estimés in situ. La direction de l'axe de contrainte principale, α (ou de l'incrément en contrainte principale, $\alpha_{d\sigma}$) et l'amplitude relative de la contrainte principale intermédiaire, liée au paramètre $b = (\sigma_2 - \sigma_3)/(\sigma_1 - \sigma_3)$ ont été choisies comme variables contrôlées liées à la contrainte. L'étude examine leur influence sur la résistance au cisaillement maximale. Il a été montré que la résistance au cisaillement est fortement anisotrope et l'on a pu observer un effet potentiel du paramètre b. Avec $b = 0,5$, par exemple, on a pu noter un minimum du pic q/p' pour α autour de 45–67° et la valeur maximale, supérieure de près de 40 %, est observée pour $\alpha = 0°$. Un ensemble de données limité, réalisé sur des spécimens plus larges, suggèrent qu'une charge à $\alpha = 90°$ puisse également conduire à une résistance au cisaillement relativement faible. Le développement de la pression hydrique interstitielle avant fracture reflète l'anisotropie des caractéristiques de déformation pour des contraintes plus faibles et contribue à l'anisotropie de la résistance au cisaillement non drainé totale. En ce qui concerne l'influence de b, la ligne du critère de rupture de Mohr–Coulomb s'ajustant aux données pour $b = 0$ ou 1 a fourni une limite inférieure de toutes les données pour une valeur donnée de α. On a pu approfondir les horizons de l'étude de l'anisotropie de la strate grâce à des essais de profil comportant une extension triaxiale et des conditions de cisaillement simples. Bien que les résultats obtenus confirment les tendances générales de l'anisotropie observée à 5,2 et 10,5 m en dessous du niveau du sol, le degré d'anisotropie est apparu comme devenant plus important avec la profondeur.

INTRODUCTION

The importance of the mechanical anisotropy of soils has long been recognised in geotechnical engineering. For example Bjerrum (1973) and Jardine & Smith (1991), among others, drew attention to principal stress axis rotation in soft clays under embankment loading, and highlighted the need to understand better shear strength anisotropy. Zdravkovic *et al.* (2002) simulated a full-scale embankment test successfully with an anisotropic constitutive model calibrated to laboratory test data, also exploring interactions between strength anisotropy, embankment geometry, ground movements and collapse heights.

However, similar advances do not appear to have been made yet with stiff overconsolidated plastic clays. One of the major hurdles in interpreting and simulating failure of stiff clays, in addition to sampling disturbance, is their markedly progressive failure processes, deriving from their brittle constitutive relations (Burland, 1990; Burland *et al.*, 1996; Jardine *et al.*, 2004). The progressive mechanisms, with particular reference to the London Clay, were explored in numerical analysis by Potts *et al.* (1997) and Kovacevic *et al.* (2004). Noting these and other areas of uncertainty, the potential effects of shear strength anisotropy have been largely neglected. The currently available reliable experimental datasets regarding the shear strength anisotropy of stiff plastic clays are severely limited and do not offer sufficient evidence to develop and apply anisotropic constitutive models in practice. The research work undertaken in the 1960s on the

Manuscript received 5 May 2006; revised manuscript accepted 14 November 2006.
Discussion on this paper closes on 1 July 2007, for further details see p. ii.
[*] Imperial College, London, UK.
[†] Atkins Ltd; formerly Imperial College, London, UK.

natural London Clay (Ward *et al.*, 1965; Bishop, 1966; Bishop & Little, 1967; Agarwal, 1968) forms an important part of the available database. However, these studies had to rely on triaxial tests on diagonally-cut cylindrical specimens that are subject to several inherent difficulties, as recognised in later years. Such disadvantages include the generation of bending moments in soil specimens (e.g. Saada, 1970; Saada & Townsend, 1981; Molenkamp, 1998) and the infeasibility of applying the in situ K_0 stress conditions.

Investigations connected with the Heathrow Airport Terminal Five (T5) project offered the opportunity to apply more advanced laboratory testing techniques to high-quality block samples of the natural London Clay from three different depths, involving stress-path triaxial cells and hollow cylinder apparatus' (HCAs). Multiple block samples from two key horizons allowed suites of tests to examine anisotropy by imposing σ_1 in different directions, while other conditions such as the reconsolidation regime and the intermediate principal stress ratio, $b = (\sigma_2 - \sigma_3)/(\sigma_1 - \sigma_3)$, applied during shearing were held constant, so allowing assessment of the net influence of anisotropy. The investigation was extended to deeper horizons by means of multiple triaxial extension and simple shear tests on rotary cores recovered from 6 m to 35 m below ground level. Referring to existing triaxial compression test data, the shear strength profiles in these shear modes were used to gauge how anisotropy developed over the main thickness of the stratum.

This paper focuses mainly on the peak shear strength anisotropy, and the influence of the intermediate principal stress on the effective and total-stress shear strengths available on shearing from in situ stress conditions. The authors' experiments investigated the clay's behaviour over a wide strain range. The anisotropic and non-linear pre-failure deformation characteristics are reported and discussed in a companion paper (Gasparre *et al.*, 2007b). The clay's geology, composition and structure, which are the causes of its anisotropy, are discussed in a second companion paper (Gasparre *et al.*, 2007a), which also gives details of index properties. Hight *et al.* (2007) synthesise the authors' findings with other field and laboratory investigations, and refer to the potential practical application of the results.

APPARATUS AND MATERIAL
Hollow cylinder apparatus (HCA)
The boundary forces and pressures applied to a hollow cylindrical specimen and the stresses and the strains induced in its elements are illustrated in Fig. 1. The most attractive feature

of an HCA is the four degrees of freedom in stress and strain control attained through combinations of axial force, torque and inner and outer cell pressures. The four stress components induced in a specimen, σ_z, σ_r, σ_θ and $\tau_{z\theta}$, were assessed as outlined by Hight *et al.* (1983), and $\tau_{zr} = \tau_{r\theta} = 0$ was assumed. They are conveniently converted into an equivalent set of four stress-related parameters, p, q, α and b, where

$$p = \frac{\sigma_1 + \sigma_2 + \sigma_3}{3} \tag{1}$$

$$q = \sigma_1 - \sigma_3 \tag{2}$$

$$\alpha = \frac{1}{2}\tan^{-1}\left(\frac{2\tau_{z\theta}}{\sigma_z - \sigma_\theta}\right) \tag{3}$$

$$b = \frac{\sigma_2 - \sigma_3}{\sigma_1 - \sigma_3} \tag{4}$$

where α is the angle between the major principal stress direction and the vertical. The effects of α on material behaviour are therefore direct indices of anisotropy. The quantity b is called the intermediate principal stress ratio, and varies between 0 and 1.

Two different computer-controlled HCAs were employed: the ICRCHCA (Porovic, 1995; Nishimura, 2006) and the ICHCA II (Jardine, 1996; Minh, 2006), which are illustrated in Figs 2(a) and 2(b) respectively. Both apparatus sets have the ability to perform experiments under mixed stress and strain control, depending on the particular test requirements. The nominal outer diameters, inner diameters and heights of test specimens were 70 mm, 38 mm and 170–190 mm in the ICRCHCA, and 100 mm, 60 mm and 200 mm in the ICHCA II. The ICRCHCA was equipped with a Hardin oscillator, with which torsional resonant column tests were performed to obtain dynamic shear moduli. The sample deformation was measured globally, apart from the semi-local (platen-to-platen) torsional rotation with a proximity transducer system. A full suite of high-resolution local transducers was available for a part of the test programme conducted with the ICHCA II. The axial displacement and torsional rotation were measured with a system consisting of enhanced electrolevel inclinometers, and changes of the inner and outer radii were measured with a linear variable displacement transducer (LVDT) and a set of three proximity transducers respectively. From these boundary displacements, the average strains induced in a soil specimen were calculated according to the equations given by Hight *et al.* (1983). Most of the triaxial extension tests were performed on 38 and 100 mm diameter specimens with the locally instrumented hydraulic stress-path triaxial cells described by Gasparre (2005).

Natural London Clay samples
Hand-trimmed block samples and triple-tube rotary cores were retrieved from the T5 site, at the locations identified by Hight *et al.* (2007). The blocks were taken from excavation bases formed at 1·0 m, 5·2 m and 10·5 m below ground level, with the latter two depths being chosen as key horizons for the detailed investigation of anisotropy. The rotary cores were retrieved from a location where the London Clay was overlain by a 6 m thick Quaternary gravel layer; this layer had been removed decades earlier at the block sampling location. The cores from 6 m below ground level (and hence the top of the London Clay) to 35 m below the ground level were tested in the present study. As described by Hight *et al.* (2007), this range covers lithological units A3, B2 and C, established by King (1981); unit B1 was not recovered successfully.

Table 1 summarises the index properties of the samples from 5·2 m and 10·5 m bgl. The standard deviations recorded indicate that the samples were sufficiently uniform at

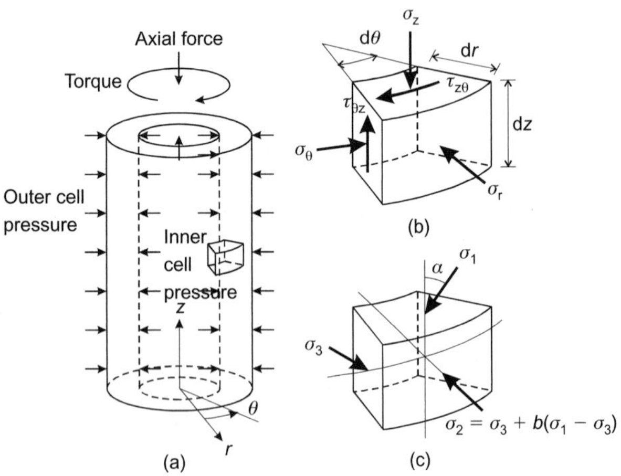

Fig. 1. Definition of coordinate system, stresses, strains and stress-related variables

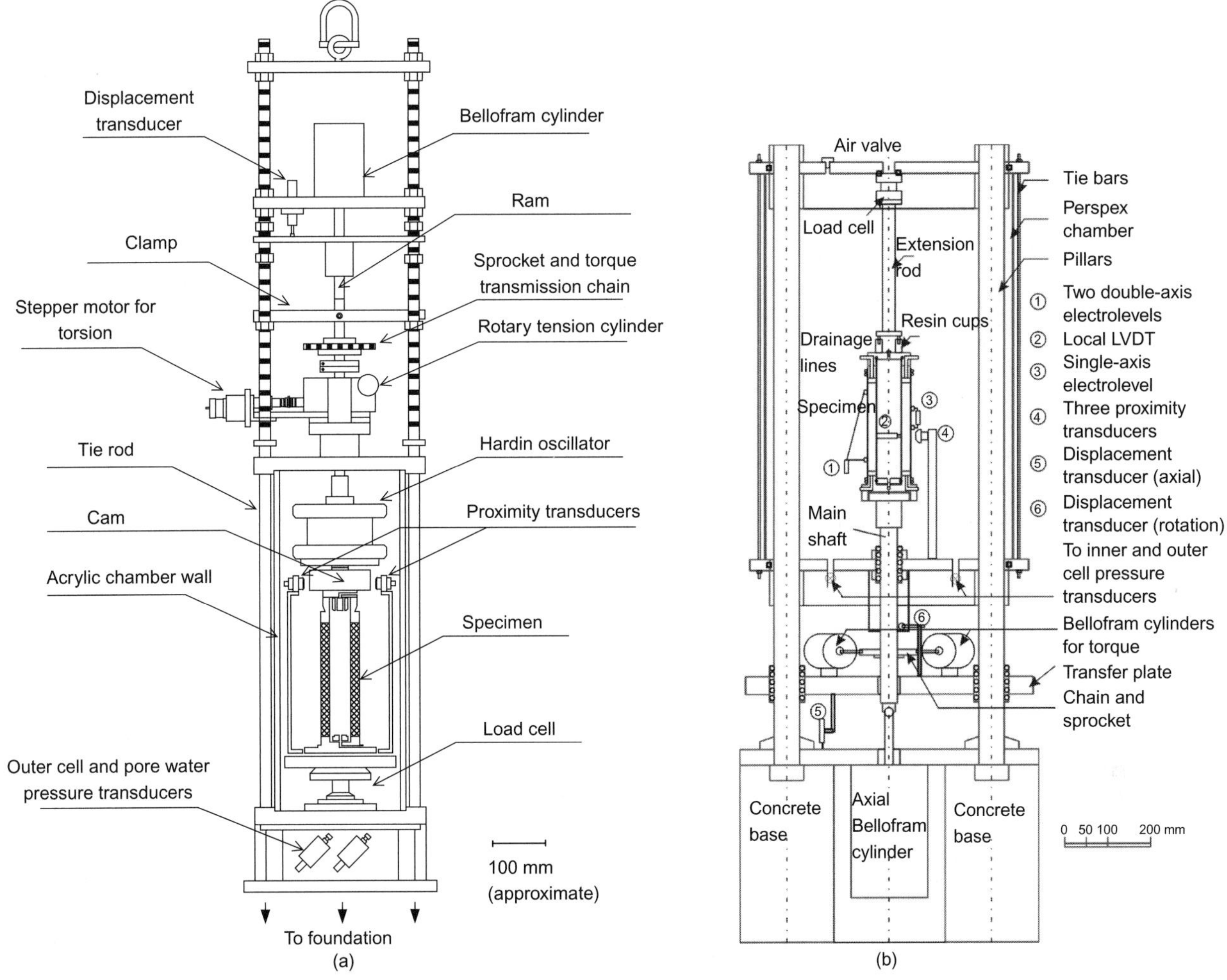

Fig. 2. Schematic illustration of HCAs: (a) ICRCHCA (Porovic, 1995; Nishimura, 2006); (b) ICHCA II (Jardine, 1996; Minh, 2006)

Table 1. Index and other properties of block samples

	5·2 m bgl (12·5 mOD)		10·5 m bgl (7·2 mOD)	
	Average	Std dev.	Average	Std dev.
Natural water content, w: %	23·7	0·3	24·9	0·5
Plastic limit	25	0·3	26	0·8
Liquid limit	69	0·3	70	0·9
Specific gravity, G_s	2·77	0·01	2·77	0·01
Clay fraction: %	51	1·3	55	1·0
Sand content: %	3·7	1·1	1·7	0·3
Unit weight, γ: kN/m³	19·9	0·2	19·9	0·3
$G_{z\theta}$ after consolidation: MPa	–	–	92·5	3·6

each depth in terms of index properties. Also tabulated is the range of shear moduli $G_{z\theta}$ obtained from 22 resonant column tests at $p' = 323$ kPa at the end of isotropic reconsolidation. The relatively small deviation suggests that variable disturbance imposed during the sampling and preparation was minimised.

One of the dominant features in the London Clay tested was the macrofabric represented by natural discontinuities. These include fissures and joints, although no particular distinction between them is made here. As noted by Skempton *et al.* (1969), many ran in horizontal to sub-horizontal and vertical to sub-vertical directions. Although their spacing was difficult to quantify, most of the visible discontinuities appeared to be separated by 2 to 5 cm. The

blocks from 5·2 m bgl were more heavily fissured than those from 10·5 m bgl, although the upwardly intensifying trend reported by many researchers (e.g. Ward *et al.*, 1965; Chandler & Apted, 1988) was not recognised in samples from the gravel-capped rotary borehole locations.

HCA TESTING PROCEDURES AND SCHEMES

The HCA stress-path tests in which α (or $\alpha_{d\sigma}$; defined later) and b were varied as controlled parameters are described in detail in this section; briefer comments are also given on the simple shear and triaxial extension tests.

The driving systems in both the ICRCHCA and ICHCA II incorporated both stress- and strain-controlling devices. Axial displacements and torsional rotations can be driven by stepper-motor-controlled, fluid-filled pumps with the ICHCA II. The ICRCHCA has a stepper-motor-driven torsional rotation drive, but relies on a pneumatic stress-controlling Bellofram cylinder for axial loading. A robust feedback system and algorithm was developed that could maintain prescribed axial strain rates and avoid post-peak instability with the potentially brittle London Clay. Satisfactorily linear strain–time relationships were obtained with all types of servo system when operating under strain control. Nishimura (2006) gives data demonstrating these features, along with a flowchart detailing the ICRCHCA control algorithm.

The simple shear tests were conducted in the ICRCHCA by testing hollow cylindrical specimens (as described by Shibuya and Hight, 1987, and Pradhan *et al.*, 1988) to overcome some of the drawbacks inherent in conventional simple shear apparatus. Strain-controlled loading was achieved by applying torsional rotation through the stepper motor and gearing arrangement while prohibiting mechanically axial movement of the ram and a volume change of the inner cell and the sample. The mechanical axial restraint was further assisted by a servo-system control to counteract the apparatus compliance. The undrained triaxial tests were conducted under axial strain control in the computer-controlled stress-path apparatus described by Gasparre (2005), working mainly with specimens of 200 mm height and 100 mm diameter; reference is also made to extension tests on smaller samples, 76 mm by 38 mm.

Sample preparation

All samples were trimmed to the specified outer diameter and height with a wire saw in a rotary soil lathe. Hollow cylinder samples were formed by placing the solid cylinders in a metal mould that was transferred to a metal working lathe. While the specimen was rotated at 190 rev/min, a set of six drill bits was advanced from one end. The process involved gradually increasing the bit diameters until the desired inner diameter was achieved. For drier samples (including all the block samples and deeper rotary core samples), the friction between the inner clay surface and the larger-diameter drills tended to open natural discontinuities and disturb the specimen. For such samples, the final drilling stages were replaced by careful point-edge reaming to reach their final diameter.

Reconsolidation scheme

After being set up and saturated to an acceptable B value, all test samples were reconsolidated to isotropic or anisotropic effective stress states similar to the estimated in situ conditions. The final reconsolidation stresses were $p' = 323$ kPa and $(\sigma_z - \sigma_\theta)/2$ $(= (\sigma_z - \sigma_r)/2) = -83$ kPa for the 10·5 m bgl HCA samples (test Series AC), corresponding to $K = \sigma_\theta'/\sigma_z' = 1\cdot78$. For the 5·2 m bgl HCA samples (test Series AM), a wider range of p' and $(\sigma_z - \sigma_\theta)/2$ values was employed, as shown in Table 2. The estimated OCRs under these conditions are 9–12. Isotropic reconsolidation was first conducted to reach the desired p' values, and then $(\sigma_z - \sigma_\theta)/2$ was reduced, keeping p' constant, to reach the eventual reconsolidation points. The IC and IM series of tests involved only isotropic reconsolidation, and the subsequent undrained shearing was initiated from $p' = 323$ kPa and $(\sigma_z - \sigma_\theta)/2 = 0$ kPa in Series IC and from $p' = 280$ kPa and $(\sigma_z - \sigma_\theta)/2 = 0$ kPa in Series IM. In test NC4505, the

Table 2. Loading conditions in HCA stress-path test series

Series	Test name	α_{peak}*: degrees	$\alpha_{d\sigma}$: degrees	b	Series	Test name	p_0': kPa	$(\sigma_z - \sigma_\theta)_0/2$: kPa	α: degrees	b
AC	AC0000	0	0	0	AM	AM0000	280	−55	0	0
	AC0005	0	0	0·5		AM4500	280	−55	45	0
	AC2305	23	15	0·5		AM0003	280	−76	0	0·3
	AC4505	48	30	0·5		AM0003B	280	−80	0	0·3
	AC6705	67	55	0·5		AM3003	280	−82	30	0·3
	AC9005	90	90	0·5		AM0005	312	−80	0	0·5
	AC4507	45	30	0·7		AM3005	312	−80	30	0·5
	AC4510	46	30	1		AM5005	312	−80	50	0·5
	AC6710	72	55	1		AM6005	332	108	60	0·5
	AC9010	90	90	1		AM9005	312	−80	90	0·5
IC	IC0000	0	0	0		AM4510	280	−78	45	1·0
	IC2300	23	23	0		AM6010	280	−70	60	1·0
	IC4500	45	45	0		AM9010	280	−75	90	1·0
	IC4503	45	45	0·3	IM	IM0005	280	0	0	0·5
	IC0005	0	0	0·5		IM9005	280	0	90	0·5
	IC2305	23	23	0·5						
	IC4505	45	45	0·5						
	IC6705	67	67	0·5	Series		p_0': kPa	$(\sigma_z - \sigma_\theta)_0/2$: kPa	$\tau_{z\theta0}$: kPa	
	IC9005	90	90	0·5						
	IC9005B	90	90	0·5	AC		323	−83	0	
	IC9010	90	90	1	IC		323	0	0	
					NC		323	0	83	
NC	NC4505	45	45	0·5	AM		See above table		0	
					IM		See above table		0	

*The values at maximum q/p'.
p_0', $(\sigma_z - \sigma_\theta)_0/2$ and $\tau_{z\theta0}$ are p', $(\sigma_z - \sigma_\theta)/2$ and $\tau_{z\theta}$ at final reconsolidation points respectively.

sample was reconsolidated to $p' = 323$ kPa, $(\sigma_z - \sigma_\theta)/2 = 0$ kPa and $\tau_{z\theta} = 83$ kPa (i.e. $\alpha = 45°$).

The undrained triaxial extension and undrained simple shear tests were performed following a reconsolidation scheme similar to that of the HCA samples, with the eventual K_0 reconsolidation effective stress states simulating the conditions expected in situ at each sample depth (Nishimura, 2006).

Shear scheme

Slightly different shearing procedures were applied in the HCA test series. In each case sufficient time was allowed after the end of the consolidation stages for creep to stabilise. Typically this involved a pause of several days, waiting for the rates of strain components to fall below 0·002%/h. In the series starting from isotropic stress conditions (Series IC and IM), undrained shearing was performed applying a nominal shear strain rate of $d(\varepsilon_1 - \varepsilon_3)/dt = 4\%$ per day while maintaining constant p, α and b values through the servo-control arrangements. In the anisotropically consolidated Series AC, b was changed from its initial value of 1 $(K_0 > 1)$ to a specified value under undrained conditions, keeping p, q and α constant. After allowing 12 h for creep, further undrained shearing was initiated with constant p, b and $\alpha_{d\sigma}$ values, where $\alpha_{d\sigma}$ is the major principal axis direction of stress increment and defined in a similar way to equation (3). It follows that α varied during undrained shearing in the Series AC tests except in the tests with $\alpha_{d\sigma} = 0°$ and $90°$, as illustrated in the inset in Fig. 3(a). The combinations of $\alpha_{d\sigma}$ and b employed are listed in Table 2, which also shows the α values recorded at the eventual peak shear stresses. In the other anisotropically consolidated Series, AM, the reconsolidation stage was followed by a drained change of b and then by an undrained change of α, except in tests AM0003, AM0003B and AM3003, in which the α-change stage was replaced by an undrained q reduction back to isotropic stress. Further undrained shearing was conducted with constant p, α and b and increasing q. The selected combinations are shown in Table 2. These shear schemes are illustrated in Fig. 3(a) for Series AC, IC and NC and in Fig. 3(b) for Series AM and IM. Note that the conditions $(\alpha, b) = (0°, 1)$ and $(90°, 0)$, and their vicinities are the undesirable 'no-go' zones identified by Symes (1983), where significant stress non-uniformity is expected across an HCA specimen. Note also that the apparatus employed had wall thicknesses and height-to-diameter ratios designed to minimise stress non-uniformity (Rolo, 2003).

The triaxial extension and simple shear tests employed reconsolidation schemes, creep rate criteria and shearing strain rates similar to those outlined above. One difference that applied to the 100 mm diameter undrained triaxial extension tests was the lower axial strain rate (0·02%/h) applied over the early stages of the test to help define the small-strain response. However, the axial strain rate was increased to 0·1%/h (equivalent to $d(\varepsilon_1 - \varepsilon_3)/dt = 3\cdot6\%$ per day) long before failure developed.

ANISOTROPY OF LONDON CLAY FROM 10·5 m AND 5·2 m BGL

General behaviour and anisotropy of deformation characteristics

Gasparre *et al.* (2007a, 2007b) and Hight *et al.* (2007) describe the characterisation of the full Heathrow T5 profile. The following sections refer to intensive HCA studies made on samples from two levels; a later section relates these results to behaviour observed at other depths in simple shear, triaxial compression and triaxial extension tests.

The general shear strength characteristics, and the deformation characteristics observed in HCA tests run at medium to large strains on samples from 5·2 to 10·5m bgl, are briefly described below, presenting the key results. Because of the four degrees of stress and strain freedom provided by the tests, a variety of 2D and 3D plots are required to fully describe the observed effective stress paths and stress–strain relationships. Nishimura (2006) and Minh (2006) provide complete datasets for the tests on 10·5 m and 5·2 m bgl samples respectively. However, it is necessary to concentrate here on a restricted set of plots, covering the elements that are essential to the subsequent discussion.

It is worth noting that the onset of rupture and shear localisation could be detected during testing by inspecting the deformation of a grid pattern drawn on the outer membrane. Owing to the optical magnification effect caused by the cell water, a shear displacement of less than 0·2 mm could be resolved by eye along any grid line, allowing the point of shear band formation to be identified before the shear strain had grown by 0·1%, or less, provided that continuous observations were made. Matching this potential strain lag with the tangential stress–strain data suggests that q at rupture should not be overestimated by more than 10 kPa in most cases. Continuous optical observations were made in Series AC, IC and NC, whereas inspection of the shear surface formation was less frequent in the other series. The test data before and after the onset of localisation of shear deformation, or visible displacement along natural discontinuities (both are simply called rupture hereinafter), are therefore distinguished only for Series AC and IC.

The stress ratio–shear strain relationships, $q/p'-(\varepsilon_1 - \varepsilon_3)$, observed during the undrained shear stages of the HCA tests are shown in Fig. 4. Considering the data from Series AC and IC, it can be seen that rupture generally preceded the peak points in q and q/p', whereas these two peaks were practically concurrent in most of the tests (see Fig. 5). This feature may be absent with ductile materials but is often seen in with brittle soils. Torsional shear modes (where $30° < \alpha < 60°$) provoked shear discontinuity formation at smaller shear strains than other shearing types. In many tests, the post-peak behaviour did not exhibit a marked displacement-softening, although the latter trend is common in triaxial tests on highly structured or plastic soils. The 'post-rupture' strengths (Burland, 1990) were not well defined in these HCA tests, nor were residual strength conditions approached. It should be borne in mind that an HCA test is in fact a non-uniform boundary value problem, as are most types of soil shear testing. The behaviour observed after shear surface formation is a consequence of interactions between the intact soil, the soil within shear bands, and the apparatus.

The trajectories of the effective stress paths observed in the above tests are shown in Fig. 5 in the $(\sigma_z - \sigma_\theta)/2-p'$ plane. In Fig. 5(b), the undrained effective stress paths followed during the α-change stage are also included. Note that the final reconsolidation effective stress in AM6005 $(\alpha = 60°)$ was slightly different from that of the rest of the dataset, and its effective stress path does not join the others. The effective stress paths exhibited a common inclination $d(\sigma_z - \sigma_\theta)/d(2p')$ up to the final dilative stage near failure, suggesting that the apparent 'dilatancy' is predominantly a response to the $(\varepsilon_z - \varepsilon_\theta)$-mode deformation, and that $\gamma_{z\theta}$-mode deformation has comparatively little effect. Such deformation characteristics are similar to those of a cross-anisotropically elastic material with the axis of material symmetry set in the z-direction. Estimates for the effective stress-path inclinations, $d(\sigma_z - \sigma_\theta)/d(2p')$, were computed by Nishimura (2006) by applying cross-anisotropic elastic theory and small-strain stiffness parameters obtained experimentally by Gasparre (2005). The calculated inclinations were

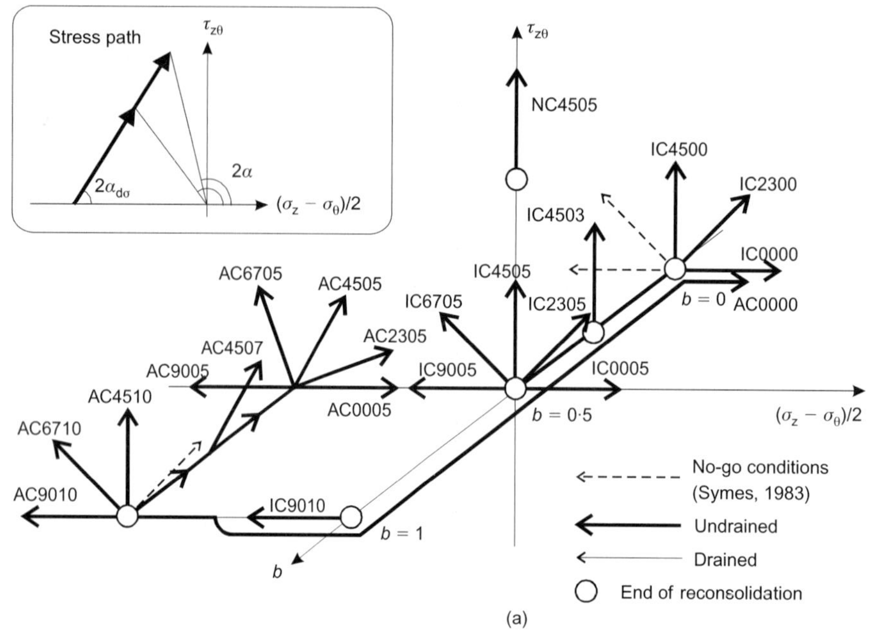

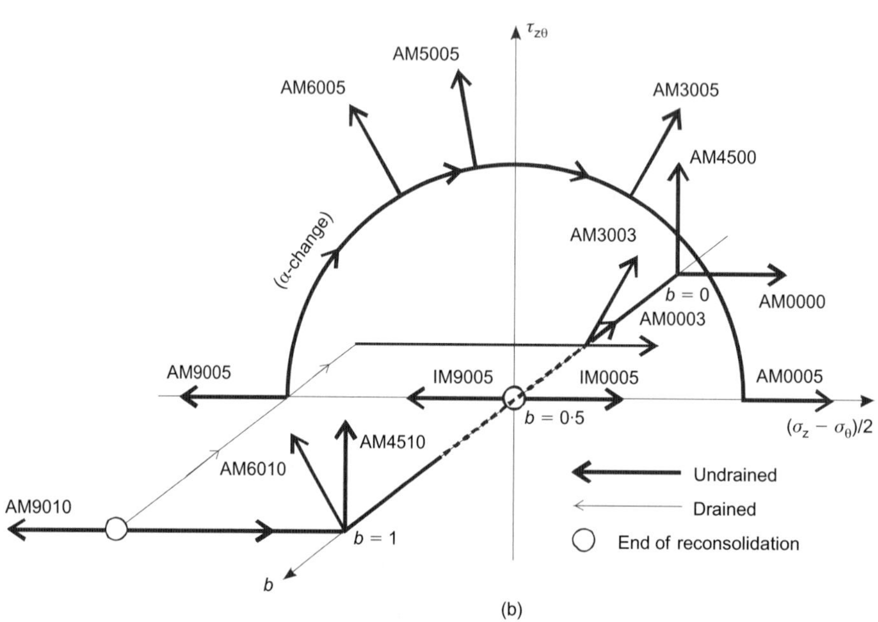

Fig. 3. Undrained shear schemes in HCA stress-path test series: (a) Series AC, IC and NC; (b) Series AM and IM (end of reconsolidation points not the same for all tests; see Table 2)

found to be consistent, if not always quantitatively coincident, with the observed values. The general trend of pre-failure 'dilative' behaviour for extensional loading ($\alpha_{d\sigma} > 45°$) and 'contractive' behaviour for compressional loading ($\alpha_{d\sigma} < 45°$) is one of the causes for the anisotropy of the clay's undrained shear strength, as discussed below. The above observations applied irrespective of the particular b value applied in each test.

Anisotropy in undrained shearing resistance

Accurate representation of the observed shear strength in terms of effective stress was hindered by the fact that samples failed at different p' values in different tests, with the p' values at peak strength varying systematically with consolidation K value, b and α. Earlier triaxial testing (e.g.

Bishop *et al.*, 1965; Hight & Jardine, 1993) and the T5 data presented by Gasparre *et al.* (2007a) and Hight *et al.* (2007) indicate curved peak q–p' envelopes and hence a peak q/p' ratio as a function of p'. Although similar trends may well apply to the HCA experiments, the authors' programmes did not include multiple tests where non-triaxial α and b values were maintained while the p' levels were varied. Noting that the range of p' values at which the HCA specimens failed is relatively narrow (for example, 338 kPa $\pm$ 23% in Series AC and 361 $\pm$ 20% in Series IC), one approach would be to neglect potential envelope curvature (and perhaps true effective cohesion) and denote the shear strength by q/p'. In order to avoid a biased interpretation caused by this simplification, the data are also reported in total stress terms by plotting peak S_u against α. These two parameters, q/p' and S_u, represent the shear strength characteristics of two types of

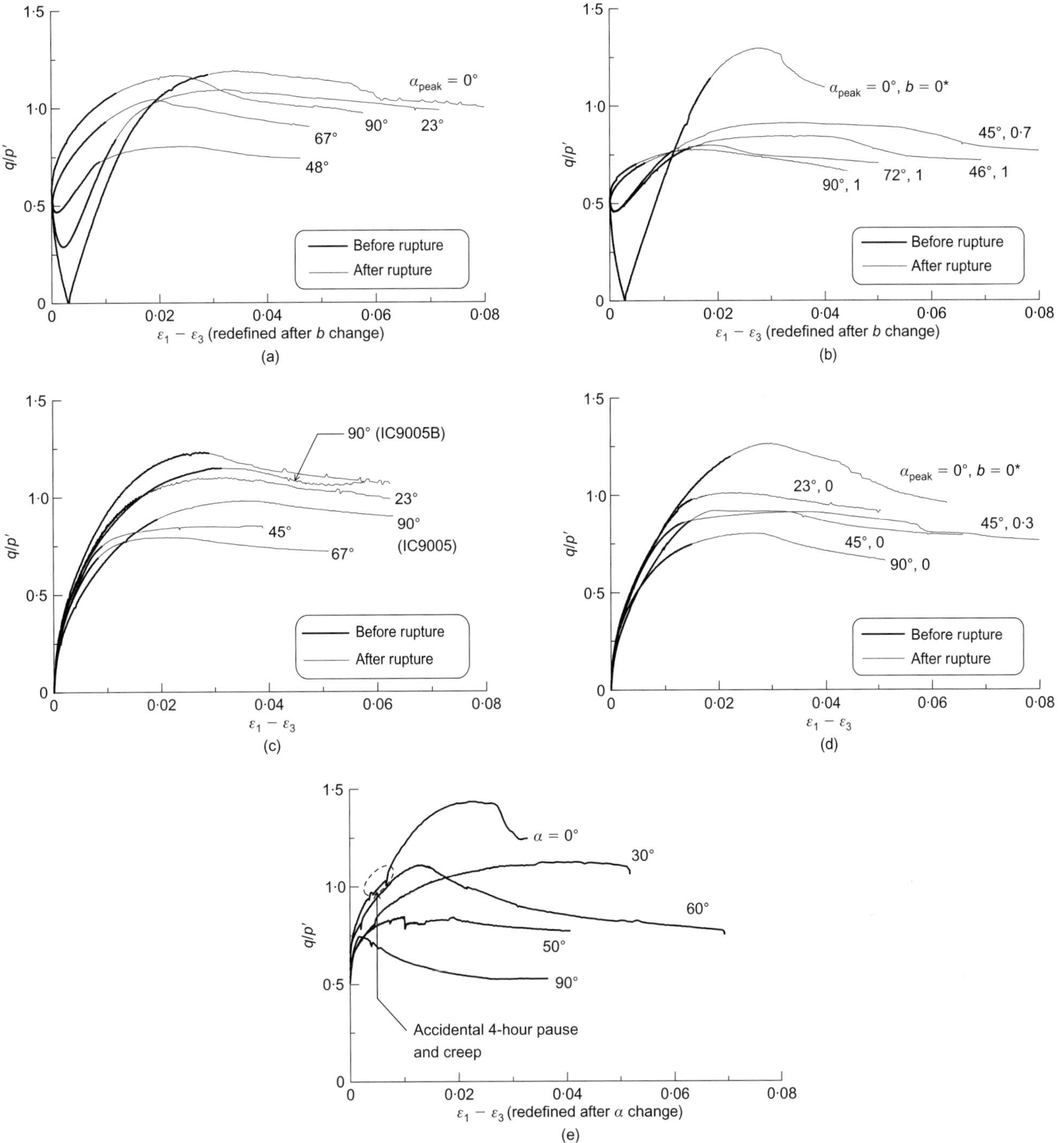

Fig. 4. Observed stress ratio–shear strain relationships: (a) 10·5 m bgl, Series AC, $b = 0·5$; (b) 10·5 m bgl, Series AC, other b values (b value changed from 1 to 0 stepwise during shear); (c) 10·5 m bgl, Series IC, $b = 0·5$; (d) 10·5 m bgl, Series IC, other b values; (e) 5·2 m bgl, Series AM, $b = 0·5$

idealised material: namely, a purely frictional material with constant ϕ' (for any given b and α), and a purely cohesive material where shear strength is independent of p' respectively. These two perspectives bracket the available possibilities when assessing the anisotropic shear strength trends; the true (effective) stress-dependence of anisotropy remains to be established.

The peak stress ratios, q/p', observed for $b = 0·5$ are plotted in Figs 6(a) and 6(b) against the α values applying at failure for 10·5 m and 5·2 m bgl respectively. The shear strength anisotropy is clearly recognisable in each case. The AC and IC series tests on 10·5 m bgl samples exhibit close agreement for $0° < \alpha < 45°$, with peak q/p' decreasing by

some 40% as α increases from 0° to 45°. A similar trend is followed by the 5·2 m bgl samples for this α range, but with higher effective stress ratios. The slightly higher q/p' ratios developed by the shallower samples for compressional loading modes are compatible with the general failure envelope curvature, shown by the London Clay at T5, reflecting variations with depth of OCR, and possibly of lithology (Gasparre et al., 2007a; Hight et al., 2007). For $45° < \alpha < 90°$, the 10·5 m bgl samples exhibited peak q/p' rising again at larger α values, but this pattern was not established from the limited number of tests on 5·2 m bgl samples. The low shear strengths recorded for $\alpha = 90°$ in Series AM and IM may be due partly to the larger specimen size in the ICHCA

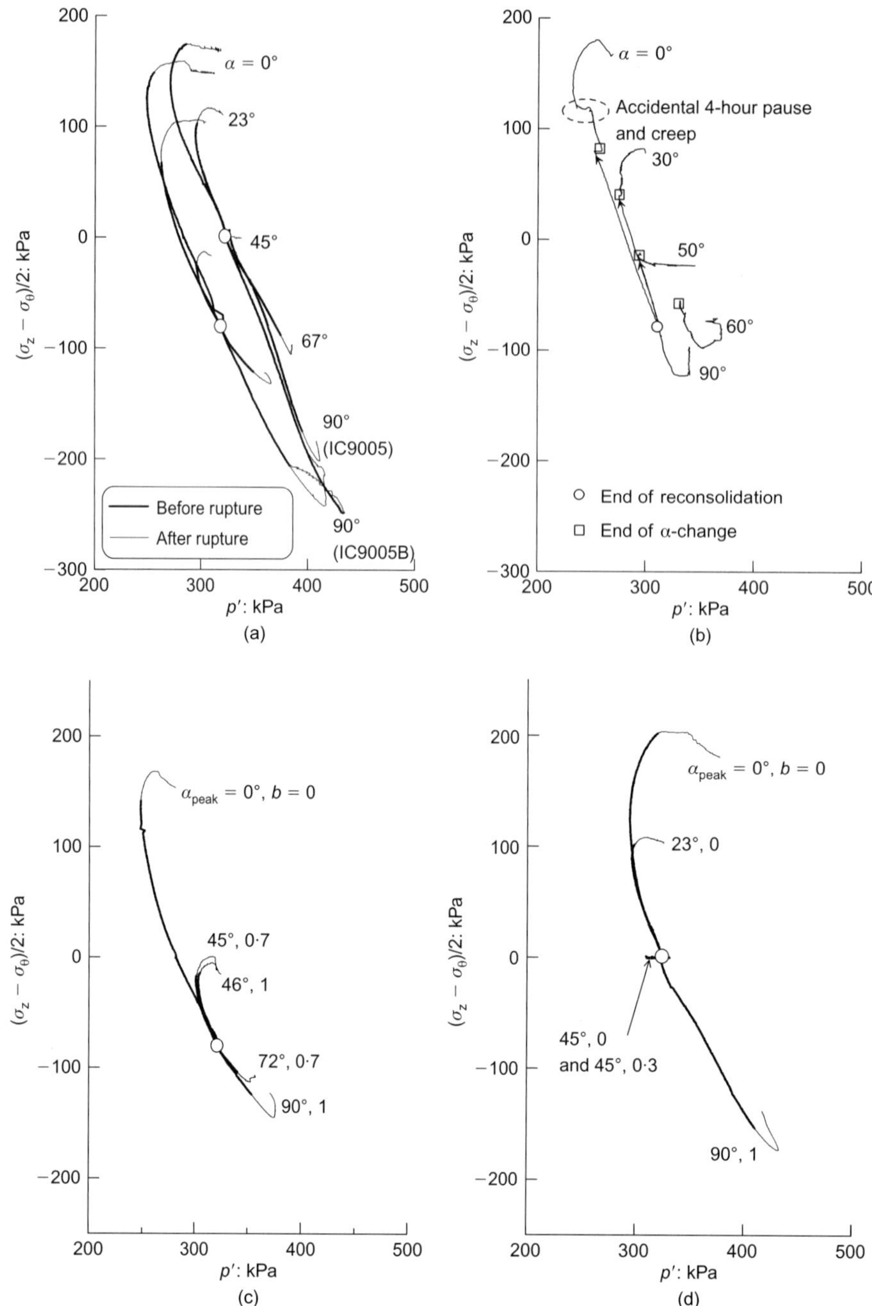

Fig. 5. Observed undrained effective stress paths: (a) Series AC and IC, $b = 0·5$; (b) Series AM, $b = 0·5$; (c) Series AC, other b values; (d) Series IC, other b values

II and hence the inclusion of more complete discontinuity network within the specimen. A sample-size effect might also have led to the apparently lower shear strengths at high α values of the larger specimens (Series AM) tested with $b = 1$, shown in Fig. 6(c), but any such effect was absent in the tests with $\alpha < 45°$ at all b values. Such a directionally dependent sample-size effect might result from the abundant sub-horizontal natural discontinuities, which were often activated as part of the torsional and extensional shear failure modes. It is worth noting that the test IC9005B ($\alpha = 90°$ and $b = 0·5$) developed a single spiral shear plane upon shearing, indicative of bifurcation in intact clay, whereas the equivalent tests on 5·2 m bgl samples (AM9005 and IM9005) simply failed along a natural discontinuity. Nishimura (2006) and Minh (2006) made careful observations of the patterns of discontinuities that existed in the

natural samples and of how these developed during shearing, showing in many cases a correlation between the original discontinuity patterns, the final shear mechanisms and the shear strength.

It is noticed in Fig. 6(c) that test AM4500 gave extraordinarily high shear strength in relation to the other data points. The reason for this was not established by re-examination of the physical properties, the test specimen itself and the failure mode. The only possible explanation at the moment is that the sample may have contained an untypical local concentration of cementing and/or a lower degree of chemical alteration by pyritisation.

Figures 6(d), 6(e) and 6(f) show the anisotropy of the undrained shear strength S_u. In interpreting these results it should be borne in mind that, with brittle soils such as the London clay, the peak undrained shear strengths tend to be

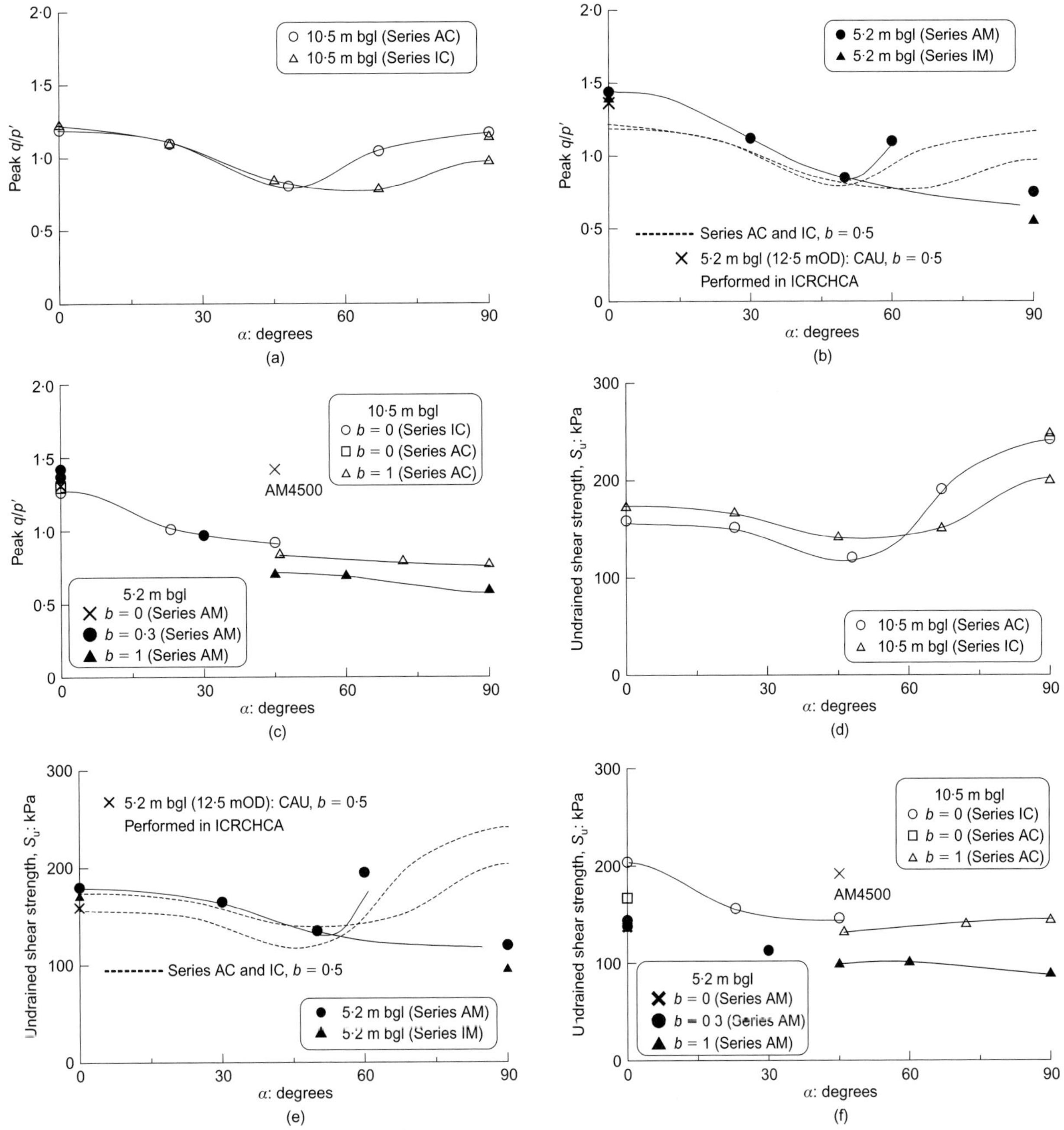

Fig. 6. Anisotropy of peak shear strength: (a) $b = 0.5$, 10·5 m bgl; (b) $b = 0.5$, 5·2 m bgl; (c) Other b values; (d) $b = 0.5$, 10·5 m bgl; (e) $b = 0.5$, 5·2 m bgl; (f) Other b values

reached shortly after engaging their failure envelopes without showing extended periods of dilative behaviour. Hence S_u depends on the effective stress state under which the envelope is first engaged, and is not uniquely related to water content or void ratio (e.g. Jardine *et al.*, 2004). Therefore a soil with a fixed failure envelope and a fixed water content could exhibit a variety of S_u values, depending on its initial effective stress states and the effective stress paths followed towards failure. Considering this fact, the anisotropically reconsolidated test results (i.e. Series AC and AM) are the most relevant for estimating the in situ undrained shear strength anisotropy.

As noted above, extensional loading with $\alpha > 45°$ resulted in a more dilative pore water pressure response, which affects the S_u anisotropy and leads to a significantly different pattern from the q/p' trend. It is interesting that all the data for $b = 1$ show little variation in peak q/p' or S_u in the case where $\alpha > 45°$.

Influence of reconsolidation conditions on shear strength anisotropy

Jardine *et al.* (1997) and Zdravkovic & Jardine (2001) demonstrated that the shear strength anisotropy of low-OCR granular soils depends strongly on their consolidation history, especially when this involved rotation of the σ_1 axis. It was important to assess whether similar features applied to the London Clay. Figs 7(a) and 7(b) show the rupture (i.e. onset of shear strain localisation) and failure envelopes obtained from test Series IC, AC and NC with $b = 0.5$ plotted in $(\sigma_z - \sigma_\theta)/2p'-\tau_{z\theta}/p'$ space. Such a space described by tensorial stresses offers a simpler and clearer

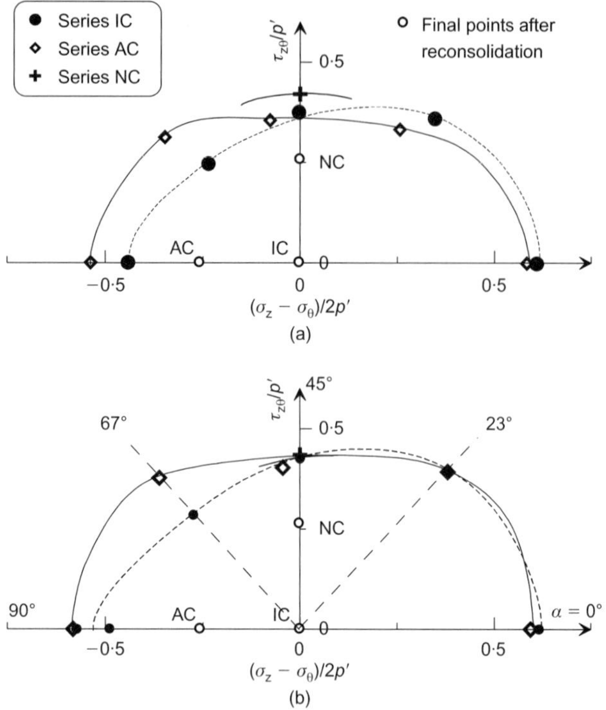

Fig. 7. Influence of reconsolidation regimes on development of shear strength: (a) rupture; (b) peak shear strength

view of the distortion and kinematics of the failure envelope than the q/p'–α plot. Distortion of the rupture envelope towards the reconsolidation effective stress points is clearly observed. However, the peak strength surface was less affected, with only two data points at $\alpha = 67°$ and $90°$, suggesting the possibility that reconsolidation affected the peak shear strength surface.

Influence of overconsolidation and natural soil structure

It is of interest to compare the anisotropy of the natural London Clay with the more familiar patterns known for low-OCR reconstituted soils. Fig. 8 shows the S_u anisotropy of the natural London Clay shown earlier for $b = 0.5$, normalised by S_u at $\alpha = 0°$ and set against a range of results from previously published tests on low-OCR (OCR = 1–4) K_0-reconstituted soils. They include clay, silt and clayey and silty sands (Whittle *et al.*, 1994; Menkiti, 1995; Zdravkovic, 1996). Low-OCR soils generally exhibit S_u falling monotoni-

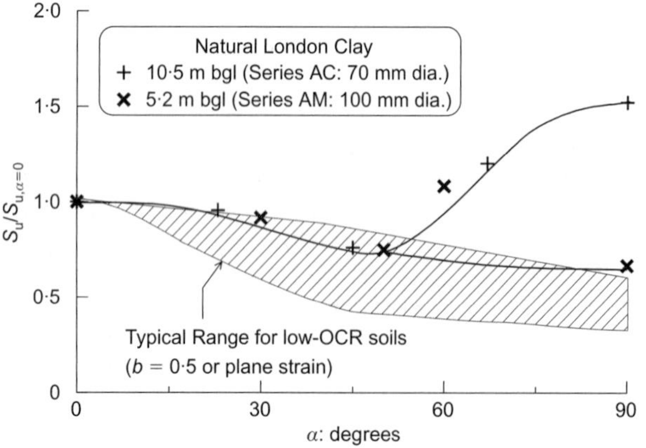

Fig. 8. Undrained shear strength anisotropy of low-OCR K_0-reconstituted soils and natural London Clay

cally with α, giving $S_{u,\alpha=90}/S_{u,\alpha=0} = 0.3$–$0.6$. However, the upper data curve for the London Clay, obtained from Series AC, follows the opposite trend, with the $S_{u,\alpha=90}/S_{u,\alpha=0}$ ratio being as high as 1.5. Whereas this upper bound is considered to reflect the influences of both the micro- and macrofabric, the lower data curve obtained from Series AM may reflect more the discontinuous (highly fissured) macrofabric. Heavy overconsolidation (i.e. OCR > 9) appears to change the microstructural anisotropy from that seen at low OCR, while discontinuities developed (most probably during or after the overconsolidation) modify the anisotropy *en masse* further. A further investigation is currently being undertaken by testing reconstituted London Clay samples at OCRs comparable to those of the natural samples.

INFLUENCE OF b ON PEAK SHEAR STRENGTH

The influence of b on the shape of the failure surface projected onto the π plane has been identified as an important but often neglected factor in geotechnical analysis (e.g. Potts, 2003). This feature for the London Clay was investigated by plotting the peak q/p' points on the π plane in Fig. 9 for groups of tests with common α ranges. Although the axes are normalised by p' here, it should be noted that the pore water pressure development was predominantly a function of $\alpha_{d\sigma}$, and was affected much less by b. As a result, p' at failure fell within a narrow range for a given α_{peak} value. For example, the failure p' range for the six data points from Series IC and AC in Fig. 9(a) was 303–343 kPa. Tests with $\alpha = 45$–$48°$ were performed covering the entire range of b, from 0 to 1. As discussed above, the peak shear strength for torsional shear may not be significantly affected by the K values applied during triaxial reconsolidation. No distinction is therefore made here between the data from Series AC and IC. Also shown as references in Fig. 9(a) are the Mohr–Coulomb and Prager–Drucker (corresponding to $(s_{ij}s_{ij})^{1/2}/p' = $ constant, where $s_{ij} = \sigma_{ij} - \delta_{ij}\sigma_{kk}/3$) envelopes, fitted to pass through the $b = 0$ test data. The former was thought to apply best to the natural London Clay by Bishop (1966), whereas the latter is preferred in some critical state soil model variants. The Mohr–Coulomb failure line for $\phi' = 23.5°$ and $c' = 0$, fitted at $b = 0$, gives a well-defined lower bound of shear strength for $\alpha = 45$–$48°$, although some data points lie well above this line. Mohr–Coulomb failure lines drawn with $c' = 0$ and passing through either the $b = 0$ or 1 data points are presented in Figs 9(b) to 9(e) for other α ranges. Although the datasets are less complete, the Mohr–Coulomb lines again provide lower bounds of shear strength, whereas $b = 0.5$ conditions tended to result in development of higher ϕ'. It is difficult, however, to propose specific curved envelopes that encapsulate the potential influence of b based on these limited data. In the absence of further data, an anisotropic Mohr–Coulomb criterion calibrated to data obtained at $b = 0$ or 1 may be regarded as a reasonably safe option, although not the best fit.

Finally, the influences of α and b on the peak shear strength, q/p', are summarised in Figs 10(a) and 10(b) for 10·5 m and 5·2 m bgl samples respectively. As discussed later, the shear strength exhibited in test AM4500 was exceptionally high, and this outlying data point is excluded from Fig. 10(b). Despite the detailed differences between the two horizons discussed above, the overall shapes are generally similar. In the range explored, the highest peak value of q/p' is available for $\alpha = 0$ with $0 < b < 0.5$, and the lowest is encountered for $\alpha = 45°$ with $b = 0.5$ or $\alpha = 90°$ and $b = 1$.

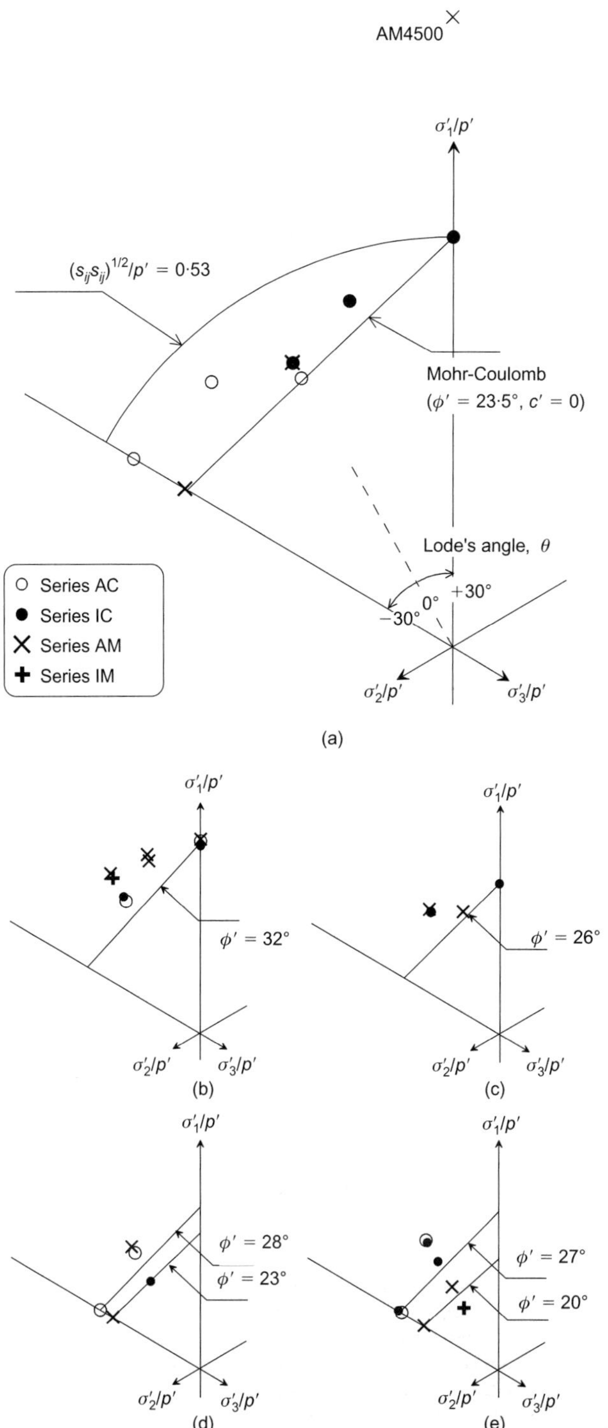

Fig. 9. Peak shear strength points plotted on π-plane for five α ranges: (a) $\alpha = 45–48°$; (b) $\alpha = 0°$; (c) $\alpha = 23–30°$; (d) $\alpha = 60–72°$; (e) $\alpha = 90°$

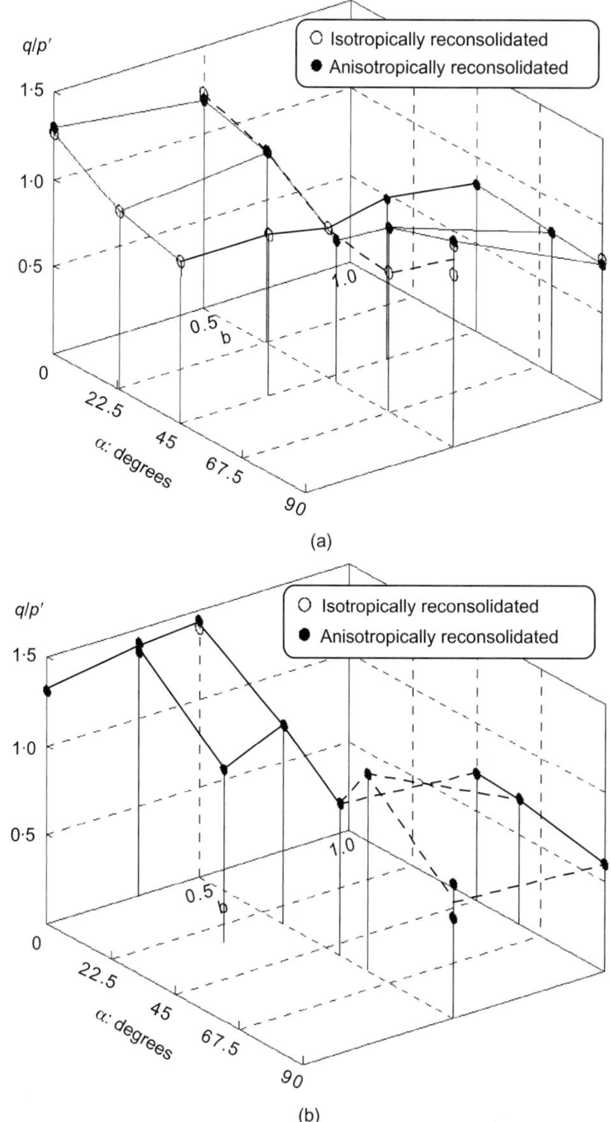

Fig. 10. Influence of α and b on peak shear strength: (a) 10·5 m bgl (7·2 mOD); (b) 5·2 m bgl (12·5 mOD)

PEAK SHEAR STRENGTH ANISOTROPY FOR 6–35 m BGL

The intensive investigation of the shear strength anisotropy developed at 5·2 m bgl and 10·5 m bgl was extended through sets of triaxial compression, triaxial extension and HCA simple shear tests conducted between 6 and 35 m bgl. Although the (α, b) combinations in triaxial compression and extension are fixed to (0°, 0) and (90°, 1) respectively, neither variable can be pre-specified in strain-controlled simple shear tests. In the present cases of the London Clay simple shear tests, the peak shear stress q was reached at $50° < \alpha < 65°$ and $0·5 < b < 0·7$ (Nishimura, 2006). The

comparison of the shear strength in these three different shear modes is therefore subject to mixed effects of α and b. Nevertheless, compilation of such test data provides useful benchmarks for the possible variations in anisotropy throughout the stratum. The water contents and index properties of the samples tested are included in the overview given in the companion paper (Hight *et al.*, 2007), and are omitted here for reasons of space.

The effective stress paths observed during simple shear and triaxial extension tests are shown in Fig. 11. Generally, p' changed little in the simple shear tests before engaging the final failure surface and experiencing accelerated p' increases, whereas the pre-failure effective stress paths were consistently 'dilative' (p' increasing) in the triaxial extension tests, features that are consistent with the small-strain stiffness anisotropy, as described earlier. The points of maximum q/p' are plotted in Fig. 12, set against the envelopes for triaxial compression originally proposed by Hight & Jardine (1993) for high-quality rotary cores of the London Clay, mainly from central London and partly from Suffolk, slightly modified here according to Gasparre (2005). The potential curvature of the failure envelopes mentioned earlier is evident here, along with variations due to changes in depth

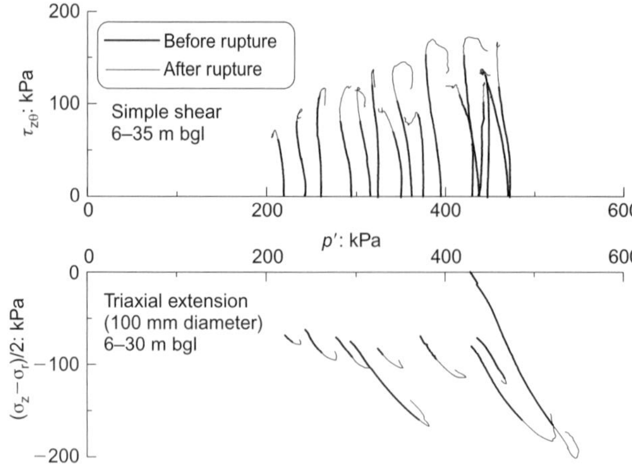

Fig. 11. Effective stress paths observed for HCA simple shear and triaxial extension tests (only data for 100 mm diameter samples are shown)

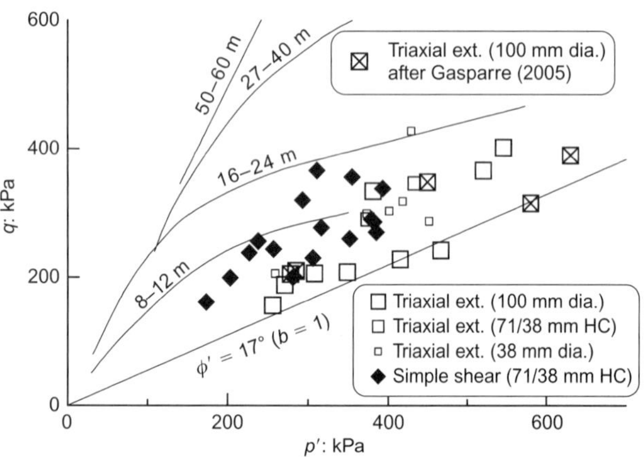

Fig. 12. Comparison of triaxial extension and simple shear strength with triaxial compression envelopes given by Hight & Jardine (1993)

and structure. Multiple en-echelon upper-bound envelopes were employed to account for changes in dilatancy and an apparently increasing degree of cementation in the deeper samples. Hight *et al.* (2003) and Gasparre (2005) found that this set of envelopes broadly applied to triaxial compression strength data for T5 samples. The fissure shear strength line ($\phi' = 17°$ and $c' = 0$) given also by Hight & Jardine (1993) provides a lower bound for all test types. The deepest samples from 35 m bgl, tested with $p' > 400$ kPa, show the largest gap between the data points and the 27–40 m bgl triaxial compression envelope, indicating that the anisotropy appears to grow with depth. This is particularly the case with the simple shear, which gave strength closer to the triaxial compression upper bound in shallower horizons. This finding may be related to the SEM observation made by Gasparre (2005) that the deeper clay had more uniform particle orientation. As discussed in the companion paper (Gasparre *et al.*, 2007b), the small-strain stiffness also exhibited stronger anisotropy in the deeper samples.

The undrained shear strength profiles obtained with the three shear modes are shown in Fig. 13. The triaxial compression data are quoted from Gasparre (2005). Reflecting the above discussion, the shear strengths in the three different shear modes are not significantly different for shallow samples, but the upper bound of the triaxial compression strength below 10 m bgl locates substantially above those for

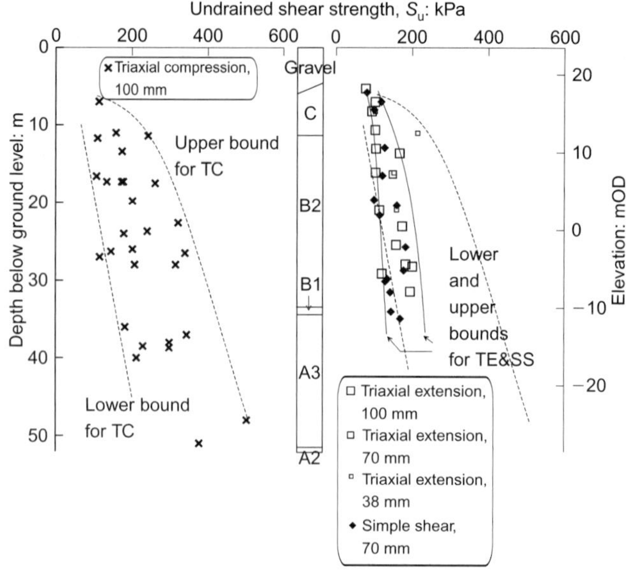

Fig. 13. Undrained shear strength profiles of three different shear modes; triaxial compression data from Gasparre (2005). The T5 lithological stratigraphy described by Hight *et al.* (2007) is shown in the central column between the two S_u profiles

the other shear modes. Considering that the simple shear tests were performed with 70 mm diameter hollow specimens, whereas the rest were typically performed with 100 mm diameter solid specimens, allowing for any sample-size effect would further accentuate the anisotropic hierarchy in S_u values.

Despite the apparent increases in shear strength anisotropy with depth, the overall hierarchy, (triaxial compression) > (simple shear) > or ≈ (triaxial extension), is fully compatible with the finding from the HCA investigation in Fig. 10.

CONCLUSIONS

The present study was undertaken to investigate the mechanical anisotropy of the natural London Clay through a comprehensive programme of laboratory tests performed in HCAs and stress-path triaxial cells. This paper focused only on the peak shear strength; a companion paper (Gasparre *et al.*, 2007b) discusses the anisotropy found at smaller strains.

The directional dependence of q/p' or S_u was evaluated at two fixed depths (5·2 m and 10·5 m below ground level) by multiple tests involving fixed values of the intermediate principal stress ratio b. Although the patterns for these two measures differ because anisotropic deformation characteristics lead to direction-dependent development of pore water pressure up to failure, both q/p' and S_u typically exhibited their maxima at either $\alpha = 0°$ or $90°$ and minima at around $\alpha = 45°$ in tests with $b = 0·5$. The effects of anisotropy led to maximum and minimum q/p' values that could differ by 40%. This feature contrasts strongly with the behaviour of low-OCR K_0-reconstituted soils, for which shear strength generally decreases monotonically with α from $0°$ to $90°$. The tests with larger samples suggested the possibility that the shear strength can be equally very low against extensional loading (i.e. $\alpha > 45°$), if samples contain an intense discontinuity network. For all the combinations of α and b, excluding the 'no-go' combinations in HCA testing, the lowest peak shear strength was found at $(\alpha, b) = (45°, 0·5)$ and $(90°, 1)$, with another possible minimum at $(90°, 0·5)$. Profiles of triaxial extension, compression and simple shear tests on rotary cores from 6 m to 35 m bgl indicated the same hierarchy of shear strengths and also suggested an increase in the degree of anisotropy with depth. The top 5 m layer of the London Clay stratum appeared to be much less anisotropic than the layers below, particularly when expressed in terms of S_u.

The influence of b on the shear strength was less well defined. For a given α value, the Mohr–Coulomb failure criterion fitted at either $b = 0$ or 1 seems to provide a reasonable lower-bound shear strength envelope, although higher strengths were generally mobilised at intermediate values of b.

ACKNOWLEDGEMENTS

The authors express their gratitude to their former colleague Dr Akihiro Takahashi (now of PWRI, Japan) and the Imperial College laboratory technicians, Steve Ackerley, Graham Keefe and Alan Bolsher. Dr Apollonia Gasparre is thanked for providing her triaxial test data and Cedric Allenou for his help with block sampling at Heathrow T5. The research was funded by EPSRC, British Airports Authority and London Underground Ltd. The second author was also funded by the Vietnamese Government Overseas Scholarship Programme Project 322.

NOTATION

B	Skempton's pore water pressure coefficient
b	intermediate principal stress ratio
c'	effective cohesion
G_s	specific gravity
$G_{z\theta}$	shear modulus
K	coefficient of earth pressure
K_0	coefficient of earth pressure at rest
p	mean normal stress
p'	mean normal effective stress
q	deviatoric stress
S_u	undrained shear strength
w	natural water content
α	angle between major principal stress axis and vertical
$\alpha_{d\sigma}$	angle between major principal stress increment axis and vertical
α_{peak}	value of α at maximum q/p'
γ	unit weight
$\gamma_{z\theta}$	shear strain in polar coordinate system
δ_{ij}	Kronecker delta
$\varepsilon_1, \varepsilon_3$	principal strains
θ	Lode's angle
$\sigma_1, \sigma_2, \sigma_3$	principal stresses
σ_{ij}	stress tensor
$\sigma_z, \sigma_r, \sigma_\theta$	normal stresses in cylindrical coordinate system
$\tau_{z\theta}$	shear stress in cylindrical coordinate system
ϕ'	angle of shearing resistance

REFERENCES

Agarwal, K. B. (1968). *The influence of size and orientation of sample on the undrained strength of London Clay*. PhD thesis, Imperial College, University of London.

Bishop, A. W. (1966). The strength of soils as engineering materials. *Géotechnique* **16**, No. 2, 91–130.

Bishop, A. W., Webb, D. L. & Lewin, P. I. (1965). Undisturbed samples of London Clay from the Ashford Common shaft: strength-effective stress relationships. *Géotechnique* **15**, No. 1, 1–31.

Bishop, A. W. & Little, A. L. (1967). The influence of the size and orientation of the sample on the apparent strength of the London Clay at Maldon, Essex. *Proceedings of the geotechnical conference*, Oslo, Vol. 1, pp. 89–96.

Bjerrum, L. (1973). Problems of soil mechanics and construction on soft clays and structurally unstable soils (collapsible, expansive and others). *Proc. 8th Int. Conf. Soil Mech. Found. Engng, Moscow* **3**, 111–159.

Burland, J. B. (1990). On the compressibility and shear strength of natural clays. *Géotechnique* **40**, No. 3, 329–378.

Burland, J. B., Rampello, V. N., Georgiannou, V. N. & Calabresi, G. (1996). A laboratory study of the strength of four stiff clays. *Géotechnique* **46**, No. 3, 491–514.

Chandler, R. J. & Apted, J. P. (1988). The effect of weathering on the strength of London Clay. *Q. J. Engng Geol.* **21**, No. 1, 59–68.

Gasparre, A. (2005). *Advanced laboratory characterization of London Clay*. PhD thesis, Imperial College London (downloadable from www.imperial.ac.uk/geotechnics).

Gasparre, A., Nishimura, S., Coop, M. R. & Jardine, R. J. (2007a). The influence of structure on the behaviour of London Clay. *Géotechnique* **57**, No. 1, 19–31.

Gasparre, A., Nishimura, S., Minh, N. A., Coop, M. R. & Jardine, R. J. (2007b). The stiffness of natural London Clay. *Géotechnique* **57**, No. 1, 33–47.

Hight, D. W., Gens, A. & Symes, M. J. P. R. (1983). The development of a new hollow cylinder apparatus for investigating the effects of principal stress rotation in soils. *Géotechnique* **33**, No. 4, 355–384.

Hight, D. W. & Jardine, R. J. (1993). Small-strain stiffness and strength characteristics of hard London tertiary clays. In *Geotechnical engineering of hard soils and soft rocks* (eds A. Anagnostopoulos, F. Schlosser, N. Kalteziotis & R. Frank), Vol. 1, pp. 533–552. Rotterdam: Balkema.

Hight, D. W., McMillan, F., Powell, J. J. M., Jardine, R. J. and Allenou, C. P. (2003). Some characteristics of London Clay. In *Characterisation of engineering properties of natural soils* (ed. T. S. Tan), pp. 851–908. Rotterdam: Balkema.

Hight, D. W., Gasparre, A., Nishimura, S., Minh, N. A., Coop, M. R. & Jardine, R. J. (2007). Characteristics of the London Clay from the Terminal 5 site at Heathrow Airport. *Géotechnique* **57**, No. 1, 3–18.

Jardine, R. J. (1996). *Development of the new HCA*, internal report. London: Imperial College.

Jardine, R. J. & Smith, P. R. (1991). Evaluating design parameters for multi-stage construction. *Proc. Int. Conf. on Geotechnical Engineering for Coastal Development, Geo-coast '91, Yokosuka*, **1**, 197–202.

Jardine, R. J., Zdravkovic, L. & Porovic, E. (1997). Anisotropic consolidation including principal stress axis rotation: experiments, results and practical implications. *Proc. 14th Int. Conf. Soil Mech. Found. Engng, Hamburg*, **4**, 2165–2168.

Jardine, R. J., Gens, A., Hight, D. W. & Coop, M. R. (2004). Developments in understanding soil behaviour. *Advances in Geotechnical Engineering, Proceedings of the Skempton Conference*, Vol. 1, pp. 103–206. London: Thomas Telford.

King, C. (1981). The stratigraphy of the London Clay and associated deposits. *Tertiary Research Special Paper*, No. 6.

Kovacevic, N., Hight, D. W. & Potts, D. M. (2004). Temporary slope stability in London Clay. *Advances in Geotechnical Engineering, Proceedings of the Skempton Conference*, Vol. 2, pp. 842–855. London: Thomas Telford.

Menkiti, C. O. (1995). *Behaviour of clayey-sand, with particular reference to principal stress rotation*. PhD thesis, Imperial College, University of London.

Minh, N. A. (2006). *An investigation of the anisotropic stress–strain–strength characteristics of an Eocene clay*. PhD thesis, Imperial College London (downloadable from www.imperial.ac.uk/geotechnics).

Molenkamp, F. (1998). Principle of axial shear apparatus. *Géotechnique* **48**, No. 3, 427–431.

Nishimura, S. (2006). *Laboratory study on anisotropy of natural London Clay*. PhD thesis, Imperial College, University of London (downloadable from www.imperial.ac.uk/geotechnics).

Porovic, E. (1995). *Investigation of soil behaviour using a resonant-column torsional shear hollow-cylinder apparatus*. PhD thesis, Imperial College, University of London.

Potts, D. M. (2003). Numerical analysis: a virtual dream or practical reality? *Géotechnique* **53**, No. 6, 535–573.

Potts, D. M., Kovacevic, N. & Vaughan, P. R. (1997). Delayed collapse of cut slopes in stiff clay. *Géotechnique* **47**, No. 5, 953–982.

Pradhan, T. B. S., Tatsuoka, F. & Horii, N. (1988). Simple shear testing on sand in a torsional shear apparatus. *Soils Found.* **28**, No. 2, 95–112.

Rolo, R. (2003). *The anisotropic stress–strain–strength behaviour of brittle sediments*. PhD thesis, Imperial College, University of London (downloadable from www.imperial.ac.uk/geotechnics).

Saada, A. S. (1970). Testing of anisotropic clay soils. *J. Soil Mech. Found. Div. ASCE* **96**, No. SM5, 1847–1852.

Saada, A. S. & Townsend, F. C. (1981). State-of-the-art: Laboratory strength testing of soils. In *Laboratory shear strength of soil*, ASTM STP 740, pp. 7–77.

Shibuya, S. & Hight, D. W. (1987). On the stress path in simple shear. *Géotechnique* **37**, No. 4, 511–515.

Skempton, A. W., Schuster, R. L. & Petley, D. J. (1969). Joints and fissures in the London Clay at Wraysbury and Edgware. *Géotechnique* **19**, No. 2, 205–217.

Symes, M. J. P. R. (1983). *Rotation of principal stresses in sand*. PhD thesis, Imperial College, University of London.

Ward, W. H., Marsland, A. & Samuels, S. G. (1965). Properties of the London clay at the Ashford Common shaft: in-situ and undrained strength tests. *Géotechnique* **15**, No. 4, 321–344.

Whittle, A. J., DeGroot, D. J., Ladd, C. C. & Seah, T.-H. (1994). Model prediction of anisotropic behavior of Boston Blue Clay. *J. Geotech. Engng Division, ASCE* **120**, No. 1, 199–224.

Zdravkovic, L. (1996). *The stress–strain–strength anisotropy of a granular medium under general stress conditions*. PhD thesis, Imperial College, University of London.

Zdravkovic, L. & Jardine, R. J. (2001). The effect on anisotropy of rotating the principal stress during consolidation. *Géotechnique* **51**, No. 1, 69–83.

Zdravkovic, L., Potts, D. M. & Hight, D. W. (2002). The effect of strength anisotropy on the behaviour of embankments on soft ground. *Géotechnique* **52**, No. 6, 447–457.

Sorensen, K. K., Baudet, B. A. & Simpson, B. (2007). *Géotechnique* **57**, No. 1, 113–124

Influence of structure on the time-dependent behaviour of a stiff sedimentary clay

K. K. SORENSEN[*], B. A. BAUDET† and B. SIMPSON‡

In this paper results from laboratory tests on London Clay and artificially cemented kaolin are presented and used to develop a preliminary framework for the time-dependent behaviour of soils, applicable to stiff clays and other soils. In the same way that the natural structure of clays has been shown to influence their monotonic behaviour, it is shown that it can also alter their response to changes in strain rate. The relative influence of the two main components of post-sedimentation structure—overconsolidation and diagenesis—on the time-dependent behaviour of London Clay was investigated in triaxial compression tests. The study was carried out in two steps, comparing first the behaviours of normally and overconsolidated reconstituted samples of London Clay subjected to stepwise changes in strain rate, and then the behaviours of overconsolidated reconstituted and undisturbed London Clay samples. The test results show that overconsolidation does not seem to affect the response of reconstituted London Clay to strain rate changes, which is consistent with published data on other stiff clays. However, intact and reconstituted overconsolidated samples show different behaviours, highlighting that it is the elements of structure resulting from diagenesis that influence the time-dependent behaviour of London Clay. Effects of cementing on strain rate sensitivity were investigated in triaxial compression of artificially cemented kaolin. The results obtained were different from the results for London Clay, suggesting that the difference in strain rate effects in the intact and reconstituted London Clay cannot be simply associated with cementing. A preliminary framework is proposed, where the time-dependent behaviour of soil depends on its particulate or continuum nature.

KEYWORDS: clays; fabric/structure of soils; laboratory tests; time dependence

Cet article présente des essais réalisés en laboratoire sur de l'argile de Londres (London Clay) et du kaolin cimenté artificiellement. Ils sont utilisés pour développer un cadre préliminaire pour le comportement dépendant du temps des sols, applicable aux argiles raides et autres sols. De la même manière qu'il a été prouvé que la structure naturelle des argiles influence leur comportement monotone, il est montré qu'elle affecte également leur réponse aux changements en terme de vitesse de déformation. Des essais de compression triaxiale ont permis d'étudier l'influence relative des deux principaux composants de structure post-sédimentation, surconsolidation et diagénèse, sur le comportement en fonction du temps de l'argile de Londres (London Clay). L'étude a été réalisée en deux temps. Tout d'abord, les comportements d'échantillons de London Clay reconstitués, normalement consolidés et surconsolidés, soumis à des changements incrémentiels de vitesse de déformation ont été comparés. La comparaison a ensuite été réalisée entre les comportements d'échantillons de London Clay reconstitués surconsolidés et non remaniés. Les résultats montrent que la surconsolidation ne semble pas affecter la réponse de l'argile de Londres (London Clay) reconstituée en terme de changements de vitesse de déformation, ce qui est cohérent avec les données publiées sur d'autres argiles raides. Cependant, les échantillons intacts et reconstitués surconsolidés affichent des comportement différents, soulignant ainsi que ce sont les éléments de structure résultant de la diagénèse qui influencent le comportement en fonction du temps de l'argile de Londres. Des essais de compression triaxiale réalisés sur du kaolin artificiellement cimenté ont permis d'évaluer les effets de la cimentation sur la sensibilité de la vitesse de déformation. Les résultats obtenus sont différents de ceux observés avec l'argile de Londres, suggérant ainsi que la différence en terme d'effets de vitesse de déformation pour les échantillons intacts et reconstitués d'argile de Londres ne peut pas être simplement associée à la cimentation. Un modèle préliminaire est ici proposé, où le comportement en fonction du temps du sol dépend de sa nature particulaire ou continue.

INTRODUCTION
The time dependence of soil behaviour plays an important role for predicting the short- and long-term performance of geotechnical structures. Various studies reported in the literature show that different soils—for example sands, weak rocks and soft clays—have different responses to changes in strain rate. Little is known of the factors that cause these differences. A difficulty in studying rate effects is that they are usually small, and unless the effects are investigated on tests where the stress or strain rate is changed in a stepwise manner it is not possible to differentiate specimen and test variability from real rate effects. There are not many reliable data available on the time-dependent behaviour of stiff clays, particularly natural stiff clays. This paper adds valuable data on the effects of changes in strain rate in a stiff sedimentary clay of marine origin—London Clay—for which no comprehensive set of data on strain rate effects is currently available, and examines whether structure is one factor that affects its time-dependent behaviour. The data presented are from rigorous varying stress or strain rate tests on both reconstituted and intact clay, which makes them a unique reliable source of reference for rate effects in London Clay.

Manuscript received 28 April 2006; revised manuscript accepted 16 October 2006.
Discussion on this paper closes on 1 July 2007, for further details see p. ii.
* Engineering College of Aarhus, Denmark; formerly University College London, UK.
† Department of Civil and Environmental Engineering, University College London, UK.
‡ Ove Arup and Partners Ltd, London, UK.

The structure of clays is usually referred to as a combination of fabric (arrangement between particles) and bonding (forces between particles) (Mitchell, 1976). Several studies on stiff clays have highlighted the differences between the behaviours of natural and reconstituted clay (e.g. Vallericca Clay, Rampello & Silvestri, 1993; Pappadai Clay, Cotecchia & Chandler, 1997; London Clay, Gasparre et al., 2007). These differences are attributed to the natural structure that developed in the soil during and/or after sedimentation. Clay structure can be unstable, with significant effects on the compressibility and strength of the soil (e.g. Pisa Clay; Callisto & Calabresi, 1998), but the existence of a macrofabric, for example in varved clays (Connecticut Valley Varved Clay; DeGroot & Lutenegger, 2003) or in layered clays (Sibari Clay; Coop & Cotecchia, 1995), can enable the clay to remain stable upon loading. By analogy several authors have proposed that for simplification the (micro-) fabric of soils could be used to describe the stable components of structure, and that bonding could be used to refer to the unstable components of structure of the soil (Coop et al., 1995; Baudet & Stallebrass, 2003). This is a convenient analogy, which will be used in this paper so that stiff sedimentary clays can be said to usually have a large overconsolidation ratio and a structure dominated by fabric.

A simple classification of structure has been proposed by Cotecchia & Chandler (2000), whereby 'sedimentation' structure refers to the elements of structure developed during sedimentation and consolidation, whereas 'post-sedimentation' structure refers to the elements of structure developed after sedimentation and burial. The geological processes that allow the development of natural structure post-sedimentation and burial can be mechanical unloading (overconsolidation), change in physical or chemical composition (diagenesis), and cementation. Usually the phenomena other than mechanical unloading are referred to as ageing effects. Ageing effects are one of the aspects of a soil time dependence. In natural soils these effects have accumulated over geological periods of time, but they can sometimes be observed in the laboratory. For example, Perret et al. (1995) carried out a drained creep test in the oedometer on a normally consolidated specimen of a sample of artificially sedimented Jonquière Clay from Canada, where the specimen was allowed to rest for 3 months under a constant vertical stress corresponding to its preconsolidation pressure. On re-loading the specimen an increase in the yield vertical effective stress was observed that could not be explained by the change in void ratio in the specimen only. This effect seems to be encountered in soils only when the rate of straining becomes slow enough to allow ageing effects to dominate the stress–strain behaviour, as was observed by Leroueil et al. (1996) in one-dimensional compression tests on the same artificially sedimented Jonquière Clay at axial strain rates of the order of 0·036%/h. In artificially cemented soils, at early stages of curing the rate of ageing (in this case curing of the cement) is very high, and ageing effects are typically found to dominate the behaviour even under high applied axial strain rates, for example greater than 2·6%/h (Sorensen et al., 2007). Older clays, such as stiff clays that have aged over a geological time for typically millions of years, have very slow rates of ageing, which so far have proved too difficult to characterise during laboratory-scale periods of time (e.g. undisturbed London Clay; Yimsiri, 2001). However, the influence of structure resulting from ageing can be investigated with laboratory tests, as will be shown in this paper.

A complicated problem to investigate is the interaction between ageing effects and strain rate effects, strain rate effects being the second aspect of time dependence of soils. Strain rate effects are usually referred to as viscous effects, as they are linked to the viscosity (dependence of shear stress on shear strain rate) of a soil. Investigating strain rate effects in stiff clays necessarily involves studying the interaction with ageing, and in particular the influence of structure on the viscous behaviour of the clay. The work presented focuses on understanding the relative influence of the two main components of post-sedimentation structure— overconsolidation and diagenesis—on the time-dependent behaviour of London Clay. The influence of structure was studied in two steps: first the influence of structure induced by mechanical unloading was identified, and then the influence of structure induced by ageing was observed. This was achieved by comparing the behaviours of normally and overconsolidated reconstituted samples of London Clay subjected to stepwise changes in strain rate in triaxial compression, and then comparing the behaviours of overconsolidated reconstituted and undisturbed London Clay samples. The tests performed to investigate time effects in London Clay were long—generally weeks, and up to 3 months in some cases—and complex. These factors increased the risk of failure of the system during testing, and meant that the tests could not be repeated.

As cement is one aspect of the structure of stiff clays, the set of tests on London Clay was complemented by a series of triaxial compression tests on artificially cemented kaolin, as a way to investigate the interaction between ageing (in this case curing of the cement) and strain rate effects in laboratory time. A preliminary framework is proposed based on these results, where the time-dependent behaviour of soil depends on its particulate or continuum nature. This paper describes the results of this research.

TESTING MATERIALS AND SAMPLE PREPARATION
London Clay
London Clay was deposited under marine conditions in the Eocene period, around 30 million years ago. The clay minerals present are illite and, to a lesser extent, kaolinite and smectite. Natural London Clay can be characterised as a very stiff and heavily overconsolidated fissured clay (Skempton & Henkel, 1957). Nominally undisturbed London Clay samples were retrieved from the ground investigation at Heathrow Airport Terminal 5 (T5), West London, by means of high-quality rotary coring. Rotary cores, 100 mm in diameter, were obtained from a depth of 13·95–15·45 m below ground level, corresponding to the top of the $B_{2(b)}$-sub-unit, and wax-sealed at the site (Gasparre, 2005). Geotechnical index properties of the London Clay samples are given in Table 1.

Within the $B_{2(b)}$-sub-unit Gasparre et al. (2007) recorded an overconsolidation ratio (OCR), defined as the ratio of the maximum (vertical) preconsolidation stress σ'_p to the current vertical stress σ'_v, of about 8·5, based on the geological history of the clay; the yield stress ratio (YSR), defined as the ratio of the gross yield stress to the current stress, was found to vary between 9 and 24. According to Cotecchia & Chandler's (2000) framework the values of OCR and YSR are similar for clays where the post-sedimentation structure results from overconsolidation only, whereas a structure arising from post-sedimentation diagenetic processes will result in a value of YSR greater than that of the OCR. For the samples investigated, Gasparre et al. (2007) showed evidence that the state boundary surface for the intact clay lies far outside that for the reconstituted clay, which suggests that diagenesis occurred throughout the whole clay deposit after deposition and burial, and created a post-sedimentation structure in the clay. Additional evidence from micrographs shows that fabric rather than cementing is responsible for the different structure in the natural clay (Gasparre et al.,

Table 1. Geotechnical index properties of testing materials

Material	w_p: %	w_L: %	PI: %	G_s	CF: %
TA kaolin (Komoto, 2004)	21	46	25	2·68	~100
London Clay, $B_{2(b)}$-sub-unit (Gasparre, 2005)	28	65	37	2·65–2·76	54

2007). In this paper, differences between the time-dependent behaviours of the overconsolidated reconstituted and undisturbed clay samples will be interpreted as differences in (micro-)structure.

The undisturbed specimens of London Clay were cut vertically from homogeneous parts of the centre of the rotary cores. Using a soil lathe placed in a simple purpose-built humidity chamber the samples were hand-trimmed into cylinders with a diameter of 38 mm and subsequently trimmed to 76 mm in length. The reconstituted specimens of London Clay were created by thoroughly mixing trimmings from the undisturbed rotary core with tap water until a homogeneous paste was obtained at a water content of about 1·6 times the liquid limit. The paste was then preconsolidated under a vertical effective stress of 90 kPa in a tall Perspex floating-ring consolidometer (inner diameter 38 mm) to make the specimens strong enough to handle. For the overconsolidated (OC) sample the preconsolidation was carried out under a vertical effective stress of about 1500 kPa in the consolidometer, so that at the start of shearing from the isotropic stress state $p' = 300$ kPa the soil had an overconsolidation ratio of about 5. After preconsolidation and extrusion the specimens were cut to about 76 mm in length before being set up in the triaxial cell.

Artificially cemented kaolin

To reproduce cementation effects similar to what is observed over many years in natural clays, artificially cemented clay was produced by mixing TA kaolin clay with a small quantity of rapid-hardening Portland (RHP) cement. The interrelationship between strain rate and cementation (ageing) effects could then be observed in the artificially cemented clay in laboratory time. TA kaolin is an inorganic high-plasticity clay, which is extracted regionally in Japan from weathered granite and has similar characteristics to Speswhite kaolin. The geotechnical index properties of TA kaolin are given in Table 1. RHP cement was chosen to achieve a rapid hydration, with the majority of the strength increase reached within a couple of days.

Specimens of cemented kaolin 50 mm in diameter and 100 mm high were prepared from a mixture of kaolin and 3% RHP cement (by weight), mixed dry in a standard kitchen blender. The soil–cement mix was pre-compressed under dry conditions in a tall split mould to a vertical effective pressure of about 600 kPa before being set-up in the triaxial cell. In the cell the specimens were initially subjected to 90 kPa suction and then slowly saturated through the top and bottom drainage leads. As the suction decreased during saturation the cell pressure was increased to keep the effective isotropic stress at about 100 kPa. The commencement of curing has been taken as the point of saturation.

TESTING APPARATUS AND TEST PROGRAMME
Triaxial tests on reconstituted and undisturbed samples of London Clay

The tests on London Clay were performed in a computer-controlled stress path triaxial apparatus (Bishop & Wesley, 1975) with strain-controlled loading. Miniature linear variable differential transformers (LVDTs) (Cuccovillo & Coop, 1997) were utilised for local strain measurements in both the axial and radial directions. Radial drains were used on the specimens, and pore water drainage was through the base.

Following saturation in the triaxial cell under a back-pressure of 300 kPa the samples were generally isotropically compressed to a mean effective stress $p' = 300$ kPa, corresponding to an OCR of about 7 for the intact samples and of about 5 for the OC reconstituted sample (refer to Table 2). The samples were then sheared either drained or undrained in triaxial compression from the initial isotropic stress state. To characterise the viscous behaviour in shearing the axial strain rate was changed in a stepwise manner. In each test the nominal axial strain rate was controlled by means of a stepping motor and varied between two to three different levels in the interval between 0·007%/h and 0·05%/h in drained tests and in the interval between 0·007%/h and 0·9%/h in undrained tests. The strains and strain rates used for analysis and plotting were measured from local transducers. Additionally, the viscous behaviour was characterised during isotropic compression of normally consolidated (NC) reconstituted London Clay by changing the confining stress rate in a stepwise manner between 1 kPa/h and 3 kPa/h.

In the following, the strain rate effects during shearing refer to axial strain rate effects. Hence, during drained shearing, the rate effects will include components of both volumetric and deviatoric strain rate effects. During undrained shearing the shear strain rate was proportional to the axial strain rate, whereas in drained shearing the radial strain rate was typically found to be minor, and the shear strain rate could also here be assumed proportional to the axial strain rate. Where possible, the influence of strain rate changes has been investigated on the relationship between shear stress and shear strain. Tables 2 and 3 summarise the tests on London Clay presented here.

Triaxial tests on cement-mixed kaolin

Tests on the cemented kaolin were performed in a stress path triaxial apparatus with a manually regulated hydraulic cell-pressure and back-pressure system and a computer-controlled mechanical axial loading system (Santucci de Magistris & Tatsuoka, 1999). The system made it possible to perform loading under constant confining pressure at very accurately controlled rates of axial strain. Axial strains were measured externally. Radial drains were used on the specimens, and pore water drainage was through both the top and the base of the specimen.

Generally, after saturation at isotropic stress state ($p' = 100$ kPa) the samples were sheared under fully drained conditions in triaxial compression. At a relatively high deviator stress of $q = 230$ kPa the samples experienced a prolonged constant effective stress creep period of 2 days. Following the creep stage the samples were sheared, drained, to failure. To characterise the viscous behaviour the axial strain rate was changed in a stepwise manner, varying between two or three different levels in the interval between 0·078%/h and 2·6%/h. No local measurements were made

Table 2. Details of triaxial compression tests performed on London Clay specimens

Sample	Description	Pre-consolidation	Shearing				
			Drainage conditions	p'_0: kPa	e_0	OCR	Nominal axial strain rates: %/h
S1LC	Undisturbed	In situ to $\sigma'_v \approx 2000$ kPa*	Undrained	300	$0.70^\dagger$	~ 7	0.05, 0.2, 0.8
S2LC			Drained		$0.72^\ddagger$		0.007, 0.05
S2LCrA5	Reconstituted	1D consolidation to $\sigma'_v \approx 1500$ kPa	Undrained		$0.75^\ddagger$	5	0.007, 0.05, 0.5
S1LCrA2		1D consolidation to $\sigma'_v \approx 90$ kPa	Undrained		$0.89^\ddagger$	1	0.05, 0.2, 0.9
S2LCrA2			Drained		$0.92^\ddagger$		0.007, 0.015, 0.05

*Estimate of maximum overburden pressure (after Skempton & Henkel, 1957).
†Based on initial water content and volumetric deformation calculated from local strain measurements.
‡Based on initial dimensions and volumetric deformation calculated from volume gauge measurements.

Table 3. Overview of isotropic compression tests on reconstituted London Clay

Sample	Pre-consolidation	Initial void ratio, $e_{0'}$	Nominal stress rates: kPa/h	Compression index, $\lambda = \Delta v/\Delta \ln p'$
S1LCrA2	1D consolidation to $\sigma'_v = 90$ kPa	1.29	3.0	0.182
S2LCrA2		1.35	1.0, 3.0	

Table 4. Overview of CD triaxial compression tests on cement-mixed kaolin

Sample	p'_0: kPa	e_0	Creep state, p'/q: kPa	Creep time: days	Nominal axial strain rates: %/h
S9cmk	100	1.30	177/230	2	0.60, 2.6
S4cmk		1.30		2	0.078, 0.60, 2.6

during testing: thus the graphs are plotted in terms of external axial strains. Table 4 summarises the tests on cement-mixed kaolin presented here.

STRAIN-RATE-DEPENDENT BEHAVIOUR OF NORMALLY CONSOLIDATED RECONSTITUTED LONDON CLAY

Samples S1LCrA2 and S2LCrA2 were reconstituted and consolidated to a normally consolidated (NC) state prior to testing. The influence of strain rate on the behaviour of the NC reconstituted stiff clay was studied for isotropic compression and shearing stress paths, drained and undrained. Fig. 1 shows the response of reconstituted London Clay to changes in stress rate during isotropic compression. Different stress rates or strain rates define different normal compression lines (NCLs) parallel to each other, with the NCL for the higher rates plotting above those for the lower rates. For a stress rate equal to zero (i.e. fixed effective stress) the clay will creep with decreasing void ratios until the strain rate becomes insignificant, defining a lower-bound NCL. Comparing with published results on other clays but from one-dimensional tests (e.g. Batiscan clay, Leroueil *et al.*, 1985; reconstituted Fujinomori clay, Acosta-Martinez *et al.*, 2003; and other soft undisturbed and reconstituted stiff clays, listed in Table 5), the general stress rate sensitivity seems to be comparable in both anisotropic and isotropic compression. Although the general response to strain rate changes appears to be independent of stress conditions, the magnitude of the

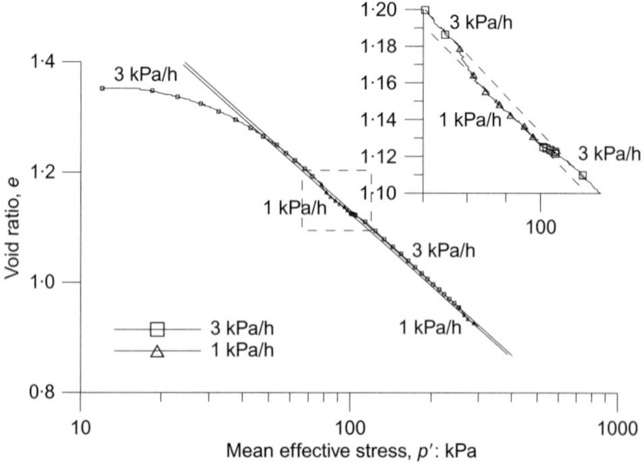

Fig. 1. Isotropic compression of NC reconstituted London Clay (sample S2LCrA2)

rate effects is likely to be influenced by the contribution to the volumetric compression from the deviator stress under anisotropic one-dimensional stress conditions; however, this is not investigated further here.

Figure 2 shows the stress–strain response of the NC reconstituted clay to drained shearing. The test failed at an early stage: thus data are available only up to 1·5% shear strain, but it is clear from the plot that at small strains the

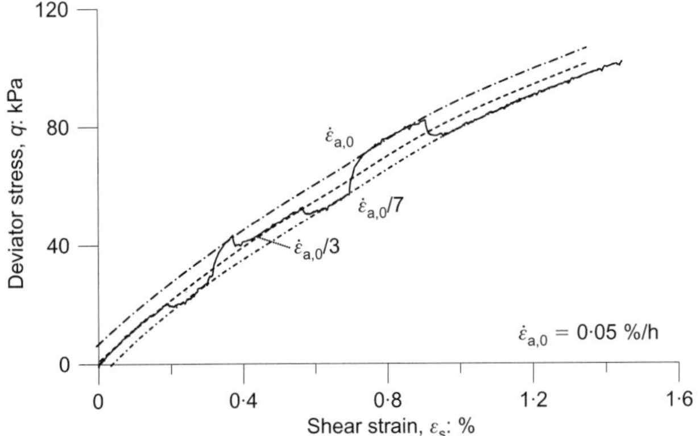

Fig. 2. Effect of stepwise change in strain rate on drained stress–strain shearing path of NC reconstituted London Clay at low stresses (S2LCrA2)

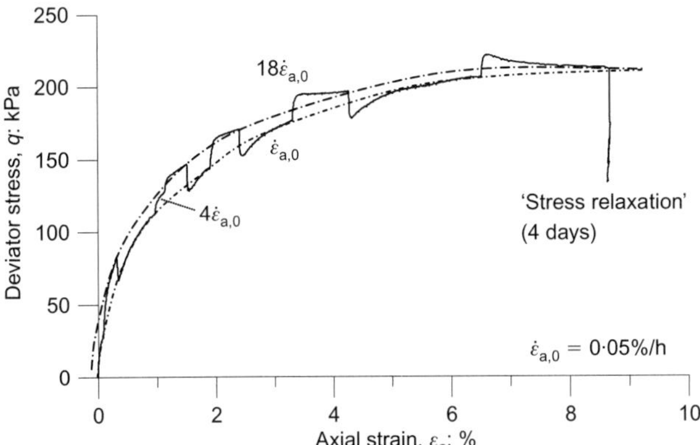

Fig. 3. Illustrating the persistent effect of stepwise change in strain rate on undrained stress–strain shearing path of NC reconstituted London Clay (S1LCrA2)

drained response of London Clay is characteristic of isotach behaviour, with different stress–strain curves for different strain rates. Fig. 3 shows the stress–strain response of the NC reconstituted clay during undrained shearing (in this test the local radial strain measurement failed, thus the graphs are plotted in terms of axial strains). For undrained loading, the response to strain rate changes seems to depend on the strain level, being characteristic of isotach behaviour at low strain levels (pre-peak strength) and changing to become more temporary with increasing strain level. The temporary response to strain rate changes has been defined by Tatsuoka *et al.* (2002) as TESRA (Temporary Effect of Strain Rate and strain Acceleration), a term that has been widely accepted and will be used here. At low strain levels for each increase in strain rate the stress–strain curve shows a stiff response (near vertical) and strain-hardens to reach the unique stress–strain curve for that specific strain rate. When the clay reaches higher strain levels and approaches its peak strength, the effects of changing strain rate become more temporary. For each increase in strain, after a stiff response the clay strain-softens to reach a persistent stress–strain curve above the curves defined by the strain rates at lower strain levels. When extrapolating the persistent response for the different strain rates it is observed that at all strain levels the points fall on unique stress–strain curves. As shown in the following, the strain rate effects give rise to an effective increase in stress at a given strain level, and hence a similar increasingly temporary behaviour can be expected for the

NC sample in drained shearing at higher strain levels (>1·5% axial strain), even though this cannot be extrapolated from the presented tests, as these were terminated at low strain levels (<1·5% axial strain).

The pore water pressure response for sample S1LCrA2 during undrained shearing is shown in Fig. 4. Some persistent effects of strain rate may be seen at low strains, but they are rather insignificant. On approaching peak each strain rate change is accompanied by a stiff response of the pore water pressure, but this is only temporary, and the pore water pressure path soon joins a unique line. This indicates that in NC-reconstituted London Clay the pore water pressure can be considered to be independent of axial strain rate.

Similarly, the relationship between the volumetric strain and axial strain is found to be independent of axial strain rate during the drained shearing of sample S2LCrA2, indicating that the strain rate changes affected the volumetric and axial strain in equal proportions. It was assumed that the test was fully drained, based on the fact that during the isotropic test the pore water pressure became significant only for stress rates greater than 3 kPa/h, corresponding to strain rates greater than approximately 0·1%/h. Furthermore, several checks, performed by closing the drainage valve for short periods up to 10 min during drained shearing of sample S2LCrA2, indicated insignificant build-up of excess pore water pressure (less than 2% of the current mean effective stress) under the highest applied rate of axial strain of 0·05%/h. The stress paths for the drained and undrained tests have been normalised for volume with respect to an equivalent pressure for a given void ratio on the isotropic NCL determined for the reconstituted clay for the reference stress rate of 3 kPa/h, with a view to determining the state boundary surface (see Fig. 5). The undrained stress path for the NC clay should follow the Roscoe–Rendulic surface and define a yield envelope or local boundary surface (LBS). The intrinsic LBS* defined by Gasparre *et al.* (2007) is also shown for comparison. This LBS* was determined using a range of strain rates, from 0·004%/h for drained tests to rates higher than 0·1%/h for the undrained tests. There are clear effects of changes in strain rate on the local boundary surface before peak, which cannot be attributed to the pore water pressure. This axial strain rate sensitivity of the undrained stress–strain path and insensitivity of the pore water pressure suggest that viscosity is associated with the soil matrix behaviour independently of drainage. This throws new light on the mechanisms governing rate effects in stiff clays.

The axial strain rate dependence of the local boundary surface LBS is consistent with that observed of the isotropic

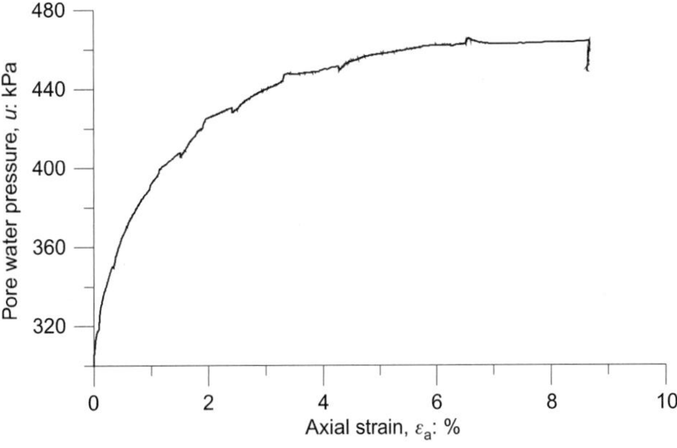

Fig. 4. Effect of strain rate on accumulation of pore water pressures during undrained shearing of NC reconstituted London Clay (S1LCrA2)

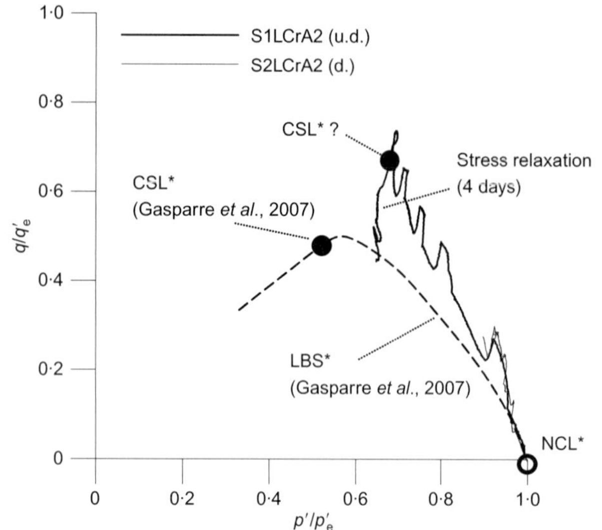

Fig. 5. Effect of stepwise change in strain rate on normalised stress paths of NC reconstituted London Clay

NCL. A comparison of Fig. 1 (isotropic compression) and Fig. 5 (undrained shearing) shows that the shift in the boundary surface is of similar magnitude in both types of test. At high shear strains the immediate effects of strain rate changes dissipate and the normalised undrained stress path converges towards a single critical state point. That point is different from the normalised CSL* determined by Gasparre *et al.* (2007), possibly because of the variability of London Clay, but the effective critical state angle of shearing resistance was found to be close to the value of 21° determined by Gasparre *et al.* (2007). The results suggest that while all effects of strain rate on the compression and shearing behaviour of London Clay are included in the size of the state boundary surface, as was shown by Tavenas *et al.* (1978) for undisturbed St Alban Clay, the critical state line in the q–p' plane is unique.

The influence of changes in strain rate on the intermediate- to large-strain stiffness of the clay is demonstrated in Fig. 6. The undrained tangent Young's modulus E_u was calculated by linear regression through the data points obtained from the LVDTs. At constant strain rate (here $\dot{\varepsilon}_{a,0}$

$= 0·05\%$/h) E_u decreases with strain level following a unique line, the S-shape stiffness degradation curve usually determined for clays. On increasing the strain rate by a factor of 18 the undrained Young's modulus increases suddenly. This abrupt increase in stiffness reflects the acceleration of strains in the sample, and is associated with the elastic stiffness. When the strain rate reaches its target value, the acceleration stops and the value of stiffness reduces with strain level, as is usually found. Here, with a target strain rate higher than the nominal rate $\dot{\varepsilon}_{a,0}$ the stiffness curve reduces to reach a curve which plots below the curve defined for $\dot{\varepsilon}_{a,0}$ (strains between 1% and 10%). Upon deceleration (decrease in strain rate by a factor of 18 to a target strain rate equal to $\dot{\varepsilon}_{a,0}$) the sample experiences a sudden fall in stiffness before rejoining the stiffness curve defined for the nominal rate. The events of acceleration and deceleration were repeated to confirm the two parallel curves for the two strain rates. The data show a clear influence of strain rate on the persistent stiffness curve at intermediate to large strains: the higher the strain rate the lower the stiffness at a given strain level.

INFLUENCE OF POST-SEDIMENTATION STRUCTURE DUE TO MECHANICAL UNLOADING ONLY ON STRAIN RATE EFFECTS IN LONDON CLAY

Mechanical overconsolidation (OC) typically results in a breakdown of the sedimentation structure (Skempton & Northey, 1952). The breakdown of structure with swelling was demonstrated by Gasparre *et al.* (2007) with results from tests on London Clay samples from the same site (T5). It is unclear, however, whether it is the sedimentation structure that degrades during unloading and, if so, by how much. According to Cotecchia & Chandler (2000) a clay with a post-sedimentation structure due to mechanical unloading will yield on its sedimentation compression line only if re-compressed. This suggests that the structure acquired during sedimentation has remained intact despite the overconsolidation. In fact, a heavily overconsolidated clay would be expected to have a strong fabric, but to have lost some of its bonding during overconsolidation. The difference observed by Gasparre *et al.* (2007) between OCR and YSR for undisturbed samples of London Clay suggests that additional bonding was created as a result of ageing after the significant unloading event(s) and/or that the sedimentation and post-sedimentation structure created prior to the significant overconsolidation remained intact to some degree. Fabric is usually associated with stable elements of structure, so that the behaviour of a clay with fabric-dominated structure is similar to that of the reconstituted clay, but with a larger state boundary surface (e.g. Ingram, 2000). In the following, results from tests on NC and OC samples of London Clay are analysed. The OC sample was preconsolidated to a significantly higher pressure than that experienced by the NC samples. The structure in the OC sample is therefore a result of the preconsolidation to high stress and unloading. By comparing the response to strain rate changes of NC and OC samples of London Clay it is expected to highlight the effect of structure due to overconsolidation.

Figure 7 shows the response to strain rate changes during undrained shearing on OC reconstituted London Clay. Similarly to the NC sample, the effect of strain rate on the stress–strain curve seems to depend on strain level. At low strain levels pre-peak strength the stress–strain curve is characteristic of isotach behaviour. When approaching the peak strength, the behaviour changes to become more temporary, but again the extrapolated persistent responses for given strain rates seem to define unique stress–strain curves. The pore water pressure response to changes in axial strain

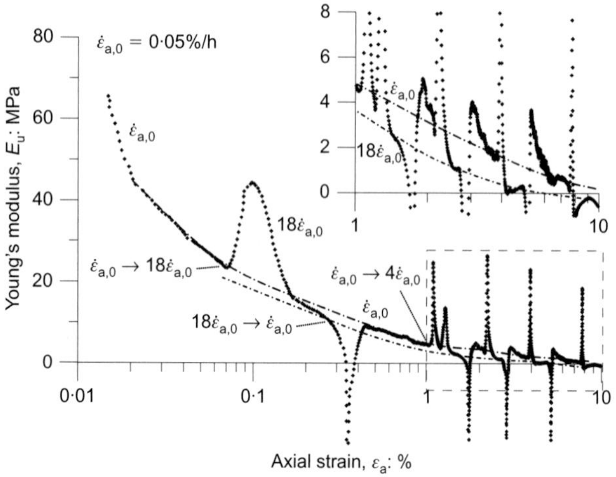

Fig. 6. Effect of stepwise change in strain rate on small- to large-strain stiffness during undrained shearing of NC reconstituted London Clay (S1LCrA2)

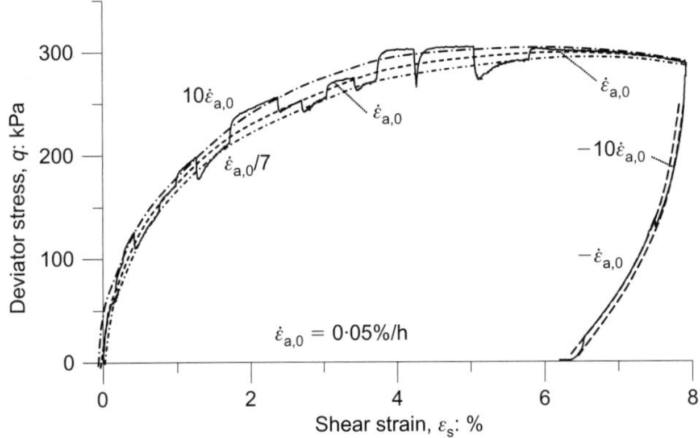

Fig. 7. Illustrating the persistent effect of stepwise change in strain rate on undrained stress–strain shearing path of OC reconstituted London Clay (S2LCrA5)

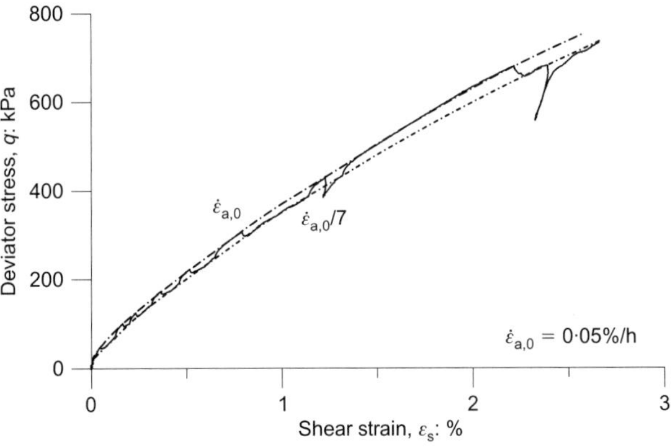

Fig. 9. Effect of stepwise change in strain rate on drained stress–strain shearing path of OC intact London Clay (S2LC)

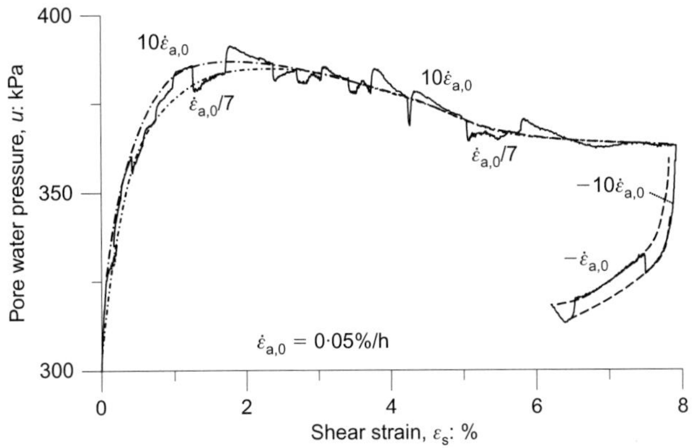

Fig. 8. Effect of strain rate on accumulation of pore water pressures during undrained shearing of OC reconstituted London Clay (S2LCrA5)

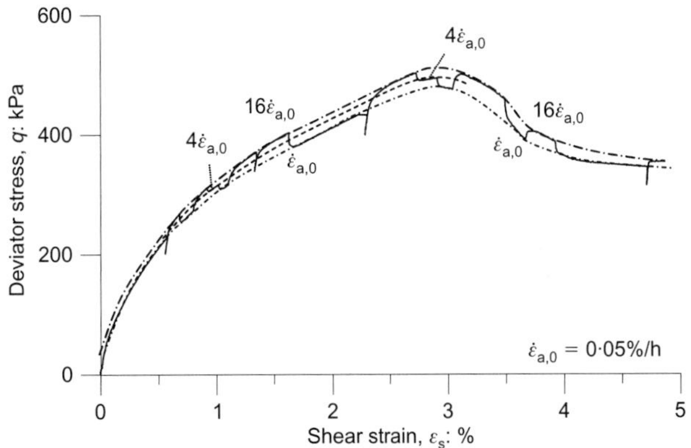

Fig. 10. Effect of stepwise change in strain rate on undrained stress–strain shearing path of OC intact London Clay (S1LC)

rate, shown in Fig. 8 for sample S2LCrA5, is at first persistent but rapidly becomes predominantly temporary, indicating that it is independent of the axial strain rate at large strain levels. A similar temporary response of the pore water pressure to changes in axial strain rate was found in the NC sample but to a lesser extent.

The comparison of the two samples of reconstituted London Clay indicates that the structure resulting from over-consolidation did not affect the overall mechanical response to axial strain rate changes significantly; however, it did affect the magnitude of the effect. Similar results were found for a reconstituted Pleistocene marine clay from Japan (Fukakusa Clay; Oka *et al.*, 2003), which was tested in NC and OC states and subjected to stepwise changes in strain rate. In that clay the general response of the NC and OC clay samples to strain rate changes was identical, showing isotach behaviour at low strains and stiff response with overshooting towards peak strength (see Table 6). Results by Sheahan *et al.* (1996) on Boston Blue Clay also showed that for a similar range in strain rates (0·05–0·5%/h) the axial strain rate sensitivity of the undrained shear strength did not vary much for overconsolidation ratios between 1 and 8.

INFLUENCE OF POST-SEDIMENTATION STRUCTURE DUE TO DIAGENETIC PROCESSES OTHER THAN OVERCONSOLIDATION ON STRAIN RATE EFFECTS IN LONDON CLAY

Figures 9 and 10 show the stress–strain response to changes in strain rate of the undisturbed London Clay during drained and undrained shearing. The stress–strain data for the drained test (Fig. 9) were measured up to 2·6% shear strain only because of a failure in the data acquisition system, but despite this the data clearly show that at low strains the response is typical of isotach behaviour. The undrained test (Fig. 10) also shows an isotach behaviour at low strains that continues even after peak. It cannot be extrapolated whether similar results would have been obtained for the drained test. The effect of strain rate changes on the pore water pressure response during undrained shearing, shown in Fig. 11 for low strains, seems to be persistent. A failure in the pore pressure transducer meant that readings were not available at axial strains greater than about 1·5%: hence it cannot be extrapolated whether the pore water response would show similar temporary characteristics at high strain level towards the peak strength, as was seen for the OC reconstituted clay. The persistent increase (see Fig. 10) of the undrained shear strength due to changes in axial strain rate that is observed in the stress–strain curve cannot, however, be attributed to the pore water pressure. Generally both drained and undrained tests suggest that the rate effects in London Clay are governed by the soil matrix independently of drainage. The stiffness of the undisturbed clay appeared to be unaffected by strain rate, but it is not clear (Fig. 12). Fig. 13 shows the stress paths for the OC reconstituted and undisturbed London Clay, normalised for volume with respect to an equivalent pressure on the isotropic NCL determined for the reconstituted clay. The path for the OC reconstituted sample fails close to the CSL* determined by Gasparre *et al.* (2007), as would be expected,

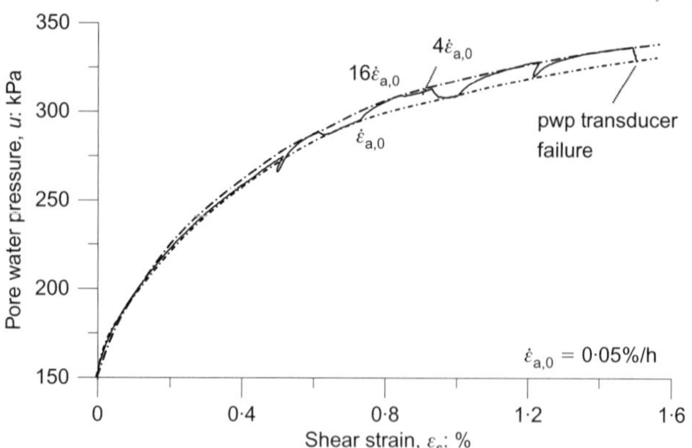

Fig. 11. Effect of strain rate on accumulation of pore water pressures during undrained shearing of OC intact London Clay (S1LC)

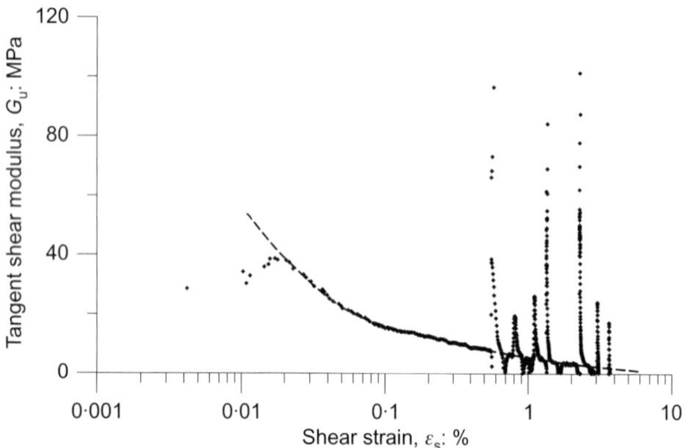

Fig. 12. Effect of stepwise change in strain rate on small- to large-strain stiffness during undrained shearing of OC intact London Clay (S1LC)

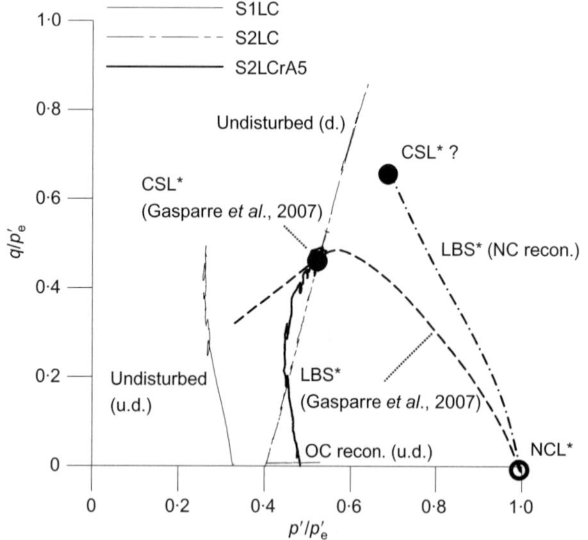

Fig. 13. Effect of stepwise change in strain rate on normalised stress paths of OC intact London Clay

but the path for the undisturbed clay lies significantly outside the LBS*, highlighting the influence of the natural structure. It appears from the data for the reconstituted soil that the critical state is not affected by strain rate, but nothing can be said from these data on whether the critical state is affected by structure.

The viscous effects on the stress–strain response of a soil can be quantified by the stress jump in deviatoric stress seen immediately after a change in the strain rate, and calculated using the definition illustrated in Fig. 14. Fig. 15 shows the influence of stress level and ratio of strain rate change (noted as 9x for an increase by 9 times) on the total immediate deviatoric stress jump experienced by the recon-

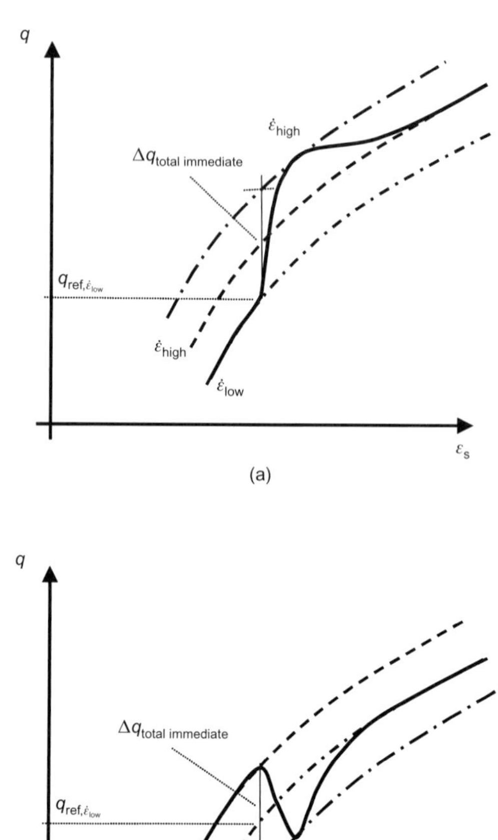

Fig. 14. Definition of stress jump due to step changes in strain rate: (a) step increase; (b) step reduction

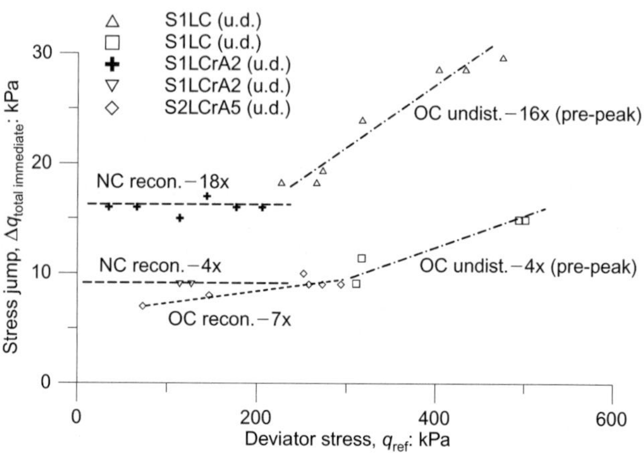

Fig. 15. Comparing influence of strain rate change and stress level on total immediate viscous stress jump during shearing of London Clay

stituted and undisturbed London Clay samples. Generally, for the OC samples, both undisturbed and reconstituted, the stress jump increases with an increase in the ratio of strain rate change, and seems to be influenced by the deviatoric stress level at which the increase in strain rate is performed. For the NC sample, however, the stress jump remains constant for a given ratio of strain rate change, independently of stress level. Strain rate effects, when quantified by the total immediate stress jump after changes in strain rate, are seen to be noticeably lower in the OC reconstituted clay than in the NC reconstituted clay and in the OC undisturbed clay, when comparing results for similar ratios of strain rate change. This may indicate that mechanical unloading did in fact alter the structure in the reconstituted clay, and that the natural structure in London Clay is responsible for the different strain-rate-related behaviours of the undisturbed and reconstituted OC samples.

The isotach behaviour pre- and post-peak, as seen for undisturbed London Clay, has been found in other undisturbed stiff clays, for example Kitan clay from Japan (Komoto *et al.*, 2003) and Vallericca Clay (Tatsuoka *et al.*, 2000): see Table 7. The difference in behaviour between the OC reconstituted and undisturbed London Clay must be due to the natural structure that was created post-sedimentation. In order to understand better how the structure type (fabric-dominated or bonding-dominated) influences the time dependence, some tests were performed on samples of artificially cemented kaolin. These tests highlight the effects of bonding on the strain-rate-related response of a clay.

INFLUENCE OF CEMENTING ON STRAIN RATE EFFECTS

One of the aspects of structure in stiff clays is cementing. By adding cement to kaolin it is intended to simulate bonding created in natural clays due to diagenetic processes but at a much faster rate. For comparison, effects of strain rate changes on the behaviour of pure kaolin need to be considered. Results obtained recently by Tatsuoka *et al.* (2002) from undrained shearing tests on normally consolidated pure kaolin show that it follows an isotach behaviour from low strains to critical state (Fig. 16), similar to what is usually found in other soft clays (e.g. Belfast clay and Winnipeg clay; Graham *et al.*, 1983) and interestingly also soft rocks (e.g. soft sandstone; Tatsuoka *et al.*, 2002) (Table 7). In contrast, results presented here show that the effects of changes in strain rate on the stress–strain response of cemented kaolin during drained shearing become rapidly temporary on approaching peak strength (Fig. 17). This is rather similar to the response observed on other cemented soils (e.g. cemented gravel; Kongsukprasert & Tatsuoka, 2003) and reconstituted stiff natural clays such as reconstituted Fujinomori Clay (Tatsuoka *et al.*, 2000) or reconstituted London Clay, normally consolidated or overconsolidated (Table 6). Dominating temporary (TESRA) effects of strain rate are also usually encountered in granular materials (e.g. Hostun sand; Tatsuoka *et al.*, 2000) (Table 6). One possible explanation for the behaviour of the cemented kaolin could be that when sheared to high strains the cementing starts breaking, forming aggregates of cement-mixed kaolin, and that these aggregates contribute to changing the soil behaviour from that of a continuum to that of a granular (particulate) material. It could be interpreted that in natural clays there exist microscopic aggregates of particles that cannot be destroyed by reconstitution (or perhaps these are made up by interparticle bonding immediately after reconstitution), so that reconstituted samples made from natural samples behave similarly to 'destructured' cement-mixed kaolin samples. This series of tests on simple ce-

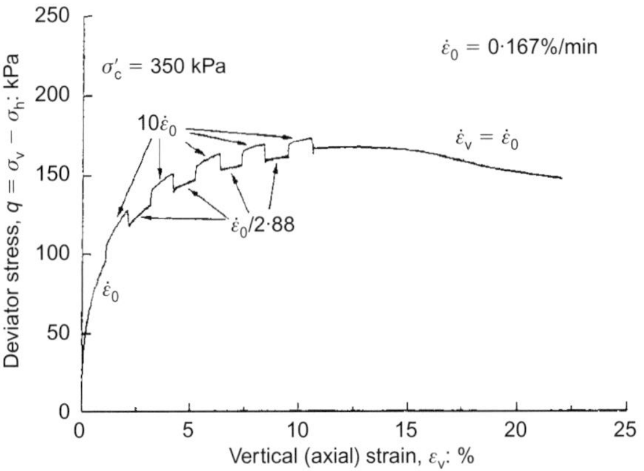

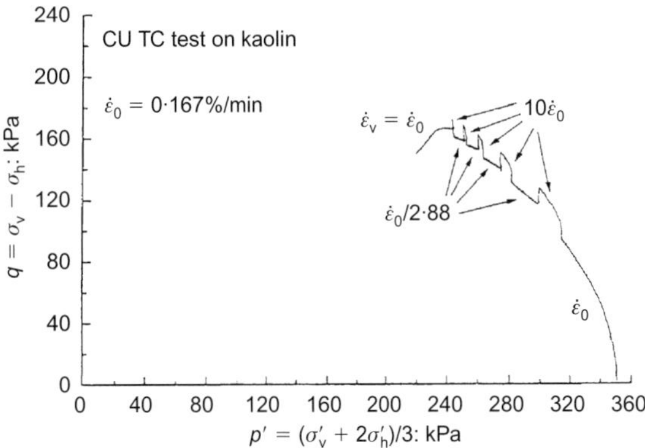

Fig. 16. CU TC tests on reconstituted NC kaolin (Tatsuoka *et al.*, 2002)

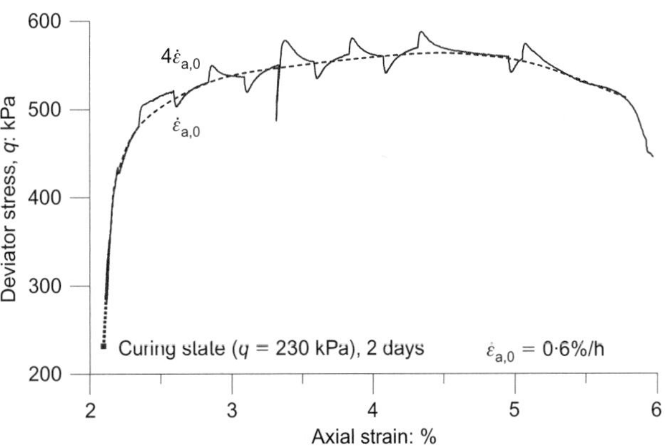

Fig. 17. Persistent effects of stepwise change in strain rate on drained stress–strain shearing path of cement-mixed kaolin after 2 days of curing at high deviator stress (S9cmk)

mented clay showed the effect that a simple structure, rather than the more complex one of London Clay, might have on the viscous behaviour of clays. This suggests that in London Clay the observed effect of post-sedimentation structure on the response to changes in axial strain rate cannot be associated with simple cementing between particles.

Based on the above, it seems that the behaviour of reconstituted stiff clays including London Clay is typical of that of a continuum at low strains, becoming increasingly temporary towards peak strength and thus more typical of that of cemented kaolin at high strains and granular materials. The behaviour of undisturbed London Clay, however, is more typical of that of soft rocks, and a parallel cannot be established between cemented kaolin and undisturbed London Clay. This is in agreement with the evidence shown on micrographs by Gasparre *et al.* (2007) that the structure in the natural clay is not due to bonding but primarily due to fabric. It seems that soils behaving like a continuum, because their particles are bonded together either through 'flexible' forces (for example electrical and chemical forces in young normally consolidated clays, such as normally consolidated samples of pure kaolin) or by heavy pressures (for example soft rocks or perhaps undisturbed London Clay), have an isotach behaviour, whereas in soils behaving like granular (particulate) materials, for example sands or 'broken' cemented kaolin, strain rate effects are only temporary. One can imagine that in a particulate material the deformation is more localised at interparticle contact points, whereas in a continuum the behaviour may be associated with the deformation of the entire volume of soil. The nature of the research, where 'macroscopic' samples are tested, means that only strain rate effects at the macroscopic scale can be determined. It could therefore be speculated that the soils defined as 'continua' are found to show a greater degree of viscous effect than the particulate soils because common laboratory tests characterise macro-behaviour and would miss the viscous effects at interparticle contact. This would be in agreement with the fact that clays, classified here as continua, generally show significantly greater creep deformation than granular soils, which are particulate in nature.

CONCLUSIONS

The data presented in this paper form a unique set of data for the axial strain-rate-dependent behaviour of London Clay. Not too surprisingly, the results fit reasonably well with the existing data base for different soils (Tables 5–7). New results on pore water pressure and volumetric response during shearing throw new light on the mechanisms governing viscous effects in London Clay.

(*a*) Overconsolidation does not seem to affect the general axial strain-rate-dependent behaviour of reconstituted London Clay, but it affects the magnitude of the effect. Similar results have been seen in tests on other reconstituted stiff clays.

(*b*) Reconstituted London Clay has an isotach response to isotropic compression and shearing at low strain levels, but at higher strain levels towards peak strength strain rate effects become increasingly temporary in shearing. This is similar to the behaviour of other reconstituted stiff clays and rather similar to the observed behaviour of cemented soils. It was also found that the pore water pressure during undrained shearing and the relationship between volumetric and axial strain during drained shearing are independent of the axial strain rate: this, together with the observed strain rate dependence of the stress path, suggests that the viscous effects in London Clay are governed by the soil matrix only, independently of the pore water pressure response.

(*c*) Undisturbed London Clay shows an isotach behaviour both pre- and post-peak during shearing. This is similar to the behaviour of soft rocks, soft clays and other undisturbed stiff natural clays.

(*d*) Cemented kaolin shows temporary effects of strain rate during shearing in the peak and post-peak region. This is similar to the behaviour of granular soils and other cemented soils post-peak.

The factors defining why the reconstituted London Clay behaves like a granular material cannot be explained here, and further research involving microscopy studies should be planned to clarify it.

ACKNOWLEDGEMENTS

The authors would like to thank Imperial College for providing the rotary cores of undisturbed London Clay, without which this research could not have been done, and particularly Dr Coop for his guidance and help with the experimental testing at the start of the research. The research was made possible through funding from EPSRC's Cooperation Awards in Science and Engineering (CASE) in collaboration with Ove Arup and Partners. The tests on cement-mixed kaolin were performed at the Institute of Industrial Science, University of Tokyo, under the guidance of Professor Tatsuoka and Professor Koseki. The authors are grateful for the funding provided by the Royal Academy of Engineering, the University College London Graduate School, and the Department of Civil Engineering, University of Tokyo, which enabled the study visit to Tokyo.

APPENDIX 1. OBSERVED VISCOUS BEHAVIOUR OF DIFFERENT SOILS IN ONE-DIMENSIONAL, ISOTROPIC AND TRIAXIAL COMPRESSION

Table 5. Observed viscous behaviour in one-dimensional and isotropic compression for different clays

Soil name	Reference	Description	PI	Observed viscous behaviour		
				Isotach	Combined TESRA and Isotach	TESRA
Batiscan clay	Leroueil *et al.* (1996)	Sensitive natural	21	1D compression	–	–
Fujinomori clay	Acosta-Martinez *et al.* (2003)	Reconstituted, NC	33		–	–
Oimachi clay	Acosta-Martinez *et al.* (2003)	Reconstituted, Pleistocene	20		–	–
Kitan clay	Acosta-Martinez *et al.* (2003)	Undisturbed, stiff, Pleistocene	22–62		–	–
London Clay	This study	Reconstituted, NC	37	Isotropic compression	–	–

Table 6. Observed viscous behaviour in triaxial compression for sands, gravels, cement-mixed soils and reconstituted clays

Soil name	Reference	Description	PI	Observed viscous behaviour		
				Isotach	Combined TESRA and Isotach	TESRA
Hostun	Tatsuoka et al. (2000)	Quartz-rich sub-angular, poorly graded medium-sized sand		–	–	Until peak
Toyoura sand	Tatsuoka et al. (2000)			–	–	Until peak
Chiba gravel	Tatsuoka et al. (2000)	Very dense, crushed sandstone		–	Pre-peak and post-peak	–
Metramo silty sand	Tatsuoka et al. (2000)	Dense, crushed granite, fines content 16%	14	–	Pre-peak?	Post-peak
Cement-mixed sand	Tatsuoka et al. (2000)	4% cement		–	Pre-peak and post-peak	post-peak?
Cement-mixed gravel	Kongsukprasert & Tatsuoka, 2003	Chiba gravel mixed with 2·5% cement		Pre-peak	–	Post-peak
Cement-mixed kaolin	This study	3% RHP cement		?	?	Post-peak
Lower Mississippi River Valley clay	Richardson & Whitman (1963)	Reconstituted slightly organic back-swamp alluvial deposit	38	Possibly at low stresses	Evidence at higher stresses pre-peak	Possibly at peak state
Fukakusa clay	Oka et al. (2003)	Reconstituted, NC and OC Pleistocene marine clay	27	At low stresses	High stresses pre-peak and post-peak	–
Fujinomori clay	Tatsuoka et al. (2000)	Reconstituted, NC	33	At low stresses	High stresses pre-peak and post-peak	–
Kitan clay	Komoto (2004)	Reconstituted, Pleistocene	22–62	At low stresses	Intermediate stresses pre-peak	High stresses and peak state
Kitan clay	Komoto (2004)	Reconstituted; aged for one year	22–62	At low stresses	Intermediate stresses pre-peak	High stresses and peak state
Oimachi clay	Komoto et al. (2003)	Reconstituted, Pleistocene	20	At low stresses (<3% strain)	?	?
London Clay	This study	Reconstituted, NC	37	At low stresses	Intermediate stresses pre-peak	High stresses and peak state
London Clay	This study	Reconstituted, OC	37	At low stresses	High stresses pre-peak and post-peak	–

Table 7. Observed viscous behaviour in triaxial compression for reconstituted and undisturbed soft clays, soft rocks and undisturbed stiff clays

Soil name	Reference	Description	PI	Observed viscous behaviour		
				Isotach	Combined TESRA and Isotach	TESRA
St Alban clay	Tavenas et al. (1978)	Undisturbed, OC	17–23	Until peak	–	–
Haney clay	Vaid & Campanella (1977)	Undisturbed sensitive marine, NC	18	Until peak	–	–
Vallerica clay[10]	Tatsuoka et al. (2000)	Undisturbed stiff marine clay, OC	26	At low stresses	Possibly at high stresses pre-peak and post-peak	–
Belfast clay	Graham et al. (1983)	Undisturbed organic estuarine, lightly OC	50–70	Pre-peak and post-peak	–	–
Winnipeg clay	Graham et al. (1983)	Undisturbed plastic lacustrine, lightly OC	35–55	Pre-peak and post-peak	–	–
Kaolin clay	Tatsuoka et al. (2002)	Reconstituted, NC	42	Until peak	–	–
Kazusa sedimentary rock	Tatsuoka et al. (2000)	Rotary core sedimentary soft sand/silt stone	–	Until peak	–	–
Kitan clay	Komoto (2004)	Undisturbed, stiff, Pleistocene	22–62	Until peak	–	–
Oimachi clay	Komoto et al. (2003)	Undisturbed, stiff, Pleistocene	20	At low stresses (<3% strain)	?	?
London Clay	This study	Undisturbed, stiff, heavily OC	37	Pre-peak and post-peak	–	–

NOTATION

B pore pressure coefficient
e void ratio
e_0 void ratio at start of shearing
$e_{0'}$ void ratio at start of compression
E_u undrained tangent Young's modulus $= \Delta\sigma_a/\Delta\varepsilon_a$
G_s specific gravity of solids
OCR   overconsolidation ratio $= \sigma'_p/\sigma'_v$
p' current mean effective stress $= \frac{1}{3}(\sigma'_a + 2\sigma'_r)$
p'_0 mean effective stress at start of shearing
p'_e equivalent pressure on the intrinsic NCL
PI   plasticity index
q deviator stress, $= \sigma'_a - \sigma'_r$
q_{ref} reference deviator stress level on stress–strain curve for lowest applied rate of strain
$q_{ref}, \dot\varepsilon_{low}$ total immediate stress jump after a step change in strain rate
u pore water pressure
v specific volume $= 1 + e$
w_L liquid limit
w_p plastic limit
ε_a axial strain
ε_r radial strain
ε_s shear strain $= \frac{2}{3}(\varepsilon_a - \varepsilon_r)$
$\dot\varepsilon_{a,0}$ reference axial strain rate
λ compression index/gradient of NCL $= \Delta v/\Delta\ln p'$
σ'_a axial effective stress (principal)
σ'_p vertical preconsolidation pressure, maximum past pressure
σ'_r radial effective stress (principal)
σ'_v vertical effective stress

REFERENCES

Acosta-Martinez, H. E., Tatsuoka, F. & Li, J.-Z. (2003). Viscosity in one-dimensional deformation of clay and considerations for its modelling. *Proc. 38th Jpn Nat. Conf. Geotech. Engng, Akita*, 255–256.

Baudet, B. A. & Stallebrass, S. E. (2003). Modelling effects of fabric and bonding in natural clays. *Proceedings of the international workshop on geotechnics of soft soils*, Noordwijkerhout, 167–174.

Bishop, A. W. & Wesley, L. D. (1975). A hydraulic triaxial apparatus for controlled stress path testing. *Géotechnique* **25**, No. 4, 657–670.

Callisto, L. & Calabresi, G. (1998). Mechanical behaviour of a natural soft clay. *Géotechnique* **48**, No. 4, 495–513.

Coop, M. R., Atkinson, J. H. & Taylor, R. N. (1995). Strength and stiffness of structured and unstructured soils. *Proc. 11th Eur. Conf. Soil Mech. Found. Engng* **1**, 55–62.

Coop, M. R. & Cotecchia, F. (1995). The compression of sediments at the archeological site of Sibari. *Proc. 11th Eur. Conf. Soil Mech. Found. Engng, Copenhagen* **8**, 19–26.

Cotecchia, F. & Chandler, R. J. (1997). Influence of structure on the pre-failure behaviour of a natural clay. *Géotechnique* **47**, No. 3, 523–544.

Cotecchia, F. & Chandler, R. J. (2000). A general framework for the mechanical behavior of clays. *Géotechnique* **50**, No. 4, 431–447.

Cuccovillo, T. & Coop, M. R. (1997). The measurement of local axial strains in triaxial tests using LVDTs. *Géotechnique* **47**, No. 1, 167–171.

DeGroot, D. J. & Lutenegger, A. J. (2003). Geology and engineering properties of Connecticut Valley Varved Clay. *Proceedings of the international workshop on characterisation and engineering purposes of natural soils*, Singapore, Vol. 1, pp. 695–724.

Gasparre, A. (2005). *Advanced laboratory investigation of London Clay*. PhD thesis, Imperial College of Science Technology and Medicine, University of London.

Gasparre, A., Nishimura, S., Coop, M. R. & Jardine, R. J. (2007). The influence of structure on the behaviour of London Clay. *Géotechnique*, **57**, No. 1, 19–31.

Graham, J., Crooks, J. H. A. & Bell, A. L. (1983). Time effects on the stress–strain behaviour of natural soft clays. *Géotechnique* **33**, No. 3, 327–340.

Ingram, P. J. (2000). *The application of numerical models to natural stiff clays*. PhD thesis, City University, London.

Komoto, N. (2004). *Experimental study on ageing effect using cement-mixed clay*. MSc thesis, University of Tokyo (in Japanese).

Komoto, N., Tatsuoka, F. & Nishi, T. (2003). Viscous stress–strain properties of undisturbed Pleistocene clay and its constitutive modelling. *Proc. 3rd Int. Symp. Deformation Characteristics of Geomaterials, IS Lyon*, 579–587.

Kongsukprasert, L. & Tatsuoka, F. (2003). Viscous effects coupled with ageing effects on the stress–strain behaviour of cement-mixed gravel. *Proc. 3rd Int. Symp. Deformation Characteristics of Geomaterials, IS Lyon*, 569–577.

Leroueil, S., Kabbaj, M., Tavenas, F. & Bouchard, R. (1985). Stress–strain-strain rate relationship for the compressibility of sensitive natural clays. *Géotechnique* **35**, No. 2, 159–180.

Leroueil, S., Perret, D. & Locat, J. (1996). Strain rate and structuring effects on the compressibility of a young clay. *ASCE Geotechnical Special Publication* **61**, 137–150.

Oka, F., Kodaka, T., Kimoto, S., Ishigaki, S. & Tsuji, C. (2003). Step-changed strain rate effect on the stress–strain relations of clay and a constitutive modelling. *Soils Found.* **43**, No. 4, 189–202.

Perret, D. (1995). *Diagénèse mécanique précoce des sédiments fins du Fjord du Saguenay*. PhD thesis, Université Laval, Ste-Foy, Québec, Canada.

Rampello, S. & Silvestri, F. (1993). The stress–strain behaviour of natural and reconstituted samples of two overconsolidated clays. *Proc. Int. Symp. Geotech. Engng Hard Soils – Soft Rocks, Athens* **1**, 769–778.

Richardson, A. M. & Whitman, R. V. (1963). Effect of strain rate upon undrained shear resistance of a saturated remoulded fat clay. *Géotechnique* **13**, No. 4, 310–324.

Santucci de Magistris, F. & Tatsuoka, F. (1999). Time effects on the stress–strain behaviour of Metramo silty sand. *Proc. 2nd Int. Symp. Pre-failure Deformation Characteristics of Geomaterials, IS Torino* **1**, 555–564.

Skempton, A. W. & Henkel, D. J. (1957). Tests on London Clay from deep borings at Paddington, Victoria and the South Bank. *Proc. 4th Int. Conf. Soil Mech. Found. Engng, London* **1**, 100–106.

Skempton, A. W. & Northey, R. D. (1952). The sensitivity of clays. *Géotechnique* **3**, No. 3, 30–53.

Sorensen, K., Baudet, B. & Tatsuoka, F. (2007). Coupling of ageing and viscous effects in an artificially structured clay. *Proceedings of the geotechnical symposium on soil stress–strain behavior: measurement, modeling and analysis*, Rome. In press.

Tatsuoka, F., Ishihara, M., Di Benedetto, H. & Kuwano, R. (2002). Time-dependent shear deformation characteristics of geomaterials and their simulation. *Soils Found.* **42**, No. 2, 103–129.

Tatsuoka, F., Santucci de Magistris, F., Hayano, K., Koseki, J. & Momoya, Y. (2000). Some new aspects of time effects on the stress–strain behaviour of stiff geomaterials. *Proc. 2nd Int. Symp. Hard Soils – Soft Rocks, Napoli*, 1285–1371.

Tavenas, F., Leroueil, S., La Rochelle, P. & Roy, M. (1978). Creep behaviour of an undisturbed lightly overconsolidated clay. *Can. Geotech. J.* **15**, No. 3, 402–423.

Vaid, Y. P. & Campanella, R. G. (1977). Time-dependent behaviour of undisturbed clay. *J. Geotech. Div. ASCE* **103**, No. GT7, 693–709.

Yimsiri, S. (2001). *Pre-failure deformation characteristics of soils: anisotropy and soil fabric*. PhD thesis, University of Cambridge.

Gens, A., Vaunat, J., Garitte, B. & Wileveau, Y. (2007). *Géotechnique* **57**, No. 2, 207–228

In situ behaviour of a stiff layered clay subject to thermal loading: observations and interpretation

A. GENS[*], J. VAUNAT[*], B. GARITTE[*] and Y. WILEVEAU[†]

The paper presents an interpretation of an in situ heating test carried out on Opalinus clay in the Mont Terri underground laboratory. Opalinus clay is a stiff, strongly bedded, Mesozoic clay of marine origin. When subjected to thermal loading, saturated stiff clays exhibit a strong pore pressure response that significantly affects the hydraulic and mechanical behaviour of the material. The observations gathered in the in situ test have provided an opportunity to examine the integrated thermo-hydro-mechanical (THM) response of this sedimentary clay. Coupled THM numerical analyses have been carried out to provide a structured framework for interpretation, and to enhance understanding of THM clay behaviour. Numerical analyses have been based on a coupled theoretical formulation that incorporates a constitutive law especially developed for this type of material. The law includes degradation of bonding by damage. By performing three-dimensional computations, it has been possible to incorporate anisotropy of material parameters and of in situ stresses. The 3D simulation has proved able to furnish a satisfactory representation of the development of the in situ test and of the main observed patterns of behaviour. A sensitivity analysis has also been carried out to examine the potential effect of various key or uncertain parameters. The critical examination of test observations and the results of the numerical analyses have allowed the classification, by differing degrees of significance, of the various coupled phenomena present in the problem.

KEYWORDS: anisotropy; clays; in situ testing; numerical modelling; radioactive waste disposal; temperature effects

Cet article présente une interprétation d'un essai de chauffage in situ sur une argile Opalinus dans le laboratoire sous-terrain de Mont Terri. L'argile Opalinus est une argile mésozoïque ferme et fortement stratifiée d'origine marine. Lorsque les argiles fermes saturées sont soumises à une charge thermique, on observe une forte réponse de pression interstitielle qui affecte significativement le comportement hydraulique et mécanique du matériau. Les observations recueillies dans l'essai in situ ont permis d'examiner la réponse thermo-hydromécanique (THM) intégrée de cette argile sédimentaire. On a effectué des analyses numériques THM couplées pour fournir un cadre structuré à l'interprétation et pour améliorer la compréhension du comportement THM de l'argile. Les analyses numériques sont basées sur une formulation théorique couplée qui incorpore une loi constitutive spécialement développée pour ce type de matériau. La loi inclut la dégradation de la cimentation par détérioration. Grâce à des calculs tridimensionnels, on a pu incorporer l'anisotropie des paramètres matériels et des contraintes in situ. La simulation tridimensionnelle a permis de fournir une représentation satisfaisante du développement de l'essai in situ et des principaux modes de comportement observés. On a également effectué une analyse de sensibilité pour examiner l'effet potentiel de divers paramètres clés ou incertains. Grâce à l'examen critique des observations des essais et aux résultats des analyses numériques, on a pu effectuer la classification des divers phénomènes couplés présents, par divers degrés de signification.

INTRODUCTION

Stiff sedimentary clays provide the geological background to many civil engineering projects. In recent years, interest in these types of material has increased, because they are being considered as potential host geological media for underground repositories of high-level radioactive waste (Gens, 2003). They exhibit favourable characteristics, such as low permeability, a degree of self-healing capacity when fractured, significant retardation properties for solute transport, and no foreseeable economic value. Possible shortcomings are the likely need for support of the excavated openings, and sensitivity to chemical actions and to desaturation caused by ventilation. High-level radioactive waste is heat-emitting. Therefore the use of stiff sedimentary clays in this type of application brings to the fore the thermal response of this type of material and, especially, the interaction of

thermal phenomena with hydraulic and mechanical behaviour.

The possible use of these types of clay as geological hosts for radioactive waste has prompted the construction of several underground laboratories. They include: the Hades laboratory in Mol (Belgium), excavated in Boom clay; the Meuse/Haute Marne laboratory of Eastern France, sited in a thick stratum of Callovo-Oxfordian mudstone; and the Mont Terri laboratory of northern Switzerland, constructed in Opalinus clay. Underground laboratories allow, by the performance of appropriate in situ tests, observation of the clay response in complex situations that mimic some of the conditions likely to be encountered in a deep geological repository. In particular, special attention is paid to the coupled thermo-hydro-mechanical (THM) behaviour of the potential host clay.

The existence of underground laboratories offers both an opportunity and a challenge. The opportunity lies in the availability of extensive experimental studies, both in the laboratory and in situ, that provide a level of characterisation of the stiff sedimentary clay much higher than that normally available in a conventional civil engineering project. The challenge lies in the need to achieve a good understanding of the generalised behaviour of these clays when subjected

Manuscript received 9 May 2006; revised manuscript accepted 8 January 2007.
Discussion on this paper closes on 1 August 2007, for further details see p. ii.
[*] Department of Geotechnical Engineering and Geosciences, Technical University of Catalonia, Spain.
[†] ANDRA, Chatenay Malabry, France.

to complex demands in a natural environment. Generally, numerical analyses are an essential tool towards achieving such an understanding.

Various in situ heating tests have been performed involving the observation of the response of natural sedimentary clay. For instance, in the Hades laboratory the following experiments have been performed: the CACTUS test, the ATLAS test and the CERBERUS test (Picard et al., 1994; Bernier & Neerdael, 1996; De Bruyn & Labat, 2002). The results obtained in these tests have led to a better characterisation of the THM properties of the Boom clay formation. The CERBERUS test has also included radioactive material in the heating source to study the additional effects of radiation on material behaviour. In the Opalinus clay of the Mont Terri laboratory, the test HE-D has been performed, the results of which are examined and interpreted in this paper. Finally, in the Meuse/Haute Marne laboratory, an in situ heating test (TER) is under way at present. In all cases the sedimentary clays react to the temperature increase with a rise of pore water pressures and associated mechanical effects, although the specific features of the response naturally depend on the particular clay being tested. A good understanding of the resulting coupled THM behaviour is therefore necessary for a rational design of potential repositories.

This paper concentrates on the response of Opalinus clay, a stiff, layered Mesozoic clay, to thermal loading in the context of the HE-D in situ heating test performed in the Mont Terri underground laboratory. Special attention is given to the interplay between the thermal, hydraulic and mechanical aspects of the clay behaviour. Some geological information on Opalinus clay is given first, together with a summary of the properties relevant to the case considered. Opalinus clay exhibits a strongly bedded structure that results in a distinct anisotropy of several THM properties. The in situ stresses in the massif are also anisotropic. After the description of the material, an outline of the theoretical formulation used in subsequent analyses is presented. The in situ heating test is then described. The observations are interpreted using a 3D coupled thermo-hydro-mechanical (THM) analysis that accounts for the main features of the generalised clay behaviour, including its anisotropy. Associated 2D axisymmetric computations have been used as reference. Finally, a sensitivity analysis (using the axisymmetric model) has been carried out to achieve an enhanced understanding of the clay behaviour and a more specific identification of critical parameters and interacting phenomena.

CHARACTERISTICS OF OPALINUS CLAY

Opalinus clay is a stiff overconsolidated clay of Lower Aalenian age, corresponding to the Middle Jurassic. It is found in the Jura Mountains of Northern Switzerland. Its mineralogy consists mainly of sheet silicates (illite, illite–smectite mixed layers, chlorites, kaolinites), framework silicates (albites, K-feldspar), carbonates (calcite, dolomite, ankerite and siderite), and quartz (Bossart et al., 2002). There are three slightly different facies containing different mineral proportions: a shaly facies in the lower part of the deposit; a 15 m thick sandy-silty facies in the centre; and a sandy facies interstratified with the shaly facies in the upper part. The content of clay minerals may range from 40% to 80%, depending on the facies. The clay was sedimented in marine conditions. Its pore water is highly mineralised, with total dissolved salts up to 20 g/l; this water contains a significant content of seawater millions of years old. Total thickness is about 160 m. To place Opalinus clay in the context of the sedimentary clays present in other underground laboratories, a comparison of the typical values of some basic properties is presented in Table 1. Boom clay is a soil-like material with a relative large porosity in spite of the important burial depth. Opalinus clay and the Callovo-Oxfordian mudstone are indurated clays of low porosity and higher strength. Naturally, permeability is very low in all cases.

Opalinus clay behaviour has been intensely studied by means of laboratory and in situ experimental programmes. A synthesis of the main physical and geotechnical parameters is reported in Bock (2001). Based on this information, the reference values of a series of parameters together with their likely ranges have been listed in Table 2. Note that some of the parameters have different values depending on the orientation of the material, reflecting the anisotropy caused by the intense bedding of the clay. When information is scarce, no range estimate has been quoted. The following additional remarks can be made.

(a) Reference elastic parameters have been based on measurements made in triaxial and uniaxial compression tests and on the results of field dilatometer tests.

(b) The reference shear strength parameters have been derived from laboratory triaxial tests. As strength depends on water content, only strength data for which the moisture content was within the range of the natural water content were considered.

(c) Permeability measurements were made on laboratory specimens and using in situ boreholes.

(d) Thermal dilation coefficient and Biot's coefficient are quite uncertain because the data on those parameters are limited.

The reference values have been used as a basis for estimating the parameters required in the numerical analyses reported below. However, alternative values have been adopted if information referring more specifically to the location of the in situ heating test was available.

Table 1. Some typical basic properties of Opalinus clay, Callovo-Oxfordian mudstone and Boom clay (Horseman et al., 1987; Baldi et al., 1991; Bernier et al., 1997; Harrington et al., 2001; Croisé et al., 2004; ANDRA, 2005; Coll, 2005; Wileveau, 2005; Bastiaens et al., 2006).

	Opalinus clay	Callovo-Oxfordian mudstone	Boom clay
Dry density: g/cm^3	2·22–2·33	2·21–2·33	1·61–1·78
Calcite content: %	6–22	23–42	0–3
Porosity: %	13·5–17·9	< 13	> 30
Water content: %	4·2–8·0	< 5·5	> 9·5
Young's modulus: MPa	4000–10 000	4000–5600	200–400
Unconfined compression strength: MPa	4–22	20–30	2
Permeability: m/s	1–5 × 10^{-13}	1–5 × 10^{-13}	2–5 × 10^{-12}

Table 2. Reference parameters for Opalinus clay (Wileveau, 2005)

	Parameter	Orientation*	Reference value	Range
Mineralogy	Clay content: %	–	62	44–80
	Carbonate content: %	–	14	6–22
	Quartz content: %	–	18	10–27
Petrophysical properties	Density, ρ: g/cm^3	–	2·45	±0·03
	Water content, w: %	–	6·1	±1·9
	Porosity, ϕ: %		15·7	±2·2
Mechanical properties	Uniaxial compression strength, R_c: Mpa	Parallel	10	±6
		Perpendicular	16	±6
	Tensile strength, R_t: Mpa	Parallel	1	–
		Perpendicular	0·5	–
	Elastic modulus, E: Mpa	Parallel	10 000	±3700
		Perpendicular	4000	±1000
	Poisson's ratio, ν	Parallel	0·24	–
		Perpendicular	0·33	–
	Shear strength parameters			
	c': MPa	Parallel	2·2	–
		Perpendicular	25	–
	Φ': degrees	Parallel	5·0	–
		Perpendicular	25	–
Thermal and thermomechanical properties	Thermal conductivity, λ: W/(m K)	Parallel	2·1	±10%
		Perpendicular	0·995	±10%
	Heat capacity of dry material at 20°C, E_s: J/(kg K)	–	800	±10%
	Coefficient of linear thermal expansion, α: K^{-1}	–	$2·6 \times 10^{-5}$	–
Hydraulic and hydromechanical properties	Water permeability of sound clay, k: m/s	–	10^{-13}	10^{-12}–10^{-14}
	Biot's coefficient, b	–	0·6	0·42–0·78

* Orientation to bedding.

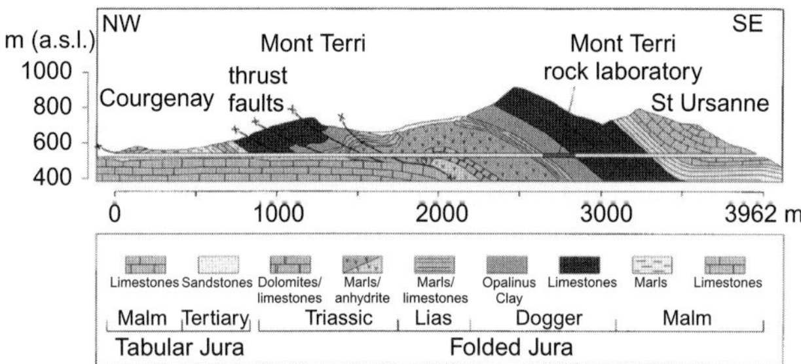

Fig. 1. Geological profile in vicinity of Mont Terri underground laboratory (Schaeren & Norbert, 1989; Thury & Bossart, 1999)

Fig. 2. View of Opalinus clay in an excavation at Mont Terri underground laboratory

In the location of the Mont Terri laboratory (Fig. 1), overburden varies between 250 m and 320 m (it is estimated that overburden reached at least 1000 m in the past), and the closely spaced bedding dips at an angle of approximately 45°. Fig. 2 presents a picture of the Opalinus clay appearing on one side of a niche excavation, and shows clearly the strong layering of the material.

A significant number of measurements of the in situ stress have been made using different procedures (borehole slotter, undercoring, and hydraulic fracturing). They have been supplemented by geological observations and back-analysis of instrumented excavations. A synthesis of the information available is reported in Wermeille & Bossart (1999) and in Martin & Lanyon (2003). It appears that the most reliable data are provided by the undercoring method, yielding the estimation presented in Table 3. The following comments can be made.

(a) The major principal stress is subvertical, and corresponds approximately with the overburden weight.

(b) The magnitude of the intermediate principal stress obtained from the undercoring technique is consistent with the results from hydraulic fracture tests.

(c) The value of the minor principal stress is quite low, and is probably controlled by the presence of a deep valley to the SW of the laboratory. A low value of the minor

Table 3. Estimated in situ stress system

Principal stress	Orientation	Azimuth: degrees	Dip: degrees	Value: MPa
Major, σ_1	Subvertical	N210	70	6·0–7·0
Intermediate, σ_2	Subhorizontal	N320	10	4·0–5·0
Minor, σ_3	Subhorizontal	N50	20	2·0–3·0

principal stress is consistent with the small number of breakouts observed in vertical boreholes.

THEORETICAL FORMULATION
Coupled phenomena and THM formulation

When a thermal load is applied to a saturated porous material, hydraulic and mechanical effects ensue. A pore pressure rise will generally result from an increase of temperature due to the differential rate of expansion of the clay skeleton, solid phase and water. Indeed, in stiff materials, the increase in pore pressure may be very large and approach or even reach hydraulic fracture conditions. Material stresses and strains will be generated by both the change in temperature (thermal expansion) and the variation of pore pressure. Conversely, volume changes will affect the hydraulic regime via the well-understood consolidation process. Coupling towards the thermal problem also occurs via energy transport by water convection (hydraulic), and by energy loss associated with deformation processes (mechanical). Naturally, the strength and significance of each coupling relationship vary widely (Bai & Abousleiman, 1997; Zimmerman, 2000).

Observations made during the test did not indicate any desaturation of the clay; it is presumed that the material has remained saturated throughout. Coupled THM formulations for saturated porous media have been proposed by several authors (e.g. Booker & Savvidou, 1985; Katsube, 1988; Kurashige, 1989; Wang & Papamichos, 1999; Kanj & Abousleiman, 2005). They are usually applied to the derivation of analytical solutions and to the solution of simplified problems. Applications to field engineering cases are rare.

The theoretical THM formulation used herein is a particular case of the general formulation presented in Olivella et al. (1994) for saturated and unsaturated media. For space reasons, the formulation is only outlined in this section. Two phases are considered, solid (s) and liquid (l), corresponding to the two species mineral and water (w). Although, in this case, phases and species coincide, it is convenient to maintain the distinction for future generalisation.

The solution of a coupled THM problem requires the simultaneous solution of the following balance equations.

Balance of solid:

$$\frac{\partial}{\partial t}[\rho_s(1-\phi)] + \nabla \cdot (\mathbf{j}_s) = 0 \tag{1}$$

Balance of water mass:

$$\frac{\partial}{\partial t}(\rho_l \phi) + \nabla \cdot (\mathbf{j}_l) = f^w \tag{2}$$

Balance of internal energy:

$$\frac{\partial}{\partial t}[E_s\rho_s(1-\phi) + E_l\rho_l S_l\phi] + \nabla \cdot (\mathbf{i}_c + \mathbf{j}_{Es} + \mathbf{j}_{El}) = f^Q \tag{3}$$

Equilibrium:

$$\nabla \cdot \boldsymbol{\sigma} + \mathbf{b} = \mathbf{0} \tag{4}$$

where ϕ is porosity; ρ is density, $\mathbf{j}$ is total mass flux; $\mathbf{u}$ is the solid displacement vector; $\boldsymbol{\sigma}$ is the stress tensor; $\mathbf{b}$ is the body forces vector; E is the specific internal energy; $\mathbf{i}_c$ is the conductive heat flux; and $\mathbf{j}_E$ is the energy flux due to mass motion.

Using the definition of material derivative,

$$\frac{D_s(\bullet)}{Dt} = \frac{\partial(\bullet)}{\partial t} + \frac{d\mathbf{u}}{dt} \cdot \nabla(\bullet) \tag{5}$$

equation (1) becomes

$$\frac{D_s\phi}{Dt} = \frac{1}{\rho_s}\left[(1-\phi)\frac{D_s\rho_s}{Dt}\right] + (1-\phi)\nabla \cdot \frac{d\mathbf{u}}{dt} \tag{6}$$

Now, the solid mass balance (equation (6)) can be eliminated by introducing it into the water mass balance relationship (equation (2)). Making use of the material derivative definition again, the following equation results:

$$\phi\frac{D_s\rho_w}{Dt} + \frac{\rho_w}{\rho_s}(1-\phi)\frac{D_s\rho_s}{Dt} + \rho_w\nabla\frac{d\mathbf{u}}{dt} + \nabla(\rho_w q_l) = 0 \tag{7}$$

The first two derivatives of this expression can be developed further, taking into account the dependences of the liquid and solid densities on temperature, solid pressure and pore pressure, as follows:

$$\rho_w = \rho_{w0} \exp\left[\beta_w(p_l - p_{l0}) + b_wT\right] \tag{8}$$

$$\rho_s = \rho_{s0} \exp\left[\beta_s(p_s - p_{s0}) + 3b_s(T - T_{\text{ref}})\right] \tag{9}$$

where β_w and β_s are the water and solid compressibilities respectively, and b_w and b_s are the volumetric and linear thermal expansion coefficients for water and the solid grain respectively. Expanding the first two derivatives of equation (7) results in

$$[\phi b_w + (1-\phi)3b_s]\frac{D_sT}{Dt} + \phi\beta_w\frac{D_sp_w}{Dt}$$
$$+(1-\phi)\beta_s\frac{D_sp_s}{Dt} + \nabla\frac{d\mathbf{u}}{dt} + \frac{\nabla(\rho_w q_l)}{\rho_w} = 0 \tag{10}$$

Equation (10) contains the THM couplings that explain the variation of pore pressure when a temperature change is applied to the clay. The first term expresses the differential thermal expansion of the solid and liquid phases. The second and third terms are the volume changes of water and solid phase water associated with a pore pressure change; the fourth term is the volume change of the material skeleton (includes contributions from stresses, pore pressures and temperature); and the fifth term is the volume change associated with the flow of water in or out of the element considered. The pore pressure generated will be the result of the interplay of all these terms in each particular case.

The formulation must be completed with various constitutive laws that describe the various phenomena under consideration. They have been presented and discussed elsewhere (e.g. Gens & Olivella, 2000). The main ones correspond to the flow of heat by conduction, the advective flow of water, and the mechanical constitutive law for mechanical behaviour.

Heat conduction is governed by Fourier's law:

$$\mathbf{i}_c = -\lambda \nabla T \tag{11}$$

where λ is the coefficient of thermal conductivity.

Water flow is controlled by Darcy's law:

$$q_l = -K_l(\nabla p_l - \rho_w g) = -\frac{k}{\mu_l}(\nabla p_l - \rho_w g) \tag{12}$$

where K_l is the liquid permeability and k is the intrinsic permeability. Liquid permeability depends on temperature through the variation of water viscosity, and intrinsic permeability depends on porosity. In non-isothermal problems it is more convenient to characterise the hydraulic conductivity of a material using the concept of intrinsic permeability.

Composite mechanical constitutive law

A special mechanical constitutive law is adopted for the description of the stress–strain behaviour of the stiff sedimentary clay. It has been presented in Vaunat & Gens (2003), but a brief description is given here for completeness. It considers the material as a composite, accounting separately for the clay matrix and the bonding (Fig. 3). For simplicity, the equations are expressed here in triaxial space only.

When a load is externally applied to the medium, part of the stresses will be carried by the bonds and part by the matrix. The two materials will then experience different local values of stresses and strains. These values are constrained by the condition that local strains must be compatible with externally applied deformations, by the stress–strain relationships of the matrix and the bonds, and by the fact that local stresses must be in equilibrium with the external load. The model must therefore include a constitutive model for the matrix, a constitutive model for the bonds, and a stress-partitioning criterion to specify the way in which the applied stresses are shared.

Because of the characteristics of the test analysed, matrix strength does not play a significant role, and matrix behaviour is assumed to be elastic. Non-linearity arises from the behaviour of the bonds, defined by means of a damage model that provides a relationship between bond stresses and bond strains. More specifically, the damage model established by Carol et al. (2001) has been selected, in which a logarithmic damage measure is proposed, as follows:

$$L = \ln\left(\frac{1}{1-D}\right) \tag{13}$$

Equations defining this law are

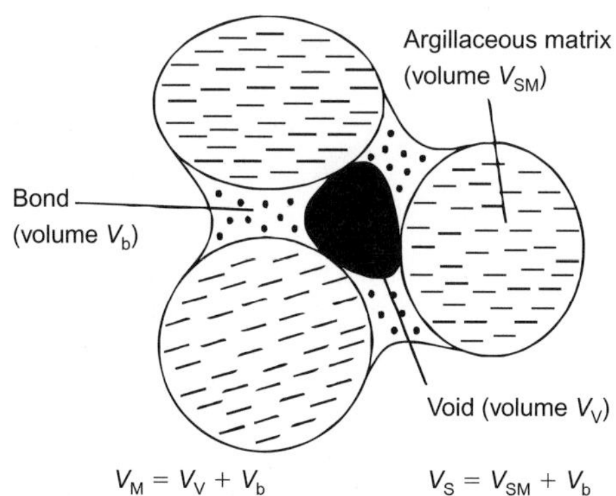

Fig. 3. Conceptual scheme underlying the constitutive law. Material is composed of matrix and bonding

$$p_b = (1-D)K_{b0}\varepsilon_{vb} = e^{-L}K_{b0}\varepsilon_{vb} \tag{14}$$

$$q_b = (1-D)G_{b0}\varepsilon_{qb} = e^{-L}G_{b0}\varepsilon_{qb} \tag{15}$$

where p_b and q_b are the mean and deviatoric bond stress, and ε_{vb} and ε_{qb} are the volumetric and shear strain (triaxial space).

D is a measure of damage or fissuring of the material, and is equal to the ratio of bond fissures over the whole area of bonds. Fissures are assumed to have null stiffness, and bond material between the fissures is considered as linear elastic, with bulk and shear moduli K_{b0} and G_{b0}. When $D = 0$ the material is intact, and bond stiffness is determined by K_{b0} and G_{b0}. As D increases, fissures develop and material stiffness reduces progressively. When $D = 1$, no more resisting area exists inside the bonding, and bond stiffness is equal to 0. In this modelling framework, bond response is totally determined if K_{b0}, G_{b0} and the evolution of D with load are known. According to Carol et al.'s (2001) proposal, change of D is linked to the energy increment input to the bonds du_b (equal to $(p_b - p_{b0})d\varepsilon_{vb} + (q_b - q_{b0})d\varepsilon_{qb}$ in triaxial conditions). The following relationship has been used:

$$r(L) = r_0 e^{r_1 L} = u_b \tag{16}$$

The current bond damage locus is defined in the stress–strain space as a threshold of equal energy r, corresponding to the maximum energy input to the bond during its history. This condition draws an ellipse in p_b–q_b space. For a stress state moving inside the ellipse, no further damage develops. When the ellipse is reached by the current stress state, damage occurs.

Any load applied to an element of cemented material after the time of bond deposition will distribute itself between the soil matrix and the bonding according to a ratio that depends on the geometric arrangement of both components. Cordebois & Sidoroff (1982) proposed using the energy equivalence principle that establishes the equality between the energy of the composite material and the sum of energies for all components. In this case, this principle leads to the expressions

$$p = p_M(1 + \chi_v) + \chi_v(p_b - p_{b0}) \tag{17}$$

$$q = q_M(1 + \chi_q) + \chi_q(q_b - q_{b0}) \tag{18}$$

where p_M and q_M are the clay matrix stresses. χ_v and χ_q define the part of load (p, q) carried respectively by bonds and matrix. With the assumption that the strains prevailing in the unfissured part of the bonds are equal to $\sqrt{1-D}$ times the bond strains (ε_{vb}, ε_{qb}), χ_v and χ_q can be rewritten as

$$\chi_v = \chi_q = \chi = \sqrt{1-D}\chi_0 \tag{19}$$

where χ_0 is a coefficient related to initial bonding intensity.

According to equation (19), χ_v and χ_q evolve from χ_0 to 0 during the process of bond damage. This mechanism is accompanied by a destructuration of the material and a progressive transfer of load from bonds to clay matrix. The damage parameter of the bonding component provides a useful variable for relating changes of permeability to increasing degradation of the material. In this case this is achieved by assuming an increase of porosity with damage in accordance with the expression

$$\phi' = \phi + c_b\sqrt{D} \tag{20}$$

DESCRIPTION OF THE IN SITU TEST

The in situ heating test (HE-D) is located in the shaly facies of Opalinus clay, which contains a higher proportion of clay minerals. To perform the experiment, a niche was excavated from the main laboratory tunnel, from which a 30 cm diameter borehole (D0) has been drilled horizontally with a total length of 14 m. In the section close to the end of the borehole, two heaters have been installed. The heaters are 2 m long, and were pressurised to 1 MPa to ensure a good contact with the clay. The separation between heaters is 0·8 m. In addition, various auxiliary boreholes have been constructed to install a variety of instruments for monitoring the test. Fig. 4 shows a top view of the test area. A horizontal test layout was selected in order to have a largely uniform lithology.

Temperatures were measured along two boreholes (D1 and D2) drilled from the niche HE-D. However, perhaps the most relevant observations were those combining measurements of temperatures and pore pressures at the same point in order to relate the two variables directly. This was achieved in borehole D3 (drilled parallel to the heater borehole) and in a series of small-diameter boreholes (D7 to D17) drilled from the MI niche. The pore pressure measurements of sensors located below the main borehole were quite successful, but the pore pressure probes located above the main borehole exhibited a rather slow response, attributed to difficulties encountered in de-airing the sensor area. Finally, sliding micrometer tubing was installed in boreholes D4 and D5 to measure incremental deformations at 1 m intervals. Special care was taken to ensure accuracy in the direction and length of the instrumentation boreholes to guarantee the correct location of the sensors. All instruments were in place before the drilling of the main borehole containing the heaters. In this way hydro-mechanical effects during excavation could also be recorded.

Approximately one month after installation and pressurisation, the heaters were switched on with a total power of 650 W (325 W per heater). The heaters were then left under constant power for 90 days. Afterwards the power was increased threefold, to 1950 W (975 W per heater), and maintained at that level for a further 248 days. At the end of this second heating stage, the heaters were switched off and the clay was allowed to cool down. Temperatures, pore pressures and deformations were measured throughout. In Fig. 5 the position of the main temperature and pore pressure sensors with respect to heater axis are indicated, together with the main orientation of the bedding planes. Heater 2 became depressurised because of a leak 109 days after the start of the heating. This event did not noticeably affect the development of the test.

NUMERICAL SIMULATION AND INTERPRETATION OF TEST RESULTS
Features of analysis and clay properties

Interpretation of the patterns of behaviour exhibited by the test observations will be aided by the results of a finite

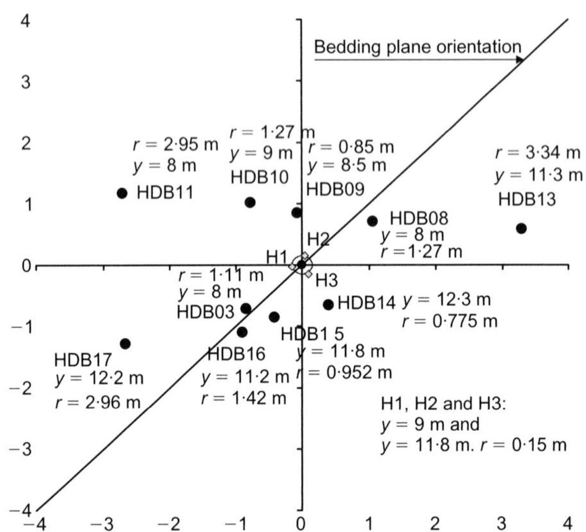

Fig. 5. Location of main observation points in HE-D in situ test. Coordinate r is distance to axis of main borehole D0 containing the heaters. Coordinate y represents distance to HE-D niche. For reference, mid-point between the two heaters lies at coordinate $y = 10·4$ m; centres of two heaters lie at values of y of 9 m and 11·8 m respectively

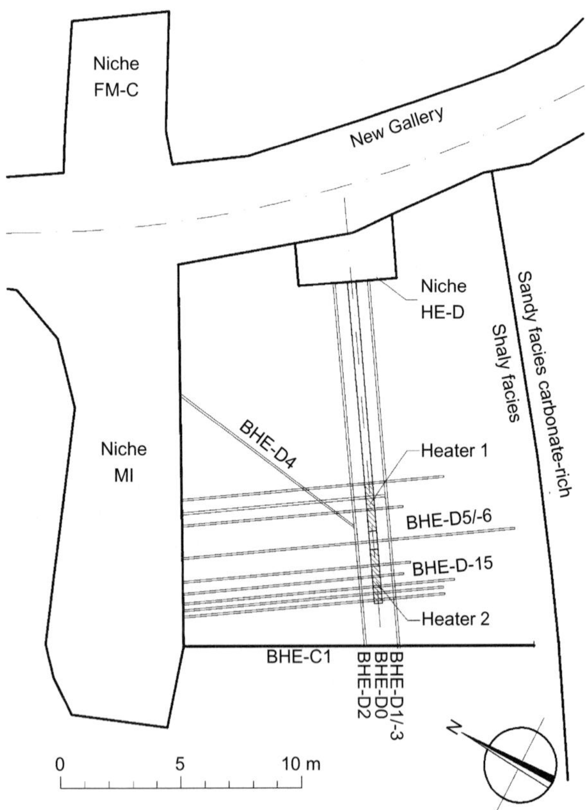

Fig. 4. Schematic layout of in situ test HE-D. Locations of observation boreholes drilled from niches HE-D and MI are shown

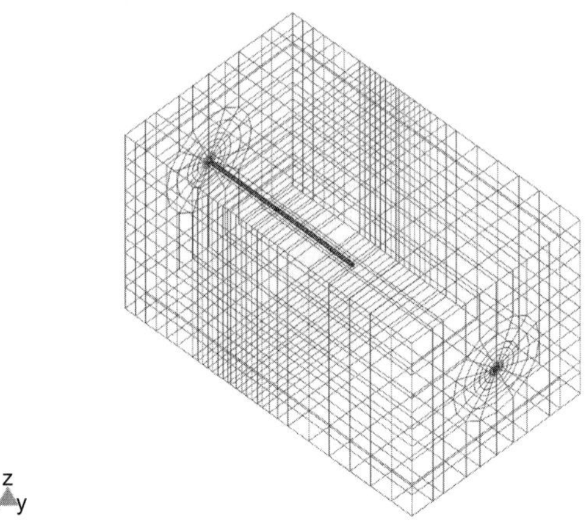

Fig. 6. Finite element mesh three-dimensional analysis. Dimensions are 16 m × 16 m × 28 m. Computation domain centred on axis of main borehole

element simulation of the experiment. A three-dimensional coupled THM analysis has been performed in order to incorporate the anisotropic properties of the clay and the anisotropy of the in situ stress state. The domain modelled and the finite element mesh used are depicted in Fig. 6. The dimensions are 16 m × 16 m × 28 m and the computation domain is centred on the axis of the main borehole. The distance to the boundary is 8 m, which corresponds to the distance from the borehole axis to the niche MI. Different boundary conditions are selected for the two lateral surfaces. A constant stress boundary is assumed on the surface that corresponds to the cavity MI (thereby allowing some lateral displacement), whereas a condition of constrained lateral displacements is adopted for the opposite lateral boundary.

The initial water pressure was set to the field-measured value of 0·9 MPa, lower than the value typical of the massif because of the influence of the various tunnels and niches present in the area. Values of 7 MPa, 5 MPa and 3 MPa as major, intermediate and minor initial principal stresses were assigned to the domain, consistent with the information in Table 3. The pore pressures on the borehole boundary are set to atmospheric. Initial temperature is 15°C. The various stages of the experiment were mimicked by the various phases of the numerical analyses. They are listed in Table 4. The start of heating is adopted as the origin of the time-scale.

Because the aim of the analysis was interpretation and not prediction, some of the clay parameters that were deemed especially important were determined from field observations. For instance, a proper interpretation of the thermo-hydraulic coupling requires a good representation of the thermal field throughout the domain. To achieve this, it is necessary to know, as precisely as possible, the value of the thermal conductivity of the clay. A three-dimensional back-

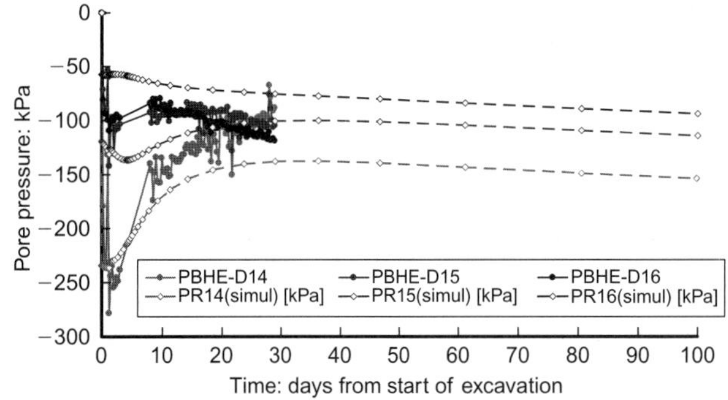

Fig. 7. Evolution of pore pressures during excavation of main boreholes: observed values and computed results using back-calculated parameters

Table 4. Phases of the analyses

Phase	Start: day	End: day	Duration: days	Activity
1	−29	−27·75	1·25	Excavation
2	−27·75	−23·75	4	Placement of heaters
3	−23·75	0	23·75	Heaters pressurised (1 MPa)
4	0	91	91	First heating phase (325 W/heater)
5	91	339	248	Second heating phase (975 W/heater)
6	339	518	179	Cooling phase

Note: On day 109 the heater pressure is set to zero to simulate an accidental pressure loss of the heaters that actually occurred during the in situ experiment.

Table 5. Clay properties adopted in the 3D THM analysis

	Parameter	Orientation*	Value	
Mechanical properties	Young's modulus, E: MPa	Parallel	9300	Back-analysis of borehole excavation. Anisotropic stiffness ratio from Bock (2001)
		Perpendicular	5800	
	Poisson's ratio, ν	–	0·295	Reference value
H properties	Intrinsic permeability, k_l	–	$5 \cdot 10-m^2$	Back-analysis from borehole excavation
	Biot coefficient, b	–	0·6	Reference value
TM properties	Linear thermal expansion of solid, α_T: K^{-1}	Parallel	$1\cdot7 \times 10^{-5}$	Value and anisotropic ratio from Auvray (2004)
		Perpendicular	$1\cdot1 \times 10^{-5}$	
	Linear thermal expansion of the solid grain, b_s: K^{-1}	–	$1\cdot4 \times 10^{-5}$	Auvray (2004)
T properties	Thermal conductivity, λ: W/(m K)	Parallel	2·8	Anisotropic back-analysis of thermal field
		Perpendicular	1·6	
	Specific heat capacity of the solid, c_s: J/(kg K)	–	840	Auvray (2004)
Petrophysical properties	Solid compressibility, β_s: MPa^{-1}	–	$2\cdot5 \times 10^{-5}$	From mineral composition
	Specific weight, ρ_s: kg/m³	–	2·7	Reference value
	Porosity, ϕ	–	0·137	Martin & Lanyon (2003)

* Orientation to bedding.

analysis exercise has been performed to find the value of thermal conductivity that achieves the closest agreement with measured temperatures, according to the least-squares criterion. This has required the performance of several 3D purely thermal analyses. The following values of thermal conductivity were obtained: 2·8 W/(mK) in the direction parallel to the bedding, and 1·6 W/(mK) in the direction perpendicular to the bedding.

The stiffness (Young's modulus) and permeability of the clay have been obtained from a back-calculation of the evolution of pore pressures measured in the clay when the main borehole was excavated. Because of the large number of coupled HM analyses to be performed, an axisymmetric model was used for this purpose. Therefore only averaged isotropic parameters could be determined in this way. The parameter values obtained were $E = 7500$ MPa and intrinsic permeability $k = 5 \times 10^{-20}$ m^2 (approximately 5×10^{-13} m/s for water at 15°C temperature). The agreement reached can be seen in Fig. 7. The pore pressure peak observed is quite sensitive to the value of E, whereas the rate of decay depends mainly on permeability. Thus the two parameters can be identified in a relatively independent way.

A list of the basic clay parameters used in the analyses is given in Table 5, with an indication of the source used to derive some of the values. It can be observed that many of the properties exhibit anisotropy. No clear evidence of anisotropy of permeability has been obtained (Bock, 2001; Croisé et al., 2004; Coll, 2005), so the single back-calculated value has been used. In contrast, an anisotropic Young's modulus has been adopted, consistent with the average back-calculated value but keeping the measured

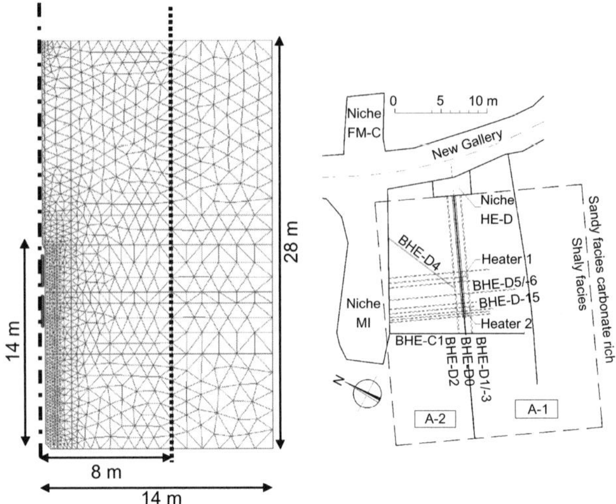

Fig. 8. Domain modelled and finite element mesh used in axisymmetric analyses A1 and A2

ratio of stiffness normal and parallel to the bedding direction.

Model parameters for the mechanical constitutive model (Table 6) have been based mainly on the experimental database of the material. Some mineral physical data have also been used to characterise bond stiffness. Finally, the combination of the Young's moduli of bonding and matrix has been selected in such a way that the global values of the Young's moduli derived from the observations during the

Table 6. Parameters of the composite mechanical constitutive model used in the 3D THM analysis

Parameter	Orientation*	Value
Bonding Young's modulus, E_{b0}: MPa	Parallel	18 150
	Perpendicular	11 315
Matrix Young's modulus, E_{m}: MPa	Parallel	3720
	Perpendicular	2320
Bonding Poisson's ratio, ν_b	–	0·295
Matrix Poisson's ratio, ν_m	–	0·295
Structuration parameter, χ	–	1·2
Bonding concentration, c_b	–	0·3
Damage threshold, r_0: MPa.m/m	–	1×10^{-3}
Damage evolution, r_1	–	1·8

* Orientation to bedding.

Table 7. Clay properties used in the axisymmetric analyses; the values used in the 3D simulation are presented for comparison

	Parameter	Axisymmetric analysis	3d analysis
Mechanical properties	Young's modulus, E: Mpa	7500	9300// 5800⊥
	Bonding Young's modulus, E_{b0}: Mpa	14 640 Mpa	18 150// 11 315⊥
	Matrix Young's modulus, E_{m}: Mpa	3000	3720// 2320⊥
TM properties	Linear thermal expansion of solid, α_T: K^{-1}	$1·4 \times 10^{-5}$	$1·72 \times 10^{-5}$// $1·1 \times 10^{-5}$⊥
T properties	Thermal conductivity, λ: W/(m K)	2·2	2·8// 1·6⊥

// Parallel to bedding.
⊥ Perpendicular to bedding.

excavation of the borehole are recovered. Nevertheless, there is considerable uncertainty in the values of some of the parameters because of lack of sufficient laboratory or field evidence. The parameter r_0, which controls the damage threshold, and the parameter r_1, which defines the damage evolution, are examples of this. The values adopted have been derived by analogy from tests performed in another stiff sedimentary clay: the Callovo-Oxfordian clay from the Meuse/Haute Marne laboratory.

For reference, two axisymmetric analyses have been performed and the results compared with those of the 3D simulation. The domain modelled and the finite element meshes adopted are depicted in Fig. 8. In the analysis (A1) the domain has a length of 28 m and a width of 14 m. A second analysis (A2) has been run with a width of 8 m only, to examine the possible effect of the presence of Niche MI 8 m away from the main borehole. Naturally the symmetry axis coincides with the axis of the main borehole D0 containing the pressurised heaters. The in situ stress has been represented by a normal stress of 4·2 MPa applied to the boundary. This value is an approximate average of the anisotropic stress state values. Material parameters of the axisymmetric analyses are the same as in the 3D computation, except for the properties where anisotropy has been assumed (Table 7). A preliminary axisymmetric analysis of this test using a simplified model and different material parameters has been reported in Gens et al. (2006).

Thermal results

The observed evolution of temperatures at the interface between heaters and clay is shown in Fig. 9 together with the computed results from the 3D analysis. The two heating stages and the cooling phase can be clearly identified. Maximum temperatures reach values just above 100°C at the end

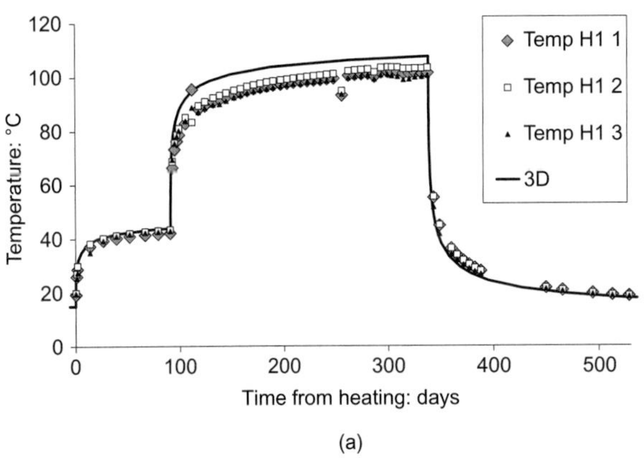

(a)

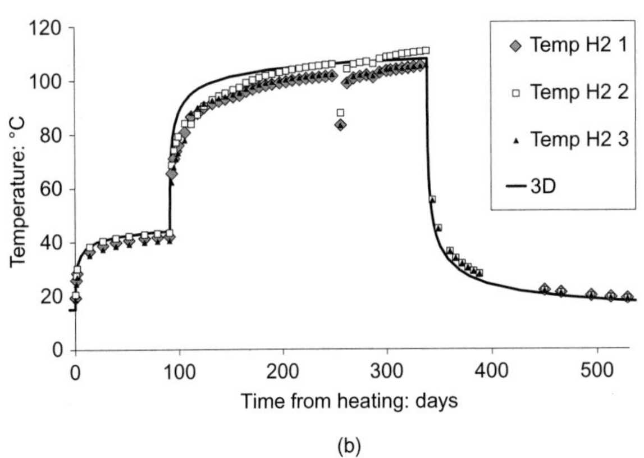

(b)

Fig. 9. Evolution of temperatures on heater/clay interface surface throughout the test: (a) Heater H1; (b) Heater H2

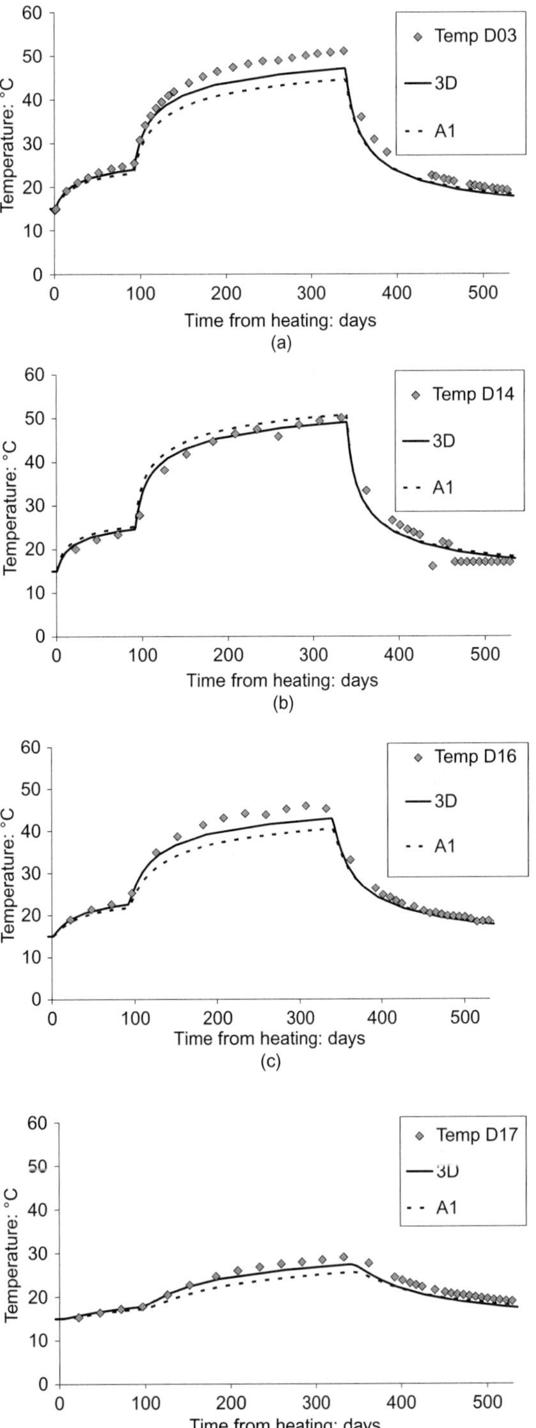

(a)

(b)

(c)

(d)

Fig. 10. Evolution of temperatures at various points in the clay: observed and computed results

of the second heating stage. There were three sensors in each heater, set at different orientations with respect to bedding. However, anisotropic effects are hardly noticeable; very similar temperatures are recorded at all three points. The 3D analysis captures the temperature variation well; of course, thermal conductivity has been back-calculated, and conduction is the dominant form of heat transport. In any case, the whole evolution of temperatures is correctly reproduced. In this instance, the axisymmetric results coincide exactly with the 3D computations and are not plotted.

Anisotropic effects are more noticeable in the temperatures measured in the clay mass (Fig. 10). Take, for instance,

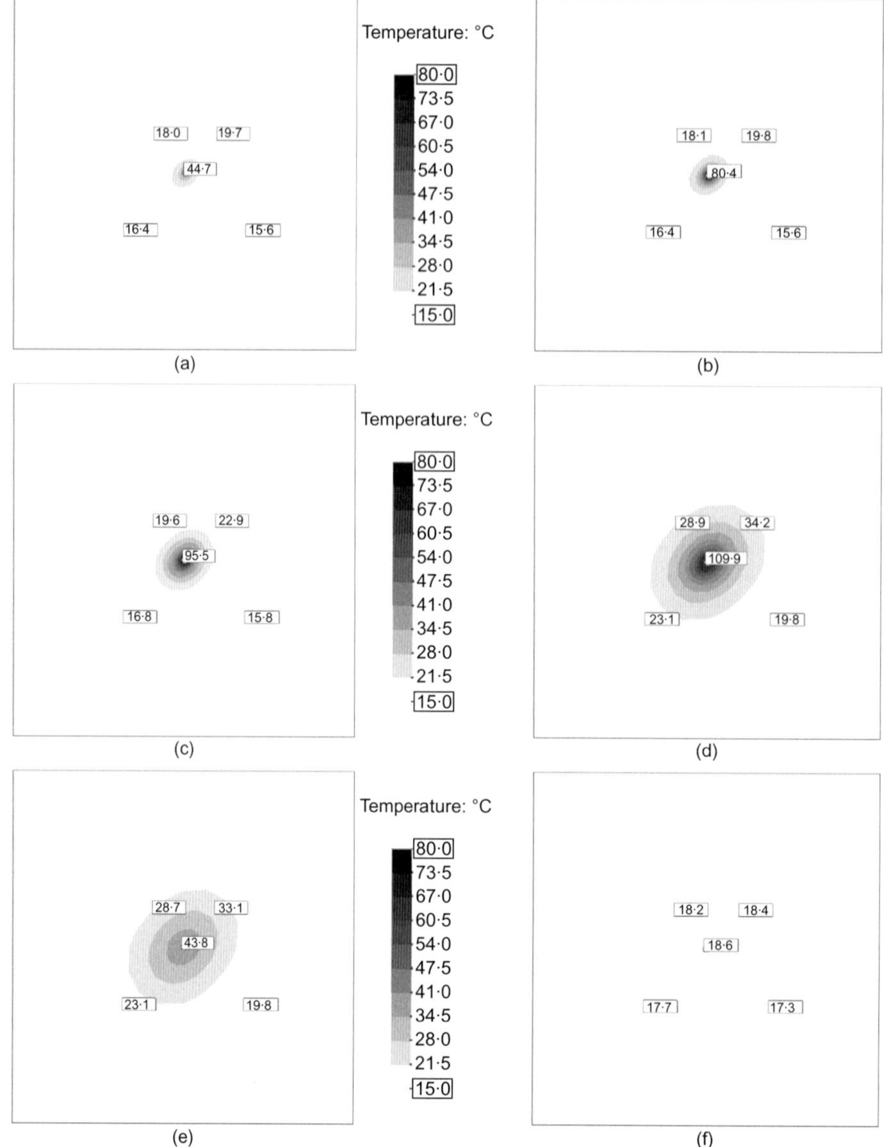

Fig. 11. Computed contours of equal temperature (°C) in a cross-section across Heater 2
for: (a) 90 days; (b) 93 days; (c) 110 days; (d) 338 days; (e) 346 days; (f) 501 days

points D03 and D14: the first is oriented parallel to the bedding with respect to the main borehole, and the second is perpendicular to the bedding. They reach a similar temperature, about 50°C, although they are located at different distances from the heater axis. It can be seen that the results of the 3D analysis (which include thermal conductivity anisotropy) are closer to the observed temperatures, although the differences from the axisymmetric computations are not large. Only axisymmetric analysis A1 is plotted, because the results of analysis A2 are the same.

More information from the analyses is given in Fig. 11 in terms of contours in a section across Heater 2 at various relevant stages. The following times have been selected for plotting: 90 days (end of the first heating stage); 93 days (just after the start of the second stage of heating); 110 days (time at which maximum pore pressures are achieved); 338 days (at the end of the second stage of heating); 346 days (during the cooling phase); and 501 days (towards the end of cooling phase). The anisotropic temperature distribution is apparent, with higher temperatures being reached in the direction of the bedding planes. Similar information is presented in Fig. 12 in a three-dimensional view.

The distributions of temperatures on the section across Heater 2 are plotted in Fig. 13 for the same times as selected before. The progressive rise of temperature can be readily observed, as well as the subsequent temperature reduction during the cooling period. It is interesting to note that the region with a significant temperature increase appears to be limited to a radius of about 5 m around the main borehole.

Pore pressures

As expected, temperature increases triggered a large upsurge of pore pressures. A typical set of observations is presented in Fig. 14, where the temperature and pore pressure recorded in borehole D03 are plotted. It is noticeable that the pore pressure responds immediately to the rise of temperature, and the larger the change of temperature is, the stronger is the response. The maximum pore pressure increase in this case is 2·5 MPa, a significant magnitude. It can also be observed that the pore pressure evolution does not exactly match that of temperature: at some time the pore pressure stops rising, even though temperatures increase

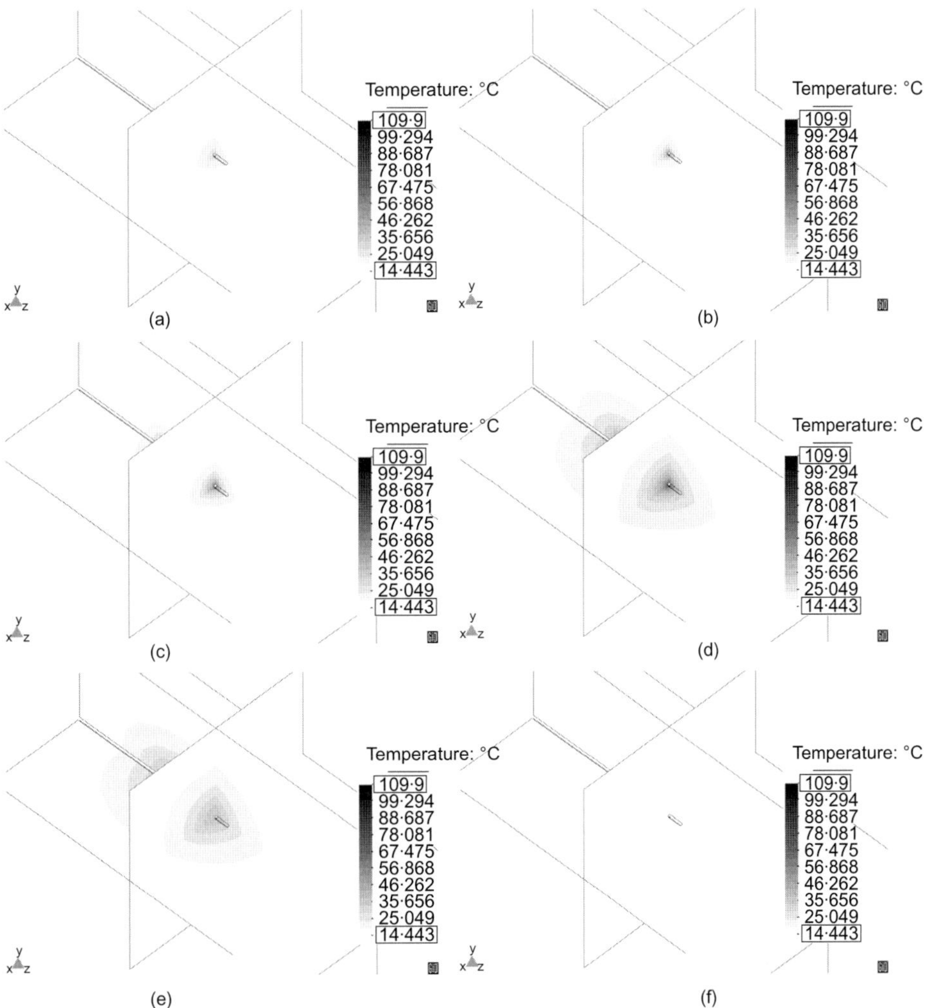

Fig. 12. Three-dimensional view of computed contours of equal temperature (°C) for: (a) 90 days; (b) 93 days; (c) 110 days; (d) 338 days; (e) 346 days; (f) 501 days

further. Obviously, dissipation by liquid flow overcomes the thermal effect. This phenomenon is especially noticeable in the second stage of heating.

A comparison between the results of the 3D analysis and observations, in terms of pore pressure increases, for various points of the clay mass is presented in Fig. 15. The results of the two axisymmetric analyses have also been added for reference. It is apparent that the maximum pore pressure increase is reasonably well captured with the formulation and parameters used. The evolution of pore pressure is also reproduced well, with the noticeable exception of borehole D17, located further away from the heaters. The pore pressure rise in the first heating stage seems to be sharper in the computations than in the observations. This could be due to a slow response of the pore pressure sensors; it is significant that in the second stage the observed and computed rates of pore pressure increase appear to be the same.

In Fig. 15, the time at which the maximum pore pressure increase is calculated is indicated for each measurement point. The time becomes larger as the distance to the main borehole increases. This results from the combined effect of the movement of temperature rise outwards and of pore pressure dissipation from the inner zones. This evolution can easily be observed in Fig. 16, where two pore pressure distributions on a section perpendicular to the main borehole and located between the two heaters are plotted for various significant times. It can be seen that the cooling associated

with the switching off of the heaters induces a pore pressure reduction, the counterpart of the pore pressure increase during heating. The two distributions plotted correspond to two orthogonal directions, one along the bedding plane and the other normal to the bedding. Some differences between the two sets of distributions can be observed, especially for shorter times. These differences do not arise from hydraulic effects (permeability has been assumed isotropic) but from the differential temperature rise caused by thermal conductivity anisotropy.

Anisotropy of generated pore pressures can also be observed in the contours of equal pore pressure presented in Figs 17 and 18. The displacement of the maximum pore pressure away from the main borehole as time goes on is also readily apparent. In addition, it is interesting to note (Figs 15, 16, 17(f) and 18(f)) that, at the end of the experiment, pore pressures have not yet returned to equilibrium levels. Close to the heater, pore pressures are depressed owing to the reduction brought about by cooling. In contrast, some positive excess pore pressures prevail away from the heater, where temperature changes are more gradual.

Displacements

The combined effects of temperature changes and pore pressure generation and dissipation cause mechanical effects.

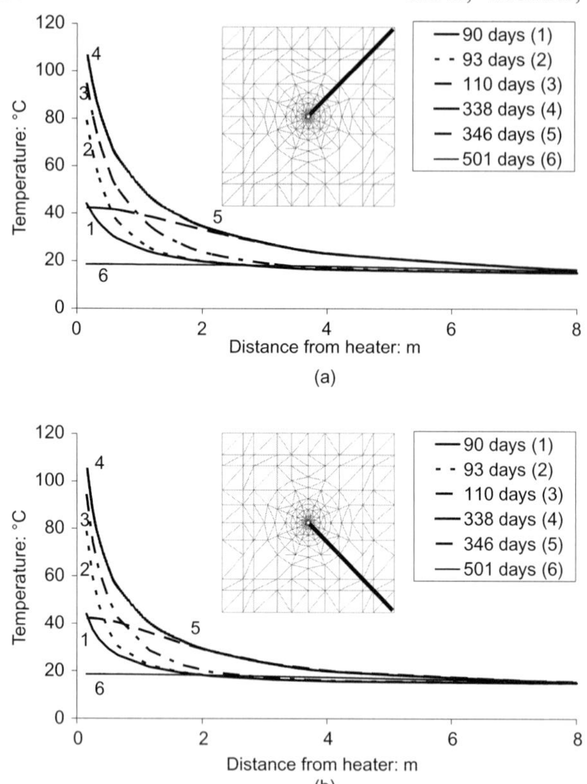

Fig. 13. Computed temperature distributions at various times on section across Heater 2: (a) bedding plane direction; (b) perpendicular to bedding plane direction

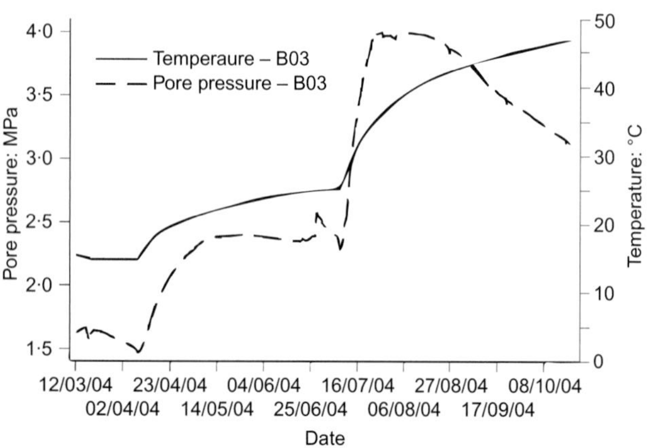

Fig. 14. Evolution of temperature and pore pressure in borehole D3 during in situ test

Those effects were monitored during the performance of the in situ test by measuring relative displacements in boreholes D4 and D5 using a sliding micrometer (Fig. 19). Fig. 20 shows the movements measured along borehole D5 at various times. This borehole was drilled from Niche MI at a direction approximately normal to the main borehole. It can be observed that, in the region around the heaters, extension deformations occur, but they become compressive strains at locations further away from the heater. The dilation of the clay close to the heater, driven by thermal expansion, is partially taken up by the compression of the outer zones.

The evolution of strains of some selected intervals is shown in Figs 21 and 22 (boreholes D4 and D5 respectively) together with the computed results from the 3D analysis. Again, the results from the two axisymmetric computations

are added for comparison. A first observation is the very small magnitude of the strains (and hence displacements) measured and computed, consistent with the high stiffness exhibited by the clay. It can be seen that, in borehole intervals close to the heater (e.g. intervals 6–7, 7–8 and 8–9 of borehole D-5), dilatant strains are observed in response to the temperature increase; indeed, the two heating stages can be easily identified. In contrast, in borehole intervals away from the heater (e.g. intervals 11–12 and 12–13), a compressive strain is measured initially, as a result of the expansion of the inner regions of the test. When the thermal field extends with time, the material expands, and compressive strains become dilatant strains. In spite of some scatter of the observations (a consequence of the small magnitudes being measured), the numerical analyses satisfactorily capture the main patterns of behaviour.

The interplay between compressive and dilatant strains can also be examined by reference to Fig. 23, where the distributions of computed radial displacements at various times are presented. Displacement differences can be observed in the computed distributions when plotted parallel and perpendicular to the bedding direction resulting from the combined effect of the anisotropy considered for stiffness and thermal dilation properties. It can also be seen that radial displacements exhibit a peak. Clay expansion occurs before the maximum of displacements (which are always directed outwards), and the material contracts in the zone after the peak. The transition between the two strain regions migrates slowly outwards. By reference to Fig. 13, it can be seen that the region of dilatant strains coincides with the zone in which temperature increases are significant, and compressive strains prevail where the temperature rise is small or negligible. This phenomenon is also directly illustrated in Fig. 24, where the distributions of computed radial displacements and temperatures for a particular time and direction are plotted together.

Damage zone

The mechanical constitutive model used in the 3D analysis (and also in the axisymmetric analyses reported so far) incorporates the concept of a bonding subject to damage as a constituent part of the material. The onset and development of damage degrades the elastic parameters in accordance with equations (14) and (15). Although there is significant uncertainty regarding the parameters defining the damage threshold and evolution, it is of interest to investigate the extent to which, according to the model, the damage zone has spread during the in situ test. Fig. 25(a) shows the distributions, at various times, of the unloading elastic modulus (the main parameter affected by the degree of damage) along a horizontal direction in a section perpendicular to Heater 2. The model reproduces the possible damage of the clay due to thermal loading in the initial stages of heating. Afterwards, the damaged zone becomes larger during pore pressure dissipation because of the consequent increase of effective stress. During the cooling period, damage does not develop further.

During the dismantling of the in situ test, a borehole was drilled from the MI niche, perpendicular to the main borehole. The unconfined Young's moduli of a series of samples taken from this borehole at different distances from the heaters were determined by means of measurement of the compression wave velocity. The measured moduli have been added to the plot of Fig. 25(a). The values obtained appear to confirm the existence of a damaged zone close to the heaters but, perhaps, not extending back into the clay as much as shown by the computations at later times. The amount of independent experimental data, however, is not

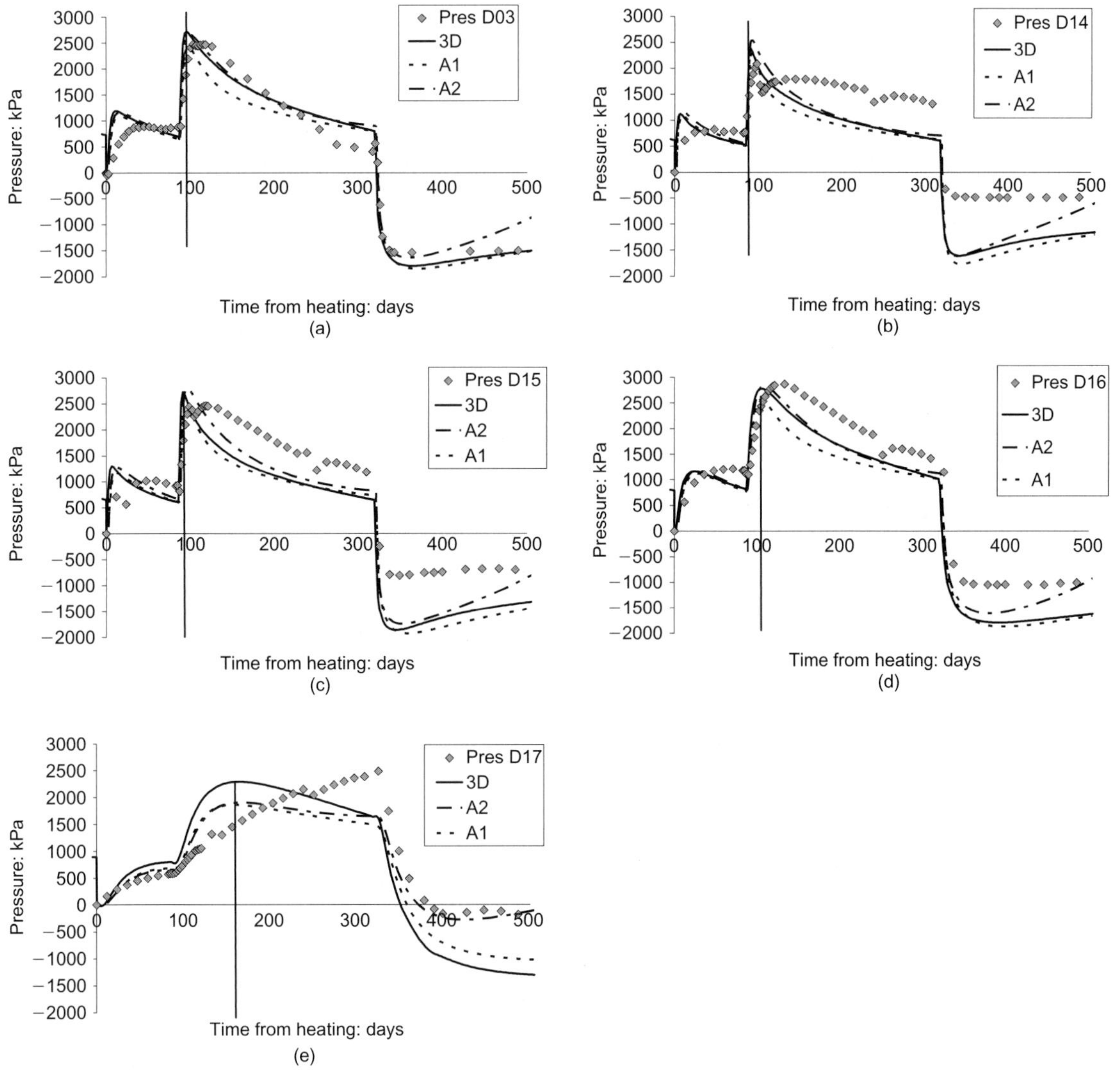

Fig. 15. Evolution of pore pressure increments at various points in the clay: observed and computed results

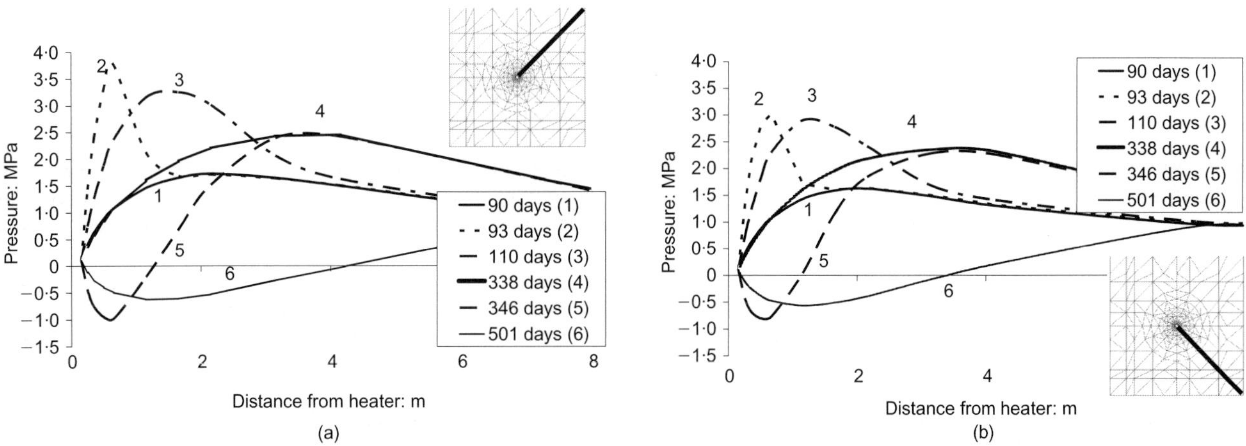

Fig. 16. Computed pore pressure increment distributions at various times on section across Heater 2: (a) bedding plane direction; (b) perpendicular to bedding plane direction

sufficient yet for drawing firm conclusions in this regard. Fig. 25(b) shows the distributions of permeability on the same section derived from the amount of damage computed in the analysis (equation (20)). It is very interesting to note that, in the laboratory tests on damaged Opalinus clay samples reported by Coll (2005), the values of measured

intrinsic permeability were in the range of 10^{-17} to 10^{-19} m^2, similar to that predicted by the analysis. The extent and evolution of the damaged zone obtained in the 3D calculations can be observed directly by plotting contours of the damage parameter L at different times (Figs 26 and 27). The samples extracted at the end of the experiment

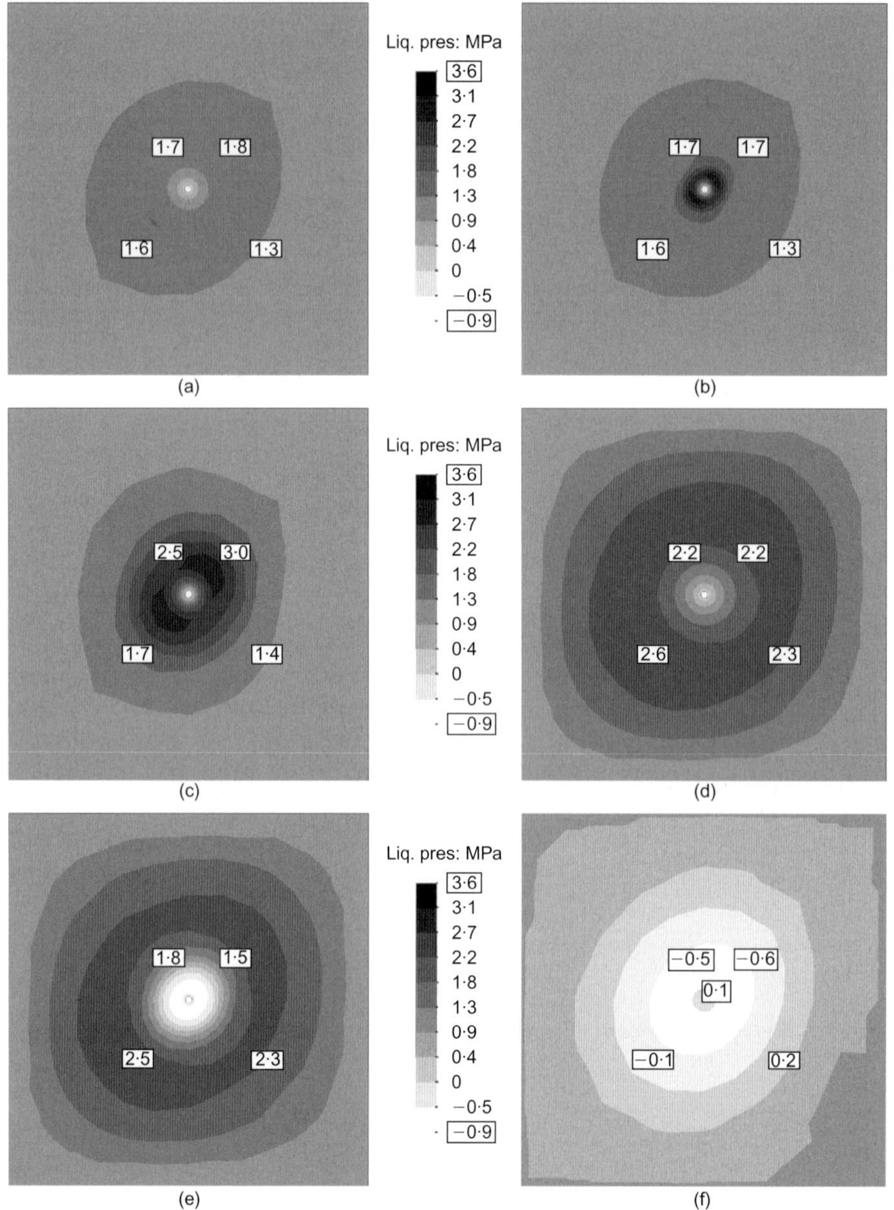

Fig. 17. Computed contours of equal pore pressure increase (MPa) in cross-section across
Heater 2 for: (a) 90 days; (b) 93 days; (c) 110 days; (d) 338 days; (e) 346 days; (f) 501 days

would correspond to the states of the damage zone shown in Figs 26(d) and 27(d).

SENSITIVITY ANALYSIS

A sensitivity analysis has been performed to examine the effect of various parameters and, in this way, improve the understanding of the system subject to thermo-hydro-mechanical perturbations. Because many fully coupled THM computations had to be performed, axisymmetry conditions have been accepted, and mechanical non-linearity has been avoided by using a linear mechanical constitutive model as the base case. For space reasons, the effect of variation of individual parameters will be demonstrated by reference to the pore pressure increment evolution in borehole D03 and strain variation of interval 8–9 of borehole D4, the closest to the heater. The thermal field is very similar in all analyses except when thermal conductivity changes are considered.

In the previous section, it has been shown that, although anisotropy effects are certainly noticeable, they are moderate

and do not significantly affect the overall behaviour of the experiment examined in this parametric study. Similarly, the chief effect of introducing damage in the constitutive law is a strain increase in the zone close to the heater as a result of stiffness degradation occurring on that region. But, again, the main features of THM behaviour remain largely unchanged. This can be checked in Fig. 28, where the results of analysis A1 (including damage in the constitutive law) and analysis A1-E (linear behaviour) are compared. Thus the use of analysis A1-E as base case is justified.

A selection of the sensitivity analyses performed will be presented, focusing on the most important or uncertain parameters: Young's modulus E; Biot coefficient b; solid compressibility β_s; thermal expansion coefficient of solid grains and skeleton b_s; thermal conductivity λ; and intrinsic water permeability k.

The value of *Young's modulus* naturally affects deformations, but also the magnitude of the pore pressure generation due to temperature changes. Analyses have been performed using values of E equal to 11 500 MPa and 2000 MPa, which go beyond the expected range of reference values (Table 1;

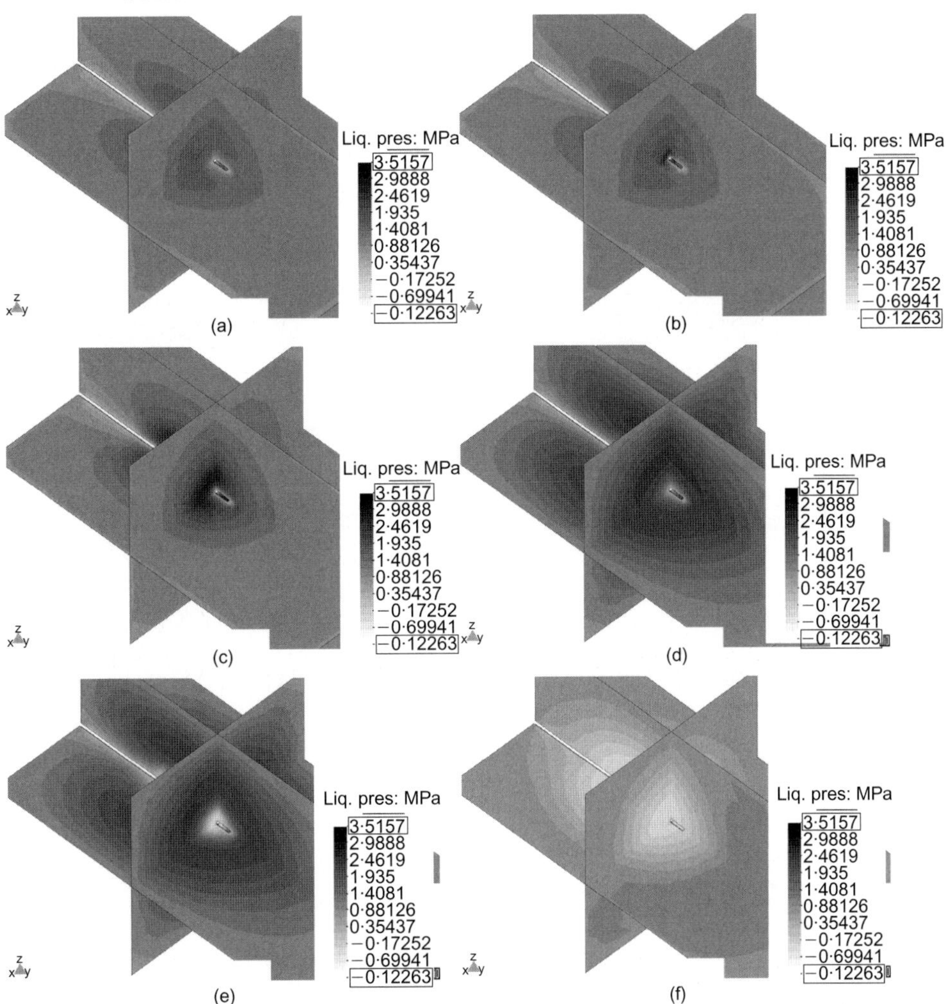

Fig. 18. Computed contours of equal pore pressure increase (MPa) for: (a) 90 days; (b) 93 days;
(c) 110 days; (d) 338 days; (e) 346 days; (f) 501 days

Wileveau, 2005). The results are shown in Fig. 29. As expected, a lower Young's modulus results in a reduced pore pressure generation and larger displacements, but the general pattern of results remains unchanged. Opposite effects are observed when using a larger value of Young's modulus.

As indicated above, there are major uncertainties regarding the actual value of *Biot's coefficient b* for Opalinus clay. Consequently, additional analyses have been run using values of b equal to 0·4 and 1. The latter case would, of course, correspond to the assumption of Terzaghi's definition for effective stress. As Fig. 30 shows, the effect of a variation of Biot's coefficient is noticeable but not large. This result also indicates that the major contributor to strains and displacements is thermal expansion.

Solid compressibility β_s had to be estimated from mineral composition and compressibility values for individual minerals obtained from the literature. To check on its possible influence on the results two further analyses have been carried out, one with a totally rigid solid mineral and a second one with a higher compressibility. As Fig. 31 indicates, the effects of the variation of this parameter are slight.

The *thermal expansion* of solid grains (b_s) and clay skeleton (α_T) is another uncertain parameter, as laboratory experimental results from different sources have provided a wide range of results (e.g. Gens, 2000; Auvray, 2004; Wileveau, 2005). No significant structural rearrangement is expected during limited thermal expansion, and therefore

both parameters are assumed equal. Thermal expansion properties obviously affect thermal strains, but they also influence pore pressure generation. Consequently, Fig. 32 shows a small effect on computed pore pressures but a very large effect on the computed strains. It is more evidence of the dominance of thermally generated strains and displacements.

The low permeability of the clay ensures that heat transport by advection is negligible, and therefore the only significant heat transport mechanism is heat conduction. Consequently, a variation in *thermal conductivity λ* must affect the temperature field directly. This is immediately apparent in the results shown in Figs 33(a) and 33(b). The temperature in the clay/heater interface varies significantly with the value of λ (Fig. 31(a)), the base case parameter giving a good estimation of the recorded temperature. The computed effects on pore pressure (Fig. 33(c)) and strains (Fig. 33(d)) are a consequence of the differences in temperature.

The mechanism underlying the hydraulic behaviour in the clay is a competition between the generation of pore pressures by the differential thermal expansion of liquid and solid and the dissipation of pore pressures, the rate of which is controlled mainly by the value of water permeability. Therefore variations of *intrinsic water permeability k* are bound to have a profound effect on test behaviour, especially concerning the hydraulic aspects. Fig. 34 shows the results of sensitivity analyses where the values of intrinsic per-

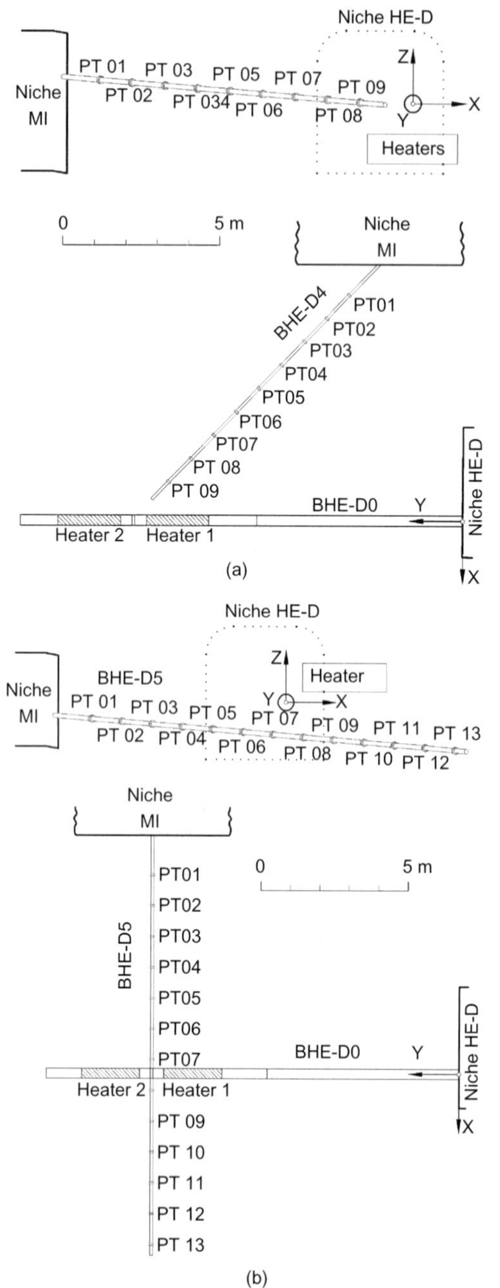

Fig. 19. Plan and elevation of (a) borehole D4 and (b) borehole D5 for displacement measurements

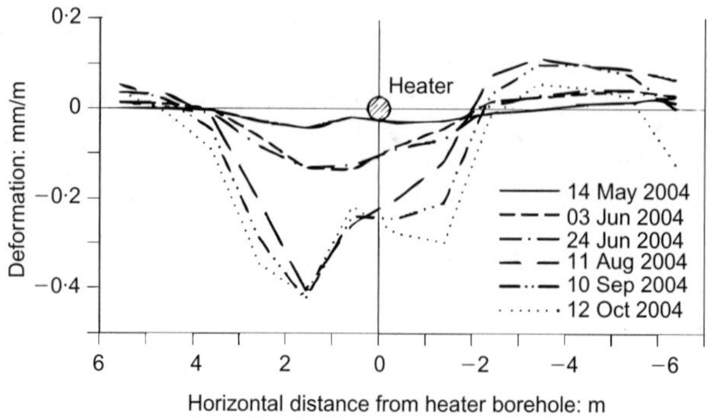

Fig. 20. Distributions of deformation measured at different times in borehole D5, drilled approximately perpendicular to main borehole

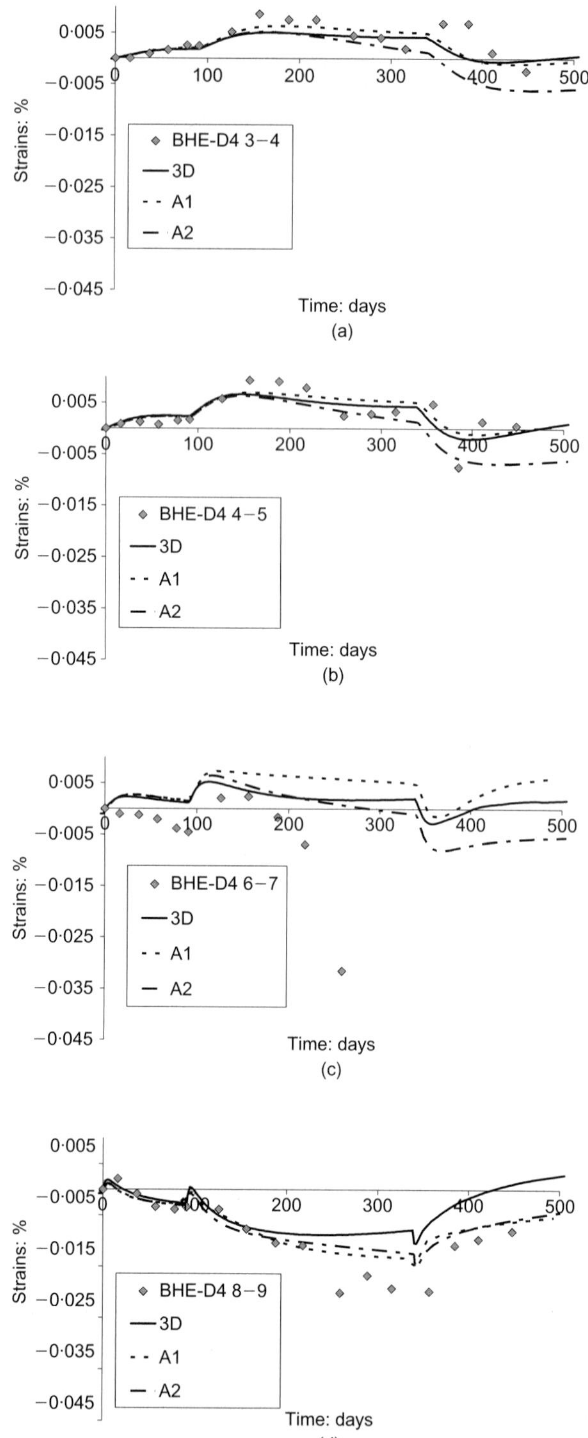

Fig. 21. Evolution of strain increments at various points in the clay: observed and computed results. Borehole D4

meability have been varied by an order of magnitude (from $10^{-19}\,\mathrm{m}^2$ to $10^{-20}\,\mathrm{m}^2$), staying within the range of reference values of Table 1. It is apparent that the effect on generated pore pressures is very large, and their subsequent dissipation also affects the predicted strains considerably. Because of the importance of this parameter in the global behaviour of the test, a further two analyses have been performed to represent practically undrained and drained conditions. To this end intrinsic permeabilities of $10^{-23}\,\mathrm{m}^2$ and $10^{-16}\,\mathrm{m}^2$ have been used, well beyond the range of measured permeabilities. The results are shown in Fig. 35. In the very low-

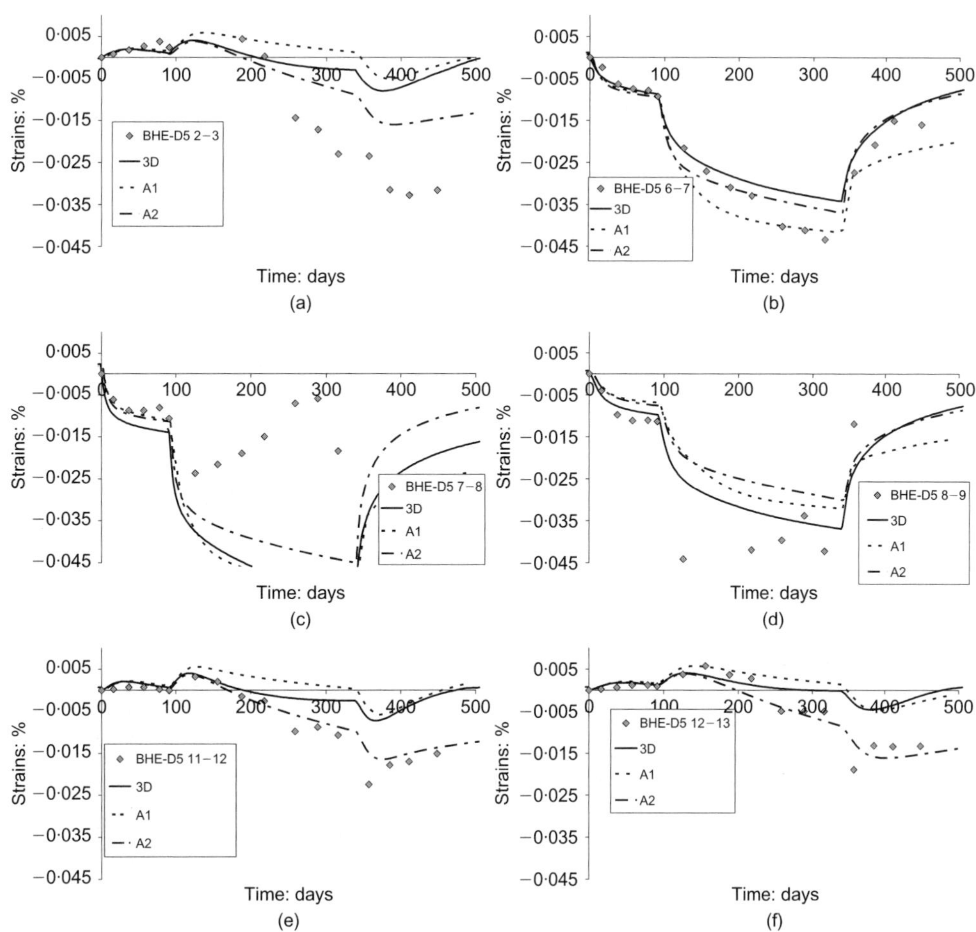

Fig. 22. Evolution of strain increments at various points in the clay: observed and computed results. Borehole D5

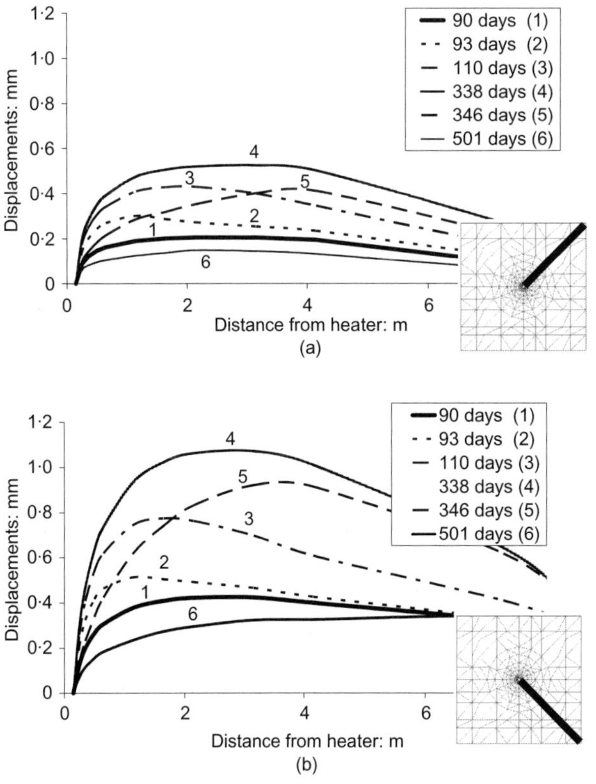

Fig. 23. Computed radial displacement distributions at various times on a section between the two heaters: (a) bedding plane direction; (b) perpendicular to bedding plane direction

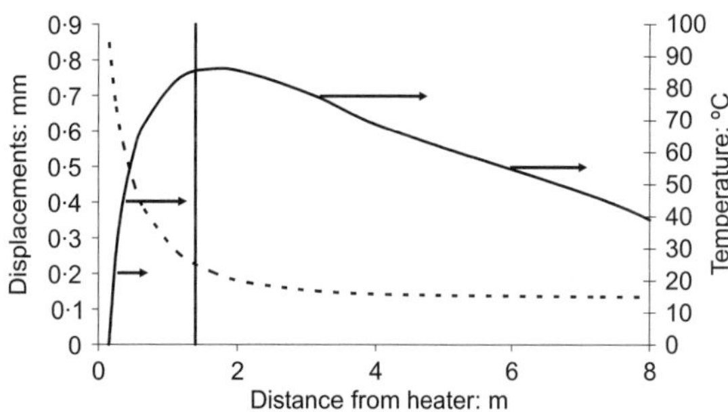

Fig. 24. Computed displacement and temperature distributions at 110 days in cross-section of Heater 2, in direction perpendicular to bedding

permeability case there is an extremely large rise of pore pressures, which follows the temperature evolution. Dissipation is practically negligible: that is, undrained conditions prevail. By contrast, in the very large-permeability case no pore pressure generation is observed, as pore pressure dissipation dominates throughout: that is, drained conditions occur. The different hydraulic conditions result in noticeable differences of strain development.

CONCLUSIONS

Opalinus clay is a stiff layered Mesozoic clay of marine origin. When saturated stiff clays are subjected to thermal loading, they may develop a strong pore pressure response.

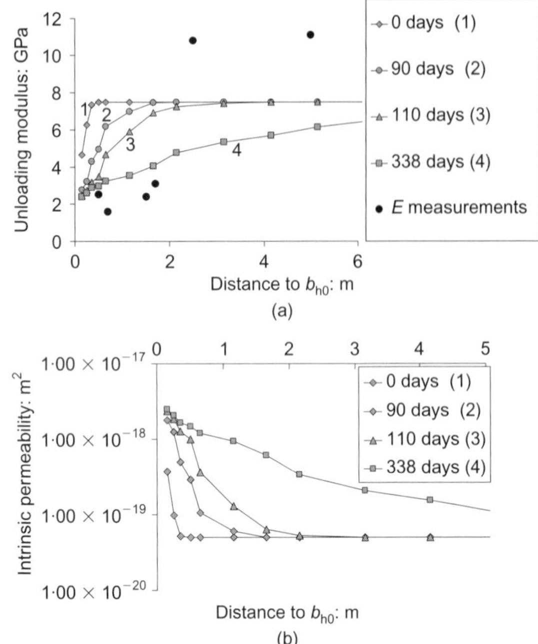

Fig. 25. (a) Distribution of unloading elastic modulus at various times on section perpendicular to Heater 2; elastic moduli measured on samples taken at end of test are indicated. (b) Distribution of computed permeability at various times on section perpendicular to Heater 2

In turn, the generated pore pressures will impact on subsequent thermo-hydro-mechanical behaviour. The performance of a heating test in the Mont Terri underground laboratory has provided the opportunity to observe in situ the development of the coupled THM behaviour in this type of material. *A priori*, the strongly bedded nature of the clay suggested that anisotropy effects could be significant.

Performance of three-dimensional coupled THM analysis has proved very useful in providing a structured framework for interpretation. A 3D simulation has allowed the incorporation of anisotropic material properties and anisotropic in situ stress state. By comparing the results of the 3D computations with those of axisymmetric analyses, it is possible to conclude that the anisotropic effects are certainly noticeable although not large. The mechanical behaviour of the material has been described by a constitutive model that explicitly considers bonding and its subsequent damage. It is of interest to note that the predicted damaged zone by thermal effects appears to be quite consistent with independent clay stiffness determinations performed after the test. Overall, the theoretical formulation adopted and the 3D analysis performed have been able to provide a satisfactory reproduction of in situ test observations, even from a quantitative point of view.

A sensitivity analysis has also been carried out to assist in the interpretation of the test and to achieve an enhanced understanding of the phenomena involved. The intensity of cross-coupling between the different aspects of the problem

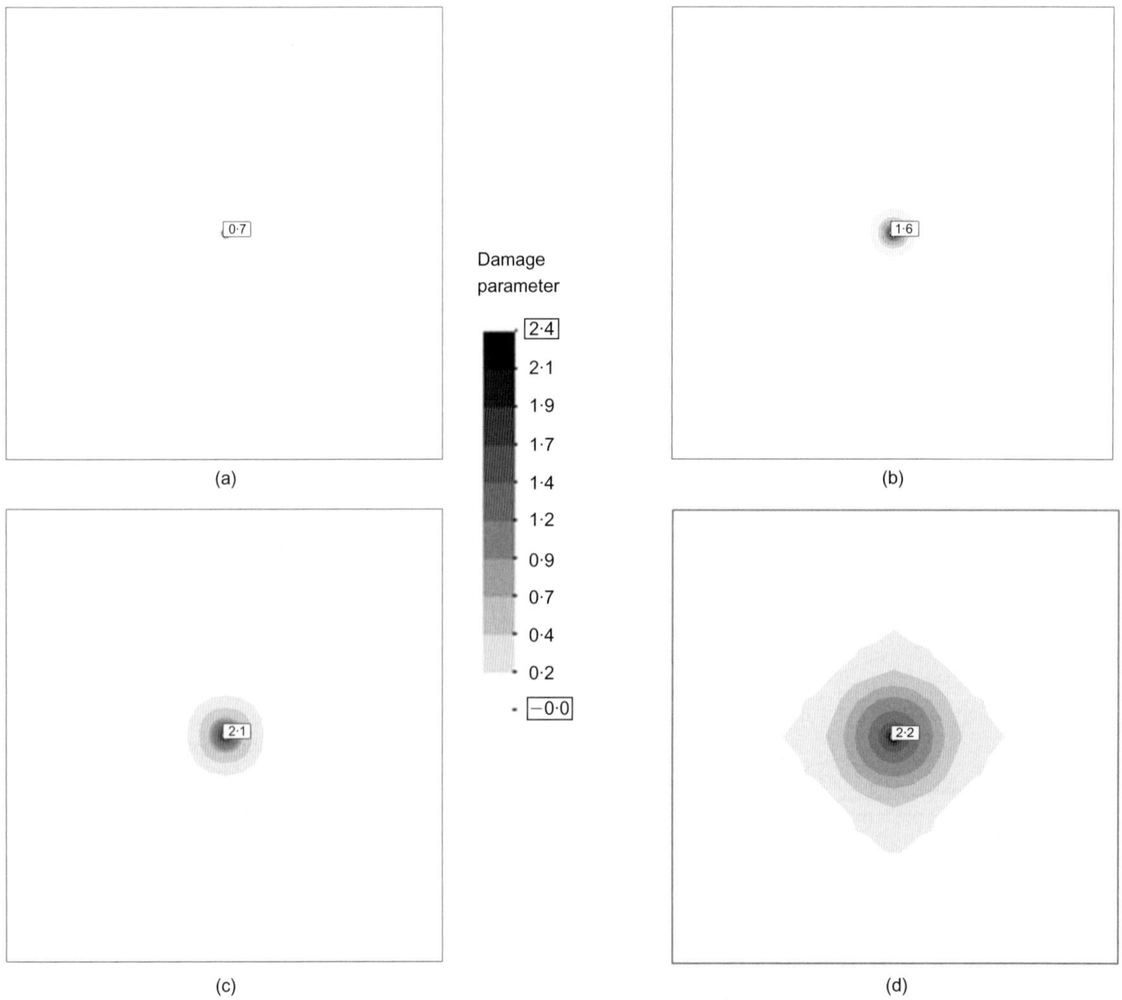

Fig. 26. Contours of damage parameter L in cross-section across Heater 2 at various times: (a) before heating starts (day 0); (b) end of first stage of heating (day 90); (c) pore pressure maxima (110 days); (d) end of test (508 days)

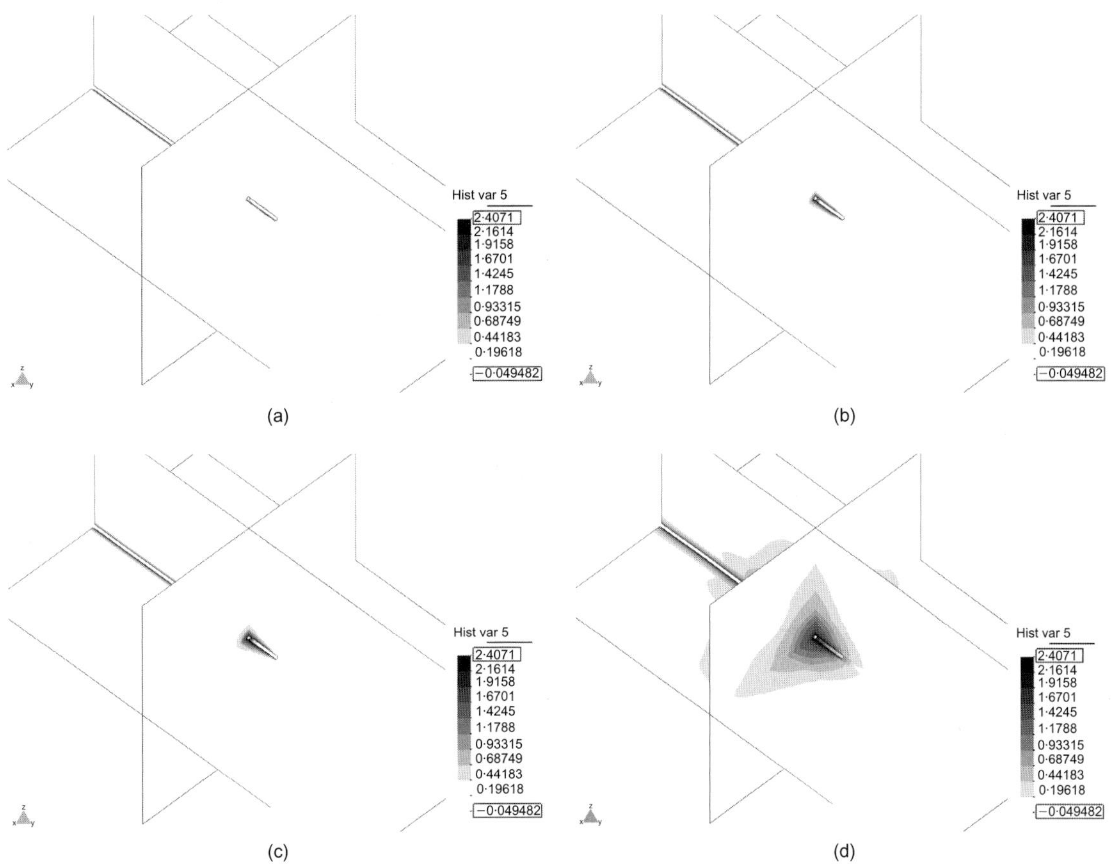

Fig. 27. Three-dimensional view of contours of damage parameter L at various times: (a) before heating starts (day 0); (b) end of first stage of heating (day 90); (c) pore pressure maxima (110 days); (d) end of test (508 days)

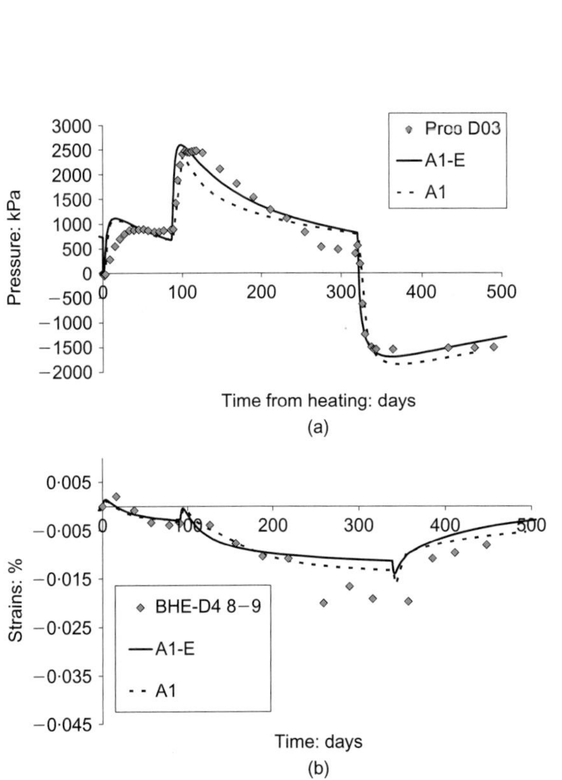

Fig. 28. Comparison of computed results of analysis A1 (considering damage) and A1-E (linear clay behaviour) referring to: (a) pore pressure evolution; (b) strain evolution

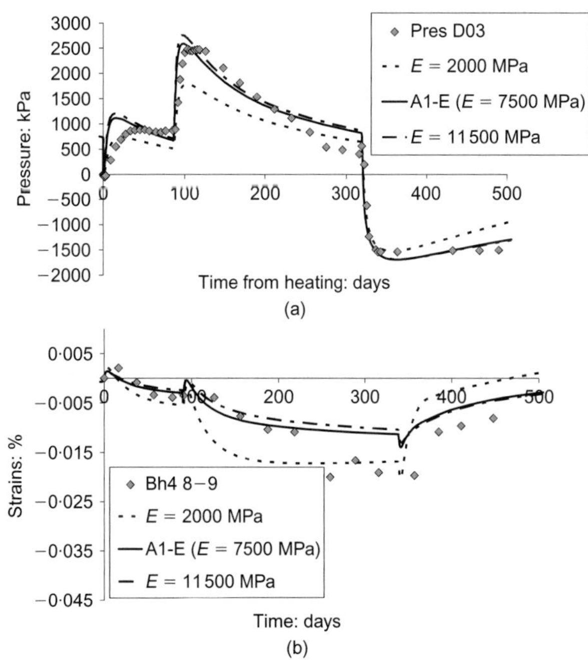

Fig. 29. Effect of variation of Young's modulus on: (a) pore pressure evolution; (b) strain evolution

turns out to be very variable, and can be organised in a hierarchical manner, as follows.

(*a*) The strongest coupling is found from thermal to hydraulic and mechanical behaviour. Pore pressure

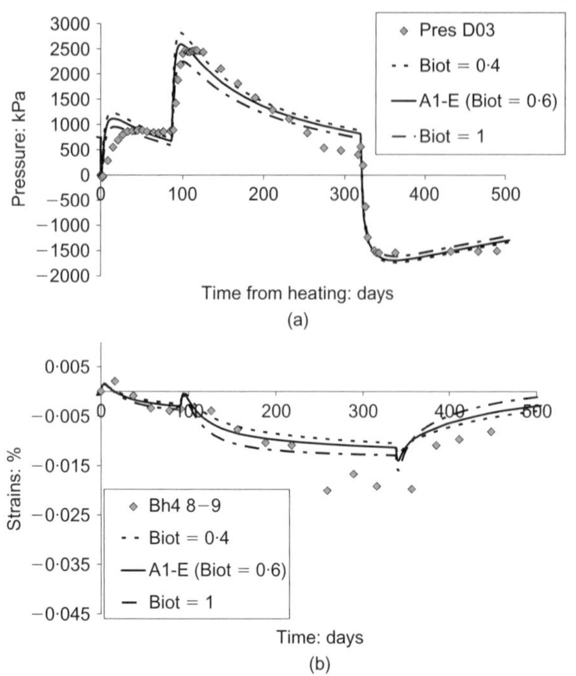

Fig. 30. Effect of variation of Biot's coefficient on: (a) pore pressure evolution; (b) strain evolution

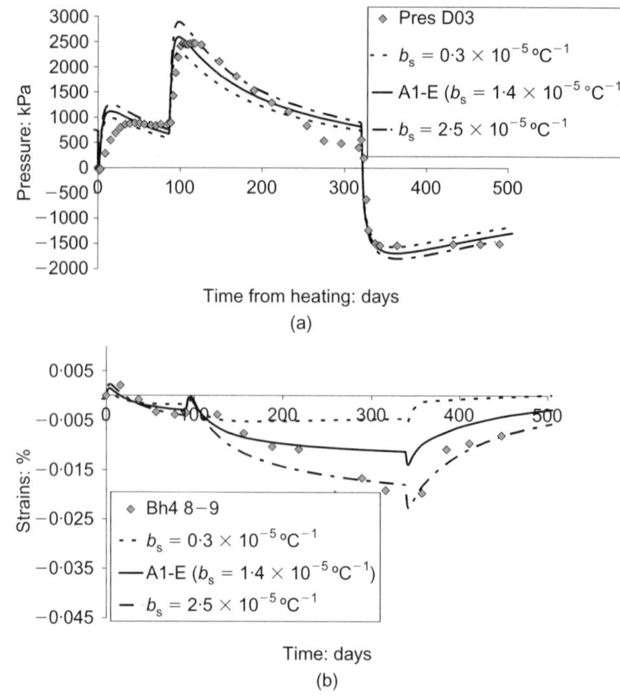

Fig. 32. Effect of thermal expansion coefficient on: (a) pore pressure evolution; (b) strain evolution

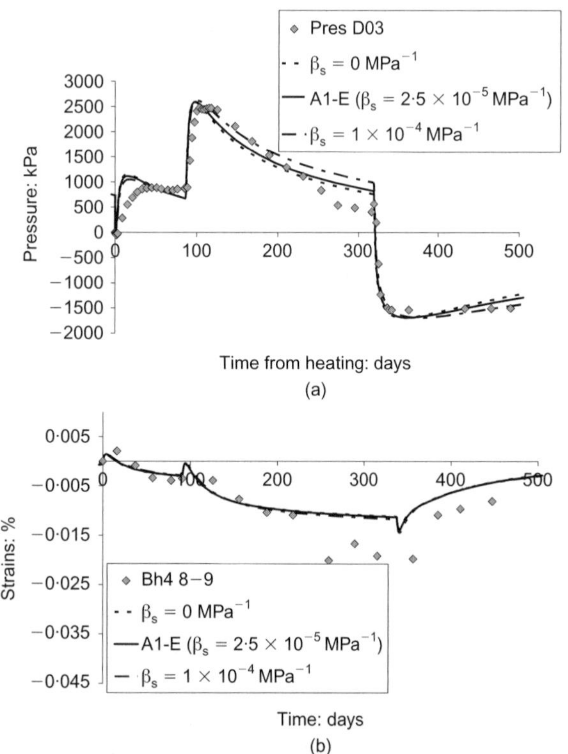

Fig. 31. Effect of variation of solid phase compressibility on: (a) pore pressure evolution; (b) strain evolution

generation is controlled primarily by temperature increase, and the largest contributor to deformation and displacements is thermal expansion.

(*b*) Significant but more moderate effects are identified from the coupling of hydraulic to mechanical behaviour. The dissipation of pore pressures induces additional displacements and strains, but, because of the large clay stiffness, they are smaller than thermally induced deformations.

(*c*) In principle, mechanical damage could impact on the hydraulic results because of the development of higher permeability owing to material damage. However, the damaged zone appears to be very limited in this case, and the effects do not seem to be very significant.

(*d*) There is no noticeable coupling from hydraulic to thermal behaviour. Practically all heat transport is by conduction, and the thermal conductivity of the material does not change, as the material remains saturated throughout.

(*e*) Coupling from mechanical to thermal behaviour is negligible. The changes of clay porosity are so slight that they do not affect thermal conductivity. In addition, mechanical energy dissipation is insignificant in this non-isothermal case.

ACKNOWLEDGEMENTS

The work presented in this paper was financially and technically supported by the Agence Nationale pour la gestion des déchets radioactifs (ANDRA) of France. The authors also want to acknowledge the support of the European Commission via a Marie Curie Fellowship awarded to Benoit Garitte within the framework of the Mechanics of Unsaturated Soils for Engineering Research Training Network. The contribution of the Spanish Ministry of Education and Science through research grant BIA2005-05801 is also acknowledged. The authors are grateful to J. Pineda for providing the data from the experimental determination of Opalinus clay stiffness.

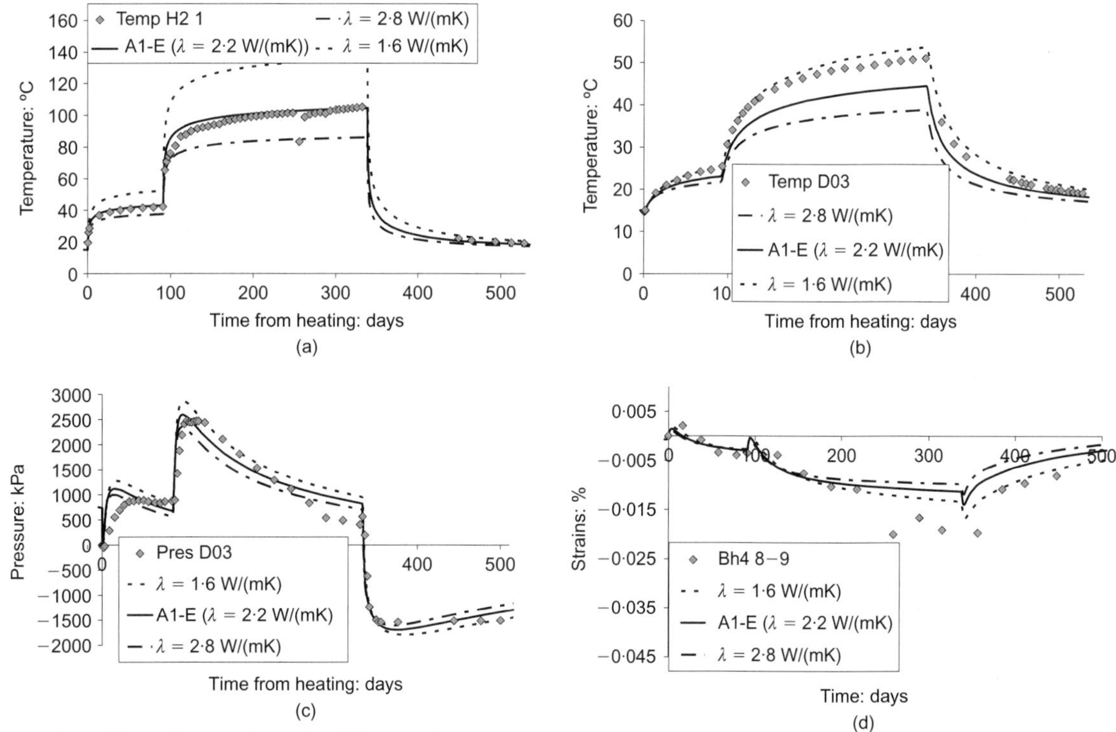

Fig. 33. Effect of thermal conductivity on: (a) temperature evolution on the clay/heater interface; (b) temperature evolution of borehole D03; (c) pore pressure evolution of borehole D03; (d) strain evolution of interval 8–9 of borehole D4

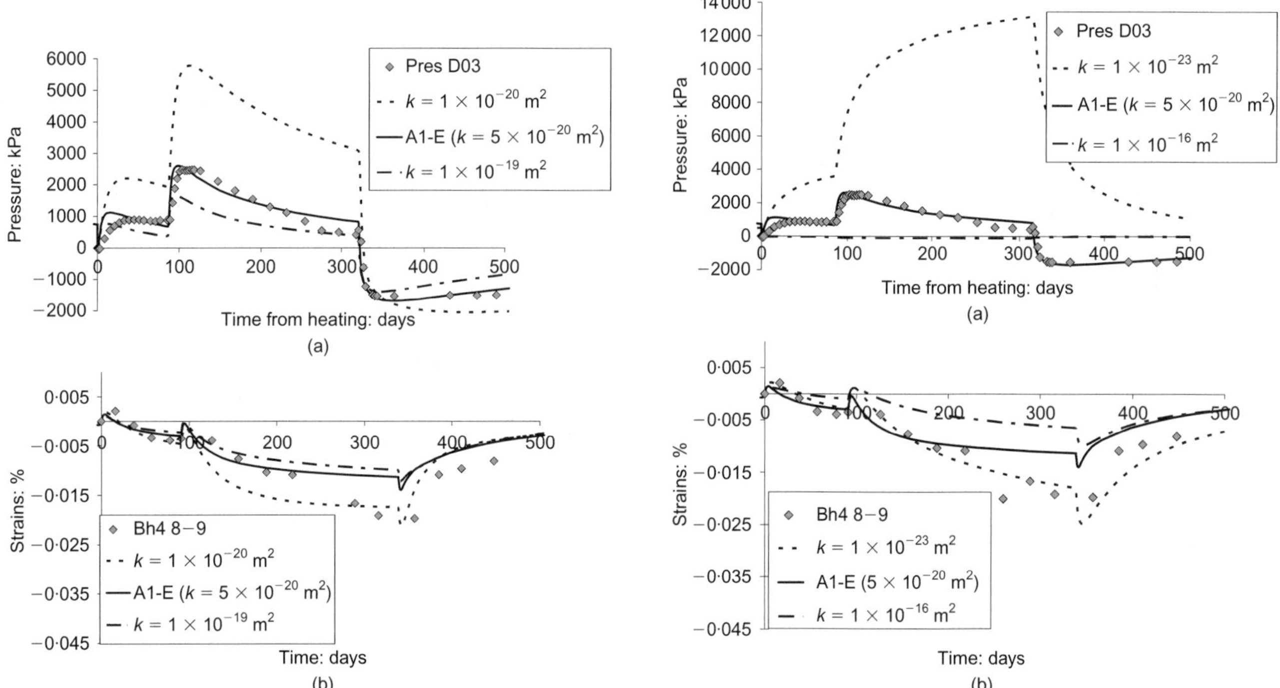

Fig. 34. Effect of intrinsic permeability on: (a) pore pressure evolution; (b) strain evolution. Permeability parameters in the range of reference values

Fig. 35. Effect of intrinsic permeability on: (a) pore pressure evolution; (b) strain evolution. Undrained and drained conditions

REFERENCES
ANDRA (2005). *Dossier 2005, Référentiel du site Meuse/Haute Marne, Tome 2*, Report CRPADS040022_B. Chatenay-Malabry: ANDRA.
Auvray, C. (2004). *Thermomechanical tests on Opalinus clays of the Mont Terri*, ANDRA Report C.RP.OENG.04–0239. Chatenay-Malabry: ANDRA.

Bai, M. & Abousleiman, Y. (1997). Thermoporoelastic coupling with application to consolidation. *Int. J. Numer. Anal. Methods Geomech.* **21**, No. 2, 121–132.
Baldi, G., Hueckel, T., Peano, A. & Pellegrini, R. (1991). *Developments in modelling of thermo-hydro-geomechanical behaviour of Boom Clay and clay-based buffer materials*. ISMES, Final report, EUR 13365/1. Luxembourg: ISMES.

Bastiaens, W., Bernier, F. & Li, X. L. (2006). An overview of long-term HM measurements around HADES URF. In *EUROCK 2006: Multiphysics coupling and long-term behaviour in rock mechanics* (eds Van Cotthem, A., Charlier, R., Thimus, J. F. and Tshibangu, J. P.), pp. 15–25. London: Taylor & Francis.

Bernier, F. & Neerdael, B. (1996). Overview of in situ thermo-mechanical experiments in clay: concept, results and interpretation. *Engng Geol.* **41**, Nos 1–4, 51–64.

Bernier, F., Volckaert, G., Alonso, E. & Villar, M. (1997). Suction-controlled experiments on Boom Clay. *Engng Geol.* **47**, No. 4, 325–338.

Bock, H. (2001). *RA experiment. Rock mechanics analyses and synthesis: Data report on rock mechanics*, Technical Report 2000-02. Mont Terri Project.

Booker, J. R. & Savvidou, C. (1985). Consolidation around a point heat source. *Int. J. Numer. Anal. Methods Geomech.* **9**, No. 2, 173–184.

Bossart, P., Meier, P. M., Moeri. A., Trick, T. & Mayor, J.-C. (2002). Geological and hydraulic characterisation of the excavation disturbed zone in the Opalinus clay of the Mont Terri rock laboratory. *Engng Geol.* **66**, Nos 1–2, 19–38.

Carol, I., Rizzi, E. & Willam, K. (2001). On the formulation of anisotropic elastic degradation. I. Theory based on a pseudo-logarithmic damage tensor rate. *Int. J. Solids Struct.* **38**, No. 4, 491–518.

Coll, C. (2005). *Endommagement des roches argileuses et perméabilité induite au voisinage d'ouvrages souterrains.* Doctoral thesis, Université Joseph Fourier, Grenoble, France.

Cordebois, J. P. & Sidoroff, F. (1982). Endommagement anisotrope en élasticité et plasticité. *J. de Mécanique Théorique et Appliquée*, Numéro Spécial, 45–60.

Croisé, J., Schlickenrieder, L., Marschall P., Boisson J.-Y., Vogel, P. & Yamamoto, S. (2004). Hydrogeological investigations in a low permeability claystone formation: the Mont-Terri Rock Laboratory. *Phys Chem. of the Earth* **29**, No. 1, 3–15.

De Bruyn, D. & Labat, S. (2002). The second phase of ATLAS: the continuation of a running THM test in the HADES underground research facility at Mol. *Engng Geol.* **64**, Nos 2–3, 309–316.

Gens, A. (2000). *HE experiment: Complementary rock laboratory tests*, Technical Note TN 2000-47. Mont Terri Project.

Gens, A. (2003). The role of geotechnical engineering in nuclear energy utilisation (Special Lecture). *Proc. 13th Eur. Conf. Soil Mech. Geotech. Engng, Prague* **3**, 25–67.

Gens, A. & Olivella, S. (2000). Non isothermal multiphase flow in deformable porous media: Coupled formulation and application to nuclear waste disposal. In *Developments in theoretical geomechanics: The John Booker Memorial Symposium* (eds D. W. Smith and J. P. Carter), pp. 619–640. Rotterdam: Balkema.

Gens, A., Vaunat, J., Garitte, B. & Wileveau, Y. (2006). Response of a saturated mudstone under excavation and thermal loading. In *EUROCK 2006: Multiphysics coupling and long-term behaviour in rock mechanics* (eds Van Cotthem, A., Charlier, R., Thimus, J. F. and Tshibangu, J. P.), pp. 35–44. London: Taylor & Francis.

Harrington, J. F., Horseman, S. T. & Noy, D. J. (2001). Swelling and osmotic flow in a potential host rock. *Proc. 6th Int. Workshop on Key Issues in Waste Isolation Research*, Paris, 169–188.

Horseman, S. T., Winter, M. G. and Entwistle, D. C. (1987). *Geotechnical characterization of boom clay in relation to the disposal of radioactive waste*, EUR 10987. Luxembourg: Commission of the European Communities.

Kanj, M. & Abousleiman, Y. (2005). Porothermoelastic analyses of anisotropic hollow cylinders with applications. *Int. J. Numer. Anal. Methods Geomech.* **29**, No. 2, 103–126.

Katsube, N. (1988). The anisotropic thermo-mechanical constitutive theory for fluid-filled porous materials with solid/fluid outer boundaries. *Int. J. Solids Struct.* **24**, No. 4, 375–380.

Kurashige, M. (1989). A thermoelastic theory of fluid-filled porous materials. *Int. J. Solids Struct.* **25**, No. 9, 1039–1052.

Martin, C. D. & Lanyon, G. W. (2003) Measurement of in situ stress in weak rocks at Mont Terri Rock Laboratory, Switzerland. *Int. J. Rock Mech. Mining Sci.* **40**, Nos 7–8, 1077–1088

Olivella, S., Carrera, J., Gens, A. & Alonso, E. E. (1994). Non-isothermal multiphase flow of brine and gas through saline media. *Trans. Porous Media* **15**, No. 3, 271–293.

Picard, J. M., Bazargan, B. & Rousset, G. (1994). *Essai thermo-hydro-mécanique dans une argile profonde: Essai CACTUS*, CEC Report EUR 15482 FR. Luxembourg: CEC.

Schaeren, G. & Norbert, J. (1989). *Tunnel du Mont Terri et du Mont Russelin. La traversée des roches à risques: marnes et marnes à anhydrite*, SIA-Dokumentation D037, La traversée du Jura – Les projets des nouveaux tunnels. Zürich: Swiss Society of Engineers and Architects (SIA).

Thury, M. & Bossart, P. (1999) The Mont Terri rock laboratory, a new international research project in a Mesozoic shale formation, in Switzerland. *Engng Geol.* **52**, Nos 3–4, 347–359.

Vaunat, J. & Gens, A. (2003). Numerical modelling of an excavation in a hard soil/soft rock formation using a coupled damage/plasticity model. *Proc. 7th Int. Conf. on Computational Plasticity COMPLAS 2003, Barcelona* (CD-ROM).

Wang, Y. & Papamichos, E. (1999). Thermal effects on fluid flow and hydraulic fracturing from wellbores and cavities in low permeability formations. *Int. J. Numer. Anal. Methods Geomech.* **23**, No. 15, 1819–1834.

Wermeille, S. & Bossart, P. (1999). In situ *stresses in the Mont Terri Region: Data compilation,* Technical Report 99-02. Mont Terri Project.

Wileveau, Y. (2005). *THM behaviour of host rock (HE-D) experiment: Progress report. Part 1*, Technical Report TR 2005-03. Mont Terri Project.

Zimmerman, R. W. (2000). Coupling in poroelasticity and thermoelasticity. *Int. J. Rock Mech. Min. Sci.* **37**, Nos 1–2, 79–87.

Delage, P., Le, T.-T., Tang, A.-M., Cui, Y.-J. & Li, X.-L. (2007). *Géotechnique* **57**, No. 2, 239–244

TECHNICAL NOTE

Suction effects in deep Boom Clay block samples

P. DELAGE*, T.-T. LE*, A.-M. TANG*, Y.-J. CUI* and X.-L. LI†

KEYWORDS: clays; laboratory tests; radioactive waste disposal; sampling; suction

INTRODUCTION

Various investigations have been carried out on Boom Clay, a stiff clay from Belgium, in the context of research into deep radioactive waste disposal (ONDRAF/NIRAS, 2001). Investigations are linked to research conducted at the Mol Underground Research Laboratory (URL), excavated in a layer of Boom Clay at a depth of 223 m by SCK-CEN, the Belgian organisation for nuclear studies, near the city of Mol.

Most research has been carried out on triaxial specimens that were trimmed from blocks extracted during excavation sequences in the URL. Various questions arose concerning the quality of the specimens and the best procedure to adopt prior to running triaxial tests (Winter & Horseman, 1993; Coll, 2005). Special concern related to the great depth at which the blocks were extracted, to the corresponding stress release, and to the resulting suction (Skempton, 1961; Skempton & Sowa, 1963; Doran *et al.*, 2000), which is clearly larger than in common geotechnical engineering practice. The swelling observed in Boom Clay specimens when releasing suction by putting them in contact with saturated porous stones in the triaxial cell also caused some concern because of possible alteration of the natural initial microstructure (Sultan, 1997), with consequences for the mechanical properties of the clay, and more particularly the overconsolidation ratio.

In this paper suction effects in deep Boom Clay block samples are investigated through characterisation of the water retention and swelling properties of the clay. The relationship between suction and stress changes during loading, and unloading sequences are also examined by running oedometer tests with suction measurements. Finally, the effect of suction release under an isotropic stress close to the estimated suction is investigated.

MATERIAL AND EXPERIMENTAL METHODS

The study was carried out on intact Boom Clay specimens extracted at a depth of 223 m in the URL of Mol, Belgium. The Boom Clay formation belongs to the Rupelian geological period in the Tertiary sub-era, which dates from 36 to 30 million years before the present. Its thickness is around 102 m in the Mol area, and the layer dips gently ($\pm 1°$) towards the north-north-east (Mertens *et al.*, 2003). As noted by Horseman *et al.* (1987), cited in Burland (1990), the clay is geologically lightly overconsolidated, but the yield stress

of Boom Clay may be larger than the preconsolidation pressure owing to mechanisms such as creep and diagenesis, resulting in an overconsolidation ratio $R_{OC} = 2\cdot4$.

Table 1 presents the mineralogical composition of Boom Clay, taken from various sources (Decleer *et al.*, 1983; Al-Mukhtar *et al.*, 1996; ONDRAF/NIRAS, 2001). The clay fraction is dominant (50–60% $< 2\,\mu$m), but some differences exist in the smectite fraction between Al-Mukhtar *et al.* and Decleer *et al.* on the one hand (33%) and ONDRAF/NIRAS on the other (17%). Other differences are observed in the kaolinite and illite fractions. These differences can be due to the natural variability of Boom Clay. They show, however, that some confirmation of the mineralogical composition of Boom Clay is still needed. The geotechnical properties of Boom Clay are shown in Table 2. In situ water content measurements were made on excavated blocks during excavation in the URL (connecting gallery, excavated between 23 January and 23 April 2002). Average values between 24·5% and 25·5% were obtained at that time.

Table 1. Mineralogical composition of Boom clay according to Decleer *et al.* (1983), Al-Mukhtar *et al.* (1996) and ONDRAF/ NIRAS (2001)

	Decleer *et al.* (1983): %	Al-Mukhtar *et al.* (1996): %	ONDRAF/ NIRAS (2001): %
Clay minerals	50	62	52
Illite	12	16	28
Kaolinite	5	13	6
Smectite	33	33	17
Chlorite			3
Glauconite			3
Quartz	35	20–25	20
Calcite, dolomite	1		1–5
Pyrite	1	4–5	1–5
Feldspar			5–10
Microcline	9	4–5	
Plagioclase	4	4–5	
Organic material			1–3

Table 2. Geotechnical properties of Boom clay according to Belanteur *et al.* (1997) and Dehandschutter *et al.* (2005)

Property	Belanteur *et al.* (1997)	Dehandschutter *et al.* (2005)
Unit mass of solid: Mg/m^3	2·67	
Unit mass: Mg/m^3		1·9
Liquid limit, w_L	59–76	70
Plastic limit, w_P	22–26	25
Plastic index, I_P	37–50	45
Water content: %		25–30
Natural porosity: %		35
Poisson's ratio		0·4
Internal friction angle: degrees		18
Permeability: m/s		10^{-12}

Manuscript received 25 April 2006; revised manuscript accepted 29 January 2007.
Discussion on this paper closes on 1 August 2007, for further details see p. ii.
* École Nationale des Ponts et Chaussées (CERMES, Institut Navier), Paris, France.
† EURIDICE Group, Mol, Belgium.

Excavated blocks were immediately vacuum-packaged in reinforced aluminium foil and thermo-welded. Blocks were stored in a room with temperature varying between 15°C and 20°C and an average relative humidity of 45%. The laboratory water content determination that was made during this study (in 2005) gave smaller values of between 20·2% and 21·6%. The difference in water content between the two measurements is related to drying that may have occurred during sample storage between 2002 and 2005. As compared with the initial suction at the time when the sample was excavated, this drying obviously corresponds to an increase in suction, which will be further commented on based on the water retention properties of the sample.

The water retention properties were determined on both rectangular clay samples (approximately 30 mm × 30 mm × 10 mm) and cylindrical oedometer samples ($d = 70$ mm, $h = 20$ mm). Starting from initial water contents close to 21%, the rectangular clay samples were dried at controlled values of suction by using the vapour equilibrium method (e.g. Delage *et al.*, 1998) with five saline solutions: $CuSO_4$ ($s = 2·8$ MPa); K_2SO_4 ($s = 4·2$ MPa); KNO_3 ($s = 8·5$ MPa); NaCl ($s = 37·8$ MPa); and $MgCl_2$ ($s = 152·8$ MPa). Along the drying path, three specimens were used at each suction level to determine the water content at equilibrium. Along the wetting path, the three oedometer specimens were smoothly wetted by putting them in contact with humid filter papers, and the resulting suction was subsequently measured by using a tensiometer. The volume changes of the oedometer samples were determined with a precision caliper. The volume changes of the rectangular specimens were determined by hydrostatic weighing after having immersed the samples in a non-aromatic hydrocarbon liquid called Kerdane.

An oedometer cell equipped with a high-suction tensiometer (capable of measuring to 1500 kPa) was used to measure suction changes during oedometer compression, as suggested by Dineen & Burland (1995) (Fig. 1). The principle of the high-suction tensiometer was first reported by Ridley & Burland (1993). The tensiometer used in this work was built at ENPC-CERMES. As seen in the figure, the tensiometer is located within the base of the oedometer cell in contact with the lower face of the sample, whereas the upper face of the sample is in contact with a porous stone that is fixed to the piston.

Finally, an isotropic compression test was carried out in a high-pressure triaxial cell (Delage *et al.*, 2000) on a standard triaxial specimen ($d = 38$ mm, $h = 76$ mm). The confining pressure and the back-pressure were applied by two GDS volume/pressure controllers. During the test, the sample volume changes were measured by monitoring the volume of water in the cell through the pressure/volume controller used for the confining pressure. As will be seen later, this was

necessary: (a) to control the volume changes during the compression sequence with dry porous stones (i.e. with no water permitted to infiltrate inside the sample); and (b) to control the swelling when water was allowed to hydrate the sample through the porous stones. In high-pressure metal cells, this method appeared to give satisfactory results (Delage *et al.*, 2000).

EXPERIMENTAL RESULTS
Water retention curve and swelling behaviour

Figure 2 shows some aspects of the water retention properties of Boom Clay, presented in terms of changes in water content (w) and degree of saturation (S_r) as a function of the logarithm of suction ($\log s$). As described previously, starting from initial water contents of 20·2–21·6%, three points were obtained along a wetting path (with measured suctions equal to 180, 280 and 600 kPa respectively) and five suctions were used along a drying path (with imposed suctions equal to 2·8, 4·2, 8·5, 37·8 and 152·8 MPa and with three specimens at each suction level). The data obtained along the drying path show good compatibility between the various points obtained under the same suction, in terms both of water content and of degree of saturation.

The S_r–$\log s$ plot shows that the two points obtained along the drying path at the two low suctions (2·8 and 4·2 MPa) indicate that the samples remained saturated. Desaturation starts above 4·2 MPa, and the degree of saturation at a suction of 8·5 MPa is 90%. As shown in the figure, the air entry value of Boom Clay can be estimated at approximately 5 MPa. At the highest suction (152·8 MPa), the degree of saturation is equal to 31%.

Along the wetting path, the curve shows that, curiously, the degree of saturation of the oedometer samples decreases below 100%, with values lying between 90% and 100%. As the volume change measurements carried out by hydrostatic weighing on the samples subjected to suctions of 2·8 and 4·5 MPa showed that the sample remained saturated, it is believed that the initial state of the sample close to $w = 21\%$ was saturated (unfortunately, the degree of saturation was not determined in the initial state). Hence the samples subjected to suctions lower than the initial one that reached higher values of water content (22·5%, 29% and 29·5%) should not desaturate. The value of degree of saturation reported in the figure along the wetting path hence appears

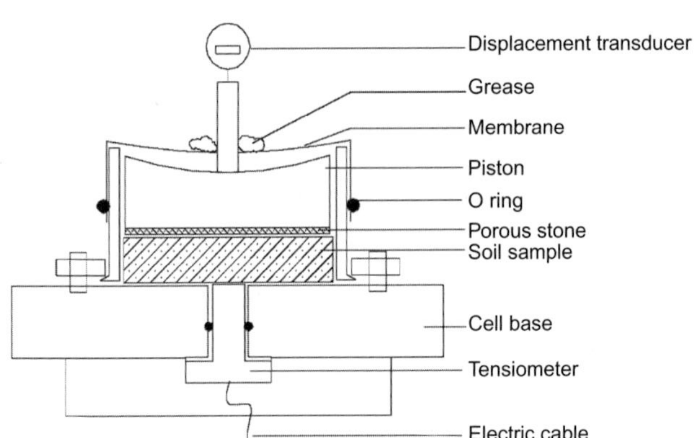

Fig. 1. Oedometer cell equipped with a tensiometer at the basis

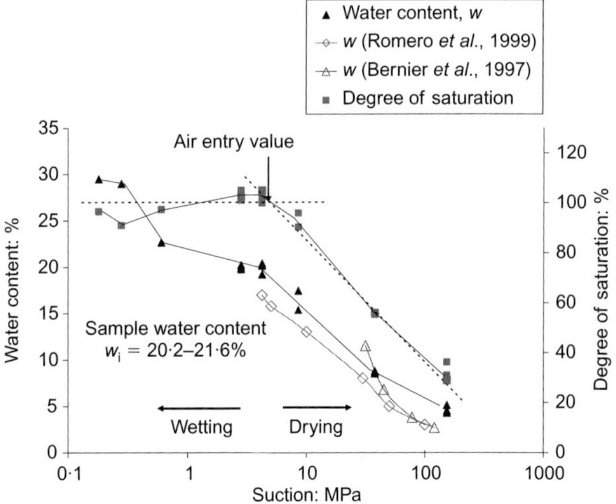

Fig. 2. Water retention properties of Boom Clay sample in terms of changes in water content and degree of saturation against suction (wetting and drying paths followed respectively starting from a 21% initial water content)

to show some discrepancy due to the lack of precision of the precision caliper volume measurements, as compared with hydrostatic weighing. Under the hypothesis of a saturated state, the increase in water content obtained along the wetting path corresponds to some swelling, which will be further investigated later. Conversely, up to the air entry value pressure (5 MPa), drying occurs with some shrinkage under a saturated state. The curve follows the main drying path at suction higher than 5 MPa, when the sample starts desaturating. At a suction as high as 152·8 MPa, Boom Clay is able to retain 5% water content. Referring to Table 1, this could indicate a high smectite content, probably closer to 33% than 17%.

Water retention data obtained by Bernier *et al.* (1997) and Romero *et al.* (1999) on compacted Boom Clay samples at a dry unit mass of 1·7 Mg/m^3 are also presented for comparison. The data show that the curve of Romero *et al.* (1999) is parallel, with less water being retained by the compacted sample at the same suction. The curve of Bernier *et al.* (1997) is similar to that of Romero *et al.* (1999) at high suction.

Figure 3 shows the volume changes with respect to suction that correspond to the drying and wetting paths of Fig. 2. A significant swelling of 18% is observed when suction is reduced to 180 kPa. A shrinkage of 15% is observed at a suction of 152·8 MPa. The slope that characterises swelling (average slope $\Delta e/\Delta \log s = -0·5$) is larger than the shrinkage slope (average slope $\Delta e/\Delta \log s = -0·1$). Bernier *et al.* (1997) found a similar trend on compacted Boom Clay specimens subjected to change in suction under a small vertical load in the oedometer.

It is difficult to provide a value for the initial suction of the sample just after extraction, based on the data from the curve of Fig. 2. The drying that reduced the water content from an average in situ value close to 25% down to an average value of 21% after the storage period obviously increased the sample suction. The suction given by the figure at a water content close to the in situ water content (25%) is equal to 400 kPa.

As quoted by various authors, including Skempton & Sowa (1963), Bishop *et al.* (1975) and Doran *et al.* (2000), there is a relation between the suction of a saturated sample and the state of stress at the depth at which the sample has been extracted. In the case of the 'perfect sampling' (Bishop *et al.*, 1975) of an isotropic elastic sample, the suction of the sample is equal to the mean effective stress supported by the sample prior to extraction. Doran *et al.* (2000) showed, however, that this was no longer true in anisotropic elastic soils, even after 'perfect sampling' due to coupling of volumetric strain with deviator stress during unloading. Because of the depth at which it has been sampled (223 m), and the high corresponding in situ vertical effective stress

(estimated at 2·45 MPa with an average soil unit mass $\rho = 2·1$ Mg/m^3 at a depth of 223 m), Boom Clay can certainly be considered as a cross-anisotropic material. This is also suggested by the scanning electron microscope observations carried out by Dehandschutter *et al.* (2005), in which bedding planes were clearly observed. Note also that the Boom Clay layer dips gently ($\pm 1°$) towards the north-north-east (Mertens *et al.*, 2003).

Based on in situ stresses estimations, Horseman *et al.* (1993) calculated for Boom Clay a K_0 value equal of 0·8. The equivalent effective mean stress (p'), derived from the vertical load (σ'_v), is

$$p' = \tfrac{1}{3}(\sigma'_v + 2\sigma'_h) = 0·87\sigma'_v = 2·12\,\text{MPa} \tag{1}$$

This value is significantly higher than the suction of 400 kPa that could be estimated at $w = 25\%$ in Fig. 2. The 400 kPa value seems to be too small to be explained by possible effects related to non-perfect sampling or to cross-anisotropy, and a deeper examination of the water retention properties (Fig. 2) appears to be necessary. In this regard, Fig. 4 presents a more detailed view of the water retention curve, in which an estimated initial state has also been plotted by adopting the mean effective stress calculated above (2·12 MPa) as an approximate suction at a water content of 25%. Based on this hypothesis, the drying process that occurred during storage can be interpreted. The figure shows that the position of the initial point is reasonably compatible with the drying curve of the water retention curve. Note that, given the shape of the curve, this is also true for suction values down to 1 MPa. Also, the shape of the wetting path determined by the three points obtained at low suction in the oedometer illustrates a hysteresis effect typical of clays, as observed for instance by Croney (1952) on a heavy clay soil ($w_p = 26\%$, $w_L = 78\%$) that had been subjected to a suction cycle starting from its initial undisturbed state. In other words, at the same water content, the suction reached by a sample after a drying–wetting cycle can be significantly lower than the initial suction.

An estimation of the sample suction with a value of water content close to 21% can now be made by considering the estimated drying path of Fig. 4. Accounting for the uncertainty concerning the initial state, the figure shows that this value should be between 2 and 3 MPa. This range would also have been obtained by starting from an initial value of 1 MPa.

Oedometer tests
The three samples used for determining the wetting path were then used for oedometer testing, with initial suction

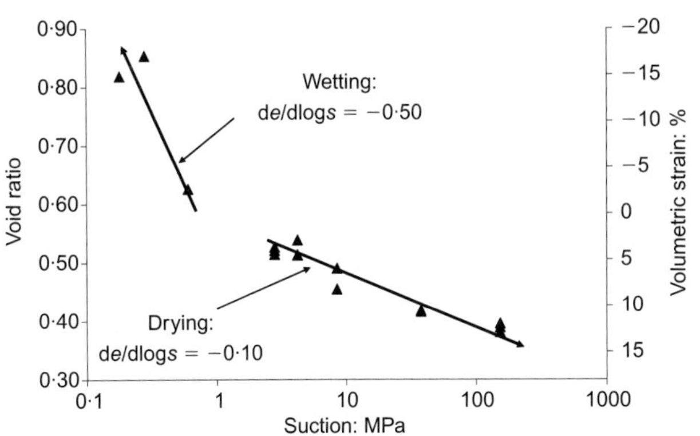

Fig. 3. Volume changes of Boom clay samples under suction changes during drying and wetting

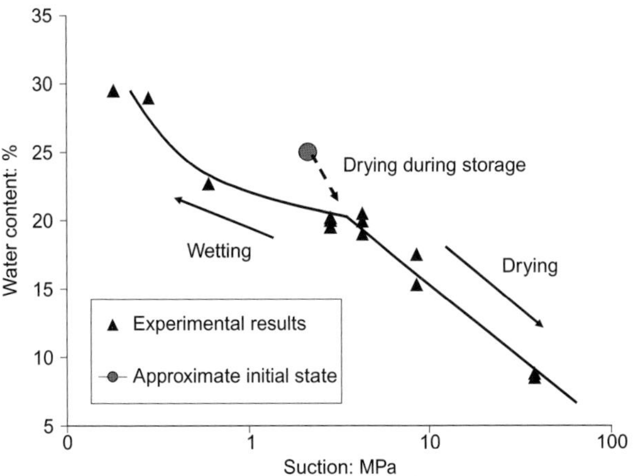

Fig. 4. Hysteresis effects in the lower suction range

values of 180, 280 and 600 kPa respectively (Fig. 2). The results of the oedometer compression test with suction measurement carried out with an initial suction of 280 kPa are presented as a function of time in Fig. 5 as follows

(a) loading sequence (standard oedometer step loading with loading stages generally close to 24 h)
(b) vertical displacement
(c) pore water pressure changes, starting from an initial value of -280 kPa.

Figure 5(b) shows that each loading step induces an instantaneous settlement followed by an equilibration stage. Interestingly, instantaneous coupled peaks in positive pressure are observed at each loading step (Fig. 5(c)), even when starting from an initial negative pore pressure. For vertical stresses smaller than 800 kPa, measured pore water pressures come back and stabilise at negative values. The preservation of a suction state within the sample means that no water has been expelled from the sample, resulting in a constant water content condition.

The transition from negative to positive pressure was observed when loads greater than 800 kPa were applied. Water pressure stabilised at 0 kPa after the 800 and 1600 kPa loading stages. This shows that the suction/pressure gauge worked properly in the positive pressure range, describing a standard positive pore pressure dissipation process and a decrease in water content of the sample with water expelled in the porous stone located above the sample. Similarly, instantaneous coupled pressure decreases are observed during the unloading step, with suction as high as 700 kPa reached when passing from 1600 kPa to 800 kPa vertical stress. Note that this instantaneous suction value is close to the load release (800 kPa), which is reasonably compatible with saturated 'inverse' consolidation.

The results of the three oedometer tests are presented in Fig. 6 in diagrams giving the changes in vertical stress with respect to suction changes, once equilibrium has been reached. These diagrams show how suction is reduced by loading and then increased by unloading, starting from three different initial suctions.

During compression, a linear relation between changes in suction and vertical load is observed with slopes $ds/d\sigma_v$ that vary between -0.45 and -0.56. Compression paths also show that zero suction is reached at smaller loads when starting from smaller initial suctions, s_i

(a) $\sigma_{vn} = 1200$ kPa for $s_i = 600$ kPa
(b) $\sigma_{vn} = 600$ kPa for $s_i = 280$ kPa
(c) $\sigma_{vn} = 450$ kPa for $s_i = 180$ kPa.

The slopes obtained during unloading were higher, and varied from -0.59 to -0.80. The variability between the different slopes measured in Fig. 6 is related to the different initial void ratios of the samples given in Fig. 3. Both samples hydrated at smaller initial suction (180 and 280 kPa) exhibited around 18% swelling, with similar void ratios, equal to 0.86 and 0.87 respectively. For both samples the slopes in the unloading stage (-0.74 at 180 kPa and -0.80 at 280 kPa) are larger than in the loading stage (-0.45 at 180 kPa and -0.53 at 280 kPa), with slopes

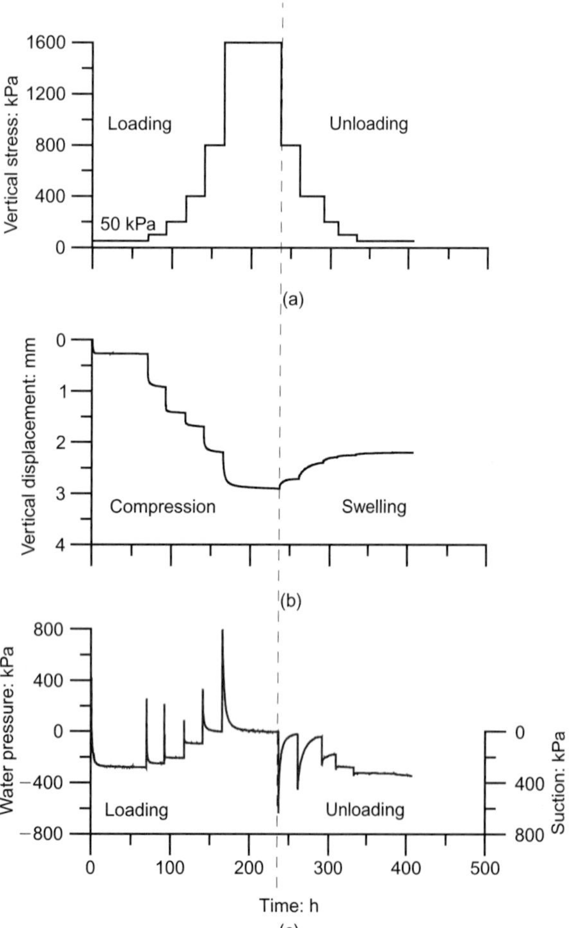

Fig. 5. Oedometer compression test with suction measurement. (a) Vertical stress, (b) vertical displacement and (c) suction changes given as a function of elapsed time (initial suction 280 kPa)

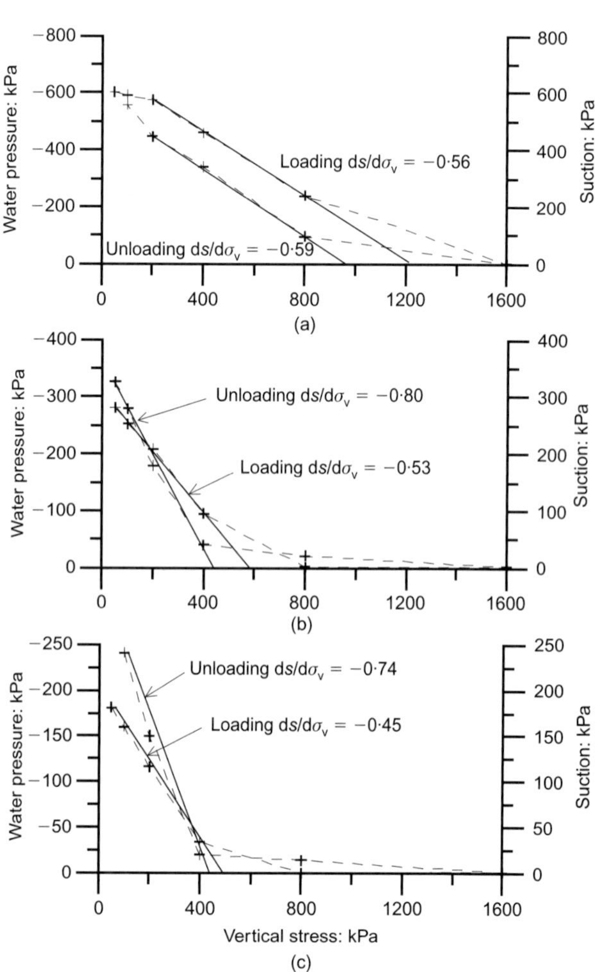

Fig. 6. Suction variation under vertical stress change during oedometer compression for three tests with initial suction s_i equal to: (a) 600 kPa; (b) 280 kPa; (c) 180 kPa

slightly smaller at lower suction. The 600 kPa suction sample, which exhibited a significantly smaller swelling (4%, leading to a void ratio of 0·64), has the highest value of slope during the loading phase (−0·56) and the lowest in the unloading phase (−0·59). The variability of the slopes in the loading phase (between −0·56 and −0·45) is less than in the unloading phase (between −0·59 and −0·80).

The range of the slopes determined during the unloading sequences (between −0·59 and −0·80) gives an idea of the changes in suction that occur in Boom Clay samples when releasing the vertical stress. They can be compared by the value of s/σ_v' ranging from 0·35 to 0·75 obtained by Bishop *et al.* (1975) for natural soils, in which s is the sample suction and σ_v' is the in situ vertical effective stress. Based on the sample suction obtained here (between 2 and 3 MPa), values of s/σ_v' between 0·94 and 1·42 are obtained. These much higher values show the effects of drying. The resulting uncertainty makes it difficult to use for a precise determination of suction changes induced by stress release.

Suction release under isotropic stress

While wetting Boom Clay sample at low suctions under zero stress, a swelling of around 18% due to suction release has been observed (Fig. 3). This swelling is consistent with the 11–14% swelling found by Coll (2005) under a low effective pressure in the triaxial apparatus. This saturation procedure and the resulting swelling are likely to alter the initial state of the soil and to lead to unexpectedly low values of the yield stress, as shown by Horseman *et al.* (1993) and Sultan (1997). After Graham *et al.* (1987), during saturation under the in situ state of stress, suction reduces to zero and no further swelling should occur, making it possible to put the sample in contact with water and apply a back-pressure in a standard manner with no significant modification of the initial microstructure.

A test in which suction was released under a constant isotropic stress was carried out on a sample with an initial water content of 21·6% (Fig. 7). In order to minimise the volume change during soaking, an isotropic stress close to the estimated initial sample suction was chosen. Based on the range of suctions estimated from the water retention curve (2–3 MPa), it was decided to adopt an isotropic stress of 2·5 MPa. This value is larger than the in situ mean effective stress estimated previously (2·12 MPa). The sample was isotropically compressed from 0·1 MPa to 2·5 MPa while keeping the porous stones dry. The porous stones were then soaked, and a low back-pressure (50 kPa) was applied to the soil sample while measuring volume changes. The confining pressure and the back-pressure were afterwards simultaneously increased in order to keep the effective stress equal to 2·5 MPa, the final back-pressure being equal to

1 MPa. Finally, the soil sample was compressed up to an effective pressure of 10 MPa with a constant low pressure-change rate (0·5 kPa/min). Sultan *et al.* (2002) showed that this rate was slow enough to ensure drained conditions in Boom Clay. The corresponding experimental data are shown in Fig. 7. Under 2·5 MPa, a slight swelling is observed during saturation ($\varepsilon_v = 1·7\%$), which corresponds to an increase in void ratio from 0·582 to 0·608. This swelling is much smaller than that (14%) obtained after saturation at low confining pressure by Coll (2005). Furthermore, the yield stress (p_y') obtained in Fig. 7 is close to 5 MPa, giving an overconsolidation ratio $R_{OC} = 2·1$, in good agreement with the data of Horseman *et al.* (1993; $R_{OC} = 2·4$) and Coll (2005; $R_{OC} = 2·2$). Note, however that, as noted by Horseman *et al.* (1987) cited in Burland (1990), based on geological evidence the yield stress of Boom Clay may be larger than the preconsolidation pressure owing to mechanisms such as creep and diagenesis.

By saturating the sample under low effective pressure Sultan (1997) obtained an underestimate value $p_y' = 0·4$ MPa. This showed that swelling may alter the natural microstructure of a swelling clay and erase the memory of the stress history by significantly reducing the overconsolidation ratio.

The data of the test of Fig. 7 confirm, as suggested by Graham *et al.* (1987), that some precautions have to be taken before releasing the suction of a natural sample. Note, however, that the slight swelling observed during the test indicates that the initial suction might be higher than 2·5 MPa.

CONCLUSIONS

Suction effects were investigated in intact Boom Clay block samples extracted at great depth (223 m) in the underground research laboratory of Mol (Belgium). Suction effects in deep intact samples are considered to be significant because of the relationship between suction and the in situ stress state of the sample, as shown by various authors (Skempton, 1961; Skempton & Sowa, 1963; Doran *et al.*, 2000).

Suction effects were investigated through characterisation of the water retention and swelling properties of intact Boom Clay. Some drying that occurred during the storage period between the block extraction and the present experimental investigation was interpreted in terms of hysteresis effects. The sample suction was estimated to be between 2 and 3 MPa. The swelling-shrinkage behaviour under changes in suction was investigated, and an air entry value of 5 MPa was determined.

The relationship between suction and stress changes during loading and unloading sequences was also examined by running oedometer tests with suction measurements. Slopes $ds/d\sigma_v$ between −0·59 and −0·80 were obtained in the unloading phases, to compare with the range of values of s/σ_v' (0·35–0·75) proposed by Bishop *et al.* (1975) for natural soils. The s/σ_v' values obtained with the block sample used in this work were in the range 0·94–1·42. These much higher values showed the negative effects of drying.

Finally, the effect of suction release under an isotropic stress close to the estimated suction (2·5 MPa) was investigated. A slight swelling (1·7%) was observed, and a further compression sequence showed that a satisfactory overconsolidation ratio (2·1) was obtained. This result confirmed the importance of taking some precautions before putting a swelling soil in contact with water prior to triaxial testing, as suggested by Graham *et al.* (1987).

These data confirmed the importance of suction and

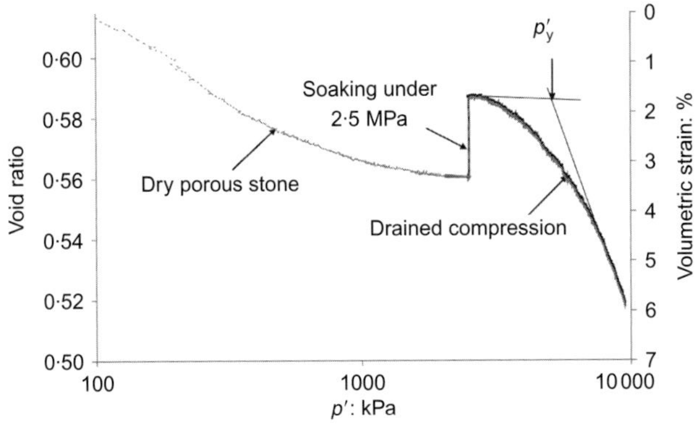

Fig. 7. Isotropic compression test

suction release effects, particularly in deep swelling samples of stiff clay. Obviously, some of the conclusions drawn here should be confirmed or improved by running similar tests on fresh samples just after extraction, in order to eliminate drying effects.

ACKNOWLEDGEMENTS

EURIDICE (European Underground Research Infrastructure for Disposal of nuclear waste In Clay Environment, Mol, Belgium) is gratefully acknowledged for funding the work presented in this paper. This work is part of the PhD thesis prepared at ENPC Paris by the second author. The financial support of ENPC is also acknowledged. The authors are also grateful to the MUSE European Research and Training Network (MRTN-CT-2004-506861). They also thank the referee and assessor, whose comments greatly helped in improving the paper.

NOTATION

d sample diameter
e void ratio
h sample height
K_0 coefficient of earth pressure at rest
p'_y yield stress
R_{OC} overconsolidation ratio
S_r degree of saturation
s suction
s_i initial suction
w water content
w_i initial water content
ε_v volumetric deformation
σ_h horizontal stress
σ_v vertical stress
σ'_v vertical effective stress
σ_{vn} vertical stress when suction decreases to 0

REFERENCES

Al-Mukhtar, M., Belanteur, N., Tessier, D. & Vanapalli, S. K. (1996). The fabric of a clay soil under controlled mechanical and hydraulical stress states. *Appl. Clay Sci.* **11**, Nos 2–4, 99–115.

Belanteur, N., Tacherifet, S. & Pakzad, M. (1997). Étude des comportements mécanique, thermo-mécanique et hydro-mécanique des argiles gonflantes et non gonflantes fortement compactées. *Rev. Fr. Géotech.* **78**, 31–50.

Bernier, F., Volckaert, G., Alonso, E. & Villar, M. (1997). Suction-controlled experiments on Boom clay. *Engng Geol.* **47**, No. 4, 325–338.

Bishop, A. W., Kumapley, N. K. & El Ruwayih, A. (1975). The influence of pore water tension on the strength of a clay. *Phil. Trans. Royal Soc. London* **278**, No. 1286, 511–554.

Burland, J. (1990). On the compressibility and shear strength of natural clays. *Géotechnique* **40**, No. 3, 329–378.

Coll, C. (2005). *Endommagement des roches argileuses et perméabilité induite au voisinage d'ouvrages souterrains.* PhD thesis, Université Joseph Fourier-Grenoble 1, France.

Croney, D. (1952). The movement and distribution of water in soils. *Géotechnique* **3**, No. 1, 1–16.

Decleer, J., Viane, A. & Vandenberghe, N. (1983). Relationships between chemical, physical and mineralogical characteristics of the Rupelian Boom Clay, Belguim. *Clay Miner.* **18**, No. 1, 1–10.

Dehandschutter, B., Vandycke, S., Sintubin, M., Vandenberghe, N. & Wouters, L. (2005). Brittle fractures and ductile shear bands in argillaceous sediments: inferences from Oligocene Boom Clay (Belgium). *J. Struct. Geol.* **27**, No. 6, 1095–1112.

Delage, P., Howat, M. D. & Cui, Y. J. (1998). The relationship between suction and swelling properties in a heavily compacted saturated clay. *Engng Geol.* **50**, Nos 1–2, 31–48.

Delage, P., Sultan, N. & Cui, Y. J. (2000). On the thermal consolidation of Boom clay. *Can. Geotech. J.* **37**, No. 2, 343–354.

Dineen, K. & Burland, J. B. (1995) A new approach to osmotically controlled oedometer testing. *Proc. 1st Int. Conf. on Unsaturated Soils, Paris,* 459–465.

Doran, I. G., Sivakumar, V., Graham, J. & Johnson, A. (2000). Estimation of in situ stresses using anisotropic elasticity and suction measurements. *Géotechnique* **50**, No. 2, 189–196.

Graham, J., Kwok, C. K. & Ambroise, R. W. (1987). Stress release, undrained storage, and reconsolidation in simulated underwater clay. *Can. Geotech. J.* **24**, No. 2, 279–288.

Horseman, S. T., Winter, M. G. & Entwistle, D. C. (1987). *Geotechnical characterisation of Boom clay in relation to disposal of radioactive waste.* Luxembourg: Office for Official Publications of the European Communities.

Horseman, S. T., Winter, M. G. & Entwistle, D. C. (1993). Triaxial experiments on Boom clay. In *The engineering geology of weak rock* (ed. J. C. Cripps), pp. 36–43. Rotterdam: Balkema.

Mertens, J., Vanderberghe, N., Wouters, L. & Sintubin, M. (2003). The origin and development of joints in the Boom clay formation (Rupelian) in Belgium. In *Subsurface sediment mobilization* (eds P. Van Rensbergen, R. R. Hillis, A. J. Maltman and C. K. Morley), pp. 61–71. Geological Society of London, Special Publication 158, pp. 61–71.

ONDRAF/NIRAS (2001). *Aperçu technique du rapport SAFIR 2. Safety assessment and feasibility interim report 2.* Publication NIROND 2001–05 F, p. 280.

Ridley, A. M. & Burland, J. B. (1993). A new instrument for the measurement of soil moisture suction. *Géotechnique* **43**, No. 2, 321–324.

Romero, E., Gens, A. & Lloret, A. (1999). Water permeability, water retention and microstructure of unsaturated compacted Boom clay. *Engng Geol.* **54**, Nos 1–2, 117–127.

Skempton, A. W. (1961). Horizontal stresses in an over-consolidated Eocene clay. *Proc. 5th Int. Conf. Soil Mech. Found. Engng, Paris,* 351–357.

Skempton, A. W. & Sowa, V. A. (1963). The behaviour of saturated clays during sampling and testing. *Géotechnique* **13**, No. 4, 269–290.

Sultan, N. (1997). *Etude du comportement thermo-mécanique de l'argile de Boom: expériences et modélisation.* PhD thesis, Ecole Nationale des Ponts et Chaussées, Paris.

Sultan, N., Delage, P. & Cui, Y. J. (2002). Temperature effects on the volume change behaviour of Boom clay. *Engng Geol.* **64**, Nos 2–3, 135–145.

Winter, M. G. & Horseman, S. T. (1993). Specimen preparation technique for a very stiff clay:a technical note. In *The engineering geology of weak rock* (ed. J. C. Cripps), pp. 83. Rotterdam: Balkema.

Piriyakul, K. & Haegeman, W. (2007). *Géotechnique* **57**, No. 2, 245–248

TECHNICAL NOTE

Void ratio function for elastic shear moduli for Boom Clay

K. PIRIYAKUL* and W. HAEGEMAN†

KEYWORDS: anisotropy; clays; elasticity; shear modulus; void ratio

INTRODUCTION

The initial shear modulus, G_0, is widely considered to be a fundamental soil stiffness property, and is a parameter for practical geotechnical problems, both in earthquake engineering and in the prediction of soil–structure interaction. The shear modulus can be measured in triaxial testing with the bender element technique. This paper presents a possibility for evaluating the initial shear moduli $G_{0(ij)}$ of Boom clay material at very small strains by measuring G_{vh}, G_{hh} and G_{hv} on the same sample at any stress state with independent control of the vertical and horizontal stresses. Hardin & Blandford (1989) presented the possibility of expressing the dependence of the initial shear modulus $G_{0(ij)}$ on the current state of a clay in the relationship

$$G_{0(ij)} = S_{ij}F(e)(\text{OCR})^k p_a^{(1-ni-nj)}(\sigma_i')^{ni}(\sigma_j')^{nj} \tag{1}$$

where σ_i' and σ_j' are the effective principal stresses acting on the plane in which G_0 is measured (in the case of seismic body waves the i and j directions correspond to the propagation and particle motion directions respectively); k is an empirical exponent that depends on the plasticity index I_p of the clay; S_{ij} is a non-dimensional material constant of a given soil that also reflects its fabric; ni and nj are empirical exponents; p_a is atmospheric pressure; and $F(e)$ is the void ratio function.

Jamiolkowski *et al.* (1995) evaluated the constants for a number of clays and found that $k = 0$. Therefore OCR is not an independent variable, and does not influence the magnitude of the small-strain elastic shear modulus, $G_{0(ij)}$. The multidirectional shear moduli G_{vh}, G_{hh} and G_{hv} of clay soils at very small strains are expressed as

$$G_{vh} = S_{vh}e^{-x}p_a^{(1-nv-nh)}(\sigma_v')^{nv}(\sigma_h')^{nh} \tag{2}$$

$$G_{hh} = S_{hh}e^{-x}p_a^{(1-2nh)}(\sigma_h')^{nh}(\sigma_h')^{nh} \tag{3}$$

$$G_{hv} = S_{hv}e^{-x}p_a^{(1-nh-nv)}(\sigma_h')^{nh}(\sigma_v')^{nv} \tag{4}$$

LABORATORY EQUIPMENT
Triaxial testing with multidirectional bender elements

Two sets of triaxial apparatus are used for measuring the shear wave velocity. A conventional triaxial cell for consoli-

dation under isotropic stress conditions and a stress path cell for testing under K_0- and K-consolidation are equipped with multidirectional bender elements and local strain devices. In both cases, vertical bender elements as proposed in Brignoli *et al.* (1996) are used. Within the framework of this research, a horizontal bender element technique is also installed. This technique uses friction in order to generate shear waves, as explained in detail in Fioravante (2000) and Fioravante & Capoferri (2001). Therefore the multidirectional bender element technique offers the possibility of obtained multidirectional shear moduli (G_{vh}, G_{hv} and G_{hh}) in the same soil sample under identical stress conditions, as seen in Fig. 1.

Processing shear wave velocities

The shear wave velocity is measured in two ways. In the first method it is calculated as the ratio of the time required by the shear wave to cover the distance between the two transducers to the tip-to-tip distance between these transducers. This time delay is assessed by a visual interpretation of the sending and receiving signals (first arrival method). In this method different waveforms can be used, such as a single- or multiple-period sinusoidal signal.

The second method was developed by Blewett *et al.* (1999). A continuous sine wave is applied to the transmitter

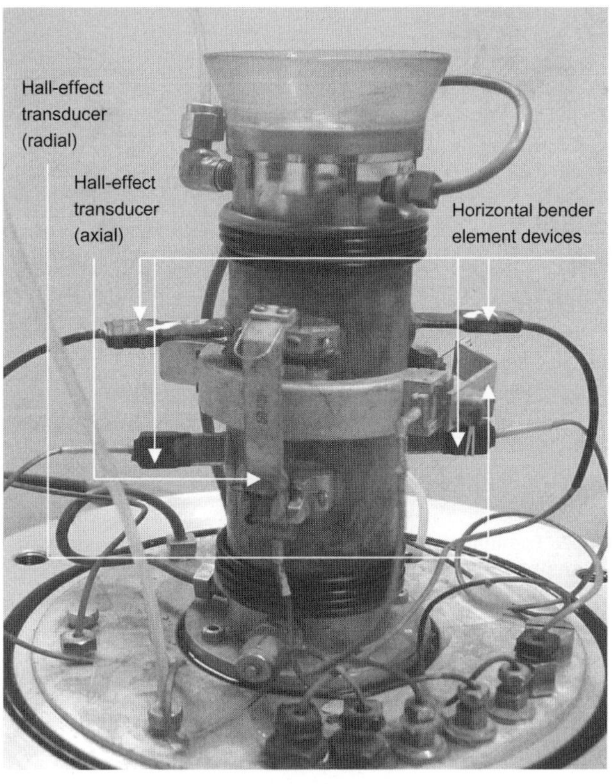

Fig. 1. Set-up of bender elements and local strain devices on soil sample in triaxial test

Manuscript received 26 April 2006; revised manuscript accepted 29 November 2006.
Discussion on this paper closes on 1 August 2007, for further details see p. ii.
* Department of Civil Engineering, King Mongkut's Institute of Technology, North Bangkok, Thailand.
† Laboratory of Geotechnics, Department of Civil Engineering, Ghent University, Belgium.

transducer at a low frequency. The output from the receiver transducer and the cross-correlation function between the sending and receiving signals are displayed at the same time. The frequency of the input signal is then gradually increased until the output signal comes into phase (cross-correlation time = 0). At this point a shear wave with one full wavelength is developed between the bender element transducers. Then the input frequency is gradually increased again until a multiple number of wavelengths (2, 3, 4, …) is obtained, and the in-phase frequency is noted. The shear wave velocity V_s can be calculated as

$$V_s = \frac{L}{N} f \qquad (5)$$

where N is the number of wavelengths, and f is the in-phase frequency of the signal. As an example, Fig. 2 shows the transmitted and received signals of a continuous sinusoidal waveform with a frequency of 9012 Hz together with the in-phase cross-correlation. Using equation (5), a shear wave velocity of 189·16 m/s is calculated, since in this test the number of wavelengths is four and the tip-to-tip distance is 83·96 mm.

Table 1 presents the shear wave velocities on a Boom clay sample at 5 m depth consolidated to an isotropic stress of 100 kPa. The calculation of the shear wave velocity is improved if a sufficiently high frequency is used and the number of wavelengths is at least four, as near-field effects are then reduced.

Table 2 shows the calculated shear wave velocity and shear modulus from the first arrival method using a single or four-period sinusoidal signal combined with the results from the cross-correlation technique with continuous sine wave. Two conclusions can be drawn from these data: first, there is no significant effect of waveform on the determination of the shear wave velocity in the first arrival method; and, second, the interpretation in the cross-correlation method is as good as that in the first arrival method.

Table 1. V_s from continuously sinusoidal shear wave measurements with different frequencies

Number of wavelengths	Frequency: Hz	Shear wave velocity: m/s
1	2 050	172·12
2	3 915	164·22
3	6 150	172·12
4	9 012	189·16
5	11 050	185·55

SAMPLES AND TEST PROCEDURES
Undisturbed Boom clay

Undisturbed Boom clay is sampled at the research site Sint-Katelijne-Waver, where the Boom clay formation outcrops. Haegeman & Mengé (2001) reported that the geological condition at this site consists of a homogeneous layer of Boom clay to a large depth. In its upper part the Boom clay exhibits horizontal layering and has a medium to high degree of fissuring. Therefore the Boom clay in its upper part has to be described as a stiff, fissured, layered overconsolidated clay. Fig. 3 provides boring logs along with SCPT test data obtained by Karl (2005) at this research site. The natural water table is found at 2·8–3·3 m below ground level. In this research two sampling techniques are used for sampling the overconsolidated stiff clay: borehole drilling with a thin-walled tube sampler, and rotary core drilling with a triple-tube wire line coring sampling system. Table 3 reports the engineering properties of the Boom clay at the

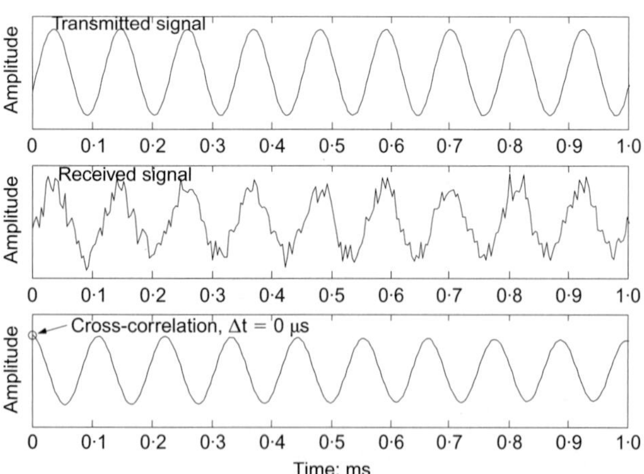

Fig. 2. Example of measurement of continuously sinusoidal shear wave velocity in Boom clay

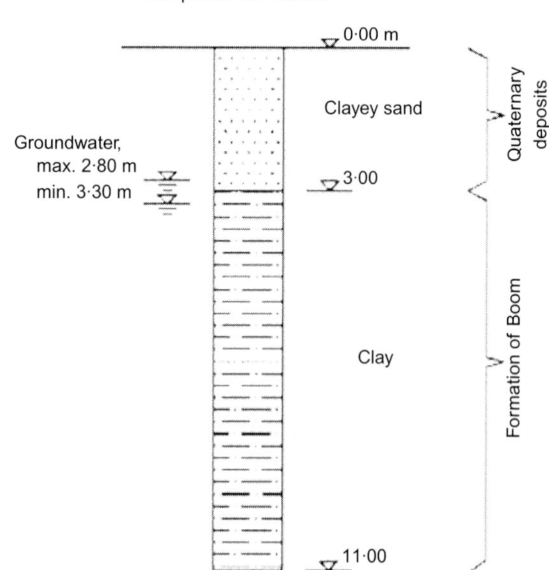

Fig. 3. Summary of test results at Sint-Katelijne-Waver (after Karl, 2005)

Table 2. Shear moduli of Boom clay using different waveforms

Methods	V_s: m/s	G: MPa
Single-period sinusoidal waveform	185·30	66·92
Four-period sinusoidal waveform	185·31	66·93
Continuously sinusoidal waveform	189·16	69·74

site determined on samples at depths of 5·0 m and 8·0 m. The water content of the samples from the triple-tube sampling technique (C-samples) appears to be higher. This can possibly be explained by the flushing process used in this sampling technique. Piriyakul & Haegeman (2005) studied the possible soil disturbance due to the two sampling techniques, and confirmed that both techniques fulfil the requirement of category class A sampling; however, the triple tube technique is preferred for its fast operation and the continuity of sampling.

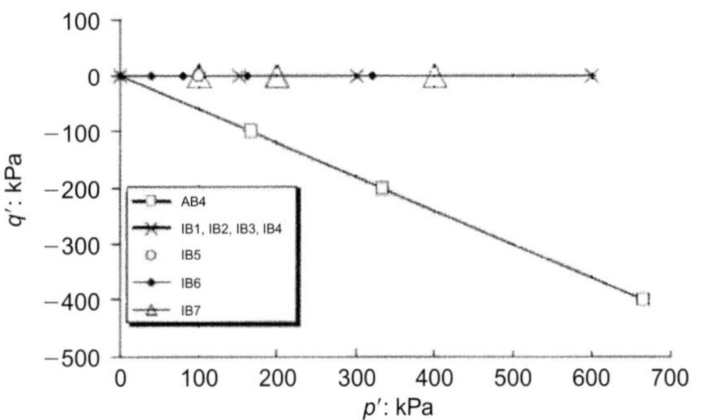

Fig. 4. Stress path of triaxial tests

Triaxial testing programme

Several tests (IBx) were performed on the undisturbed Boom clay C5/C8 and D8, consolidating the samples under isotropic conditions. In a similar way, tests ABx were performed on the undisturbed Boom clay C8, consolidating the samples under anisotropic conditions ($K = 2·0$). Fig. 4 summarises the stress paths followed in these tests.

VOID RATIO FUNCTION

The void ratio function $F(e)$ is adopted according to Lo Presti (1989) and Jamiolkowski *et al.* (1991) as given in equations (2) to (4): $F(e) = e^{-x}$ A value x of about 1·30 is proposed by Lo Presti (1995) and Jamiolkowski *et al.* (1995). Using this value, a regression analysis is performed including all data from the isotropic tests IB1, IB2, IB4 and IB7 and the anisotropic test AB4. The normalised values of $G_{0(ij)}$ at different stress states are shown in Fig. 5. The data points for the consolidated samples fall close to a single straight line, showing the values of S_{ij} and nj.

A void ratio function with the proposed value $x = 1·30$ fits the test data for the Boom clay less well, because the values of nv and nh are not equal. Therefore the test data are re-analysed in order to evaluate the void ratio function of the Boom clay. Based on the assumption that the principal stress exponent $nv = nh = n$, as in Jamiolkowski *et al.* (1995), the void ratio function with $x = 1·15$ leads to a unique value of n equal to 0·21, as seen in Fig. 6.

Furthermore, this research analyses the test data without fixing the value of $n = nv = nh$ or fixing the value of $x = 1·30$ for the void ratio function. So the values of nv, nh and x are free during regression analysis. Finally $x = 1·21$ is found for the void ratio function; S_{vh}, S_{hh}, S_{hv}, nv and nh are obtained as given in Fig. 7, and summarised in Table 4. The above-mentioned parameters are consistent with the values for six Italian clays, as seen in Table 5. It should be noted that no cross-anisotropy ($G_{vh} = G_{hv}$) is measured. It is assumed that, because of anisotropic fissuring and layering of this overconsolidated Boom clay, the axes of testing in the triaxial cell are not equal to the axes of cross-anisotropy in situ, giving rise to differences between G_{vh} and G_{hv}.

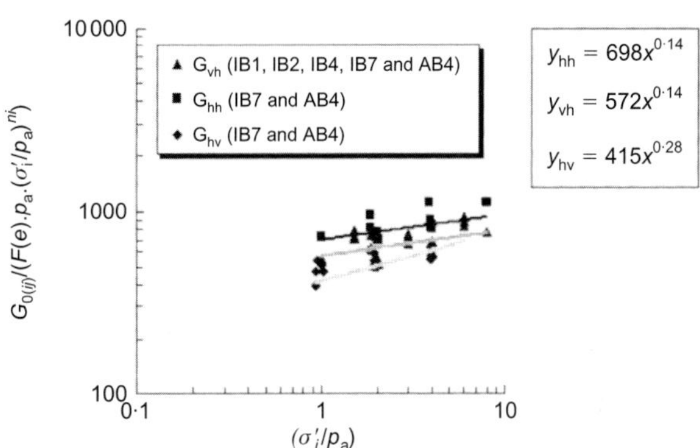

Fig. 5. Normalised shear moduli against effective consolidation stresses; $x = 1·30$

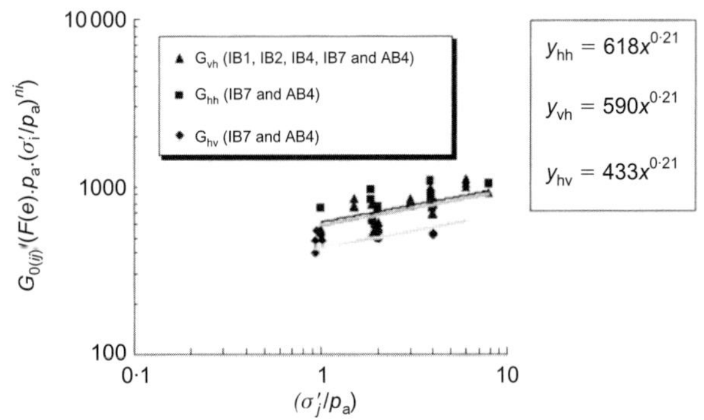

Fig. 6. Normalised shear moduli against effective consolidation stresses; $x = 1·15$

ACKNOWLEDGEMENTS

The authors are grateful to the Fund for Scientific Research – Flanders (FWO) for Grant B/02232 under which this study was carried out.

Table 3. Summary of properties of Boom clay

Parameter	Sample			
	C5	D5	C8	D8
Liquid limit, w_{LL}: %	65·38	65·95	57·51	54·12
Plastic limit, w_{PL}: %	22·23	23·21	21·02	22·89
Plasticity index, I_p: %	43·15	42·74	36·49	31·23
Specific gravity, G_s	2·71	2·69	2·70	2·71
Water content, w: %	29·42	24·00	28·25	23·32
Density, ρ: kg/m^3	2030	2036	2042	2034

Table 4. Parameters of Boom clay

Method	nv	nh	x	S_{vh}	S_{hh}	S_{hv}
Fixing $x = 1·30$	0·28	0·14	1·30	572	698	415
Fixing $n = nv = nh$	0·21	0·21	1·15	590	618	433
Free x and n	0·23	0·19	1·21	579	629	424

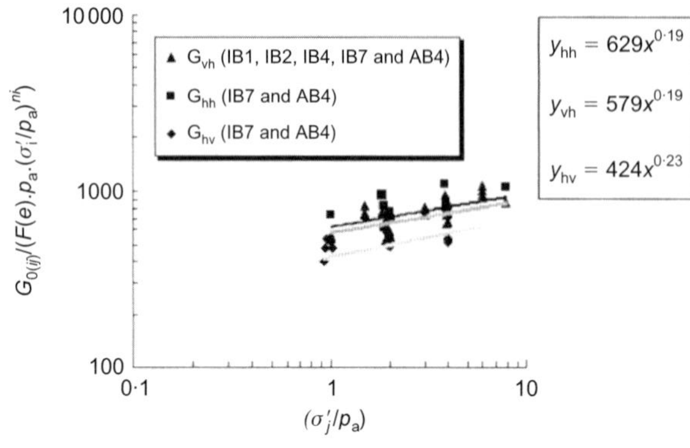

Fig. 7. Normalised shear moduli against effective consolidation stresses; $x = 1·21$

Table 5. Parameters of six Italian clays (after Jamiolkowski et al. 1995)

Soil	$nv = nh$	x	S_{vh}
Panigaglia	0·25	1·30	520
Pisa	0·22	1·43	640
Garigliana	0·29	1·11	560
Fucino	0·20	1·52	640
Montalto di Castro	0·20	1·33	632
Avezzano	0·23	1·27	810

REFERENCES

Blewett, J., Blewett, I. & Woodward, P. (1999). Measurement of shearwave velocity using phase-sensitive detection techniques. *Can. Geotech. J.* **36**, No. 5, 934–939.

Brignoli, E. G. M., Gotti, M. & Stokoe, K. H. (1996). Measurement of shear waves in laboratory specimens by means of piezoelectric transducers. *Geotech. Test. J.* **19**, No. 4, 384–397.

Fioravante, V. (2000). Anisotropy of small strain stiffness of Ticino and Kenya sands from seismic wave propagation measured in triaxial testing. *Soils Found.* **40**, No. 4, 129–142.

Fioravante, V. & Capoferri, R. (2001). On the using of multi-directional piezoelectric transducers in triaxial testing. *Geotech. Test. J.* **24**, No. 3, 243–255.

Haegeman, W. & Mengé, P. (2001). In situ and laboratory shear wave velocity measurements of the Boom clay. *Proc. 15th Int. Conf. Soil Mech. Geotech. Engng, Istanbul*, 415–418.

Hardin, B. & Blandford, G. (1989). Elasticity of particulate materials. *J. Geotech. Engng Div. ASCE* **115**, No. 6, 788–805.

Jamiolkowski, M., Leroueil, S. & Lo Presti, D. (1991). Design parameter from theory to practice. Proc. *GEO-COAST '91, Yokohama*, 877–917.

Jamiolkowski, M., Lancellotta, R. & Lo Presti, D. (1995). Remarks on the stiffness at small strains of six Italian clays. *Proc. 1st Int. Symp. on Prefailure Deformation Characteristics of Geomaterials, Sapporo*, 817–836.

Karl, L. (2005). *Dynamic soil properties out of SCPT and bender element tests with emphasis on material damping.* PhD thesis, Ghent University.

Lo Presti, D. (1989). Proprieta dinamiche dei terreni. *XIV Conferenza Geotechnica di Torino*, 1–62.

Lo Presti, D. (1995). Measurement of shear deformation of geomaterials in the laboratory. *Proc. 1st Int. Symp. on Prefailure Deformation Characteristics of Geomaterials, Sapporo*, 1067–1088.

Piriyakul, K. & Haegeman, W. (2005). Soil disturbance assessment in soil sampling. *Can. Geotech. J.* (submitted).

Stallebrass, S. E., Atkinson, J. H. & Mašín, D. (2007). *Géotechnique* **57**, No. 2, 249–253

Manufacture of samples of overconsolidated clay by laboratory sedimentation

S. E. STALLEBRASS*, J. H. ATKINSON* and D. MAŠÍN[†]

KEYWORDS: clays; compressibility; fabric/structure of soils; laboratory tests; sedimentation; stiffness

INTRODUCTION

The behaviour of a given soil depends on its state, defined in terms of stresses and specific volume or water content, and structure, which may be due to fabric and/or bonding. Stiff sedimentary clays will have a structure resulting from their deposition by sedimentation and in some cases from post-depositional processes. An important effect of this structure is to change the limiting combinations of stress state and specific volume at which such clays can exist.

Soil samples for testing are usually either undisturbed from the ground or reconstituted in the laboratory as slurry and then subjected to one-dimensional consolidation. Samples reconstituted in the laboratory will not have the same structure as the natural sedimentary clay. In addition, because of the effect of structure on volumetric states it will not always be possible to achieve the same specific volume and stress state as occurs in the natural clay or to do so by following the same stress history. This problem is illustrated schematically in Fig. 1, with reference to one-dimensional compression and swelling curves for undisturbed samples of clay with a stable structure and clay reconstituted in the laboratory. Undisturbed and reconstituted samples at A have the same specific volume and effective vertical stress but different overconsolidation ratios.

The advantage of testing reconstituted samples of soils is that it is easy to manufacture essentially identical samples, whereas two natural samples are rarely identical. For stiff sedimentary clays, this is particularly important in detailed studies of pre-failure deformation, which can be influenced by a wide range of factors (Hight *et al.*, 2003). The aim is to build up a framework to characterise this behaviour and in particular the influence of the depositional and post-depositional structure.

This note describes a process developed to create samples of overconsolidated sedimented clay in the laboratory, which re-created, as far as possible, the structure in natural stiff clay resulting from sedimentation, compression and swelling under one-dimensional stress conditions. Evidence for the presence of this structure in the sedimented clay is presented, together with preliminary data comparing small-strain stiffness and pre-failure deformation measurements for samples at overconsolidated states.

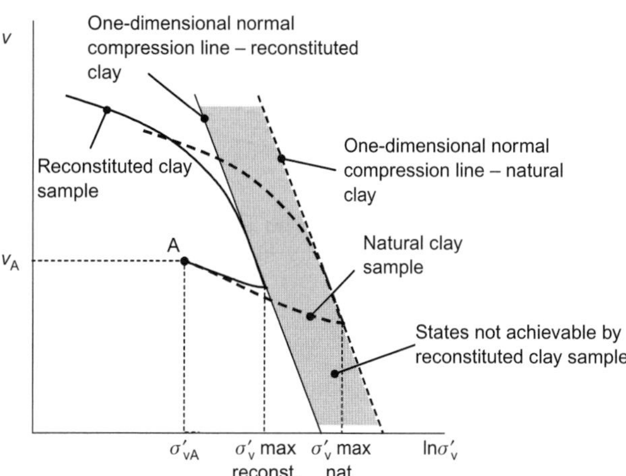

Fig. 1. Diagram showing differences in stress state, specific volume and stress history for natural and reconstituted overconsolidated clay samples

Laboratory sedimentation is not a new process, but it has not previously been used as part of a procedure to create overconsolidated clay samples. Bjerrum & Rosenqvist (1956) and Locat & Lefebvre (1985, 1986) were primarily using sedimentation to investigate the effect of the salinity of the pore water on the volumetric compression of highly sensitive natural marine clays, whereas Olson (1962), Monte & Krizek (1976) and Ting *et al.* (1994) have examined how the salinity of the pore water affects flocculation or dispersion of clay particles for specific clay minerals such as illite and kaolinite and subsequent volumetric compression. Studies on sedimentation by Been & Sills (1981) and Edge & Sills (1989) used natural soils to create sediments, but were concerned mainly with examining the transition from suspension to soil.

In this study reconstituted and sedimented samples were made from the same bulk sample of unweathered London Clay. It was obtained from the site of the former Knightsbridge Crown Court, from a horizon about 52 m above the base of the Formation. Typical Atterberg limits for this horizon were $I_l = 75$ and $I_p = 30$.

The principal purpose of the study was to investigate the differences in samples made by sedimentation and made in the conventional manner by reconstitution and compression from a slurry. The laboratory samples were sedimented through salt water to represent, so far as possible, the natural sedimentation process of the London Clay.

Manuscript received 5 May 2006; revised manuscript accepted 15 August 2006.
Discussion on this paper closes on 1 August 2007, for further details see p. ii.
* Geotechnical Engineering Research Centre, City University, London, UK.
† Charles University, Prague, Czech Republic.

PREPARATION OF LABORATORY-SEDIMENTED CLAY

The key principles of the sedimentation process are:

(*a*) reconstitution of natural clay into a dilute slurry

(*b*)   sedimentation through a column of water
(*c*)   one-dimensional consolidation and swelling.

The structure formed by laboratory sedimentation may be affected by the pore water chemistry, as noted above, but also by whether slurry is introduced into the column of water in a continuous or discontinuous process and at what rate. This determines the layering in the sedimented sample, which is an important aspect of fabric, as reported by Coop & Cotecchia (1995).

Reconstituted slurry used for sedimentation

Slurry with a water content of about 1250% was prepared by combining London Clay at its natural water content with salt water that had a salinity of 3·51%. In these initial tests the slurry contained about 300 g of London Clay. This slurry was then poured into the sedimentation column (Fig. 2(a)), which already contained about 12 l of salt water, resulting in slurry with an initial water content of approximately 5800%. Using salt water ensured that the sedimentation procedure was representative of conditions likely to occur during the sedimentation of the natural clay. The salinity in the pore water acted as a flocculating agent, which should create a more random and open structure within the sedimented sample (Locat & Lefebvre, 1985). It also ensured that sedimentation was relatively quick. After self-weight sedimentation was complete when no change in the position of the slurry water interface was observed, usually after 3 days, further slurry manufactured as described above was added. This process was repeated three times, resulting in clay sediment containing four layers.

Sedimentation column

The equipment is described in Fig. 2. It consists of a purpose-built sedimentation column shown in Fig. 2(a), made from a 2 m high Perspex tube of internal diameter 94·2 mm. The tube may be split into two parts, which facilitates the application of a vertical stress to the clay sediment during the second stage of preparation of the samples. After sedimentation a stress is applied through a submersible piston (Fig. 2(b)), which, when fitted with its metal rod and top platen, has positive buoyancy. This made it possible to apply very small stresses to the sediment at the start of loading. On the face of the piston, and in contact with the soil, there is a porous metal plate that allows drainage around the piston. There is also drainage at the base of the column, which is closed off during sedimentation and opened during the compression of the clay sediment. Fig. 2(c) shows a detail of the arrangement at the base of the column, which includes an aluminium cap that can be removed to allow thin-walled sampling of the compressed sediment.

One-dimensional compression

The clay sediment was first compressed by applying dead weights to the submersible piston. The weights were gradually increased until the total axial stress was 70 kN/m². When consolidation was complete, the base of the sedimentation column was opened and three 38 mm diameter samples were obtained using thin-walled samplers (see Fig. 3). The height of the clay sediment at this stage was approximately 140 mm. Two of the samples obtained were immediately compressed further in a stress path triaxial apparatus and the third was sealed with wax for use at a later date.

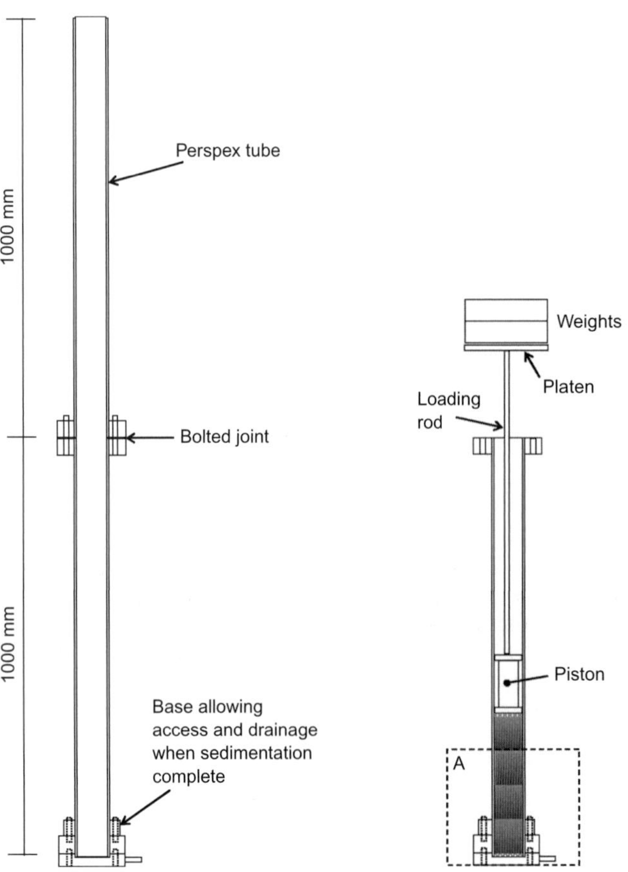

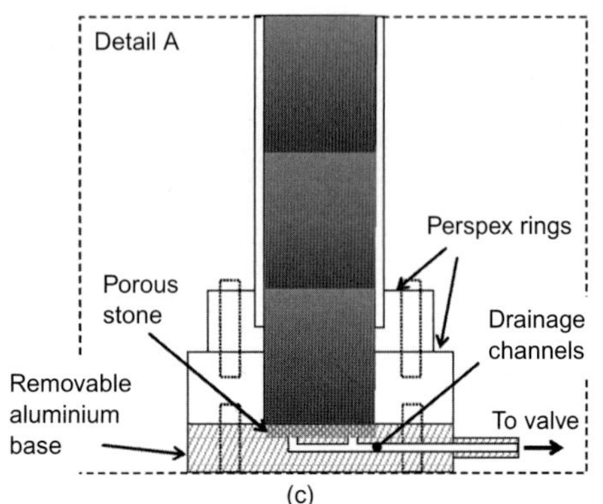

Fig. 2 Sedimentation apparatus

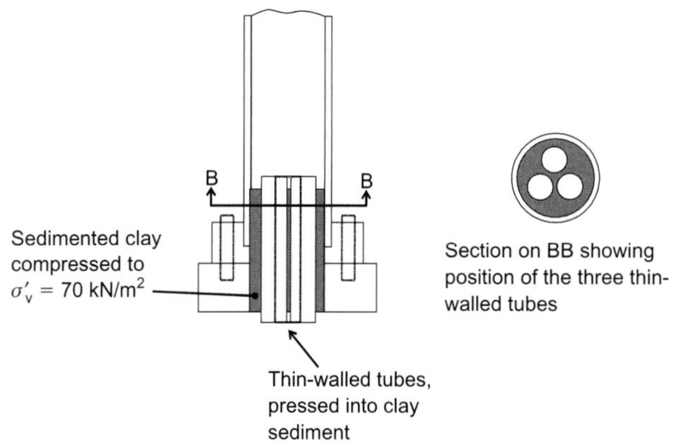

Fig. 3. Method used to obtain samples of clay sediment from sedimentation column

Fig. 4. Photograph of two split samples of London Clay, one sedimented and the other reconstituted

The samples were one-dimensionally compressed to higher stresses and then one-dimensionally swelled in a stress path triaxial apparatus fitted with local axial and radial gauges and bender elements. Samples taken from the sedimentation column were trimmed to about 91 mm in height and placed in the triaxial cell. Local axial and radial strain gauges were glued directly to the membrane. Careful measurements of the mass and dimensions of the sample were taken at this point, which were used with the final water content to determine the specific volume of the sample. The samples were one-dimensionally compressed to an effective vertical stress of 400 kN/m² and swelled one-dimensionally to various overconsolidation ratios. Bender element measurements of shear stiffness were taken at intervals during this process. Samples were then sheared under a range of stress paths.

EVIDENCE FOR THE DEVELOPMENT OF STRUCTURE IN LABORATORY-SEDIMENTED SAMPLES

The sedimented clay samples are compared with samples prepared conventionally from London Clay reconstituted to a slurry with a water content of 125% and consolidated in floating ring consolidation tubes to a final vertical effective stress of 70 kN/m². These samples were made both with distilled water and with salt water of the same salinity as that used in the samples of sedimented clay. In this way it was possible to establish the effect of both the sedimentation process and the pore water chemistry on the mechanical response of the clay. These samples were also one-dimensionally compressed and swelled in stress path triaxial cells to the same vertical effective stress using exactly the same procedures as have been described for the samples of sedimented clay.

Figure 4 shows split samples from clay sediment and conventionally reconstituted clay taken before one-dimensional compression in a stress path cell. The samples were air-dried until they could be conveniently split, after about 24 h. Both samples are intact with no open fissures; the photograph shows the surface texture/fabric where the samples were split. The reconstituted sample, on the right, has a random fabric with a few scattered larger-sized particles, whereas the sedimented sample on the left has a clearly layered fabric in the upper section and a layer of larger silt particles at the base, which represents the interface between one layer of slurry and the layer below. The sedimented sample broke apart at this silt/clay interface. The segregation of larger particles during sedimentation was also observed by Edge & Sills (1989) and creates a layered fabric in the

samples. At this point the layers are approximately 30 mm thick.

As might be expected, the sedimentary layered fabric, clearly visible in Fig. 4, causes the one-dimensional compression curves for the sedimented samples, plotted as $\ln v : \ln p'$, to be distinctly different from the one-dimensional curves for reconstituted samples. This is illustrated in Fig. 5 by data taken from Mašín et al. (2003), which shows one-dimensional compression data for seven tests carried out on both sedimented and reconstituted samples. All the curves are approximately parallel at higher stresses, but fall into two distinct groups, indicating that the sedimentation has created a stable structure or fabric, which can be characterised by values of sensitivity S_t, as defined by Cotecchia & Chandler (2000), which vary between 1·5 and 1·8. These data and others presented in Mašín et al. (2003) also demonstrate that the pore water chemistry has little effect on the mechanical behaviour of the reconstituted clay. It is thought that variations in the initial states of the sedimented samples are due to variations in the water content of the sedimented sample, which had an average specific volume of 2·93 in the bottom third and 2·70 in the top third. This is probably due to side friction in the sedimentation column. The specific volume of the samples after testing in the stress path triaxial cells is much more uniform, changing throughout the sample by less than 0·02.

PRE-FAILURE DEFORMATION OF SEDIMENTED CLAY

Figure 6 shows very-small-strain stiffness measured from bender elements during one-dimensional compression and swelling to overconsolidated states. There is no apparent difference in G for sedimented and reconstituted clay samples, confirming that G_{max} for both reconstituted and sedimented samples of clay is dependent on current state and overconsolidation ratio (Viggiani & Atkinson, 1995) and is not affected by the sedimentation structure or fabric in these clays. Rampello et al. (1994) found similar results for bender element tests, and among others Coop et al. (1995) and Cotecchia (1996) presented data from small-strain stiffnesses obtained from triaxial tests, which also support this result. This may not be the case at larger strain levels, as is evidenced by data from tests on two samples, one sedimented and the other reconstituted, which after compression to $\sigma'_v = 300$ kN/m² were swelled one-dimensionally to $\sigma'_v = 100$ kN/m², recompressed one-dimensionally to $\sigma'_v = 266$ kN/m², and then sheared at constant p' in extension.

Fig. 7 shows the variation in tangent shear stiffness G with strain during constant-p' shearing for these samples. Although the small-strain stiffness is very similar for both samples and close to the value obtained from bender element

tests, at shear strains of around 0·0002 the stiffness of the reconstituted sample is about half that of the sedimented sample. Further work is necessary to investigate these phenomena in more detail.

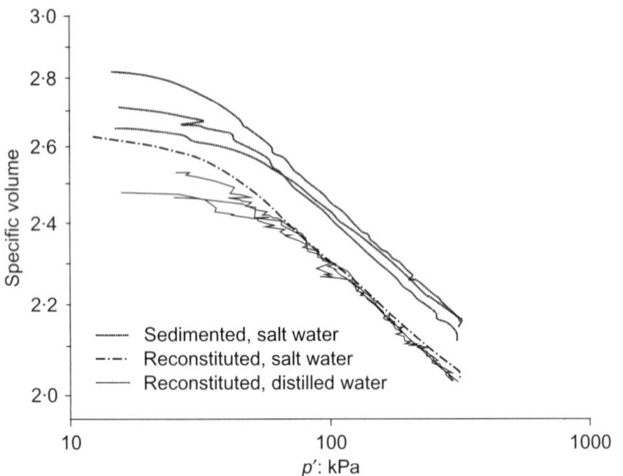

Fig. 5. One-dimensional compression data for reconstituted and sedimented samples (after Mašín *et al.*, 2003)

CONCLUSIONS

A method of creating samples of sedimented clay in the laboratory has been described. Visual comparison of split samples of sedimented and conventionally reconstituted clay has shown that the sedimented samples have a layered structure both within the fine-grained material and due to segregation of larger particles during sedimentation. This sedimented structure is also manifest in the large-strain response of the samples, as reported in Mašín *et al.* (2003) and illustrated here by a comparison of one-dimensional compression curves from sedimented and reconstituted samples. The value of sensitivity S_t, a measure of the effect of structure, for the sedimented clay varies from 1·5 to 1·8, compared with the value of 2 estimated by Cotecchia (1996) for natural London Clay. For stiff sedimentary clays prefailure deformations are of particular significance. Small-strain stiffness measured from bender elements was not affected by the structure in the sedimented clays, but there is some evidence that shear stiffness measured at intermediate strains, say 0·0001 to 0·01, may be influenced by the sedimentary structure.

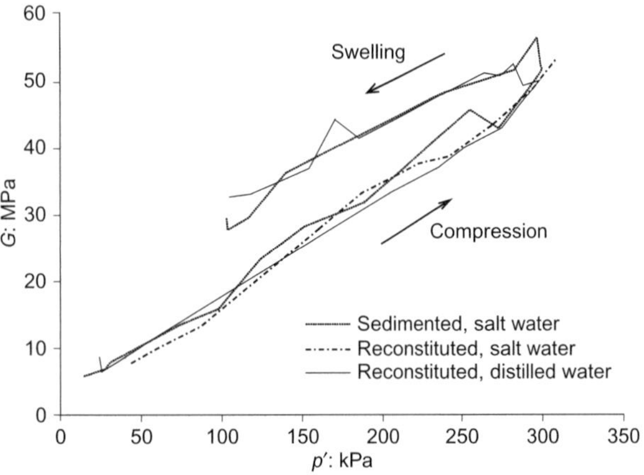

Fig. 6. Shear stiffness obtained from bender element tests during one-dimensional compression and swelling of reconstituted and sedimented samples

NOTATION

G tangent shear stiffness
G_{max} shear stiffness measured in bender element tests
I_l water content at the Liquid Limit
I_p water content at the Plastic Limit
p' mean effective stress
S_t sensitivity of a natural clay, a measure of the effect of structure.
v specific volume
ε_s deviatoric shear strain
σ'_v vertical effective stress

REFERENCES

Been, K. & Sills, G. C. (1981). Self-weight consolidation of soft soil: an experimental and theoretical study. *Géotechnique* **31**, No. 4, 519–535.

Bjerrum, L. & Rosenqvist, I. T. (1956). Some experiments with artificially sedimented clays. *Géotechnique* **6**, No. 6, 124–136.

Coop, M. R. & Cotecchia, F. (1995). The compression of sediments at the archaeological site of Sibari. *Proc. 11th Eur. Conf. Soil Mech. Found. Engng, Copenhagen* **8**, 19–26.

Coop, M. R., Atkinson, J. H. & Taylor, R. N. (1995). Strength and stiffness of structured and unstructured soils. *Proc. 11th Eur. Conf. Soil Mech. Found. Engng, Copenhagen* **1**, 55–62.

Cotecchia, F. (1996). *The effects of structure on the properties of an Italian Pleistocene clay*. PhD thesis, University of London.

Cotecchia, F. & Chandler, J. (2000) A general framework for the mechanical behaviour of clays. *Géotechnique* **50**, No. 4, 431–447.

Edge, M. J. & Sills, G. C. (1989). The development of layered sediment beds in the laboratory as an illustration of possible field processes. *Q. J. Engng Geol.* **22**, No. 4, 271–279.

Hight, D. W., McMillan, F., Powell, J. J. M., Jardine, R. J. & Allenou, C. P. (2003) Some characteristics of London Clay. In *Characterisation and engineering properties of natural soils* (eds Tan, T. S., Phoon, K. K., Hight, D. W. and Leroueil, S.) 851–908. Lisse: Swets & Zeitlinger.

Locat, J. & Lefebvre, G. (1985). The compressibility and sensitivity of an artificially sedimented clay soil: the Grande-Baleine marine clay, Quebec, Canada. *Mar. Geotechnol.* **6**, No. 1, 1–28.

Locat, J. & Lefebvre, G. (1986). The origin of structuration of the

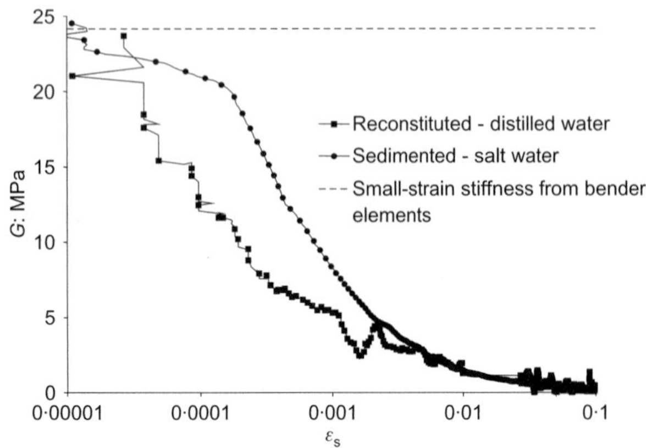

Fig. 7. Variation of shear stiffness with strain during constant-p' extension for reconstituted and sedimented samples

Grande-Baleine marine sediments, Quebec, Canada. *Q. J. Engng Geol.* **19**, No. 4, 365–374.

Mašín, D., Stallebrass, S. E. & Atkinson, J. H. (2003). Laboratory modelling of natural structured clays. *Proc int. workshop on geotechnics of soft soils: theory and practice*, VGE, The Netherlands, pp. 491–496.

Monte, J. L. & Krizek, R. J. (1976). One-dimensional mathematical model for large strain consolidation. *Géotechnique* **26**, No. 3, 495–510.

Olson, R. E. (1962). The shear strength properties of calcium illite. *Géotechnique* **12**, No. 1, 22–43.

Rampello, S., Viggiani, G. M. B. & Silvestri, F. (1994) The dependence of small strain stiffness on stress state and history for fine-grained soils: the example of Vallericca clay. In *Prefailure deformation of geomaterials* (eds Shibuya, S., Mitachi, T. and Miura, S.), pp. 273–278. Rotterdam: Balkema.

Ting, C. M. R., Sills, G. C. & Wijeyesekera, D. C. (1994). Development of K_0 in soft soils. *Géotechnique* **44**, No. 1, 101–109.

Viggiani, G. M. B. & Atkinson, J.H. (1995). Stiffness of fine-grained soil at very small strains. *Géotechnique* **45**, No. 2, 245–265.

INFORMAL DISCUSSION

Session 2
Laboratory and In Situ Techniques and their Interpretation

CHAIR: DR DAVID NASH AND DR SARAH STALLEBRASS

Points for discussion:

1. How robust are determination of stiffness parameters, both elastic and at intermediate strains?
2. Do drained laboratory tests have any value in engineering practice?
3. How can we measure values of permeability which will be useful in design?

Dr Matthew Coop, *Imperial College London*

In his introductory overview, David Nash raised a question about interpretation of bender elements. At Imperial College we use time domain and frequency domain techniques and we get similar values from both methods on samples of London Clay. The frequency domain measurements were made using the π point method and were within a few percent of what we got using the time domain approach. In other soils there may be typically 10 or 20% difference between the time domain and the frequency domain values of shear wave velocity from bender element measurements.

Dr David Nash *Chairman*

Dr Coop would you also like to comment on the use of axial bender elements? I notice from your papers that you weren't using the axial benders to measure the VH shear wave velocity.

Dr Matthew Coop, *Imperial College*

Yes, that was a pragmatic choice, simply because the samples were quite heavy. We were testing 100 mm diameter samples. Placing them on axial bender elements was considered to be too awkward to be practical.

Dr Kenny K. Sorensen, *Engineering College of Aarhus*

I would like to comment on the analysis of bender element tests. I have also done tests with bender elements in a triaxial cell on London Clay and compared various methods for obtaining shear wave velocities; time domain, frequency domain, single pulse and multiple pulse signals. I consistently found when comparing frequency domain and time domain techniques that the frequency domain gave higher velocities—typically around 10% higher. The method I used was multiple pulse or a continuous signal sent at the sample resonance. With the peak to peak correlations in time domain analyses, there can be problems if a signal is used that's not at a resonant frequency because you have a transformation of the signal. I came to the conclusion that the best approach would be to send a multiple pulse or a continuous signal at the resonant frequency thereby avoiding transformation of the signal. I found this to be a very consistent technique.

Dr Apollonia Gasparre, *Geotechnical Consulting Group*

Just to comment on the bender element tests reported in our papers, we found that the signal of the bender elements in London Clay was very clear. I think one of the advantages for us was that we were using relatively large samples and therefore the travel path was long. With greater separation between the elements it was quite clear where the first arrival was. The agreement obtained between the measured and calculated small strain stiffness parameters in the triaxial tests demonstrates the applicability of the cross-anisotropic model in London Clay. In the field tests we assumed that G_{vh} was equal to G_{hv}, which was a pragmatic assumption–it allowed us to calculate other independent parameters that could not be calculated in triaxial tests. However the field data showed that there are some differences between G_{vh} and G_{hv}. This indicates that the soil is not wholly cross anisotropic. We were able to compare the cross anisotropic stiffness parameters that I could calculate from my triaxial tests with parameters that we could measure directly in the hollow cylinder. Again we obtained good agreement between the calculated and measured values. So I consider that although the cross anisotropic model is an idealisation it provides a good approximation to the behaviour of London Clay.

Dr David Nash *Chairman*

And a very good set of data has been obtained as a result of adopting the cross-anisotropic model.

Professor Richard Jardine, *Imperial College London*

I would like to respond first to David Nash's comment on the normalised summary London clay stiffness-strain curves shown in Fig 15 by Hight *et al.*, where we reported that trends from our recent high resolution triaxial tests fall below the 'bounds' previously estimated for London Clay. It is important to bear in mind the history of the latter 'bounds'. While the local strain sensors first deployed in our laboratories, around 30 years ago, provided a huge improvement on previous practice, we recognised (Jardine *et al.*, 1984, *Géotechnique*) that triaxial stiffnesses defined at $10^{-2}\%$ were reliable to only +/- 10%. Speculative curves were plotted to marginally lower strains, but we recognised that these could diverge widely—and that we could not define any linear range for soils such as London clay. Techniques have improved since. By the early 1990s, Professor Tatsuoka's high resolution LDT devices could resolve linear behaviour in triaxial tests at strains of $10^{-4}\%$ to $10^{-3}\%$. By the time of the 1997 *Géotechnique* Symposium in Print, Kuwano and Jardine (1997) could report similar measurements, as could colleagues at City University and elsewhere. This had become possible, with some difficulty, in research but has not been generally achievable in commercial practice. Reports continue to show a wide spread of results at small strains that may be interpreted selectively to give curves that may be unrealistically stiff at very small strains. The importance of time and rate effects has also been recognised: Clayton and Heymann (2001) showed that stiffness curves vary tremendously depending on creep pause durations and the potential interactions between creep and the strain rates mobilised in stress probing tests. Our awareness of this problem led us, from the 1980s, to specify procedures designed to isolate creep and stress probing response. For practical reasons these were specified in terms of external strain rates, so potential difficulties remained if apparatus compliance was significant. All of the above factors led to potential bias in the trends established from databases assembled from either commercial testing or research conducted without the highest resolution equipment. Most of the data contributing the 'previously established' bounds plotted in Hight et $al.$'s Figure 15 would have been affected by the above discussed limitations.

A further point to consider in relation to the Figure is its normalisation by p_0'. It appears from our recent research, and that by Pantelidou and Simpson (2007), that normalisation by the p_0' value imposed in the laboratory immediately prior to shearing can give disperse the results. It is more satisfactory to normalise by the in situ, geologically established, p_0' value, even though this depends on having a reliable estimate for K_0. If this is done, then the disparities seen in Fig 15 are less marked.

My second comment is to emphasise that the strong stiffness anisotropy of London Clay, and most other natural geomaterials, makes it inappropriate to combine data obtained in tests that impose different shearing paths (and rates) into a single stiffness-strain plot—as has been common in the past. For example, resonant column, bender element and field geophysical values of G_{vh} or G_{hh} cannot be plotted as joining onto the left hand side of non-linear stiffness curves from drained or undrained triaxial tests: we can't mix these disparate types of moduli.

Dr Minh Nguyen Anh, *Atkins*

Just want to make two comments to support Dr Gasparre's points. Firstly, bulk modulus measurements are notoriously difficult to carry out especially at very small strains. In order to do so, local strain instrumentation is required and this often provides data at only one section. To apply these data to the sample as a whole, reasonably uniform properties must be assumed. Secondly, the accuracy of the Poisson's ratio determination needs to be considered. These values are calculated as the ratios of two strain components at very low strain levels. This places high demands on the measurement equipment and procedures. The last point that I want to mention is the stiffness anisotropy of London Clays. My work was based on hollow cylinder tests, which allow us to check the symmetric condition of the elastic stiffness matrix. We found that cross anisotropy elasticity was a reasonable model at very small strains but the symmetry of the matrix was not always fully satisfied.

Mr K.C. Law, *Edge Consultants UK Ltd*

Just a couple of general questions more than comments. Firstly, I would like to ask about the value of sophisticated laboratory tests such as bender elements in practice—they are very expensive and time consuming. Is it worth doing them or not? Are samples you get from standard sampling of boreholes adequate for such tests. I picked up from one of the previous speakers that sampling techniques have not been improved over the last 20 years or so. My second question is about in situ testing. The Symposium papers put a lot of emphasis on laboratory testing. However, for coastal engineering and near shore engineering, we use CPT test results to provide a range of design parameters. There is little mention in the Symposium of CPTs in stiff clays. Are such in situ tests still considered to be useful?

Dr John Powell, *BRE*

Just before I respond to the question on CPT, perhaps I could pick up on two other matters. I think it is very encouraging to see that laboratory testing is now resolving anisotropic stiffness values which match very closely what is seen in the field measurements. I would encourage the use of in situ measurements of V_{hh}, V_{hv} and V_{vh} in the field. The field values act as a check on laboratory measurements which may be affected by the applied consolidation stresses and creep on test samples. Secondly, I notice that in the data from the Imperial College tests that the resonant column stiffness values are higher than those found using bender elements. At BRE we have found a similar trend in our test data. Have similar trends also been observed between other types of stiffness measurement?

Regarding the use of CPT in stiff clays, one of the best correlations that we have got from the CPT in London clay is between q_t and the G_{hh} value. We can profile pretty reliably the small strain G_{hh} from the CPT. There have been attempts to correlate q_t with G_{vh} but this relationship is much less convincing.

Dr Matthew Coop, *Imperial College*

I would like to respond to the question regarding the cost of small strain shear stiffness measurements. Bender elements are not expensive—a few tens of pounds. In order to make the measurements a computer card is required. These cost a few hundred pounds—and for this modest outlay very accurate measurements of stiffnesses can be made. Most of the cost is in the time and effort required to obtain good data.

Professor Richard Jardine, *Imperial College London*

In response to John Powell's question about possible disagreements between different types of geophysical laboratory and field G_{vh} measurements with other soil types, we do have a limited comparative database. We have both resonant column and bender measurements for Dunkirk and Ham River sands, with field seismic CPT profiles also for Dunkirk. In addition, some tests were made recently for Florence silt, as part of the investigation for the high speed train link and station, which have been published by our colleagues at Arup Geotechnical (Hocombe et $al.$, 2007). Field geophysical measurements were made at another one of our research sites, Pentre in Shropshire. In this thick profile of low OCR glaciolacustrine clay-silts, the laboratory resonant column G_{vh} values fell marginally (around 10 to 20%) below the seismic cone test profile found by BRE (see Chow, 1997), while at Dunkirk bender element and resonant column tests both match the BRE's seismic cone profile reasonably well (Kuwano, 1999). While for Ham River Sand the resonant column tests reported by Porovic (1995) fell below the bender element data gathered by Kuwano (1999) by 0 to 30%, depending on the effective stress level and test interpretation technique, other tests by Connolly gave closer agreement (Kuwano, 1999).

Dr Angus Skinner, *Engineering Consultant*

David Nash pointed out that the stiffness of samples as measured in triaxial tests is very dependent on rate. In bender element tests, stiffnesses are measured at very high strain rates. The stress paths imposed by engineering construction in the field are followed at very slow strain rates. I'd like to know what the authors think about the effects of strain rate on stiffness. Given that stiff clays are anisotropic, is it possible that the strain rate effects are different for the different components of stiffness or for different stress paths? The second point is to do with the discontinuities which develop when loading triaxial samples. Is it possible that discontinuities were initiated and propagated in the hollow cylinder tests reported in the Symposium.

Mr Howard Roscoe, *Atkins*

Just moving on to the question of measuring values of permeability in stiff clays. The lesson from our work at Ashford was that the values of in situ permeability measurements were very significant and affected many aspects of the construction. They affected the construction of the piles, (using the permeability values we designed the walls on a drained basis). They affected the calculated pressures on retaining walls, the temporary excavation sequences, and demonstrated that there was a requirement for long term drainage and passive relief wells. We found a strong contrast between the Atherfield and Weald Clays. The Atherfield Clay doesn't have much fabric while the Weald Clay has a strong horizontal fabric which gives anisotropic and comparatively high values of permeability.

Regarding the question of how to measure permeability– no laboratory determinations of permeability were attempted. There were a number of measurements undertaken in boreholes during the site investigation which were misleading. It was when we were able to get onto the site and do pumping tests and dewatering trials that we actually got to grips with realistic values of permeability. There were fairly complicated pumping tests done from trial wells but the technique that gave us the greatest help was some ejector well trials. The Contractors undertook these trials to determine whether or not we could effectively lower pore pressures in the Weald Clay to give us an advantage with the temporary works. We found that by imposing a big suction on the Weald Clay using ejector wells we got quite surprising and informative responses with draw downs over quite a wide area. The drainage we were able to achieve in the Weald Clay provides an example of what may be achieved in some other stiff clay formations.

Dr Kenichi Soga, *Cambridge University*

In response to that contribution, the measurement of reliable values of permeability in stiff clays is an important and complex issue. At Cambridge University, we've been working on self boring permeameter testing for some time. One of the techniques that we have been looking at is to increase the pocket size and then determining the effect of changes of scale on the measured values of permeability. These data were presented in a paper in *Géotechnique* about a year ago. We plotted size of the test pocket versus permeability and showed that the apparent permeability was increasing as the scale of the test pocket increased. I would also like to respond to the question about the value of drained laboratory tests–do they have any value in engineering practice? The answer depends on the nature of the engineering problem. For example, the paper we wrote for this Symposium was looking at long term effects of tunnelling in stiff clay and we needed drained properties. But also we need undrained properties, and partially drained to drained properties. The question for the modeller is whether their models are calibrated to characterise the stiff clay in undrained, partially drained and drained conditions.

Mr Sunday Obeahon, *Jacobs UK Ltd*

I would also like to raise a question that is related to the relevance of drained laboratory tests in clay soil.

In a recent Jacobs project that I was involved in, an existing slope cut at 60 degrees has been standing for 15 years. The soil was stiff clay. If we carry out a slope stability analysis using strength parameters, we would find that the slope shouldn't be standing. We actually had to carry out a back-analysis to try to determine the engineering parameters. So the question is why is it standing? How long does it take for drained conditions to be established in stiff clays and can this time be calculated reliably? If this is not possible, is it not better to use undrained parameters instead of drained parameters for the analysis of such slopes?

Dr David Nash *Chairman*

Without knowing the details it is hard to comment but it does strike me that steep slopes in clay, such as one gets in eroding cliffs, are very often held up by the presence of suction. I recall taking students to Allum Bay and discussing with them the very steep slopes in the Reading Beds that are present there on the Isle of Wight. We noted that although the drained strength of the Reading Beds would suggest that the slope angle should be quite shallow the cliff is actually standing quite steeply and it must be because there is not full drainage. Pore suctions must be present and these are the likely explanation for the observed temporary stability of stiff clay cliffs.

Session 3

Characterisation of Formations and Material Models

Hight, D. W., Gasparre, A., Nishimura, S., Minh, N. A., Jardine, R. J. & Coop, M. R. (2007). *Géotechnique* **57**, No. 1, 3–18

Characteristics of the London Clay from the Terminal 5 site at Heathrow Airport

D. W. HIGHT*, A. GASPARRE†, S. NISHIMURA‡, N. A. MINH§, R. J. JARDINE‡ and M. R. COOP‡

The innovative engineering approach adopted for the new Terminal 5 at Heathrow Airport called for an advanced investigation of the London Clay strata, including detailed in situ profiling and stress path laboratory testing on high-quality rotary-cored samples. Although the scope of the investigations exceeded that normally specified for conventional design, questions relating to the structure and anisotropy of stiffness and strength of the clay remained that could not be answered. Further research was required, and the deep excavations at the site provided the opportunity for a team from Imperial College to take multiple block samples from three depths, supplemented by two additional dedicated rotary-cored boreholes. Intensive research was performed at Imperial College on these samples, as described in three companion papers by Gasparre *et al.* and Nishimura *et al.* This overview paper integrates the findings from the recent research with those from the commercial investigation and with earlier studies to extend our understanding of the geology and key characteristics of this stiff clay, and their variation with depth. The influence of lithology, structure and destructuring is examined, and the practical implications of the work are discussed.

KEYWORDS: anisotropy; clays; compressibility; permeability; shear strength; stiffness

L'approche technique innovatrice adoptée pour le nouveau Terminal 5 de l'aéroport d'Heathrow demandait la réalisation d'une étude approfondie de la strate de London Clay (argile de Londres), comprenant un profil in situ détaillé et des essais de cheminement de contrainte effectués en laboratoire sur des échantillons de carottes de haute qualité extraites parla méthode de forage rotary. Bien que la portée des ces travaux dépassaient celle normalement spécifiée dans le cadre de la conception géotechnique conventionnelle, des questions relatives à la structure, l'anisotropie de la rigidité et la résistance de l'argile restaient sans réponse. Il s'avérait donc nécessaire de poursuivre la recherche dans ce domaine. Les excavations profondes sur le site ont donné à une équipe de l'Imperial College l'opportunité de prélever des échantillons de blocs multiples, issus de trois profondeurs, complétés par deux trous de forage dédiés creusés au moyen de la technique rotary. L'Imperial College a réalisé des travaux d'étude intensifs sur ces échantillons, comme décrit dans trois articles associés de Gasparre *et al.* et Nishimura *et al.* Cet article de revue intègre les observations de travaux de recherche récents avec celles de l'étude commerciale, ainsi que les études précédentes, pour approfondir notre compréhension de la géologie, des caractéristiques clés de cette argile raide et de leur variation en fonction de la profondeur. L'influence de la lithologie, de la structure et de la déstructuration y est examinée et les conséquences pratiques de ce travail sont discutées.

INTRODUCTION

Ground investigations at the site of London's Heathrow Airport's newest terminal, Terminal 5 (T5), were carried out in various phases and included contamination studies, the site being previously occupied by the Perry Oaks Sewage Works, and archaeological studies. Phase 4 of the ground investigations was aimed at providing the geotechnical parameters required to make predictions of ground movements that would accompany the various supported excavations and predictions of the stand-up time of the temporary slopes to deep open-cut excavations (Kovacevic *et al.*, 2007). The Phase 4 investigation comprised a cluster of three boreholes at the location of the main terminal building and a rotary-cored borehole at the site of each other main excavation. The cluster of three boreholes (406, 407 and 408 in Fig. 1) comprised

(*a*) a continuously sampled rotary-cored borehole, which was used for a detailed examination of the lithology and fabric of the London Clay

(*b*) a second rotary-cored borehole from which high-quality samples were taken for laboratory testing, from depths selected on the basis of the first rotary-cored hole

(*c*) a borehole from which pushed thin-walled tube samples were taken and on which suction measurements were made on-site to assist in the estimate of in situ stresses.

Crosshole and downhole seismic wave measurements were made using the three co-linear boreholes. Laboratory testing was carried out in commercial laboratories and included the measurement of small-strain stiffness on locally instrumented 100 mm diameter rotary-cored samples.

Results from the commercial investigation have been published by Hight *et al.* (2002) and set in the context of the data obtained from other sites involving state-of-the-art commercial investigations of the London Clay. Attempts were made in the interpretation of the data to assess the significance of variations between the mechanical properties of the lithological units identified in the London Clay by King (1981). Restrictions on the availability of the necessary testing equipment in commercial laboratories meant that the investigations into the anisotropy of strength and stiffness of the clay were incomplete, as was the link between these properties and the lithological units. British Airports Authority generously provided access to the T5 site to allow a team of researchers from Imperial College (IC) to take block samples from one of the deep excavations, and provided samples of London Clay over its full thickness from two

Manuscript received 16 May 2005; revised manuscript accepted 14 November 2006.
Discussion on this paper closes on 1 July 2007, for further details see p. ii.
* Geotechnical Consulting Group, London, UK.
† Geotechnical Consulting Group; formerly Imperial College, London, UK.
‡ Imperial College, London, UK.
§ Atkins Ltd; formerly Imperial College, London, UK.

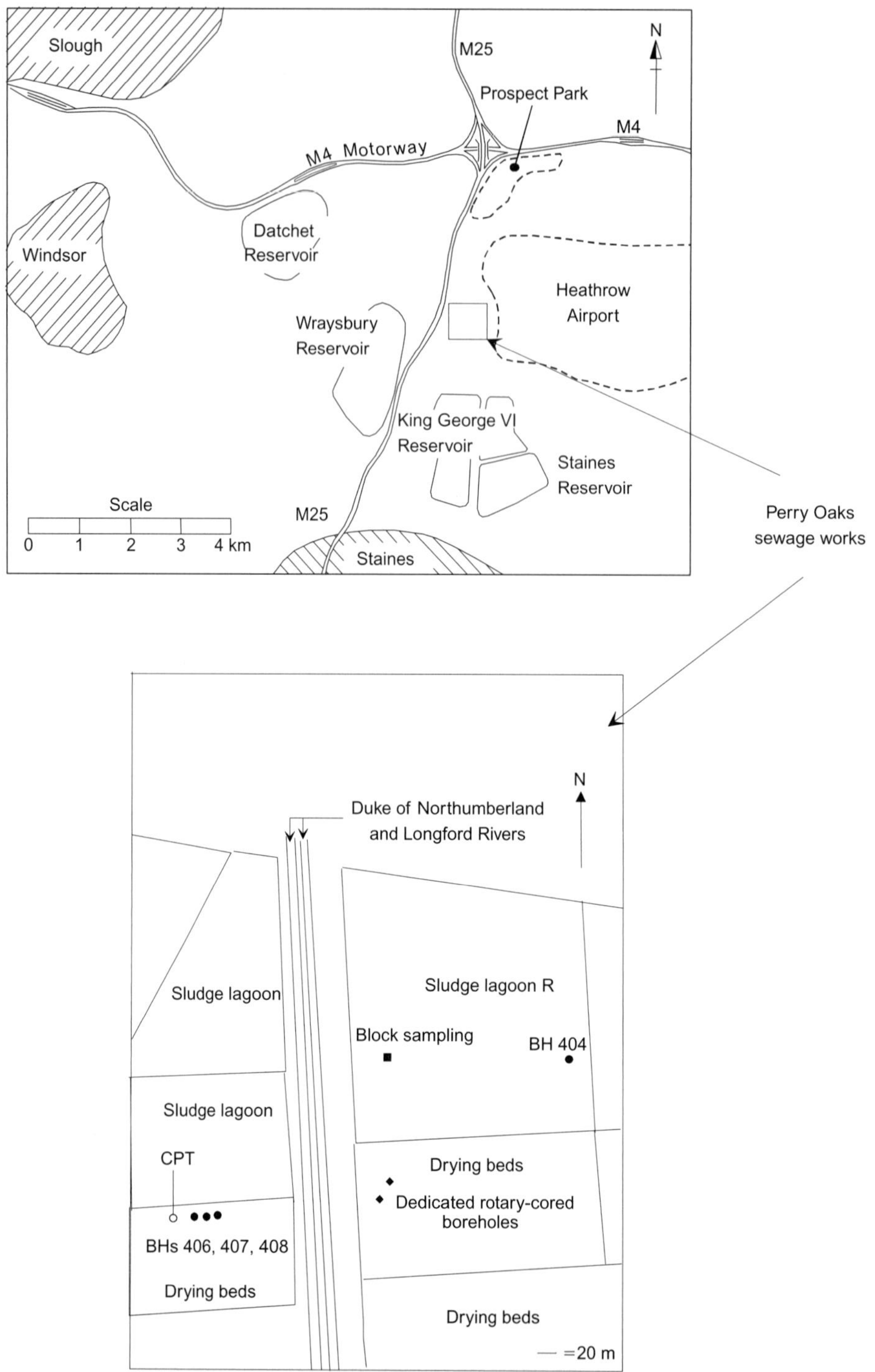

Fig. 1. Layout of sampling locations in Perry Oaks sewage works

adjacent rotary-cored boreholes, both advanced using a triple-tube core barrel with a plastic liner and a natural polymer mud flush technique. The samples were retrieved from their liners as soon as they were recovered from the borehole, and an outer annulus of around 5 mm thickness was trimmed off before preserving the samples in layers of clingfilm (plastic wrap) and wax. Block samples of 300 mm cube were taken from the benches of an excavation about 140 m away; the blocks were cut at depths below the top of the clay of 1·2, 5·2 and 10·5 m. The locations of the additional boreholes and of the block sampling are shown in Fig. 1 in relation to the commercial boreholes from which data will be drawn. The rotary cores were retrieved from a location where the London Clay was overlain by a 6 m thick Quaternary gravel layer; this layer had been removed ap-

proximately 25 years earlier at the block sampling location to create one of the lagoons in the sewage works.

The research carried out at IC, which was also supported by London Underground Limited and the EPSRC, involved testing of the block samples in two types of hollow cylinder apparatus, advanced triaxial stress path testing on the block and rotary-cored samples, and the use of dynamic testing techniques. The results of the research are described by Gasparre *et al.* (2007a, 2007b) and by Nishimura *et al.* (2007), where details of the tests, apparatus and techniques can be found. The aim of this paper is to integrate the results of the research with the results of the earlier investigations, to explore in more detail the differences in mechanical behaviour between the lithological units, and to highlight the practical significance of the findings.

BACKGROUND TO THE LITHOLOGY OF LONDON CLAY

The following summary is based on King (1981) and King (pers. comm.).

London Clay was deposited in marine conditions in the Eocene (30 million years BP) in a basin that was subsiding, and in which sea levels were rising. Bioturbation is common, taking the form of *Chondrites* burrows of 1–3 mm in diameter, infilled with grey silt (Fig. 2). Pyrite occurs as small aggregates, often infilling burrows and fossils. Phosphatic and claystone nodules are common. Very fine sand and silt dustings, partings and lenses are frequent in the siltier clays, and thicker sand layers occur in the sandy clayey silts.

Central London was near enough to the basin margin for the sedimentation to be affected by sea level changes. Thus a fall in sea level was associated with a coarsening of the material being deposited. The inferred cycles of sea level change and accompanying changes in sedimentation provide the basis for King's division of the London Clay into lithological units, based on the content of fossil fauna and partly on the contents of sand and silt (King, 1981). Five major transgressive-regressive cycles are recognised within the London Clay. The cycles are used to define five divisions in the clay (divisions A to E). Each cycle ideally marks the base of a coarsening-upward facies sequence.

The lowest unit, A2, which is approximately 12 m thick, is poorly sorted with a high percentage of silt, and occasional wood fragments and pyrite nodules. Within A2 there are several alternations of sandy clays and silty clays with diffuse boundaries, reflecting minor sea level changes (King, 1981). Partings and lenses of silt and fine sand are numerous. Unit A2 is generally non-calcareous and contains few claystones. It is often referred to as the basal beds.

Unit A3 has an overall thickness of 12 m. At its base is the first silty clay layer, which is homogeneous and slightly calcareous. The first main claystone layer is also encountered near the base of A3 and is more or less continuous. There are a further three or four claystone layers in close succession. Silt and sand partings become more common in the silty clay towards the top of the unit, and impersistent thin claystone layers occur.

The boundary between units A and B is marked by a 1 m thick sandy clay (Unit B1), which is glauconitic. Unit B2 comprises silty clays with weak silt and sand partings and numerous claystones, the lowest of which is the most

prominent and continuous. Sedimentary cycles (up to six) are weakly discernible within Unit B2. The total thickness of unit B is 25 m.

Units C1, C2, C3, D1, D2 and E are recognised in the full sequence of London Clay, which is up to 150 m in South Essex (King, 1981). At T5 only the lower part of the sequence is present, comprising units A2, A3, B1, B2 and the lower part of C; unit B1 was not recovered successfully in the recent sampling operation.

The increasing depth of deposition in moving from west to east across the London Basin, which was responsible for the increasing plasticity from west to east identified by Burnett & Fookes (1974), means that the lithological units are likely to become less distinct moving in this direction.

Dr King inspected the continuous core from borehole 404 at the T5 site and identified the units, based largely on the texture of the clay. Dr King's log of borehole 404 is presented in Fig. 3. Standing (pers. comm.) had found that the units that Dr King identified for him at St James's Park in Central London could also be identified on the basis of water content profiles. Water content profiles have been used successfully at T5 and other sites in Central London to identify the units (Hight *et al.*, 2002), and have confirmed the consistent thicknesses of the main units. The water content profiles at T5 (Fig. 4) suggested that further subdivisions of engineering importance might be possible within the main sub-units (Hight *et al.*, 2002). Fig. 4 shows on the left the unit boundaries identified by Dr King on the basis of his inspection of the cores and on the right the subdivisions of unit B(2) proposed by Hight *et al.* (2002) on the basis of water content. Unit B(2) has now been separated into three sub-units on the basis of their biostratigraphy by de Freitas & Mannion (2007). In the following presentations the boundaries proposed by King and de Freitas & Mannion have been adopted.

STRESS HISTORY

The soils that constitute the London Clay Formation in Central London were deeply buried as a result of continuing deposition. They were later subjected to mild tectonic loading during the Alpine orogeny forming the syncline of the London Basin. Erosion in late Tertiary and Pleistocene times has removed all the overlying deposits, and, especially in the Thames Valley, much of the London Clay itself. The amount of erosion has been estimated to range from about 150 m in Essex (Skempton, 1961) to 300 m in the Wraysbury district (Bishop *et al.*, 1965); these estimates were based on an interpretation of the vertical yield stress in oedometer tests on high-quality samples but ignored the effects of structure. Nevertheless, Chandler (2000) has shown from geological evidence that the amount of overburden removed is likely to have been of the order of 200 m. The clay at T5 is therefore heavily overconsolidated. The deposition of the late Quaternary gravel sheets along the Thames Valley followed the last part of this erosional process. At T5 the remaining thickness of the London Clay is 52 m and the gravels are up to 6 m thick. The gravels have protected the underlying clay from the effects of weathering, and alteration of the clay is restricted to approximately the top 1 m.

FABRIC OF THE LONDON CLAY

Various studies of the micro- and macro-fabric of the London Clay have been carried out. Micro-fabric studies, including those by Gasparre *et al.* (2007a), have confirmed the expected increasingly compact structure and increasing level of preferred orientation of the clay and clay aggregates with depth and maximum past pressure. This is likely to lead

Fig. 2. Bioturbation by *Chrondrities* (photo courtesy of Dr F. C. Chow)

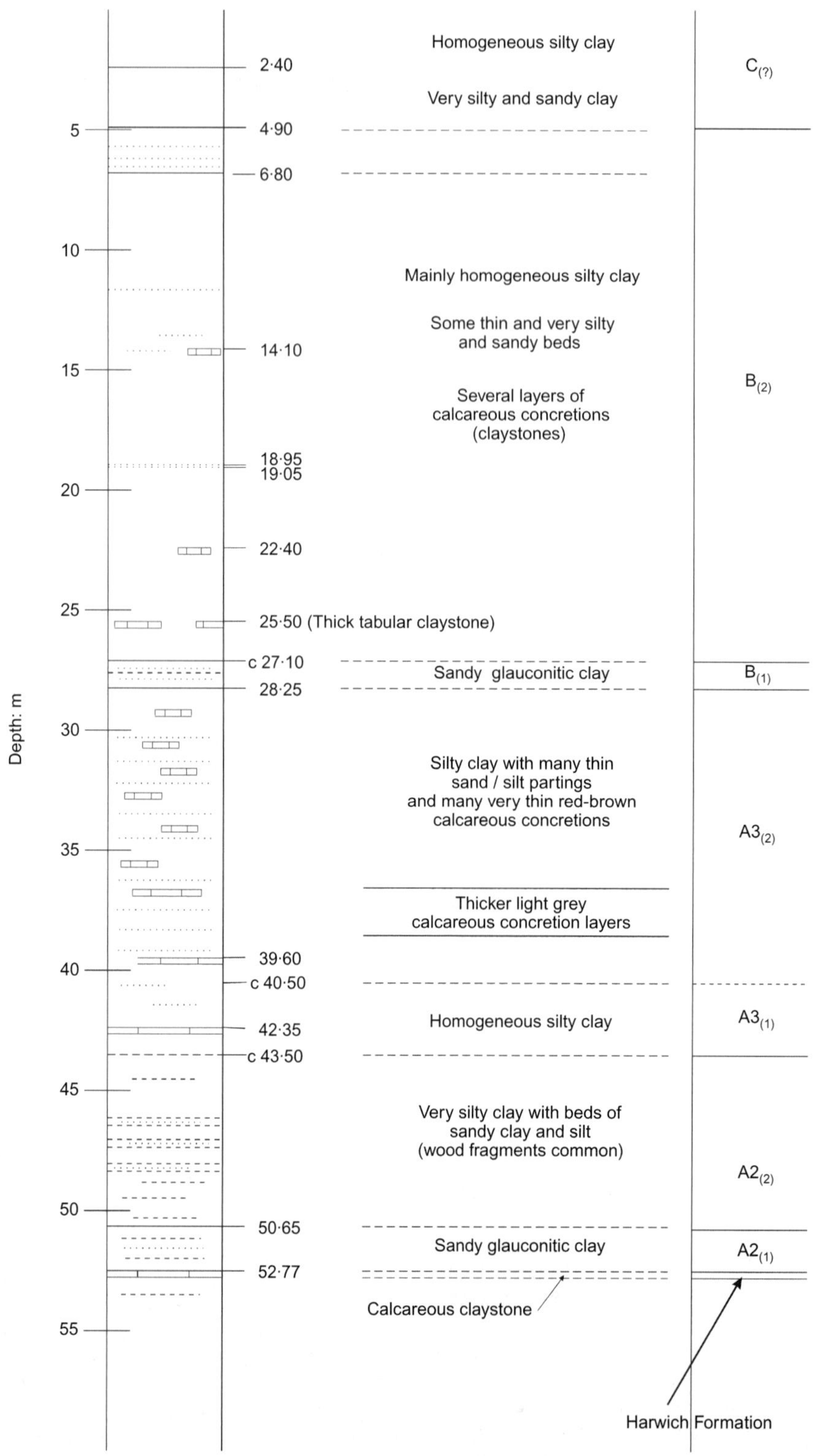

Fig. 3. Borehole log for BH 404 prepared by Dr King, identifying lithological units

to differences in behaviour between the units, with unit C having the most open and least anisotropic fabric and units A2 and A3 having the most compact and anisotropic fabrics. Dewhurst *et al.* (1998) illustrated the importance of the silt content in disrupting the orientation of the clay particles and aggregates in the deeply buried clay. On this basis, and the increasing fineness with depth in a unit, one would expect increasing levels of anisotropy with depth within a unit.

London Clay is heavily fissured and jointed, with joints often infilled with silts: examples from the T5 excavations are presented in Figs 5 and 6. The fissuring and jointing in London Clay was painstakingly investigated at the nearby site of the Wraysbury reservoir (Skempton *et al.*, 1969). The fissures are not polished but, as suggested by Hight (1998), appear to represent surfaces of separation between compressed and aligned clay aggregates: an example of a fissure observed by SEM is presented in Fig. 7 and shows a gently undulating surface with some pyrite infill.

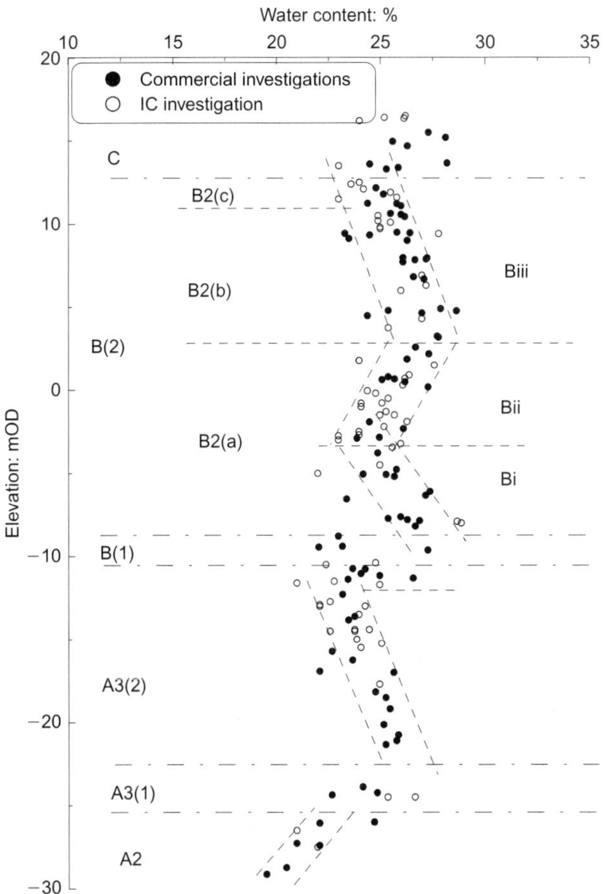

Fig. 4. Water content profile with lithological unit and sub-unit boundaries

Fig. 5. Fissure pattern at shallow depth at T5

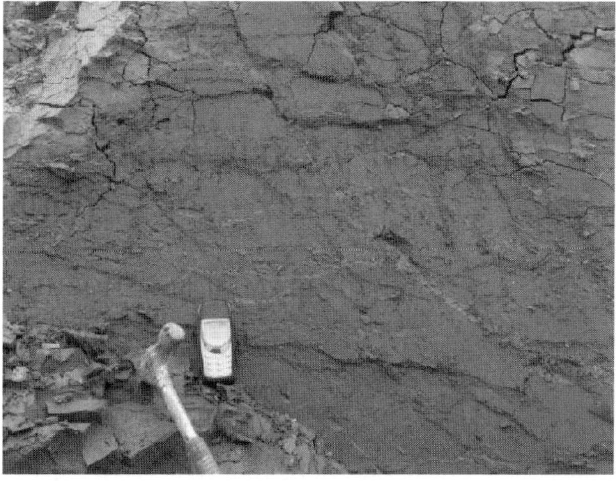

Fig. 6. Silt infill to discontinuities

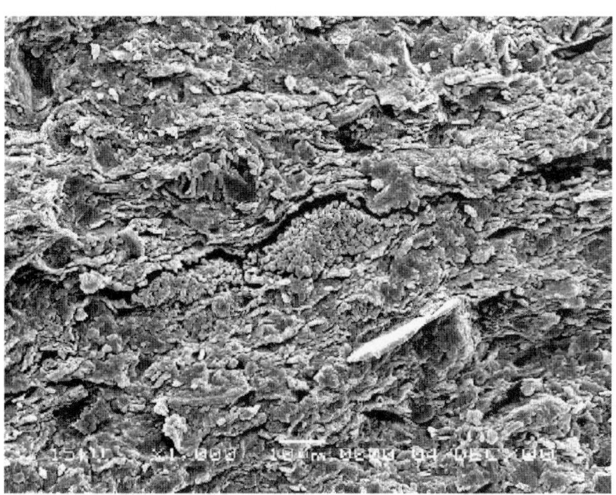

Fig. 7. SEM of fissure in London Clay (photo courtesy of Professor J. Locat)

Fig. 8. Shallow slope failures in T5 excavations

Observations from block sampling and sample preparation in the laboratory revealed that the fissures tend to have a variable spacing from a few centimetres to a few tens of centimetres, tend to be discontinuous, and extend up to approximately 15 cm, with orientations that are horizontal to sub-horizontal, and vertical to sub-vertical. The blocks from the 5·2 m depth below the top of the clay were more heavily fissured than those from 10·5 m depth, which is consistent with the observations reported by Skempton *et al.* (1969) that fissure intensity reduces with increasing depth.

The fissures are known to have an important effect on the mass behaviour of the clay, particularly at shallow depths. An illustration of this is provided in Fig. 8, which shows a typical shallow slope failure in the T5 excavations, with the clay separating on the fissures and joints into gravel and cobble-size blocks.

A key observation in relation to fissures and lithological units is that unit A2 is not fissured, presumably because of its larger sand content. In addition, in tests on samples on unit A3, Gasparre (2005) reports that the failure of only one of the eight samples was influenced by a pre-existing fissure, a much lower proportion than in the overlying units,

TECTONIC SHEARS

The tectonic forces associated with the Alpine orogeny produced discontinuities of the form described by Chandler *et al.* (1998) and observed at the nearby Prospect Park (see Fig. 1 for location). A major effort was made in the ground investigations at T5 to identify the locations at which these discontinuities might exist because of their potential influence on the stability of the temporary slopes. Surfaces were identified in core on which there was evidence of movement, and these were assumed to be tectonic shears. At the depths where these surfaces were found, 'disturbed zones' had been recorded in several boreholes, with occasional driller's descriptions of 'broken core and smeared clay'.

INDEX PROPERTIES

Profiles of Atterberg limits, combining the data from the commercial and research investigations, are presented in Fig. 9, with the lithological unit boundaries superimposed. Comparing Figs 4 and 9 it is apparent that the lithological boundaries coincide more clearly with breaks in the trends in the water content data than in the plasticity data. However, the coarsening upwards is reflected by the trend for a reducing liquid limit in moving up through units A3 and B2(b). Unit B2 and its sub-units are generally more plastic than units A3 and A2.

STRUCTURE

Gasparre *et al.* (2007a) discuss the structure of London Clay in detail by comparing the behaviour of the intact samples from the different lithological units with that of reconstituted samples formed by mixing the trimmings from intact samples at water contents about 1·25 times the liquid limit. A key comparison between the one-dimensional compression of the intact and reconstituted clay from the different units is presented in Fig. 10, in terms of void ratio (Fig. 10(a)) and void index (Fig. 10(b)), where the values of void index for the natural samples were determined directly by comparison with the reconstituted data for the same unit. The locations of the intrinsic compression and swelling lines (ICLs and ISLs) in e–σ'_v space vary with unit, the more plastic units from B2 lying above those from the slightly less plastic A3 and A2 units. The slopes of the ICLs and ISLs scatter around similar means for all units, with average values of 0·386 and 0·184 for C^*_c and C^*_s respectively. Gasparre *et al.* (2007a) report that, for the reconstituted soil, the separation of the ICL and critical state line is similar for each unit, with the triaxial compression critical state q/p' ($= M$) values all falling around 0·85 (ϕ'_{cs} of 21·3°).

Superimposing the data from the intact samples on Fig. 10(b) it is evident that the yield of the natural London Clay is poorly defined, with compression paths that continue to diverge from the ICL even at high stresses. The lack of convergence of the intact and reconstituted samples at high

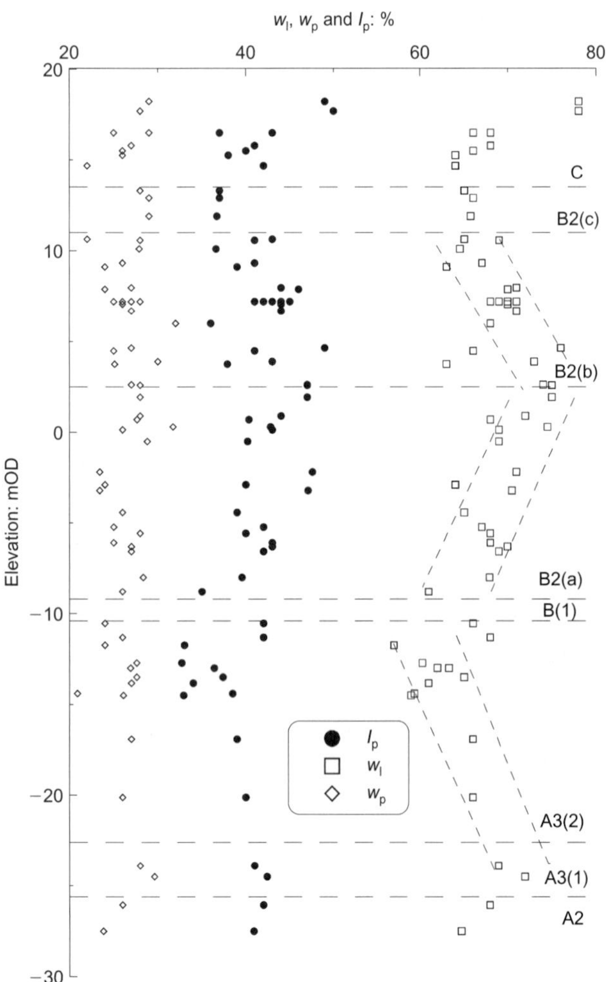

Fig. 9. Profile of index properties with lithological unit and sub-unit boundaries

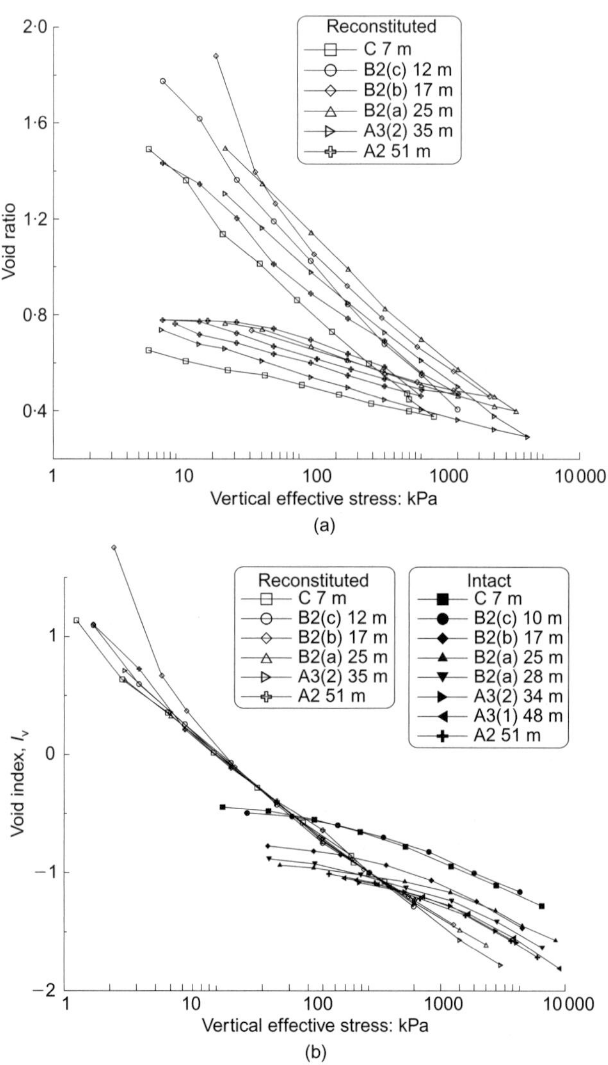

Fig. 10. One-dimensional compression of reconstituted and intact London Clay: (a) void ratio; (b) void index

stresses may be a reflection of the strength of the clay aggregates in the natural clay, which are more easily broken down by the reconstitution process than by one-dimensional compression to high stresses.

Gasparre *et al.* (2007a) found that there were clear differences in structure between the lithological units, with the compression paths for intact samples from the shallower units reaching states further outside the ICL than the deeper units. However, tests on samples from different depths within the same unit showed less variation, both in the location of the ICL and in the compression paths of the intact samples, than had been expected from the coarsening upward sequences within each unit.

IN SITU STRESSES

As a result of its heavy overconsolidation it is known that high horizontal effective stresses are likely to exist in the London Clay at T5. It is also thought that values of K_0 are reduced near the surface of the clay, where it has been reloaded in the past by the deposition of the Terrace Gravel or by old construction (Burland *et al.*, 1979). Estimates of K_0 at T5 were made on the basis of suction measurements, made on-site with the Imperial College suction probe (Ridley & Burland, 1993) on thin-wall tube samples, which were extruded immediately after their retrieval and from which the outer 10 mm was trimmed. (The outer 10 mm contains clay that has been sheared as a result of pushing in the sampling tube and in which negative excess pore pressures will have been generated in this heavily overconsolidated clay. It is removed immediately to reduce the effect of this zone increasing the effective stress in the remainder of the sample.) It was assumed that the mean effective stress in the sample was equal to the mean effective stress in the ground. The K_0 profiles estimated from the suction measurements are shown in Fig. 11 for two locations, namely the lagoon, where the gravel had previously been removed, and the drying bed, where the gravel remains. The K_0 profile for the lagoon lies above that for the drying bed, as would be expected because K_0 has been increased by the effects of unloading caused by removal of the gravel. Although K_0 is higher below the lagoon, the in situ horizontal effective stresses are lower than below the drying beds, where the gravel is in place. Also shown in Fig. 11 are the values of K_0 estimated from the block samples taken at Ashford Common and quoted by Webb (1964): these are in close agreement with the values estimated for T5.

SEISMIC WAVE MEASUREMENTS

Measurements of shear and compression wave velocities in both downhole and crosshole tests were made by the Building Research Establishment (BRE), using the borehole array 406, 407, and 408. V_{hv} and V_{hh} were each measured with the seismic source in borehole 408, with receivers in boreholes 407 and 406, and with the source in borehole 406 and the receivers in boreholes 407 and 408. Thus measurements were made using the true interval method. V_{vh} was measured in downhole tests in borehole 407 with the shear waves polarised in a north–south direction and east–west direction. Values of G_{vh}, G_{hv} and G_{hh} were derived from the two sets of shear wave velocities, and the averages of the two are plotted against elevation in Fig. 12.

Repeatability of the V_{hv} and V_{vh} measurements was reasonable above an elevation of $-10\,\mathrm{mOD}$; repeatability of the V_{hh} measurements was poorer, and below an elevation of $0\,\mathrm{mOD}$ there was much scatter in the data. Almost certainly the anomalously high values of G_{hh} in unit B2(a) were the result of the horizontal shear waves passing through the

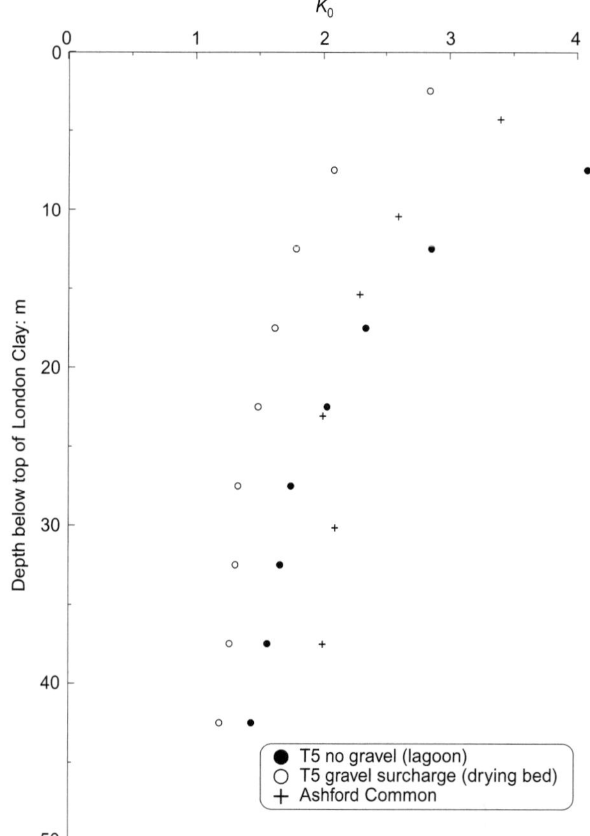

Fig. 11. K_0 profile estimated on basis of suction measurements

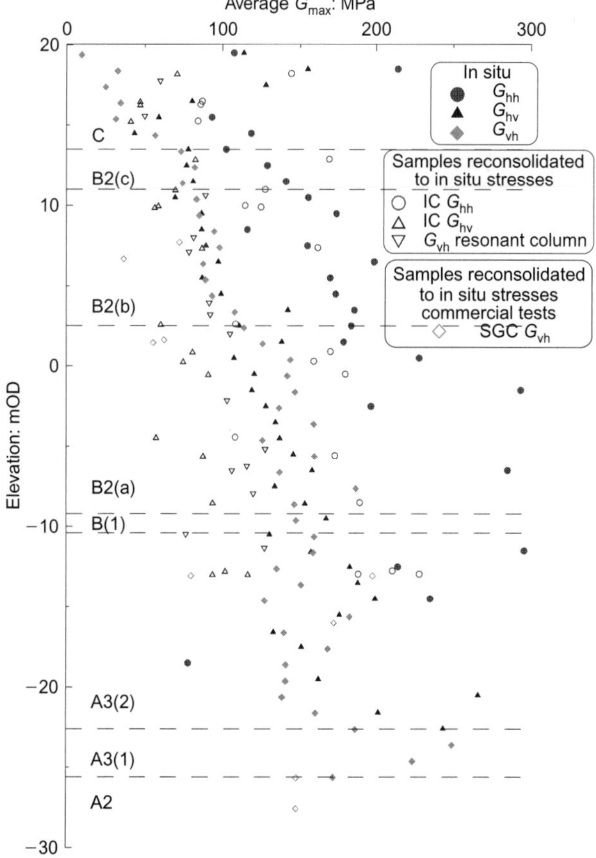

Fig. 12. In situ and laboratory measurements of dynamic shear moduli

stiffer sub-layers provided by the claystones and other nodules when these lay close to the level of the energy source in the borehole.

Despite the scatter, the overall difference between the derived values of G_{hh} and G_{vh} or G_{hv} is clear and consistent with other published data (e.g. Hight *et al.*, 2002), confirming the strong anisotropy of small-strain stiffness in situ. In situ values of G_{hv}/p_0' and G_{vh}/p_0', where p_0' is the mean effective stress in situ, have an average value of approximately 360 whereas in situ values of G_{hh}/p_0' have an average value of approximately 700. (Although these values can serve as a useful first indicator of in situ values, it should not be assumed that normalisation of these small-strain stiffnesses by p_0' is a rational choice.)

In the commercial laboratory tests, measurements were made of V_{vh}, using bender elements, after reconsolidating the triaxial samples to in situ stresses. In the IC study, measurements were also made after reconsolidating samples to in situ stresses of both V_{hv} and V_{hh} using bender elements and measurements of V_{vh} using a resonant column apparatus. The values of G_{vh}, G_{hv} and G_{hh} derived from these measurements have been added to Fig. 12. It is apparent that in units C, B2(c) and B2(b) the agreement between measured field and laboratory values is reasonable. In unit B2(a), where the influence of the frequent claystone layers on the in situ measurements has already been noted, and in unit A3, the laboratory values are lower than the in situ values. The ground level at the location where the in situ measurements were made was approximately 3 m lower than the level at which the rotary-cored boreholes were put down and the level on which the reconsolidation stresses were based. Allowing for this difference in p_0' values will increase any divergence between the two sets of measurements, although, based on the finding of Gasparre *et al.* (2007a) that the effect of stress level on dynamic G values in the intact clay was limited, the increased divergence will be small. It can be seen in Fig. 12 that G_{vh} values derived from the resonant column apparatus are slightly higher than those derived from bender elements.

Although there are differences between the measured in situ and laboratory values of dynamic stiffness, because of scale effects and differences in wavelength, the anisotropy in situ and in the laboratory samples, expressed as the ratio of G_{hh}/G_{hv}, is similar (see Fig. 13), the ratio having an average value of 2 based on the laboratory measurements, but a much wider scatter based on the in situ measurements.

STRESS PROBES

Gasparre *et al.* (2007b) describe the results of small-strain stress probes aimed at establishing the drained stiffness anisotropy of the natural London Clay. The limits to the kinematic Y_1 surface that encompasses the elastic range were established. When normalised by p_0' a common surface was found for all tested units; behaviour within the relatively small Y_1 regions was essentially cross-anisotropic elastic, although the compliance matrix terms varied with depth and effective stresses. Fig. 14 presents the profiles of E_{hmax}' and E_{vmax}' applying at in situ stresses, together with the data on G_{hv} and G_{hh} from the bender element tests. The corresponding Poisson's ratios (v_{vh}', v_{hh}' and v_{hv}') also showed strong anisotropy, having values far from those routinely assumed in foundation analyses. The Poisson's ratios were subject to scatter (up to ± 0.15), and differences existed between statically and dynamically determined values. However, the overall trends were: for v_{vh}' to fall between 0 and 0.3; for v_{hh}' to be small and negative (0 to -0.2); and for v_{hv}' to be positive, larger (around 0.4 to 0.8), and to increase with depth.

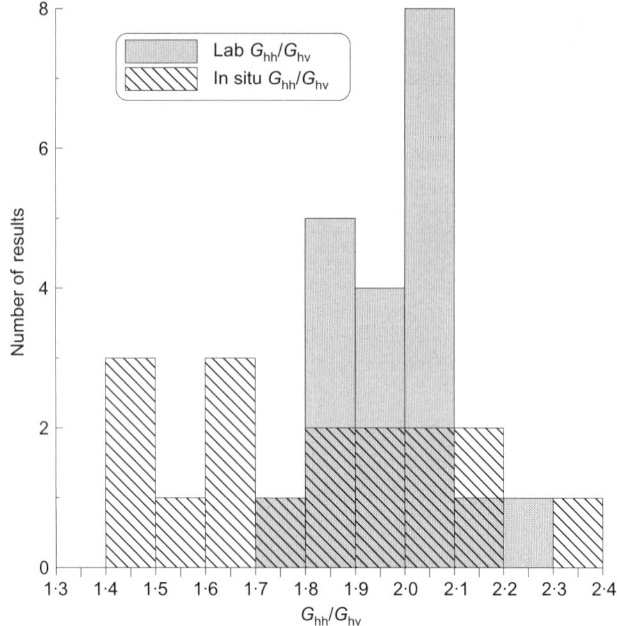

Fig. 13. Comparison of in situ and laboratory measurements of dynamic stiffness anisotropy

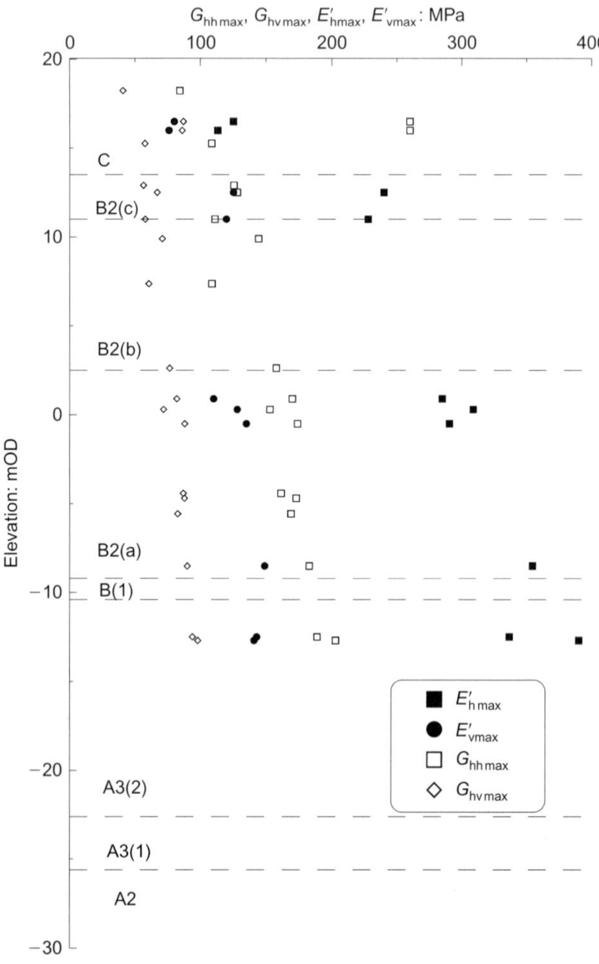

Fig. 14. Profiles of dynamic and static small-strain stiffnesses

The probing tests also identified the second Y_2 kinematic surface that surrounded Y_1 and defined the point at which (a) strain increment directions would change from their initial elastic patterns and (b) the soil microstructure would be altered by continued straining. The sizes of the Y_2 kinematic surfaces also grew in proportion to p_0'.

MODULUS DECAY CURVES

The modulus decay curves for the London Clay were measured in commercial tests in four pairs of CAU tests. Each pair of tests comprised an undrained triaxial compression test and an undrained triaxial extension test on rotary-cored samples taken from similar depths. The reconsolidation paths differed between pairs of tests, in terms of whether or not there was a reversal in stress path direction when starting the undrained shear stage. The results from each pair of tests are presented in Fig. 15 in terms of the undrained secant modulus, E_u, normalised by the value of the mean effective stress prior to the start of undrained shear, p_0', plotted against axial strain on a log scale. The normalised data are compared in these figures with a previously established database of normalised modulus decay for London Clay that has been used successfully in predictions of ground movements.

The normalised stiffness data obtained by Gasparre (2005) and by Nishimura (2006) from CAU tests in the IC study has been added to Fig. 15. These normalised stiffnesses are significantly lower than some of the measurements made in the commercial tests, and lower than the database values. The differences between the new IC tests and the tests for which the database has been established are as follows.

(a) The rates of stress change on the approach path to the in situ stress are lower in the new tests.
(b) The pause periods between the end of reconsolidation and the start of undrained shear were of the order of 7 days, considerably longer than in earlier tests.
(c) The rate of undrained shearing (0·02%/h) was 10 times slower than in the previous tests.
(d) The samples were stored for longer periods.
(e) The methods of local strain measurement differed.

The suction values in the rotary cores tested at IC were similar to those measured on site, immediately after taking the samples: this, combined with the reasonable agreement between laboratory and in situ measurements of dynamic moduli in units C, B2(c) and B2(a), gives confidence that the samples were of a high quality, and that there was not an effect of sample storage. The effect of the difference in rate of undrained shearing will have been relatively minor at the very small strains, so that the most likely causes of the lower normalised stiffnesses are the extended drained pause periods before the start of undrained shearing and the slower rates of stress change along the approach path to the in situ stresses. Gasparre *et al.* (2007b) have confirmed the findings of Clayton and Heymann (2001) that there are no effects of recent stress history in tests on London Clay that involve comparably short approach paths and long creep periods: provided identical samples are tested, degradation curves from undrained triaxial compression and extension tests should start at the same elastic stiffnesses, but diverge with increasing strain.

With insufficient pause periods there is interaction between creep and renewed shearing, with the effect depending on whether or not there is a reversal in stress direction. In tests in which the direction of undrained shearing is a continuation of the direction of the approach stress path, and of the direction of ongoing creep, the true stiffness will be underestimated as a result of the creep strains, whereas in tests in which the undrained shearing involves a reversal of the loading direction the true stiffness will be overestimated as a result of the creep strains.

The effect is graphically illustrated by the results of the commercial tests, which were meticulously carried out to a specification that required the creep rate to be 100 times slower than the subsequent undrained shearing rate. Those tests that involved a reversal of stress direction between undrained loading and the final reconsolidation approach path (tests marked as 'reversal' in Fig. 15) resulted in much higher stiffnesses, whereas the tests in which the direction of undrained shearing was a continuation of the direction of the final reconsolidation approach path (tests marked as 'no reversal' in Fig. 15) resulted in lower stiffnesses. In addition, each pair of modulus decay curves for compression and extension were separated at small strains as a result of ongoing creep. The problem arises from the fact that, although the undrained stages of the triaxial tests were run with a nominal external rate of strain of 0·004%/min, the internal strain rate was significantly slower during the early stage of the tests as a result of compliance and bedding, allowing additional creep strains to develop.

The differences between the modulus decay curves from the new tests and those from the earlier commercial tests can be explained by the extended drained pause periods before the start of undrained shearing and the slower rates of stress change along the approach path to the in situ stresses in the new tests, but it remains to explain why the database of higher modulus decay curves has been successful when used in finite element analyses to predict ground movements in London Clay (e.g. Jardine *et al.*, 1991; Hight *et al.*, 1993; St John *et al.*, 1993; Jardine *et al.*, 2005). These earlier predictions modelled the stress–strain non-linearity of the clay, but assumed it to be isotropic, so that factors relating to anisotropy may help to explain the apparent anomaly.

The cross-anisotropic effective stress stiffness parameters assessed for London Clay in the present study can be rearranged (following, for example, the approaches set out by Lings, 2001, or Nishimura, 2006) to derive equivalent undrained relationships. Application to the data presented in Fig. 14 indicates some interesting results. For example, uniaxial horizontal undrained loading (which cannot be performed in a standard triaxial cell) can be shown to give E_{uh} values over the limited elastic range greater than the E_{uv} values applying to undrained vertical compression or extension tests, with the ratio E_{uh}/E_{uv} ranging between 1·5 and 2·0. In the same way, the octahedral elastic shear stiffnesses developed in undrained plane strain (active or passive) loading are found to be significantly (30% or more) greater than those applying in undrained triaxial tests, while plane strain pure shear (applied on horizontal planes) may invoke a similar global stiffness to the latter. In many cases, therefore, the effects of anisotropy, which have not been taken into account, may have compensated for the effects of ongoing creep during the early stages of undrained shear discussed above, and therefore the effects of differences in strain rate. The degree of compensation would have varied according to the boundary conditions involved in each type of problem.

On the basis of Fig. 15 there is no major difference between the normalised decay curves for the different units, except for the sandier unit A2, which, consistent with findings from other sites, has a higher normalised stiffness.

BULK STIFFNESS

Data on the decay of bulk modulus with volumetric strain has been obtained from the recompression stages of the triaxial tests, during reconsolidation to in situ stresses. The bulk moduli, normalised by the current value of the mean effective stress during swelling, p', are plotted against volumetric strain in Fig. 16.

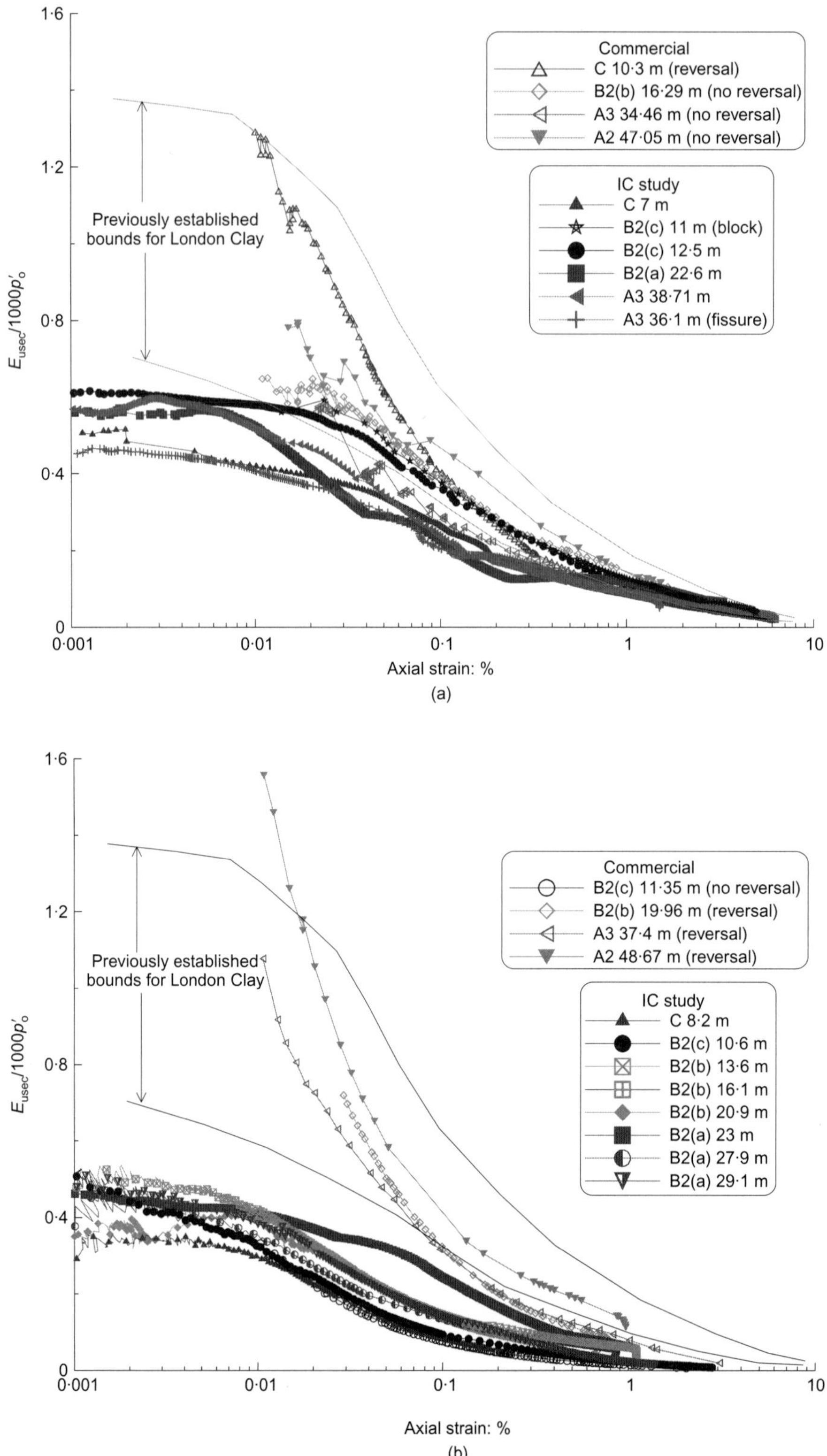

Fig. 15. Normalised modulus decay curves for: (a) undrained triaxial compression; (b) undrained triaxial extension

Measurement of bulk stiffness is notoriously difficult, being affected by creep and inaccuracies in volume strain measurements: external measurements of volume change are unreliable, and internal assessments of volume change are usually based on the measurement at only one diameter. Although these problems contribute to the scatter shown in Fig. 16, much of the scatter is between the different lithological units.

EFFECTIVE STRESS STRENGTH PARAMETERS

The results of the undrained triaxial compression and extension tests on the rotary-cored and block samples are shown in Fig. 17 for the different lithological units in terms of their effective stress paths using coordinates of $(\sigma_v - \sigma_h)/2$ and $(\sigma'_v + \sigma'_h)/2$. In the tests carried out at IC it was noted that in compression a shear zone formed immediately pre-peak, confirming the findings of Sandroni (1975),

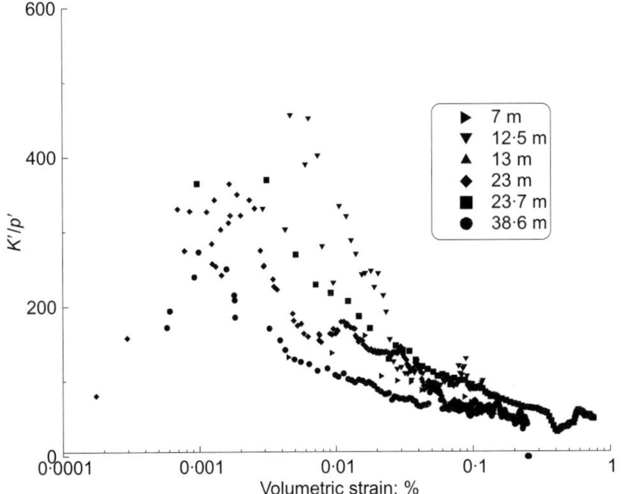

Fig. 16. Normalised bulk modulus decay curves

whereas in extension the failure of all the samples tested by Gasparre (2005) was influenced by a pre-existing fissure.

In units B2(c), B2(b), B2(a) and A3 there are sufficient data to define two failure envelopes in triaxial compression and extension, one representing a peak envelope to the stress paths and one representing the post-rupture strength (Burland, 1990). A linear approximation to a curved peak envelope has been adopted. The definition of each envelope is somewhat subjective, particularly when there is limited data in triaxial extension and the identification of post-rupture is not clear-cut. The failure envelopes are compared in Fig. 18, which includes indicative envelopes for units C and A2.

It can be seen from Fig. 18 that peak failure envelopes for compression and extension loading rise with increasing depth and therefore differ between lithological units. It would appear that a significant cohesion intercept applies to the deeper units, B2(a), A3 and A2, reflecting increasing structure. The increasing lithification with depth is shown clearly in Fig. 17(e) where the data for unit C, the uppermost unit at T5, may be compared with those from unit A2, the deepest.

In triaxial extension the post-rupture envelopes are similar between all the units, with ϕ' lying between $16°$ and $18°$. In triaxial compression the post-rupture envelopes for each sub-unit from B2 are also effectively the same, when one allows for the difficulty in defining that strength and for the uncertainties in making area and membrane corrections. It can also be seen that there is no difference in the post-rupture strengths of the initially intact samples and those with pre-existing fissures, indicating that the natural fissures have not been subjected to large movements.

Nishimura et al. (2007) report the results of a series of simple shear tests carried out in the hollow cylinder apparatus. The results are presented in Fig. 19 in terms of stress paths given by $(\sigma_1 - \sigma_3)/2$ and $(\sigma_1' + \sigma_3')/2$. In the hollow cylinder apparatus it is not possible to strain the samples sufficiently to define a post-rupture strength. However, it is apparent that the failure of several samples was influenced by the presence of fissures, and a failure envelope to these particular samples is given by ϕ' of $19.5°$. Although there is evidence from the simple shear tests of a higher peak failure envelope for unit B2(a) than for B2(b), both envelopes are significantly lower than found for triaxial compression, suggesting a more pervasive influence of the fissuring for this shearing mode.

UNDRAINED STRENGTH

The undrained strength of the London Clay is determined by the effective stress strength parameters applicable to the particular volume of clay and direction of shearing, and by the slope of the effective stress path, which is itself determined by the direction of shearing and the stiffness anisotropy in the clay. The effect of shearing direction is illustrated in Fig. 20 using the results of a set of undrained hollow cylinder tests on block samples of the clay from an elevation of 7 mOD reported by Nishimura et al. (2007). In this test series b was fixed at 0·5, and the major principal stress direction to the vertical during shear was $0°$, $23°$, $45°$, $67°$ and $90°$. As the major principal direction is rotated from the vertical $(\alpha = 0°)$ both the relevant effective stress strength failure envelope and the direction of the effective stress path change. Superimposed on Fig. 20 are the peak and post-rupture failure envelopes for unit B2(b) determined from the triaxial compression tests. The peak in the hollow cylinder apparatus (HCA) test with $b = 0.5$ and $\alpha = 0$ is close to the peak envelope found in triaxial compression ($b = 0$, $\alpha = 0$), whereas the peak in the HCA test with $b = 0.5$ and $\alpha = 67$ is coincident with the post-rupture failure envelope.

As a result of the anisotropy of the effective stress failure envelopes and the anisotropy of stiffness, the undrained strength of the London Clay is strongly anisotropic. Nishimura et al. (2007) illustrate this by presenting both undrained strength profiles for different shearing modes and the variations in undrained strength with principal stress direction at two levels.

BRITTLENESS

Both the drained and undrained brittleness, i.e. the drop from peak to post-rupture strength, of the clay increases with depth, and therefore varies with unit. Brittleness is reduced by fissures that are aligned so that they are influential in the shearing process. Unit A2, which appears not to be fissured and is the deepest unit, shows extreme brittleness (see Fig. 17(e)).

DESTRUCTURING BY SWELLING

The swelling behaviour of the London Clay was of particular importance for the T5 project, because it determined the behaviour of the piles supporting the main terminal building, the construction of which resulted in a significant net unloading of the ground, and it was influential in determining the stand-up time of the temporary slopes. Tests carried out on rotary-cored samples in the commercial investigation revealed the major increase in expansibility of the clay when swelled below effective stresses of approximately 75 kPa (Fig. 21). It was expected that swelling below this threshold would lead to destructuring of the clay and to a lowering of its failure envelope. Tests reported by Gasparre et al. (2007a) confirmed that the swelling had a significant effect on both initial stiffness and yield in compression. However, as illustrated in Fig. 22, the swelling did not appear to modify the peak failure envelope.

PERMEABILITY

Self-boring permeameter tests were carried out using a technique recently introduced by Cambridge Insitu in which a modified self-boring pressuremeter is advanced 1·2 m beyond the base of a borehole and then withdrawn a distance of 0·5 m to create a pocket in which the first constant-flow permeability test is carried out. Subsequently the pressuremeter is readvanced in stages, reducing the length of the pocket in which the next constant-flow test is

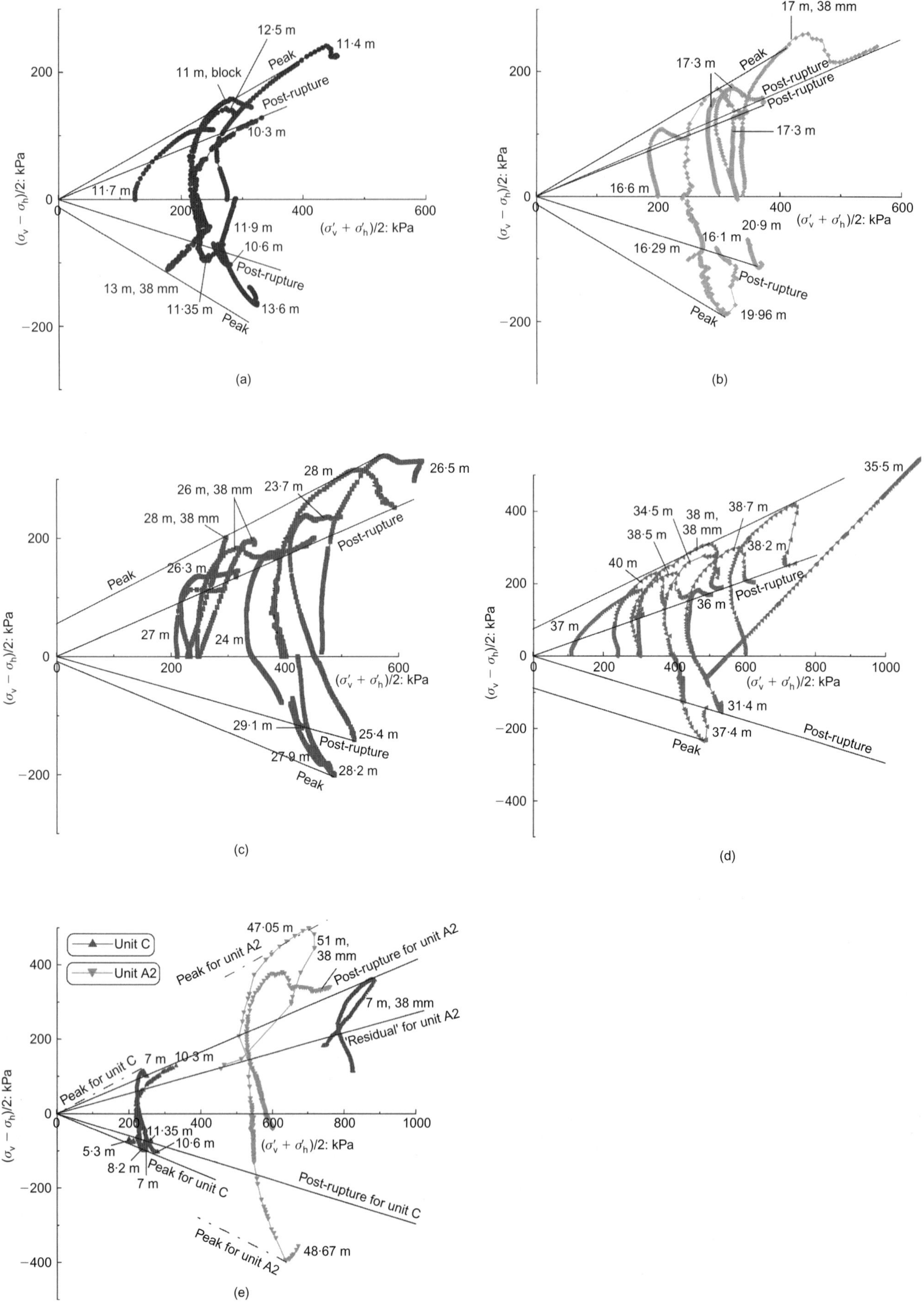

Fig. 17. Effective stress paths in undrained triaxial compression and extension (100 mm diameter rotary-cored samples except where noted): (a) unit B2(c); (b) unit B2(b); (c) unit B2(a); (d) unit A3; (e) units C and A2

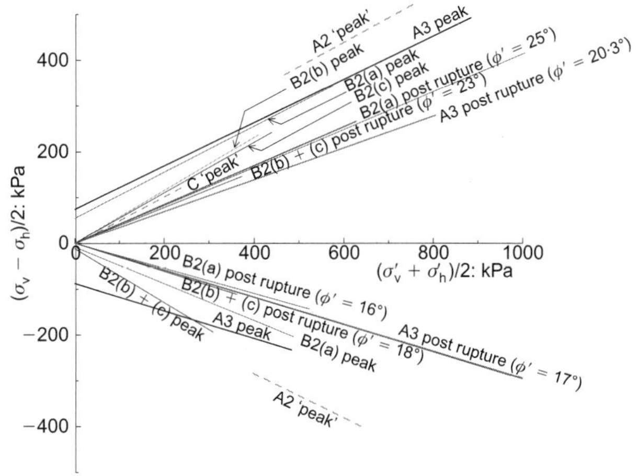

Fig. 18. Peak and post-rupture failure envelopes for different lithological units

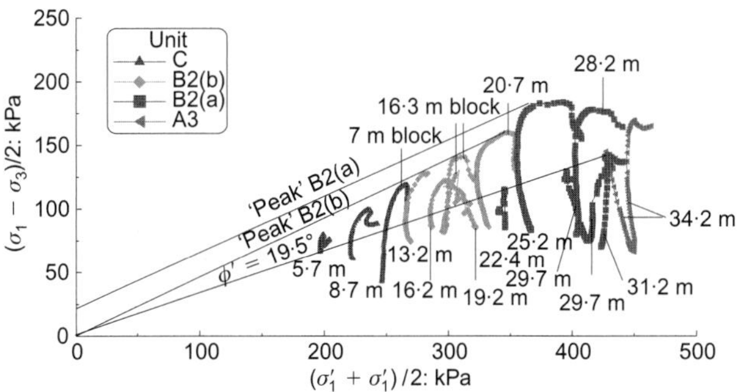

Fig. 19. Effective stress paths in simple shear tests

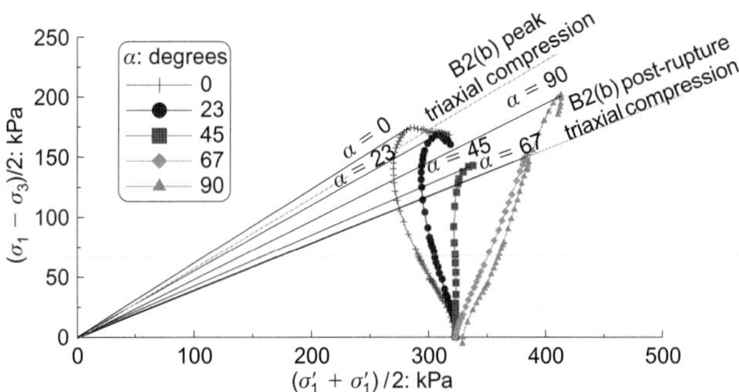

Fig. 20. Effect of principal stress direction on stress path inclination and on failure envelope in HCA tests

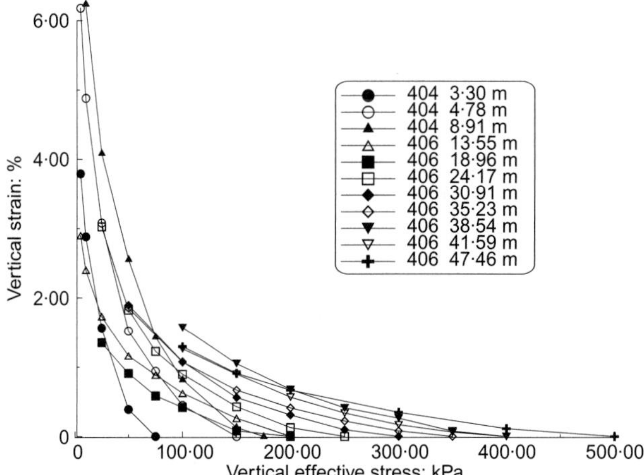

Fig. 21. One-dimensional swelling of intact London Clay

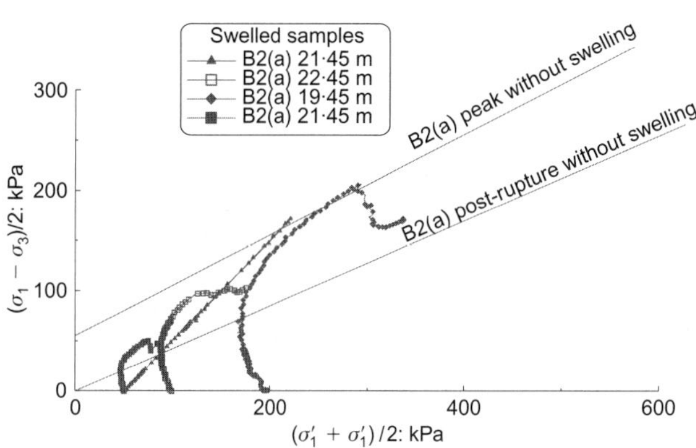

Fig. 22. Effective stress paths of samples swelled to 10 kPa, reconsolidated and sheared undrained in triaxial compression

carried out. The permeability values determined from the first stage of each self-boring permeameter test are plotted against depth in Fig. 23 and compared with other reliable in situ measurements made at sites in London Clay. The T5 data fit within the broad trends of the existing data; values in unit B2(a) are noticeably higher, presumably being influenced by the silt and sand seams.

DISCUSSION
Lithological units

Taking into account the subtle changes in depositional environment and resulting variations in composition and fabric, it is reasonable to expect vertical variations in the mechanical properties of the London Clay, over and above those due to differences in stress state, and for these to correlate with the lithological units defined by King (1981). In contrast, only a gradual variation in properties can be expected laterally, unless there has been faulting or other tectonic events.

Research and commercial testing described in this paper, and in the companion papers of Gasparre *et al.* (2007a, 2007b) and Nishimura *et al.* (2007), has confirmed that there are important differences in the behaviour between the lithological units. Units B2(a), A3 and A2 are clearly more structured than the overlying units, having significant cohesive components of strength. Unit A2 is the most brittle, but is non-fissured. Anisotropy increases with the depth of unit and varies within units. Except for unit A2, there are no major differences between the modulus decay curves of the different units.

Establishing the elevation of the boundaries between these units is therefore a prerequisite in any advanced investigation of London Clay sites. As the units appear to be of consistent thickness, it is sufficient to establish the base of the London Clay at a site in order to anticipate the position of the lithological units within the profile.

Assessment of strength

For clay that has not been subject to previous shearing there are two bounds to the effective stress strength parameters of the London Clay: an upper bound given by the parameters defining the peak failure envelope for the intact clay, that is, without fissures; and a lower bound given by the parameters defining the strength on a fissure. The fissure strength was determined by Skempton *et al.* (1969) to correspond to $c' = 0$ and $\phi' = 20°$. It has been shown that these parameters also apply approximately to the post-

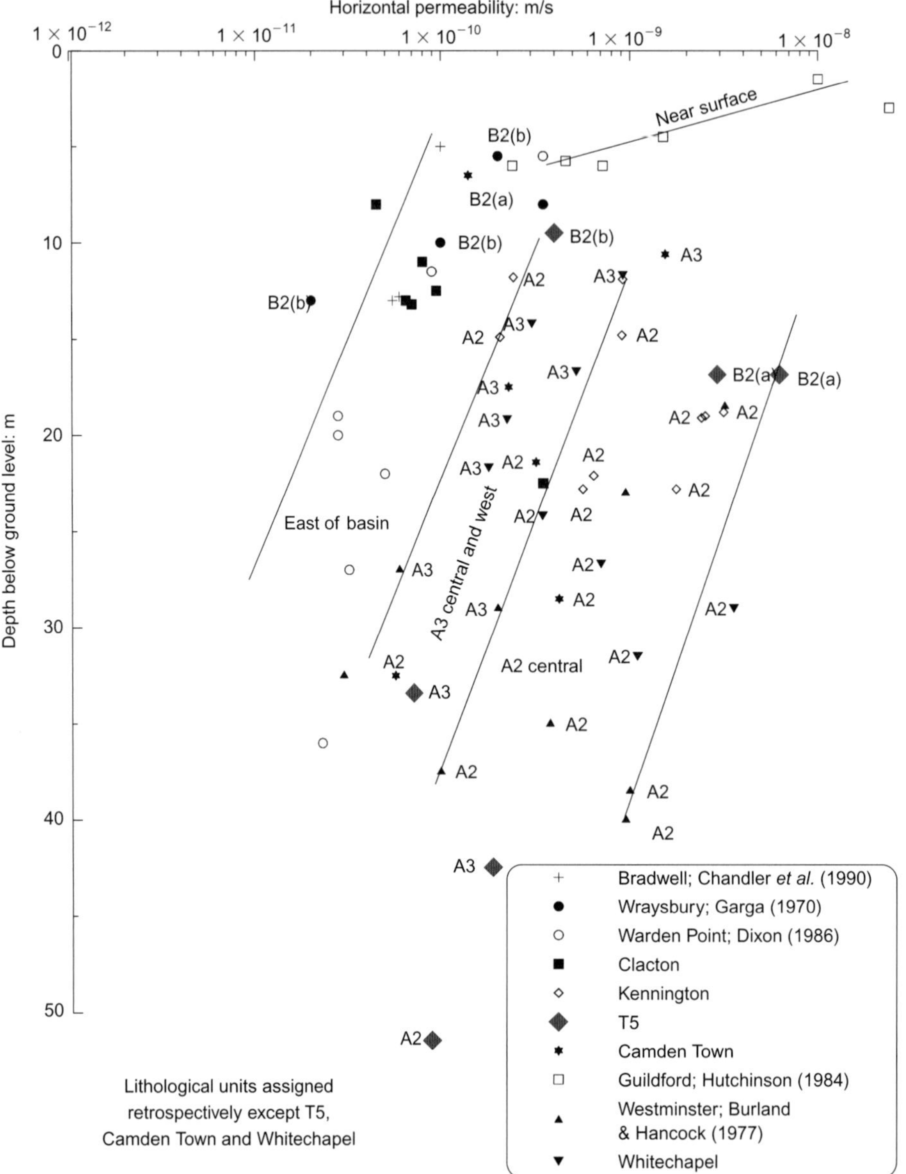

Fig. 23. Permeability measurements at T5 using self-boring pressuremeter compared with database

rupture strength in triaxial extension and in simple shear, and to the intrinsic strength of the clay.

The effective stress strength parameters that apply to a particular volume of clay will lie between these two bounds and will be determined by the spacing, extent and orientation of the fissures, by the direction of shearing, and by the kinematic constraints—that is, whether potential shear surfaces can seek out the lower-strength fissures. It is not necessarily the case that a small sample will provide the intact strength, although this is more likely; nor is it likely that any sample will fail wholly on a fissure and so define the fissure strength. For a fixed sample size and for a fixed direction of shearing, fissures, because of their varying extent and location, introduce major natural variability.

The preferred orientation of the fissures in London Clay means that fissures are more likely to be influential for certain directions of loading. In its simplest form this is demonstrated by a comparison in Fig. 24 between the effective stress strength values found in triaxial compression and in triaxial extension. It is evident that extension strengths are likely to be closer to the fissure strength than compression strengths. Adding the failure points from the simple shear tests carried out in the hollow cylinder apparatus

reinforces the importance of shearing direction and the preferred orientation of the fissures, with several of the failure points falling on or close to the fissure strength envelope. While fissures were influential in all the triaxial extension tests reported by Gasparre et al. (2007), this is not always the case, as shown in Figs 17 and 24. The similarity between the fissure strength and post-rupture strengths in triaxial extension and simple shear is another indicator of the influence of the fissures on these modes of shearing.

For clays that have previously been sheared there is a bound below the fissure strength that is determined by the amount of displacement on the shear surface and by whether that movement was on a pre-existing fissure or not. Chandler et al. (1998) report the strength on a tectonic shear to be 13·5 degrees, which, because of the limited movement on the shear surface, is slightly higher than the ultimate lower bound, namely the residual strength, which lies between 10 and 12 degrees.

Clearly an assessment of strength requires knowledge of the fissure spacing, extent and orientations, in relation to the direction of shearing, as well as the parameters defining the intact and fissure strengths. If the clay has been sheared previously then knowledge is required of the location of the

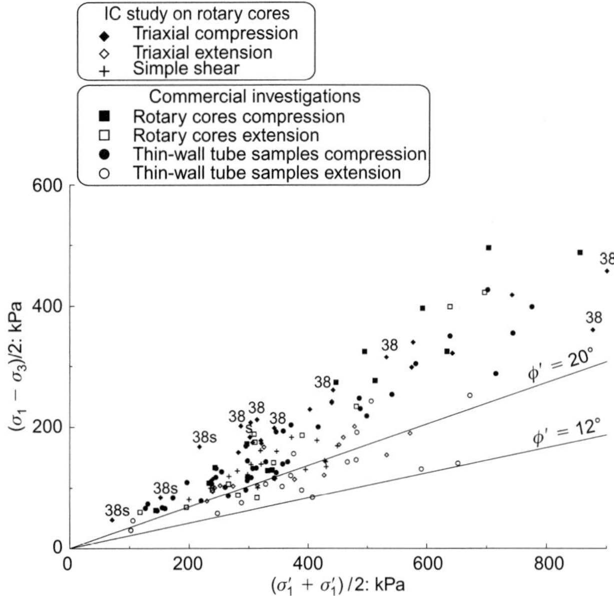

Fig. 24 Effective stress failure points for different types of sample in undrained triaxial compression, triaxial extension and simple shear. 100 mm samples except where noted; s = swelled prior to undrained shear

existing shear surfaces and the amount of movement on them. If the brittleness of the clay is not modelled explicitly in analyses then allowance must be made for the effects of progressive failure as the strength drops from peak to post-rupture to residual.

Assessment of stiffness

It is clear that reliable measurements of modulus decay in the laboratory require longer drained pause periods than have been applied in the past, and the requirement for a 100-fold difference between ongoing creep rate and subsequent externally applied undrained shearing rate needs to be increased to allow for compliance and bedding effects. Preferably strain rates should be controlled on the basis of feedback from local measurements of axial strain.

Claystone layers and other nodule bands complicate the interpretation of in situ measurements of dynamic stiffness, particularly in unit B2(a) and below.

Sampling effects

The new research, and earlier work, has shown that rotary-cored samples are generally of a similar quality to block samples, and has highlighted potential problems with pushed thin-wall tube samples for the measurement of stiffness and strength. Unless a residual surface is perfectly aligned in a triaxial sample it is unlikely that residual strengths will be mobilised in triaxial tests. However, the data presented in Fig. 24 show that strengths close to the residual have been observed in triaxial extension tests on 100 mm samples taken by pushed thin-wall tube sampling. This can possibly be explained by movement on the fissures during the tube sampling process and by the need for only small movements on a fissure to drop towards residual strength. It follows from this that the presence of fissures during sampling can be influential in terms of sample disturbance. Mismatching of fissure surfaces, which would affect small-strain stiffness, is possible if suctions are not sustained across the surfaces.

All types of sample can be adversely affected by the presence of the sand and silt seams, which allow water content redistribution, and which appear to be more pervasive towards the top of each lithological unit

CONCLUSIONS

It has been known for many years that the shear strength of natural London clay is profoundly affected by fissuring, and that its shear stiffness is highly non-linear. The overriding influence of fissuring in natural London Clay has been reaffirmed in this study, and advanced HCA testing has shown that it contributes to a marked anisotropy in shear strength. The clay's stiffness anisotropy has also been explored at all strain levels. Strong elastic anisotropy has been established within kinematic Y_1 yield surfaces that have been established over the depth profile, together with a second set of Y_2 kinematic yield surfaces. Stiffness anisotropy, which is related to preferred orientation of the clay aggregates, determines the generation of pore pressures during undrained shear and contributes, with the preferred orientation of the fissures, to the substantial undrained strength anisotropy of the clay.

The study has emphasised the correlation between engineering properties and the different lithological units of the London Clay. Significantly different behaviours arise both from changes in intrinsic behaviour and also from differences in structure. These latter are particularly evident in the peak shear strength characteristics and compression behaviour, while permeability is seen to reduce sharply with depth, following a pattern that also correlates well with the lithological units.

The new findings regarding permeability, stiffness and shear strength have many implications for geotechnical engineering in deposits such as London Clay, affecting the timing of works and the assessment of stability as well as the magnitudes, patterns and rates of ground movements.

NOTATION

b — intermediate principal stress ratio
c' — apparent cohesion
C_c^* — intrinsic compression index
C_s^* — intrinsic swelling index
e — void ratio
E'_{hmax}, E'_{vmax} — maximum drained Young's moduli for horizontal and vertical loading
E_u — undrained Young's modulus
E_{uh}, E_{uv} — undrained Young's moduli for horizontal and vertical loading
G_{hh} — shear modulus in the horizontal plane
G_{hv}, G_{vh} — shear modulus in the vertical plane
G_{hhmax}, G_{hvmax} — maximum value of G_{hh}, G_{hv}
G_{max} — maximum shear modulus
I_p — plasticity index
I_v — void index
K' — drained bulk modulus
K_0 — coefficient of earth pressure at rest
M — q/p' at critical state in compression
p' — mean effective stress during swelling
p'_0 — mean effective stress prior to start of undrained shear
q — deviatoric stress $(\sigma_1 - \sigma_3)$
V_{hh} — shear wave velocity of horizontally polarised shear wave travelling horizontally
V_{hv} — shear wave velocity of vertically polarised shear wave travelling horizontally
V_{vh} — shear wave velocity of horizontally polarised shear wave travelling vertically
w_l, w_p — liquid and plastic limits
α — angle of inclination of major principal stress to vertical

ν'_{hh} drained Poission's ratio for horizontal strains due to horizontal strain

ν'_{hv} drained Poission's ratio for horizontal strains due to vertical strain

ν'_{vh} drained Poission's ratio for vertical strains due to horizontal strain

σ_h, σ_v horizontal and vertical total stresses

σ'_h, σ'_v horizontal and vertical effective stresses

σ_1, σ_3 major and minor principal total stresses

σ'_1, σ'_3 major and minor principal effective stresses

ϕ' angle of shearing resistance

ϕ'_{cs} critical state angle of shearing resistance

REFERENCES

Bishop, A. W., Webb, D. L. & Lewin, P.I. (1965). Undisturbed samples of London Clay from Ashford Common shaft: strength-effective stress relationship. *Géotechnique* **15**, No. 1, 1–31.

Burland, J. B. (1990). On the compressibility and shear strength of natural soils. *Géotechnique* **40**, No. 3, 329–378.

Burland, J. B. and Hancock, R. J. R. (1977). Underground car park at the House of Commons, London; geotechnical aspects. *The Structural Engineer* **1**, 13–29.

Burland, J. B., Simpson, B. & St John, H. D. (1979). Movements around excavations in London Clay. *Proc. 7th Eur. Conf. Soil Mech., Brighton* **1**, 13–20.

Burnett, A. D & Fookes, P. G. (1974). A regional engineering geological study of the London Clay in the London and Hampshire basins. *Q. J. Engng Geol.* **7**, No. 3, 257–295.

Chandler, R. J. (2000) Clay sediments in depositional basin: the geotechnical cycle. *Q. J. Engng Geol. Hydrogeol.* **33**, No. 1, 7–39.

Chandler, R. J. Leroueil, S. & Trenter, N. A. (1990). Measurements of London Clay using a self-boring permeameter. *Géotechnique* **40**, No. 1, 113–124.

Chandler, R. J., Willis, M. R., Hamilton, P. S. & Andreou, I. (1998) Tectonic shear zones in the London Clay formation. *Géotechnique* **48**, No. 2, 257–270.

Clayton, C. R. I. & Heymann, G. (2001). Stiffness of geomaterials at very small strains. *Géotechnique* **51**, No. 3, 245–255.

de Freitas, M. H. & Mannion, W. (2007). A biostratigraphy for the London Clay in London. *Géotechnique* **57**, No. 1, 91–99.

Dewhurst, D. N., Aplin, A. C., Sarda J.-P. & Yang, Y. (1998). Compaction-driven evolution of porosity and permeability in natural mudstones: an experimental study. *J. Geophys. Res. Solid Earth* **103**, No. B1, 651–661.

Dixon, N. (1986). *The mechanics of coastal land slides in London Clay at Warden Point, Isle of Sheppey.* PhD thesis, Kingston Polytechnic.

Gasparre, A. (2005). *Advanced laboratory characterisation of London Clay.* PhD thesis, Imperial College, London.

Gasparre, A., Nishimura, S., Coop, M. R. & Jardine, R. J. (2007a). The influence of structure on the behaviour of London Clay. *Géotechnique* **57**, No. 1, 19–31.

Gasparre, A., Nishimura, S., Minh, N. A., Coop, M. R. & Jardine, R. J. (2007b). The stiffness of natural London Clay. *Géotechnique* **57**, No. 1, 33–47.

Hight, D. W. (1998). Soil characterisation: the importance of structure and anisotropy. 38th Rankine Lecture, 18 March 1998, British Geotechnical Society, London. *Géotechnique*, in preparation.

Hight, D. W., Pickles, A. R., De Moor, E. K., Higgins, K. G., Jardine, R. J. & Potts, D. M. (1993). Predicted and measured tunnel distortions associated with construction of Waterloo International Terminal. *Proceedings of the Wroth Memorial Symposium on Predictive Soil Mechanics*, Oxford, pp. 317–338.

Hight, D. W., McMillan, F., Powell, J. J. M., Jardine, R. J. & Allenou, C. P. (2002). Some characteristics of London Clay. In *Characterisation and engineering properties of natural soils* (eds T. S. Tan, K. K. Phoon, D. W. Hight and S. Leroueil), pp. 851–908. Rotterdam: Balkema.

Hutchinson, J. N. (1984). Landslides in Britain and their countermeasures. *J. Japan Landslide Soc.* **21**, No. 1, 1–21

King, C. (1981). The stratigraphy of the London Basin and associated deposits. *Tertiary Research Special Paper* **6**, Rotterdam: Backhuys.

Kovacevic, N., Hight, D. W. & Potts, D. M. (2007). Predicting the stand-up times of temporary London Clay slopes at Terminal 5, Heathrow Airport. *Géotechnique* **57**, No. 1, 63–74.

Jardine, R. J., St John H. D., Hight, D. W. & Potts, D. M. (1991) Some applications of a non-linear ground model. *Proc. 10th Eur. Conf. Soil Mech. Found. Engng, Florence* **1**, 223–228.

Jardine, R. J., Standing, J. R & Kovacevic, N. (2005). Lessons learned from full scale observations and the practical application of advanced testing and modelling: Keynote paper. *Proceedings of the international symposium on deformation characteristics of geomaterials*, Lyon, Vol. 2, pp. 201–245.

Lings, M. L. (2001). Drained and undrained anisotropic elastic stiffness parameters. *Géotechnique* **51**, No. 6, 555–565.

Nishimura, S. (2006). *Laboratory study on anisotropy of natural London Clay.* PhD thesis, Imperial College, London.

Nishimura, S., Minh, N. A. & Jardine, R. J. (2007) Shear strength anisotropy of natural London Clay. *Géotechnique* **57**, No. 1, 49–62.

Ridley, A. M. & Burland, J. B. (1993). A new instrument for measuring soil moisture suction. *Géotechnique* **43**, No. 2, 321–324.

Sandroni, S. S. (1975). *The strength of London Clay in total and effective stress terms.* PhD thesis, Imperial College, London.

Skempton, A. W. (1961). Horizontal stresses in an overconsolidated Eocene clay. *Proc. 5th Int. Conf. Soil Mech. Found. Engng, Paris* **1**, 351–357.

Skempton, A. W., Schuster, R. L. & Petley, D. J. (1969) Joints and fissures in the London Clay at Wraysbury and Edgware. *Géotechnique* **19**, No 2, 205–217.

St John, H. D., Potts, D. M., Jardine, R. J. & Higgins, K. G. (1993). Prediction and performance of ground response due to construction of a deep basement at 60 Victoria Embankment. *Proceedings of the Wroth Memorial Symposium on Predictive Soil Mechanics*, Oxford, pp. 581–608.

Webb, D. L. (1964). *The mechanical properties of undisturbed samples of London Clay and Pierre Shale.* PhD thesis, University of London.

Atkinson, J. (2007). *Géotechnique* **57**, No. 2, 127–135

Peak strength of overconsolidated clays

J. ATKINSON*

The peak strength of overconsolidated clay soil is conventionally related to the effective stress by a linear Mohr–Coulomb criterion in which the parameters c'_p and ϕ'_p depend on the nature of the soil grains, on the water content, and on the range of stress over which measurements are made. For soils that have zero or very small interparticle bonding the failure line must pass through or close to the point $\tau = 0$ and $\sigma' = 0$, and it must meet the critical state line, so a linear Mohr–Coulomb criterion cannot satisfactorily describe the peak strength of soil except over relatively small ranges of effective stress. The paper reports the results of special triaxial tests in which the peak strengths of samples of stiff clays from southeast England were measured at small effective stresses and large overconsolidation ratios. For each soil the peak strength, with a suitable normalising parameter to take account of water content, was related to the effective stress by a non-linear power law criterion similar to that familiar in rock mechanics. In addition, the peak stress ratios were linked to a state parameter through a relationship similar to that for ordinary Cam clay. For each clay tested the material parameters that describe the peak strength were found to be only loosely related to their plasticity.

KEYWORDS: clays; laboratory tests; shear strength; stress paths

La résistance de pic d'un sol argileux surconsolidé est reliée conventionnellement à la contrainte effective par un critère linéaire de Mohr-Coulomb dans lequel les paramètres c'_p et ϕ'_p dépendent de la nature des grains de sol, du contenu en eau et du type de contrainte utilisé pour faire les mesures. Pour les sols qui ont une cimentation interparticulaire nulle ou très petite, la ligne de rupture doit passer à travers ou près du point $\tau = 0$ et $\sigma' = 0$ et doit rencontrer la ligne d'état critique en sorte qu'un critère de Mohr–Coulomb linéaire ne puisse décrire la résistance de pic du sol de manière satisfaisante, excepté pour des valeurs rapprochées de contrainte effective. Cet article fait part des résultats des essais triaxiaux spéciaux dans lesquels on a mesuré les résistances de pic d'échantillons d'argiles fermes du sud-est de l'Angleterre avec de petites contraintes effectives et de grands rapports de surconsolidation. Pour chaque sol, la résistance de pic, avec des paramètres de normalisation appropriés pour prendre en compte le contenu en eau, a été reliée à la contrainte effective par un critère de loi de puissance non linéaire semblable à celui utilisé en mécanique rocheuse. De plus, les rapports de contrainte de pic étaient reliés à un paramètre d'état par une relation semblable à celle existant pour une argile de Cam ordinaire. Pour chaque argile mise à l'essai, on a trouvé que les paramètres matériels décrivant la résistance de pic n'étaient que vaguement reliés à la plasticité.

THE MOHR–COULOMB STRENGTH CRITERION FOR UNBONDED SOIL

The effective stress strength of soil is normally represented by the linear Mohr–Coulomb equation

$$\tau = c' + \sigma' \tan\phi' \tag{1}$$

where c' is a cohesion intercept and ϕ' is a friction angle. Values for c' and ϕ' depend on, among other things, which strength is being described (peak, critical state or residual). The critical state strength is given by equation (1) with $c' = 0$ and $\phi = \phi'_c$, and this is the double line OC in Fig. 1. (The critical state line (CSL) is conventionally shown as a double line; Schofield & Wroth, 1968.)

If soil has a peak strength it must, by definition, lie above the critical state line at a point such as P in Fig. 1(a). The strength of an unbonded soil is $\tau = 0$ when the effective normal stress is $\sigma' = 0$, so the peak strength line must pass through the point O in Fig. 1(a). At large effective stresses, where the state is on the wet side of critical, soils do not have a peak strength larger than the critical state strength, and so the peak strength line must terminate at a point such as C in Fig. 1(a). The only smooth line that can join the points O, P and C is a curve such as that shown in Fig. 1(a): a straight line cannot represent the peak strength of

unbonded soil over the range of effective stress from zero to the critical state.

The linear Mohr–Coulomb failure criterion is also intrinsically unsafe when the normal effective stresses in the ground are greater or smaller than the normal effective stresses applied in the tests. Fig. 1(b) shows the curved peak failure line from Fig. 1(a) with peak strengths P_1 to P_3 measured in tests: the broken line is a best-fit line for the data. The strength of the soil is less than that given by the linear Mohr–Coulomb criterion except where the broken line is below the solid line. There are many combinations of c' and ϕ' that describe the peak strengths of the same soil over different ranges of normal effective stress.

A curved peak failure line can be expressed as a power law of the form (de Mello, 1977)

$$\tau = A\sigma'^b \tag{2}$$

The implications of curved failure lines for analyses of slope stability were investigated by Charles & Watts (1980) and by Charles (1982). Power law failure criteria are common in rock mechanics, and have been extended to include bonding (Hoek & Brown, 1980).

LABORATORY TESTS ON OVERCONSOLIDATED CLAYS

Laboratory tests were carried out on reconstituted samples of kaolin clay and on reconstituted and intact samples of several of the plastic clays found in south-east England. The intact samples were trimmed from tube samples taken from

Manuscript received 5 May 2006; revised manuscript accepted 9 November 2006.
Discussion on this paper closes on 1 August 2007, for further details see p. ii.
* School of Engineering and Mathematical Sciences, City University, London, UK.

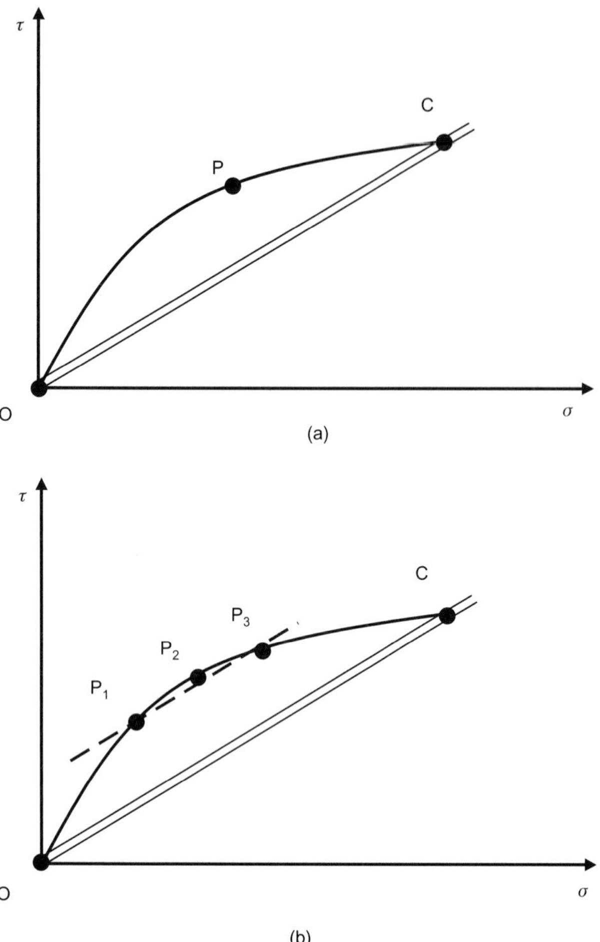

Fig. 1. Peak strength envelopes for soil: (a) general curved envelope; (b) linear envelope fitted to test measurements

compacted motorway embankment slopes. They were prepared for testing, saturated, compressed and swelled to their initial states as described below. The tests on reconstituted kaolin clay were carried out to establish the basic nature of the peak strength of clays. The tests on plastic clays were undertaken as part of a programme of research into the stability of motorway slopes carried out by the Transport and Road Research Laboratory (Atkinson & Crabb, 1991). The locations on the UK motorway network from which the samples were obtained are given in Table 1.

All test samples were 38 mm diameter and were tested in a hydraulic stress path triaxial apparatus similar to that described by Bishop & Wesley (1975). Cell pressure and pore pressure were measured using conventional transducers, but, because many of the samples were failed at relatively low effective stresses, they were also measured on a differential pressure transducer. This measured the difference between the cell pressure and the pore pressure at the base of the sample, and hence gave a direct reading of radial effective stress. Particular care was taken with calibration of the

Table 1. Locations from which samples were taken

Soil	Location of sampling site
Gault clay	M25: Dunton Green
Kimmeridge clay	M4: Swindon
London clay	M4: Wokingham
Oxford clay	M1: Junction 13
Reading clay	M11: Stansted
Weald clay	M23: Redhill

internal axial load cell, and the usual corrections (Bishop & Henkel, 1957) were applied to determinations of deviator stress.

All the reconstituted samples were made from slurry with a water content about twice the liquid limit by initial one-dimensional consolidation in a thick-wall tube to a vertical effective stress of about 100 kPa. Intact samples were trimmed from 38 mm diameter thin-wall tube samples. After they were set up in the triaxial cell the reconstituted samples were isotropically compressed to an effective stress of 300 kPa and the intact samples were isotropically compressed to an effective stress of 100 kPa. All samples were then swelled to effective stresses in the range 15–50 kPa, where they were on the dry side of the critical state (Schofield & Wroth, 1968). In all cases the pore pressure during initial compression and swelling was 150 kPa, and the value of B before testing was larger than 0·96. The samples were then sheared in compression following one of the stress paths illustrated in Fig. 2. Path OA is simple drained triaxial compression with constant cell pressure and constant pore pressure. This path is inconvenient for investigating failure at small effective stress as its gradient takes it away from the small stress region. Path OB has constant axial stress and constant pore pressure, and the radial stress decreases. Path OCD starts with simple drained triaxial compression along OC to a convenient point at C. The test then proceeds with the total stresses held constant as the pore pressure is raised. (This path corresponds to increasing pore pressure in a slope.) In all these tests stresses or pore pressures were changed at a rate of 1 kPa per hour. This rate of loading was calculated from the consolidation characteristics of the samples using the methods described by Cherrill (1990) to give excess pore pressures in the sample less than 1 kPa. Close to failure in stress-controlled tests the samples became unstable, and their peak states were identified as the points at which strains suddenly accelerated.

Additional sets of drained triaxial compression tests at a constant strain rate were carried out on reconstituted and normally consolidated samples of kaolin clay and on each of the clays listed in Table 1 to determine their basic isotropic compression and critical state parameters. These samples all reached well-defined critical states.

PEAK AND CRITICAL STATES OF KAOLIN CLAY

In Fig. 3 the solid Mohr's circles represent effective stress at the peak state in triaxial tests on samples of reconstituted kaolin clay, all swelled from a maximum isotropic effective stress of 300 kPa. The broken Mohr's circles represent effective stresses at the critical state in triaxial tests on samples of normally compressed kaolin clay. The overconsolidated

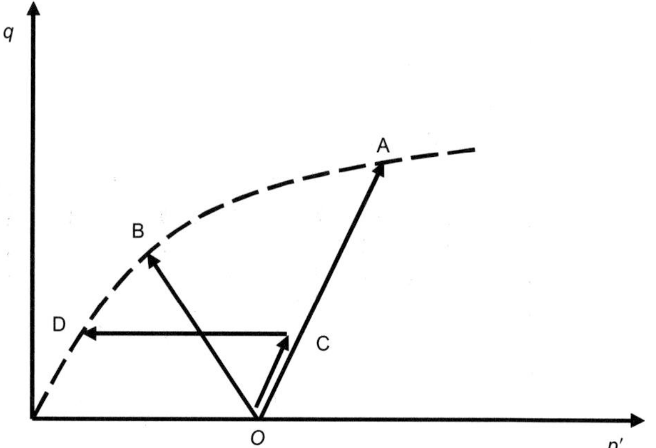

Fig. 2. Stress paths in triaxial compression tests

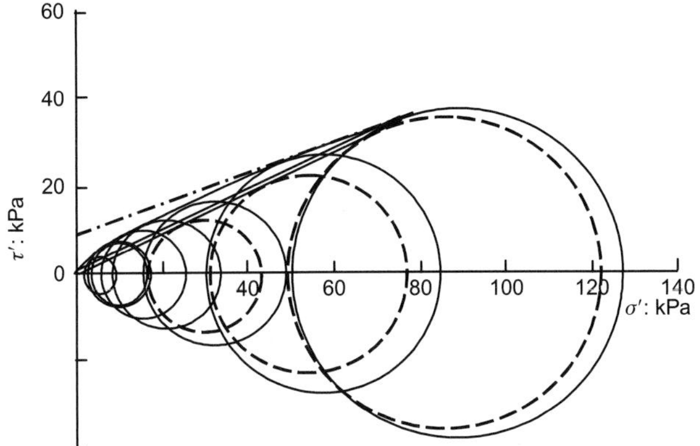

Fig. 3. Mohr's circles of peak and critical states of kaolin clay

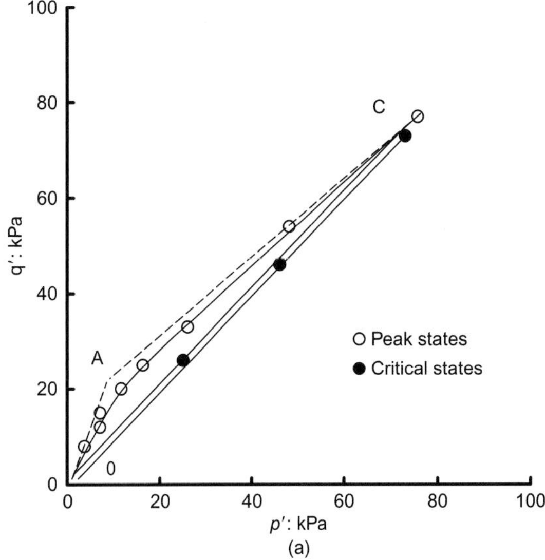

Fig. 4(a)

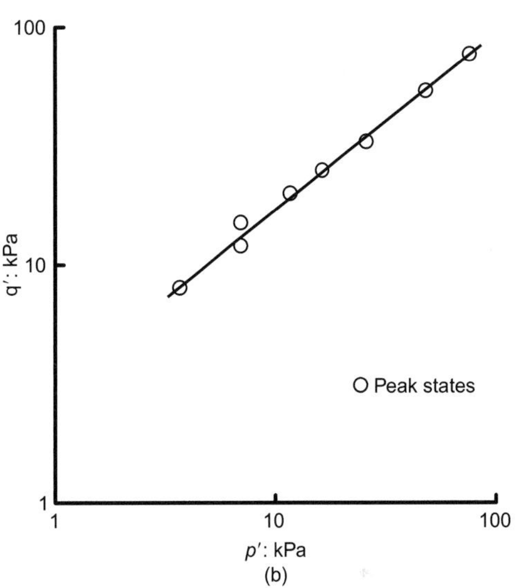

Fig. 4(b)

Fig. 4. Peak and critical states of kaolin clay

samples developed clear slip planes as they failed at their peak states in stress-controlled tests. The normally consolidated samples did not develop slip planes in strain-controlled tests, and they reached clearly defined critical states. Although the stress paths applied to the overconsolidated samples approached the peak strength envelope from different directions, all the samples failed in triaxial compression with $\sigma'_a > \sigma'_r$, so the differences in the stress paths should not influence the peak strengths.

In Fig. 3 a smooth curve has been drawn that approximately limits the Mohr's circles for the peak states and which passes through the origin. (Owing to scatter in the data, some circles just cross the smooth curve whereas others are just inside it.) The two test results with normal effective stresses greater than about 50 kPa can be approximated by a linear Mohr–Coulomb criterion shown as a chain dotted line with $c'_p = 9$ kPa and $\phi'_p = 20°$. For smaller normal effective stresses the peak strength line is distinctly curved, and the Mohr's circles are well below the linear Mohr–Coulomb line. The double line in Fig. 3 is tangent to the Mohr's circles for the critical states. This is the critical state line given by equation (1) with $c' = 0$ and $\phi' = 25\frac{1}{2}°$.

Figure 4(a) shows the same data as in Fig. 3 but plotted with axes $q = \sigma'_a - \sigma'_r$ and $p' = \frac{1}{3}(\sigma'_a + 2\sigma'_r)$. This is more convenient for analysis of triaxial test data because it avoids the need to draw Mohr's circles and locate points of limiting stress ratio. There is a critical state line given by

$$q = Mp' \tag{3}$$

with $M = 1.0$ and a curved line representing the peak states. The linear broken line AC is through the two peak points with the largest effective stresses, and the broken line OA represents the no-tension cut-off (Atkinson & Bransby, 1978), for which $\sigma'_r = 0$. (These correspond to the chain dotted line and the τ axis in Fig. 3.)

Figure 4(b) shows the peak strength points from Fig. 4(a) plotted with logarithmic scales. These are close to a straight line given by

$$\log q = \log A + b \log p' \tag{4}$$

and the curved peak failure line in Fig. 4(a) can be represented by a simple power law,

$$q = Ap'^b \tag{5}$$

where A and b are peak strength parameters. The parameter A is similar to a friction coefficient, and the parameter b governs the degree of curvature. For the kaolin clay samples tested, which had all been previously compressed to a maximum stress of 300 kPa, $A = 1.8$ and $b = 0.69$.

Figure 5 shows the initial and critical states of kaolin clay

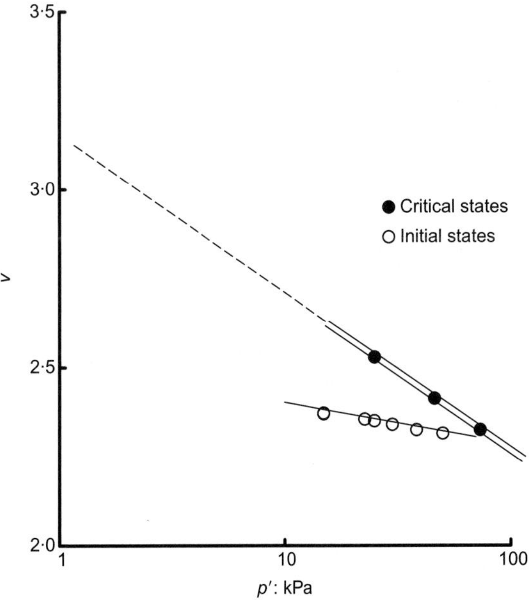

Fig. 5. Initial and critical states of kaolin clay

with axes v and p' plotted to a logarithmic scale. The critical state line is given by

$$v = v_\Gamma - \lambda \ln p' \tag{6}$$

with $\lambda = 0.19$ and $v_\Gamma = 3.14$. The swelling line corresponds to a maximum pre-consolidation pressure of $p' = 300$ kPa, and its gradient is $\kappa = 0.05$. These, together with $M = 1.0$ from Fig. 4(a), are the basic critical state parameters for the kaolin clay.

NORMALISING TO TAKE ACCOUNT OF PRECONSOLIDATION PRESSURE AND SPECIFIC VOLUME

The data for kaolin clay shown in Fig. 4 are for samples compressed to 300 kPa, swelled to different stresses and sheared following different stress paths: consequently they all have different overconsolidation ratios and specific volumes, both at the start of shearing and at their peak states. If the samples had all been previously compressed to different effective stresses their peak states would have fallen close to another peak state line (Atkinson & Little, 1988). These differences can be taken into account by normalising the peak and critical states using appropriate state parameters.

Figure 6 shows a peak state at P described by its specific volume v_p and mean effective stress p'_p. The state at P is conveniently defined by the distance it is from the critical state line (CSL) either as $v_\Gamma - v_\lambda$ or as p'_p/p'_c. The critical pressure p'_c is obtained from

$$v_p = v_\Gamma - \lambda \ln p'_c \tag{7}$$

and the value of v_λ is obtained from

$$v_p = v_\lambda - \lambda \ln p'_p \tag{8}$$

Figure 7 shows the same test data as those shown in Fig. 4. The mean stresses have been normalised with respect to p'_c and the deviator stresses have been normalised with respect to Mp'_c. The critical state data are all at a single point C where $p'/p'_c = 1.0$ and $q'/Mp'_c = 1.0$. The peak state points are close to a smooth curve. The broken line OA is the no-tension cut-off (Atkinson & Bransby, 1978) corresponding to $\sigma'_r = 0$. The broken line AC is a linear envelope through the peak state points at the two largest values of p'/p'_c. This is the Hvorslev surface (Atkinson & Bransby, 1978). At smaller values of p'/p'_c the peak state points fall well below the line AC. (If the isotropic normal consolidation line is at $p'/p'_c = 2$ a value of $p'/p'_c = 0.5$ corresponds to an overconsolidation ratio of 4.)

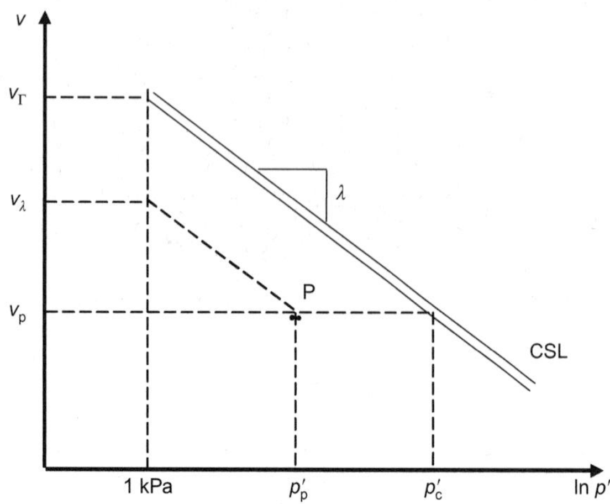

Fig. 6. State parameters p'_c and v_λ

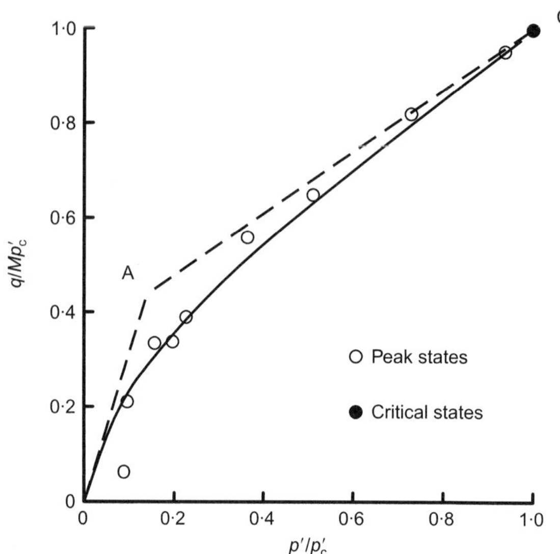

Fig. 7. Normalised peak and critical states of kaolin clay

Figure 8(a) shows the data from Fig. 7 plotted with logarithmic axes. The data are close to a line given by

$$\log\left(\frac{q}{Mp'_c}\right) = \log \alpha + \beta \log\left(\frac{p'}{p'_c}\right) \tag{9}$$

which can be written

$$\frac{q}{Mp'_c} = \alpha\left(\frac{p'}{p'_c}\right)^\beta \tag{10}$$

The peak strength line in Fig. 8(a) meets the critical state point where $p'/p'_c = 1.0$ and $q/Mp'_c = 1.0$. Hence $\alpha = 1$, and equation (9) becomes

$$\frac{q}{Mp'_c} = \left(\frac{p'}{p'_c}\right)^\beta \tag{11}$$

in which the parameter β describes the degree of non-linearity. From the data in Fig. 8(a) the value of $\beta = 0.63$. This and the values $\lambda = 0.19$, $v_\Gamma = 3.14$ and $M = 1.0$ define the peak strength of kaolin clay.

Figure 8(b) shows values of q/Mp' at the peak and critical states plotted against the state parameter $v_\Gamma - v_\lambda$. (As before, the parameter M has been included, so the critical state is at $q/Mp' = 1.0$ and $v_\lambda = v_\Gamma$.) The data fall close to a straight line through the critical state point given by

$$\frac{q}{Mp'} = 1 + \chi(v_\Gamma - v_\lambda) \tag{12}$$

where χ is its gradient. From the data in Fig. 8(b) the value of χ is 2·8.

In equations (11) and (12) the specific volume v_p and the mean stress p'_p at the peak state are included in the state parameters p'_c and v_λ through equations (7) and (8), and so the parameters M, β and χ are material parameters: they depend only on the nature of the grains and not on the current state. Values for these and the parameters λ, κ and v_Γ for kaolin clay are summarised in Table 2.

Equations (11) and (12) are two different relationships for normalised peak states of kaolin clay. Mathematically these are not equivalent. The power law relationship given in equation (11) is similar to that proposed by de Mello (1977) and used routinely in rock mechanics (Hoek & Brown, 1980). Equation (12) is similar to the relationship between stress ratio and state parameter proposed by Been & Jeff-

tuted normally consolidated samples, and in each case the peak strength line found from tests on overconsolidated samples was distinctly curved. Values for the Atterberg limits and the critical state parameters M, λ, κ and v_Γ are summarised in Table 2.

Figures 9(a) to 14(a) show peak and critical states plotted with axes q/Mp'_c and p'/p'_c, both with logarithmic scales. In

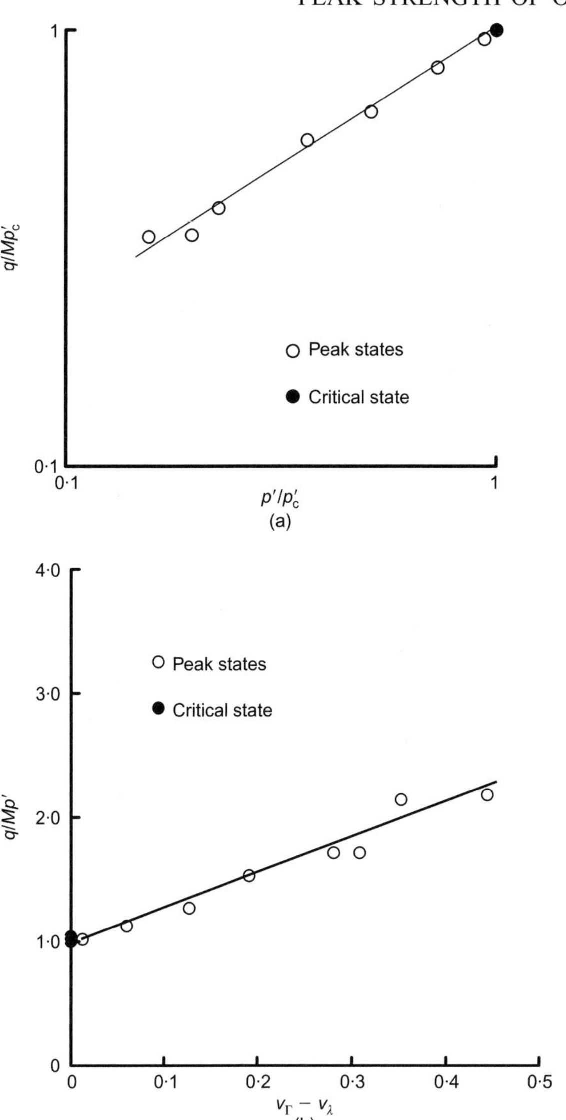

Fig. 8. Critical and peak states of kaolin clay normalised with respect to a state parameter

eries (1985). The data in Figs 8(a) and 8(b) are not sufficiently precise to distinguish which of the two relationships fits the data best.

PEAK AND CRITICAL STATES OF SOME PLASTIC CLAYS

Stress path triaxial tests similar to those described above were carried out on samples of several different plastic clays shown in Table 1. Qualitatively the behaviour of these soils was the same as that of kaolin clay. In each case there were clearly defined critical states found from tests on reconsti-

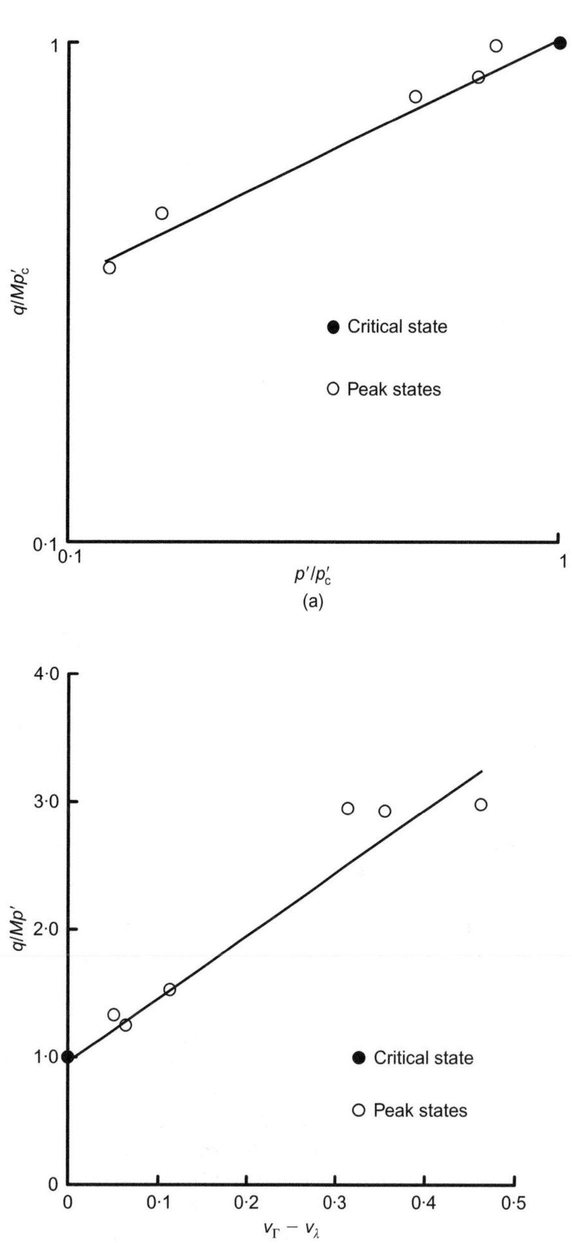

Fig. 9. Critical and peak states of Gault clay normalised with respect to a state parameter

Table 2. Parameters for peak and critical state strengths of some UK plastic clays

Soil	PI	λ	κ	v_Γ	M	β	χ	$1/(\lambda - \kappa)$
Kaolin clay	30	0·190	0·050	3·14	1·00	0·63	2·8	7·1
Gault clay	30	0·168	0·032	2·58	0·94	0·47	4·8	7·4
Kimmeridge clay	27	0·192	0·032	2·68	0·84	0·64	3·4	6·3
London clay	27	0·168	0·050	2·50	1·00	0·33	6·8	8·5
Oxford clay	28	0·191	0·027	2·65	1·00	0·48	4·8	6·1
Reading clay	32	0·230	0·073	2·72	0·76	0·17	6·0	6·4
Weald clay	23	0·148	0·026	2·34	0·97	0·73	3·7	8·2

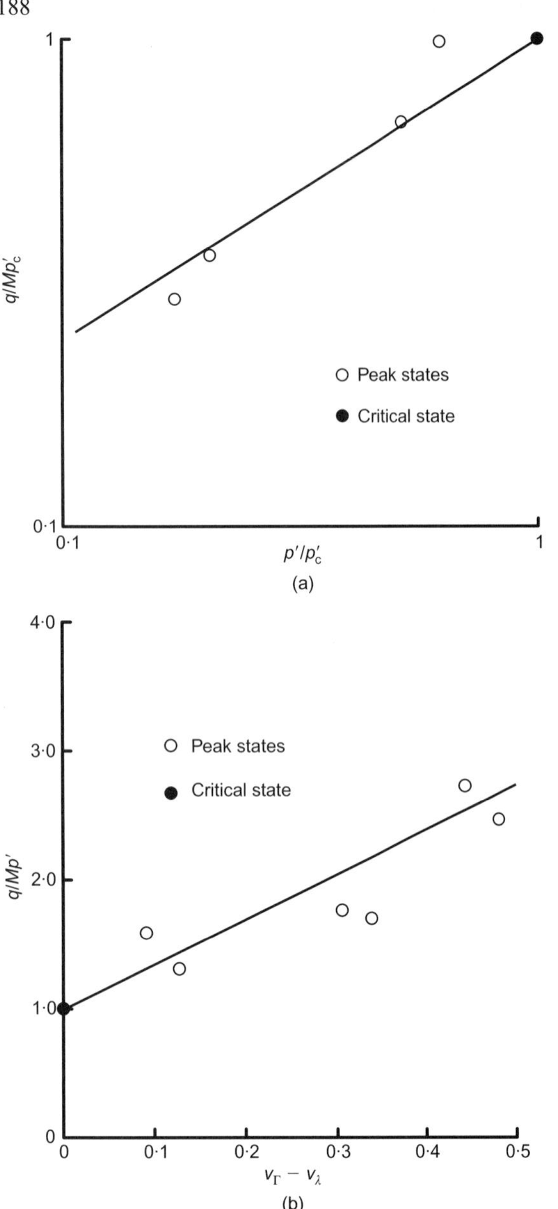

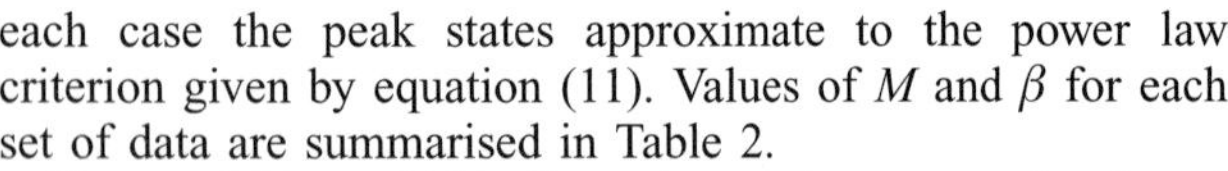

(b)

Fig. 10. Critical and peak states of Kimmeridge clay normalised with respect to a state parameter

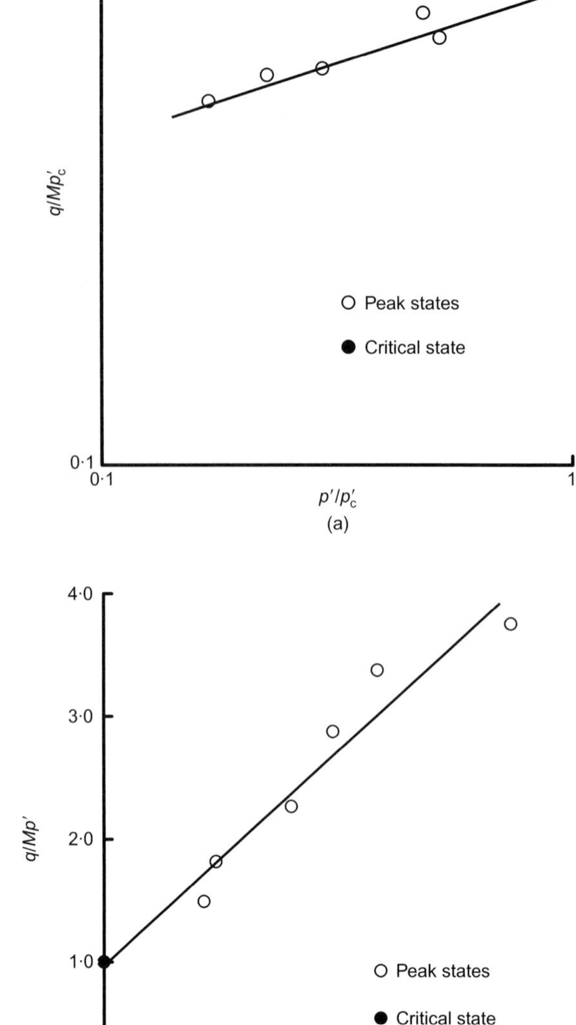

Fig. 11. Critical and peak states of London clay normalised with respect to a state parameter

each case the peak states approximate to the power law criterion given by equation (11). Values of M and β for each set of data are summarised in Table 2.

Figures 9(b) to 14(b) show the peak stress ratio plotted with axes q/Mp' and $(v_\Gamma - v_\lambda)$. In each case the peak states approximate to a linear relationship given by equation (12). Values for χ for each set of data are summarised in Table 2.

The equation describing the state boundary surface of ordinary Cam clay (Schofield & Wroth, 1968) is

$$v_p = v_\Gamma + \lambda - \kappa - \lambda \ln p'_p - \frac{(\lambda - \kappa)q_p}{Mp'_p} \tag{13}$$

where v_p, q_p and p'_p are the peak states on the dry side of the critical state. Substituting v_p from equation (8)

$$\frac{q_p}{Mp'_p} = 1 + \frac{1}{(\lambda - \kappa)}(v_\Gamma - v_\lambda) \tag{14}$$

and, from equation (12)

$$\chi = \frac{1}{\lambda - \kappa} \tag{15}$$

Table 2 shows values of $1/(\lambda - \kappa)$ for each of the soils tested. Except for the data for Reading clay, the values of χ obtained from the test data shown in Figs 8(b) to 14(b) are significantly smaller than those calculated from the Cam clay equation with values of λ and κ found from the gradients of the critical state line and a swelling line. The peak state lines in Figs 8(b) to 14(b) are approximately linear. This indicates that the general form of the Cam clay equation is a reasonable representation of soil behaviour, although the measured peak states are generally smaller than those predicted from equation (15).

If the parameters λ, κ, M, v_Γ, β and χ are material parameters—that is, they do not depend on the state of the soil—they should be related to some parameter that describes the nature of the soil grains. Fig. 15 shows values for the critical state parameters λ, M and v_Γ plotted against the plasticity index PI. There are reasonable correlations between these. Fig. 16 shows values for the peak state

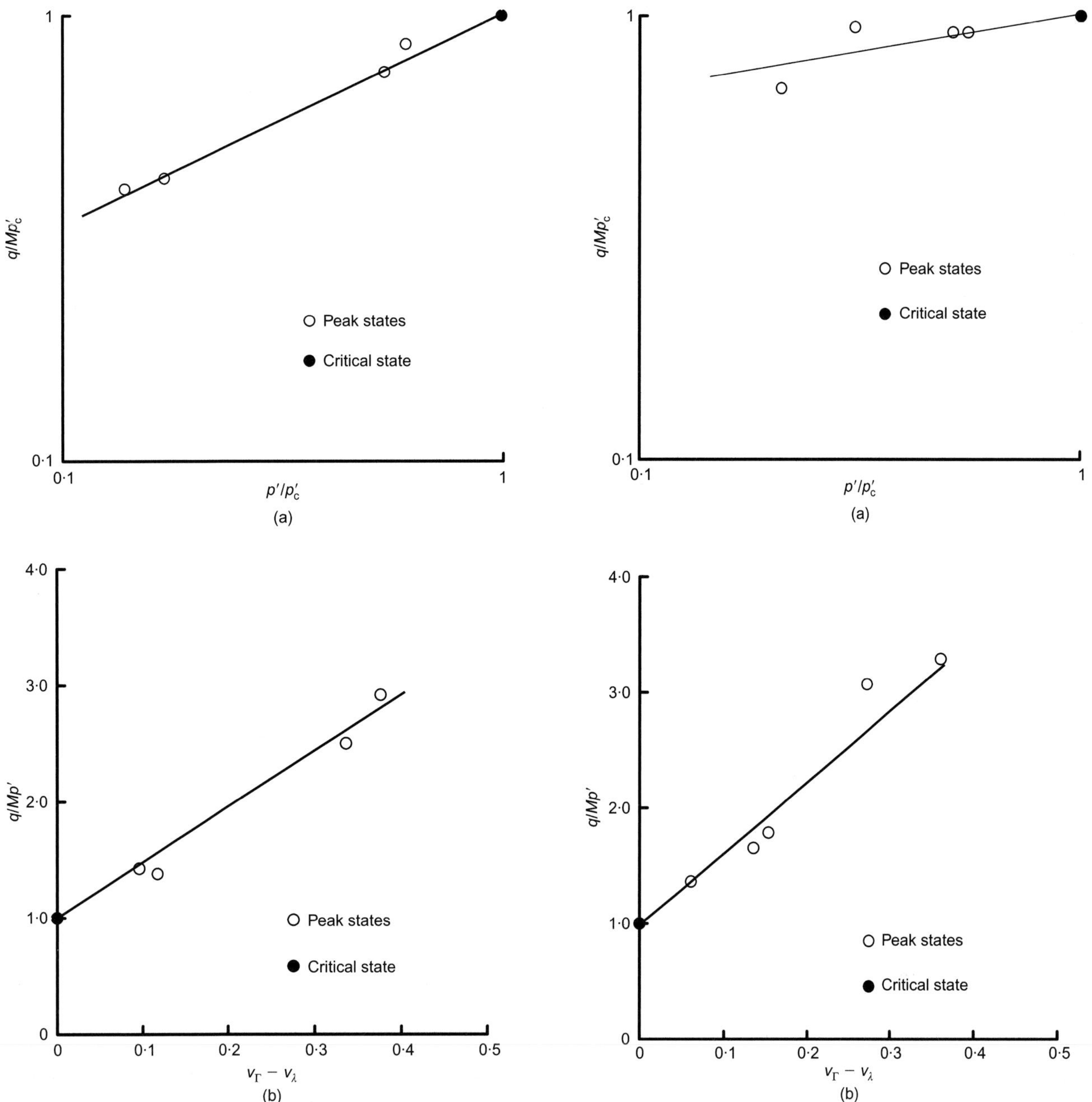

Fig. 12. **Critical and peak states of Oxford clay normalised with respect to a state parameter**

Fig. 13. **Critical and peak states of Reading clay normalised with respect to a state parameter**

parameters β and χ plotted against PI. There is a trend of decreasing β with increasing plasticity, but there is no clear correlation between χ and plasticity.

PEAK STRENGTH IN LABORATORY TESTS AND IN THE GROUND

The peak strength envelope of kaolin clay shown in Fig. 7 is markedly non-linear, particularly for overconsolidation ratios greater than about 4 (corresponding to values of p'/p_c' smaller than 0·5). The value of β, which describes the degree of non-linearity, given in Table 2 is 0·63. (If the envelope is linear, $\beta = 1$.) Values of β for the plastic clays given in Table 2 are mostly smaller than that for kaolin clay, indicating that, for these clays, the degree of non-linearity is greater than that for kaolin clay. In addition, many natural stiff clays in the ground have apparent preconsolidation pressures in excess of 1 MPa, and for these soils the effec-

tive stress below which non-linear peak strength becomes significant would be about 250 kPa.

In most geotechnical structures stresses vary from place to place: at depth and near foundations and walls they can be relatively large, but elsewhere they will be small. It is often difficult to choose a range of stresses (or overconsolidation ratios) to apply in laboratory tests such that a linear Mohr–Coulomb envelope represents the non-linear peak strength everywhere in the ground.

SUMMARY

Special stress path triaxial tests were carried out on reconstituted samples of a variety of plastic clays to examine their peak and critical states. In all cases the critical states were close to lines given by

$$q = Mp' \qquad (3)$$

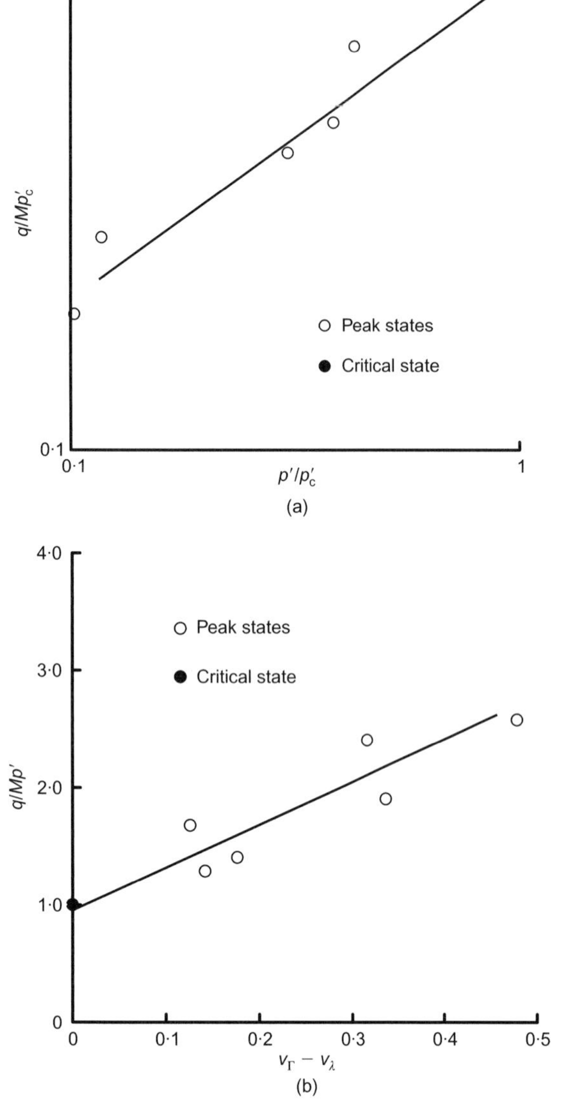

Fig. 14. Critical and peak states of Weald clay normalised with respect to a state parameter

and

$$v = v_\Gamma - \lambda \ln p' \tag{6}$$

with material parameters M, λ and v_Γ that were related to the plasticity index of the soil.

In all cases the peak states were related to effective stresses by a power law better than by a linear relationship. The peak strengths of the soils tested could be represented by a power law of the form

$$\frac{q}{Mp'_c} = \left(\frac{p'}{p'_c}\right)^\beta \tag{11}$$

where p'_c is the critical pressure that takes account of differences in specific volume, and β is a material parameter that describes the degree of curvature of the peak strength line.

Alternatively the peak strength of the soils tested can be related to their peak state parameter $v_\Gamma - v_\lambda$ by

$$\frac{q}{Mp'} = 1 + \chi(v_\Gamma - v_\lambda) \tag{12}$$

where χ is a material parameter that describes the relationship between the peak stress ratio and the distance of the

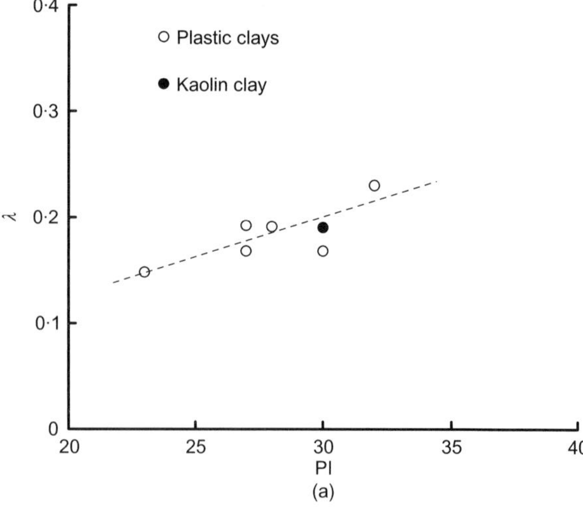

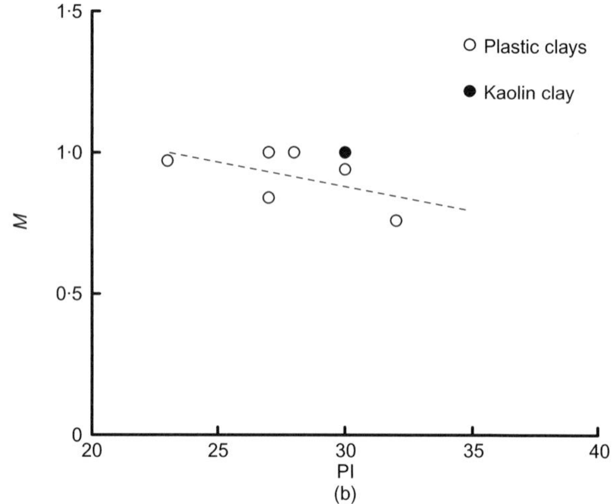

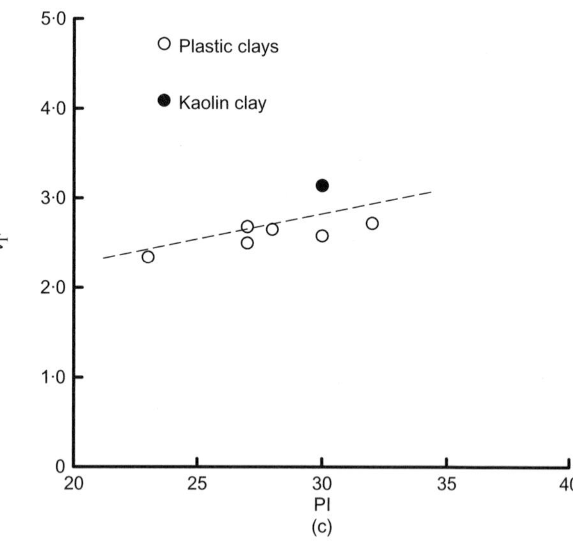

Fig. 15. Critical state parameters: (a) λ; (b) M; (c) v_Γ

peak state from the critical state. The failure criterion given by equation (12) is equivalent to the ordinary Cam clay state boundary surface with $\chi = 1/(\lambda - \kappa)$, although the values of χ and the peak strengths measured in the laboratory tests were generally smaller than those predicted from the critical state parameters λ and κ.

The peak strength criteria given by equations (11) and (12) are not mathematically equivalent, and the present

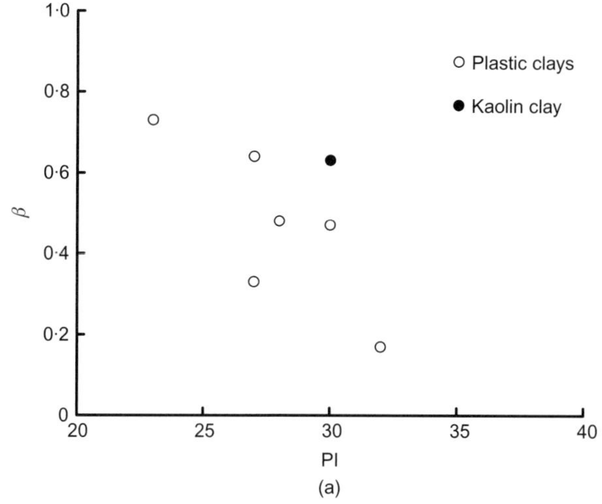

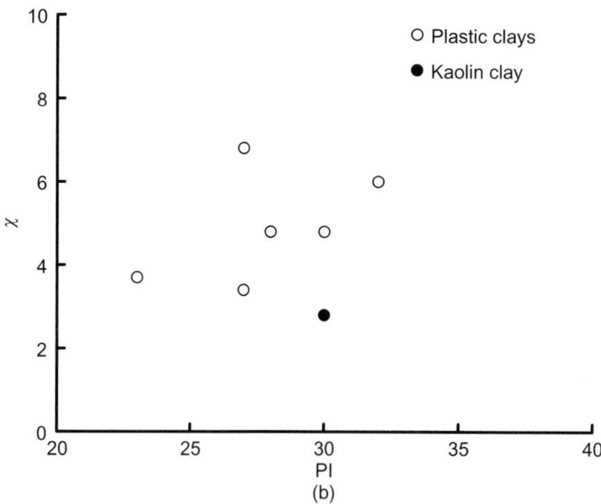

Fig. 16. Peak state parameters: (a) β; (b) χ

experimental data are not sufficiently precise to distinguish which represents the data better. The peak state parameters β and χ obtained from the present tests do not appear to be simply related to the plasticity of the soil.

Use of a linear Mohr–Coulomb failure criterion for the peak strength of soil is potentially unsafe. For the soils tested, the peak strength envelope becomes significantly non-linear at overconsolidation ratios in excess of about 4.

NOTATION

A parameter in a power law strength envelope in equation (5)
b parameter in a power law strength envelope in equation (5)
c' cohesion intercept in a general Mohr–Coulomb strength envelope
c'_p cohesion intercept in a Mohr-Coulomb envelope for peak states
M value of q'/p' at the critical state
p' mean effective stress $= \frac{1}{3}(\sigma'_a + \sigma'_r)$
p'_c critical pressure defined in equation (7)
p'_p value of p' at the peak state
q deviatoric stress $= (\sigma'_a - \sigma'_r)$
v specific volume
v_p value of v at the peak state
v_Γ value of v on the critical state line at $p' = 1$ kPa
v_λ normalising parameter defined by equation (8)
α parameter in a power law strength envelope in equation (10)
β parameter in a power law strength envelope in equation (10)
κ gradient of a swelling line
λ gradient of the critical state line
σ' effective normal stress
σ'_a axial effective stress
σ'_r radial effective stress
τ shear stress
ϕ' angle of friction in a general Mohr-Coulomb strength envelope
ϕ'_p angle of friction in a Mohr-Coulomb envelope for peak states
χ gradient of a peak strength envelope defined in equation (12)

REFERENCES

Atkinson, J. H. & Bransby, P. L. (1978). *The mechanics of soils.* London: McGraw-Hill.

Atkinson, J. H. & Crabb, G. I. (1991). Determination of soil strength parameters for the analysis of highway slope failures. *Proceedings of the ICE international conference on slope stability engineering developments and applications*, London, pp. 11–16.

Atkinson, J. H. & Little, J. A. (1988). Undrained triaxial strength and stress–strain characteristics of a glacial till soil. *Can. Geotech. J.* **25**, No. 3, 428–439.

Been, K. & Jefferies, M. G. (1985). A state parameter for sands. *Géotechnique* **35**, No. 2, 99–12.

Bishop, A. W. & Henkel, D. J. (1957). *The measurement of soil properties in the triaxial test.* Edward Arnold.

Bishop, A. W. & Wesley, L. D. (1975). A hydraulic triaxial apparatus for controlled stress path testing. *Géotechnique* **25**, No. 4, 657–670.

Charles, J. A. (1982). An appraisal of the influence of a curved failure envelope on slope stability. *Géotechnique* **32**, No. 4, 389–392.

Charles, J. A. & Watts, K. S. (1980). The influence of confining pressure on the shear strength of compacted rockfill. *Géotechnique* **30**, No. 4, 355–367.

Cherrill, H. E. (1990). *The influence of loading rate on excess pore pressures in triaxial tests.* PhD thesis, City University, London.

de Mello, V. F. B (1977). Reflections on design decisions of practical significance to embankment dams. *Géotechnique* **27**, No. 3, 279–355.

Hoek, E. & Brown, E. T. (1980). *Underground excavation in rock.* London: Institution of Mining and Metallurgy.

Schofield, A. N. & Wroth, C. P. (1968). *Critical state soil mechanics.* Cambridge: Cambridge University Press.

Pinyol, N., Vaunat, J. & Alonso, E. E. (2007). *Géotechnique* **57**, No. 2, 137–151

A constitutive model for soft clayey rocks that includes weathering effects

N. PINYOL*, J. VAUNAT* and E. E. ALONSO*

Microstructural and mineralogical observations of clay-stones reveal the presence of clay aggregates bonded by a skeleton of inert minerals such as calcium carbonate. The behaviour of these natural materials evolves from a rock-like behaviour, when undisturbed, to clay-like soil when weathered or when subjected to straining. A model has been developed to simulate the constitutive behaviour of these transitional materials. Following the proposal made by Vaunat & Gens, the material is conceived as a composite medium made of a clay matrix and a quasi-brittle bonding microstructure. The basic model has been modified and extended to reproduce the expansive behaviour of the clay matrix, a fundamental aspect to simulate weathering effects, induced largely by drying–wetting cycles. The clay matrix reacts to stress and suction changes, whereas the bond component is not affected by suction changes. An elasto-plastic double-structure model for expansive clay soils describes the clay matrix. The bonding structure follows a damage model. The inter-action between the two constitutive models derives from strain compatibility conditions and energy considerations. The model performance is first illustrated by means of a sensitive analysis that explores the effect of initial bond strength, bond damage rate, bonding concentration, and wetting–drying cycles. Simple stress paths (uniaxial deformation; triaxial compression) are used to highlight the features of the formulation and the role of significant parameters. Some published tests are also reproduced with the model. They have been selected to show some relevant features of evolving soft clay rocks: stiffness and strength degradation during loading and enhanced rebound during unloading, and the effect of drying–wetting cycles on subsequent stiffness and strength degradation.

KEYWORDS: constitutive relations; expansive soils; plasticity; soft rocks; suction

Les observations microstructurales et minéralogiques de l'argilite ont révélé la présence d'agrégats argileux cimentés par un squelette de minéraux inertes tels que le carbonate de calcium. Le comportement de ces matériaux naturels évolue, d'un comportement proche d'une roche (lorsqu'ils laissés tels quels) au comportement d'un sol de type argileux (lorsqu'ils sont dégradés ou soumis à des contraintes). On a développé un modèle pour simuler le comportement constitutif de ces matériaux de transition. Suite à la proposition de Vaunat & Gens, on considère que le matériau est un milieu composite fait d'une matrice argileuse et d'une microstructure de cimentation quasi-fragile. Le modèle de base a été modifié et étendu pour reproduire le comportement expansif de la matrice argileuse, un aspect fondamental pour simuler les effets de dégradation essentiellement induits par les cycles de séchage-mouillage. La matrice argileuse réagit aux changements de contrainte et d'aspiration tandis que la composante de cimentation n'est pas affectée par les changements d'aspiration. On peut décrire la matrice argileuse des sols argileux expansifs à l'aide d'un modèle à double-structure élasto-plastique. La structure de cimentation, quant à elle, utilise un modèle de détérioration. L'inter-action entre les deux modèles constitutifs provient des conditions de compatibilité de contrainte et des considéra-tions énergétiques. La performance des modèles est d'abord illustrée par l'intermédiaire d'une analyse de sensibilité qui explore l'effet de la résistance de la cimenta-tion initiale, le taux de détérioration de la cimentation, la concentration de la cimentation et les cycles de séchage-mouillage. On a utilisé des chemins de contraintes simples (déformation uniaxiale, compression triaxiale) pour sou-ligner les caractéristiques de la formulation et le rôle des paramètres significatifs. Outre le modèle, cet article repro-duit des essais publiés. Ceux-ci ont été sélectionnés pour montrer certaines caractéristiques pertinentes des roches argileuses fermes évolutives : dégradation de la fermeté et de la résistance lors de la compression et augmentation lors de la décompression par effet de rebond ainsi que l'effet des cycles de séchage-mouillage sur la dégradation ultér-ieure de la fermeté et de la résistance.

INTRODUCTION

'Clayey rocks' is a general term for a wide variety of materials that include clay shales, marls, sulphate-bearing claystones, and many types of indurated overconsolidated clay of pre-Pleistocene age. They are 'soft' rocks in the sense that their unconfined compression strength is moderate or low, typically smaller than 30 MPa. They are common in nature, and are often present in civil engineering works.

Examples of distribution of clay shales in North America and Italy are given by Hsu & Nelson (1993) and Bertuccioli & Lanzo (1993). A characteristic feature of these materials is that they weather or degrade when they are exposed to stress or environmental conditions that differ from their undisturbed state. The term 'weathering' implies mechanical, physical and chemical changes, whereas 'degradation' usual-ly refers to a loss of mechanical competence. In their undisturbed state the type of clayey rocks considered in this work are in an intermediate stage in the diagenetic processes that change clay sediments into indurated rocks. Processes leading to rock formation include the increase in confining stress due to increased depth of burial, and the cementation of minerals that precipitate from solution in the pore water. Mineral changes may also occur. The precipitated minerals

Manuscript received 31 May 2006; revised manuscript accepted 14 December 2006.
Discussion on this paper closes on 1 August 2007, for further details see p. ii.
* Department of Geotechnical Engineering and Geosciences, UPC, Barcelona, Spain.

add strength and stiffness to the material. These precipitated minerals are conceived as relatively rigid and brittle materials if they are compared with the mechanical properties of the clay matrix.

The two components of the clayey rock just mentioned— a rigid but brittle bonding and a more deformable clay matrix—are the starting point for the constitutive model described in the paper. The geometrical arrangement of clay matrix components (clay platelets tend to stack in larger units or aggregates), cementing minerals such as carbonates, and residual pore space provides interesting information for modelling purposes, because it may orient some decisions concerning the basic hypothesis of the constitutive framework, as mentioned later.

A conceptual model of the microstructural nature of a mudstone from eastern France is discussed by Sanmartino (2001), on the basis of the analysis of SEM micrographs. Two stratigraphic levels, one rich in carbonates and the other dominated by clay minerals, were examined. The microstructure of the clay-rich horizon showed a clay matrix that surrounds isolated quartz grains and carbonate crystals. In the carbonate-rich zone the carbonate minerals are uniformly present within the clay matrix. The observation that increasing carbonate content leads to increased strength is quite common: one example is provided by Hsu & Nelson (1993) when they report the strength of Taylor and Austin chalk formations in Texas.

Engineering works such as excavations have two distinct effects on the exposed materials: a reduction in confining stress (and possibly an increase in deviatoric stress components), and a cyclic change in humidity due to atmospheric interaction. Field profiles of the water content of exposed clayey rock formations often show the presence of a well-defined 'weathering front'. Below the front the water content remains at its undisturbed low value. Above the front the material evolves rapidly into a clay formation with a substantial increase in water content. Increasing the water content (which is also an indication of increased porosity) leads to a rapid reduction in mechanical properties. Two examples are given in Figs 1 and 2. Fig. 1 shows the unconfined compression strength measured in rock cores taken in boreholes located in the floor of Lilla Tunnel (Tarragona, Spain). Lilla Tunnel is excavated in early Eocene argillaceous rock containing anhydrite and gypsum veins. The plot shows the rapid reduction in strength as the water content increases.

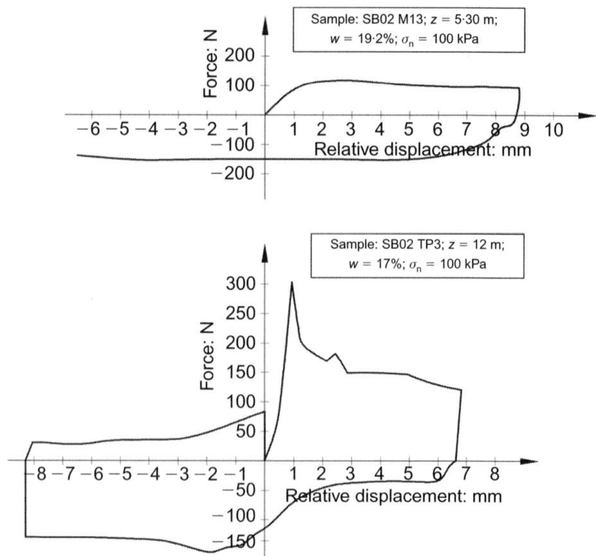

Fig. 2. Curves of force against relative displacement measured in two cores from El Bierzo tertiary clay formations

Visual identification of cores as 'fresh' or weathered is also shown in the figure. A description of this case is given in Alonso & Berdugo (2005).

The loss in strength is associated with a parallel reduction of brittleness. This is shown in Fig. 2, which compares two slow direct shear tests performed on cores recovered from red continental low-plasticity Tertiary clays from El Bierzo, Spain. The two specimens were tested under identical confining stresses (100 kPa). The unweathered core, with a water content of 17%, was taken at a depth of 12 m, and exhibited a marked brittle behaviour. A shallower specimen, having a small increase in water content (2·2%), reacted as a ductile material. Its shear strength was close to the 'residual' strength determined in the shear box, after a full shearing cycle. The strength envelopes determined for this material are shown in Fig. 3. Peak strengths show a considerable scatter. 'Residual values' as determined after a cycle of shear stress applications show a marked loss in strength, but the point to note here is that there are two alternative circumstances leading to the low 'residual' strength: either a shear straining of the undisturbed material, or a natural weathering, which resulted in a (relatively small) increase in water content. The identification of these apparently separate mechanisms is not so simple in this case, however, because

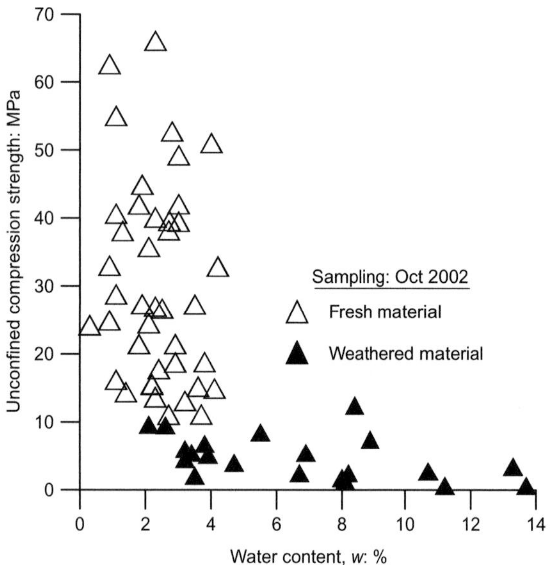

Fig. 1. Variation of unconfined compression strength with water content of Lilla claystone

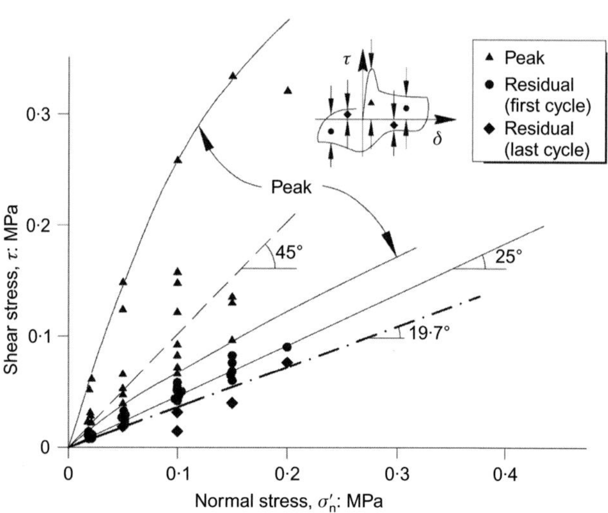

Fig. 3. Strength envelopes for El Bierzo tertiary clay

the cores were recovered from a natural slope with symptoms of shallow instability, and therefore some previous natural straining of the upper core specimen cannot be disregarded in this case.

This paper describes a constitutive model inspired by the observed features of clayey rocks, as outlined above. Specific attention has been given to the simulation of weathering effects and, in particular, to two important mechanisms that are believed to control, in practice, the loss of mechanical properties of these materials: the straining associated with stress changes, and the wetting–drying cycles imposed by atmospheric action. Other effects mentioned in the literature (Mitchell, 1976) that contribute to weathering (thermal straining, crystal dissolution or precipitation, chemical reactions) are not considered here. Before the model is described, some additional features of clay rock behaviour, which justify some modelling hypotheses, are described in the next section. In a second part of the paper, the model performance is presented and compared with some selected laboratory tests.

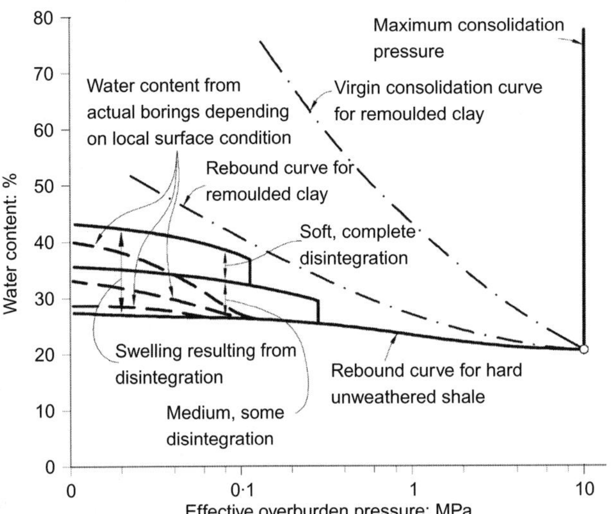

Fig. 4. Geological history of Bearpaw clay shale (Bjerrum, 1967)

ADDITIONAL FEATURES OF CLAYEY ROCK BEHAVIOUR

The consequences of bonding under stress

Bjerrum (1967), in his study of montmorillonitic Bearpaw clay shales ($w_L = 90–120\%$), provided an interpretation for the observed water content of recovered specimens. When water content is plotted against the calculated effective overburden stress for each specimen, a 'rebound' or 'unloading' curve should be obtained (Fig. 4), because the region was subjected to high (10 MPa) preconsolidation stress before being unloaded.

Specimens from different locations were plotted on different rebound curves. Bjerrum attributed this behaviour to a varying degree of shale 'disintegration'. Hard unweathered specimens maintain the lowest water contents. Increasing degradation (rebound curves are shown in Fig. 4) lead to increasing water contents. The loading–unloading curve for the remoulded clay is also shown in Fig. 4. The field curve of soft clay shale tends toward the unloading or swelled curve of the remoulded structure-free specimen. Bjerrum (1967) attributed the swelling of the unloaded shale to the release of the 'locked-in strain energy'. The expression provides a clue to the reason for swelling and damage of clay shales. If a clay soil, in the process of diagenesis, becomes bonded when under stress, it has the potential to swell if the stress is removed. It all depends on the relative intensity of the stress changes induced by unloading, and on the internal bond strength.

The idea is illustrated in Fig. 5. A clay matrix, organised in aggregates, is first compressed and then cemented under stress. The associated changes in void ratio are sketched in the same figure. Then stress changes induced by unloading damage the cementing bonds to some extent, and allow clay aggregates to expand. Further reloading may contribute to additional bond damage.

Swelling and swelling-induced damage are not the exclusive behaviour of clay rocks that contain active clay minerals (such as Bearpaw shale). In fact, clays of any plasticity may behave in a qualitatively similar manner. Even granular soils cemented under load may exhibit swelling if unloaded. The energy stored in hertzian contacts between grains may be released if the bonds are broken. Such behaviour was observed in some elegant tests reported by Fernández & Santamarina (2001).

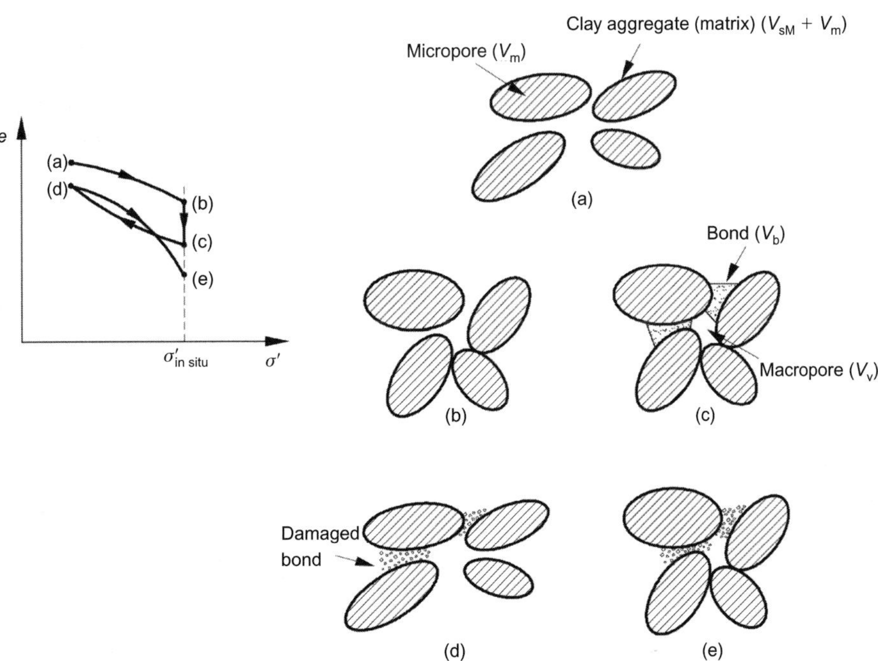

Fig. 5. Interpretation of diagenesis process and structural effects of an unloading–reloading cycle on a clayey rock

The effect of loading and unloading cycles

Besides the natural loading and unloading cycle implied by overconsolidation, an additional unloading is normally associated with core boring operations. When load is then imposed in a test, the material has already experienced a particular stress path history. In practice, excavation followed by foundation construction imposes an unloading–loading cycle. Tunnelling excavation and lining leads, in a similar way, to stress reversals. It is therefore appropriate to investigate the response of bonded clay shales to loading–unloading cycles. One interesting example is the behaviour of Laviano 'scaly' clay shale, a highly tectonised material with moderate expansive behaviour, described by Picarelli (1991). A sequence of applied loading–unloading cycles under oedometric conditions led to a progressive loss of material stiffness during cycle applications. This behaviour is consistent with the concept of destructuration suffered by the clay when loaded above a certain limit and the subsequent release of swelling potential during unloading. According to Kavvadas (2000), the swelling index is an indicator of the amount of structure in soils, because the reconstituted material is expected to swell more than the natural one.

A different behaviour was observed for Todi clay (Calabresi & Scarpelli, 1985), a heavily overconsolidated and intensively fissured lacustrine clay. The reported behaviour (Fig. 6) will later be compared with model calculations. During first loading, a compression curve similar to that of Laviano clay was observed. However, during the subsequent unloading, a very high swelling strain occurred and the void ratio increased above its initial value. During the second loading cycle, the sample experienced first a stiff response similar to that of the intact material, but it apparently reached a yield point at a lower value, indicating that some destructuration had previously occurred. The slope of the NC line is now higher than the slope of the intact material, but still lower than the slope of the reconstituted clay. The

swelling coefficient during the second unloading path lies between values of the intact and the destructured material. This response is attributed to a partial damage of bonds, which removes the interparticle forces that prevent swelling and releases the expansive potential of clay minerals.

Drying–wetting cycles

Changes in relative humidity (RH), as well as rainfall and evapotranspiration phenomena, lead, in many engineering situations, to the application of suction cycles to the exposed material. Even in tunnelling works, protected from rainfall and transpiration phenomena, changes in RH are significant. Records of relative humidity measured in Lilla Tunnel (data on the variation of strength of Lilla claystone with water content were given in Fig. 1) over 8 months indicated RH values ranging between 93% and 20%. Similar records for the Orange-Fish tunnel in South Africa have been given by Olivier (1987). RH changed from full saturation to values close to 55%. In this case, the rock excavated was a carboniferous expansive mudrock. Olivier (1987) reported the strong effect of full wetting–air drying cycles on unconfined specimens. Some specimens were described as broken after two wetting–drying cycles.

Wong (1998) performed drained triaxial tests on specimens of intact La Biche shale, a medium-plasticity ($w_\mathrm{L} = 44\cdot6$–$45\cdot8\%$) compact shale from western Canada, previously subjected to free or semi-confined swell. He also investigated the effect of swelling under different salt concentrations of the water added in swelling tests. Fig. 7(a) illustrates the behaviour of the intact material. Increasing the confining stress leads to a progressive ductility of the shale. Once the specimens are swelled under a 50 kPa vertical

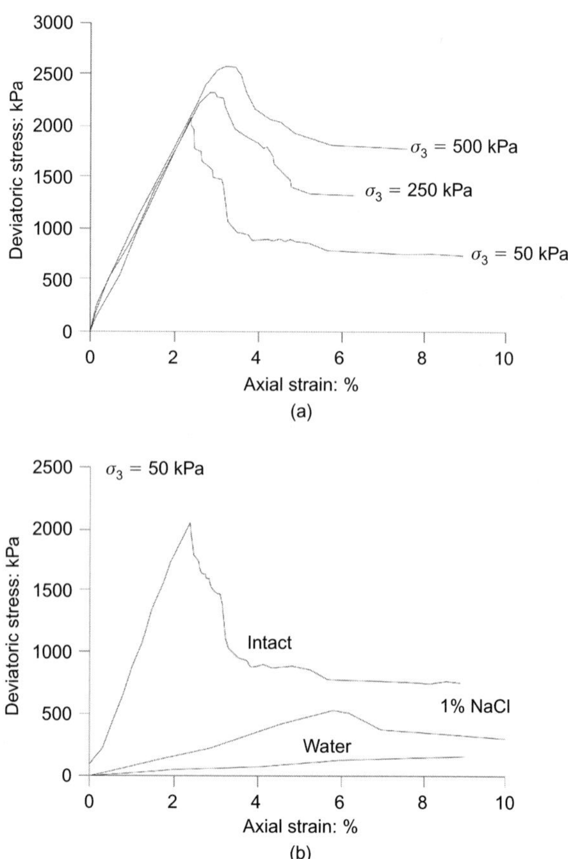

Fig. 7. Drained triaxial tests on: (a) intact La Biche shale: (b) swollen specimens at a confining stress of 50 kPa (after Wong, 1998)

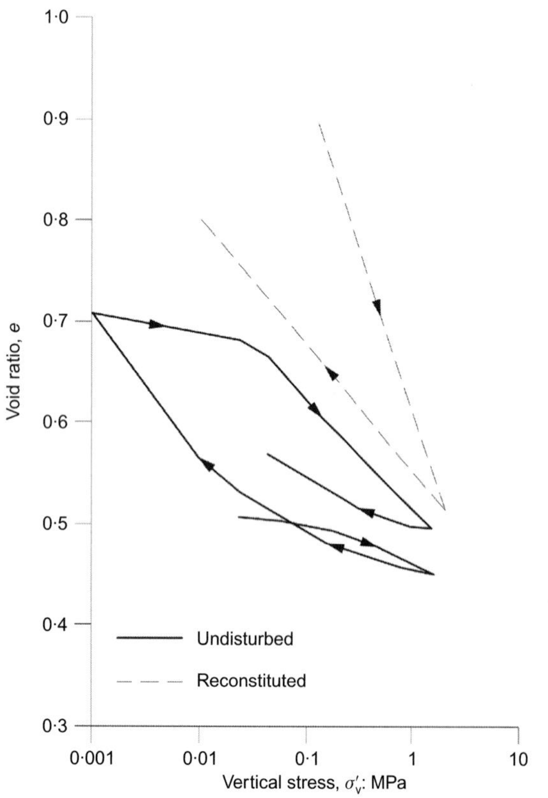

Fig. 6. Compression oedometer tests on Todi clay (Calabresi & Scarpelli, 1985)

stress, a profound degradation in strength and stiffness is observed in the subsequent triaxial test (Fig. 7(b)). Application of pure water (during swelling) leads to the maximum structural breakdown. When swelling is restricted by using a salt concentration in the water, or by imposing a constant vertical stress during wetting, the degradation is also limited, and the shale may retain some brittleness under the set of confining stresses investigated.

Similar results were obtained by Rampello (1991), who investigated the effect of a single desiccation–resaturation cycle on the subsequent triaxial response under triaxial testing of Todi Clay.

In all the tests mentioned, the strength of the weathered material falls well below the 'end of test' strength values recorded on undisturbed specimens.

Summary of results
The described features can be summarised as follows:

(*a*) When bonding occurs under confining stress, a potential expansibility is built into clayey rocks.

(*b*) Changes in stress may damage the bonding and allow the release of the swelling potential. Unloading–loading cycles are common in engineering works, and they contribute to accumulated bond damage and soil swelling. Some experiments indicate that the damaged material progressively approaches the state of the remoulded material, although this is not always the case.

(*c*) Wetting–drying cycles are also expected in exposed surfaces of clayey rocks. Experiments indicate that suction cycles are capable of inducing substantial damage in bonded clayey rocks.

MODEL DESCRIPTION

The clayey soft rock is conceptualised as a composite material made of a clay matrix and a distributed network of bonds, which partially occupy the macropores of the clay matrix (Vaunat & Gens, 2003). The clay matrix has its own structure. Following accepted ideas of clay particle association in clays (Gens & Alonso, 1992), clay platelets are stacked in aggregates that may swell or shrink. A description of clay stacks in shaly rocks and clay shales is given by Pye & Krinsley (1983) and Seedsman (1987). Volume changes of these stacks (which may integrate thousands of individual clay sheets) are thought to disrupt the bonded structure of the rock. The structure of the clay matrix is conveniently described as a micro–macro structure. 'Micro' refers to the porosity inside the clay aggregates; 'macro' describes the void space between clay aggregates. The 'macro' space includes open voids available for fluid transfer, and the volume occupied by the cementation minerals (bonds). Two types of pore coexist in the medium: the small pores inside the aggregates; and the relatively large pores between aggregates. A sketch of the model components is given in Figs 5(a) and 5(c).

Topologic considerations and strain compatibility
In the reference volume (Fig. 5), the following volumetric variables may be defined.

V total volume
V_{sM} solid volume of (clay) matrix
V_b solid volume of bond material (it is assumed to be non-porous)
V_v volume of macropores
V_m volume of micropores

The following relation holds:

$$V = V_{sM} + V_b + V_v + V_m \tag{1}$$

The total volume of solids, V_s, is

$$V_s = V_{sM} + V_b \tag{2}$$

It is convenient, in order to define later strains, to consider the existing volume between clay aggregates, V_M^*. It includes the bond volume and the volume of macropores:

$$V_M^* = V_b + V_v \tag{3}$$

A change in V_M^* is a measurement of the relative motion of the clay aggregates. The asterisk (*) is introduced to indicate that V_M^* may be considered as a 'fictitious' macropore space because it includes real voids as well as the bond volume.

In a general situation the bond mass may increase or decrease as a result of precipitation or dissolution of minerals. Only the solid volume of minerals included in the clay matrix is considered fixed. Therefore V_{sM} is adopted as the reference volume to define void ratios, as follows.

Void ratio of macropores

$$e_v = \frac{V_v}{V_{sM}}$$

Void ratio of bonds

$$e_b = \frac{V_b}{V_{sM}}$$

Void ratio of micropores

$$e_m = \frac{V_m}{V_{sM}}$$

Void ratio of pores between clay aggregates

$$e_M^* = \frac{V_M^*}{V_{sM}} = \frac{V_v + V_b}{V_{sM}}$$

First, volumetric strain relationships will be first written; then the results will be generalised to triaxial conditions.

A global (externally observed) volumetric strain will have two components: a change in the topological configuration of the aggregates, and the volume change of the aggregates themselves. The first component introduces the following volumetric strain.

$$\varepsilon_{vol}^M = -\frac{\Delta V_M^*}{V} = -\frac{\Delta V_v}{V} - \frac{\Delta V_b}{V} \tag{4}$$

Note that a volumetric strain ε_{vol}^M (in compression, for instance) implies a reduction in pore space and a reduction of bond volume. The asterisk (*) is dropped in the remaining paragraphs to simplify the notation.

The second component, the volumetric strain of micropores, is written

$$\varepsilon_{vol}^m = -\frac{\Delta V_m}{V}$$

The volumetric strain of bonding material is defined as

$$\varepsilon_{vol}^b = -\frac{\Delta V_b}{V_b}$$

The straining of bonds is defined locally. The equivalent strain for the representative volume V will be

$$-\frac{\Delta V_b}{V} = -\frac{\Delta V_b}{V_b}\frac{V_b}{V} = \varepsilon_{vol}^b \, C_b \tag{5}$$

where C_b is the volume concentration of bonding material.

From equations (4) and (5), the volumetric strain associated with the change in position of aggregates will be

$$\varepsilon_{\text{vol}}^{M} = \varepsilon_{\text{vol}}^{V} + C_{b}\,\varepsilon_{\text{vol}}^{b} \tag{6}$$

Finally, the measurable 'external' volumetric strain, $\varepsilon_{\text{vol}}^{\text{ext}}$, will be obtained by adding equation (6) and the micro-component.

$$\begin{aligned}\varepsilon_{\text{vol}}^{\text{ext}} &= \frac{\Delta V_{v} + \Delta V_{b} + \Delta V_{m}}{V} \\ &= \varepsilon_{\text{vol}}^{V} + C_{b}\,\varepsilon_{\text{vol}}^{b} + \varepsilon_{\text{vol}}^{m} \\ &= \varepsilon_{\text{vol}}^{M} + \varepsilon_{\text{vol}}^{m}\end{aligned} \tag{7}$$

If triaxial conditions are considered the deviatoric strain states are written as

$$\varepsilon_{q}^{\text{ext}} = \varepsilon_{q}^{V} + C_{b}\,\varepsilon_{q}^{b} = \varepsilon_{q}^{M} \tag{8}$$

Note that in terms of deviatoric strain the basic model for a non-expansive soft rock is recovered.

Constitutive model: matrix behaviour

The matrix reacts to changes in stresses (σ^{M}) and to suction changes. For expansive materials, an explicit consideration of the two structural levels defined for the matrix (micro and macro) was found convenient to describe the features of expansive soil behaviour. A qualitative framework for the model was first described by Gens & Alonso (1992). The model was later formulated in more detail in Alonso *et al.* (1999). Within the context of unsaturated soil mechanics, the matrix stress should more properly be defined as a 'net' matrix stress (the excess of total stress over air pressure). 'Matrix stress' will be the term used here for simplicity.

A brief description of the volumetric part of the model is given here. As clay expansion is interpreted as a local hydration phenomenon in which clay aggregates 'capture' the water available in macropores, it is reasonable to assume that, away from equilibrium, two different suction values exist in the soil: a microstructural suction, s^{micro}, prevailing inside the aggregates, and a macrostructural suction existing in macropores. The microstructural behaviour is assumed to be volumetric and non-linear elastic. Mean stress and s^{micro} control their volumetric response. It is accepted that the mean stress in the aggregates is in equilibrium with the externally applied mean stress on the matrix, p^{M}. Suction in macropores, s^{macro}, will not be necessarily in equilibrium with s^{micro}. Local suction gradients will result in local fluid transfer. However, in the applications presented later, equilibrium will be assumed, and only the drained response of the material will be discussed ($s^{\text{macro}} = s^{\text{micro}} = s$)

As clay aggregates are supposed to remain saturated, in view of the very small spacing between clay platelets, the effective stress defined by the sum ($p^{M} + s^{\text{micro}}$) holds, and a simple constitutive law may be assumed for the volume change of aggregates.

$$\mathrm{d}\varepsilon_{\text{vol}}^{m} = \frac{\kappa_{m}}{1 + e_{m}}\,\frac{d\left(p^{M} + s\right)}{p^{M} + s} = \frac{d\left(p^{M} + s\right)}{K_{m}} \tag{9}$$

where κ_{m} is a (constant) compressibility coefficient and K_{m} is the bulk modulus of clay aggregates.

Macrostructural behaviour (defined as the rearrangement of the clay aggregates) includes the competition of two mechanisms: the response of aggregates to load and suction, as it is considered in a non-expansive material; and the response induced by swelling and shrinkage of the aggre-

gates. The first mechanism is modelled through the Barcelona Basic Model (Alonso *et al.*, 1990). The second mechanism is introduced by means of an additional component of volumetric plastic strain that depends on the volumetric change experienced by the microstructure and the distance of the current stress state to the collapse yield loci. The onset of this coupled plastic strain is defined by two additional yield loci: the SD (suction decrease) yield surface, which marks the development of irreversible swelling strains; and the SI (suction increase) surface, which defines the development of irreversible shrinkage.

The elastic response of the arrangement of aggregates (macro) is defined by an expression similar to equation (9), but now the single effective stress principle does not hold, and two separate contributions associated with the matrix stress and the (macro) suction lead to the elastic strain.

$$\begin{aligned}\mathrm{d}\varepsilon_{\text{vol}}^{Me} &= \frac{\kappa}{1 + e_{M}^{*}}\,\frac{\mathrm{d}p^{M}}{p^{M}} + \frac{\kappa_{s}}{1 + e_{M}^{*}}\,\frac{\mathrm{d}s}{\left(s + p_{\text{atm}}\right)} \\ &= \frac{\mathrm{d}p^{M}}{K_{M}} + \frac{\mathrm{d}s}{K_{s}}\end{aligned} \tag{10}$$

where κ and κ_{s} are the elastic macro-stiffness parameters for changes in mean matrix stress and in macro-suction respectively; K_{M} and K_{s} are the bulk elastic moduli of the clay aggregate arrangement against changes in matrix stress and (macro) suction; and p_{atm} indicates the atmospheric pressure.

A representation of yielding states of the model (for isotropic stress states) is given in Fig. 8 in a plane defined by the mean matrix stress (p^{M}) and the suction s.

When SI is activated, an additional irreversible strain component equal to the volumetric microstrain $\mathrm{d}\varepsilon_{\text{vol}}^{\text{micro}}$ multiplied by an interaction function $f_{I} = f_{I0} + f_{I1} \times (p^{M}/p_{0}^{M})^{n_{I}}$ is added to the volumetric matrix strain. When SD is activated, the additional plastic strain component reads $\mathrm{d}\varepsilon_{\text{vol}}^{m}[f_{D0} + f_{D1}(1 - p^{M}/p_{0}^{M})^{n_{D}}]$, where p_{0}^{M} represents the yield point on the LC surface at the suction under consideration. Therefore the plastic straining associated with basic (elastic) mineral expansion or contraction is controlled by the current overconsolidation ratio. Note also that p_{0}^{M} is also a measure of the matrix density. Expressions for the interaction functions f_{I1} and f_{D1} are given in Alonso *et al.* (1999). The remaining symbols (f_{I0}, n_{I}, f_{D0}, n_{D}) are scalar material parameters.

Irreversible plastic straining induced by suction changes modifies the yield stress, p_{0}^{M}. This link helps to model the common observation that swelling strains decrease with in-

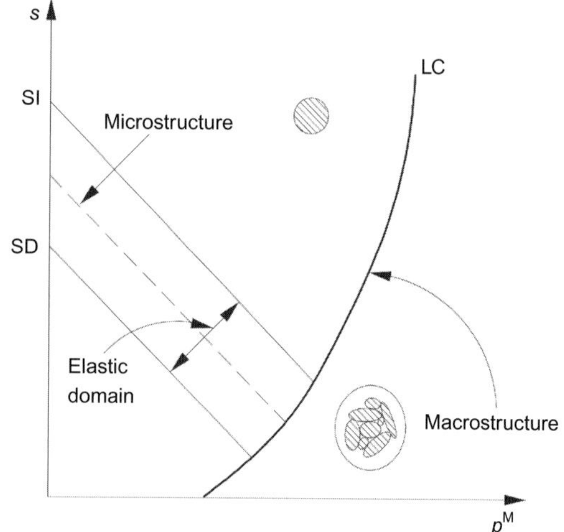

Fig. 8. Elastic zone and yield loci for matrix expansive model (volumetric behaviour)

creasing confining stress. The formulation of the plastic straining associated with suction changes in terms of two yield limits, SI and SD, helps to reproduce the evolving deformation of clays when several wetting–drying cycles are applied.

Constitutive model: bond behaviour
Selecting an appropriate constitutive model for the bonding is not an easy task. Its effect on the overall behaviour does not depend only on the properties of the bond material itself or on bond concentration. Probably as important is the three-dimensional geometry of the bonding network and its links with other mineral constituents, namely the clay matrix. It may be expected that, for high bond concentrations, the nature of the bond material will dominate the bond constitutive behaviour. For low to moderate concentrations, which is the case for hard and cemented clay formations, the nature of the bond material has probably a more limited effect. Test results discussed before suggest that bonding introduces a stiff response before yielding, and that it also provides a distinct brittleness. The interpretation of test results suggests also that bonds may fail in compression, in shear or in extension. However, bonding cannot be directly tested, and its properties can only be inferred from global material behaviour. The idea favoured here is that bonding becomes damaged when it is subjected to a given amount of strain energy, irrespective of the stress path applied. This hypothesis is also consistent with the idea that bonding is established once the matrix is stressed. Bond strain energy takes, as a reference zero state, the initial stress of the investigated stress path.

Bond behaviour is modelled through isotropic damage theory. A linear elastic response is assumed for the bond in an undamaged state. Under triaxial conditions, the stress–strain relationship is simply

$$p^b = K_{b0}\varepsilon^b_{vol} \tag{11a}$$

$$q^b = G_{b0}\varepsilon^b_q \tag{11b}$$

where K_{b0} and G_{b0} are the bulk and shear moduli of the bond in an intact state.

The isotropic damage is introduced by means of the following simple form.

$$p^b = (1 - D)K_{b0}\varepsilon^b_{vol} = K_b\varepsilon^b_{vol} \tag{12a}$$

$$q^b = (1 - D)G_{b0}\varepsilon^b_q = G_b\varepsilon^b_q \tag{12b}$$

D is the scalar damage parameter, which represents the damaged area of bonds over the total area of bond. It takes the value 0 when the material is intact, and 1 when all the bond is damaged. $K_b = (1 - D)K_{b0}$ and $G_b = (1 - D)G_{b0}$ are the bulk and shear modulus of the bond in the current damaged state.

As proposed by Carol *et al.* (2001), an alternative damage variable is used,

$$L = \ln\left(\frac{1}{1 - D}\right) \tag{13}$$

which varies from 0 to ∞.

Using the logarithmic damage variable, the bond bulk and shear moduli are written as

$$K_b = e^{-L}K_{b0} \tag{14a}$$

$$G_b = e^{-L}G_{b0} \tag{14b}$$

The evolution of L is required to define the degradation of bond stiffness totally, and to determine the bond behaviour.

Carol *et al.* (2001) proposed that damage and subsequent changes of L should be linked to the increments of energy stored in the bonds per unit of volume. This energy is defined as the elastic secant energy that would be recovered upon unloading ($u_b = \frac{1}{2}(p^b\varepsilon^b_{vol} + q^b\varepsilon^b_q)$)

With this formulation, the current bond damage locus is defined by a threshold (r_0) of the secant elastic energy that can be represented by an ellipse in (p^b, q^b) space. Inside this ellipse, the bond stiffness remains constant. When changes in the tensional state of the bond lead the secant elastic energy to reach the threshold, damage takes place (L increases) and stiffness reduces. The damage evolution is determined by means of the function suggested by Carol *et al.* (2001):

$$r(L) = r_0 e^{r_1 L} = u_b \tag{15}$$

where r_1 determines the damage rate. The consistency condition associated with this energy locus allows the determination of L.

Strain partition
An assumption is now made concerning the form of the bond strain ε^b.

$$\varepsilon^b_{vol} = \chi\varepsilon^v_{vol} \tag{16a}$$

$$\varepsilon^b_q = \chi\varepsilon^v_q \tag{16b}$$

where $\chi(\chi_0, L)$ is assumed to depend on an initial reference value χ_0, and on the logarithmic damage variable L. Equations (16) explain that in a highly damaged material (χ is small) the local bond deformation will be small when the macrovoids are deformed (Fig. 5(d) illustrates this idea). Conversely, the local bond strain will increase when damage is reduced, because the bond stiffness is high in comparative terms.

Therefore, from equations (16), (7) and (8),

$$\varepsilon^b_{vol} = \frac{\chi}{1 + \chi C_b}\varepsilon^M_{vol} \tag{17a}$$

$$\varepsilon^b_q = \frac{\chi}{1 + \chi C_b}\varepsilon^M_q \tag{17b}$$

The following relationship is adopted for χ:

$$\chi = \chi_0 e^{-L/2} \tag{18}$$

where parameter χ_0 provides the strain partition before any bond damage has occurred. This is a constitutive parameter— the bonding deformation parameter for the undamaged state.

Stress partition: conjugate variables
Two stress components perform work: a stress associated with the matrix aggregates (p^M and q^M), and a stress resisted by the bonds (p^b and q^b). p^M and q^M deform the clay aggregates and modify its relative position, creating additional straining. Therefore the matrix stress acts on the external strain ε^{ext}, the only one being measured. The bond stress performs work when the bond deforms ($C_b\varepsilon^b$).

In order to derive an expression for the external stress σ^{ext} in terms of the two defined stress components, the principle of virtual work is applied as follows.

$$p^{ext}d\varepsilon^{ext}_{vol} + q^{ext}d\varepsilon^{ext}_q = p^M d\varepsilon^{ext}_{vol} + q^M d\varepsilon^{ext}_q \\ + p^b C_b d\varepsilon^b_{vol} + q^b C_b d\varepsilon^b_q \tag{19}$$

Equation (19), valid for any external change in strain, provides a procedure to find the external stress associated with any change in external strain, as illustrated in the following simplified cases.

Infinitely rigid clay aggregates and no damage The soil does not exhibit any expansion or shrinkage phenomena. The material is now a cemented granular soil. The bond strain, in view of equations (7), (8) and (17) will be

$$\varepsilon_{\text{vol}}^{\text{b}} = \frac{\chi}{1 + \chi C_{\text{b}}} \varepsilon_{\text{vol}}^{\text{ext}} \tag{20a}$$

$$\varepsilon_{\text{q}}^{\text{b}} = \frac{\chi}{1 + \chi C_{\text{b}}} \varepsilon_{\text{q}}^{\text{ext}} \tag{20b}$$

Substituting these values in equation (19) and assuming no damage (χ constant),

$$p^{\text{ext}}\Delta\varepsilon_{\text{vol}}^{\text{ext}} + q^{\text{ext}}\Delta\varepsilon_{\text{q}}^{\text{ext}} = p^{\text{M}}\Delta\varepsilon_{\text{vol}}^{\text{ext}} + q^{\text{M}}\Delta\varepsilon_{\text{q}}^{\text{ext}}$$

$$+ \frac{C_{\text{b}}\chi}{1 + \chi C_{\text{b}}}\left(p^{\text{b}}\Delta\varepsilon_{\text{vol}}^{\text{ext}} + q^{\text{b}}\Delta\varepsilon_{\text{q}}^{\text{ext}}\right) \tag{21}$$

and as equation (21) is valid for any $\Delta\varepsilon^{\text{ext}}$,

$$p^{\text{ext}} = p^{\text{M}} + \frac{C_{\text{b}}\chi}{1 + C_{\text{b}}\chi} p^{\text{b}} \tag{22a}$$

$$q^{\text{ext}} = q^{\text{M}} + \frac{C_{\text{b}}\chi}{1 + C_{\text{b}}\chi} q^{\text{b}} \tag{22b}$$

For a given C_{b}, χ determines the redistribution of external stress among the bond and the matrix. When damage occurs and χ decreases, stress is progressively transferred from bond to matrix. When a fully destructured state is reached, the bond has lost its stiffness and strength, and the external stress is fully applied to the clay matrix. When the bond concentration vanishes ($C_{\text{b}} = 0$) the external stress is also fully applied to the clay matrix.

Elastic behaviour of matrix and no damage (at constant suction) The bond stress is given by

$$\Delta p^{\text{b}} = K_{\text{b}}\Delta\varepsilon_{\text{vol}}^{\text{b}}$$

$$= K_{\text{b}}\frac{\chi}{1 + C_{\text{b}}\chi}\Delta\varepsilon_{\text{vol}}^{\text{M}} \tag{23a}$$

$$= \frac{\chi}{1 + C_{\text{b}}\chi}\frac{K_{\text{b}}}{K_{\text{M}}}\Delta p^{\text{M}}$$

$$\Delta q^{\text{b}} = G_{\text{b}}\Delta\varepsilon_{\text{q}}^{\text{b}}$$

$$= G_{\text{b}}\frac{\chi}{1 + C_{\text{b}}\chi}\Delta\varepsilon_{\text{q}}^{\text{M}} \tag{23b}$$

$$= \frac{\chi}{1 + C_{\text{b}}\chi}\frac{G_{\text{b}}}{G_{\text{M}}}\Delta q^{\text{M}}$$

where equations (9) and (11) have been used.

If the matrix constitutive law is now invoked,

$$\Delta p^{\text{M}} = K_{\text{M}}\Delta\varepsilon_{\text{vol}}^{\text{M}}$$

$$= K_{\text{M}}\left(\Delta\varepsilon_{\text{vol}}^{\text{ext}} - \Delta\varepsilon_{\text{vol}}^{\text{m}}\right) \tag{24a}$$

$$\Delta q^{\text{M}} = G_{\text{M}}\Delta\varepsilon_{\text{q}}^{\text{M}} = G_{\text{M}}\Delta\varepsilon_{\text{q}}^{\text{ext}} \tag{24b}$$

where the micro volumetric strain is given by

$$\Delta\varepsilon_{\text{vol}}^{\text{m}} = \frac{\Delta p^{\text{M}}}{K_{\text{m}}}$$

Then

$$\Delta p^{\text{M}} = \left(\frac{1}{K_{\text{M}}} + \frac{1}{K_{\text{m}}}\right)^{-1}\Delta\varepsilon_{\text{vol}}^{\text{ext}} = K_{\text{Mext}}\Delta\varepsilon_{\text{vol}}^{\text{ext}} \tag{25a}$$

$$\Delta q^{\text{M}} = G_{\text{M}}\Delta\varepsilon_{\text{q}}^{\text{ext}} \tag{25b}$$

where K_{Mext} is the average bulk modulus of the clay matrix. Therefore

$$\Delta p^{\text{b}} = \left(\frac{\chi}{1 + C_{\text{b}}\chi}\right)\frac{K_{\text{b}}}{K_{\text{M}}} K_{\text{Mext}}\Delta\varepsilon_{\text{vol}}^{\text{ext}} = K_{\text{bext}}\Delta\varepsilon_{\text{vol}}^{\text{ext}} \tag{26a}$$

$$\Delta q^{\text{b}} = \left(\frac{\chi}{1 + C_{\text{b}}\chi}\right)G_{\text{b}}\Delta\varepsilon_{\text{q}}^{\text{ext}} \tag{26b}$$

where a bulk modulus for the bond stress, K_{bext}, has been defined in terms of the external volumetric strain.

The principle of virtual work (equation (19)), valid for any external strain, is now written as

$$p^{\text{ext}}\Delta\varepsilon_{\text{vol}}^{\text{ext}} = p^{\text{M}}\Delta\varepsilon_{\text{vol}}^{\text{ext}} + C_{\text{b}}p^{\text{b}}\Delta\varepsilon_{\text{vol}}^{\text{b}} \tag{27a}$$

$$q^{\text{ext}}\Delta\varepsilon_{\text{q}}^{\text{ext}} = q^{\text{M}}\Delta\varepsilon_{\text{q}}^{\text{ext}} + C_{\text{b}}p^{\text{b}}\Delta\varepsilon_{\text{q}}^{\text{b}} \tag{27b}$$

where

$$\Delta\varepsilon_{\text{vol}}^{\text{b}} = \frac{\chi}{1 + C_{\text{b}}\chi}\Delta\varepsilon_{\text{vol}}^{\text{M}} = \frac{\chi}{1 + C_{\text{b}}\chi}\frac{K_{\text{Mext}}}{K_{\text{M}}}\Delta\varepsilon_{\text{vol}}^{\text{ext}} \tag{28a}$$

$$\Delta\varepsilon_{\text{q}}^{\text{b}} = \frac{\chi}{1 + C_{\text{b}}\chi}\Delta\varepsilon_{\text{q}}^{\text{M}} = \frac{\chi}{1 + C_{\text{b}}\chi}\Delta\varepsilon_{\text{q}}^{\text{ext}} \tag{28b}$$

Therefore, from equations (27) and (28),

$$p^{\text{ext}} = p^{\text{M}} + p^{\text{b}}\frac{\chi}{1 + C_{\text{b}}\chi}\frac{K_{\text{Mext}}}{K_{\text{M}}} \tag{29a}$$

$$q^{\text{ext}} = q^{\text{M}} + q^{\text{b}}\frac{\chi}{1 + C_{\text{b}}\chi} \tag{29b}$$

These equations provide the expression for the external stresses in terms of bond and matrix stresses. In a general case, the stress partition cannot be expressed in a closed form. The constitutive equation has to be integrated numerically. Again, an increment of external strain is applied to the model. The first step is to find the macro and micro strain components ($\Delta\varepsilon^{\text{ext}} = \Delta\varepsilon^{\text{M}} + \Delta\varepsilon^{\text{m}}$). As each strain component is associated with the matrix stress, the expansive elasto-plastic model provides the changes in (p^{M}, q^{M}) stress (and allows computation of the current macro stress (p^{M}, q^{M})). The incremental bond strain is then calculated in terms of $\Delta\varepsilon^{\text{m}}$ and the current value of χ and bond concentration (C_{b}) through equation (20). Integrating the damage model, the incremental bond stress (Δp^{b}, Δq^{b}) is calculated and (p^{b}, q^{b}) is updated. Finally, a direct use of the principle of virtual work equation provides the value of (p^{ext}, q^{ext}).

FEATURES OF THE MODEL THROUGH SIMULATED TESTS

A few simple tests have been run to illustrate the model capabilities in a parametric study. Synthetic oedometric and triaxial tests have been simulated on two hypothetical non-expansive bonded clays. A third case deals with the degradation induced by suction changes.

Oedometric behaviour: effect of initial damage state (r_0) and damage rate (r_1)

A material defined with the parameters presented in Table 1 (column 1) has been chosen to illustrate the oedometric response of a bonded soil. The results, plotted in Figs 9 and 10, illustrate the effect of varying the initial damage state

Table 1. Material parameters for cases shown

Symbol	Definition of parameter	Value					Units
		1	2	3	4	5	
		Oedometer	Triaxial	Suction cycles	La Biche shale	Todi clay	
ν_M	Matrix Poisson's ratio	0·3	0·3	0·3	0·3	0·3	–
	Expansive matrix						
κ	Elastic macro stiffness parameter for changes in net mean stress	0·05	0·005	0·01	0·01	0·008	–
κ_s	Elastic macro stiffness parameter for changes in suction	0·005	0·005	0·01	0·001	0·01	–
κ_m	Elastic micro stiffness parameter for changes in effective mean stress	Non-expansive matrix	Non-expansive matrix	Analysed parameter	0·012	0·01	–
$\lambda(0)$	Macro stiffness parameter for virgin states in saturated conditions	0·25	0·074	0·054	0·054	0·04	–
$r,\ \beta$	Parameters that describe the rate of changes of virgin compressibility parameter with macro suction	0·75, 0·1	0·75, 0·1	0·75, 0·1	0·75, 0·1	0·75, 0·1	–, MPa^{-1}
M	Slope of critical-state line	0·8	1·0	1·0	0·5	1·0	–
k_s	Parameter that controls the increase of cohesion with suction	0·0	0·0	0·0	0·0	0·0	–
p_0^{M*}	Mean yield stress for saturated conditions	0·011	3·0	3·0	2·5	6·0	MPa
p_c	Reference stress	0·042	0·042	0·01	0·01	0·042	MPa
$s_I,\ s_D$	Hardening parameters for SI and SD yield surfaces respectively	Non-expansive matrix	Non-expansive matrix	10·05, 10·05	0·55, 0·45	0·02, 0·02	MPa
$f_{I0},\ f_{I1},\ n_I$	Parameters micro–macro coupling functions when SI is active	Non-expansive matrix	Non-expansive matrix	0·0, 1·2, 1·0	0·0, 1·2, 1·0	0·45, 3·0, 1·5	–
$f_{D0},\ f_{D1},\ n_D$	Parameters micro–macro coupling functions when SD is active	Non-expansive matrix	Non-expansive matrix	0·5, 1·2, 1·0	1·0, 1·2, 1·0	1·0, 2·5, 5·5	–
	Bond						
C_b	Bond concentration	0·4	Analysed parameter	0·6	0·7	0·1	–
K_{b0}	Bond undamaged bulk modulus	10	3000	100·0	130	110	MPa
G_{b0}	Bond undamaged shear modulus	3·85	1154	38	50	42	MPa
r_0	Initial elastic energy of bond that defines initial bond damage state	Analysed parameter	0·1	0·001	0·02	0·001	MPa
r_1	Bond damage rate	Analysed parameter	0·05	0·05	−0·3	0·15	–
	Coupling						
χ_0	Bonding deformation parameter for undamaged state	1·0	Analysed parameter	1·0	5·0	1·0	–

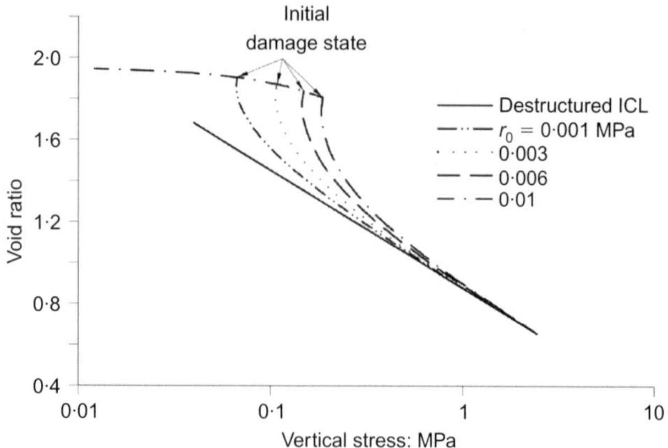

Fig. 9. Set of simulated oedometer tests for a bonded material: effect of initial damage state r_0

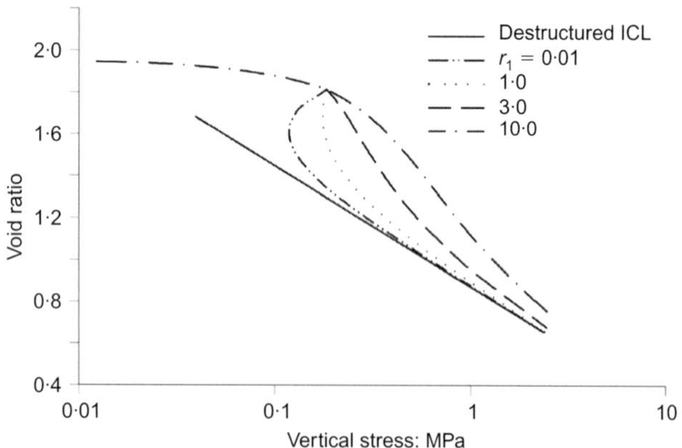

Fig. 10. Set of simulated oedometer tests for a bonded material: effect of damage rate r_1

(r_0) and damage rate (r_1). For a given bond concentration C_b and a given bonding deformation parameter χ_0, the initial stiff response of the material corresponds to the undisturbed state. The slope of the first part of the compression curve is identical for all the simulated tests. The initiation of damage is controlled by the initial size of the bond energy locus (r_0). As load is increased, the induced damage results in a sharp transition towards a softer response (Fig. 9). As strain accumulates, the material evolves toward the matrix behaviour. The reason for this is that, as bonding degrades, the stress is progressively transferred to the matrix. The elasto-plastic matrix response (a continuous line in Fig. 9) has been computed with the same parameters of the undisturbed bonded material but with $C_b = 0$ imposed.

The evolution of damage is controlled by the rate parameter r_1 (Fig. 10): the higher r_1 is, the smoother is the transition from an undisturbed to a damaged state.

Triaxial behaviour: effect of bond deformation parameter (χ) and bond concentration (C_b)

To illustrate the behaviour under triaxial conditions, a hypothetical soft rock has been defined. Material parameters are given in column 2 of Table 1. Compared with the bonded soil described in the previous section, the higher bond concentration and stiffer bond moduli define a more indurated material. The bond strength is controlled partially by the r_0 value, because it marks the maximum (secant) strain energy accepted by the bonds before the degradation process is initiated.

The bonding deformation parameter χ_0 also controls the

initial stiffness and peak strength of the composite material, because it provides the amount of straining arriving at the bonds (see equations (17) and (18)). χ_0 is a constitutive variable that defines the effectiveness of the bond to limit the changes in macroporosity when stress is applied. A high value ($\chi_0 = 5$) indicates that porosity changes are essentially described by bond deformation. In the opposite extreme ($\chi_0 = 0$), the bond material does not oppose any restriction to the matrix deformation. Isolated bodies of bond material, scattered in a clay matrix, provide a conceptual model for a $\chi_0 = 0$ situation (Fig. 5(d)).

Figure 11 shows the effect of χ_0 on the triaxial stress response ($\Delta\sigma_3 = 0$; $\Delta\sigma_1 > 0$) of the composite material. In all cases a concentration $C_b = 0.6$ is assumed. When $\chi_0 = 0$, the matrix behaviour is recovered. For any strain, an effective bonding ($\chi > 0$) results in a higher mobilised deviatoric stress. Accumulation of strain destroys the effectiveness of bonds, and the (residual) strength of the material approximates the matrix behaviour.

The effect of bond concentration (when $\chi_0 = 1$) is analysed in Fig. 12 for the same material. Bond concentration C_b is assumed to change from 80% to 0%. The initial stiffness now remains fairly constant, but the peak strengths change substantially with the amount of bonding. The bond degradation law provides a typical strain-softening behaviour. However, in some tests, when the bond concentration is low, the matrix behaviour at large strains provides a slow, ductile gain in strength. This effect could be removed by changing some of the model parameters, but it shows the capabilities of linking two different basic material behaviours (bond and matrix).

Degradation induced by wetting–drying cycles

In the previous examples, no reference to suction was made. Now, the effect of a common situation in exposed

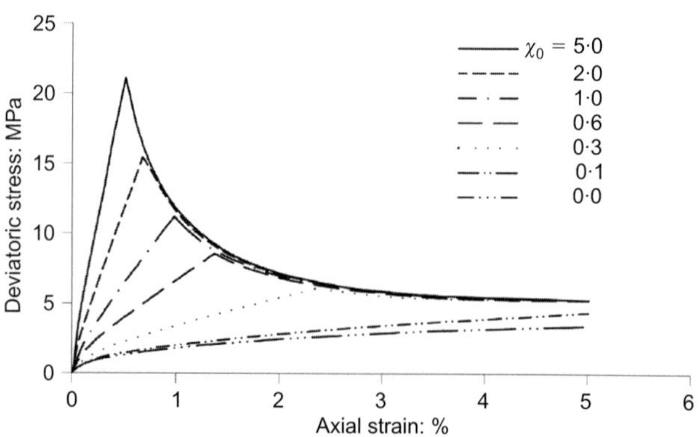

Fig. 11. Set of simulated triaxial tests for a bonded material: effect of initial bonding deformation parameter χ_0

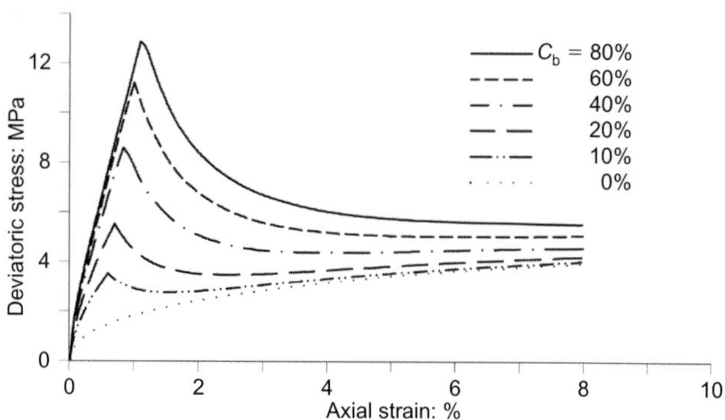

Fig. 12. Set of simulated triaxial tests for a bonded material: effect of bond concentration C_b

materials, namely the application of weather-induced suction cycles, will be explored. This is believed to be an important ultimate reason for rock degradation.

The matrix now reacts to suction changes and, as irreversible volumetric swelling accumulates, bonding is degraded. Material parameters are given in Table 1 (column 3). The values of the coefficient of the interaction functions lead to an accumulation of plastic expansion as cycles are imposed. The imposed suction cycles are schematically given in Fig. 13 (during cycles, suction ranges from 100 MPa to 0·5 MPa). Three clay materials are considered, each characterised by a different expansion potential. Aspects controlling the expansion potential are the nature of the clay minerals, the density of the clay matrix, the confining stress, and the specific arrangement or geometry of clay aggregates in the clayey rock. In the model, the nature of the clay minerals dictates the basic microstructural response (equation (9)); the density of the clay matrix is reflected in the position of the LC yield locus (Fig. 8); the effects of the confining stress are included in the non-linear basic law of microstructured response (equation (9)), and they are explicitly introduced in the structure of the interaction macro–micro functions f_I or f_D. The effectiveness of mineral expansion to induce macrostructural irreversible deformations is also included in f_I and f_D.

The expansion potential of the three materials considered

has been simply represented by a different elastic stiffness of the microstructure, $\kappa_m = 0·001$, $0·004$ and $0·009$. These three materials are identified as having a low, a moderate, and a high expansive potential respectively. Numerical tests were run under isotropic stress conditions and a confining stress of 0·05 MPa. The calculated evolution of the deformation parameter is given in Fig. 13. Degradation in χ occurs rapidly, once the bond limiting surface is reached. As χ evolves, the material degrades and the volumetric strains increase, as shown in Fig. 14. The accumulated irreversible volumetric strains during drying–wetting induce bond damage, and the full potential of the clay matrix is progressively released. 'Tests' were stopped when some effective bonding was still available in the three specimens.

MODELLING SOME REPORTED EXPERIMENTS
The behaviour of La Biche shale (Wong, 1998)

The particular triaxial tests reproduced in Fig. 7(b) were selected for the comparison exercise. The triaxial response of the intact rock specimen, under a confining stress of 50 kPa, is compared with the triaxial response of the same material once it was subjected to a drying–wetting cycle. No details of the suction reached during drying are given in Wong's paper, but as the specimen was exposed to the atmosphere of the laboratory, a high suction (100 MPa) was applied in the simulation. The subsequent wetting is simulated by reducing suction to a low value (0·5 MPa). Then the damaged specimen was sheared under a 50 kPa confining stress. The stress paths of the two specimens to be compared are given in Fig. 15.

The selection of model parameters was facilitated by some additional data given in Wong's paper. It was reported that he measured the free swelling strain of the wetted specimen at around 15%. For the simulation as performed, the calculated shrinkage and swelling strains are given in Fig. 16(a). A final swelling strain close to 13% was obtained in the model. The set of material parameters selected for La Biche shale are given in Table 1 (column 4).

The experimental and calculated triaxial responses of the two specimens are compared in Fig. 17. The model captures well the extreme degradation experienced by this material during the swelling cycle. The deviatoric response is reasonably well reproduced. The volumetric response shows the correct trends: the undisturbed specimen dilates strongly

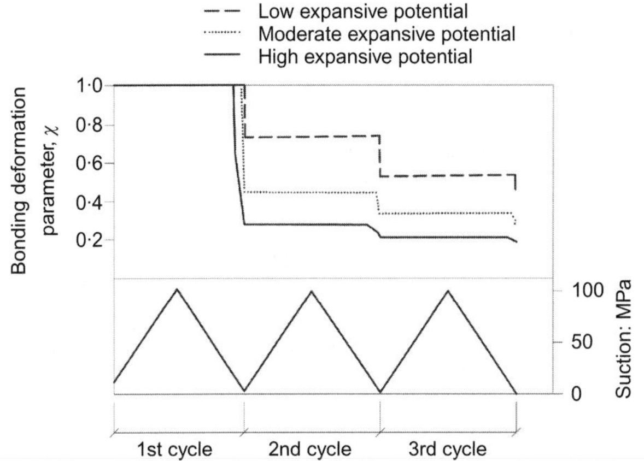

Fig. 13. Applied suction cycles and evolution of deformation parameter χ

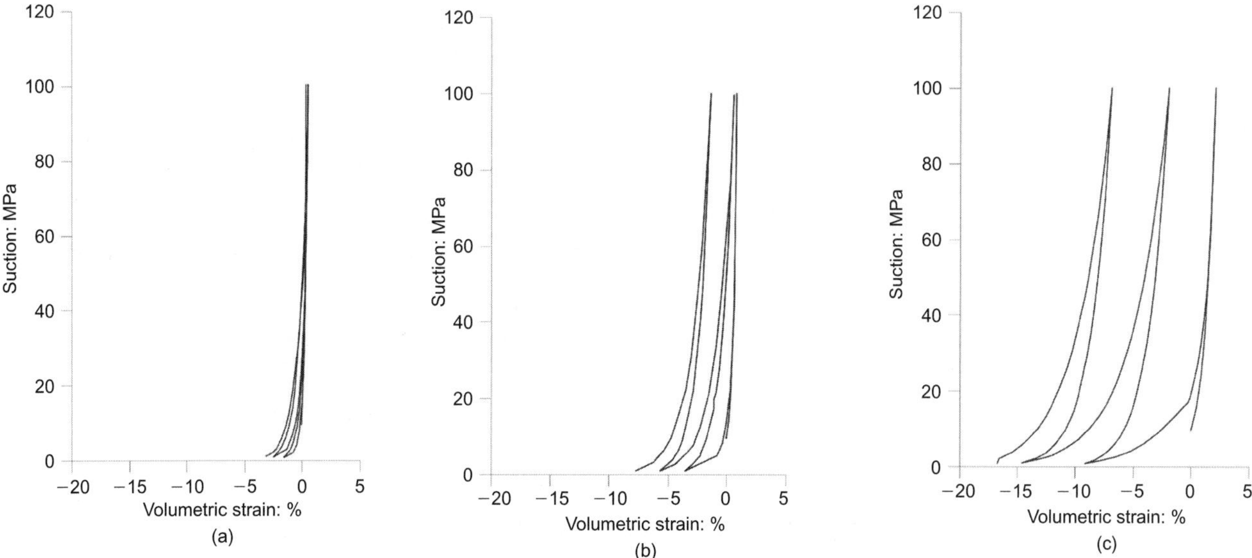

Fig. 14. Calculated expansion of three simulated tests under suction cycles: (a) low, (b) moderate and (c) high expansion potential of clay matrix

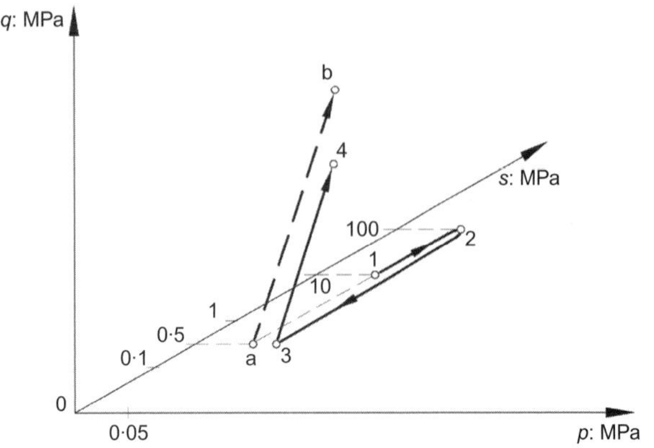

Fig. 15. Stress paths of La Biche shale tests: a–b, intact specimen; 1–2–3–4, specimen preconditioned by a drying–wetting cycle

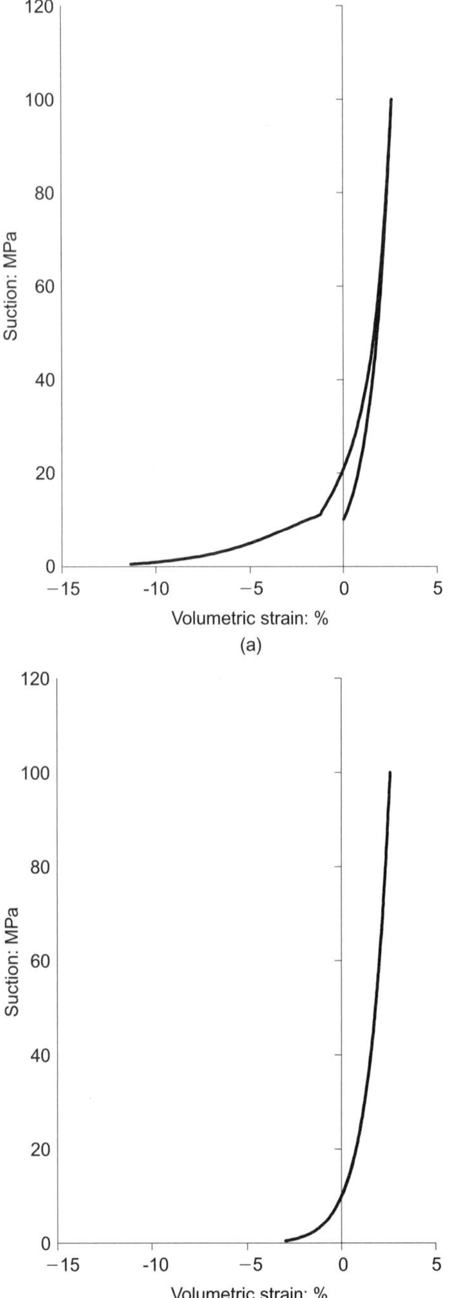

Fig. 16. La Biche shale model. Calculated swelling strains during the drying–wetting cycle: (a) expansive matrix; (b) non-expansive matrix

after peak strength. However, the expanded material shows a strong compressive behaviour during shear, probably explained by the extreme degradation of the shale after several days of unconfined soaking, which is not well reproduced by the model.

Further insight into the developed model is obtained by comparing the performance of a swelling and a non-swelling clay matrix. If material parameters are maintained, except for the microstructural volumetric swelling strains (equation (9)), which are now removed, the model response is as shown in Figs 16(b) and 17(b). Final wetting strains are now much smaller, and they scarcely damage the bond component of the model at all. The subsequent deviatoric response is almost identical to the intact behaviour. Note that, when the microstructural swelling is eliminated, the plastic shrinkage and swelling strains associated with the SD and SI yield loci are also zero, and they do not contribute to damage to the bonds.

Volumetric behaviour of Todi clay under two loading–unloading cycles

The behaviour of Todi clay was presented in Fig. 6. These results are a challenge to constitutive modelling because of the rather singular response of the soil. The initial loading response is quite rigid, a result that may be attributed mainly to the bonding response. Then the bond is damaged, and the soil expands substantially. However, upon reloading, once the bonding has essentially been destroyed, the material maintains a stiff behaviour, which has to be attributed to the clay matrix. These considerations helped in selecting the set of constitutive parameters given in Table 1, column 5.

A more detailed interpretation of the test results, accepting the model developed as a suitable representation of the test, is now given. Reference is made to the numbers that identify the stress–strain response of the test (Fig. 18(a)). Initially (step 1–2) the stiffness of the existing bonds determines the response of the material. Damage starts at point 2, and it increases until point 4. During the first unloading step (3–4) the rebound behaviour is defined by the elastic stiffness of matrix and the already highly damaged stiffness of the bond due to the previous loading path. At point 4, the yield surface SD is reached and an accumulated irreversible expansion is calculated. These plastic swelling strains explain the large recorded final void ratio when unloading is completed. The stress acting on the bonds is now very small. The state of damage at the end of the first loading–unloading cycle implies a new stress partition between matrix and bond. In fact, during the second cycle, the behaviour of the matrix component is dominant in the global response. During step 5–6 the SI yield surface is reached, but its effect in the compressibility of the matrix is small. Owing to the interaction defined between the LC and SI yield surfaces, and because during this second cycle the matrix component receives more load, the apparent preconsolidation stress is reduced and yielding conditions are reached at a low stress level (point 6). Beyond this point, the clay responds in an elasto-plastic manner following the matrix model. During the second unloading (step 7–9) the response is initially elastic, and the slope of the curve increases when the SD yield surface is reached again (point 8). In this case, the expansive potential of the material is lower because the stress point is now close to the current LC yield locus (lower overconsolidation ratio).

It is again interesting to compare the described response with the behaviour of a 'regular', non-expansive bonded material subjected to the same stress path, as in the previous case. The response is given in Fig. 18(b). The first compression steps (1–2–3) are identical to the previous case. The

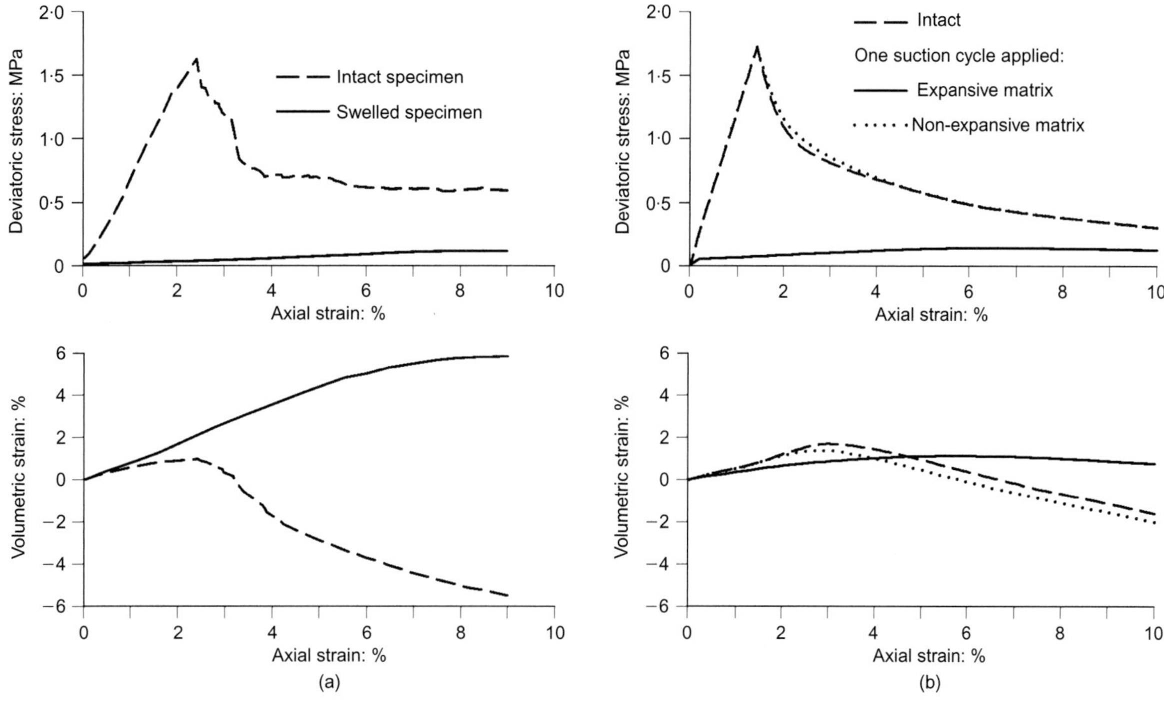

Fig. 17. Triaxial tests on La Biche shale: (a) experimental results (Wong, 1998); (b) computed results

bond also becomes damaged at point 2, and damage progresses thereafter. Upon unloading (3–4), the material follows the unloading stiffness, but it no longer yields (in expansion) at point 4. The similarities between the two cases end here. The regular bonded material remains essentially elastic for the subsequent loading–unloading cycles.

CONCLUSIONS

Clay shales are commonly found in engineering works, and they are known in practice for the evolving nature of their mechanical properties. Stress changes, as well as environmental actions (notably the drying–wetting cycles experienced by the material exposed to atmospheric action), often lead to material degradation.

Two features are particularly relevant to explain the nature of weathering: the cementation of the clay matrix under stress, and the swelling/shrinkage behaviour of the clay aggregates when subjected to suction cycles. Both aspects are covered by the model presented in the paper.

Two constitutive models have been integrated into a common framework: an elasto-plastic model for expansive materials, and a damage model for the bond material. The integration follows some rules imposed by internal strain compatibility conditions. Stress partitioning between the two constituent materials (expansive clay matrix and bond) is achieved by means of the virtual work principle.

The resulting model has many features that appear to follow experimental observations. The stiff bond response explains the linear deviatoric stress–strain response often observed in triaxial tests before the peak strength is reached. Bond degradation contributes to the marked strain-softening observed after peak. The formulation ensures that the fully damaged material follows the behaviour of the clay matrix. A relevant feature of the model is the ability to incorporate damage induced by wetting and drying. This is a consequence of the elasto-plastic formulation of the clay matrix, which includes suction effects by means of a double-structure framework. The model is capable of reproducing the effect of cyclic drying–wetting, a fundamental aspect of weathering. When suction cycles are capable of irreversibly

deforming the clay matrix, they may induce bond damage, which facilitates further straining. The material evolves towards new states characterised by decreasing strength, decreasing initial stiffness, increasing ductility and reduced dilatancy. These comments are illustrated in a few examples developed to show the capabilities of the model.

The ability of the model to achieve good quantitative results was shown by comparing model calculations with the reported response of two natural materials. One of the cases presented involved a comparison of the triaxial response of a shale, in both the unweathered and the fully weathered state (induced by full swelling when soaking the specimen in water). The two stress-suction paths were reproduced, and the calculated results were compared with the experiment. In a second case, the remarkable behaviour of an expansive structured clay subjected to a double loading–unloading cycle under oedometric conditions was reproduced.

More important than achieving a good representation of a given experiment, which typically requires a trial and error process in order to define a set of material parameters, is obtaining a set of consistent results for a variety of stress paths. In this sense, the developed model may help in gaining a more fundamental insight into the nature of weathering, an important practical aspect that is often approached in qualitative, descriptive and markedly empirical terms.

ACKNOWLEDGEMENTS

The authors wish to acknowledge the support provided by the Departament d'Educació í Universitats de la Generalitat de Catalunya and the European Social Fund during the development of the work reported in the paper.

NOTATION

C_b band concentration
D scalar damage variable
e_b void ratio of bonds
e_M^* void ratio of pores between clay aggregates
e_m void ratio of micropores
e_v void ratio of macropores

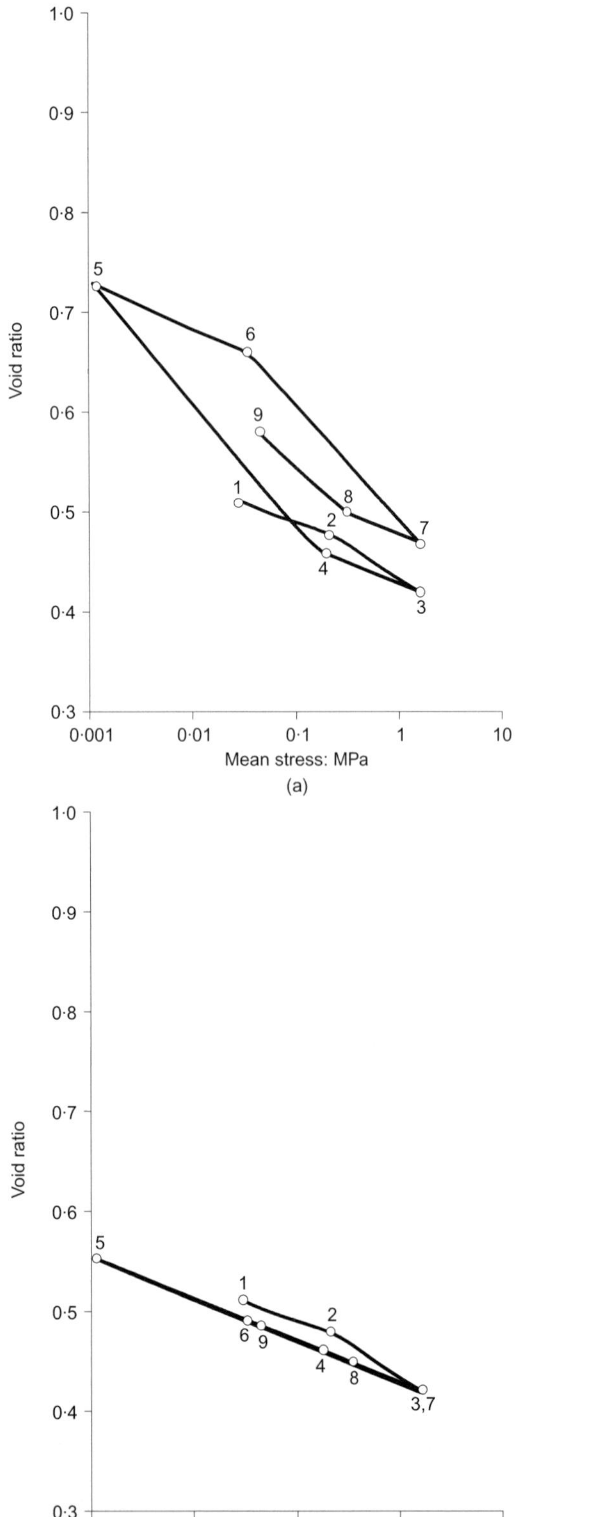

Fig. 18. Calculated response of Todi clay under two loading–unloading cycles: (a) expansive matrix; (b) non-expansive matrix

changes in external mean stress in terms of external shear strain

G_{b0} undamaged bond shear modulus for changes in bond deviatoric stress

G_M macro shear modulus (of arrangement of clay aggregates) for changes in matrix deviatoric stress

G_{Mext} shear modulus of clay matrix (considering micro and macro effects) in terms of external shear strain

K_b bond bulk modulus at current damage state for changes in bond mean stress

K_{bext} bond bulk modulus at current damage state for changes in external mean stress in terms of external volumetric strain

K_{b0} undamaged bond bulk modulus for changes in bond mean stress

K_M macro bulk modulus (of arrangement of clay aggregates) for changes in matrix mean stress

K_m micro bulk modulus (of clay aggregates) for changes in matrix mean stress

K_{Mext} bulk modulus of clay matrix (considering micro and macro effects) in terms of external volumetric strain

K_s macro bulk modulus (of arrangement of clay aggregates) for changes in macro suction

L logarithmic damage variable

p_0^M mean yield stress at current suction

p_0^{M*} mean yield stress for saturated conditions

r_0 initial damage state

r_1 bond damage rate

s^{macro} macrostructural suction

s^{micro} microstructural suction

u_b secant elastic energy stored by bonds per unit of volume

V total volume

V_b solid volume of bond material (assumed to be non-porous)

V_M^* volume between clay aggregates

V_m volume of micropores

V_s total volume of solids

V_{sM} solid volume of clay matrix

V_v volume of macropores

$\boldsymbol{\varepsilon}^{ext};\ \varepsilon_{vol}^{ext},\ \varepsilon_q^{ext}$ external strain; volumetric component, deviatoric component

$\boldsymbol{\varepsilon}^b;\ \varepsilon_{vol}^b, \varepsilon_q^b$ strain of bond; volumetric component, deviatoric component

$\boldsymbol{\varepsilon}^M;\ \varepsilon_{vol}^M, \varepsilon_q^M$ strain between aggregates; volumetric component, deviatoric component

$\boldsymbol{\varepsilon}^m;\ \varepsilon_{vol}^m$ strain of micropores; volumetric component

$\boldsymbol{\varepsilon}^V;\ \varepsilon_{vol}^V,\ \varepsilon_q^V$ strain of macropores; volumetric component, deviatoric component

κ elastic macro stiffness parameter for changes in mean stress

κ_m elastic micro stiffness parameter for changes in mean stress and micro suction

κ_s elastic macro stiffness parameter for changes in macro suction

$\boldsymbol{\sigma}^{ext}, p^{ext}, q^{ext}$ external stress, external mean stress, external deviatoric stress

$\boldsymbol{\sigma}^M, p^M, q^M$ matrix stress, matrix mean stress, matrix deviatoric stress

$\boldsymbol{\sigma}^b, p^b, q^b$ bond stress, bond mean stress, bond deviatoric stress

χ bonding deformation parameter

χ_0 bonding deformation parameter associated with undamaged state

f_{I0}, f_{I1}, n_I parameters of micro–macro function when SI is active

f_{D0}, f_{D1}, n_D parameters of micro–macro function when SD is active

G_b bond shear modulus at current damage state for changes in bond deviatoric stress

G_{bext} bond shear modulus at current damage state for

REFERENCES

Alonso, E. E., Gens, A. & Josa, A. (1990). A constitutive model for partially saturated soils. *Géotechnique* **40**, No. 3, 405–430.

Alonso, E. E., Vaunat, J. & Gens, A. (1999). Modelling the mechanical behaviour of expansive clays. *Engng Geol.* **54**, Nos 1–2, 173–183.

Alonso, E. E. & Berdugo, I. (2005). Expansive behaviour of sulphate-bearing clays. *Proceedings of the international conference on problematic soils*, Famagusta, pp. 1–25

Bertuccioli, P. & Lanzo, G. (1993). Mechanical properties of the Italian structurally complex clay soils. *Proc. Symp. Geotech. Engng Hard Soils—Soft Rocks, Athens*, 383–389.

Bjerrum, L. (1967). Progressive failure in slopes of overconsolidated plastic clay and clay shales. *J. Soil Mech. Found. Div. ASCE* **93**, No. SM5, 3–49.

Calabresi, G. & Scarpelli, G. (1985). Effects of swelling cause by unloading in overconsolidation clays. *Proc. 11th Int. Conf. Soil Mech. Found. Engng, San Francisco* **1**, 411–414.

Carol, I., Rizzi, E. & Willam, K. (2001) On the formulation of anisotropic elastic degradation. I. Theory based on a pseudologarithmic damage tensor rate. *Int. J. Solids Struct.* **38**, No. 4, 91–518.

Fernández, A. & Santamarina, C. (2001). The effect of cementation on the small strain parameters of sand. *Can. Geotech. J.* **38**, No. 1, 191–199.

Gens, A. & Alonso, E. (1992). A framework for the behaviour of unsaturated expansive clays. *Can. Geotech. J.* **29**, No. 6, 1013–1032.

Hsu, S. C. & Nelson, P. P. (1993). Characterization of Cretaceous clay shales in North America. *Proc. Symp. Geotech. Engng Hard Soils—Soft Rocks, Athens* **1**, 139–146.

Kavvadas, M. J. (2000). General report. Modelling the soil behaviour: selection of soil parameters. *Proc. 2nd Int. Symp. Geotech. Hard Soils—Soft Rocks*, **2**, 1441–1481.

Mitchell, J. K. (1976). *Fundamentals of soil behaviour.* New York: Wiley.

Olivier, H. J. (1987). Some aspects of the influence of mineralogy and moisture redistribution on the weathering behaviour of mudrock. *Proc. 4th Int. Conf. Rock Mech.*, Montreaux, **3**, 467–474.

Picarelli, L. (1991). Discussion: 'The general and congruent effects of structure in natural soils and weak rocks' by S. Leroueil & P. R. Vaughan. *Géotechnique* **40**, No. 2, 281–284.

Pye, K. & Krinsley, D. H. (1983). Interlayered clay stacks in Jurassic shales. *Nature* **304**, No. 5927, 618–620.

Rampello, S. (1991). Some remarks on the mechanical behaviour of stiff clays: the example of Todi clay. *Proceedings of the international workshop on experimental characterization and modelling of soils and soft rocks*, Napoli, pp. 131–190.

Sanmartino, S. (2001). *Construction d'un modèle conceptuel de la porosité et de la minéralogie dans les argiles du site de Bure.* Internal report D.RP.OCPE.03. 001, Paris: ANDRA.

Seedsman, R. W. (1987). Strength implications of the crystalline and osmotic swelling of clays in shales. *Int. J. Rock Mech. Min. Sci. Geomech. Abstr.* **24**, No. 6, 357–363.

Vaunat, J. & Gens, A. (2003). Bond degradation and irreversible strains in soft argillaceous rock. *Proc. 12th Panam. Conf. Soil Mech. Geotech. Engng, Boston* **1**, 479–484.

Wong, R. C. K. (1998). Swelling and softening behaviour of La Biche shale. *Can. Geotech. J.* **35**, No. 2, 206–221.

Amorosi, A. & Rampello, S. (2007). *Géotechnique* **57**, No. 2, 153–166

An experimental investigation into the mechanical behaviour of a structured stiff clay

A. AMOROSI* and S. RAMPELLO†

In recent years, fundamental research has been carried out into the properties of some natural stiff clays and the corresponding reconstituted materials, highlighting the role of microstructural features in the observed differences. In this paper the results of an experimental investigation into the mechanical behaviour of an Italian stiff clay of marine origin are presented. Medium-pressure and high-pressure stress-controlled triaxial cells were used in which natural samples underwent isotropic and anisotropic compression and swelling before drained or undrained shearing. Comparison of soil behaviour observed after different compression histories up to different values of maximum effective stress allowed the following aspects to be discussed: the effects of the initially structured state on the medium to large strain response and shear strength characteristics of the soil; the relevance of volumetric and deviatoric plastic strain to the structure degradation; the role and implications of the imposed non-isotropic stress histories; the permanent differences between reconstituted samples and fully de-structured natural samples; and the uniqueness of the critical state condition.

KEYWORDS: clays; compressibility; fabric/structure of soils; laboratory tests; shear strength

Récemment, des recherches fondamentales ont été effectuées sur les propriétés de certaines argiles fermes naturelles et sur celles de matériaux reconstitués correspondants : ces recherches ont mis en évidence le rôle des caractéristiques microstructurales dans les différences observées. Cet article présente les résultats d'une recherche expérimentale sur le comportement mécanique d'une argile ferme italienne d'origine marine. On a utilisé des cellules triaxiales de pression moyenne et de forte pression dans lesquelles des échantillons naturels ont subi des compressions isotropes et anisotropes et un gonflement puis un cisaillement drainé ou non drainé. En comparant le comportement du sol observé après différentes séquences de compression avec des valeurs différentes de contrainte effective maximale, on pu discuter les aspects suivants : les effets de l'état structuré initial sur la réponse des déformations moyennes ou grandes et les caractéristiques de la résistance au cisaillement du sol ; la pertinence de la déformation plastique volumétrique et déviatorique sur la dégradation de la structure ; le rôle et les implications des séquences de contrainte non isotropes imposées ; les différences permanentes entre les échantillons reconstitués et les échantillons naturels entièrement déstructurés ; et le caractère unique de la condition d'état critique.

INTRODUCTION

In recent years, experimental work has highlighted the role of microstructural features in the compressibility and shear strength of natural stiff clays, often comparing such properties with those observed on the corresponding reconstituted material (e.g. Horseman *et al.*, 1987; Rampello, 1989; Burland, 1990; Leroueil & Vaughan, 1990; Coop & Cotecchia, 1995; Coop *et al.*, 1995; Burland *et al.*, 1996; Kavvadas & Anagnostopoulos, 1998; Cotecchia & Chandler, 2000; Rampello *et al.*, 2002). In these cases the reconstituted clay was used as a reference for understanding and interpreting the behaviour of the natural soil. The observed differences were attributed to structure, defined as the combination of particle arrangement (fabric) and interparticle bonding (Mitchell, 1976).

Interparticle bonding can be related to several different processes that might occur in a clayey deposit on a geological timescale, such as percolation of calcium carbonate, or weak lithification. Its main engineering effect is enhancement of the initial stiffness and strength of the material. Upon increase of the applied loads, the stresses at certain interparticle contacts can be thought to reach the bond strength, and a mechanical bond degradation (de-bonding)

process is initiated. Bond degradation is an irreversible phenomenon that, experimentally, appears to be controlled by plastic strain accumulation (e.g. Smith *et al.*, 1992; Lagioia & Nova, 1995; Callisto, 1996). It is characterised by an initially more intense effect, which reduces as more irreversible strains are accumulated, leading asymptotically towards a final fully de-bonded state.

The macroscopic effects of different fabrics have been highlighted experimentally by comparing the mechanical behaviour of the same material (usually a reconstituted soil), compressed and unloaded along different radial stress paths prior to shearing (e.g. Henkel & Sowa, 1963; Ladd, 1965; Parry & Nadarajah, 1973; Amerasinghe & Parry, 1975; Gens, 1982). The results indicate that the position and shape of the yield surface or limit state surface are modified, depending on the applied virgin consolidation path, leading to anisotropic and non-coaxial responses for non-isotropic paths. This macroscopic behaviour can be qualitatively related to the clay microstructure. Based on this idea, and adopting electron micrograph images, Anandarajah *et al.* (1996) have observed that, for a reconstituted clay, the fabric evolves towards different stable configurations depending on the direction of the imposed virgin consolidation path. The results mentioned above indicate that the mechanical behaviour of clays depends not only on the overconsolidation ratio, but also on the loading path direction imposed during virgin consolidation (e.g. Topolnicki *et al.*, 1990). Consistently, the fabric evolution appears to be controlled by the amounts of volumetric and deviatoric plastic strain, and by their ratio.

In this paper, the main results of an experimental investi-

Manuscript received 21 June 2006; revised manuscript accepted 11 December 2006.
Discussion on this paper closes on 1 August 2007, for further details see p. ii.
* Technical University of Bari, Italy.
† University of Roma 'La Sapienza', Italy.

gation into the mechanical behaviour of an Italian stiff cohesive soil, Vallericca clay, are presented.

The mechanical properties of Vallericca clay have been studied at the University of Rome over the last decade. Available data come from isotropic and anisotropic triaxial tests carried out on natural samples. The former were run under effective stresses lower than the yield stress, whereas the latter were extended to stress levels well above yield stress. Triaxial tests on normally consolidated and overconsolidated reconstituted samples were also available.

Isotropic test results have been published to discuss the effect of initial microstructure and of structure degradation on the compressibility and shear strength of the clay (Rampello *et al.*, 1993; Burland *et al.*, 1996; Callisto & Rampello, 2004). Anisotropic test data have been used for evaluating structure degradation induced by compression to high effective stress (Amorosi, 1996; Amorosi & Rampello, 1998), and to calibrate a constitutive model developed for structured soils (Kavvadas & Amorosi, 2000; Amorosi & Kavvadas, 2002).

In this work, additional high-pressure triaxial tests carried out under isotropic stress conditions are presented, together with additional oedometer tests, and the whole set of experimental data is reinterpreted comparing isotropic and anisotropic test results performed under medium to high stress levels. This allowed the initially structured state and related de-structuring process to be studied with reference to different loading paths undergone by the samples before drained or undrained shearing.

In the following, a phenomenological definition for structured clay is suggested: the material can be defined as structured when its behaviour deviates from that exhibited by its reference version, here assumed to be that obtained by isotropically compressing the intact clay up to a sufficiently high level of pressure to largely erase its initial microstructural features.

In this definition no reference is made to the reconstituted soil, based on the experimental evidence exhibited by several natural stiff clays that, when subjected to various laboratory tests, do not achieve the combination of stress and volumetric states that characterises the corresponding reconstituted materials, differently from what is typically observed for soft natural clays (e.g. Burland *et al.*, 1996; Cotecchia, 1996).

MATERIAL TESTED

Vallericca clay is a stiff overconsolidated clay deposit of Plio-Pleistocene age that was deposited in a neritic marine environment in the depression that currently coincides with the valley of the Tiber river.

To minimise the effects of sampling disturbance, large block samples were extracted from vertical faces of deep cuts. The blocks of Vallericca clay were taken in different years from two adjacent brick pits a few kilometres north of Rome. In the following sections, reference will be made to two batches of samples: one retrieved from a first brick pit, and referred to as batch 1, and a second, retrieved from the same deposit but a different brick pit, identified as batch 2.

Table 1 gives the average index properties and natural water content of the blocks from the two batches. The soil is a medium plasticity and activity clay with a calcium carbonate content of about 30%; higher values of W_0 were observed on samples taken from batch 1.

Tests for the intrinsic properties of Vallericca clay, identified by an asterisk in the following sections, were carried out on specimens trimmed from blocks of reconstituted soil. The blocks were formed by mixing the natural clay into a slurry at a water content of about 1·5 times its liquid limit. The slurry was one-dimensionally compressed by incremental loading in a large oedometer up to a vertical effective stress $\sigma'_v = 300$ kPa.

The initial microstructure of undisturbed and reconstituted samples of Vallericca clay was studied by Sciotti (1992) using a scanning electron microscope. The surfaces to be studied were obtained by the fracture procedure. The specimens were then dried in the atmosphere, and the surfaces to be observed were cleaned by peeling; a thin film of a conductor (200–300 Å thick) was finally painted onto the surfaces to avoid concentration of electrical charges and loss of resolution.

Figure 1 shows vertical sections of the natural clay. A prevalence of edge-to-face associations is observed, and two classes of irregular aggregations may be distinguished. Aggregates of the first class, 5–10 μm in size, consist of packets of partially iso-oriented clay particles. They associate by face-to-face and edge-to-face contacts to form the second class of aggregates, 40–60 μm in size. The average intra-aggregate pore space is 1–3 μm, and inter-aggregate pores are 3–6 μm in size. The presence of rounded silt particles surrounded by platy clay minerals is also observed. The many microfossils observed in the specimens justify the relatively high $CaCO_3$ content, which should not be thought as being responsible for cemented interparticle bonding. Fig. 2 shows a vertical section of a reconstituted specimen. This time, face-to-face contacts prevail, with more closely spaced clay packets.

TEST EQUIPMENT AND PROCEDURES

Oedometer tests and triaxial compression tests were carried out to study the mechanical properties of the clay.

Oedometer tests on natural samples were performed in cells with a reduced diameter of 35·7 mm in order to achieve a vertical stress as high as 12·8 MPa. The apparatus compliance was accounted for when calculating the axial displacement.

In the triaxial apparatus both isotropic and anisotropic compression was carried out on specimens of intact clay, 38 mm in diameter and 76 mm high, cut from the block samples. Triaxial compression tests were carried out at the end of the compression or swelling stages, when the rate of volume strain was lower than 0·02% per day. Undrained and drained shearing was carried out at constant rates of axial strain $\varepsilon_a = 4·5\%$ and 1·5% per day respectively.

Two different triaxial equipments were used to perform the medium-pressure (MP) and high-pressure (HP) tests. Both apparatuses consist of computer-controlled stress path cells, with cell pressure capacities of 3·2 and 14 MPa respectively; they are similar to the triaxial system described by Taylor & Coop (1993) and Cuccovillo & Coop (1999).

A mid-height pore water pressure probe (Hight, 1982) was

Table 1. Index properties of the tested material

Batch	G_s	CF: %	W_L: %	I_P: %	W_0: %	I_L
1	2·78	42	59·2	31·6	28·6	0·03
2	2·75	47	53·9	29·2	26·4	0·06

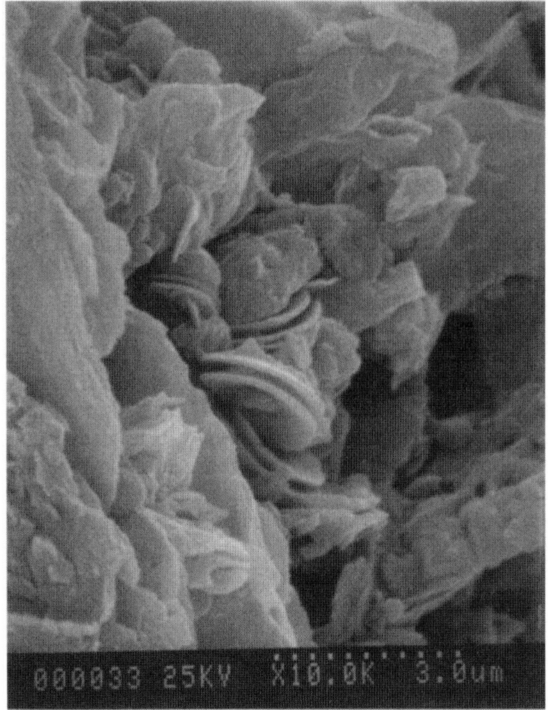

Fig. 2. Photograph of reconstituted Vallericca clay: vertical section (Sciotti, 1992)

A limited number of tests were carried out using a high-pressure triaxial cell equipped with a pair of miniaturised submersible linear variable differential transducers (LVDTs), which measured the axial displacement over the central length of the sample (Cuccovillo & Coop, 1997).

TESTING PROGRAMME

In the following, triaxial tests are distinguished according to the maximum value of the mean effective stress p'_{max} or the vertical effective stress σ'_{vmax} applied during isotropic or anisotropic compression respectively. Tests will be referred to as medium-pressure (MP) tests for values of p'_{max} or σ'_{vmax} lower than the mean effective yield stress p'_y or the vertical effective yield stress σ'_{vy} obtained from the isotropic or one-dimensional compression curves of natural samples; they will be referred to as high-pressure (HP) tests when $p'_{max} > p'_y$ or $\sigma'_{vmax} > \sigma'_y$. Concerning the different kind of compression history applied prior to shearing, in the following CI stands for isotropically consolidated samples and CA stands for anisotropically consolidated samples.

The testing programme summarised in Tables 2–4 consisted of three sets of triaxial tests carried out on batch 2 of the clay. The tables summarise the maximum applied effective stresses (p'_{max}, q_{max}) and the values of void ratio (e_0), effective stress (p'_0, q_0) and overconsolidation ratio attained prior to shearing.

A first set of data refers to high-pressure isotropically compressed samples (CI-HP). Two samples were isotropically compressed to values of $p'_{max} = 7073$ and $11\,000\,\text{kPa}$

Fig. 1. Photographs of natural Vallericca clay: vertical sections (Sciotti, 1992)

mounted on natural samples tested in the MP apparatus, and both external and internal load cells were used in the HP apparatus. An externally mounted displacement transducer was used to measure the axial strain.

Table 2. Isotropic compression: high-pressure (CI-HP) triaxial tests

Test		$p'_{max} = p'_y$: kPa	p'_0: kPa	e_0	p'_y/p'_0
HVr8	CIU	11 000	11 000	0·479	1
HVr16	CIU	7 073	7 073	0·546	1
HVr24	CIU	7 073	1 753	0·580	4

Table 3. Anisotropic compression: high-pressure (CA-HP) triaxial tests

Test		p'_{max}: kPa	q_{max}: kPa	p'_0: kPa	q_0: kPa	e_0	OCR
HVr4	CAU	4630	3160	4630	3160	0·571	1
HVr6	CAU	4630	3159	3159	1389	0·581	1·65
HVr10	CAU	4635	3172	2429	556	0·583	2·41
HVr13	CAU	4635	3172	1756	5	0·588	3·83
HVr15	CAU	4635	3172	1003	−364	0·612	8·88

Table 4. Anisotropic compression: medium-pressure (CA-MP) triaxial tests

Test		p'_{max}: kPa	q_{max}: kPa	p'_0: kPa	q_0: kPa	e_0	OCR
Vr2	CAD	1748·8	1195·7	1748·8	1195·7	0·695	1
Vr19	CAD	1764·0	1205·8	1203·6	489·6	0·715	1·68
VrL4	CAU	1777·0	1208·0	1777·0	1208·0	0·710	1
Vr11	CAU	1773·8	1227·0	1221·9	523·7	0·718	1·65
Vr14	CAU	1763·0	1205·7	942·0	234·6	0·719	2·34
Vr17	CAU	1796·0	1224·0	695·8	8·8	0·728	3·72
VrL3	CAU	1748·0	1210·0	390·0	−150·0	0·748	8·80

respectively, and then sheared undrained. A third sample was first compressed to p'_{max} = 7073 kPa and then unloaded to 1753 kPa to achieve an overconsolidation ratio equal to 4 before undrained shearing.

In the other two series of tests, anisotropic compression and swelling paths were applied to the natural samples before shearing (Fig. 3). In the first series, named CA-HP, anisotropic compression was carried out up to a vertical effective stress 2·5 times greater than the yield stress evaluated in the oedometer tests. Specifically, values of mean effective stress and deviatoric stress, $p'_{max} \cong 4630$ kPa and $q_{max} \cong 3170$ kPa, were achieved, corresponding to a maximum vertical effective stress σ'_{vmax} = 6750 kPa. Samples were then sheared or anisotropically swelled back to different OCRs prior to shearing. The second series of anisotropic triaxial tests, named CA-MP, was carried out reaching maximum levels of stress only slightly lower than the yield stress observed in oedometer tests (p'_{max} = 1766 kPa, q_{max} = 1207 kPa and σ'_{vmax} = 2570 kPa). At this stage, samples were either sheared starting from about a normally consolidated state, or anisotropically unloaded to achieve, prior to shearing, values of OCR equal to those imposed in the CA-HP tests.

Anisotropic compression or swelling was obtained by applying discrete increments of axial and radial stresses followed by consolidation. The stress paths were selected to minimise the amount of radial deformation experienced by the samples. Specifically, values of stress ratio

$$\eta = \frac{q}{p'} = 3\left(\frac{1 - K_0}{1 + 2K_0}\right) \tag{1}$$

with

$$K_0 = K_0^{oc} = (1 - \sin \phi'_{cv}) \cdot OCR^{\sin \phi'_{cv}} \tag{2}$$

or

$$K_0 = K_0^{nc} = 1 - \sin \phi'_{cv} \tag{3}$$

were applied for overconsolidated or normally consolidated states respectively, using a constant-volume angle of shearing resistance $\phi'_{cv} = 28°$.

In the following, reference is also made for comparison purposes to isotropic compression–medium pressure (CI-MP) triaxial test results obtained on natural and reconstituted samples retrieved from batch 1 of Vallericca clay (Burland *et al.*, 1996; Rampello *et al.*, 1993). The testing programme of the CI-MP series is summarised in Table 5.

Interpretation criteria

In the triaxial compression tests, rupture of the intact samples was often associated with the formation of a single well-defined slip surface or of a system of conjugate slip surfaces. In such conditions, forces and displacements measured at the sample boundaries cannot be directly interpreted as stresses and strains in the sample, because of the non-homogeneity of the deformation. It is then most important to recognise the development of the slip surfaces through the samples to interpret the stress–strain behaviour correctly. The criteria adopted for this purpose varied according to the instrumentation installed in the triaxial systems used to run the tests.

For the high-pressure triaxial cell the development of the slip surfaces was recognised by comparing the axial stress measured by the external and internal load cells. An example of such a comparison is shown in Fig. 4 for a compressed anisotropically–undrained (CAU) test in which conjugate slip surfaces developed at failure. The external and internal readings are nearly coincident until the post-

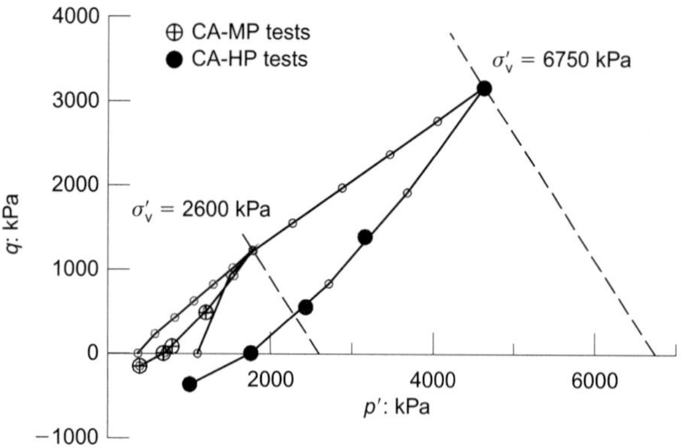

Fig. 3. Anisotropic compression of natural samples before shearing

Table 5. Isotropic compression: medium-pressure (CI-MP) triaxial tests*

Test		$p'_0 = p'_{max}$: kPa	E_0	p'_y / p'_0
Vr–60	CIU	60	0·840	31·7
Vr–200	CIU	200	0·813	9·5
Vr–412	CIU	412	0·788	4·6
Vr–428	CIU	428	0·790	4·4
Vr–620	CIU	620	0·791	3·1
Vr–817	CIU	817	0·774	2·3
Vr–1600	CIU	1600	0·756	1·2
Vr–2400	CIU	2400	0·715	1·0
Vr–3200	CIU	3200	0·682	1·0
Vr–53	Const. p'	53	0·893	35·8
Vr–101	Const. p'	101	0·843	18·8
Vr–148	Const. p'	148	0·836	12·8
Vr–50	CID	50	0·868	38·0
Vr–100	CID	100	0·829	19·0
Vr–200	CID	200	0·815	9·5
Vr–400	CID	400	0·790	4·8
Vr–800	CID	800	0·788	2·4
Vr–1200	CID	1200	0·775	1·6

* Burland *et al.* (1996) and Rampello *et al.* (1993).
† $p'_y = 1900$ kPa for $p'_y > p'_0$

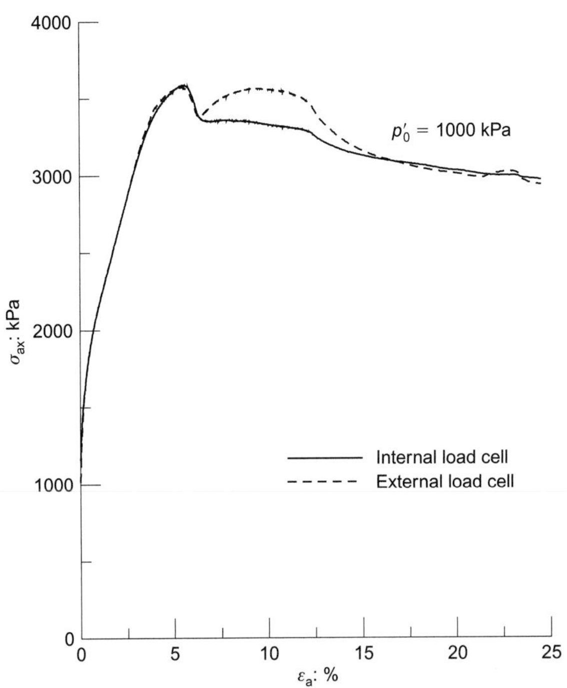

Fig. 4. CA-HP HVr15 test : internal and external measurements of axial load

with the hypothesis that the failure surface intersects the sample boundaries at that point of the test, and that hydraulic connection exists between the slip surface and the base of the sample (Viggiani *et al.*, 1993, 1995).

Finally, local strain transducers permitted the development of the slip surface to be detected in a CI-HP test in which the sample was sheared undrained from a state wet of critical and the slip surface passed outside the transducer pins. To this purpose, two indexes of strain homogeneity were used, which allow the onset of the slip surface and its complete development through the sample to be detected (Calabresi *et al.*, 1990; Rampello, 1991). Fig. 5 shows the homogeneity indexes plotted against the notional overall axial strain ε_{aext}. The first index is defined as the difference between the axial strain measured by each single transducer (Fig. 5(a)); the second index, denoted by the ratio $\Delta(\varepsilon_{aL1} - \varepsilon_{aL2})/\Delta\varepsilon_{aext}$, represents the ratio of the relative displacement of the sliding bodies to the external measurement of axial displacement (Fig. 5(b)). By definition, a homogeneous strain distribution is characterised by zero

peak load is reached ($\varepsilon_a \cong 6\cdot5\%$); thereafter they first diverge, the external load cell measuring a greater axial stress, and then merge again at $\varepsilon_a \cong 16\%$ up to the end of the test. Horizontal components of the load in the ram and, in turn, friction errors in external measurements of the axial stress can be associated with the development of a slip surface; once the conjugate slip surface develops, the friction errors reduce once more and the external and internal readings merge again.

The mid-height pore water pressure probe installed in the medium-pressure triaxial apparatus was also helpful in recognising the development of the slip surface. In fact, when the probe was close to the slip surface, probe and base measurements of pore water pressure were seen to equalise soon after the peak strength was reached. This is consistent

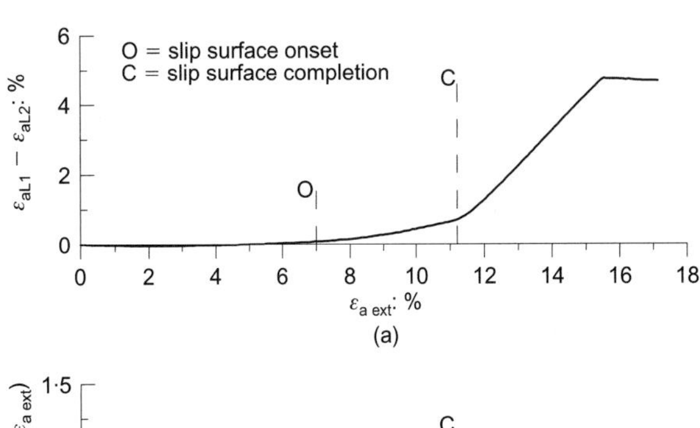

(a)

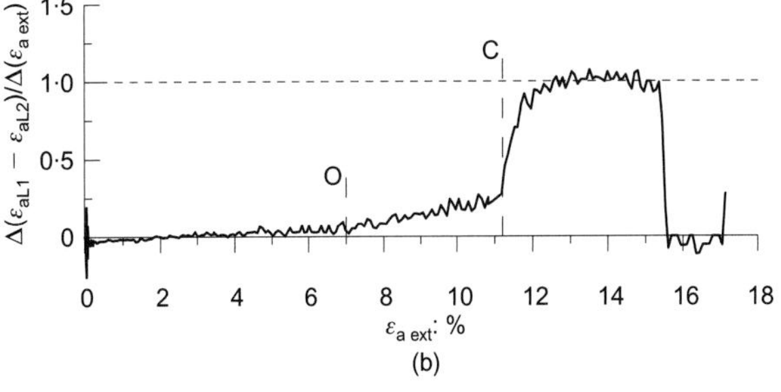

(b)

Fig. 5. Onset and development of the slip surface : (a) first and (b) second indices of strain homogeneity

values of both indexes, whereas the second index reaches a constant value of 1 once the slip surface has completed its development through the sample. Data shown in Fig. 5 indicate that strain localisation starts at an axial strain of about 7%, and the slip surface develops through the sample completely at $\varepsilon_a \approx 11\%$. Fig. 6 compares the stress conditions corresponding to slip surface onset and development for samples sheared from states dry and wet of critical. Although further experimental observations are needed, it seems that, contrary to what was observed for samples sheared from states dry of critical (Viggiani *et al.*, 1993), for the sample sheared from a state wet of critical the slip surface developed completely when the peak stress ratio q/p' was reached.

In the following sections, according to the criteria mentioned above, the stress–strain data are shown by a full line for states of continuous straining, and a dotted line is used from values of ε_a from which discontinuous slipping is inferred. Data in this latter strain range are not actually used in the interpretation of experimental results.

COMPRESSIBILITY

Figure 7 shows the oedometer compression curves, plotted

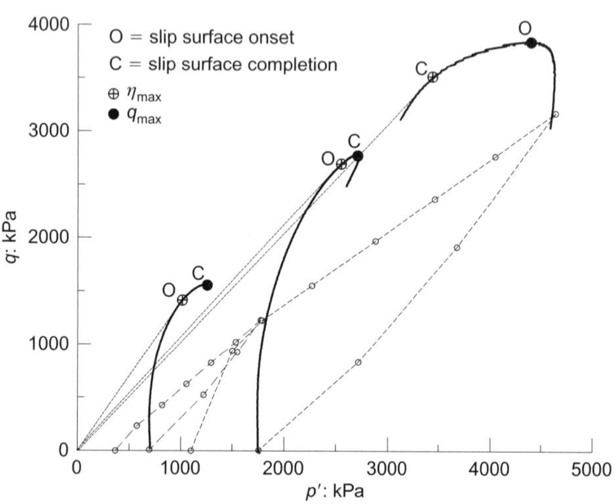

Fig. 6. Onset and development of the slip surface as detected by the local transducers

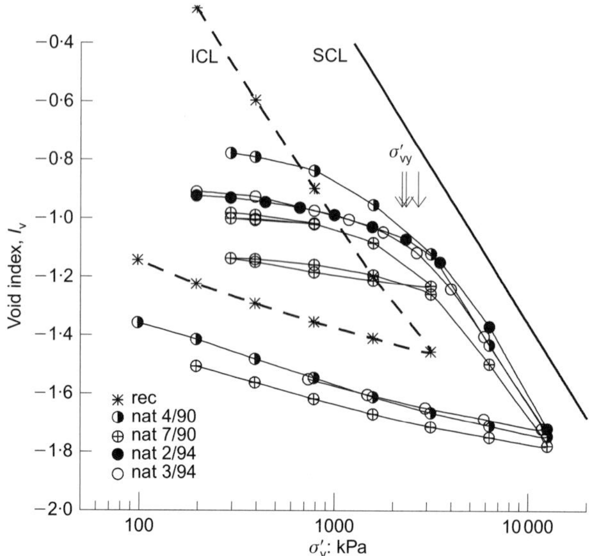

Fig. 7. Oedometer test results

in terms of the void index I_v, for tests carried out on a reconstituted sample and on four natural samples of Vallericca clay. The corresponding compression characteristics are summarised in Table 6; values of the swelling index in the table were evaluated at an overconsolidation ratio OCR = 10, and values of the yield stress were obtained using graphs of $\log(1 + e)$ against $\log\sigma'_v$. The sedimentation compression line SCL is also plotted in the figure (Burland, 1990). The intrinsic compression and swelling indexes are $C_c^* = 0.405$ and $C_s^* = 0.085$ respectively, the latter being evaluated at OCR = 10.

The compression curves for the intact samples cross the intrinsic compression line (ICL) and bend down when significant structural degradation begins, before reaching the SCL. After yielding has fully developed the states of the natural samples appear to follow normal compression curves about parallel to the ICL.

In Table 7 the swelling index C_s is compared with C_s^*. Increasing values of C_s are observed with increasing maximum vertical stress σ'_{vmax} applied before unloading. The ratio C_s^*/C_s decreases from about 5 for an applied stress less than σ'_{vy} ($\sigma'_{vmax}/\sigma'_{vy} = 0.34$) to about 1 when $\sigma'_{vmax}/\sigma'_{vy} \approx 6.0$.

In Fig. 8 the compression and swelling curves obtained from the triaxial tests are shown in the $\log p'$–v plane, where $v = (1 + e)$ is the specific volume. The full symbols refer to high-pressure isotropic and anisotropic compression tests (CI-HP and CA-HP) and the open-cross circles refer to medium-pressure anisotropic compression tests (CA-MP), both carried out on natural samples retrieved from batch 2. In addition to the CI-HP tests listed in Table 2, an isotropic compression curve is shown in Fig. 8 in which a void ratio $e = 0.302$ was attained at the maximum applied mean effective stress $p' = 47.6$ MPa. Also in the figure, the specific volume at the end of medium-pressure isotropic compression (CI-MP) tests is shown by asterisks for the reconstituted samples and by open circles for the natural samples taken from batch 1 (Rampello *et al.*, 1993). Samples Vr-2400 and Vr-3200 were compressed to values of p' higher than the mean effective yield stress $p'_y = 1900$ kPa evaluated for samples taken from batch 1. After isotropic compression, their state plots on the normal compression line defined by the CI-HP tests.

The compression parameters listed in Table 8 refer to the $\ln p'$–v plane, although data are plotted on a $\log p'$–v scale; N is the specific volume at a reference stress $p'_r = 1$ kPa, and λ and κ are the slopes of the virgin compression curves and the swelling curves respectively. The difference in the values of p'_y relevant to the CI-MP and the CI-HP tests (Table 8, Fig. 8) is consistent with the different values of natural water content observed in the samples retrieved from the two batches of the clay.

The isotropic and anisotropic virgin compression curves of the natural samples are about parallel to the isotropic intrinsic compression line, and are characterised by a single value of λ; a single value of κ was also found to be in fair agreement with the average slope of the natural swelling curves.

For samples taken from batch 2, values of p'_y evaluated under anisotropic compression in triaxial tests and of σ'_{vy} determined in oedometer yield a value of $K_0^{nc} = 0.546$, which corresponds to $\phi'_{cv} = 27°$.

To check whether K_0 conditions were maintained during anisotropic compression in triaxial tests, the increments of axial strain $\delta\varepsilon_a$, measured after isotropic compression at the initial effective stress of the samples, p'_k, are plotted against the increments of volumetric strain $\delta\varepsilon_v$ (Fig. 9). The straight line corresponding to zero increments of radial strain ($\delta\varepsilon_v = \delta\varepsilon_a$) is also shown in the figure. For samples tested in the

Table 6. Compression characteristics in oedometer tests

Batch	Sample	e_0	C_c	C_s	σ'_{vy}: MPa	σ^*_{ve}: MPa	$\sigma'_{vy}/\sigma^*_{ve}$	C_c/C^*_c
1	4/90	0·775	0·40	0·08	2·15	0·89	2·42	0·99
1	7/90	0·712	0·37	0·07	2·33	1·15	2·03	0·90
2	2/94	0·735	0·44	–	2·70	1·03	2·62	1·09
2	3/94	0·743	0·42	0·06	2·70	1·13	2·39	1·03

Table 7. Swell sensitivity in oedometer tests

Batch	Sample	σ'_{vmax}: MPa	$\sigma'_{vmax}/\sigma'_{vy}$	C_s	C^*_s/C_s
1	7/90	0·8	0·34	0·018	4·72
1	7/90	3·2	1·37	0·045	1·89
2	3/94	12·0	4·44	0·060	1·42
1	7/90	12·8	5·49	0·065	1·31
1	4/90	12·8	5·95	0·080	1·06

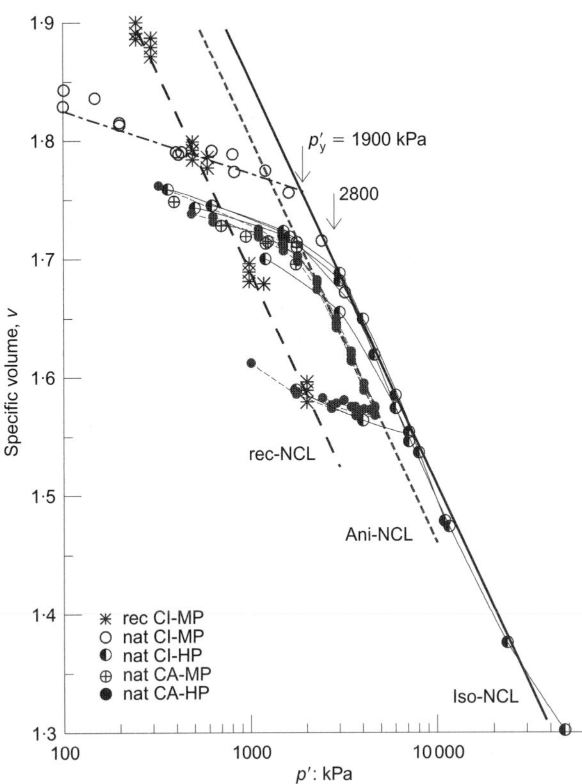

Fig. 8. Compressibility under triaxial stress conditions

medium-pressure range, data points plot closely around the $\delta\varepsilon_r = 0$ line, thus confirming the choice of $\phi'_{cv} = 28°$ to evaluate η. For samples tested in the high-pressure range, data points plot above the $\delta\varepsilon_r = 0$ line, the ratio $\delta\varepsilon_a/\delta\varepsilon_v$ being about 1·54. This implies that during anisotropic com-

pression the samples experienced an increase in diameter with values of $\delta\varepsilon_r \approx -0·18\delta\varepsilon_a$, because the applied values of η are greater than those corresponding to K_0 conditions. A slightly lower value of η and in turn of ϕ'_{cv} ($\approx 26°$) should have been used for samples tested in the high-pressure range to account for the curvature of the strength envelope.

Discussion

According to Burland (1990) and Burland *et al.* (1996), the location of σ'_{vy} relative to the intrinsic compression line is a measure of the enhanced resistance to compression caused by the natural microstructure; this is given by the ratio $\sigma'_{vy}/\sigma^*_{ve}$, where σ^*_{ve} is the equivalent pressure on the ICL corresponding to the void ratio of the natural soil at yield.

Data in Table 6 show that, for intact samples of Vallericca clay, the ratio $\sigma'_{vy}/\sigma^*_{ve}$ is about 2·4 and the ratio C_c/C^*_c is close to unity, in agreement with previous published data (Burland *et al.*, 1996). Table 7 reports decreasing values of swell sensitivity C^*_s/C_s with increasing maximum stress applied prior to unloading. Values of C^*_s/C_s approaching unity with increasing stress level suggest that the maximum loads applied in the oedometer were sufficient to substantially damage the interparticle bonding present in the natural state. However, after yield the natural compression curves in Figs 7 and 8 are seen to be permanently offset to the right of the ICL, the stress difference between the natural compression curve and the intrinsic compression line (σ'_v/σ^*_{ve}) remaining about constant with increasing stress level in a semi-logarithmic plane. If this stress difference is attributed entirely to interparticle bonding, progressive structural degradation would not be justified. Rather, the difference (σ'_v/σ^*_{ve}) can result from significant differences in fabric between the natural and reconstituted soil, which will con-

Table 8. Compression characteristics in triaxial tests

Batch	Samples	State	N	λ	κ	p'_y : MPa
1		Reconstituted	2·68	0·145	0·035	–
1	CI-MP	Natural*	2·87	0·148	0·023	1·9
2	CI-HP	Natural	2·87	0·148	0·023	2·8
2	CA-MP/HP	Natural[†]	2·82	0·148	0·023	1·9

* Rampello *et al.* (1993)
[†] Amorosi & Rampello (1998).

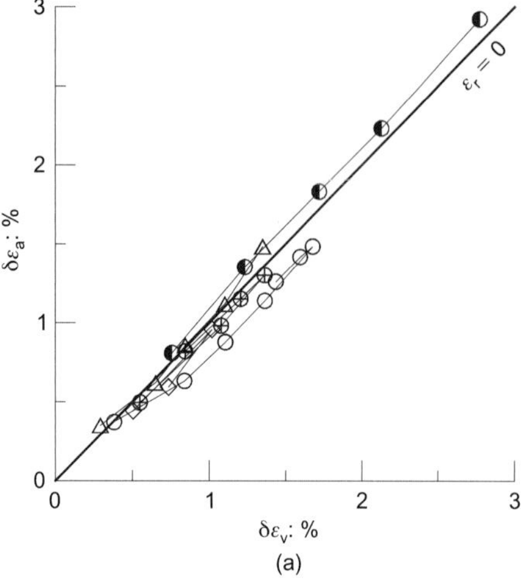

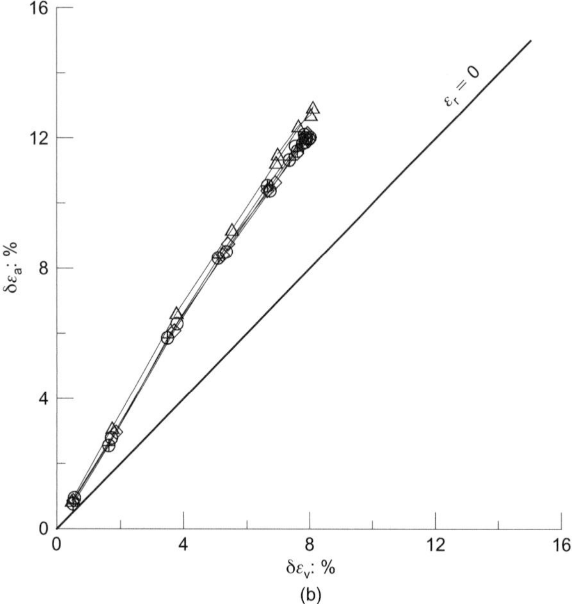

Fig. 9. Axial and volumetric strains measured during anisotropic compression : (a) CA-MP tests ; (b) CA-HP tests

tribute to these effects. In fact, there can be no doubt that increase of plastic volumetric strain, and in turn of plastic shear strain, with increasing stress level produces a progressive breakdown of interparticle bonding.

The values of the ratios $\sigma'_{vy}/\sigma^*_{ve}$, C_c/C^*_c and C^*_s/C_s proposed by Burland *et al.* (1996) as microstructure indicators seem to indicate that, for intact Vallericca clay, the main difference from the reconstituted material is due to the natural microfabric with some interparticle bonding.

The analysis of the compressibility of Vallericca clay highlights that for such a stiff material the reconstituted version may not represent a fully destructured reference state attainable along stress paths typical of civil engineering boundary value problems. This can be attributed to the relevant, and partly permanent, differences that characterise the structures of the two materials at the micro scale, leading to different asymptotic states for the same compressibility tests.

SHEAR STRENGTH

In this section the triaxial test data are gathered together according to the compression histories and the maximum stress levels applied prior to shearing.

High-pressure isotropically consolidated samples (CI-HP)

In order to complete the available set of test data obtained under isotropic stress conditions in the medium stress range (e.g. Rampello *et al.*, 1993), three triaxial tests were performed in which natural samples of Vallericca clay were isotropically compressed to high effective stress (Table 2).

Figure 10 shows the undrained stress–strain behaviour of these samples. According to the interpretation criteria specified above, stress–strain data are shown by full lines for the portion of test results for which continuous straining was inferred, and dotted lines are used from values of ε_a from which discontinuous slipping was assumed.

The normally consolidated samples show similar ductile and contractant response, with values of stress ratio $\eta = q/p'$ approaching unity at large strains. The overconsolidated sample shows a dilatant behaviour with a reduction of the excess pore water pressure Δu for values of $\varepsilon_a > 1\%$, and a larger value of the peak stress ratio for values of $\varepsilon_a \leqslant 8\%$, for which the sample can still be assumed to undergo continuous straining. At larger strains ($\varepsilon_a > 8\%$), possible development of a slip surface was detected; nonetheless, the corresponding portion of the $q/p'-\varepsilon_a$ curve tends to converge towards the same value of $\eta = 1$ attained by the normally consolidated samples.

The observed behaviour does not differ qualitatively from that typically exhibited by isotropically compressed reconstituted samples of clayey soils. In this sense it is worth noting that, owing to the high level of imposed compression stresses, these samples of Vallericca clay had experienced significant modifications of their structure prior to shearing. In fact, both bonding and fabric were likely to be altered along such extended compression paths, as discussed in the following sections.

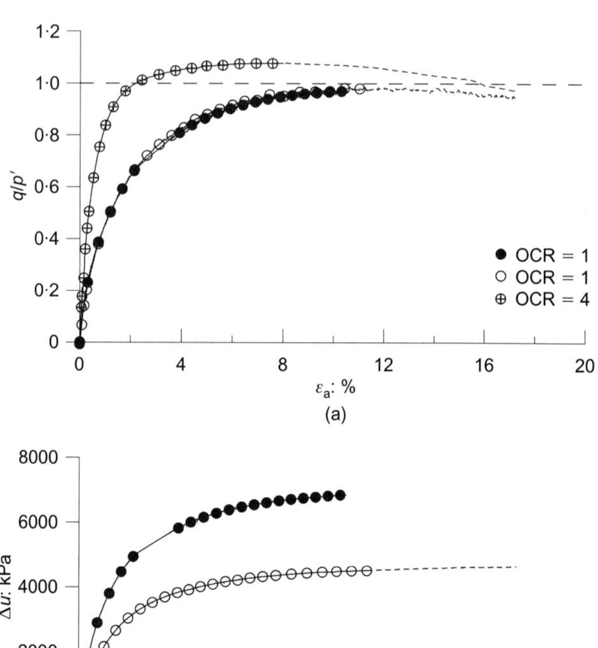

Fig. 10. CI-HP tests on natural samples : undrained stress–strain behaviour

High-pressure anisotropically consolidated samples (CA-HP)
In the CA-HP series of tests all the samples experienced maximum values of the vertical effective stress $\sigma'_{vmax} \cong$ 6750 kPa, which is about 2·5 times the vertical effective yield stress σ'_{vy} observed in oedometer tests. Table 3 summarises the maximum applied effective stresses and the values of void ratio, effective stress and overconsolidation ratio attained prior to shearing.

Figure 11 shows the undrained stress–strain behaviour of the CA-HP samples. The overconsolidated samples exhibited brittle stress–strain behaviour, with values of ε_a at peak strength increasing and post-peak reduction decreasing as the overconsolidation increases (Amorosi & Rampello, 1998). For low values of OCR (< 3), stress–strain data show values of q/p' approaching unity as ε_a increases, and a peak in the $q/p'-\varepsilon_a$ curves is observed for OCR > 3. The excess pore water pressure Δu attains its maximum value prior to peak strength; thereafter it decreases to a minimum value, which remains about constant for samples sheared from high OCRs, but gradually increases again after its minimum for samples tested at low OCRs.

For the normally consolidated sample, the contractant behaviour resulted in values of pore water pressure that increase with ε_a towards a stationary value; a value of $\eta = 1$ is again approached with increasing axial strain.

Medium-pressure anisotropically consolidated samples (CA-MP)
This series of tests was performed on natural samples anisotropically re-compressed to achieve an OCR of about 1 and then, when necessary, anisotropically unloaded under K_0 conditions. In so doing, the samples were assumed not to experience much structure modification prior to shearing, in terms of either fabric or bonding. On samples with OCR = 1 and 1·7 drained and undrained shearing was carried out

respectively, and undrained shearing was performed on the remaining samples.

The stress–strain behaviour observed for drained and undrained shearing is shown in Fig. 12 with full and open symbols respectively. Undrained tests on overconsolidated samples show a stress–strain behaviour similar to that observed for the CA-HP tests, peak deviatoric stress being attained for values of axial strain increasing with OCR.

The drained and undrained tests on normally consolidated samples show a contractant behaviour with volumetric strain and excess pore water pressure continuously increasing towards stationary values as ε_a increases.

A final value of $q/p' \approx 1$ is attained by both the normally consolidated samples, whereas all the others develop discontinuities well before the achievement of a stationary value for η.

DISCUSSION OF THE TRIAXIAL TEST RESULTS

In order to highlight the role of natural microstructure on the mechanical behaviour of Vallericca clay it is worth comparing the three sets of data presented above and interpreting the differences in terms of different particle arrangement and interparticle bonding.

Figure 13 shows the stress paths followed by the natural samples tested in the medium and high-pressure ranges after anisotropic compression histories. Undrained stress paths of normally consolidated and slightly overconsolidated samples reach peak strength with small changes in mean effective stress p'. Soon after the peak strength is attained, the stress path bends to the left, showing decreasing values of q and

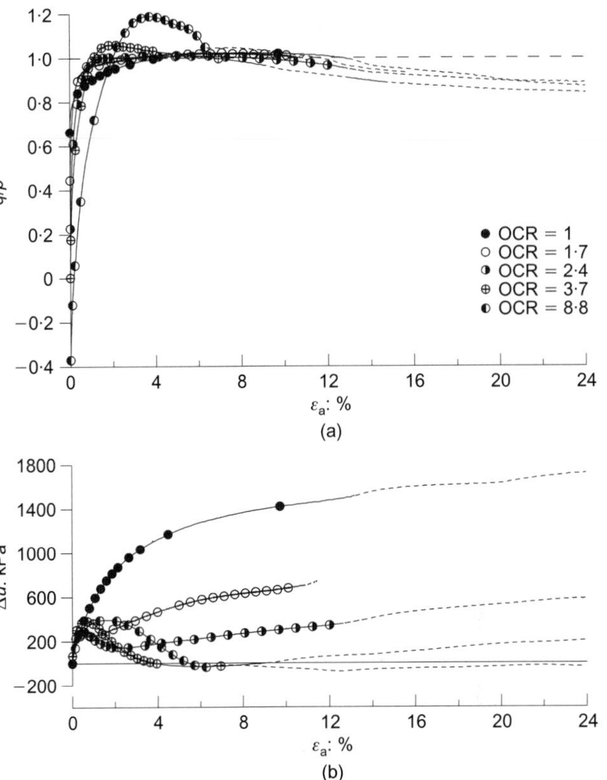

Fig. 11. Undrained stress–strain behaviour from CA-HP tests on natural samples: (a) stress ratio; (b) excess pore water pressure against axial strain

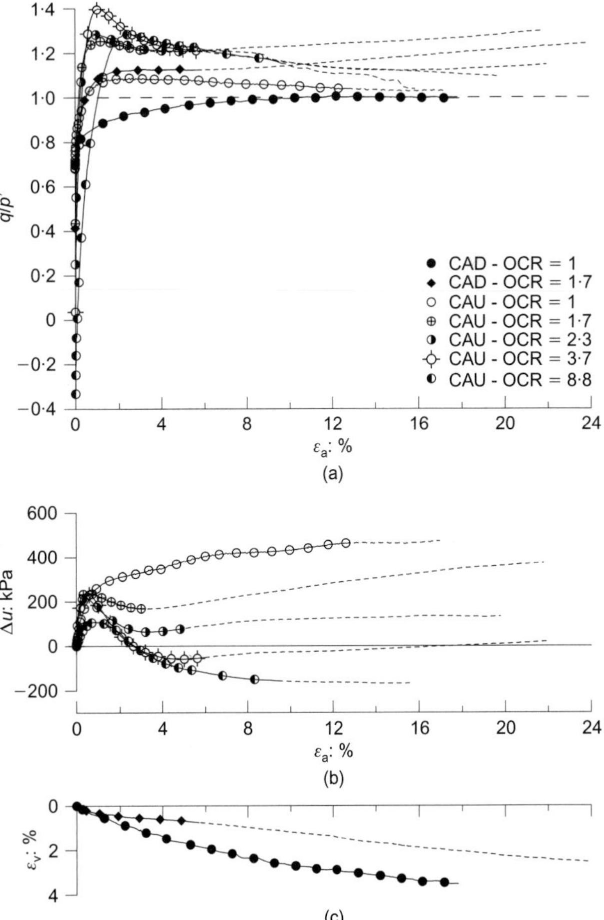

Fig. 12. Drained and undrained stress–strain behaviour from CA-MP tests on natural samples: (a) stress ratio; (b) excess pore water pressure; (c) volumetric strain against axial strain

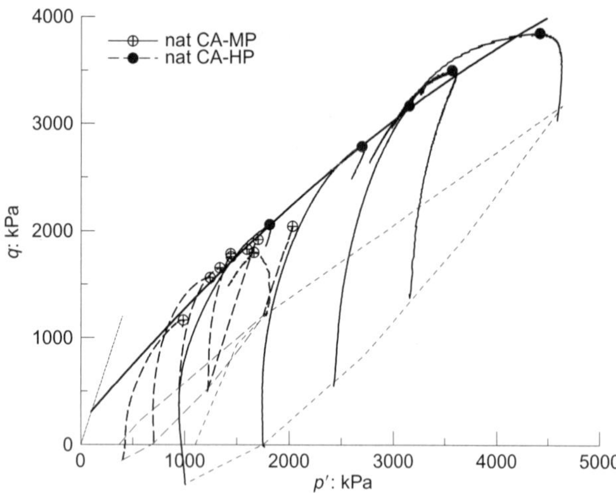

Fig. 13. Stress paths of CA-MP and CA-HP tests on natural samples

p', though increasing values of q/p' are described for a significant portion of shearing. Overconsolidated samples exhibit dilatant behaviour, with stress paths bending to the right until peak strength is reached on the failure envelope. The observed behaviour is qualitatively similar to that observed for anisotropically compressed reconstituted samples of different clayey soils (e.g. Gens, 1982; Rossato & Jardine, 1990). Similarities are observed for the two series of test results, although the maximum effective stresses experienced by the CA-MP and CA-HP samples differ substantially.

The peak strengths of natural samples sheared after either isotropic or anisotropic compression are plotted in Fig. 14 in the q–p' plane. The open symbols refer to samples tested in the range of medium pressure, and the full symbols are for high-pressure tests. The curved failure envelopes obtained for the natural samples from the two batches merge with the envelope for normally consolidated samples at different values of p' consistent with the different values of the mean effective yield stress observed during compression (Fig. 8).

The experimental data are characterised by rather different states prior to shearing, in terms of specific volume, effective stress, previous stress history and, consequently, microstructural features. To compare the stress paths of the three sets of results it is convenient to introduce a normalisation procedure. In the following, all the stress paths are normalised with respect to the equivalent pressure, defined as

$$p'_e = \exp\left(\frac{N - v}{\lambda}\right) \qquad (4)$$

where v is the current specific volume, and the parameters $N = 2\cdot87$ and $\lambda = 0\cdot148$ are those relative to the isotropic compression line for the natural soil.

Within the critical state soil mechanics (CSSM) framework this normalisation procedure results in a bi-dimensional representation of the state boundary surface (SBS), which provides a boundary separating possible and impossible states and defines the locus of all the possible states of an element of soil undergoing plastic deformation (Schofield & Wroth, 1968). The uniqueness of the SBS in the normalised plane p'/p'_e–q/p'_e stems from the hypothesis that the hardening law of the material is isotropic and depends on plastic volumetric strain only. Thus a scalar quantity related to the soil compressibility, the equivalent pressure, can be used to normalise the observed stress paths in order to obtain a unified picture of soil behaviour.

Figure 15 shows the normalised stress paths followed under drained or undrained shearing by the natural samples after either isotropic or anisotropic compression histories: open symbols and dashed lines refer to medium-pressure tests, and closed symbols and full lines are for high-pressure tests.

Tests results from high-pressure tests are compared first. The normalised stress paths followed by the normally consolidated CI-HP and CA-HP samples were used to locate the wet side of the bounding surface. However, in contrast to CSSM, the state boundary surface identified by the CA-HP test plots above that identified by the CI-HP tests, showing that it is not possible to locate a single SBS on a stress plane normalised by the equivalent pressure.

To interpret this occurrence two assumptions can be made concerning the samples soil structure.

(a) Under both isotropic and anisotropic compression up to high effective stresses major and irreversible damage to the initial interparticle bonding is induced.

(b) Significant changes in the initial soil fabric are likely to occur during isotropic compression to high pressure, whereas only minor changes should be induced by anisotropic compression under nearly K_0 stress conditions.

In light of these hypotheses, the observed modification of the shape and position of the surface could be related to the different (isotropic versus anisotropic) high-pressure compression paths and, in constitutive terms, might be accounted for by introducing anisotropic hardening in the description

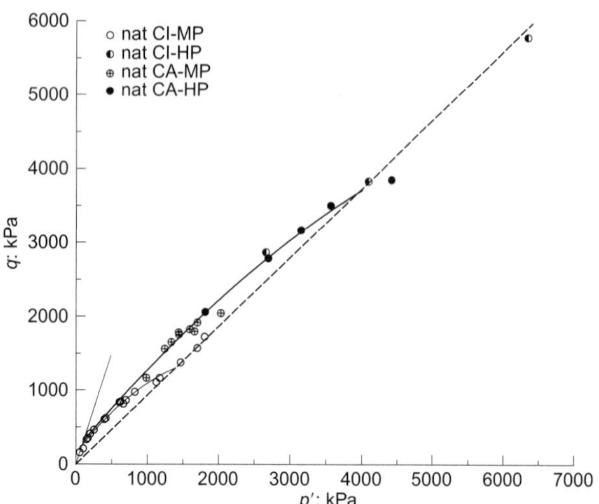

Fig. 14. Strength envelope of natural samples of Vallericca clay

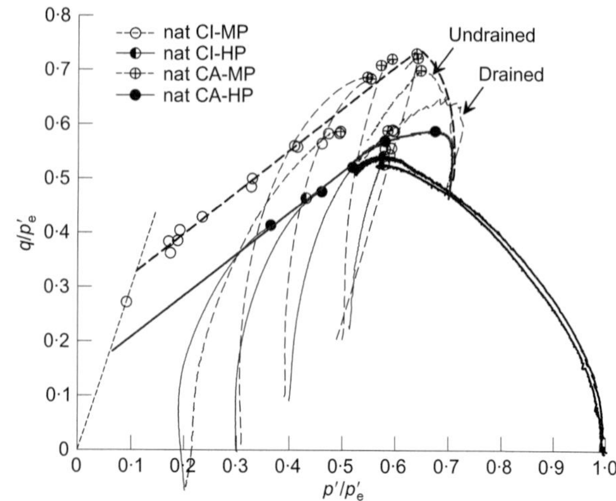

Fig. 15. Normalised stress paths of natural samples

of soil behaviour. This would imply that a tensorial quantity is likely to control soil hardening rather than a scalar one, and therefore a scalar-based normalisation procedure is, in principle, not possible.

Normalised stress paths of the overconsolidated samples were used instead to define the dry side of the state boundary surface; specifically, portions of the stress paths in between the values of η_{max} and q_{max} were used for this purpose, because post-peak discontinuous sliding causes the stress paths to deviate from the SBS. Opposite to what was observed for states wet of critical, the normalised failure point of the overconsolidated CI-HP sample plots on the same SBS defined by the overconsolidated samples that underwent anisotropic compression to high pressure. Therefore anisotropic hardening seems to have a negligible effect on the shape of the SBS for states dry of critical. However, further experimental evidence is needed to support this interpretation.

When high-pressure and medium-pressure test results are compared, it is convenient to distinguish between samples sheared from states wet of critical and those sheared from states dry of critical.

Similar to what was observed for the high-pressure tests, the normalised strengths of the overconsolidated samples tested in the medium-pressure range plot closely around a single Hvorslev strength envelope, irrespective of their isotropic or anisotropic compression history.

The wet side of the boundary surface, defined only by the normally consolidated CA-MP samples, was located by enveloping the external portions of the paths observed on samples Vr2 and VrL4, sheared drained and undrained respectively.

The normalised state boundary surfaces relevant to the CA-MP and CA-HP tests are similar in shape but do not coincide, in contrast to the CSSM framework, the SBS for samples tested in the medium-pressure range lying above that defined by the high-pressure tests.

A possible interpretation of the observed behaviour assuming an initial structured state and an induced structural degradation can be proposed, based on the following hypotheses.

(a) Samples tested in the medium-pressure range were virtually undisturbed (i.e. structured) prior to shearing, because their compression history was characterised by values of $\sigma'_{vmax} < \sigma'_{vy}$ and $\varepsilon_r/\varepsilon_a < 0.1$.

(b) Samples that underwent compression up to high effective stress were fully de-bonded before shearing.

(c) Similar soil fabric characterised samples that experienced similar anisotropic compression histories.

Consistent with the above hypotheses, CA-MP and CI-MP samples mainly underwent structural degradation during shearing. In fact, failure of overconsolidated samples was associated with early development of slip planes, and only slight structural degradation was caused by deviatoric stress paths up to failure (Callisto & Rampello, 2004). By contrast, stress paths of normally consolidated samples deviate from the state boundary surface described by the structured CA-MP samples, moving towards the critical state defined by the de-bonded CA-HP samples. The different behaviour observed during drained and undrained shearing on CA-MP normally consolidated samples indicated that both volumetric and deviatoric plastic strains control the de-structuring process. Fig. 16 shows the volumetric and deviatoric plastic strains obtained after the elastic strain components computed using a hyperelastic model were subtracted from the total strains (Borja et al., 1997). In the undrained test (VrL4) plastic deviatoric strain ε_s^p mainly develops, whereas in the drained test (Vr2) a non-negligible plastic volumetric

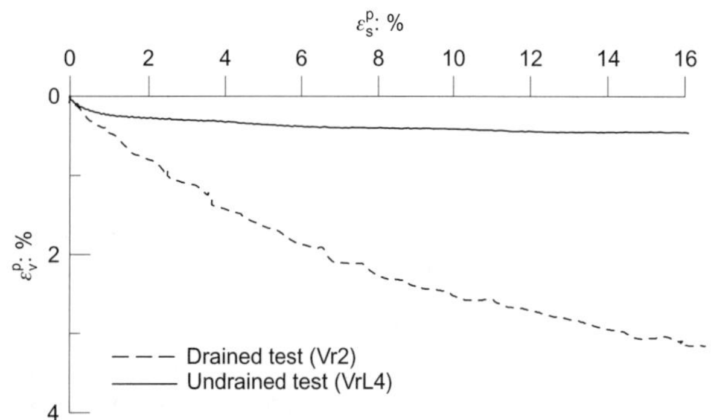

Fig. 16. CA-MP triaxial tests : volumetric against deviatoric plastic strains developed under drained (Vr2) or undrained (VrL4) shearing of normally consolidated samples

strain adds to ε_s^p, as shown in the figure. This results in a faster progressive breakdown of interparticle bonding under drained conditions, causing the normalised stress path of this test to lie below that obtained from undrained shearing.

Similarity of the normalised boundary surfaces of anisotropically compressed samples suggests that the de-bonding process mainly induces a decrease in size of the initial domain, thus indicating its isotropic character.

This interpretation is consistent with what has been proposed in several plasticity-based constitutive models in which mechanical de-bonding is assumed to depend on a combination of volumetric and deviatoric plastic strains, and is accounted for by appropriately manipulating the isotropic part of the hardening law of the de-bonded material (e.g. Gens & Nova, 1993; Lagioia & Nova, 1995; Kavvadas & Amorosi, 2000; Rouainia & Wood, 2000; Gajo & Wood, 2001; Liu & Carter, 2002; Nova et al., 2003; Amorosi & Boldini, 2003; Liu et al., 2003; Wheeler et al., 2003; Baudet & Stallebrass, 2004).

By incorporating the effect of structural degradation into the normalising parameter, it is possible to locate a single SBS on a normalised stress plane for both drained and undrained tests of a natural clay, and to evaluate the influence of volumetric and deviatoric plastic strains (Amorosi & Kavvadas, 2002; Callisto & Rampello, 2004).

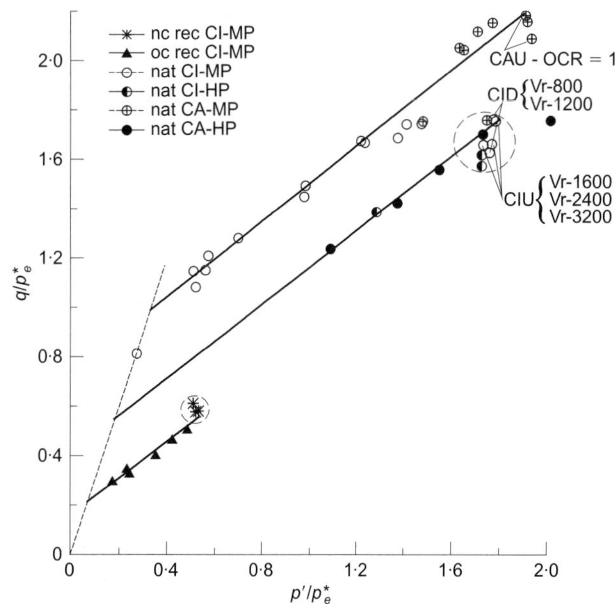

Fig. 17. Normalised strength envelopes of natural and reconstituted samples

Figure 17 shows the normalised peak strengths of the three series of triaxial tests together with data obtained from isotropically compressed samples of natural and reconstituted Vallericca clay (Rampello *et al.*, 1993). In this case the equivalent pressure is evaluated with reference to the ICL (rec-NCL in Fig. 8). Three normalised failure envelopes are shown, which are about parallel. The upper envelope is relevant to natural samples sheared after either isotropic or anisotropic compression in the medium-pressure range. The intermediate envelope is for samples that have undergone isotropic or anisotropic compression to effective stress higher than the yield stress. The lower envelope is for the reconstituted samples. It is worth noting that the normalised strength of the normally consolidated natural sample, sheared drained after anisotropic compression in the MP range, plots close to the critical state condition defined by the destructured samples. The same is observed for two drained tests and three undrained CI-MP tests in which the samples were sheared after isotropic compression to values of effective stress close to or higher than the yield stress. All these samples exhibited a contractant behaviour, and, at peak strength, continuous shearing took place at almost constant pore water pressure or volumetric strain. As a consequence, they underwent a substantial degree of structure degradation during shearing, and their normalised strengths merge with the CSL defined by the destructured samples. However, comparison of the strength envelopes of natural and reconstituted samples normalised by the intrinsic equivalent pressure shows that even the highest levels of applied effective stress were not capable of bringing the natural and the reconstituted clay together. In fact, the natural samples tend towards stationary states that lie well above those defined by the reconstituted samples.

The same occurrence can be shown by plotting the test paths of undrained and drained tests in the v–$\log p'$ plane. In Fig. 18 the end points of tests carried out on natural samples for which substantial continuous straining was observed plot closely around a single line that is parallel to the normal compression lines. A single critical state line, described by a specific volume $\Gamma = 2\cdot77$ at $p'_r = 1$ kPa and a slope $\lambda = 0\cdot148$, can then be defined for natural samples of Vallericca clay, irrespective of their isotropic or anisotropic compression histories and of the maximum effective stress experienced by the samples before shearing.

Critical states observed by Rampello *et al.* (1993) for reconstituted samples of Vallericca defined a CSL that plots below the one relevant to the natural samples, though characterised by the same slope $\lambda = 0\cdot148$: this provides further experimental evidence that, in the investigated stress range, natural Vallericca clay does not tend to the reference state defined by the corresponding reconstituted material. This could be attributed to substantial differences between the fabric of the natural and reconstituted samples, not deleted by compression to high effective stress nor by shearing.

CONCLUSIONS

Past experimental investigation into natural and reconstituted samples of Vallericca clay, together with the tests presented in this work, has led to an additional insight into the effects of structural degradation on the mechanical response of the clay. This has been achieved using experimental data from oedometer tests and triaxial compression tests, the latter being characterised by either isotropic or anisotropic compression histories and by different levels of maximum effective stress applied before shearing.

Structural degradation involves damage to interparticle bonding and changes in particle arrangement. Both of these depend on the amount of volumetric and deviatoric plastic strains. Particle arrangement is also affected by plastic strain ratio, that is, by the direction of the plastic strain path.

Compression or swelling histories characterised by values of stress ratio η similar to those experienced by the soil in situ can be assumed to produce negligible changes in soil fabric, while mainly inducing progressive breakdown of interparticle bonding as effective stress becomes larger than the yield stress. In this case, structural degradation consists mainly of a de-bonding process. When the applied compression histories are characterised by values of η different from those undergone by the soil during its past stress history, non-negligible changes in soil fabric are also induced.

Compression curves of natural samples obtained from oedometer tests and after isotropic or anisotropic compression in triaxial cells were observed to offset permanently to the right of the intrinsic compression line up to effective stress much higher than the yield stress. As substantial breakdown of interparticle bonding induced by plastic strains developed during compression to high effective stress, the observed difference may be related to significant differences in the particle arrangement of the natural and reconstituted clay, which hold even at high stress levels.

Comparing the CA-MP and the CA-HP test results, two distinct SBSs were obtained in a stress plane normalised by the natural equivalent pressure, in contrast to the CSSM framework. This occurrence has been ascribed to the de-bonding process, which affects the two sets of data differently. However, similarity in the shape of the boundary surfaces reveals the isotropic character of the de-bonding process, and makes it possible to incorporate the effect of structural degradation into a scalar hardening parameter.

When CI-HP and CA-HP tests characterised by isotropic and anisotropic compression histories are compared, changes in both fabric and bonding can be thought to contribute to structural degradation. The changes in both position and shape of the normalised SBS observed for states wet of critical can then be related to substantial changes in the particle arrangement. In this case, normalising the stress paths by a scalar hardening parameter does not bring together data from the two sets of tests, even if account is taken of structural degradation by manipulating the isotropic part of the hardening law.

Comparison of the normalised shear strength of natural

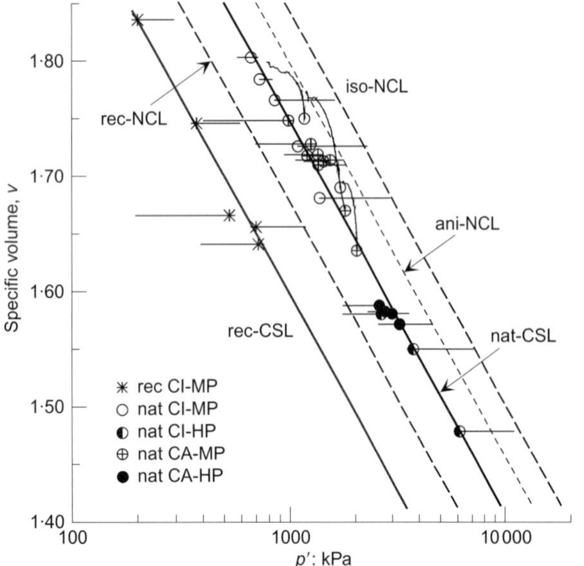

Fig. 18. Critical state conditions of natural and reconstituted samples

and reconstituted samples of Vallericca clay showed that the end points of the state paths of the natural samples define an ultimate state that lies well above that defined by the reconstituted samples, possibly because of the differences in the particle arrangement of the natural and reconstituted clay. Consistently, in the compression plane $(v-\log p')$ the natural samples define a single critical state line that plots above the intrinsic CSL and is about parallel to it. The natural CSL does not depend on the maximum effective stress applied before shearing, and is the same for both isotropic and anisotropic compression histories.

For natural Vallericca clay the reconstituted material does not represent an attainable fully destructured reference state, in that natural samples of the clay do not achieve the combination of stress and volumetric states of the reconstituted samples, even after experiencing effective stresses much higher than the yield stress.

NOTATION

C_c	compression index
C_c^*	intrinsic compression index
C_s	swelling index
C_s^*	intrinsic swelling index
CF	clay fraction
e	void ratio
G_s	specific gravity
I_L	liquidity index
I_P	plasticity index
I_v	void index
K_0	coefficient of earth pressure at rest
K_0^{nc}	coefficient of earth pressure at rest for nc states
K_0^{oc}	coefficient of earth pressure at rest for oc states
N	specific volume
OCR	overconsolidation ratio
p'	mean effective stress
p_e'	mean equivalent pressure
p_k'	initial effective stress
p_{max}'	maximum value of mean effective stress
p_r'	reference stress
p_y'	mean effective yield stress
p_e^*	intrinsic mean equivalent pressure
q	deviatoric stress
q_{max}	maximum value of deviatoric stress
v, Γ	specific volume
W_L	liquid limit
W_0	natural water content
ε_a	axial strain
$\varepsilon_{aL}, \varepsilon_{aL2}$	local axial strains
ε_{aext}	overall axial strain
ε_r	radial strain
ε_s^p	plastic deviatoric strain
ε_v	volumetric strain
ε_v^p	plastic volumetric strain
η	stress ratio
η_{max}	maximum value of stress ratio
κ	slope of swelling curve
λ	slope of virgin compression curve
σ_{ax}	axial load
σ_v'	vertical effective stress
σ_{ve}^*	intrinsic vertical equivalent stress
σ_{vmax}'	maximum value of vertical effective stress
σ_{vy}'	vertical effective yield stress
ϕ_{cv}'	constant-volume angle of shearing resistance

REFERENCES

Amerasinghe, S. F. & Parry, R. H. G. (1975). Anisotropy in heavily overconsolidated kaolin. *J. Geotech. Engng Div. ASCE* **101**, No. 12, 1277–1293.

Amorosi, A. (1996). *Il comportamento meccanico di un'argilla naturale consistente.* Doctoral thesis, University of Rome 'La Sapienza'.

Amorosi, A. & Boldini, D. (2003). Single surface hardening plasticity model for soft clays: mathematical formulation and implicit numerical integration. *Proceedings of an international workshop on geotechnics of soft soils: theory and practice*, Noordwijkerhout, pp. 153–158.

Amorosi, A. & Kavvadas, M. (2002). A critical review of the state boundary surface concept in light of advanced constitutive modelling. *Proc. 8th Int. Symp. Num. Mod. Geomech. (NUMOG), Rome*, 85–90.

Amorosi, A. & Rampello, S. (1998). The influence of natural soil structure on the mechanical behaviour of a stiff clay. *Proc. Int. Symp. Geotech. Hard Soils – Soft Rocks, Naples* **1**, 395–402.

Anandarajah, A., Kuganethira, N. & Zhao, D (1996). Variation of fabric anisotropy of kaolinite in triaxial loading. *J. Geotech. Engng. Div. ASCE* **122**, No. 8, 633–640.

Baudet, B. & Stallebrass, S. E. (2004). A constitutive model for structured clays. *Géotechnique* **54**, No. 4, 269–278.

Borja R. I., Tamagnini C. & Amorosi A. (1997). Coupling plasticity and energy-conserving elasticity model for clays. *Int. J. Geotech. Geoenviron. Engng* **123**, No. 10, 948–957.

Burland, J. B. (1990). On the compressibility and shear strength of natural clays. *Géotechnique* **40**, No. 3, 329–378.

Burland, J. B., Rampello, S., Georgiannou, V. N. & Calabresi, G. (1996). A laboratory study of the strength of four stiff clays. *Géotechnique* **46**, No. 3, 491–514.

Calabresi, G., Rampello, S. and Viggiani, G. (1990). Analisi di una prova triassiale in presenza di un fenomeno di localizzazione delle deformazioni. *Consiglio Nazionale delle Ricerche, Attività di Ricerca 1989–1990*, 111–114.

Callisto, L. (1996). *Studio sperimentale su un'argilla naturale: il comportamento meccanico dell'argilla di Pisa.* Doctoral thesis, University of Rome 'La Sapienza'.

Callisto, L. & Rampello, S. (2004). An interpretation of structural degradation for three natural clays. *Can. Geotech. J.* **41**, No. 4, 392–407.

Coop, M. R. & Cotecchia, F. (1995). The compression of sediments at the archaeological site of Sibari. *Proc. 11th Eur. Conf. Soil Mech. Found. Engng, Copenhagen* **8**, 19–26.

Coop, M. R., Atkinson, J. H. & Taylor, R. N. (1995). Strength and stiffness of structured and unstructured soils. *Proc. 11th Eur. Conf. Soil Mech. Found. Engng, Copenhagen* **1**, 55–62.

Cotecchia, F. (1996). *The effects of structure on the properties of an Italian Pleistocene clay.* PhD thesis, University of London.

Cotecchia, F. & Chandler, J. (2000). A general framework for the mechanical behaviour of clays. *Géotechnique* **50**, No. 4, 523–544.

Cuccovillo, T. & Coop, M. R. (1997). The measurement of local axial strains in triaxial tests using LVDTs. *Géotechnique* **47**, No. 1, 167–171.

Cuccovillo, T. & Coop, M. R. (1999). An automated triaxial apparatus for elevated pressures. In *Non-destructive and automated testing of soil and rock properties*, ASTM STP 1350, pp. 231–245.

Gajo, A. & Muir Wood, D. (2001). A new approach to anisotropic, bounding surface plasticity: general formulation and simulations of natural and reconstituted clays. *Int. J. Num. Anal. Meth. Geomech.* **25**, No. 3, 207–241.

Gens, A. (1982). *Stress–strain and strength characteristic of a low plasticity clay.* PhD thesis, University of London.

Gens, A. & Nova, R. (1993). Conceptual bases for a constitutive model for bonded soils and weak rocks. *Proc. Int. Symp. Hard Soils – Soft Rocks, Athens* **1**, 485–494.

Henkel, D. J. & Sowa, V. A. (1963). The influence of stress history on the stress paths in undrained triaxial tests on clays. In *Laboratory shear testing of soils*, ASTM STP 361, pp. 280–291.

Hight, D. W. (1982). A simple piezometer probe for the routine measurement of pore pressure in triaxial tests on saturated soils. *Géotechnique* **32**, No. 4, 396–401.

Horseman, S. T., Winter, M. G. & Entwistle, D. C. (1987). Geotechnical characterisation of Boom Clay in relation to disposal of radioactive waste, Report EUR 10987. Luxembourg: Commission of the European Communities.

Kavvadas, M. & Amorosi, A. (2000). A constitutive model for structured soils. *Géotechnique* **50**, No. 3, 263–273.

Kavvadas, M. & Anagnostopoulos, A. (1998). A framework for the mechanical behaviour of structured soils. *Proc. 2nd Int. Symp. Hard Soils – Soft Rocks, Naples* **2**, 603–614.

Ladd, C. C. (1965). Stress–strain behaviour of anisotropically consolidated clays during undrained shear. *Proc. 6th Int. Conf. Soil Mech. Found. Engng, Montreal* **1**, 282–290.

Lagioia, R. & Nova, R. (1995). An experimental and theoretical study of the behaviour of a calcarenite in triaxial compression. *Géotechnique* **45**, No. 4, 633–648.

Leroueil, S. & Vaughan, P. R. (1990). The general and congruent effects of structure in natural soils and weak rocks. *Géotechnique* **40**, No. 3, 467–488.

Liu, M. D. & Carter, J. (2002). A structured Cam Clay model. *Can. Geotech. J.* **39**, No. 6, 1313–1332.

Liu, M. D., Carter, J. & Desai, C. S. (2003). Modelling compression behaviour of geo-materials. *Int. J. Geomech. ASCE* **3**, No. 2, 191–204.

Mitchell, J. K. (1976). *Fundamentals of soil behaviour.* New York: John Wiley & Sons.

Nova, R., Castellanza, R. & Tamagnini, C. (2003). A constitutive model for bonded geomaterials subject to mechanical and/or chemical degradation. *Int. J. Num. Anal. Meth. Geomech.* **27**, No. 9, 705–732.

Parry, R. H. G. & Nadarajah, V. (1973). Observations on laboratory prepared, lightly overconsolidated specimens of kaolin. *Géotechnique* **24**, No. 3, 345–358.

Rampello, S. (1989). *Effetti del rigonfiamento sul comportamento meccanico di argille fortemente sovraconsolidate.* Doctoral thesis, University of Rome 'La Sapienza'.

Rampello, S. (1991). Some remarks on the mechanical behaviour of stiff clays: the example of Todi clay. *Proceedings of the workshop on experimental characterization and modelling of soils and soft rocks*, Naples, pp. 131–190.

Rampello, S., Georgiannou, V. N. & Viggiani, G. (1993). Strength and dilatancy of natural and reconstituted Vallericca clay. *Proc. Int. Symp. on Hard Soils – Soft Rocks, Athens* **1**, 761–768.

Rampello, S., Calabresi, G. & Callisto, L. (2002). Characterisation and engineering properties of a stiff clay deposit. *Proceedings of the international workshop on characterisation and engineering properties of natural soils*, Singapore, Vol. 2, pp. 1021–1045.

Rossato, G. & Jardine, R. J. (1990). *Preliminary investigations of the stress–strain properties of an artificial sandy clay*, Internal report. University of London.

Rouainia, M. & Muir Wood, D. (2000). A kinematic hardening constitutive model for natural clays with loss of structure. *Géotechnique* **50**, No. 2, 153–164.

Schofield, A. N. & Wroth, C. P. (1968). *Critical state soil mechanics.* London: McGraw-Hill.

Sciotti, A. (1992). *Compressibilità e resistenza di alcune argille sovraconsolidate dell'Italia Centrale.* Unpublished thesis, University of Rome 'La Sapienza'.

Smith, P. R., Jardine, R. J. & Hight, D. W. (1992). The yielding of Bothkennar clay. *Géotechnique* **42**, No. 2, 257–274.

Taylor, R. N. & Coop, M. R. (1993). Stress path testing of Boom clay from Mol, Belgium. *Proc. 26th Ann. Conf. Engng Group of the Geological Society, Leeds*, 77–82.

Topolnicki, M., Gudehus, G. & Mazurkiewicz, B. K. (1990). Observed stress–strain behaviour of remoulded saturated clay under plane strain conditions. *Géotechnique* **40**, No. 2, 155–187.

Viggiani, G., Rampello, S. & Georgiannou, V. N. (1993). Experimental analysis of localisation phenomena in triaxial tests on stiff clays. *Proc. Int. Symp. on Hard Soils – Soft Rocks, Athens* **1**, 849–856.

Viggiani G., Finno R. J., Rampello S. & Calabresi G. (1995). Discussion on 'Cracks, bifurcation and shear band propagation in saturated clays' by Saada, A. S., Bianchini, G. F. and Liang, L. (1994), *Géotechnique*, **45**, No. 2, 339–343.

Wheeler, S. J., Näätänen, A., Karstunen, M. & Lojander, M. (2003). An anisotropic elasto-plastic model for soft clays. *Can. Geotech. J.* **40**, No. 2, 403–418.

Bernier, F., Li, X.-L. & Bastiaens, W. (2007). *Géotechnique* **57**, No. 2, 229–237

Twenty-five years' geotechnical observation and testing in the Tertiary Boom Clay formation

F. BERNIER*, X.-L. LI* and W. BASTIAENS*

The research and development programme on geological disposal for high-level and long-lived waste (HLW) in Belgium was initiated in 1974. A deep tertiary clay formation, the Boom Clay, present under the Mol-Dessel nuclear site, was selected as a reference host formation for experimental purposes. The construction of the underground laboratory HADES (at a depth of 223 m, initiated in 1980 and extended in 2002) allowed the building of a valuable geotechnical database, and led to the development of improved excavation techniques that significantly reduce the excavation-damaged zone (EDZ). Since the operational start of HADES about 25 years ago, many geotechnical measurements have been performed around excavations. Comparison between in situ measurements and modelling results allowed a continuous improvement of our knowledge on the Boom Clay behaviour. Important issues for interpreting the measurements correctly are good control of the excavation parameters and the boundary conditions. An extensive characterisation of the hydromechanical response of Boom Clay around an excavation for short- and long-term conditions has been performed. One important finding was the occurrence of measurable hydraulic effects at a distance of about 60 m (12·5 tunnel diameters) ahead of the tunnel excavation.

KEYWORDS: clays; in situ testing; monitoring; numerical modelling and analysis; pore pressures; radioactive waste disposal

Le programme de recherche et développement sur l'évacuation géologique des déchets fortement radioactifs de haute période (HLW) en Belgique a été initié en 1974. On a sélectionné une formation argileuse tertiaire profonde, l'argile de Boom, située sous le site nucléaire de Mol-Dessel, comme formation hôte de référence pour des expériences. La construction du laboratoire sous-terrain HADES (d'une profondeur de 223m – commencé en 1980 et étendu en 2002) a permis de mettre en place une base de données géotechnique précieuse et a conduit au développement de techniques d'excavation améliorées qui réduisent significativement la zone endommagée par l'excavation (EDZ, de « excavation-damaged zone »). Depuis le début des opérations à HADES il y a environ 25 ans, on a procédé à de nombreuses mesures géotechniques autour des excavations. En comparant les mesures in situ et les résultats de la modélisation, on a continuellement amélioré nos connaissances du comportement de l'argile de Boom. Pour interpréter les mesures correctement, il est important d'avoir un bon contrôle des paramètres d'excavation et des conditions aux limites. On a procédé à une caractérisation extensive de la réponse hydromécanique de l'argile de Boom autour d'une excavation dans des conditions de court-terme et de long-terme. Une découverte importante était qu'il y a des effets hydrauliques mesurables environ 60m devant l'excavation des tunnels (diamètres des tunnels de 12,5 m).

INTRODUCTION

In all nuclear power generating countries, spent nuclear fuel and long-lived radioactive-waste management are an important environmental issue. Disposal in deep clay geological formations is one of the promising options for disposal of these wastes.

In Belgium, the research and development programme on this topic was initiated at the Belgian nuclear research centre (SCK.CEN) in 1974. A Tertiary clay formation, the Boom Clay, present between 190 m and 290 m under the Mol-Dessel nuclear site, was selected as a potential host formation for the disposal of HLW (high-level and long-lived radioactive waste). The first construction phase of the underground research facility (URF) HADES (High-Activity Disposal Experimental Site), at a depth of about 223 m, started in 1980. Since then, HADES has been expanded several times. Fig. 1 shows the construction history, and Table 1 gives basic information on each phase (Bastiaens & Bernier, 2006). The total length of HADES is about 200 m, with an average internal diameter of about 4 m. HADES is currently managed by the economic interest grouping EURIDICE (European Underground Research Infrastructure for Disposal of Nuclear Waste in Clay Environment), a joint venture between SCK.CEN and NIRAS (the National Agency for Radioactive Waste and Enriched Fissile Materials).

During the construction of HADES, and geotechnical measurements in and around HADES, experience was gained that allowed the building of a valuable geotechnical database. This knowledge led to the use of improved excavation techniques, which significantly reduce the excavation-damaged zone (EDZ). Moreover, different lining techniques and types were used: cast iron segments, steel sliding ribs and concrete segments. Both the short- and long-term behaviour of the host rock as well as its interaction with each kind of lining were studied. Comparison between in situ measurements and modelling results allowed a continuous improvement of our knowledge of the Boom Clay behaviour. This paper discusses the most relevant results concerning the hydromechanical (HM) behaviour of the Boom Clay, which cover both the short-term HM responses during gallery excavation and the long-term responses. A comparison between experimental results and numerical simulations is presented. The crack-sealing behaviour of the Boom Clay is also discussed.

Manuscript received 5 May 2006; revised manuscript accepted 23 January 2007.
Discussion on this paper closes on 1 August 2007, for further details see p. ii.
* European Underground Research Infrastructure for Disposal of Nuclear Waste in Clay Environment, ESV EURIDICE GIE, Mol, Belgium.

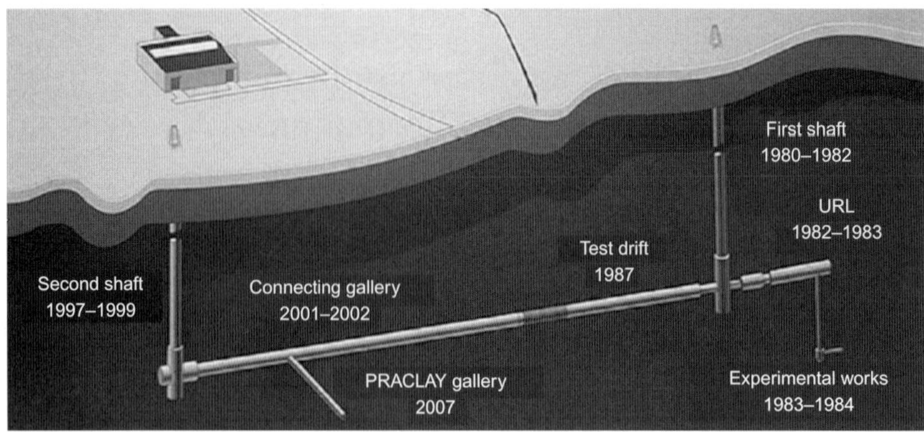

Fig. 1. Construction history of the HADES Underground Research Facility (from Bastiaens & Bernier, 2006)

Table 1. Basic information for each construction phase

Gallery	Excavation method	Lining type	Inner/outer diameter and length: m
URF 1982—1983	Frozen clay – semi-manual excavation	Cast iron segments + grout injection	3·5–4·0 m 35 m
Exploratory works 1983—1984	Semi-manual excavation	Unreinforced concrete segments + Linex plates + grout injection	1·4–2·0 m shaft: 24 m, gallery: 8·5 m
Test drift 1987	Semi-manual excavation	Unreinforced concrete segments + Linex plates + grout injection	3·5–4·7 m 65 m
Andra gallery 1987	Semi-manual excavation	Steel sliding ribs	~4·1 m 12 m
Connecting gallery 2001—2002	Shield and road-header	Unreinforced concrete segments (wedge blocks)	4·0–4·8 m 85 m

BOOM CLAY

At Mol-Dessel nuclear site, the Boom Clay is present 190–290 m below ground level. The Boom Clay layer is almost horizontal (it dips 1–2% towards the north-east), and water-bearing sand layers are situated above and below it. The total vertical stress and pore water pressure at the level of HADES are respectively some 4·5 and 2·2 MPa. There exist, however, open questions about the in situ stress state tensor. A K_0 value (ratio of horizontal to vertical effective stresses) was determined by laboratory methods and in situ investigations. The in situ investigations from HADES, using pressuremeter, dilatometer, self-boring pressuremeter (SBP), hydrofracturing tests, borehole breakouts analysis and back-analysis of the stresses in the liner, gave some scatter for K_0 values (0·3–0·9) (Bernier *et al.*, 2002). Laboratory investigations indicated values for K_0 between 0·5 and 0·8 (Henrion *et al.*, 1984, Horseman *et al.*, 1987). Deeper investigation of this subject is therefore necessary: a new set of in situ tests including SBP and hydrofracturing tests in different directions is planned for 2007. The Boom Clay is characterised by a fairly constant chemical and mineralogical composition. The overconsolidation ratio (OCR) is about 2·4 (Horseman *et al.*, 1987; Coll, 2005), and the unconfined compressive strength (UCS) is some 2 MPa.

GEOTECHNICAL INVESTIGATIONS

Since the operational start of HADES about 25 years ago, many geotechnical measurements have been performed around it to understand the hydromechanical behaviour of the Boom Clay. The geotechnical investigations include both laboratory tests (triaxial and odometer tests) and in situ testing (visual observations, dilatometer tests, self-boring pressuremeter tests, seismic measurements, etc.). More spe-

cifically, the excavations of the test drift (excavated by hand) and of the connecting gallery (excavated with a tunnel machine) were accompanied by an instrumentation programme that monitored the hydromechanical response of the Boom Clay in terms of displacement, stress and pore water pressure.

Back-analyses have enabled the hydraulic and geomechanical characteristics of the undisturbed Boom Clay to be determined (see Table 2). It is important to note, however, that the Boom Clay behaviour is characterised by a highly non-linear stress–strain response. Although there is some scatter in the results, laboratory tests showed a very clear trend of stiffness variation with strain level, as illustrated by Fig. 2: its tangent stiffness at 0·01% deformation may be one order of magnitude larger than that at 1% deformation. The drained Young's modulus in Table 2 (300 MPa) was estimated from analysis and interpretation of the in situ measurements made during excavation of the test drift (Mair *et al.*, 1992). The same value was applied to simulate some

Table 2. Hydraulic and drained geomechanical characteristics of Boom Clay

Young's modulus, E': MPa	300
Poisson's ratio, v'	0·125
Friction angle,* ϕ': degrees	18
Cohesion,* c': MPa	0·3
Dilation angle,* ψ': degrees	0–10[†]
Saturated permeability, K_w: m/s	2–4 × 10^{-12}
Porosity, n: %	39

* Averaged out over a range of mean effective stress of 2·5–4 MPa.
[†] For the numerical analysis $\psi' = 0$ has been adopted.

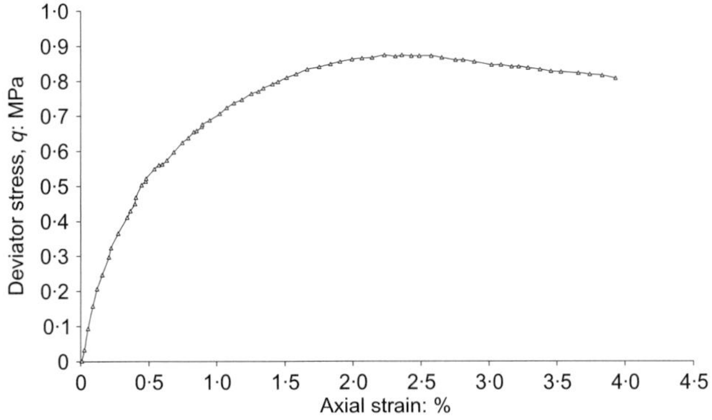

Fig. 2. An example of the stress–strain data from the drained triaxial compression tests

selected laboratory triaxial compression tests assuming linear elastic behaviour prior to reaching the maximum strength (linear-elastic perfectly plastic model), and it gave reasonably consistent results with the test stress–strain curves (Labiouse, 1997). The shearing strength characteristics (drained cohesion and internal frictional angle) correspond to the values averaged out over a range of mean effective stress of 2·5–4 MPa on the basis of the laboratory tests (Baldi *et al.*, 1987, 1991; Horseman *et al.*, 1987; Mair *et al.*, 1992; Djéran *et al.*, 1994; Labiouse, 1997).

Several laboratory and in situ experiments were also performed to study the origin and the extent of excavation-induced fractures as well as the sealing characteristics of the Boom Clay.

HYDROMECHANICAL RESPONSE AROUND EXCAVATIONS

In the Boom Clay formation, as in most soil media, excavation of galleries leads to both mechanical (convergence, redistribution of the stress) and hydraulic disturbances (pore pressure variation) in their surroundings owing to decompression of the formation, and inevitably creates an excavation-damaged zone (EDZ) in the vicinity of the excavation faces as a result of high deviator stress generation. Within the context of high-level and long-lived waste repository design, the extent and the evolution of the EDZ have to be taken into account in assessing the long-term performance of the geological disposal.

Two extreme conditions must be considered for the design of galleries in Boom Clay formation. The first concerns the immediate stability, corresponding to the gallery construction, with negative pore water pressures generated by excavation. The second is related to the long-term stability, corresponding to the situation when all excess pore water pressures have dissipated: that is, the situation after several decades.

Short-term hydromechanical response

Overall, the characterisation of the hydromechanical behaviour of the Boom Clay has been undertaken mainly using data from in situ measurements and well-controlled excavation technique. When studying the short-term hydromechanical response around excavations, it is important to have a sound knowledge of the excavation parameters and the boundary conditions. This was achieved by the use of a tunnelling machine during the construction of the connecting gallery, and the installation of a comprehensive instrumentation programme. Prior to the drilling of this tunnel, a series of instrumented boreholes A, B, C, D and E (Fig. 3) were established within the Clay Instrumentation Programme for

the Extension of an Underground Research Laboratory (CLIPEX) project (Bernier *et al.*, 2002). This suite of instrumentation allowed monitoring of the pore pressure, total stresses, and displacements around the excavation and ahead of the excavation front.

The tunnelling machine was composed of a 2·3 m long shield, a road header for excavation of the rock, and a bird-wing erector system for installation of the lining (Fig. 4). The shield was equipped with a cutting head to ensure a smooth excavation profile. This excavation technique achieved good control of the initial excavated diameter, and hence provided a good baseline for subsequent convergence measurements. This is quite important for the correct interpretation of the measurements. The total radial convergence during construction was limited to 90 mm. This value is the sum of the instantaneous radial convergence ahead of the excavated front (45 mm) and the radial convergence occurring between the excavated front and the last installed lining segment (45 mm). Owing to the very low permeability of the Boom Clay formation and the quite high excavation rate (about 3 m/day), the response of the host formation to the excavation can be considered as in the undrained condition. Therefore it was possible to dissociate the instantaneous response clearly from the delayed effects.

The measurement of the pore water pressure through piezometers has proved to be a very reliable, accurate, and mature technique. The piezometers used are metallic tubes, which are installed in a drilled borehole. At fixed locations the tubes are equipped with porous filters that allow pore pressure to be measured at those locations. The filters are sealed off by natural convergence of the borehole walls around the instrument. The water is delivered to the pressure transmitters through a twin-tube system. Displacement measurements so far have led to a good knowledge of the axial displacement along the gallery axis.

During excavation of the connecting gallery, all piezometers installed ahead of the excavation front registered a similar regular evolution of the pore water pressure with the approach of the excavation front: a progressive increase followed by a sharp drop as the excavation front approached very closely (Fig. 5(a)). The pressure response and mechanical displacement were strongly coupled, as shown in Fig. 5(b). The high decompression of the formation near the excavation face generated pore water suction (negative pore pressure), as shown in Fig. 8, showing absolute pore pressures ahead of the excavation face. It is worth noting that the range of the sensors is limited: they can measure absolute pressures only as low as 0 MPa. This corresponds to a suction of ~0·1 MPa. The actual suction was, without doubt, larger. Note that the pressure and displacement become steady as the excavation front passed, owing to the emplacement of a stiff lining support following the tunnelling machine.

An unexpectedly extended disturbed zone (both hydraulic and mechanical) due to excavation was observed. As shown in Fig. 6, the pore pressure and displacement sensors began to register regular variation when the excavation front was still more than 60 m (i.e. 12·5 tunnel diameters) distant. This far-field behaviour remains difficult to explain, and constitutes an important issue for the proper understanding of the hydromechanical behaviour of the Boom Clay. An initial attempt has been made to explain this phenomenon. It may be partly associated with the following two factors.

(*a*) The apparent increase of the excavation radius: the fractured zone observed around the connecting gallery (see below) can be seen as an apparent increase of the excavated radius, thereby extending the excavation-disturbed zone. The associated increase of the per-

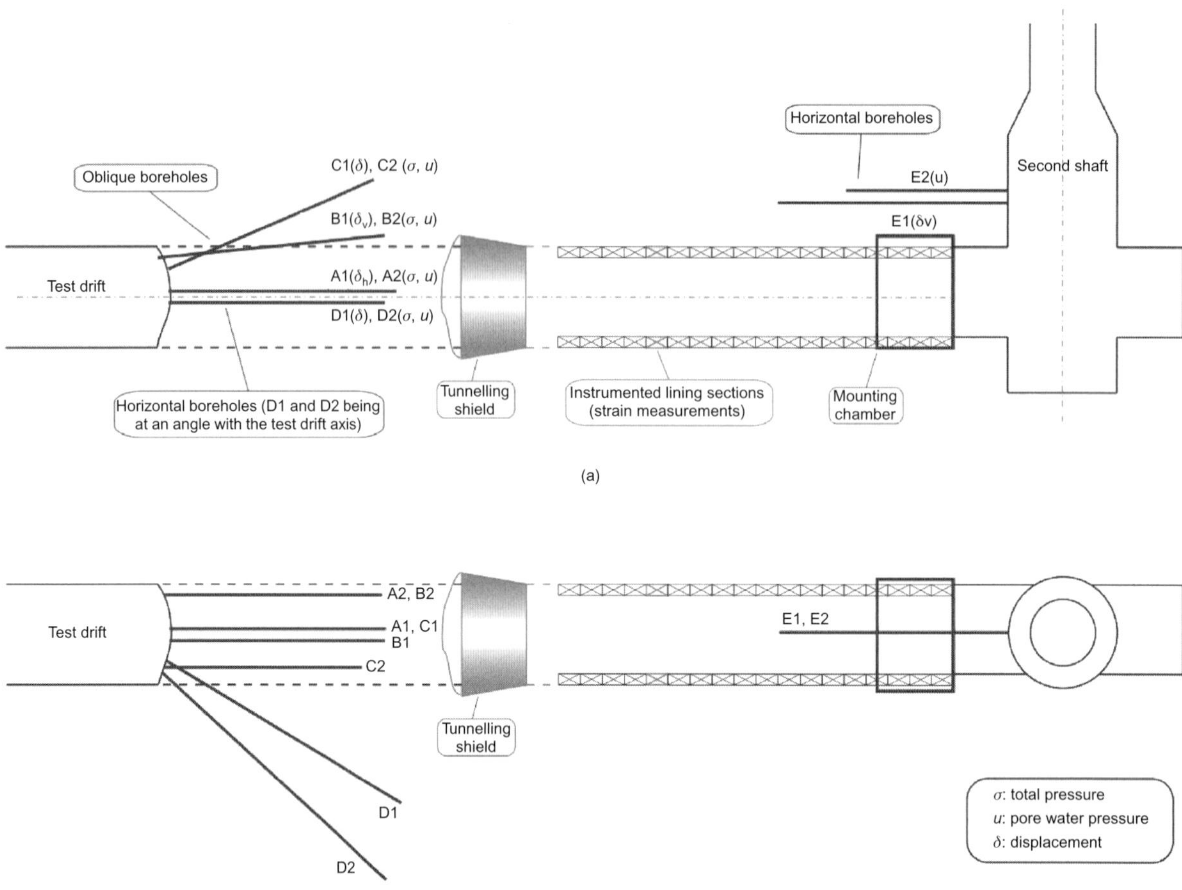

Fig. 3. CLIPEX instrumentation programme (boreholes A, B, C, D, E): (a) side view; (b) top view)

Fig. 4. The tunnel machine used for the connecting gallery (from Bastiaens & Bernier, 2006)

meability in the highly disturbed zone around the gallery reinforced this far-field phenomenon.

(*b*) The viscosity of the skeleton: preliminary numerical research indicated that the hydraulic perturbation zone is larger when using an elasto-plastic-viscoplastic model than that obtained with an elasto-plastic model (Barnichon & Volckaert, 2003). Constitutive development based on limited laboratory creep tests results has indeed shown that the viscous characteristic time for the Boom Clay is much shorter than its hydraulic characteristic time (Rousset *et al.*, 1993; Djéran *et al.*, 1994). This suggests that the viscous effect may appear very soon after the Boom Clay has been subjected to a hydromechanical disturbance, such as excavation, and thus contribute to the far-field responses. Detailed study

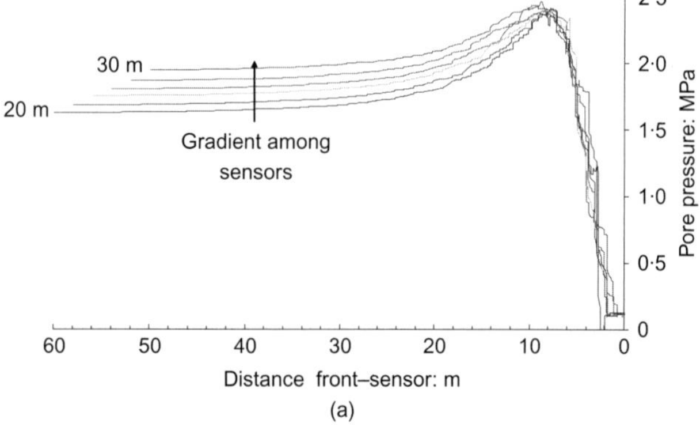

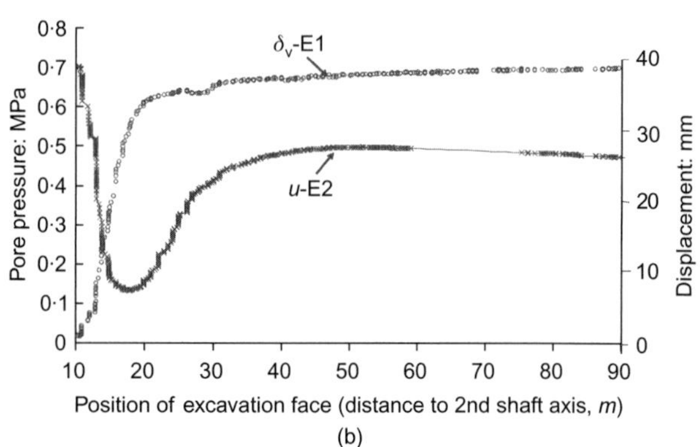

Fig. 5. (a) Pore pressure evolution in borehole B2: filters 1–6 are located at 30–20 m from the test drift face. (b) Pore pressure (*u*) and vertical displacement (δ_v) in borehole E2 and E1: sensors shown are located at ~15 m from axis of second shaft

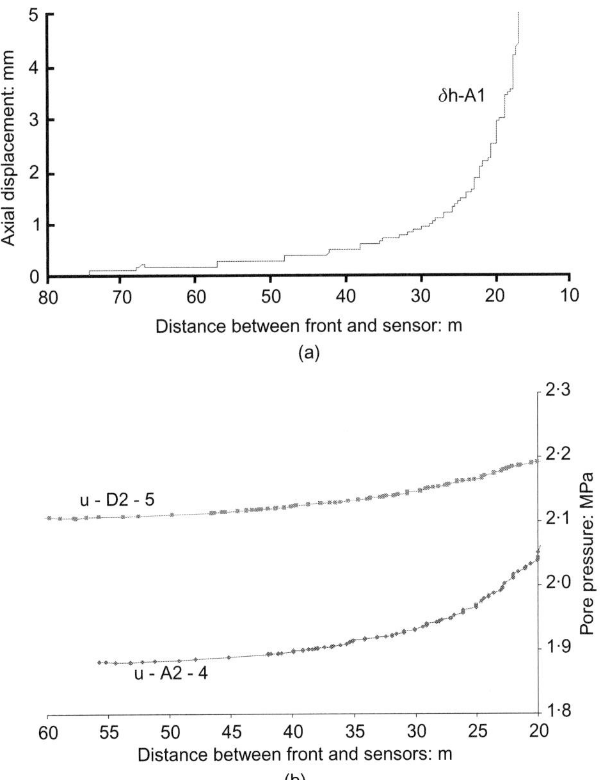

Fig. 6. Evidence of extended excavation-disturbed zone. Very early reaction of extensometers installed in: (a) borehole A1; (b) boreholes A2 and D2

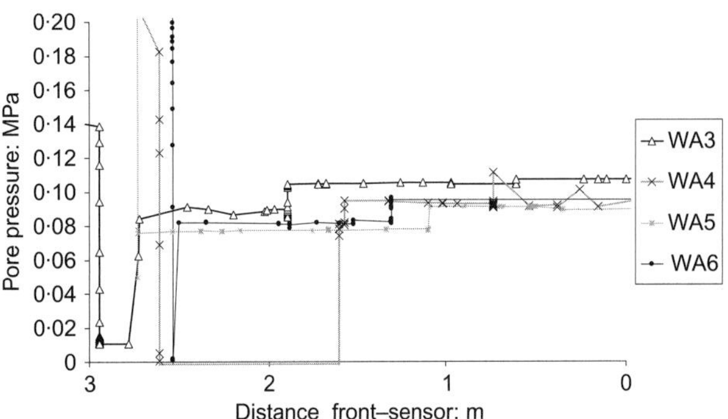

Fig. 8. Evidence of development of fractures in borehole A2 (absolute pore pressures measured ahead of excavation face)

of this subject is continuing through new laboratory tests and numerical development.

Evidence of fractures induced by the excavation has been gathered during the construction of the connecting gallery. Systematic observations of the front and side walls allowed the fracture pattern in the surrounding formation to be determined (Fig. 7). The orientation of the encountered fractures is consistent along most of the excavation. Near the end of the connecting gallery, fractures induced by excavation of the test drift 15 years ago were observed. Because of the strong hydromechanical coupling, the development of fractures was also evidenced through the pore water pressure measurements. The suction created at about 3 m ahead of the excavation front was followed by an abrupt recovery up to atmospheric pressure as the front was coming closer (Fig. 8). This sudden re-equilibrium with the atmospheric pressure indicated the opening of fractures in the formation ahead of the excavation front up to a distance of about 2–3 m along the gallery axis.

Time-dependent behaviour
The Boom Clay is characterised by elasto-viscoplastic behaviour. Viscosity of the skeleton (creep, relaxation, etc.)

and pore water pressure dissipation imply long-term effects around underground excavations.

Figure 9 shows the best estimate of the time evolution of the total stress around the gallery, measured by self-boring pressuremeter tests in horizontal boreholes at two different periods; the values are likely to be more representative of the vertical component of the in situ stress. The SBP test is a two-part process: the first part is concerned with inserting the instrument into the ground to the depth where a test is intended. The second part consists of a controlled inflation of the elastic membrane so that the instrument carries out a loading of the test cavity (Bolton & Whittle, 1999). Between the two tests, total stress close to the gallery wall appears to have risen, indicating stress build-up (re-equilibrium) around the excavation. Total stress seems to be influenced up to at least 6–8 m into the host rock. The value obtained in the far field (5–5·5 MPa) is somewhat higher than the initial in situ stress. This may be due to data interpretation, or it may be a consequence of the excavation of the gallery. Further tests using a deeper borehole are necessary to get more accurate interpretation of these likely higher values.

Figure 10 shows the results at two different times of two reference piezometers installed radially to the test drift. The test drift lining is considered to be very permeable compared with the Boom Clay. Time evolution of pore pressure is different for each orientation. It is influenced by pore pressure dissipation and the viscoplastic mechanical behaviour of the host formation.

Strain gauges were embedded in the lining segments of the connecting gallery to monitor the stresses in the lining and the pressures exerted on it by the host rock.

The convergence of the test drift has now been measured

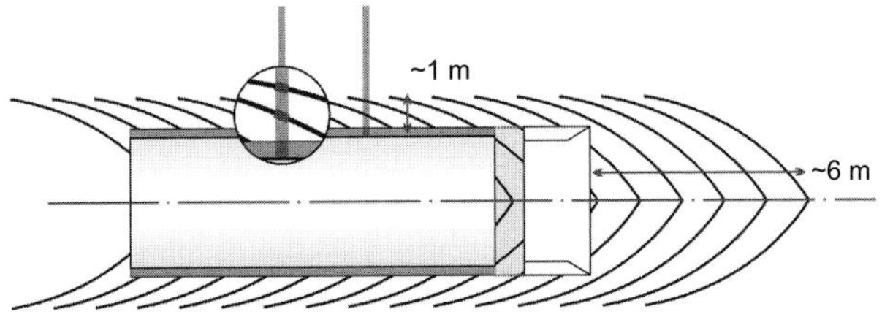

Fig. 7. Observed fracture pattern around connecting gallery (vertical cross-section) (from Bastiaens & Bernier, 2006)

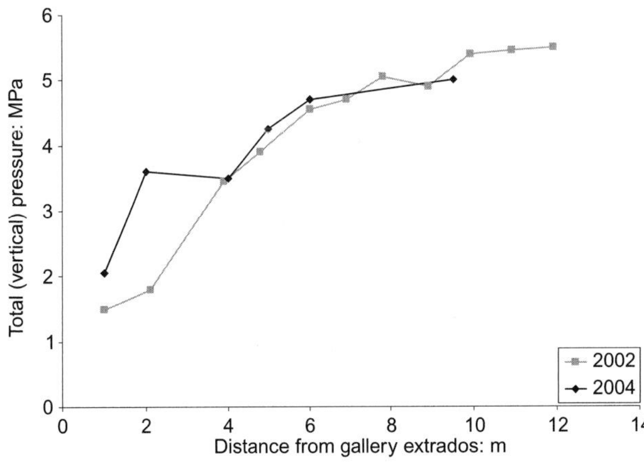

Fig. 9. Total (vertical) stress evolution based upon two series of self-boring pressuremeter tests conducted from connecting gallery in 2002 and 2004

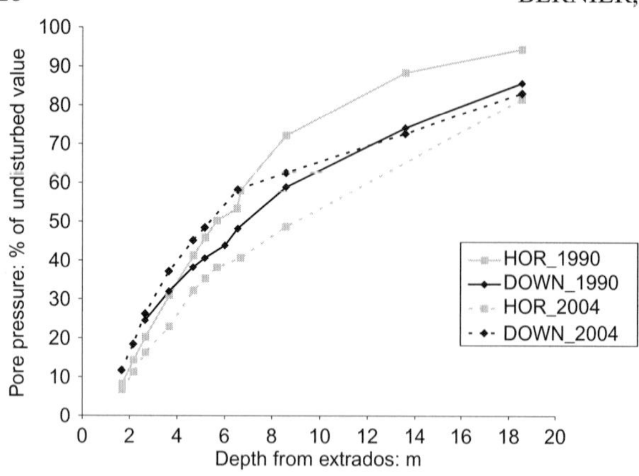

Fig. 10. Pore pressure expressed as a percentage of undisturbed in situ value of pore water pressure at each location in two reference piezometers installed in horizontal and downward boreholes. For each direction pore water pressure profiles at two dates are shown

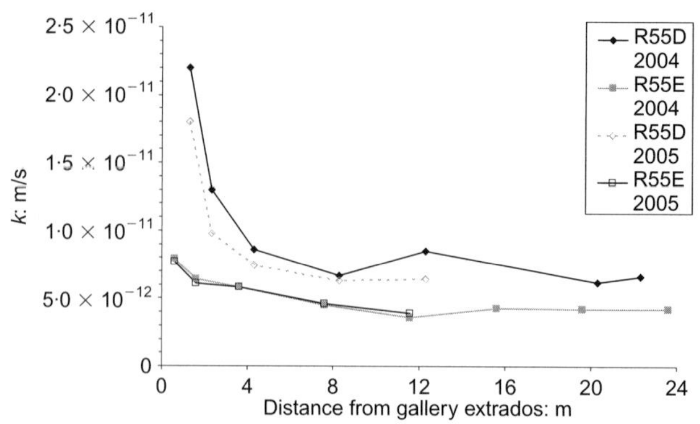

Fig. 12. Hydraulic conductivity around connecting gallery; measured on a horizontal (R55E) and a vertically downward (R55D) piezometer in 2004 and 2005

over a period of 19 years. Fig. 11 shows the diameter reductions of four sections from the central part of the test drift. The test drift was lined with unreinforced concrete blocks, 64 segments per ring, separated by compressible Linex plates. The Linex plates are 8 mm thick wooden plates with an assumed E-modulus of 100 MPa. Up to 1999 the convergence was measured by means of an INVAR wire; since 2000 optical measurements have been used. The decrease of the lining diameter was most rapid during the first year after construction. The rate of convergence has slowed progressively since the initial excavation, but convergence is still continuing, currently (some 18 years after construction) at a rate of about 0·5 mm/year. Since the measurements started, the diameter has reduced by some 60 mm.

Figure 12 shows the results of the hydraulic conductivity measurements around the connecting gallery at different periods (Bastiaens et al., 2006b). Hydraulic conductivity was determined at several filters on a horizontal and a vertical (downward) multi-port piezometer. The tests performed were single-point, steady-state measurements. An increase of hydraulic conductivity is observed up to about 6–8 m from the lining. This increase is attributed to the lower effective stress in this zone. Values of $\sim 6 \times 10^{-12}$ m/s and $\sim 4 \times 10^{-12}$ m/s were obtained outside the influenced zone on the vertical (downward) and horizontal piezometers respectively. Time evolution of the hydraulic conductivity in the vicinity of the gallery indicated clearly the time-dependent coupled hydromechanical behaviour of the Boom Clay.

In Boom Clay sealing effects are very fast, as shown, for instance, by µCT (microfocus computer tomography) images. Fig. 13 shows the µCT images of an artificially fractured clay sample taken after the artificial fracture was created (before saturation of the sample) (Fig. 13(a)) and after 4·5 h of hydraulic conductivity measurement (Fig. 13(b)). The closure of the fracture, due only to the saturation of the sample, is evidenced. Sealing was confirmed by the results of the hydraulic conductivity measurement on the sample, which were of the same order of magnitude as the undisturbed value, being $\sim 10^{-12}$ m/s (Bastiaens et al., 2006a). The EC SELFRAC project studied sealing and healing effects in clays in detail (Bernier et al., 2006).

NUMERICAL SIMULATIONS

Prior to excavation of the connecting gallery, blind predictions of the excavation HM responses of the Boom Clay were performed. The prediction results were subsequently compared with the in situ measurements. The blind predictions were carried out using the commercial finite difference program FLAC, and took into account conditions as close as possible to the reality. Numerical simulations were performed in the framework of soil mechanics for saturated porous media, taking the hydromechanical coupling into account. The Boom Clay behaviour was represented by the classical elasto-perfectly plastic Mohr–Coulomb model; the properties used are listed in Table 2.

Modelling hypotheses

The main hypotheses adopted were

(a) axisymmetric conditions
(b) non-influence of the first and second shafts
(c) perfect contact between the clay and the lining (the contact behaviour between them was not modelled)
(d) excavation under undrained conditions
(e) strong hydromechanical coupling: Biot coefficient $\alpha = 1$.

Geometry

The modelling zone is concentrated on the region around the connecting gallery and the head of the test drift. The dimension of the grid is chosen as large as possible compared with the gallery radius R_0 ($\approx$ 2·45 m), as indicated in Fig. 14, where the reference axes are also shown. The density of the mesh is increased when approaching the CLIPEX instrumentation.

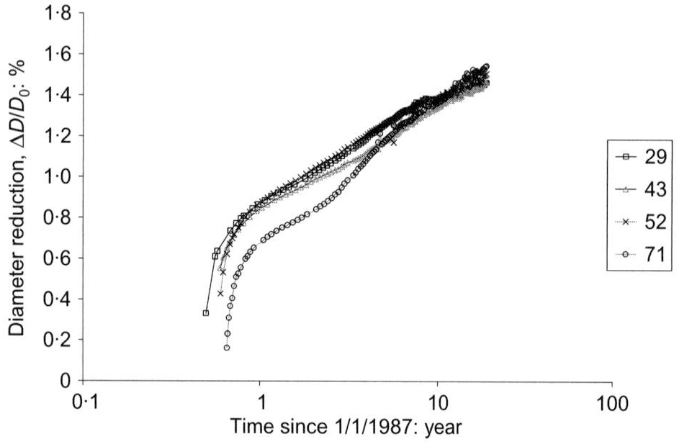

Fig. 11. Diameter reduction at four sections located in central part of the test drift (D_{ext} = 4·7 m)

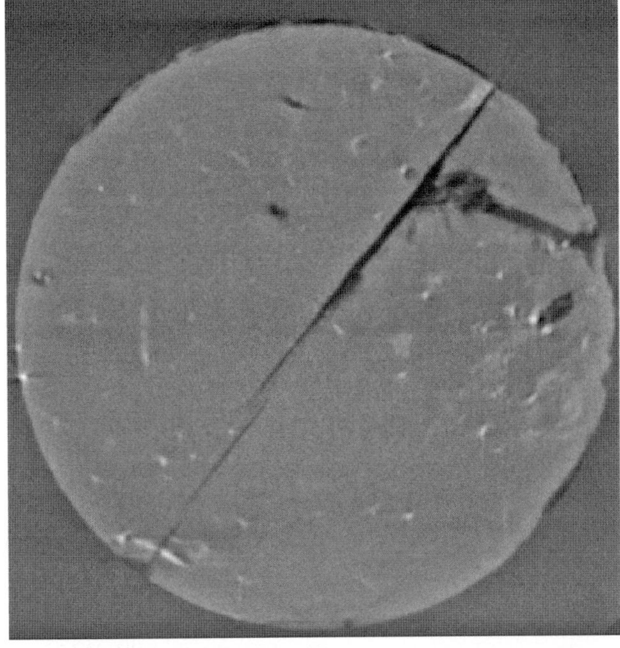

(a)

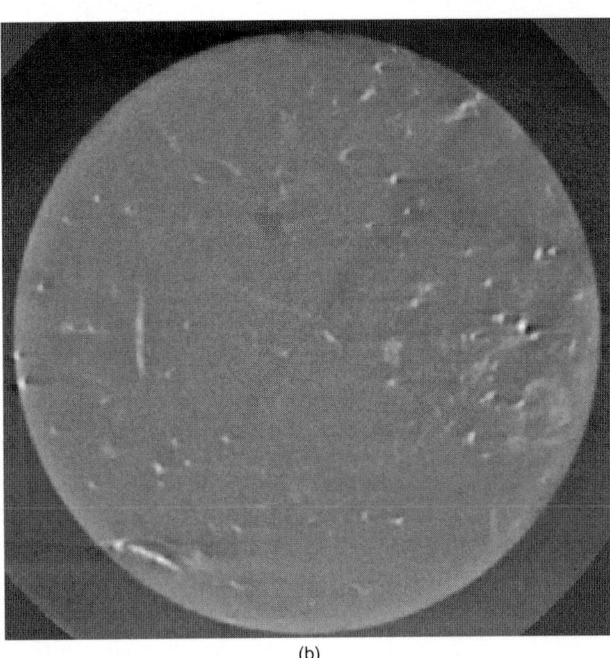

(b)

Fig. 13. μCT images of artificially fractured clay sample (30·3 mm high, diameter 38 mm): (a) taken after creation of artificial fracture; (b) taken after 4·5 h of hydraulic conductivity measurement

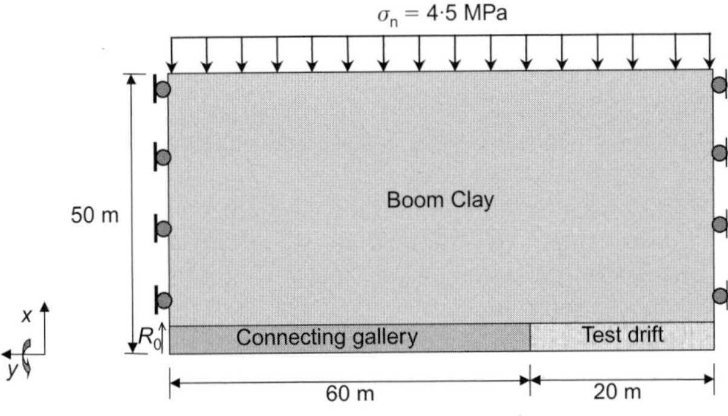

Fig. 14. Geometry of modelling zone and boundary conditions

Modelling procedure

The modelling procedure followed the construction sequence of the URF in three major stages, as follows.

(*a*) *Stage 1: Test drift excavation (only the last 20 m is considered).* Test drift excavation was modelled in order to take into account the complete unloading history of the clay formation surrounding the target instruments around the test drift.

(*b*) *Stage 2: 14-year drainage around the test drift.* The long-term hydromechanical behaviour during 1987 to 2001 (after test drift construction and before excavation of the connecting gallery excavation) was modelled by considering the drainage of the test drift.

(*c*) *Stage 3: Connecting gallery excavation.* The construction of the connecting gallery was modelled following its real construction sequence: excavation and then installation of the lining. These two steps progressed in turn successively with the advancing front. The excavation was modelled by nulling an entire section to be excavated in a single step in accompanying a linear relaxation of the boundary pressure. Because of the irregular mesh size, the length of the rounds excavated and the length of the liner varied as a function of the mesh size. The elements representing the lining were activated about 3·50 m (length of the shield plus unsupported span) behind the advancing working face. The maximum unsupported span between the rear of shield and the lining segment is 1·2 m (Fig. 15).

The convergence behind the cutting edge of the shield was modelled by fixing the radial displacement when it exceeded a prescribed amount. The lining was represented using classical elements with an elastic model. No constraints were applied to movement of the tunnel face.

Initial and boundary conditions

The clay formation is considered as homogeneous and isotropic. The initial stresses are supposed to be hydrostatic. In addition, the modelling zone is supposed to be sufficiently deep that variation of the stress and pore pressure with depth is neglected. Consequently, the initial stresses and pore pressure input to the model correspond to the values at the depth of gallery axis (Table 3). The boundary conditions are sketched in Fig. 14.

Results

Numerical prediction of the pore water pressure evolution at the points of the piezometers installed in boreholes A2 and C2 is given in Fig. 16, in which the in situ measurements are also presented for comparison. It can be seen that the pore water pressure evolution obtained by the Mohr–Coulomb model agrees well with the general tendency of the

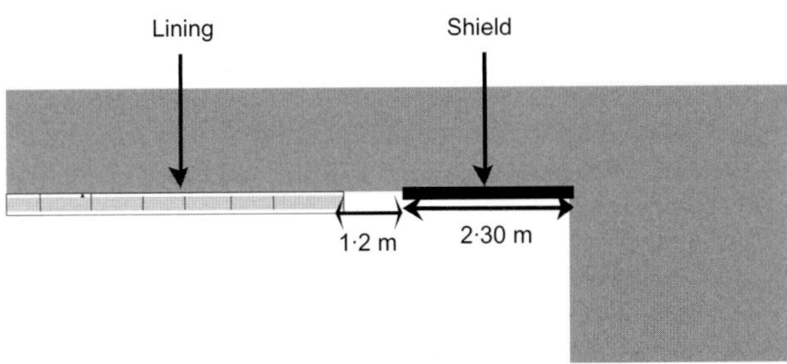

Fig. 15. Modelling sequence of connecting gallery: excavation, lining installation

Table 3. Initial stress and pore pressure in Boom Clay around gallery

Total stress, $\sigma_0 = \sigma_h = \sigma_v$: MPa	4·5
Pore pressure, u_w: MPa	2·25
Effective stress, σ_0': MPa	2·25

in situ measurements. The pore pressure increase is due to the undrained contractant behaviour of the clay (Biot coefficient $\alpha = 1$), while the drop phenomenon results from the high decompression of the formation (volumetric dilatations). However, the numerical simulations did not produce the extended hydraulic disturbed zone, and underestimated the radial variation of the pore pressure.

The measurements of the instantaneous radial convergence of the clay through the small holes in the shield indicated that the shield was in contact with the clay at its rear, over a distance of about one third to one half of its total length. They agreed very well with the modelled convergence (Fig. 17).

The modelled total axial displacement agreed with the in situ measurements (Fig. 18), but, again, the far-field behaviour was not reproduced numerically, because the sensors react very late compared with the in situ measurements (Bernier *et al.*, 2002; Li *et al.*, 2006).

CONCLUSIONS AND DISCUSSION

One important achievement in the study of the feasibility of radioactive waste disposal in Boom Clay is the demon-

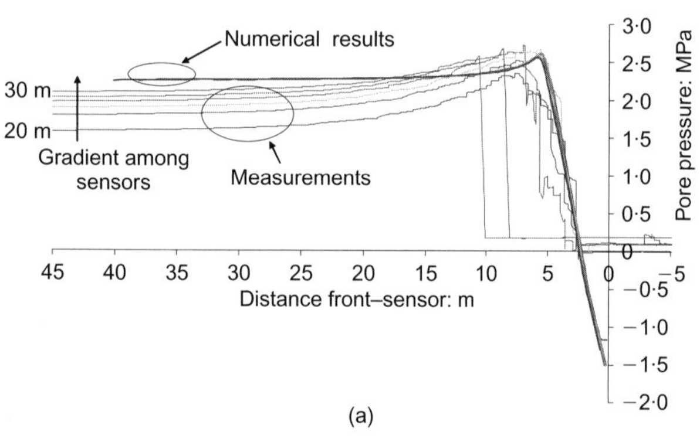

(a)

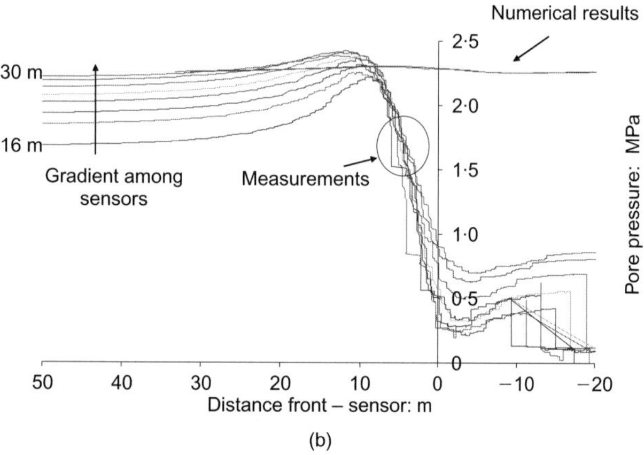

(b)

Fig. 16. Pore water pressure evolution ahead of excavation front (comparison between numerical predictions and in situ measurements). (a) Borehole A2 (installed horizontally along axis of gallery): sensors located between 20 and 30 m from test drift front. (b) Borehole C2 (installed in a vertical plan): sensors located between 16 and 30 m from test drift front

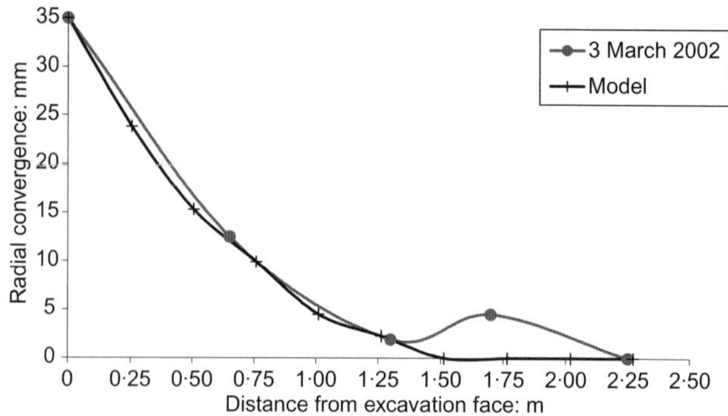

Fig. 17. Instantaneous radial convergence prediction

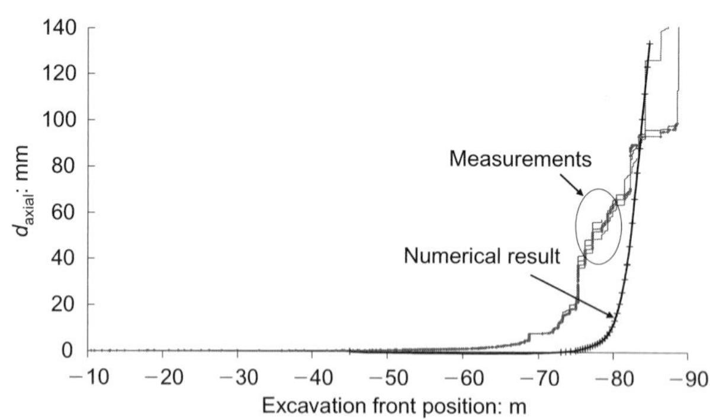

Fig. 18. Axial convergence prediction

stration that the galleries can be constructed while keeping the disturbance of the host rock at an acceptable level for the long-term safety of the disposal. Excavation in plastic clays at great depth is unusual because of the large convergences and the high confining stresses. The knowledge acquired since the beginning of the construction of the HADES underground laboratory has allowed the building of a valuable geotechnical database, and has led to the development of improved excavation techniques that significantly reduce perturbation of the digging operations in the surrounding host rock.

The instantaneous hydromechanical responses of the Boom Clay around excavation have been characterised. Convergence and pore water pressure measurements over an extended period of time allowed characterisation of its delayed effects. Comparison between in situ measurements and modelling results allowed a continuous improvement of knowledge of Boom Clay behaviour. Important issues in interpreting the measurements correctly are good control of the excavation parameters and the boundary conditions.

During excavation of a gallery, the pore pressure evolved regularly with the coming excavation front. One important finding was the unpredicted hydraulic perturbation at a distance of about 60 m from the excavation front, inside the formation. This behaviour was never observed in the past, because piezometers were not placed so far from the excavation front. Therefore this surprising behaviour could also apply for other clayey formations.

Numerical results showed that the pore pressure evolution with advancing excavation front calculated by assuming a linear elastic–perfectly plastic Mohr–Coulomb model agrees with the general tendency measured in situ, but both the extent and the magnitude ahead of the excavation front are undervalued. Numerical sensibility studies using more sophisticated constitutive laws such as Dafalias–Kaliakin (Kaliakin & Dafalias, 1990), which can take into account progressive transition of elasto-plasticity and thus the above-mentioned highly non-linear stress–strain behaviour, showed that they can actually produce a larger hydraulically influenced zone, but also larger displacements. This contradiction is still a pending issue for proper understanding of the hydromechanical behaviour of the Boom Clay. Thorough laboratory and numerical studies are continuing in order to understand this observed far-field hydromechanical behaviour better. They are concentrated on developing a hydromechanical model that considers the fracture process, suction effects, and the delayed effects through the viscosity of the clay skeleton.

The hydromechanical behaviour of the formation is influenced by fractures that developed during the excavation of the underground laboratory. Moreover, a slight desaturation was observed, and suction was created around the excavation, which made the Boom Clay stiffer. These effects show the difficulty of making a clear distinction between soft and stiff clays, as their behaviour depends strongly on the hydromechanical conditions to which they are submitted.

An increase of hydraulic conductivity was observed up to 6–8 m radially from the lining. This increase was attributed to the lower effective stress in this zone. Time evolution of the hydraulic conductivity indicated clearly the time-dependent coupled hydromechanical behaviour of the clay.

More specifically, the viscosity of the skeleton (creep, relaxation, etc.) and pore water pressure dissipation imply long-term effects. Convergence of the test drift lined with blocks segments separated by compressible Linex plates is still continuing, some 18 years after excavation, at a rate of about 0·5 mm/year.

The hydromechanical behaviour of Boom Clay is now well characterised, but understanding of the pore water variation in the far field needs more investigation. Long-term measurements and determination of the hydraulic characteristics of the host rock around tunnels are of prime importance for assessing the performance of a disposal for radioactive wastes.

ACKNOWLEDGEMENTS
Some results presented in this paper were obtained by the CLIPEX and SELFRAC projects. These projects were co-funded by the European Commission within the fourth and fifth framework programmes, key action: Nuclear Fission. This support is gratefully acknowledged.

REFERENCES
Baldi, G., Borsetto, M. & Hueckel, T. (1987). *Calibration of mathematical models for simulation of thermal, seepage and mechanical behaviour of Boom clay: Final report*, EUR 10924 EN. Luxembourg: Commission of the European Communities.

Baldi, G., Hueckel, T., Peano, A. & Pellegrini, R. (1991). *Developments in modelling of thermo-hydro-geomechanical behaviour of Boom clay and clay-based buffer materials: Final report*, EUR 13365 EN. Luxembourg: Commission of the European Communities.

Barnichon, J.-D. & Volckaert, G. (2003). Observations and predictions of hydromechanical coupling effects in the Boom Clay, Mol Underground Research Laboratory, Belgium. *Hydrogeol. J.* **11**, No. 1, 193–202.

Bastiaens, W. & Bernier, F. (2006). 25 years of underground engineering in a plastic clay formation: the HADES underground research laboratory. In *Geotechnical aspects of underground construction in soft ground* (eds K. J. Bakker, A. Bezuijen, W. Broere and E. A. Kwast), pp. 795–801. London: Taylor & Francis.

Bastiaens, W., Bernier, F. & Li, X. L. (2006a). SELFRAC: Experiments and conclusions on fracturing, self-healing and self-sealing processes in clays. *Physics and Chemistry of the Earth.* 2006, doi:10·1016/j.pce.2006·04·026, in press.

Bastiaens, W., Bernier, F. & Li, X. L. (2006b). An overview of long-term HM measurements around HADES URF. *Proceedings of the international symposium on multiphysics coupling and long term behaviour in rock mechanics*, Liège, pp. 15–26.

Bernier, F., Li, X. L., Verstricht, J., Barnichon, J. D., Labiouse, V., Bastiaens, W., Palut, J. M., Ben Slimane, J. K., Ghoreychi, M., Gaombalet, J., Huertas, F., Galera, J. M., Merrien, K., Elorza, F. J. & Davies, C. (2002). *CLIPEX: Clay Instrumentation Programme for the Extension of an Underground Research Laboratory*, EUR 20619. Luxembourg: Commission of the European Communities.

Bernier, F., Li, X. L., Bastiaens, W., Ortiz, L., Van Geet, M., Wouters, L., Frieg, B., Blümling, P., Desrues, J., Viaggiani, G., Coll, C., Chanchole, S., De Greef, V., Hamza, R., Malinsky, L., Vervoort, A., Vanbrabant, Y., Debecker, B., Verstraelen, J., Govaerts, A., Wevers, M., Labiouse, V., Escoffier, S., Mathier, J.-F., Gastaldo, L. & Bühler, Ch. (2006). *SELFRAC: Fractures and self-healing within the excavation disturbed zone in clays: Final technical report.* Luxembourg: Commission of the European Communities (to be published). Report 2007-02-13, EUR 22585.

Bolton, M. D. & Whittle, R. W. (1999). A non-linear elastic/perfectly plastic analysis for plane strain undrained expansion tests. *Géotechnique* **49**, No. 1, 133–141.

Coll, C. (2005). *Endommagement des roches argileuses et perméabilité induite au voisinage d'ouvrages souterrains.* PhD thesis, Université Grenoble I.

Djéran, I., Bazargan, B., Giraud, A. & Rousset, G. (1994). *Etude expérimental du comportement thermo-hydro-mecanique de l'argile de Boom, Rapport final*, Rapport No. 94-002. Mol: SCK.CEN.

Henrion, P. N., Monsecour, M., Fonteyne, A., Hermans, N., Manfroy, P., Neerdael, B. & de Bruyn, D. (1984). *Characterisation of the Boom Clay, Semi-annual report 18*, Contract WAS-334-83-7-B(RS), R&D Programme on Radioactive Disposal into Geological Formation (Study of a clay formation). Mol: SCK.CEN.

Horseman, S. T., Winter, M. G. & Entwistle, D. C. (1987). *Geotechnical characterisation of Boom Clay in relation to the disposal of radioactive waste: Final report*, EUR 10987. Luxembourg: Commission of the European Communities.

Kaliakin, V. N. & Dafalias, Y. F. (1990). Theoretical aspects of the elastoplastic-viscoplastic bounding surface model for cohesive clays. *Soils Found.* **30**, No. 3, 11–24.

Labiouse, V. (1997). Advanced hydro-mechanical modelling of excavations in the Boom clay formation. *Extended abstracts of the Boom clay seminar*, Alden Biesen, pp. 45–61.

Li, X. L., Bastiaens, W. & Bernier, F. (2006). The hydromechanical behaviour of the Boom Clay observed during excavation of the connecting gallery at Mol site. *Proceedings of the multiphysics coupling and long term behaviour in rock mechanics*, Liège, pp. 467–472.

Mair, R. J., Taylor, R. N. & Clarke, B. G. (1992). *Repository tunnel construction in deep clay formations: Final report*, EUR 13964. Luxembourg: Commission of the European Communities.

Rousset, G., Bazargan, B., Ouvry, J. F. & Bouilleau, M. (1993). *Etude du comportement différé des argiles profondes: Rapport final*, EUR 14438 FR. Luxembourg: Commission of the European Communities.

INFORMAL DISCUSSION

Session 3
Characterisation of Formations and Material Models

CHAIR: DR FEDERICA COTECCHIA AND DR LIDIJA ZDRAVKOVIC

Dr Lidija Zdravkovic *Chairman*

Within this discussion session, we cannot cover all the issues that emerged from the papers in the presentation. We have decided to structure our discussion giving some initial comments and then opening afterwards to your contributions. We have chosen two main characteristics of general soil behaviour that are important for geotechnical design. These are the stiffness, which is relevant when assessing serviceability limit states of structures, and strength, which comes into play when looking at ultimate limit states.

To start with the stiffness of stiff clays, the first question is related to the size of the small strain elastic region of stiff clays which has only recently been measured and reported for London Clay. Just to remind you of Dr Cotecchia's earlier presentation, what was surprising for us was the size of the elastic region in terms of strain level, which is smaller than the working strain range for typical structures. Also in terms of stresses, Dr Cotecchia has indicated that the elastic region for London Clay is limited to stress changes of less than 2 kPa. Therefore the first question we would like to pose to the authors is to give us your opinion on the size of this elastic region—why is it so small, and do they have any indication of similar behaviour from other stiff clays? This issue has clear implications for our constitutive modelling of stiff clays, geomaterials and other materials.

Dr Apollonia Gasparre, *Geotechnical Consulting Group*

I would like to address the question of the size of the elastic region. Our work was undertaken on London Clay. We undertook stress probes within the elastic region to determine whether there was a region of purely elastic response and to determine the size of that region. We found that the size of a purely elastic region is smaller than expected, it depends on the effective stress, p' and it does not exceed about 2 kPa in radius. However, athough beyond this region the strains are no longer fully recoverable there is little degradation of stiffness until a second kinematic surface, Y2, is engaged. This surface seems to be the onset of more rapid stiffness degradation and more significant changes to the clay structure and behaviour. In practical terms, therefore, the Y2 surface seems to be more relevant than the elastic region.

Professor Richard Jardine, *Imperial College*

We have made similar measurements of the Y1 'elastic' behaviour with other soil types, ranging from sands to clays, which generally show that elastic behaviour is confined to a very small strain range. However, the size of the elastic region doesn't seem to be uniquely related to p' for all materials. For example, Bothkennar Clay which is much softer than London Clay, shows a Y1 region of about the same absolute size in stress space although the consolidation

stresses involved in the tests were much lower. The extent of the Y1 zone also appears to vary with strain rate. Professor Tatsuoka, who has made a substantial contribution in this area, holds that the size of the elastic region grows when you perform a test more quickly, and will shrink if you go really slowly. In addition, if samples are left at a given stress state for a geologically significant period of time they may well undergo cementing and other changes that could modify the Y1 behaviour.

Dr Apollonia Gasparre, *Geotechnical Consulting Group*

We performed drained and undrained small strain probes using generally smaller strain rates. The drained probes were carried out at strain rates that were generally lower than the rates of undrained tests in order to ensure a fully drained behaviour of samples with such low permeability as London Clay. Within the range of strain rates that we used and within the range of strains over which the investigation was focused, we were unable to observe clear strain rate effects. This is not surprising though and there are a number of literature data that show the effects of strain rates are less recognisable at small strains and become more prominent at larger strains.

Professor Richard Jardine, *Imperial College*

Considering the stiffness behaviour developed in practical field problems, Professor John Burland and I worked with colleagues (Jardine *et al.*, 1985) to correlate our early 'small-strain' laboratory tests on London Clay with in-situ measurements made by BRE and others. The field tests involved carefully instrumented loading tests on plates, footings and piles. All employed local strain measurements, and gave results that were broadly compatible with our non-linear laboratory data, including the absence of any resolvable practical linear elastic range. Companion field seismic velocity measurements indicated dynamic stiffnesses only marginally greater than the credible 'small-strain' maxima seen in the lab and field experiments. We can now resolve the very small strain Y1 behaviour, and make successful direct comparisons between geophysical and static parameters (see Hight *et al.*, 2007). However, the main message from our earlier field tests remains that the operational elastic range is very small and is easily exceeded in a wide range of practical loading cases.

Dr Brian Simpson, *Arup*

It might be completely irrelevant but it just occurred to me that if we think of the small strain region as $10^{-3}\%$ or 10^{-5} absolute strain and multiply that by the size of the specimen we get roughly to the size of a particle. I wonder whether the size of the yield surface is in any way related to the size of the particles or to the size of the specimen.

Dr Angelo Amorosi, *Technical University of Bari, Italy*

Following Professor Potts' example I am inspired to ask a provocative question. If the true elastic range of behaviour in stiff clays is so small, why are we spending so much time and effort in trying to understand it? I also have a partial answer to my question—as I have spent some time researching in this area. In many field projects we are dealing with problems which involve undrained behaviour at stresses well below ultimate limit states. Such behaviour can readily be modelled if we use, for example, multi-surface plasticity-based models. In this case, even when we are out of the true elastic range, the response is still very much influenced by the assumptions we make in the elastic range.

Dr Lidija Zdravkovic *Chairman*

The linear small strain bubble will still exist within the large strain yield surface. The question is whether it will affect the degradation of stiffness at all if the elastic zone is so small.

Dr Matthew Coup, *Imperial College*

Dr Simpson mentioned possible particle size effects. We have done very similar tests to those discussed here but using sand. The sand has a very different particle size. The size of the Y1 yield surfaces change a little bit but not much.

Dr Lidija Zdravkovic *Chairman*

Staying with elastic stiffness but looking now at anisotropy of stiffness—this has been quite precisely measured for London Clay as reported in the Symposium papers and as Dr Cotecchia pointed out. For example, in Figures 12 and 14 of Hight *et al.*, (2007) the ratio of horizontal to vertical stiffnesses for London Clay has been measured to be 2, and even up to 2·5, which is quite different from most of the stiff clays reported in the literature where this ratio is about 1·5 or less. Do we have any answers as to why London Clay has such high stiffness anisotropy compared to other stiff clays?

Professor Richard Jardine, *Imperial College*

I was going to ask the same question regarding stiffness anisotropy of our colleagues at Bristol who made earlier studies of the Gault Clay and other more heavily over-consolidated clays.

Dr David Nash, *University of Bristol*

As I recall, the typical ratio of G_{hh}/G_{hv} for Gault Clay is about 1·6. London Clay is a younger material geologically so I suppose a possible explanation is that the fabric is very well compacted. Potentially, the particle orientation may correlate with the ratio of horizontal to vertical stiffness. But there may well be other factors.

Professor Richard Chandler, *Imperial College*

I would like to pose a question. The first yield of the soil occurs at very tiny strains. We are doing these tests on samples we have taken out of the ground where the total stress change is enormous compared to the stress changes which we are observing in the laboratory at first yield. Are we not doing tests on samples that are already irretrievably damaged by the sampling process?

I would also like to comment on observed stiffness anisotropy. I would agree with the traditional explanation that in the London Clay it is associated with relatively high horizontal in-situ stresses. In this formation, the current vertical stress is much less than the current horizontal stress so E'_h is greater than E'_v. Other stiff clays with their lower stiffness ratios are older and have been subjected to tectonic effects, which the London Clay being much younger has not experienced to the same extent. I suspect the stiffness anisotropy of the London Clay is simply a reflection of the stress history and current in situ stress anisotropy.

Dr Kenichi Soga, *Cambridge University*

I think what is interesting for the data on London Clay is that E'_h/E'_v ratio of 2 is similar to the ratio G_{hh}/G_{vh} which is also 2. I recall that Professor Houlsby published some work on cross-anisotropy in soils. I recall his suggestion that the square root of E'_h/E'_v. should be equal to G_{hh}/G_{vh}.

Dr Lidija Zdravkovic *Chairman*

I understood that these data fit very well with relationships required for a cross-anisotropic model to an accuracy of about 12%.

Dr Martin Lings, *University of Bristol*

Professor Jardine was asking about Gault Clay. Derek Pennington measured ratios of G_{hh}/G_{vh} which were about 2. The inferred ratios of E'_h/E'_v were about 4. As Professor Soga has reminded us, the approximate Graham & Houlsby model requires relationships between the parameters that happen to fit those data. However, I would like to comment on the London clay data from Gasparre *et al.* (2007) Fig. 5. The figure shows the ratio of G_{hh}/G_{vh} to be constant with depth and about 2, but the ratio of E'_h/E'_v to be markedly increasing with depth. I see no reason why the two ratios should be the same, or have a particular relationship with one another. They are after all independent stiffness parameters.

Dr David Nash, *University of Bristol*

Could I pick up Professor Chandler's point about the tortuous stress history the samples have been through. I think we can distinguish between what happens to the shear moduli and what happens to the Young's moduli. I experienced, when testing Gault Clay, that samples can be subjected to very large stress excursions and then come back to the initial stress state and still retain the initial values of shear moduli. That suggests that provided there has not been any significant gross damage caused by the sampling process, just stress changes, maybe the G values that we recover in the laboratory are not too badly affected. However, I would not have the same confidence regarding for the Young's modulus values.

Dr John Powell, *Building Research Establishment*

I would observe that we are talking about stiffness ratios determined in laboratory but measured field ratios are the same. With London Clay and Gault Clay we get observed ratios of G_{hh}/G_{hv} around 2 which corroborate the laboratory data.

Dr Apollonia Gasparre, *Geotechnical Consulting Group*

I would like to comment on the quality of our samples. We compared our shear moduli as calculated from bender elements in laboratory tests with the shear moduli measured on site and they are in good agreement. This suggests that there is no significant disturbance affecting our laboratory data. Also during the approach path followed in the laboratory to return samples to the in situ condition we checked if G was changing. We found it was not changing significantly so I do not think we disturbed the samples appreciably. Regarding stiffness degradation—I found that Y1 is very small. I think that in this area thermal effects may play a significant role in stiffness degradation. However, the Y2 boundary is probably more significant. We have shown that stress paths which move beyond Y2 seem to affect the sample's 'memory' of recent stress history events. So, if you move within Y2, the sample retains the full stress history and the stiffness does not degrade too much. It seems to me that Y2 is much more significant in terms of stiffness degradation and memory of stress history.

Dr Lidija Zdravkovic *Chairman*

I would like to extend the debate on stiffness anisotropy and pose a question as to how we take the anisotropy observed in small strains into larger strains? This is important if we are going to use anisotropic non-linear stiffness models or a kinematic yield surface models for analysing engineering problems. Do we know how these moduli decay at large strains—does the ratio of 1·5 to 2·0 for G_{hh}/G_{vh} remain constant at larger strains or does this ratio vary as strains increase?

Dr Minh Nyguyen, *Atkins*

I can comment on the stiffness anisotropy of London Clay at large strains. This behaviour can be readily investigated by the use of the hollow cylinder apparatus. The stiffness ratios certainly do change at higher strain levels.

Dr Lidija Zdravkovic *Chairman*

So you were able to measure independently the vertical and horizontal Young's moduli E'_v and E'_h with increasing strain levels?

Dr Minh Nyguyen, *Atkins*

We were able to make these measurements and found that the ratio of E'_h/E'_v decreased as the strain levels increased.

Dr Lidija Zdravkovic *Chairman*

Was the change of E'_h/E'_v a linear function of strain or was it non-linear?

Dr Minh Nyguyen, *Atkins*

We haven't done a thorough analysis to see whether the change was linear or non-linear but initial evidence suggests that it is not a linear function.

Dr Vojkan Jovicic, *IRGO Consulting*

In our paper from 1998 (Jovicic and Coop, 1998), it is demonstrated that we measured, for the first time, the in situ anisotropy of several sites in London Clay and afterwards we measured stiffnesses at very small and at larger strains of the appropriate samples in the laboratory. The ratio we measured at very small strains for E'_h/E'_v was about 1·5 in London Clay, both in situ and in the laboratory. Then we looked at the effects of plastic deformation on this anisotropy. We then compressed the soil isotropically and initially we found a stiffness ratio of about 1·5 which dropped to 1·0 at larger strains. I consider that the elastic stiffness anisotropy is not very relevant once plastic strains occur.

Dr Lidija Zdravkovic *Chairman*

So if we want to implement stiffness anisotropy in our analysis from small to large strains we have to incorporate in some way the reduction in the ratios with increasing strain.

Dr Vojkan Jovicic, *IRGO Consulting*

Yes, but I might point out that we measured the small strain shear moduli with bender elements at the same time that we imposed large triaxial strains. We found that the G_0 ratios only changed slightly with large global strain increases.

Professor Richard Jardine, *Imperial College*

The relationship between stiffness degradation and strain inevitably depends on the stress path followed. Imagine a London Clay element experiencing a high K_0 stress system. We have seen that it would both have a relatively high horizontal elastic stiffness and require only relatively small increases in horizontal effective stress to reach its failure envelope. The degradation patterns seen in our lab tests indicate stiffness reducing dramatically over relatively small strains under horizontal loading. In the reverse case, if you reduce the effective horizontal stress or increase the vertical effective stress, the stress point moves away from the failure envelope and leads to gentler stiffness-strain reductions. The experiments show that the initial stiffness anisotropy is very important in setting the initial pattern of stress-strain response, but that the stiffness-strain curve is affected by both the proximity of the boundary surface and the direction of the applied stress path. Time and rate effects are also significant, as noted earlier.

Dr Lidija Zdravkovic *Chairman*

The final question that we have in this group of questions is related to Figure 15 of Hight *et al.* (2007) for London Clay which we have already seen a couple of times today. This shows the difference in relationships of stiffness normalised for the stress level versus axial strain between good quality commercial tests and the new data from the recent

Imperial College study. The Imperial College study seems to imply lower stiffnesses than those from the previous studies, as well as a slower decay of stiffness with increasing strain. The previous stiffness curves have been widely used to predict ground movements around geotechnical structures. They have given very good predictions. The question is do we carry on using our old established database or should we start using this new data? What are the implications on predications of ground movements from numerical analyses using the new data?

Dr Benoit Jones, *Mott MacDonald*

I wonder why the apparently lower stiffness of London Clay found by Hight *et al.* couldn't be attributed to the sampling methods employed. The stiffness decay curves were based on tests on rotary cored samples, which were potentially softened by the coring process. The suction probe measurements on rotary cored samples, not given in the paper but available to those working on the T5 project, gave K_0 values less than 1·0, possibly indicating that the clay had taken on water during coring. Could this have led to the low values of stiffness measured in the laboratory? The K_0 values presented in the paper were based on suction probe measurements on thin-walled samples, which were considered too disturbed for use in stiffness and strength testing. Could this have led to an overestimation of K_0 and in turn p_0'? When the stiffness was normalised to p_0', it would have been further reduced relative to previously published data. Either one of these factors, an overestimation of K_0 due to disturbance during thin-walled sampling or an underestimation of stiffness due to softening during rotary coring, acting individually or in concert, could have caused the lower values of normalised stiffness relative to the previously published values.

Dr David Hight, *Geotechnical Consulting Group*

Before you spend too much time worrying about this, perhaps I should give you some background to Figure 15. First of all, the new lower stiffness band is significantly lower than what is measured in commercial tests. The higher band is very wide because of creep effects in standard commercial tests. These test specimens were not left long enough to reduce creep rates with the result that when the direction of shearing was reversed the stiffnesses were overestimated. The new data did not follow the same testing procedures as standard commercial tests. The IC tests were run at a strain rate of one tenth of that which was used in commercial tests. Let me also explain about the K_0 measurements. These were done on thin walled tube samples which were extruded immediately on site—the outer 10mm was stripped off and measurements made immediately using the Imperial College suction probe. There was no empirical 10% reduction of suction made when deriving K_0. It is possible that the measured suctions are a little high. Our comfort in those measurements was that they were very similar to the measurements made by Professor Bishop at Ashford Common. As to why the previous modulus decay curves have worked in the past, the curves that we used in our analyses did not take into account anisotropy. They were measured at a strain rate that doesn't apply in the field. Despite this, the stiffnesses at strain levels which generate most of the displacements in geotechnical structures have not changed very much. I think we can continue to use the analytical methods and models we have been using up to now but there is no doubt that there is a need to familiarise ourselves with the new data.

Dr Lidija Zdravkovic *Chairman*

Except that, as we said earlier, we don't know how to introduce the anisotropy of stiffness of strain level.

Dr David Hight, *Geotechnical Consulting Group*

Our approach to anisotropy of stiffness will remain empirical until we can take into account the true strain rate effect and deal with the evolution of anisotropy with strain levels.

Dr Lidija Zdravkovic *Chairman*

We will now move to our remaining group of questions on the strength of stiff clays. Most of the Symposium papers which address the strength of stiff clays relate the strength to the composition and structure of the clay. Papers on the London Clay brought into the story the effects of fissuring. In this regard Dr Cotecchia has drawn attention to a paper which illustrates the fissure pattern at various depths within the London Clay. We have seen quite a large variability of strength through the deposit which is apparently explained by the effects of fissuring. I would like to ask the authors first if they can tell us whether they were able to quantify the effects of fissuring in particular with regards to the strength anisotropy of the material. We have seen both variations of undrained strength shown in the papers and also variation in effective strength parameters with alpha. This work prompts a further question as to how significant is the interplay between the structure, composition and fissuring on the one side and the inherent anisotropy on the other side. Given the picture which is emerging on the strength anisotropy, is this an important issue to address for engineering in London Clay since we have such a range of factors affecting the strength in each location?

Professor Richard Jardine, *Imperial College*

Dr Nishimura is the best person to answer this point. His interpretation of the Hollow Cylinder tests noted that the observed strength anisotropy has at least two causes. One root is the platy micro-fabric of the material and its system of interparticle contacts. These features controlled the initial stiffness anisotropy, and the initial directions of the effective stress paths developed under undrained testing in a wide variety of modes. They would also control the shear strength anisotropy of intact, un-fissured, clay. The second clear root of the shear strength anisotropy was the effect of fissure patterns on shear band initiation and propagation. In certain shearing modes the effect of the fissures came into play earlier than in others, but it is clear that both micro-and macro-fabric factors combined to produce the anisotropic relationships reported in our paper. We also report that the intermediate principal stress had a significant effect on shear strength; the thesis by Nishimura (2006) gives far more comprehensive discussion on this point.

Dr Lidija Zdravkovic *Chairman*

We observe a similar pattern in the change of strength

with changes of b and alpha for both drained and undrained tests. However we find a few samples that do not fit that trend. What is the likely explanation for the behaviour of those samples?

Professor Richard Jardine, *Imperial College*

Our study involved a total of 58 Hollow Cylinder tests, two of which didn't fit the overall picture well. While we checked and reported these tests, we don't offer any firm explanation for their atypical behaviour—although sample scale may play a role in one case. Unfortunately, we could not repeat the two outlying experiments because we had exhausted our stock of both samples and time. Considering the heterogeneity of the clay fissuring and the difficulty of forming samples, the observed degree of deviation may not be surprising.

Professor David Potts, *Imperial College*

A possible explanation of the scatter in the data can be found if we review the nature of the samples in respect to the manner in which they are tested in the laboratory. The fissures in the clay are essentially planar features, however our standard laboratory tests (oedometer, triaxial and hollow cylinder) use cylindrical samples and assume axi-symmetric behaviour when analysing the results. In the general situation, in which some of the fissures are not horizontal, the assumption of axi-symmetric conditions is clearly wrong. This, in turn, implies that the test results will depend not only on the number of fissures in the sample, but also on their orientation and position with respect to the axis of geometric symmetry of the sample. It is therefore not surprising that we do not observe consistent behaviour between different samples.

This raises the provocative question as to whether or not it is worth performing standard laboratory tests at all. Perhaps we should be attempting to measure the behaviour of the intact soil between the fissures and that of the fissures independently. However, even if we were to achieve this, we would still be faced with the dilemma of how such data could be used to develop a constitutive model for use in the analysis of real geotechnical structures.

Dr David Hight, *Geotechnical Consulting Group*

I am not going to answer Professor Potts, other than to say that fissures are very important and that we are lucky in the London Clay that we saw at T5, the fissures having preferred orientations they are sub horizontal and sub vertical. Figure 17 of our paper shows the anisotropy of stiffness and illustrates the point that Professor Jardine was making. The anisotropy of the effective stress of a hard clay seems to determine the anisotropy of small strain stiffness. Fissures determine the location of the strength envelope. Another of our figures shows the various strength envelopes of the different geological units—these failure envelopes were based on test data from 100 mm diameter samples. We still don't know the larger scale mass strength. This is where the numerical modellers can assist. If we know the strength, orientation and length of fissures on bulk strength and we know about their orientation and we know about the behaviour of the intact material between the fissures, the material modellers should be able to tell us how the mass behaves.

James Galway, *Mott MacDonald*

The figure which illustrated fissures showed what in geology would be called joints. I'm not quite clear what the difference is between a fissure and a joint.

[Unnamed contribution]

In stiff clays such as the London Clay the joints form in two different stages. Fissures form not long after sedimentation and later are modified and extended by tectonic processes. They will form a 3-dimensional pattern of discontinuities. So it is self evident that if laboratory tests are performed on material obtained from between the fissures, the results will not be representative of the mass behaviour. There needs to be some kind of interaction with plate tests or other large scale testing to relate the lab test results to the bulk properties of the material in the field. The other point worth noting is that everywhere in the world there is an uneven stress field because the effects of plate tectonics. The most recent borehole breakout data for North West Europe indicated that the principle direction of the compressive stress field is North West to South East. This must be present in the London Clay. It's more obvious when you get borehole samples from greater depths. This must have some influence on test results which may become apparent when the tests get sufficiently sophisticated to recover stiffness and strength anisotropy of the same order of magnitude as the anisotropy of these different stress fields.

Dr Lidija Zdravkovic *Chairman*

The work presented in this Symposium has significant implications for constitutive modelling of stiff clays on the dry side of the critical state. How should we model the strength envelope and associated behaviour on the dry side? Most of our current constitutive models incorporate a symmetrical bounding surface in the shape of an ellipse. In reality stress paths on the dry side of critical state will not reach the elliptical bounding surface and will not dilate to the critical state as the stress path proceeds beyond the peak shear stress. How should we deal with destructuring from peak strength to post-peak strength—any comments?

Dr Sarah Stallebrass, *City University*

[Dr Stallebrass provided a response which indicated the drawbacks of standard models and suggested how the issue could be addressed using kinematic yield surface models on the dry side due to the developed plasticity below the bounding surface, which as a result reduces the predicted strength.]

Dr Lidija Zdravkovic *Chairman*

It seems to me that incorporating a realistic scheme for destructing behaviour on the dry side of the critical state is only part of the solution. It is not uncommon to see stress paths which do not converge to the critical state.

Dr Angeliki Grammatikpoulou, *Geotechnical Consulting Group*

In my experience it is true that kinematic hardening

models give improved predictions of the peak strength of overconsolidated clays as compared to the modified Cam Clay model. However, the stress path in these models will still travel all the way to the critical state, which is something not generally observed in stiff clays.

Dr Lidija Zdravkovic *Chairman*

At the moment we don't have these capabilities in standard models. Therefore we may conclude with the observation that there is plenty of work still to be done to improve the current material models for stiff clays.

Session 4

Engineering in Stiff Clays

Kovacevic, N., Hight, D. W. & Potts, D. M. (2007). *Géotechnique* **57**, No. 1, 63–74

Predicting the stand-up time of temporary London Clay slopes at Terminal 5, Heathrow Airport

N. KOVACEVIC[*], D. W. HIGHT[*] and D. M. POTTS[†]

The paper describes the results of finite element analyses of the temporary slope geometries in London Clay at London Heathrow Airport's Terminal 5. The aims of the analyses were to examine the times before failures developed, and to identify the failure mechanisms involved. The brittle behaviour of the London Clay was modelled, and the effects of progressive failure were taken into account. The possible presence of tectonic shears with their strength close to residual was considered by comparing analyses with and without a tectonic shear zone. The predicted time to failure of the slopes and the form of the failure, whether shallow or deep-seated, were determined by a combination of the assumed permeability profile and whether or not allowances were made for increases in permeability as the clay swelled, the average surface suction, the in situ K_0 profile, the depth of excavation, and whether or not a low-strength tectonic shear surface was present in the slope. The analyses fall into the category of Class A predictions, and were used in the assessment of how long temporary slopes to deep excavations could be left open before backfilling, and how the slopes should be monitored.

KEYWORDS: clay; excavation; numerical modelling; slopes; suction; time dependence

Cet article décrit les résultats d'analyses à éléments finis des géométries de versant temporaire dans la London Clay (argile de Londres) au Terminal 5 de l'aéroport d'Heathrow de Londres. Il modélise le comportement fragile de l'argile de Londres (London Clay) et prend en considération les effets de rupture progressive. La présence éventuelle de cisaillements tectoniques dont la force est proche de résiduelle a également été prise en compte en comparant des essais avec et sans zone de cisaillement tectonique. Le temps estimé de rupture des versants et la forme de la rupture, profonde ou peu profonde, ont été déterminés par association du profil de perméabilité supposé et la possibilité ou non d'augmentations de la perméabilité à mesure que l'argile gonfle, de la succion en surface moyenne, du profil de K_0 in situ, de la profondeur de l'excavation et de la présence ou non sur le versant d'une surface de cisaillement tectonique de faible force. Les essais entrent dans la catégorie des prédictions de Classe A. Ils ont été utilisés pour évaluer combien de temps des versants temporaires allant jusqu'à des excavations profondes peuvent rester ouverts avant de les remblayer et estimer comment contrôler ces versants.

INTRODUCTION

The majority of the below-ground construction at London Heathrow Airport's new Terminal 5 (T5) was carried out in open-cut excavations made in the London Clay Formation (hereafter London Clay) which underlies Terrace Gravels. There was a need to adopt slopes that were as steep as possible, to minimise the excavation and backfill volumes, but which would be stable for periods up to 6 months. Temporary slope geometries were selected on the basis of experience and precedents for other, usually shallower, short-term slopes cut in the London Clay (e.g. Chandler, 1984).

It is extremely difficult to assess the stability of temporary slopes cut in stiff overconsolidated plastic clays, such as London Clay, partly because they are brittle and prone to progressive failure. Nevertheless, it is still common to design these slopes using limit equilibrium (LE) methods of analysis. However, progressive failure is a strain-dependent phenomenon, and thus conventional LE stability analyses are not well suited for stability problems of this kind. In this respect finite element (FE) analyses are far superior. They do not impose a predetermined failure mechanism, and can assess operational strengths rationally on the basis of the development and distribution of strain within the slope.

Progressive failure can now be modelled using advanced numerical techniques, and these have been used to analyse the role of progressive failure in London Clay slopes in both the short and long term (Potts *et al.*, 1997; Vaughan *et al.*, 2004). The scale of the excavations at T5 (see Fig. 1) justified the use of these state-of-the-art techniques to take into account the effects of progressive failure in assessing the stand-up time of the temporary slopes.

Fig. 1. Plan view of excavations at Terminal 5

Manuscript received 5 May 2006; revised manuscript accepted 4 November 2006.
Discussion on this paper closes on 1 July 2007, for further details see p. ii.
[*] Geotechnical Consulting Group, London, UK.
[†] Imperial College, University of London, UK.

To calibrate the numerical methods, and to provide guidance on the likely hydraulic boundary conditions, two case histories of temporary slope failures local to T5 (Prospect Park and Wraysbury Reservoir) were also analysed (Kovacevic *et al.*, 2004) using the same techniques. Investigations of the temporary slope failure at Prospect Park by Chandler *et al.* (1998) revealed the presence and influence on that failure of tectonic shear surfaces, on which shear strengths were only a little above residual. As Prospect Park lies only 2 km to the north of the T5 site, identifying the possible presence of tectonic shear surfaces was an important aim of the ground investigations at T5, and allowance for their potential presence was made in the FE analyses.

The FE analyses were carried out on a 'standard' stepped section involving a series of 5 m high 1V:1H slopes in the London Clay, with 5 m wide gently sloping berms (1V:20H) between, and a maximum depth of 20 m (Fig. 2). Given the different excavation depths at T5, shallower excavation depths (approximately 15 m and 10 m) having the same section were also analysed.

GROUND CONDITIONS

Ground conditions at Terminal 5 are discussed in detail elsewhere (Hight *et al.*, 2003). A typical stratigraphical sequence is shown in Fig. 3. Approximately 4 m of Terrace Gravels overlie the London Clay. Below a depth of 45 m the London Clay is stiffer and much stronger (Unit A2).

In the ground investigations there were signs in the rotary cores of tectonic shears at depths between 13 and 15 m, and so the potential presence of a tectonic shear zone was modelled in the analyses at the depths shown in Fig. 3 (see also Fig. 2).

The best estimated profile of the coefficient of earth pressure at rest, K_0, is presented in Fig. 3. Bearing in mind the difficulties in determining K_0, the sensitivity of the predictions to its value was checked by adopting in some analyses a reduced K_0 profile, also shown in Fig. 3.

The groundwater table was in the Terrace Gravels, 2 m below the ground surface. Based on site measurements, a hydrostatic distribution of pore water pressures was adopted in the London Clay, there being no evidence of underdrainage.

SOIL MODELS AND MODEL PARAMETERS

The soil models employed in the analyses are described in some detail by Kovacevic *et al.* (2004). Whereas the Terrace

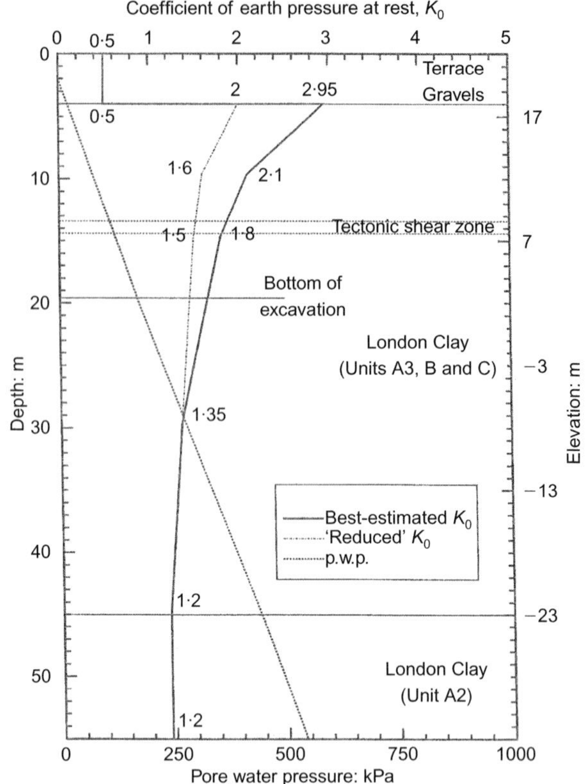

Fig. 3. Adopted soil stratigraphy, K_0 profiles and pore water pressure distribution

Gravels were modelled as a linear elastic perfectly plastic material, the London Clay was characterised by the generalised non-linear elastic strain-softening plastic model of the Mohr–Coulomb type in which effective stress shear strength parameters c' and ϕ' vary according to the plastic deviatoric strain invariant ε_D^p, defined as

$$\left(\varepsilon_D^p\right)^2 = \frac{2}{3}\left[\left(\varepsilon_1^p - \varepsilon_2^p\right)^2 + \left(\varepsilon_2^p - \varepsilon_3^p\right)^2 + \left(\varepsilon_3^p - \varepsilon_1^p\right)^2\right] \tag{1}$$

Pre-peak behaviour is formulated by a non-linear elastic model of the form described by Jardine *et al.* (1986) in which the secant shear (G^{sec}) and bulk (K^{sec}) moduli depend on both the current strain and stress levels according to

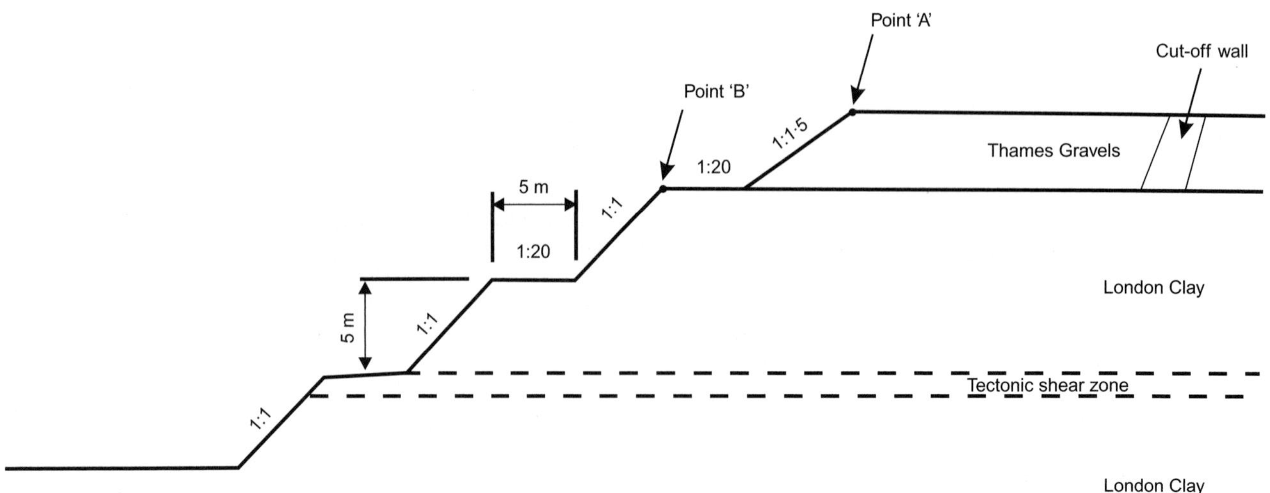

Fig. 2. 'Standard' section

$$\frac{G^{\text{sec}}}{p_0'} = A + B \cos\left\{\alpha\left[\log_{10}\left(\frac{\varepsilon_D}{3C}\right)\right]^{\gamma}\right\} \tag{2}$$

$$\frac{K^{\text{sec}}}{p'} = R + S \cos\left\{\delta\left[\log_{10}\left(\frac{\varepsilon_v}{T}\right)\right]^{\lambda}\right\} \tag{3}$$

where $p' = (\sigma_1' + \sigma_2' + \sigma_3')/3$ is the mean effective stress, ε_D is the deviatoric strain invariant defined by equation (1), $\varepsilon_v = \varepsilon_1 + \varepsilon_2 + \varepsilon_3$ is the volumetric strain, and A, B, C, R, S, T, α, γ, δ and λ are all constants.

The analyses were coupled—that is, they involved consolidation/swelling—and it was therefore necessary to specify permeability k. The database on reliably measured permeability values in London Clay (Hight *et al.*, 2003) is shown in Fig. 4(b). Overall, there is a trend for the permeability to decrease with depth, and this trend was modelled in the analyses. In the analyses of overall stability, the horizontal permeability was assumed to be three to five times higher than the vertical permeability, as shown in Fig. 4(a). More permeable zones, with even higher horizontal permeabilities, were assumed to coincide with zones where silt seams were observed. In these analyses no allowance was made for any changes in permeability of the clay during swelling, on the assumption that this effect would be restricted to a zone close to the surface of the slope, and changes in the body of the slope would be dominated by the zones of higher horizontal permeability. In contrast, for analyses of local stability, where changes in permeability close to the surface would be important, isotropic permeability was assumed and linked to the mean effective stress p' according to a non-linear relationship of the form proposed by Vaughan (1994),

$$k = k_0 e^{-bp'} \tag{4}$$

where k_0 is the permeability at zero mean effective stress (m/s), and b is a parameter that has dimensions m²/kN. The two sets of parameters adopted in the study are: (a) $k_0 = 2 \times 10^{-9}$ m/s, $b = 0{\cdot}007$ m²/kN; and (b) $k_0 = 5 \times 10^{-10}$ m/s, $b = 0{\cdot}003$ m²/kN. The resulting variations in initial permeability with depth resulting from the use of this model with the above two sets of k_0 and b are shown in Fig. 4(b).

The model parameters used in the analyses are the same as those used in the back-analyses of the failed temporary slopes at Prospect Park and Wraysbury Reservoir (Kovacevic *et al.*, 2004). For completeness, they are summarised in Tables 1 and 2. The predicted behaviour of the London Clay in drained simple shear, undrained triaxial extension[‡] and oedometer swelling tests, using the model and parameters, is shown in Figs 5, 6 and 7 respectively. Where possible, the measured response from the appropriate laboratory tests is superimposed. It can be seen that the observed behaviour was reasonably well predicted by the model used.

FINITE ELEMENT ANALYSES

The FE analyses were carried out using the computer code ICFEP. To represent the cross-section being analysed (Fig. 2), the FE mesh shown in Fig. 8 was developed.

Plane-strain eight-noded isoparametric elements with 'reduced' 2×2 integration were used. All eight nodes of an element had both displacement and pore water pressure degrees of freedom. A modified Newton–Raphson approach

---

[‡] A typical stress path during excavation is extension rather than compression.

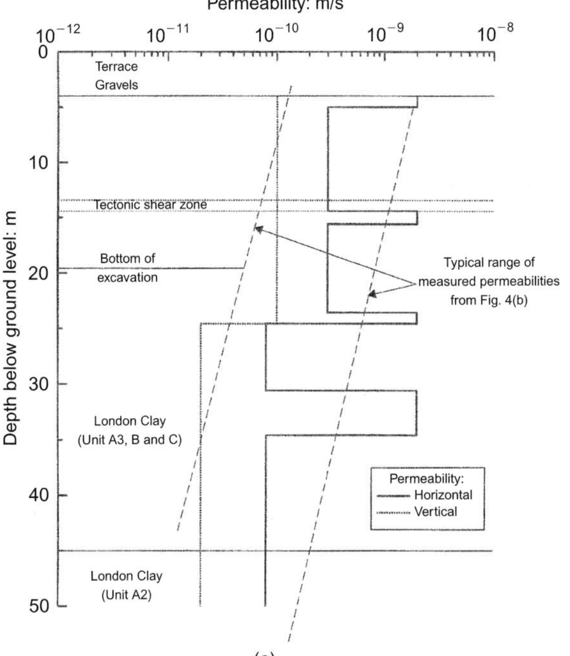

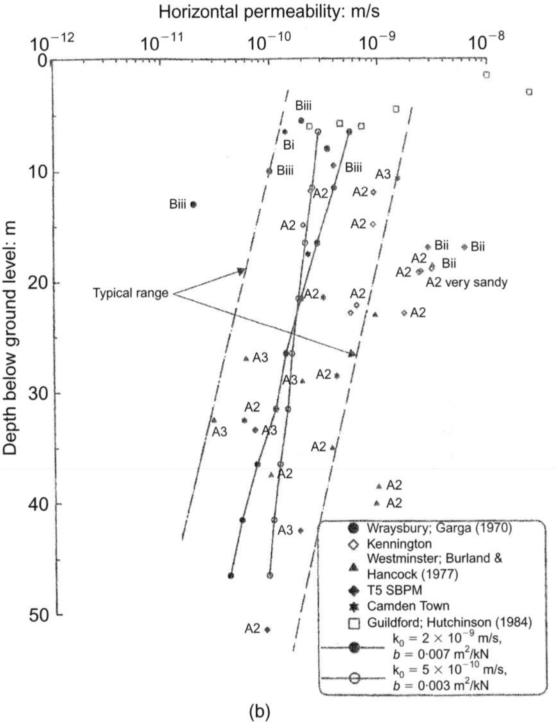

Fig. 4. Adopted permeability profiles in London Clay: (a) anisotropic permeability to investigate overall stability; (b) isotropic permeability to investigate local stability

with a sub-stepping stress point algorithm was used to solve the finite element equations (Potts & Zdravkovic, 1999).

No horizontal displacement was allowed on the vertical boundaries, whereas the bottom boundary was fixed in both the vertical and horizontal directions. The bentonite cut-off wall, assumed to be 20 m back from the top of the slope (see Fig. 2), was used to model dewatering of the Terrace Gravels. The far-end vertical boundary was assumed to be permeable (a source of water), whereas the near end vertical boundary was assumed to be the axis of symmetry, and as such was impermeable (no flow). The bottom boundary was deep (55 m), and assumed to be impermeable.

Table 1. Material properties assumed in the analyses

Property	Terrace Gravels	London Clay	Tectonic shear
Bulk unit weight, γ: kN/m^3	20	20	20
Peak strength	$c'_p = 0,\ \phi'_p = 35°$	$c'_p = 8$ kPa, $\phi'_p = 25°$	–
Residual strength	–	$c'_r = 2$ kPa, $\phi'_r = 13°$	$c'_r = 0,\ \phi'_r = 13·5°$
Plastic strain at peak, ε^p_{Dp}: %	–	2	–
Plastic strain at residual, ε^p_{Dr}: %	–	15	–
Young's modulus, E: MPa	20	*	*
Poisson's ratio, ν	0·2	*	*
Angle of dilation, ψ	17·5°	0	0
Coefficient of permeability, k	–	See Figs 4(a) and 4(b)	As for London Clay
Coefficient of earth pressure at rest, K_0	0·5	See Fig. 3	1·6[†]

*Small-strain stiffness parameters used; see Table 2.
[†]Not to violate the yield criterion.

Table 2. Coefficients and limits for non-linear elastic secant shear moduli and bulk of London Clay

Shear	A	B	C: %	α	γ	$\varepsilon_{D,min}$: %	$\varepsilon_{D,max}$: %	G_{min}: kPa
	970	890	0·001	1·47	0·7	0·00173	0·173	3333·3
Bulk	R	S	T: %	δ	λ	$\varepsilon_{v,min}$: %	$\varepsilon_{v,max}$: %	K_{min}: kPa
	772·5	712·5	0·001	2·069	0·42	0·005	0·15	4000

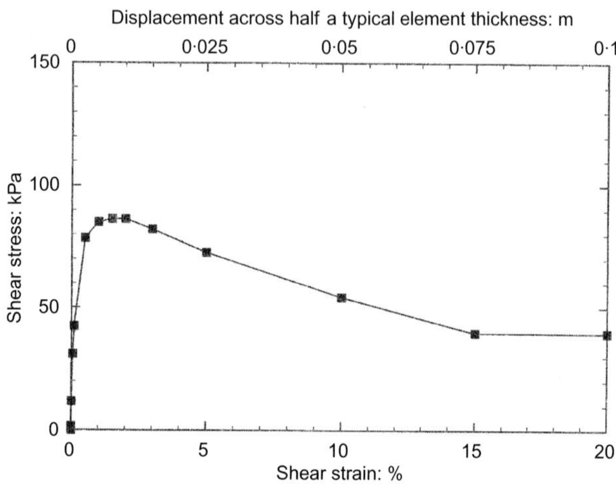

Fig. 5. **Predicted behaviour of London Clay in drained simple shear**

Excavation was simulated by removing layers of elements from the mesh at a uniform rate. Swelling of the London Clay (the Terrace Gravels were modelled as a drained material) was allowed during excavation, keeping a specified pore water pressure (suction) at the excavation boundary. Swelling was simulated in the FE analyses by applying increments of time t. The time increments during excavation were small enough (typically 0·01 years) for there to be no significant amount of swelling in materials with relatively low permeabilities (such as the London Clay). Subsequently, after excavation was completed, the time steps were increased (typically to 0·1 years). However, when collapse was approached, the time steps were reduced again.

RESULTS OF THE ANALYSES

The FE analyses of the temporary slopes at T5 were carried out in two phases. In the first phase, the aims of the analyses were to examine the time before there was overall failure and to identify the failure mechanism involved. However, prior to overall instability developing in these analyses,

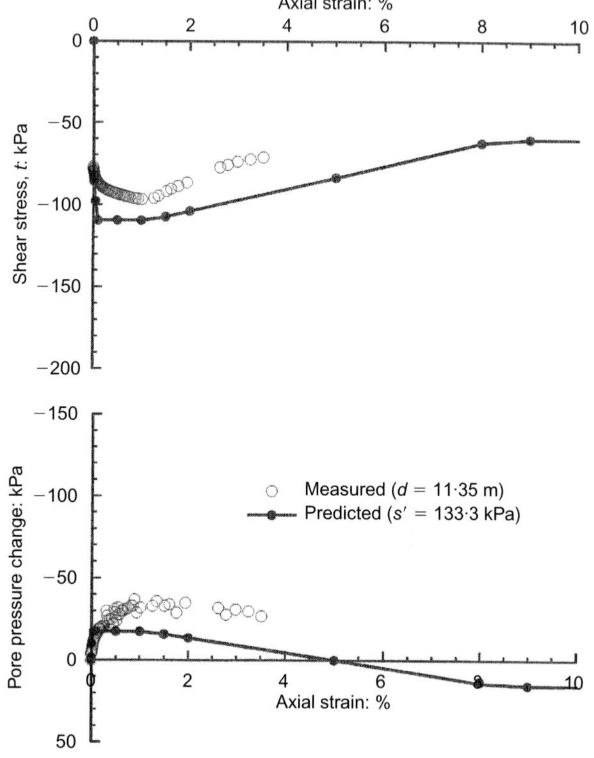

Fig. 6. **Observed and predicted response of London Clay in triaxial extension**

local instability involving the berms was predicted to occur.[§] Clearly, the local instabilities were of more immediate concern at T5 and the form and timings of these failures were investigated in the second phase.

[§] Local instabilities had to be suppressed (by allocating non strain-softening properties to the 'superficial' finite elements that were about to fail) in order to enable the overall (deep-seated) failure to occur.

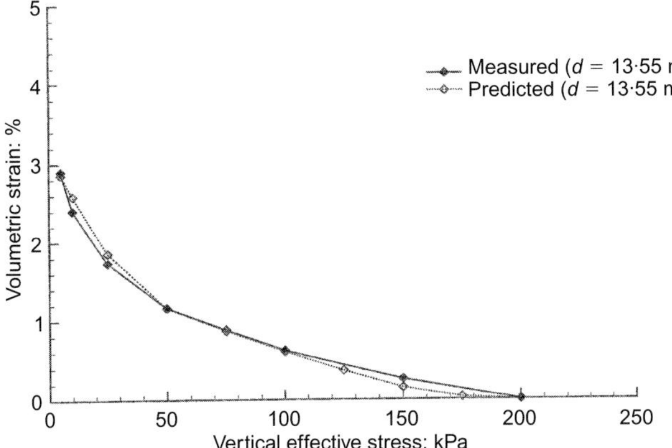

Fig. 7. Observed and predicted swelling of London Clay in oedometer test

Overall stability

The overall stability was analysed on a parametric basis. Various factors were considered, including:

(*a*) the presence of a tectonic shear zone
(*b*) the width of excavation
(*c*) the initial value of K_0
(*d*) pore pressures at the gravel/clay interface in front of the cut-off wall
(*e*) the depth of excavation.

A summary of all the analyses of overall stability performed is presented in Table 3. Note that zero pore water pressure at the excavation boundary was maintained both during excavation and subsequent swelling in all these analyses (a

conservative assumption, equivalent to a 'wet' boundary with rain infiltration in excess of evapo-transpiration).

Influence of the tectonic shear zone. None of the analyses predicted failure at the end of rapid excavation in the London Clay. The movement vectors during excavation to 20 m below ground level (bgl) are shown in Fig. 9(a) (Run 1: no tectonic shear zone present) and Fig. 9(b) (Run 3: with tectonic shear zone). The tectonic shear zone was influential during excavation, resulting in substantially greater predicted movements when it was present. This can also be seen from Fig. 10(a), which shows the development of the horizontal movements at the top of the slope (point 'A' in Fig. 2) during excavation. For the analysis with the tectonic shear zone (Run 3) the rate of movement increased after 8 m of excavation but decreased substantially once the excavation passed the depth of the tectonic shear zone (approximately 14 m bgl). With further excavation the tectonic shear zone was no longer influential. Monitoring of both surface and internal movements was recommended, therefore, to provide an indicator of whether or not a tectonic shear zone was present and likely to affect the stability.

Excavation in the London Clay gives rise to depressed pore water pressures; the amount of pore water pressure depression was larger when the tectonic shear zone was assumed to be present in the analyses (cf. Figs 11(a) and 11(b)). The analysis that modelled the tectonic shear zone predicted slope collapse to occur 7 years after excavation was complete, slightly longer than the analysis without the shear zone (6·7 years). The predicted failure mechanisms are shown in Figs 12(a) and 12(b) by means of the incremental

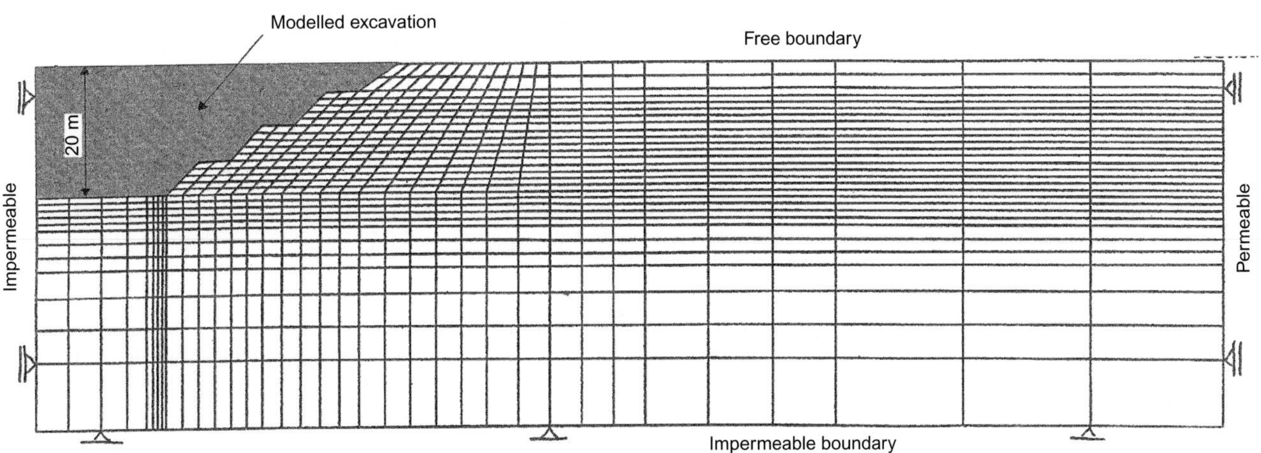

Fig. 8. Finite element mesh used in the analyses

Table 3. Summary of all analyses performed to investigate overall stability

Run	Description	Time to failure: years
1	Excavation to 20 m bgl (bottom of excavation), then pore pressure dissipation	6·7
11	Excavation to 15 m bgl (depth of third berm), then pore pressure dissipation	7·0
12	Excavation to 10 m bgl (depth of second berm), then pore pressure dissipation	10·0
2	Excavation to 20 m bgl (bottom of excavation) as in Run 1, but 'floor' width of excavation increased from 40 m (Run 1) to 80 m (Run 2)	–
3	Excavation to 20 m bgl with a tectonic shear at ~14–15 m bgl, then pore pressure dissipation	7·0
31	Excavation to 15 m bgl with a tectonic shear at ~14–15 m bgl, then pore pressure dissipation	–
4	Excavation to 20 m bgl (bottom of excavation) as in Run 1, but K_0 profile reduced according to Fig. 3	4·5
5	Excavation to 20 m bgl (bottom of excavation) as in Run 1, but increased pore water pressure at the gravel/clay interface from 0 kPa (Run 1) to 10 kPa (Run 5)	6·6

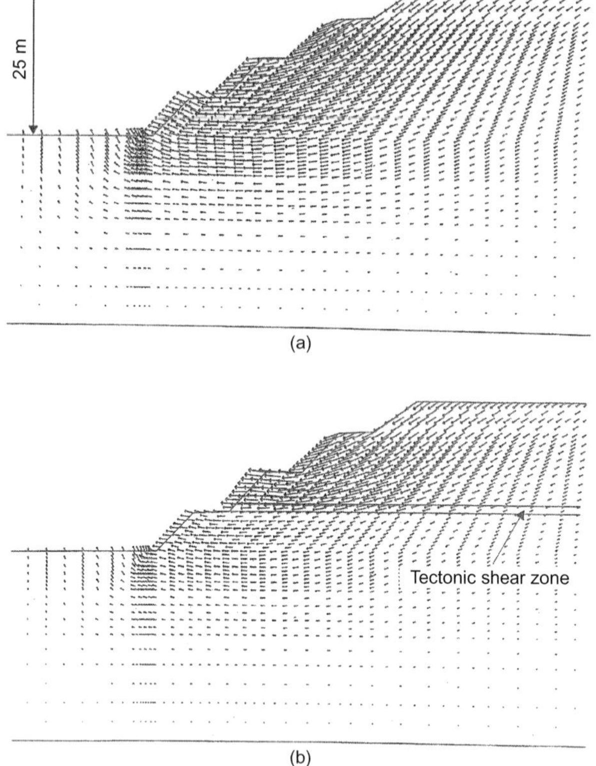

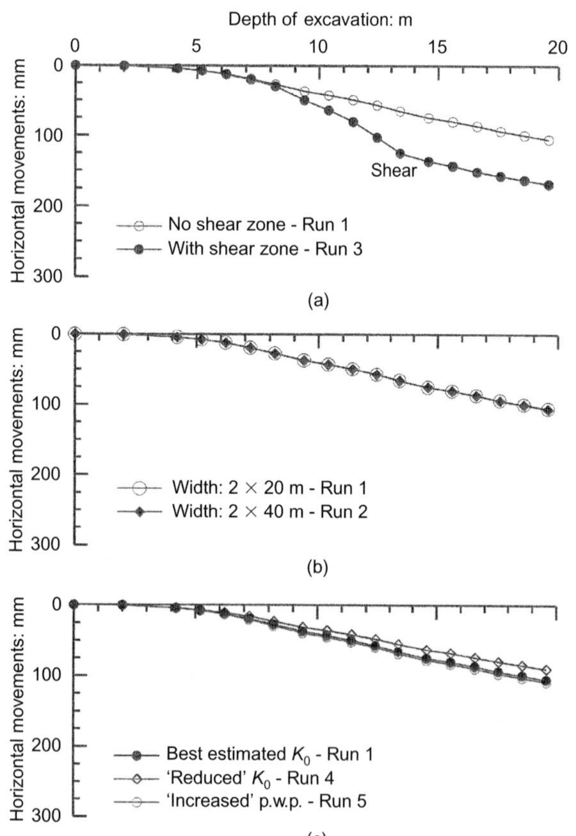

Fig. 9. Predicted displacement vectors during excavation (a) without and (b) with tectonic shear zone modelled in the analyses of overall stability

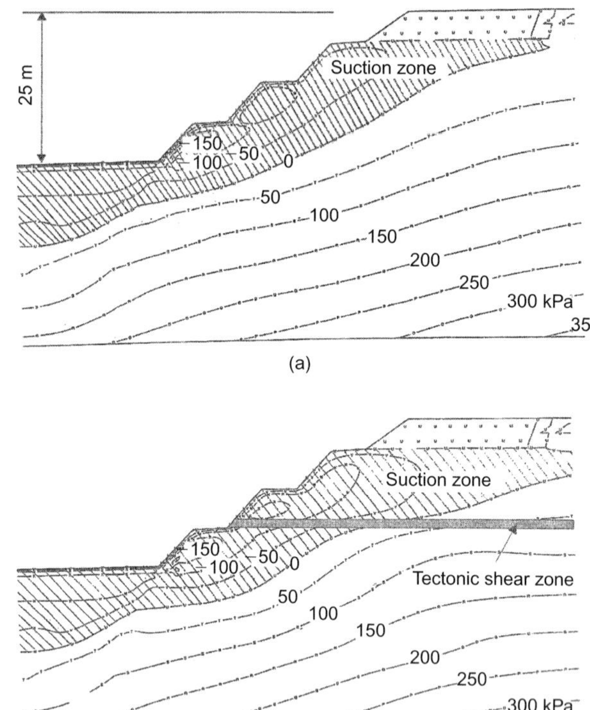

Fig. 11. Predicted contours of pore water pressure at the end excavation (a) without and (b) with tectonic shear zone modelled in the analyses of overall stability

Depth of excavation: m

Fig. 10. Predicted horizontal movements in the analyses of overall stability at the slope crest during excavation: influence of (a) tectonic shear, (b) width of excavation, and (c) initial K_0 profile and pore water pressure at the gravel/clay interface

displacement vectors[¶] just prior to collapse. The mechanisms are deep-seated and start from the toe of the slope, even when the tectonic shear zone was present in the analysis (Fig. 12(b)). In both cases the role of progressive failure was substantial, as shown by Figs 13(a) and 13(b), which indicate that the operational strength along the predicted base rupture surfaces was at residual (the plastic shear strain contour is in excess of $\varepsilon_D^p = 15\%$). The final rupture surface did not involve the modelled tectonic shear zone (see Fig. 13(b)), which seemingly caused a slightly shallower failure mechanism.

Development of the horizontal movements at the top of the London Clay (the crest of the first berm, point 'B' in Fig. 2) after excavation (i.e. excluding the movements during excavation) is shown in Fig. 14(a). The formation of the base shear zone approximately 1 year after excavation, well before eventual collapse, is evident when the tectonic shear zone was absent (Run 1). This caused larger movements to precede the collapse because of larger horizontal stress relief through the base shear zone formation, which in the case with the tectonic shear zone present (Run 3) had already partly occurred during slope excavation (see Fig. 10(a)).

Figure 15(a) shows a superficial slip in the London Clay starting from the second berm (9·2 m bgl) and propagating towards the first berm (4·2 m bgl) predicted by the analysis with no tectonic shear zone 5·3 years after excavation. A step increase in the horizontal movement at the crest of the first berm associated with the formation of this slip is clearly shown in Fig. 14(a) before the local failure was suppressed. Similarly, the analysis with the tectonic shear zone present predicted a superficial failure (see Fig. 15(b)). This local

---

¶ The absolute magnitude of the incremental displacement vectors is not of importance, because it is their relative magnitude and directions that indicate the current mechanism of behaviour.

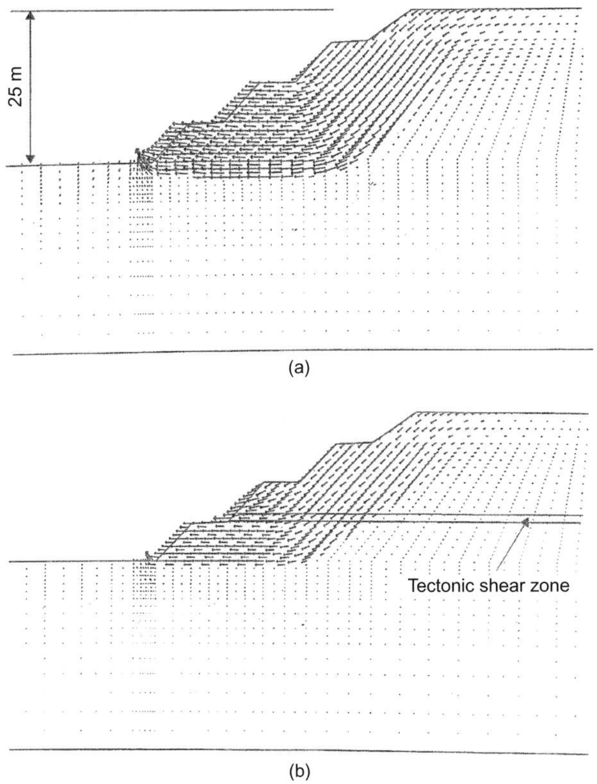

Fig. 12. Predicted incremental displacement vectors just prior to collapse (a) without and (b) with tectonic shear zone modelled in the analyses of overall stability

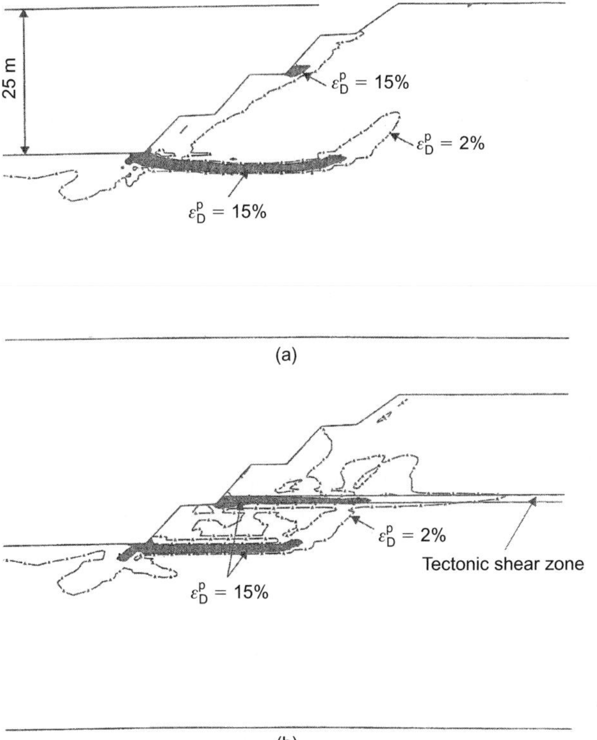

Fig. 13. Predicted contours of plastic shear strain at the end of excavation (a) without and (b) with tectonic shear zone modelled in the analyses of overall stability

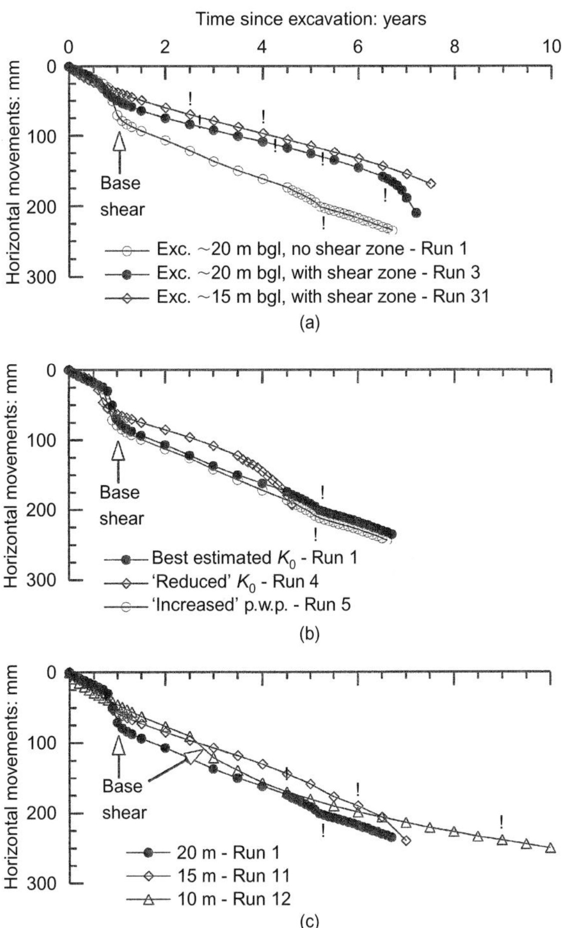

Fig. 14. Predicted horizontal movements in the analyses of overall stability at the top of London Clay after excavation: influence of (a) tectonic shear, (b) initial K_0 profile and pore water pressure at the gravel/clay interface, and (c) depth of excavation. Note: (!) denotes development of superficial failure, which was suppressed in analyses of overall stability

because they started developing from the third berm (also the position of the tectonic shear zone) and propagated towards the second berm. Note that the above times are based on a permeability that stays constant during swelling and are, therefore, overestimates of the time for superficial failures to develop.

Influence of width of excavation. To check the sensitivity of predictions to the width of the excavation, which varied considerably in the excavations at T5, the width of the floor at the end of excavation was increased from 2×20 m (Run 1) to 2×40 m (Run 2). Development of the horizontal movement at the top of the slope as excavation proceeded for the two cases considered is shown in Fig. 10(b). It can be seen that the width of the excavation has no influence on the pattern of movements during excavation. During the swelling stages of the analysis, this influence was also of no importance.

Influence of the initial K_0 profile. The previous FE study of slopes cut in the London Clay showed that the main factor that controls the behaviour of these slopes is the initial K_0 value assumed in the analyses (Potts *et al.*, 1997). This is so because progressive failure is generated primarily by the high lateral stresses (defined by the K_0 value) in the soil prior to

failure had to be suppressed at several stages during the analysis, and the exclamation marks in Fig. 14(a) indicate the timings of these events (2·5, 4·5, 5·5 and 6·6 years after excavation). It seems that all superficial failures were instigated by the presence of the modelled tectonic shear zone

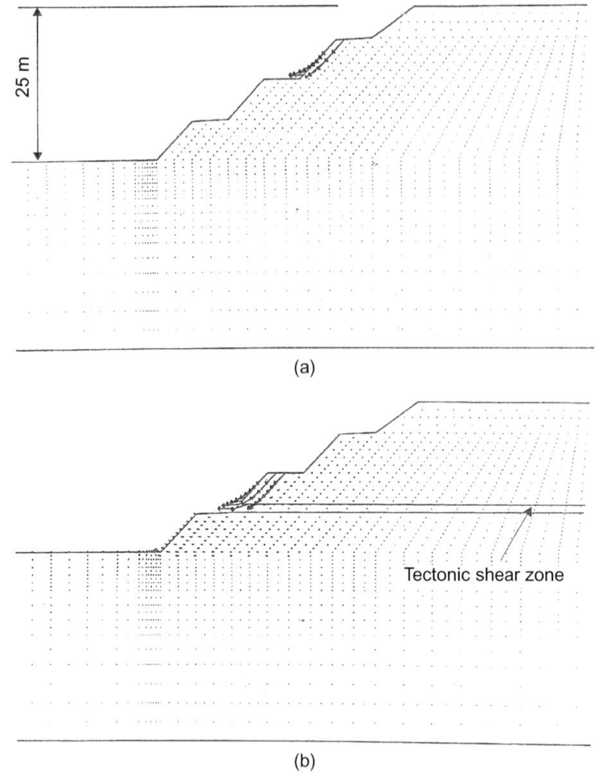

Fig. 15. Predicted superficial slips that in the analyses of overall stability were suppressed (a) without and (b) with tectonic shear zone modelled in the analyses

excavation. In particular, the value of K_0 strongly influences the location of the shear surface and the time to collapse.

To investigate the influence of this important factor on the behaviour of the temporary slopes at T5, an analysis (Run 4) was carried out with no tectonic shear present and in which the 'reduced' K_0 profile shown in Fig. 3 was assumed. Lowering K_0 values in the analysis shortened the time to failure (from 6·7 years to 4·5 years), and produced a slightly shallower slip surface and a smaller amount of shearing along the base shear zone.

The horizontal movements at the top of the slope (point 'A' in Fig. 2) during excavation (Fig. 10(c)) and at the crest of the first berm (the top of the London Clay, point 'B' in Fig. 2) after excavation was completed (Fig. 14(b)) are predicted to be smaller as a result of the 'reduced' K_0 profile. It is interesting to note that the analysis that modelled the 'reduced' K_0 profile did not predict the development of superficial failure in either of the berms.

These effects of a reduced K_0 are related to the smaller magnitude of unloading and correspondingly smaller pore pressure reduction.

Influence of pore pressures at the gravel/clay interface. The London Clay/Terrace Gravel interface was seen in excavations to undulate significantly, so that there was a risk of ponded water and of higher pore water pressure at this interface. An analysis (Run 5) was therefore performed that modelled an increased pore water pressure at the gravel/clay interface in front of the cut-off wall from 0 to 10 kPa. The tectonic shear zone was not present in this analysis.

Predicted horizontal movements during excavation and subsequent swelling from this analysis are superimposed on Figs 10(c) and 14(b) respectively. The increased pore water pressures resulted in slightly more horizontal movements, especially subsequent to excavation, during swelling (see

Fig. 14(b)). It is noteworthy that development of the base shear zone and the 'superficial' failure between the first and the second berm starts somewhat earlier. The ultimate time to deep-seated failure was also slightly reduced (from 6·7 years to 6·6 years).

Influence of the depth of excavation. Given that open-cut excavations at T5 were dug to various depths, the 'standard' section was 're-analysed', excavating to depths of the third berm ($\sim$15 m bgl, Run 11) and the second berm ($\sim$10 m bgl, Run 12). No tectonic shear zone was modelled in these analyses.

When excavation to 15 m bgl was modelled, the predicted deep-seated slip developed from the toe of the slope 7 years after excavation. The analysis that modelled the excavation to 10 m bgl predicted no deep-seated failure, even 10 years after excavation. However, the 'superficial' failures, starting from the toe of the slope and emerging at the first berm, had to be 'suppressed' on two occasions: 4·5 years and 9 years after excavation.

Development of the horizontal movements at the top of the London Clay after excavation to the different depths is shown in Fig. 14(c). As before, the exclamation signs indicate developments of the 'superficial' failures that always took place in the first berm. It can be seen that development of the base shear zones was delayed by the reduced depth of excavation. Also, the movements associated with formation of the base shear zone may well be reduced. However, the absolute magnitude of movements does not appear to be influenced by the depth of excavation.

Local stability
The analyses of overall stability of the temporary slopes at T5 showed that the slopes were stable in the short term, immediately after excavation. However, it was clear from the analyses that local instability was likely to develop before overall instability in some situations. It was probable that the time for these shallow failures to develop would depend on

(*a*) the initial permeability profile in the London Clay
(*b*) the way in which permeability changed as the clay swelled
(*c*) the pore water boundary conditions, that is, the average suction operating at the exposed surface of the London Clay.

To provide information on the likely combination of these parameters, the back-analyses of the local instability at nearby Prospect Park and Wraysbury Reservoir sites (Kovacevic *et al.*, 2004) were used. The times to failure were best matched using a non-linear permeability model (see equation (4) with $k_0 = 2 \times 10^{-9}$ m/s and $b = 0.007$ m^2/kN) and assuming a suction of 25 kPa was maintained at the London Clay surface. The same non-linear permeability model, initially with $k_0 = 2 \times 10^{-9}$ m/s and $b = 0.007$ m^2/kN and subsequently with $k_0 = 5 \times 10^{-10}$ m/s and $b = 0.003$ m^2/kN, was used in all analyses described henceforth. The sensitivity of local stability to

(*a*) the presence of the tectonic shear zone
(*b*) the surface pore pressure boundary conditions
(*c*) the initial K_0 profile
(*d*) the initial permeability profile
(*e*) the slope geometry

is described below. A summary of all the analyses of local stability performed is presented in Table 4.

It is worth mentioning that, in all the analyses that

Table 4. Summary of all analyses performed to investigate local stability

Run	Section analysed	Tectonic shear	Surface suction: kPa	K_0	Permeability	Slope	Time to failure: years
C1	Standard	Yes	0	Standard	Standard	Berms	0·17
C2	Standard	Yes	25	Standard	Standard	Berms	1·42*
C3	Standard	Yes	0	'Lower'	Standard	Berms	0·13
C4	Standard	Yes	Zero-slopes, impermeable berms	Standard	Standard	Berms	0·26
C5	Standard	Yes	0	Standard	'Lower'	Berms	1·1
D1	Standard	No	0	Standard	Standard	Berms	0·2
D1A[†]	Standard	No	0	Standard	Standard	Berms	Varied
D2	Standard	No	25	Standard	Standard	Berms	1·11*
D3	Standard	No	0	'Lower'	Standard	Berms	0·54*
D4	Standard	No	0	Standard	'Lower'	Berms	2·2*
E1	Standard	Yes	0	Standard	Standard	Uniform	0·2
E2	Standard	Yes	25	Standard	Standard	Uniform	0·57
F1	Standard	No	0	Standard	Standard	Uniform	0·43
F2	Standard	No	25	Standard	Standard	Uniform	0·59*

*Deep-seated failures predicted.
[†]Superficial failure(s) suppressed.

modelled local stability, it was assumed that the Terrace Gravels overlying the London Clay had been removed 25 years prior to the excavation of the slopes in the London Clay. This was to represent formation of the lagoons at the T5 site, which occupies the old Perry Oaks Sewage Works where lagoons had been constructed some 25 years prior to construction of T5. This was a similar situation to that at Prospect Park, where the prior removal of the gravel was shown in the back-analyses to be an important feature of the recent site history when a low-strength tectonic shear surface was present.

Influence of the tectonic shear zone with zero suction at the slope surface (Runs C1 and D1). To study the influence of the tectonic shear zone with zero suction at the slope surface, the results from Run C1 (with tectonic shear zone) and Run D1 (no tectonic shear zone) were compared. There was no influence of the tectonic shear zone for the stage of the analysis that modelled excavation of the 4 m thick Terrace Gravels 25 years before construction of T5. However, during excavation of the London Clay for T5 the presence of the tectonic shear zone caused (a) larger pore pressure changes and (b) almost twice the magnitude of ground movements. The latter can be seen in Fig. 16, which shows the development of the horizontal movements at the top of the London Clay. Predictions from the previous analyses that modelled the anisotropic (but constant) permeability, Runs 1 and 3, are also shown for comparison. Whereas there was no difference in the horizontal movements when the tectonic shear zone was absent in the analyses, its presence caused more movements in the analysis that modelled the non-linear permeability. The reason for the stronger influence of the tectonic shear zone in the latter analysis is reduction in strength on that surface during the 25-year period of swelling after removal of the gravel from the lagoons, which was modelled in the analyses of local stability.

The analysis without the tectonic shear zone (Run D1) predicted the failure of the first berm 0·2 years after excavation (see Fig. 17(a)). The tectonic shear zone caused the failure of the second berm (the berm above the tectonic shear zone) to take place 0·17 years after excavation (see Fig. 17(b), Run C1). Swelling promoted by the non-linear permeability model employed and zero suction imposed on the excavated slope surface are the main reasons for the 'quick' superficial failures predicted by these analyses.

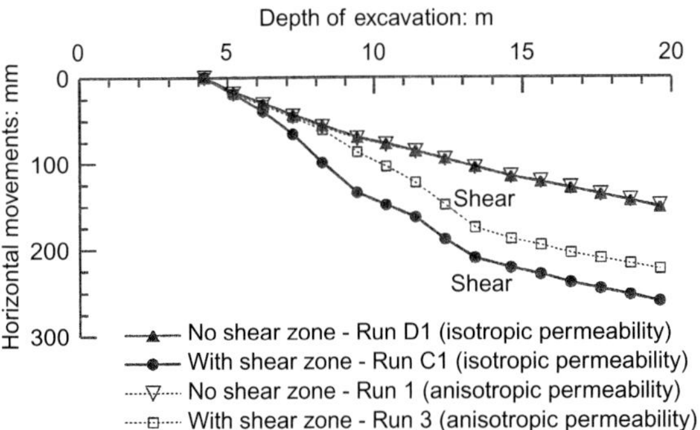

Fig. 16. Predicted horizontal movements at the top of London Clay during excavation: influence of tectonic shear and prior gravel removal

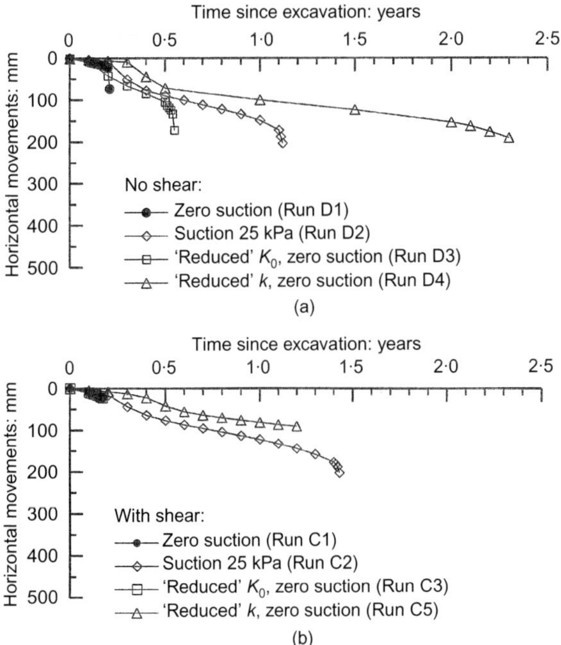

Fig. 17. Predicted horizontal movements at the top of the first berm after excavation (a) without and (b) with tectonic shear zone modelled in the analyses of local stability: influence of surface suction, initial K_0 and permeability profiles

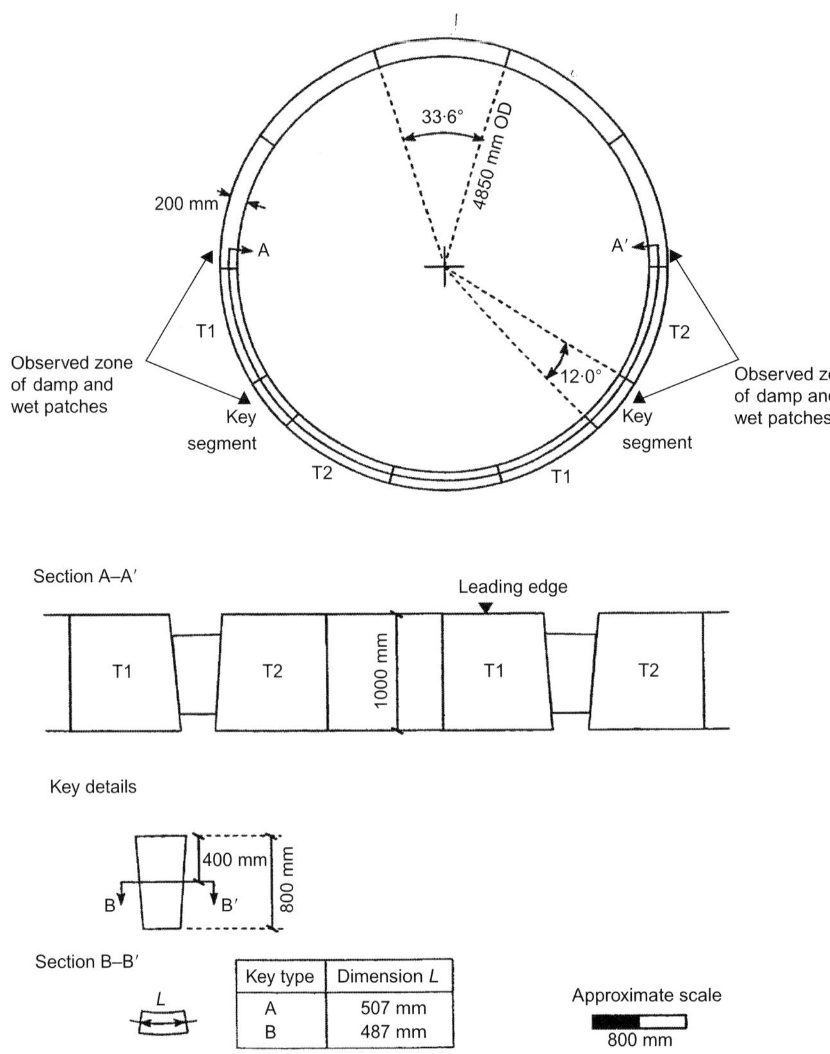

Key type	Dimension L
A	507 mm
B	487 mm

Fig. 2. General arrangement of expanded tunnel lining used under St James's Park (after Nyren, 1998)

Table 1. Summary of monitoring period timing at St James's Park (Nyren, 1998)

Period	Time span	Duration: days	Descriptions
1	1 Mar 95 to 27 Apr 95	28·4	Period prior to any tunnelling works
2	27 Apr 95 to 28 Apr 95	1·67	Construction of westbound tunnel drive
3	28 Apr 95 to 8 Jan 96	256	Rest period before eastbound excavation
4	8 Jan 96 to 10 Jan 96	4·6	Construction of eastbound tunnel drive
5	10 Jan 96 to 21 Mar 97	432	Long-term monitoring period

Note that the period is divided according to the survey number of precision levelling surveys (Nyren, 1998, table F1): period 1, survey nos 1–16; period 2, survey nos 16–29; period 3, survey nos 29–62; period 4, survey nos 62–85; period 5, survey nos 85–101.

tunnel crown and simultaneously consolidates on either sides of the tunnel in the zone close to tunnel axis level. Above these zones, the soil was settling as a rigid body, which resulted in continued settlement at the ground surface.

FINITE ELEMENT ANALYSIS

The construction of the westbound tunnel was modelled, and the subsequent ground response prior to the eastbound excavation was examined. All analyses presented in this paper were performed using the soil-fluid coupled 3D finite element (FE) method with the commercial finite element package ABAQUS™. For brevity, this numerical study utilises the data mainly obtained during Period 1 to 3 (see

Table 1) to examine the time-dependent behaviour of a single tunnel construction. This was necessary to avoid any complexity due to interactions of the two tunnels, so that a time-dependent ground deformation mechanism can be proposed based on a simple boundary condition. As future research, the numerical study can be extended to include the effects of tunnel–tunnel interaction on the long-term ground deformation behaviour.

Soil conditions and geometry

The FE model for the westbound tunnel excavation with dimensions shown in Fig. 7 was assumed to have the following soil profile:

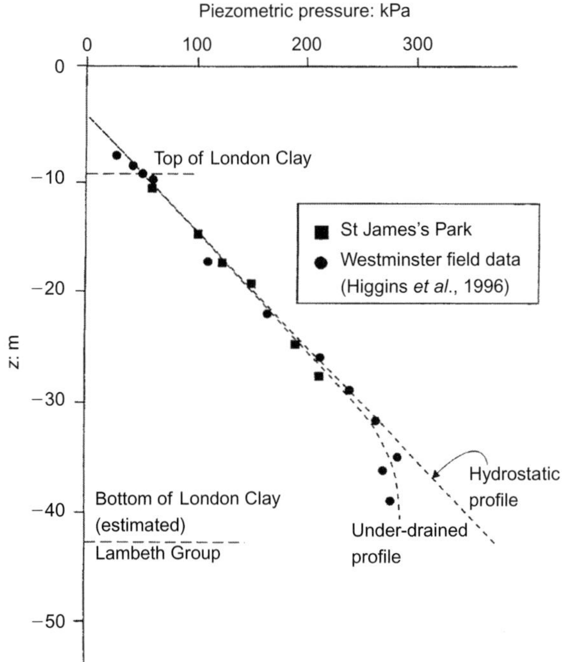

Fig. 3. Measured pore pressures at instrumented section in St James's Park and in Westminster area, London (Nyren, 1998)

(*a*) Made Ground and Alluvium, MG (0–5 m)
(*b*) Terrace Gravel, TG (5–8 m)
(*c*) London Clay, LC (8–43 m)
(*d*) Woolwich and Reading Bed Clay, WRBC (43–50 m).

It was assumed that the interface between WRB clay and underlying Basal Sand was located at a depth of 50 m (Addenbrooke, 1996; Higgins *et al.*, 1996).

To reduce analysis time, the soil was modelled with two types of element:

(*a*) 8-node trilinear displacement and pore pressure elements
(*b*) 20-node triquadratic displacement, trilinear pore pressure, reduced integration elements.

The 20-node elements were employed in the zone around the tunnel because they gave smoother pore pressure response than the 8-node elements. Tunnel lining was modelled with 8-node, doubly curved thick shell elements with reduced integration. The lining and soil elements shared the same nodes at the tunnel boundary; the interface was not modelled. The 3D model consisted of 9625 elements and 22 938 nodes. The model boundaries were set to minimise the effects on ground response (Wongsaroj, 2005).

Modelling of tunnel construction and long-term consolidation

The model had element length (size of an element in the longitudinal direction) of 2 m in the zone where the excavation was modelled. When the tip of the backhoe is fully extended, the distance from the tunnel face to the shield tail is approximately 6 m during the excavation. Assuming that there is no contact between most of the perimeter of the shield and the surrounding soil (due to over-excavation), the excavation process is modelled as a repeated sequence of

(*a*) deactivation of elements every 2 m, resulting in 6 m unsupported opening, and
(*b*) placement of 2 m long tunnel linings next to the existing tunnel linings (see Fig. 8).

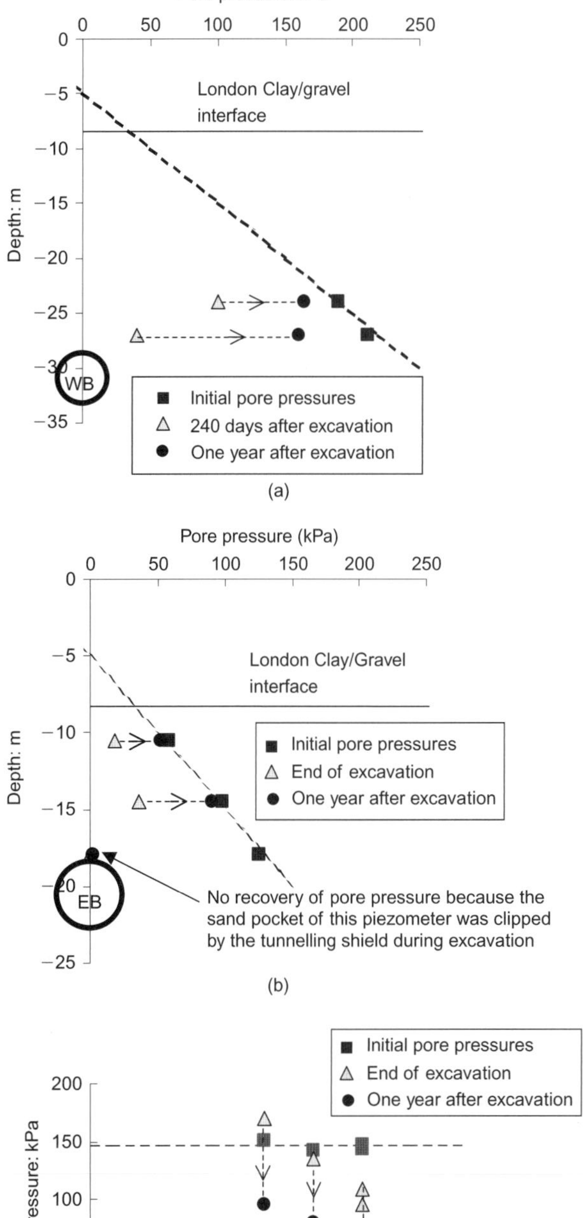

Fig. 4. Variation in pore pressures (after Nyren, 1998): (a) above westbound tunnel crown; (b) above eastbound tunnel crown; (c) at eastbound tunnel springline level

This excavation model resulted in 100% reduction of normal stress to the tunnel boundary within the 6 m unsupported length (including the tunnel face), which leads to stress redistribution and volume loss. This tunnel excavation model sequence was continued until the tunnel face reached 100 m into the model. The computed results were taken from a plane located 40 m from the start of the excavation, which is referred to as the instrument plane (see Fig. 7). It was important to model the tunnel construction process in 3D so that realistic short-term stress and pore pressure fields, together with tunnel lining loads, were obtained as the initial condition for the long-term consolidation analysis.

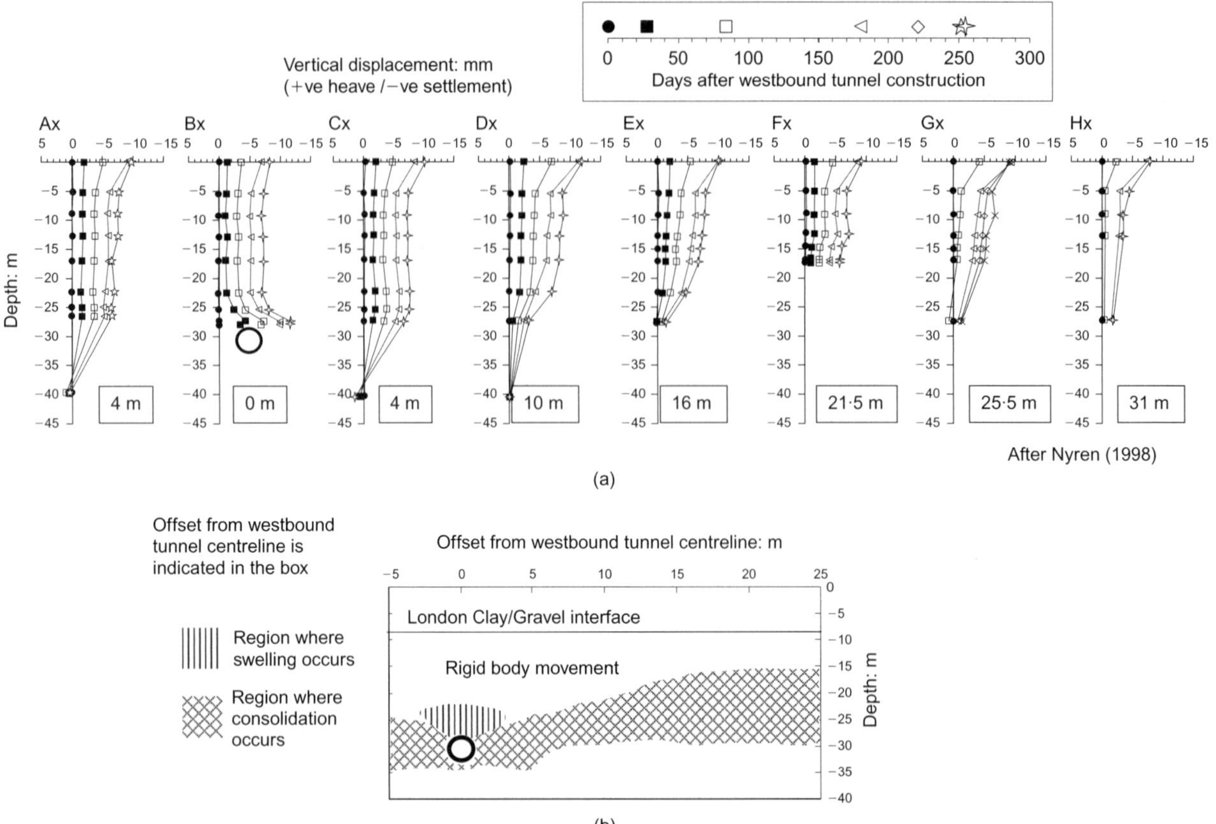

Fig. 5. (a) Field measurement of long-term ground movement after westbound tunnel construction; (b) schematic diagram indicating zones of different response during consolidation

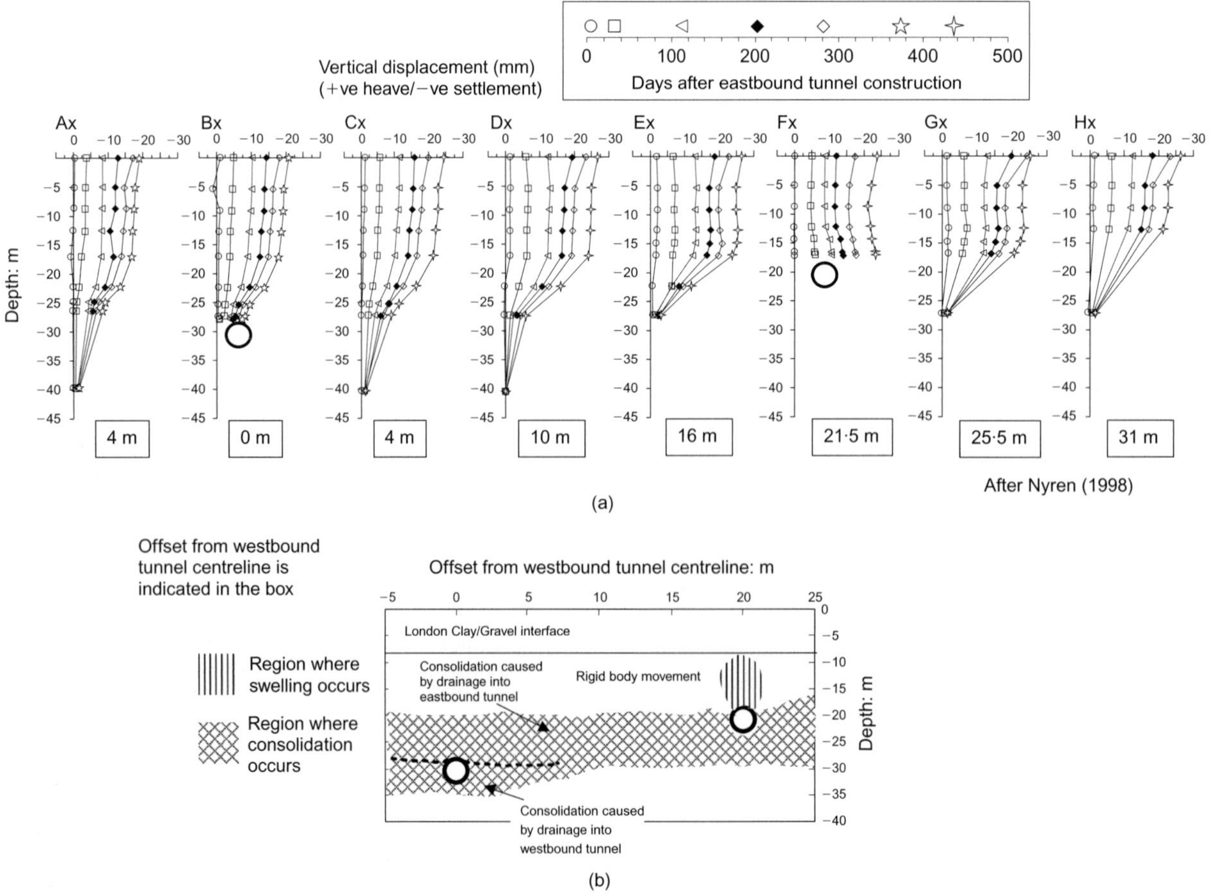

Fig. 6. (a) Field measurement of long-term ground movement after eastbound tunnel construction; (b) schematic diagram indicating zones of different response during consolidation

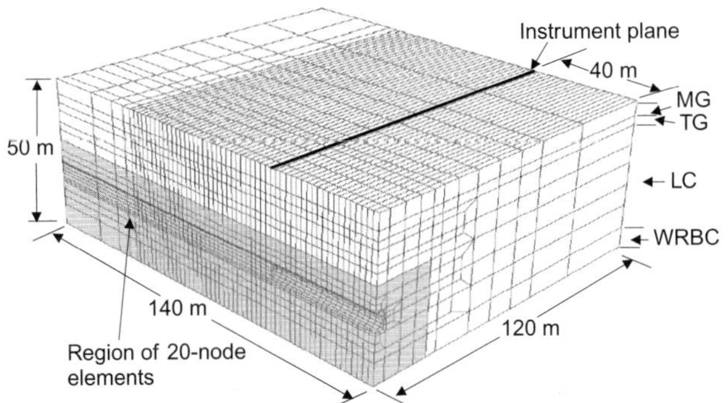

Fig. 7. Finite element model employed for analysis

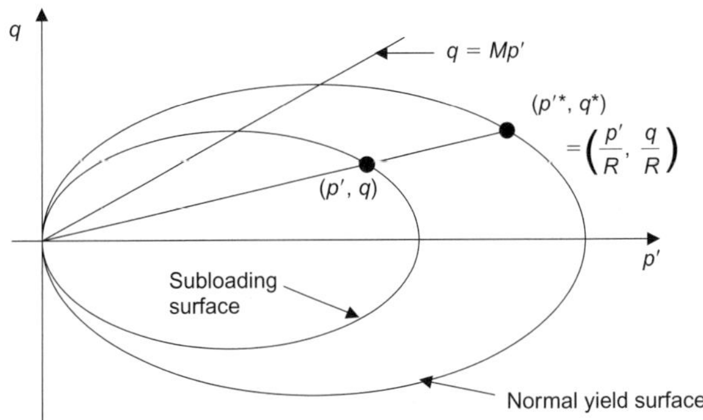

Fig. 9. Concept of subloading surface within normal yield surface and ratio R

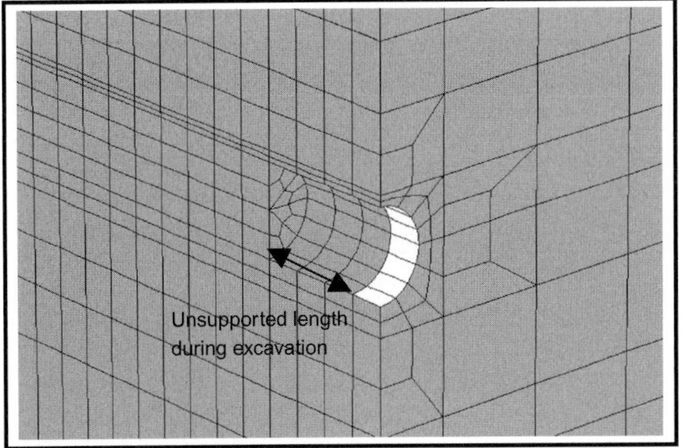

Fig. 8. Model of excavation sequence

The tunnel lining thickness of 200 mm, which is the actual thickness of the tunnel lining, was used for all simulations (neglecting the effect of the joints). The tunnel lining was modelled using a linear elastic model with a Young's modulus of 28 GPa and a Poisson's ratio of 0·15. The expansion of the tunnel lining was not modelled. The placement of the tunnel lining was simply modelled by activating the shell elements at the deformed tunnel heading. As the expansion of the lining was not modelled, the lining was not perfectly circular with the initial opening diameter of approximately 4·85 m.

When the tunnel face advanced 100 m into the model, the long-term consolidation was set to begin by assigning a drainage-only flow surface to the tunnel boundary along the entire excavated tunnel length, and consolidation was allowed until the new steady-state pore pressure regime was reached. The drainage-only flow surface controls the drainage condition at the tunnel boundary by assuming that the pore fluid velocity in the direction outwards from the soil element, v_n (m/s), is proportional to the pore pressure at the tunnel lining extrados, u_w, when it is positive. The proportional constant is termed the seepage coefficient, K_T, and is equivalent to $k_l/t_l\gamma_w$, where k_l and t_l are the permeability and thickness of the tunnel lining respectively. Additionally, no flow is allowed across the tunnel boundary when u_w is negative (i.e. water is not supplied from inside the tunnel). Hence varying the magnitude of K_T (m³/kN.s) is the same as varying the permeability of the tunnel lining for a given tunnel lining thickness, which results in different drainage conditions at the tunnel boundary. Damp patches on the tunnel lining suggest that the pore pressure in the tunnel lining segment may be negative. However, the suction within the tunnel lining is difficult to determine. Therefore atmospheric pressure was assumed inside the tunnel lining for all analyses presented in this paper.

Soil constitutive model

All soil units in this study were modelled using the non-linear elasto-plastic critical state soil model with the subloading surface concept (Hashiguchi & Ueno, 1977; Hashiguchi & Chen, 1998). The concept assumes that there is a small yield surface called the subloading surface, which always passes through the current stress point within the conventional yield surface (also termed the normal yield surface), and the ratio of their size is denoted by R, as shown in Fig. 9. The soil behaviour is elastic if the subloading surface is contracting (i.e. R is reducing), and is elasto-plastic if the subloading surface is expanding (i.e. R is increasing). The ratio R and its evolution law control the amount of plastic strain generated within the normal yield surface. Generation of plastic strains within the normal yield surface was necessary to model accurately the development of excess pore pressures during undrained shearing. The constitutive model also incorporates the following features: non-linear elasticity, anisotropic stiffness, the Matsuoka & Nakai (1985) failure criterion, and a re-invoked small-strain stiffness at stress reversal points (SRP; i.e. unloading and reloading). Further details of the model are described in Appendix 1.

The model was implemented in the ABAQUS™ user interface. The model was adopted for all soil units in the analyses. As the Made Ground is far from the tunnel, and the strains occurring within this material are small during tunnel excavation, its behaviour is more or less elastic. For the Terrace Gravel and Woolwich and Reading Beds clay, the parameters were selected such that the stiffness degradation curves are within the bounds adopted by other studies such as Addenbrooke *et al.* (1996). The critical state lines for the Made Ground and Terrace Gravel were calculated from the angles of friction given in Potts & Zdravkovic (2001). The parameters for London Clay were determined from data of undrained and/or drained triaxial tests and one-dimensional consolidation tests. In particular, emphasis was on calibrating the anisotropic model parameters so that the same model parameters can simulate data of both undrained and drained triaxial compression tests of vertically and horizontally cut London Clay specimens (Yimsiri, 2001; Wongsaroj *et al.*, 2004). The soil properties and model parameters used for the simulations are summarised in Table 2 and Table 4 in Appendix 1. In most cases, the earth pressure coefficient at rest, K_0, for London Clay was assumed to be 1·2. The preconsolidation profile adopted is shown in Fig. 10. To investigate the effects of stiffness anisotropy, simulations using a model with isotropic elastic stiffness were also performed.

Permeability model

To examine the influence on the long-term ground response of different units within London Clay and of the tunnel drainage condition, two combinations of permeability profiles and tunnel lining drainage conditions were adopted. The two permeability profiles assumed are summarised in Table 3, which are also shown in Fig. 11 along with the in situ measurement data from Burland & Hancock (1977). Although other permeability profiles were examined by Wongsaroj (2005), these two profiles were the extreme cases: profile 1 is a 'simplified' permeability model case, whereas profile 2 is a more realistic 'refined' permeability model case.

For permeability profile 1, the London Clay has a constant isotropic permeability of 10^{-9} m/s, and for these analyses the initial pore pressure (prior to tunnelling) was assumed hydrostatic with the water table at 5 m below the ground

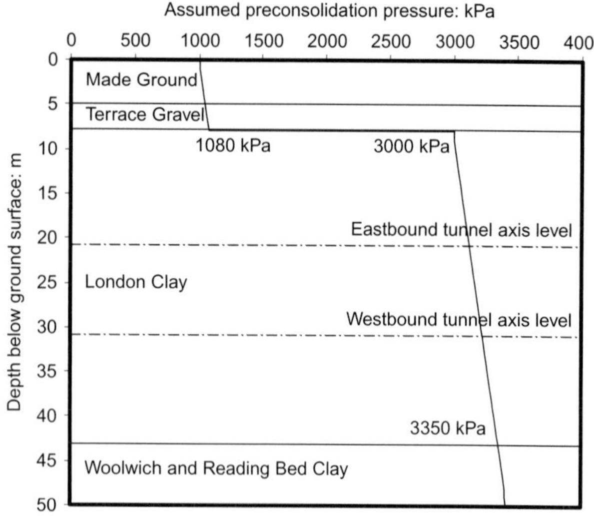

Fig. 10 **Preconsolidation pressure profile assumed for analysis**

surface. The tunnel lining was assumed to have uniform permeability.

An under-drained initial pore pressure profile was derived for the analyses that adopted permeability profile 2. Four distinct divisions (B, A3ii, A3i and A2) within the London Clay strata, established by King (1981), were identified using the detailed descriptions of split continuous U100 samples (Standing & Burland, 2006). The qualitative hierarchies of permeability suggested by Hight *et al.* (2003) and Standing & Burland (2006) are $k_{A2} > k_{A3ii(top)} > k_{A3i} \approx k_{A3ii(base)} > k_B$. Different anisotropic permeability values were assigned

Table 2. Assumed properties for all soil units

Strata	Bulk density, γ: kN/m^3	Critical angle of shearing resistance, ϕ'_{cv} : degrees	Coefficient of earth pressure at rest, K_0	Permeability, k: m/s
Made Ground	20·0	25·0	0·6	1×10^{-7}
Terrace Gravel	20·0	35·0	0·4	5×10^{-4}
London Clay	20·0	21·0	1·2*	See Table 3
Woolwich and Reading Beds Clay	20·0	27·0	1·2	See Table 3

*Only adopted for the simulations to investigate influence of permeability profiles and tunnel drainage conditions.

Table 3. Permeability profiles assumed for analyses

Soil descriptions by Standing & Burland (2006)	Permeability profile 1: m/s		Permeability profile 2: m/s	
	k_v	k_h	k_v	k_h
London Clay Division B, 8–20·5 m (95–82·5 mPD)			5×10^{-10}	10×10^{-10}
London Clay Division A3ii, 20·5–28 m (82·5–75 mPD)	1×10^{-9}		7×10^{-10} to 9×10^{-11}	45×10^{-10} to 45×10^{-11}
London Clay Division A3i, 28–35 m (75–68 mPD)			9×10^{-11} to 3×10^{-10}	18×10^{-11} to 6×10^{-10}
London Clay Division A2, 35–43m (68–60mPD)			3×10^{-10}	6×10^{-10}
Woolwich and Reading Beds Clay	5×10^{-11}		1×10^{-12}	2×10^{-12}

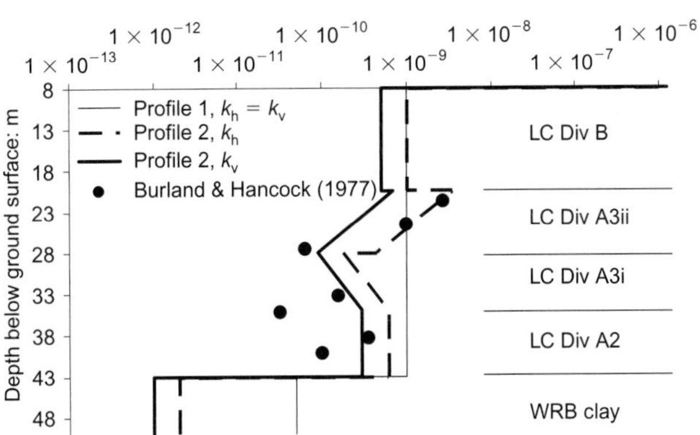

Fig. 11. Assumed permeability profiles

for these layers based on field measurement data reported in Burland & Hancock (1977) and Hight *et al.* (2003). The assumed anisotropic permeability ratio (k_h/k_v) was 2, except for division A3ii, for which the assumed k_h/k_v ratio was 5. This higher ratio was assigned because water strikes were encountered in this division (Standing & Burland, 2006). As discussed before, the tunnel lining above the springline was assumed to have finite permeability whereas the lining below the springline was assumed to be fully permeable, which is considered to be reasonably realistic based on the in-tunnel observations. The under-drained initial pore water pressure profile derived from permeability profile 2 is shown in Fig. 12.

IMPORTANCE OF SOIL STIFFNESS ANISOTROPY FOR LONG-TERM GROUND RESPONSE

It has been reported widely that stiffness anisotropy has an important influence on predicted short-term ground movements caused by tunnel excavation (e.g. Addenbrooke, 1996; Simpson *et al.*, 1996; Lee & Ng, 2002; Franzius, 2003). Similar findings were made in this study for both surface and subsurface ground movements, the use of stiffness anisotropy providing a better match to the field data on short-term ground movements (Wongsaroj *et al.*, 2004). The influence of stiffness anisotropy on long-term ground response is presented in this paper, based on results of analyses assuming $K_0 = 1\cdot5$ for London Clay. Further investigation showed that the agreement between predictions and measurements could be further improved by reducing the value of K_0 to $1\cdot2$ (Wongsaroj, 2005).

The excess pore pressure contours around the tunnel computed with the isotropic and anisotropic soil models are shown in Fig. 13. The zone of negative excess pore pressures with the isotropic model has contours that spread in the horizontal direction, whereas those computed with the anisotropic soil model give contours that spread more in the vertical direction. The computed effective stress paths at the tunnel crown and springline closely followed that of an undrained triaxial compression test on a horizontally and vertically cut sample respectively (Yimsiri, 2001; Hight *et al.*, 2003). That is, the soil adjacent to the tunnel crown extrados experiences a dilation tendency whereas the soil adjacent to the tunnel springline extrados experiences a tendency to contraction prior to dilation. This initial difference is due to the initial elastic stiffness anisotropy (e.g. Graham & Houlsby, 1983) before the plastic dilation develops at larger strains. The isotropic model gave similar stress paths above the tunnel crown and near the springline: both displayed some contraction tendency followed by dilation, defined by the isotropic soil model. Soil elements adjacent to the tunnel springline extrados experienced more shearing and greater negative excess pore pressure than soil elements adjacent to the tunnel crown extrados; this was due to the initial K_0 (>1) condition.

The computed pore pressure profile above the tunnel crown is plotted in Fig. 14. The figure shows that, at a depth of 27 m, the simulation with the anisotropic model gave a larger reduction in pore pressure than that computed with the isotropic soil model. The computed pore pressures above the tunnel crown remain positive at most depths apart from the zone very close to the tunnel crown. Based on Fig. 14, which includes field measurements of pore pressure, and the development of excess pore pressures during tunnel excavation (not shown), the trend of change in pore pressures computed with the anisotropic soil model agrees more closely with the field data than that computed with the isotropic model.

When the tunnel lining is acting as a partially permeable boundary, the pattern of excess pore pressures generated around the tunnel with the anisotropic model led to larger consolidation settlement above the tunnel centreline, causing the tunnel lining to squat in the long term. This is indicated by the increase in the horizontal diameter, as shown in Fig. 15.

The direct comparison between this prediction and field measurements cannot be made owing to the lack of field data for tunnel lining deformation during the rest period.

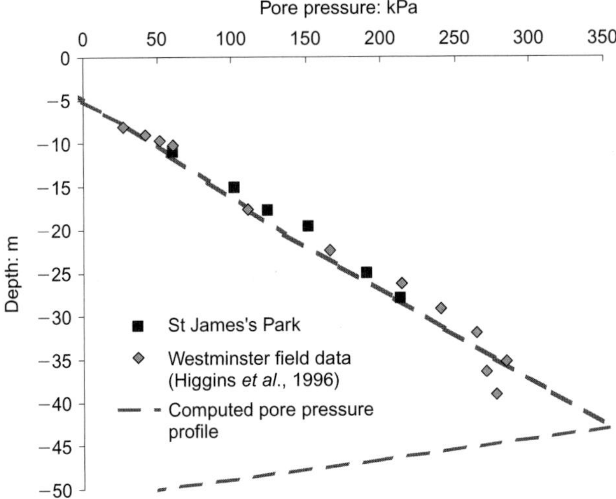

Fig. 12. Pore pressure profile derived from permeability profile 2 (see Fig. 11)

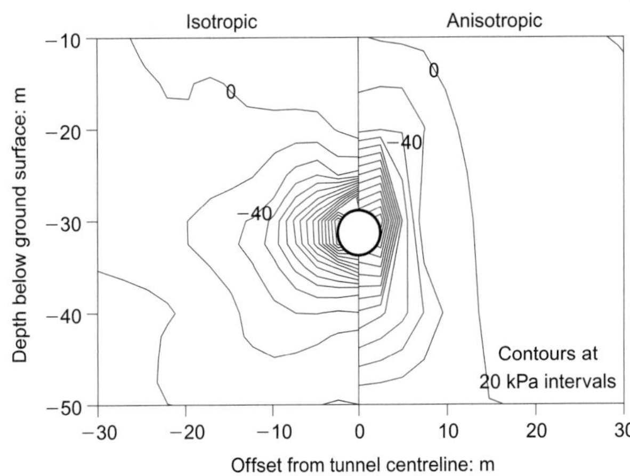

Fig. 13. Excess pore pressure contours around tunnel computed with isotropic and anisotropic soil models in 3D

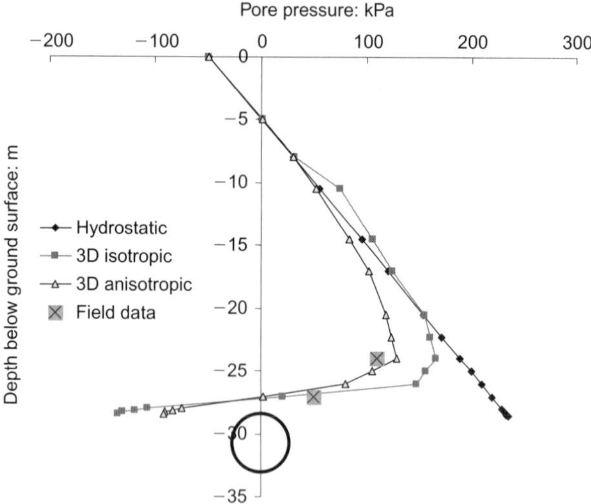

Fig. 14. Computed short-term pore pressure profile above west-bound tunnel crown

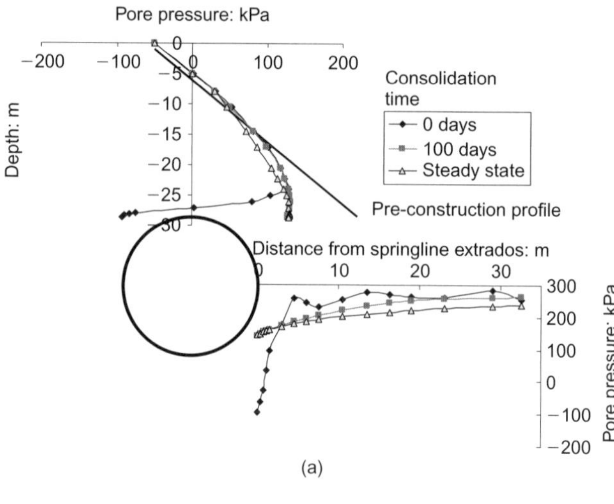

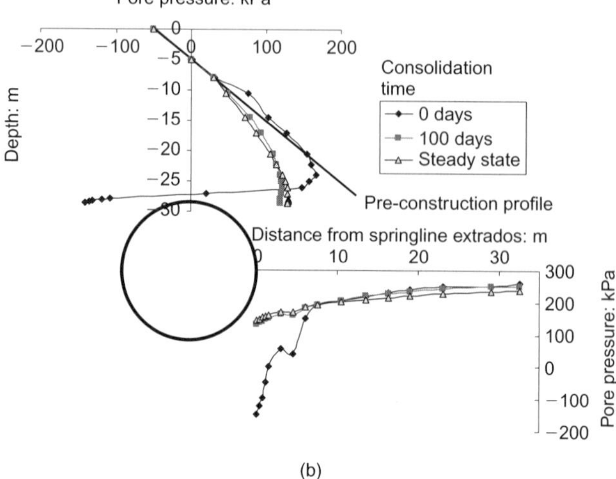

Fig. 16. Influence of soil stiffness anisotropy on long-term pore pressure profiles: (a) anisotropic soil model; (b) isotropic soil model

However, the trend of squatting with time has been observed in other tunnels in London Clay (e.g. Ward & Thomas, 1965; Nyren, 1998 (during period 5); Dimmock, 2003). The analysis with the isotropic model gave the opposite trend: that is, decreasing horizontal diameter with time, as shown in Fig. 15.

The computed pore pressure profiles immediately after, and 100 days after, the construction, and at the steady-state long term are plotted in Fig. 16; these are shown above the tunnel centreline and at the springline level. With the anisotropic model (Fig. 16(a)), there is a rise in pore pressure in the zone up to 3 m from the springline extrados. From this point the pore pressure had to decrease to reach the long-term steady state. Immediately above the tunnel crown there is a recovery of pore pressure from the tunnel crown extrados to approximately 3 m above it. In the soil within 3 m from the crown extrados there is a small reduction in pore pressure over the first 100 days of consolidation, and then there is a rise in the pore pressure to the steady state. Above the tunnel crown the soil swelled, whereas at the sides and below the tunnel the soil consolidated. The volume of consolidation was greater than that of swelling. This led to overall consolidation, which in turn decreased the thickness of the soil on both sides of the tunnel. Hence the tunnel lining squatted in the long term. This computed long-term deformation agrees with the field measurements and coincides well with the computed hoop load, which is

larger at the tunnel springline than at the crown, as reported for tunnels in London Clay (Barratt et al., 1994; Dimmock, 2003).

With the isotropic model there is a substantial increase in pore pressure during the consolidation in the zone from the springline extrados to 8 m away from it, as shown in Fig. 16(b). The majority of the soil above the tunnel crown underwent consolidation due to the reduction in pore pressure, except for the soil from the tunnel crown extrados to approximately 2 m above it, which experienced a rise in pore pressure. The rise in pore pressure at the springline and the majority of the reduction in the pore pressure above the tunnel crown caused the soil to swell at the springline and consolidate above the crown respectively. This in turn causes the tunnel lining to elongate in the vertical direction.

In summary, stiffness anisotropy is important not only to evaluate better short-term ground movements, but also to predict long-term ground and tunnel lining movements in London Clay.

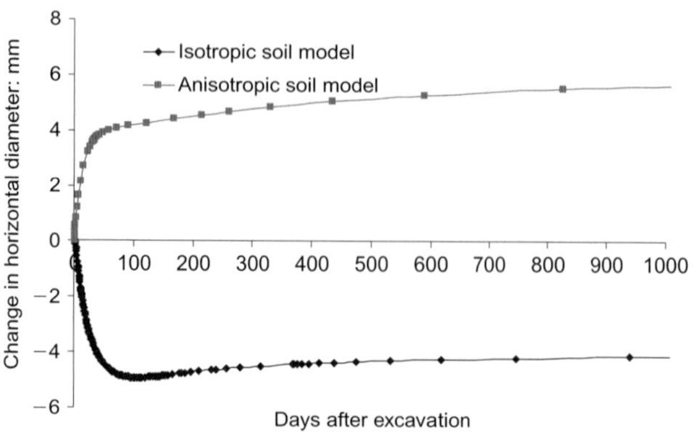

Fig. 15. Influence of stiffness anisotropy on tunnel lining deformation following tunnel excavation (increase of horizontal diameter shown positive)

IMPORTANCE OF PERMEABILITY PROFILE AND TUNNEL DRAINAGE CONDITIONS FOR LONG-TERM GROUND RESPONSE

Simplified permeability model case

The drainage condition of the tunnel boundary is often assumed to be influenced by the relative magnitude of the tunnel lining permeability and the surrounding soil permeability. In order to reasonably simulate the ground movements during the long-term consolidation in the field, the computed changes in pore pressure should have similar values to those observed in the field.

The permeability of the expanded concrete tunnel lining was evaluated by carrying out simulations of the consolidation around the westbound tunnel with different K_T values (1×10^{-14}, 1×10^{-13}, 1×10^{-12} and 1×10^{-11} m³/kN.s). For the first 300 days of consolidation (i.e. before the passage of the eastbound tunnel excavation beneath the instrumented site), the computed pore pressures at piezometers BP1 and BP2 (see Fig. 1), which are located above the westbound tunnel crown extrados, are shown in Fig. 17 along with the field measurements. The computed pore pressures at these points show an increase with time for all values of K_T. The computed response at BP2, where it is likely to be most affected by the lining-soil relative permeability, was closest to the field measurement when the value of K_T was 1×10^{-11} m³/kN.s. Non-zero pore pressures computed at the tunnel crown extrados presented in Fig. 17(c) suggest that the tunnel lining was acting as a partially permeable boundary.

The magnitude of the mass permeability of the tunnel lining can be estimated from the definition of K_T given earlier. For a K_T value of 1×10^{-11} m³/kN.s and a lining thickness of 0·2 m, the mass permeability of the tunnel lining becomes 2×10^{-11} m/s. This appears to be two orders of magnitude greater than the value of maximum concrete permeability of 1×10^{-13} m/s assumed for the design of the Jubilee Line Extension (Winterton, 1994). The larger permeability inferred from the simulation suggests that the joints between concrete segments of an unbolted expanded lining increase the mass permeability of the tunnel lining substantially, particularly in the portion above the tunnel shoulder. This may also be true of the lower part, but pore pressure measurement is required for such an evaluation.

Using the results from the analysis with the lining permeability value evaluated above, the consolidation component of vertical displacements was examined by comparing the computed values and the field measurements at various locations of extensometers (Fig. 1). Fig. 18 shows the computed subsurface vertical ground movements with depth at various extensometer locations. Despite the close matching of pore pressures near the tunnel (see Fig. 17(b)), the computed response of vertical subsurface displacements at extensometers Bx and Cx are quite different from the field measurements shown in Fig. 5(a). They show initial upward displacements and subsequent downward movements above the tunnel springline. Overall, the soil around the tunnel was predicted to swell significantly owing to the recovery of the pore pressure. Although the tunnel is in a partially drained condition, it is almost impermeable. Outside this swelling zone, the soil consolidated. With the assumed permeability of the soil being homogeneous, the consolidation is predicted to occur more or less radially, which leads to the consolidation of the London Clay even at a depth close to the clay/gravel interface (at 8 m depth).

This long-term ground movement mechanism is completely different from the mechanism proposed from the field data interpretation described earlier in the paper (see Fig. 5). The predicted swelling should occur only in a localised zone above the tunnel, and consolidation should occur at the sides of the tunnel at the tunnel axis level. The soil above the swelling and consolidating zones (see Figs 5(b) and 6(b)) settles in a rigid body manner, causing further settlement at the ground surface.

In the zone at further distance from the tunnel centreline (i.e. from the extensometer Dx outwards in Figs 18(c) and 18(d)), the analysis gives a fairly accurate consolidation

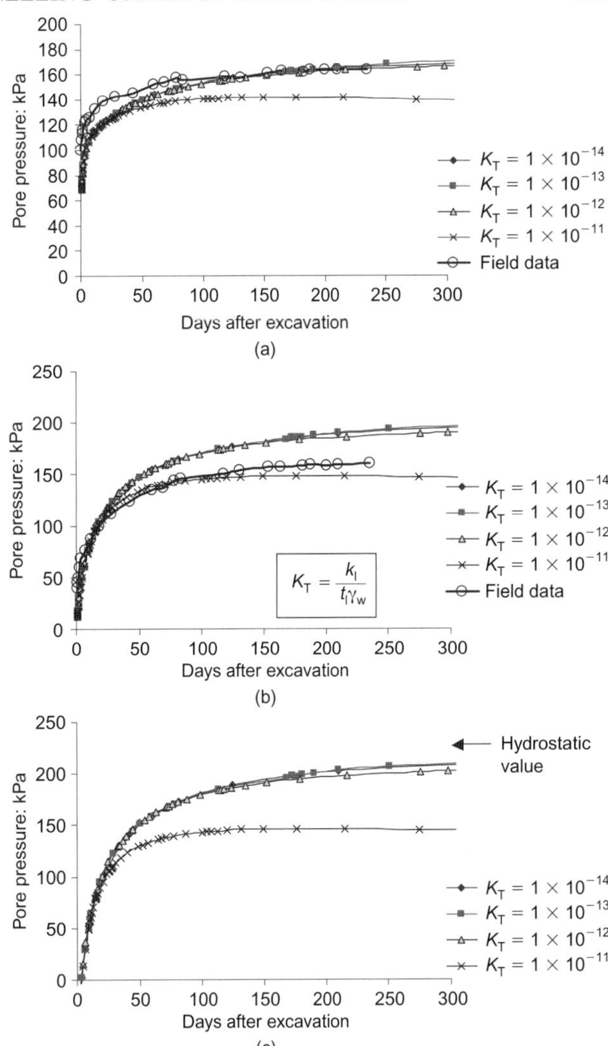

Fig. 17. Computed pore pressure response with simplified permeability model compared with field data: (a) BP1 at 24 m below ground surface; (b) BP2 at 27 m below ground surface; (c) tunnel crown extrados

response in comparison with the field data. However, field measurements indicate that consolidation appears to be more confined in the zone at the same level as the tunnel axis, whereas the analysis seems to predict consolidation that is more uniformly throughout the clay region. The confined zone of the consolidation displacements that occurs mainly close to the tunnel horizon in the field suggests that there exist layers of larger permeability within this zone of the London Clay.

The simulation results described in this section indicate that modelling the London Clay as a uniform homogeneous material with isotropic permeability with the tunnel lining as a uniform drainage boundary was not able to simulate the subsurface ground response, especially close to the tunnel boundary. A more realistic soil permeability profile and a more appropriate drainage condition for the tunnel lining are required in the analysis in order to predict the same mechanism of ground response as that interpreted from the field extensometer data. This is described in the next section.

Refined permeability model case

The in-tunnel observations by Nyren (1998) reported that there were damp and wet patches only below the springline

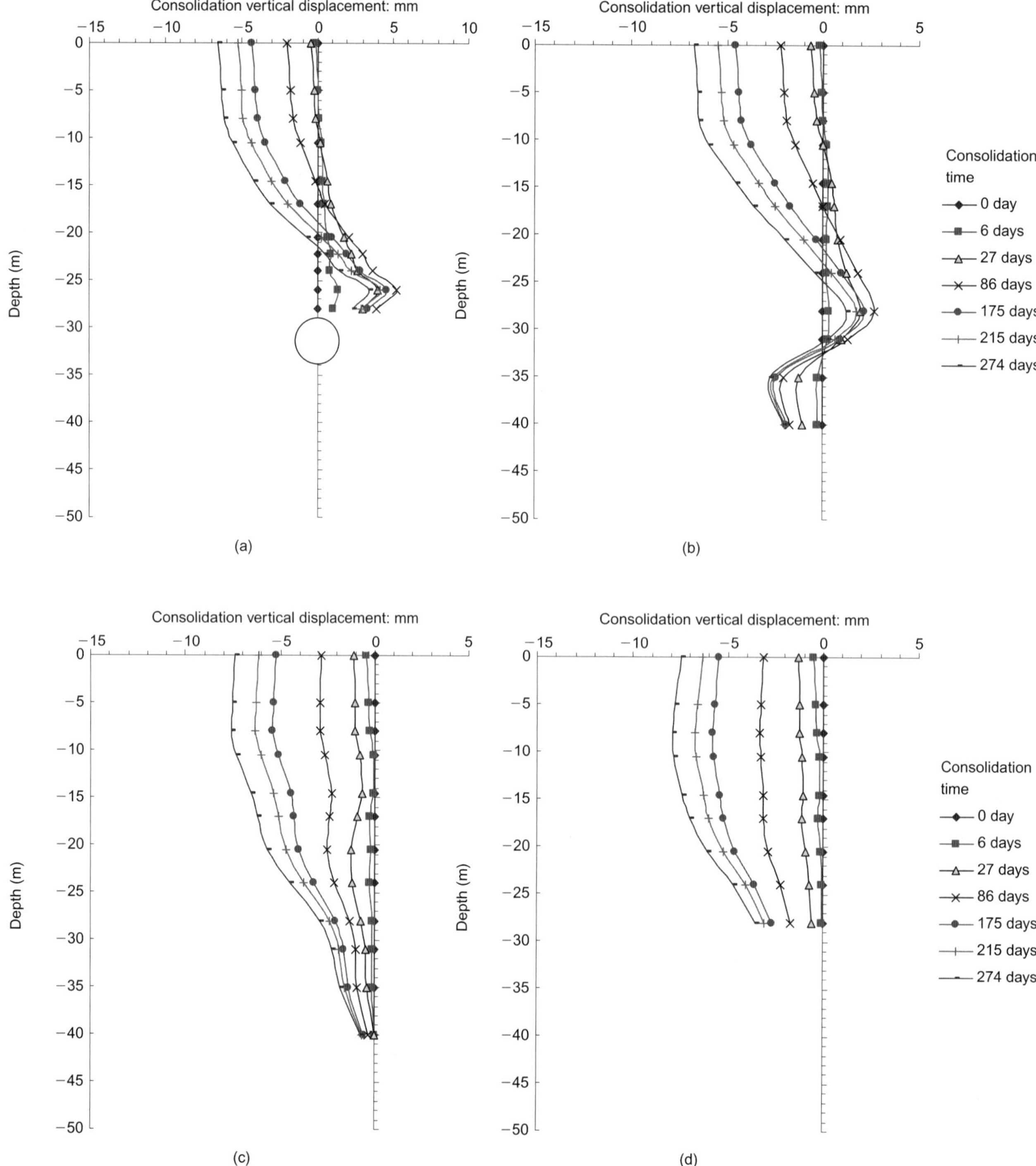

Fig. 18. Computed subsurface ground vertical movement with simplified permeability model: (a) Bx along tunnel centreline; (b) Cx 4 m from tunnel centreline; (c) Dx 10 m from tunnel centreline; (d) Ex 16 m from tunnel centreline

level. This suggests that the tunnel is significantly more permeable below the springline level: this may be associated with the key segments being smaller than the other types (see Fig. 2), because there were holes at the key locations of the tunnel. Although these holes were filled with in situ concrete, their permeability would probably be much higher than that of the precast concrete segments manufactured under controlled conditions. For demonstration purposes, the lower part of the tunnel below the springline was modelled as a permeable boundary in this analysis. To ensure the permeable condition at the tunnel below the springline, a large value of K_T relative to the soil permeability of 1×10^{-5} m^3/kN.s was assigned to this region. As a result, the tunnel can still act as an imperme-

able boundary at the top half and permeable at the bottom half. Fig. 19(a) shows the distribution of pore pressure around the tunnel extrados when K_T of the upper part was 1×10^{-12} m^3/kN.s. The partial recovery of pore pressure is computed above the tunnel springline, whereas nominally zero pore pressure is computed at the bottom half of the tunnel lining extrados.

Figures 19(b) and 19(c) show the recovery of pore pressures with time above the tunnel crown for a range of K_T values. At BP1, which is approximately 4·5 m above the tunnel crown extrados, the recovery of pore pressure seems to be less dependent on the relative lining–soil permeability. This is because the magnitude of pore pressure depends mainly on the horizontal flow of pore fluid due to the

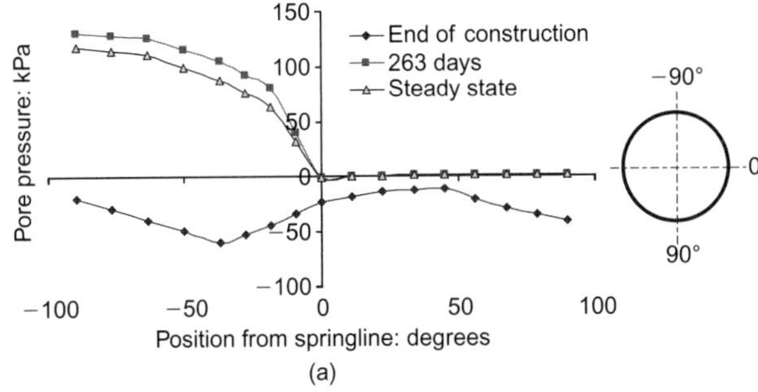

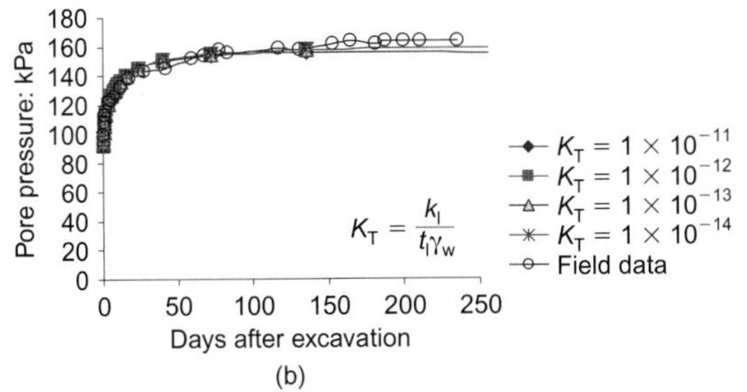

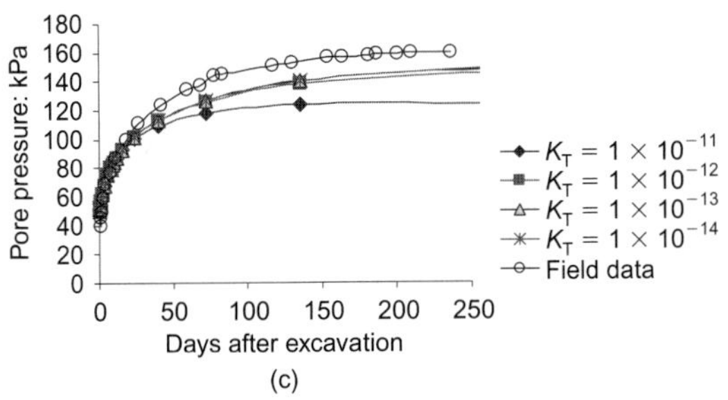

Fig. 19. Computed pore pressure response with refined permeability model compared with field data: (a) along the tunnel lining extrados; (b) BP1 at 24 m below ground surface; (c) BP2 at 27 m below ground surface

anisotropic permeability of the layer. Fig. 19(c) shows that the recovery of pore pressures at BP2, which is closer to the tunnel crown extrados than BP1, depends on the permeability of the tunnel lining. To achieve the pore pressure values close to that observed in the field, the portion of the lining above the springline must be made impermeable. Despite this, such a drainage condition causes the computed pore pressures at the tunnel crown to have slightly lower values than the field measurement. In order to achieve a better agreement with the field data, further studies on the tunnel lining and surrounding soil permeability are needed.

The computed subsurface vertical ground movements with depth at various extensometer locations are shown in Fig. 20. The rate of further settlement agrees well with the field measurement (see Fig. 5), and also the computed transverse settlement profile gives a very similar mechanism to that shown by the measurements. Non-homogeneous drainage towards the lining allowed the pore water to flow into the tunnel below the springline, causing the soil on either side of the tunnel to consolidate. Moreover, it also allowed pore pressures above the tunnel crown to partially recover, causing the soil to swell. As the water drained into the lower half of the tunnel, the increased permeability in division A3i created a slightly faster consolidation rate than the recovery of pore pressure above the tunnel. This enables the downward displacements of the soil above the tunnel crown, as shown in Fig. 20(a). The higher anisotropic permeability ratio assigned for division A3ii compared with the other layers allowed the pore pressure above the tunnel crown in

this division to recover to a higher magnitude than in the simplified model case. This is clearly shown by the larger pore pressure at BP1 computed with the permeability profile 2 (Fig. 19(b)) than that with profile 1 (Fig. 17(a)).

The findings from the results of different permeability models emphasise that appropriate tunnel drainage conditions should be assigned for a long-term analysis. The drainage conditions should take into account the presence of joints, gaps and the relatively high-permeability region in the tunnel lining, although how to account for joints and gaps is far from straightforward. Furthermore, the rate of the consolidation settlement depends principally on the permeability of the soil at the tunnel axis level, because the majority of the consolidation takes place on either side of the tunnel. This is important, particularly for the presented case where the London Clay consists of various sub-geological units (King, 1981; Hight *et al.*, 2003).

CONCLUSIONS

This study has investigated the long-term ground movements following a tunnel excavated in overconsolidated London Clay and lined with an expanded concrete lining. Based on the comprehensive set of field monitoring data obtained from St James's Park, it is found that the mechanism of long-term subsurface movements consists of a combination of swelling, consolidation, and rigid body movement. The swelling takes place in a small zone above the tunnel crown. The consolidation takes place on both sides of the tunnel, extending to a large offset from the tunnel centreline. Above the swelling and consolidation zones, the soil settles as a rigid body.

Modelling of soil layering within London Clay and application of non-homogeneous drainage conditions around the lining (i.e. with the tunnel impermeable above the springline and permeable below the springline) were necessary to accurately simulate the long-term ground response observed at St James's Park. Modelling of soil layering allows the change in pore pressures in the soil to occur at the same rate as those monitored in the field. As the majority of the consolidation takes place on either side of the tunnel, the rate of the consolidation settlement therefore depends principally on the permeability of the soil at the tunnel springline level. The application of non-homogeneous drainage conditions around the lining allows the pore water to flow into the tunnel below the springline, causing the soil on either side of the tunnel to consolidate. Moreover, it also allows the pore pressure above the tunnel crown to partially recover, causing the soil to swell. Hence it appears that appropriate tunnel drainage conditions must be assigned for successful long-term analysis. It is recognised, however, that these conclusions are based on the analyses reported in this paper of a case study with only a small number of pore pressure measurements around the tunnel. More long-term pore pressure measurement data are needed to confirm them.

Soil stiffness anisotropy is another factor affecting the long-term ground response for a high-K_0 soil. When the tunnel lining was assigned to be relatively permeable, the magnitudes of negative pore pressures were larger above and below the tunnel when the anisotropic soil model was used. This resulted in a squatting deformation of the tunnel lining in the long term, which was in good agreement with the deformations observed in the field. When the isotropic soil model was used, the opposite trend was predicted. The magnitudes of negative excess pore pressure were greater at the sides of the tunnel, which led to a decrease in horizontal diameter in the long term.

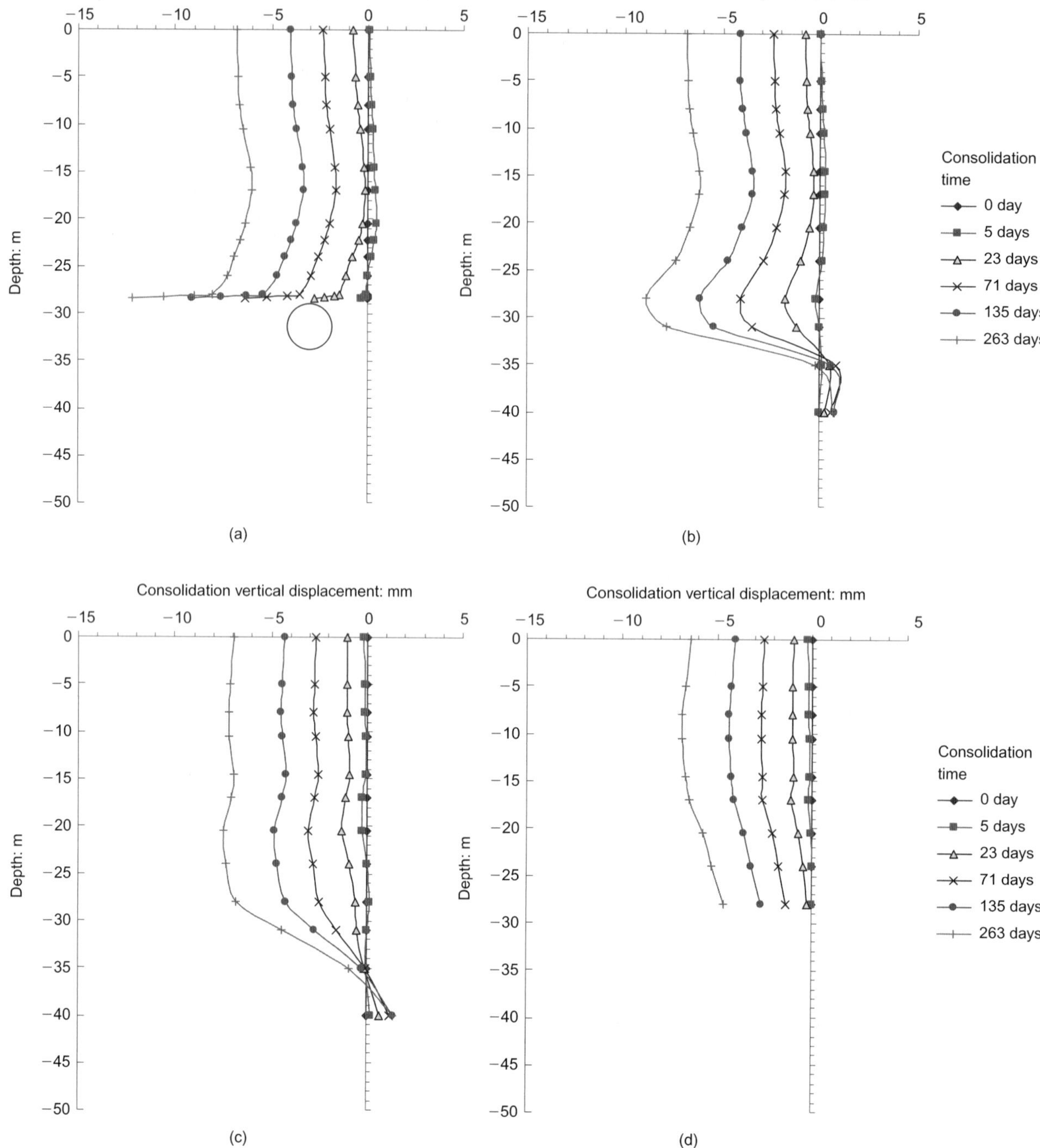

Fig. 20. Computed subsurface ground vertical movement with refined permeability model: (a) Bx along tunnel centreline; (b) Cx 4 m from tunnel centreline; (c) Dx 10 m from tunnel centreline; (d) Ex 16 m from tunnel centreline

APPENDIX 1

The normal yield surface adopted in this study is the Modified Cam Clay yield surface, which can be described in q–p' space as

$$F = q^2 - M^2(p_0' - p')p' = 0 \tag{1}$$

where M is the gradient of the critical state line and a function of the Lode angle using the Matsuoka & Nakai (1985) failure criterion, and p_0' is a constant defining the size of the normal yield surface (i.e. isotropic preconsolidation pressure). For a given value of R, the subloading surface can then be expressed as

$$F^R = q^2 - M^2(Rp_0' - p')p' = 0 \tag{2}$$

Two different evolution laws for R were proposed by Hashiguchi & Chen (1998). The one adopted in this study is expressed as

$$dR = u_1\left(\frac{1}{R^m} - 1\right)\|d\varepsilon^p\| \tag{3}$$

where u_1 and m are the material constants, and $d\varepsilon^p$ is the change in plastic strains tensor, which is obtained from the flow rule. For simplicity, the associated flow rule is assumed in the model and, therefore, the plastic potential is described by equation (2).

The conditions for the elasto-plastic process are given by

$$\left.\begin{array}{ll} R = 0: & dR = +\infty \\ 0 < R < 1: & dR > 0 \\ R = 1: & dR = 0 \\ R > 1: & dR < 0 \end{array}\right\} \quad \text{for} \quad \varepsilon^p \neq 0 \tag{4}$$

The elastic modulus of the model is assumed based on experimental data plotted in $\log_{10}e$–$\log_{10}p'$ space, where swelling behaviour is described by the gradient of the swelling line in $\log_{10}e$–$\log_{10}p'$ as (after Pestana, 1994)

$$\rho_r = \frac{1 + \omega_s\xi_s}{C_b}\left(\frac{p'}{p_a}\right)^{1/2} + D(1 - \xi^r) \tag{5}$$

Table 4. Model parameters for all soil units

Strata	M	e	u_l	m	C_b	ω_s	λ	ρ_c	D	r	ν'_{vh}	ν'_{hv}	ν'_{hh}	G_{hh}/G_{vh}
Made Ground	0·984	0·65	100	0·1	100	15	0·1	0·2476	—	—	0·2	0·2	0·2	1
Terrace Gravel	1·418	0·5	100	0·1	400	15	0·1	0·556	—	—	0·2	0·2	0·2	1
London Clay	0·814	0·7	300	0·2	200	20	—	0·3	0·05	2	0·07	0·16	0·12	1·5
Woolwich and Reading Beds Clay	1·07	0·65	100	0·1	900	50	0·15	0·37	0·05	2	0·2	0·2	0·2	1

where C_b and ω_s are material constants, which control the initial gradient of the swelling line and the non-linearity of the one-dimensional (1D) swelling line respectively; D is the gradient of the isotropic swelling line at low mean effective pressure; r controls the rate at which the isotropic swelling line reaches the gradient D; and atmospheric pressure p_a of 100 kPa is assumed for this study. ξ and ξ_s are used to re-invoke small-strain stiffness at stress reversal points (i.e. unloading and reloading). They are the dimensionless distances in space as defined by

$$\xi_s = \left\{ (\boldsymbol{\eta} - \boldsymbol{\eta}_{rev}) : (\boldsymbol{\eta} - \boldsymbol{\eta}_{rev}) \right\}^{1/2} \tag{6}$$

$$\xi = \begin{cases} p'/p'_{rev} & \text{for} \quad p' < p'_{rev} \\ p'_{rev}/p' & \text{for} \quad p' \geqslant p'_{rev} \end{cases} ; \xi \leqslant 1 \tag{7}$$

where $\boldsymbol{\eta}$ is the ratio of deviatoric stress tensor to mean effective stress, and $\boldsymbol{\eta}_{rev}$ is the value of the stress ratio at the most recent stress reversal.

The stress reversal point is defined as (Pestana, 1994)

$$\chi\dot{\chi} = \begin{cases} ^{\Delta 1}\varepsilon_p \delta\varepsilon_p & \text{for } \delta\varepsilon_p \neq 0 \\ ^{\Delta 1}\boldsymbol{\varepsilon}_s : \delta\boldsymbol{\varepsilon}_s & \text{for } \delta\varepsilon_p = 0 \end{cases} = \begin{cases} > 0 & \text{loading} \\ \leqslant 0 & \text{unloading (SRP)} \end{cases} \tag{8}$$

where $^{\Delta 1}\varepsilon$ is the accumulated strain relative to the previous stress reversal state, which can be decomposed into its volumetric and deviatoric components, $^{\Delta 1}\varepsilon_p(= \varepsilon_p - \varepsilon_{prev})$ and $^{\Delta 1}\boldsymbol{\varepsilon}_s(= \boldsymbol{\varepsilon}_s - \boldsymbol{\varepsilon}_{srev})$.

From equation (5), the tangent bulk modulus, K', during swelling is defined as

$$K' = \left(\frac{1+e}{e} \right) \frac{p'}{\rho_r} \tag{9}$$

By starting from the definition of bulk modulus and the assumption of the symmetry of the elastic D-matrix, it can be shown that

$$E'_h = E'_v \frac{\nu'_{vh}}{\nu'_{hv}} \tag{10}$$

$$E'_v = K' \left(1 - 4\nu'_{vh} + \frac{2\nu'_{vh}}{\nu'_{hv}} - \frac{2\nu'_{hh}\nu'_{vh}}{\nu'_{hv}} \right) \tag{11}$$

Under the assumption that the soil is isotropic in the plane of the deposition, the value of G_{hh} becomes

$$G_{hh} = \frac{E'_h}{2(1 + \nu'_{hh})} \tag{12}$$

The values of ν'_{vh}, ν'_{hv}, ν'_{hh} and G_{vh} are evaluated from published data.

For the present model, it is assumed that the normal consolidation line is a straight line with a gradient of ρ_c. The hardening rule therefore becomes

$$dp'_0 = \frac{1+e}{e(\rho_c - \rho_r)} p'_0 d\varepsilon_v^p \tag{13}$$

where p'_0 is the mean effective preconsolidation pressure, which controls the size of the normal yield surface, and $d\varepsilon_v^p$ is the plastic volumetric strain increment.

NOTATION

C_b parameter controlling initial gradient of swelling line or magnitude of small-strain stiffness
D gradient of 1D swelling line at low mean effective stress
E'_h drained Young's modulus in horizontal direction
E'_v drained Young's modulus in vertical direction
e void ratio
F normal yield surface function
F^R subloading surface function
G_{hh} shear modulus in horizontal plane
G_{vh} shear modulus in vertical plane
K' bulk modulus during swelling
K_T seepage coefficient
K_0 ratio of initial vertical to horizontal effective stress
k permeability of soil
k_h permeability of soil in horizontal direction
k_l permeability of tunnel lining
k_v permeability of soil in vertical direction

M gradient of critical state line in triaxial space

m, u_1 parameters controlling evolution law of R

p' mean effective stress

p_a atmospheric pressure; taken as 100 kPa

p_0' mean effective preconsolidation pressure

p_{rev}' mean effective stress at most recent stress reversal point

q deviatoric stress in triaxial space

R ratio between size of subloading surface and size of normal yield surface

r parameter controlling rate at which 1D swelling line reaches gradient D

t_l thickness of tunnel lining

u_w pore pressure at tunnel lining extrados

v_n pore fluid velocity outward from soil element

γ soil unit weight

ε^p plastic strain

ε_p volumetric strain

ε_v^p plastic volumetric strain

ε_{prev} volumetric strain at most recent stress reversal point

ε_s deviatoric strain

ε_{srev} deviatoric strain at most recent stress reversal point

$^{\Delta1}\varepsilon$ accumulated strain relative to the previous stress reversal state

η deviatoric stress tensor

η_{rev} deviatoric stress tensor at most recent stress reversal point

v_{hh}' ratio of strain in horizontal direction to other horizontal direction due to compression in horizontal direction

v_{hv}' ratio of strain in vertical direction to horizontal direction due to compression in horizontal direction

v_{vh}' ratio of strain in horizontal direction to vertical direction due to compression in vertical direction

ξ, ξ_s dimensionless distances used to re-invoke small-strain stiffness at stress reversal points

ρ_r gradient of swelling line in $\log_{10}e$–$\log_{10}p'$

ϕ_{cv}' critical angle of shearing resistance

$\chi\dot{\chi}$ parameter defining stress reversal point

ω_s parameter controlling non-linearity of one-dimensional swelling line

REFERENCES

Addenbrooke, T. (1996). *Numerical analysis of tunnelling in stiff clay*. PhD thesis, Imperial College of Science Technology and Medicine, London.

Barratt, D. A., O'Reilly, M. P. & Temporal, J. (1994). Long term measurement of loads on tunnel linings in overconsolidated clay. *Proc. Tunnelling '94, London*, 469–481.

Bowers, K. H., Hiller, M. D. & New, B. M. (1996). Ground movement over three years at Heathrow Express trial tunnel. *Proceedings of the international symposium on geotechnical aspects of underground construction in soft ground*, London, pp. 557–562.

Burland, J. B. & Hancock, R. J. (1977). Underground car park at the House of Commons, London: geotechnical aspects. *Struct. Engr* **55**, No. 2, 87–100.

Burland, J. B., Standing, J. R. & Jardine, F. M. (eds) (2001). *Building response to tunnelling*. London: CIRIA/Thomas Telford.

Dimmock, P. S. (2003). *Tunnelling induced ground and building movement on the Jubilee Line Extension*. PhD thesis, University of Cambridge.

Franzius, J. N. (2003). *Behaviour of buildings due to tunnel induced subsidence*. PhD thesis, Imperial College of Science Technology and Medicine, London.

Graham, J. & Houlsby, G. T. (1983). Anisotropic elasticity of natural clay. *Géotechnique* **33**, No. 2, 165–180.

Hashiguchi, K. & Chen, Z. P. (1998). Elastoplastic constitutive equation of soils with the subloading surface and the rotational hardening. *Int. J. Numer. Anal. Methods Geomech.* **22**, No. 3, 197–227.

Hashiguchi, K. & Ueno, M. (1977). Elastoplastic constitutive law of granular material: constitutive equations for soils. *Proc 9th Int. Conf. Soil Mech. Found. Engng, Tokyo*, Special Session 9, 73–82.

Harris, D. I. (2002). Long-term settlement following tunnelling in overconsolidated London Clay. *Proc. 3rd Int. Symp. on Geotechnical Aspects of Underground Construction in Soft Ground, Toulouse*, 393–398.

Higgins, K. G., Potts, D. M. & Mair, R. J. (1996). Numerical modelling of influence of the Westminster Station excavation on the Big Ben clock tower. In *Geotechnical aspects of underground construction in soft ground* (eds R. J. Mair and R. N. Taylor), pp. 523–530. Rotterdam: Balkema.

Hight, D. W., McMillan, F., Powell, J. J. M., Jardine, R. J. & Allenou, C. P. (2003). Some characteristics of London Clay. *Proceedings of the conference on characterisation and engineering properties of natural soils*, Singapore, Vol. 2, pp. 851–907.

King, C. (1981). *The stratigraphy of the London Basin and associated deposits*, Tertiary Research Special Paper No. 6. Rotterdam: Backhuys.

Lake, L. M., Rankin, W. J. & Hawley, J. (1992). *Prediction and effects of ground movements caused by tunnelling in soft ground beneath urban areas*, CIRIA Funders Report CP/5. London: Construction Research and Information Association.

Lee, G. T. K. & Ng, C. W. W. (2002). Three-dimensional analysis of ground settlement due to tunnelling: role of K_0 and stiffness anisotropy. *Proc. 3rd Int. Symp. on Geotechnical Aspects of Underground Construction in Soft Ground, Toulouse*, 617–622.

Matsuoka, H. and Nakai, T. (1985). Relationship among Tresca, Mises, Mohr–Coulomb and Matsuoka–Nakai failure criteria. *Soils Found.* **25**, No. 4, 123–128.

Nyren, R. (1998). *Field measurement above twin tunnel in London Clay*. PhD thesis, Imperial College of Science Technology and Medicine, London.

O'Reilly, M. P., Mair, R. J. & Alderman, G. H. (1991). Long-term settlements over tunnels: an eleven-year study at Grimsby. *Proc. Tunnelling '91, London*, 55–64.

Palmer, J. H. L. & Belshaw, D. J. (1980). Deformations and pore pressure in the vicinity of a precast, segmented, concrete-lined tunnel in clay. *Can. Geotech. J.* **17**, No. 2, 174–184.

Peck, R. B. (1969). Deep excavations and tunnelling in soft ground. *Proc. 7th Int. Conf. Soil Mech. Found. Engng, Hamburg*, State of the Art Volume, 225–290.

Pestana, J. M. (1994). *A unified constitutive model for clays and sands*. PhD thesis, Massachusetts Institute of Technology.

Potts, D. M. & Zdravkovic, L. (2001). *Finite element analysis in geotechnical engineering: application*. London: Thomas Telford, 35–72.

Simpson, B., Atkinson, A. J. & Jovicic, V. (1996). The influence of anisotropy on calculations of ground settlements above tunnels. In *Geotechnical aspects of underground construction in soft ground* (eds R. J. Mair and R. N. Taylor), pp. 591–594. Rotterdam: Balkema.

Standing, J. R. & Burland, J. B. (2006). Unexpected tunnelling volume losses in the Westminster area, London. *Géotechnique* **56**, No. 1, 11–26.

Ward, W. H. & Thomas, H. S. H. (1965). The development of earth loading and deformation in tunnel lining in London Clay. *Proc. 6th Int. Conf. Soil Mech. Found. Engng, Toronto* **2**, 432–436.

Winterton, T. R. (1994). Developments in precast concrete tunnel linings in the United Kingdom. *Proc. Tunnelling '94, London*, 601–633.

Wongsaroj, J. (2005). *Three-dimensional finite element analysis of short and long-term ground response to open-face tunnelling in stiff clay*. PhD thesis, University of Cambridge.

Wongsaroj, J., Soga, K., Mair, R. J. and Yimsiri, S. (2004). Stiffness anisotropy of London Clay and its modelling: laboratory and field. *Proceedings of the Skempton Conference on advances in geotechnical engineering*, London, Vol. 2, pp. 1205–1216.

Yimsiri, S. (2001). *Pre-failure deformation characteristic of soils: anisotropy and soil fabric*. PhD thesis, University of Cambridge.

Kovacevic, N., Higgins, K. G., Potts, D. M. & Vaughan, P. R. (2007). *Géotechnique* **57**, No. 2, 181–195

Undrained behaviour of brecciated Upper Lias Clay at Empingham Dam

N. KOVACEVIC*, K. G. HIGGINS*, D. M. POTTS† and P. R. VAUGHAN†

Empingham Dam was built in the early 1970s on a brecciated Upper Lias Clay foundation of fill derived from it. When the dam was designed, the undrained behaviour of intact stiff plastic clay in the field was poorly understood. Thus, a conservative bound to conventional strength tests was adopted to provide an undrained strength profile in the foundation. The shearing resistance of the foundation was also investigated by incorporating an instrumented 24 m high trial bank in the upstream berm of the main dam. As the trial bank was nearing completion, two slips occurred in temporary borrow pit slopes. It was considered that this field experience, together with undrained triaxial tests with pore pressure measurement on large diameter samples carried out in the laboratory, justified the original design assumptions. However, the end lengths of the main embankment on the undrained foundation started to spread laterally when the fill was some 2 m below the final level. This was in spite of the fact that the conventional factor of safety, calculated using the original design assumptions, was in excess of two. This paper describes the finite element (FE) back analysis of the observed behaviour. For this purpose two different constitutive models have been employed: a kinematically hardening 'bubble' model accounting for pre-peak plasticity, and a non-linear elastic Mohr-Coulomb plastic model accounting for strain-softening. It is demonstrated that the observed undrained behaviour of the Empingham foundation can be reproduced by FE analyses, but only if the undrained shearing resistance varies from section to section.

Le barrage d'Empingham a été construit au début des années 1970 sur une fondation bréchiforme d'argile d'Upper Lias et le remblayage utilisé provenait de la même argile. Quand le barrage a été conçu, on comprenait mal le comportement non drainé sur le terrain de l'argile plastique ferme intacte. Ainsi, pour la conception, on a adopté une limite conservatrice pour les essais de résistance conventionnels de sorte à fournir un profil de résistance non drainé dans la fondation. On a fait des recherches sur la résistance au cisaillement de la fondation en incorporant un talus d'essais de 24m de hauteur, muni d'instruments, dans la banquette en amont du barrage principal. Alors que la construction du talus d'essais était presque à son comble, il y a eu deux glissements de terrain sur les pentes de l'emprunt temporaire. On a considéré les hypothèses de conception de départ se justifiaient au vu de cette expérience sur le terrain ainsi que des essais triaxiaux non drainés effectués en laboratoire sur des échantillons de grand diamètre, avec mesure de la pression interstitielle. Il a été surprenant de constater que la largeur des extrémités du remblai sur la fondation non drainée commençait à augmenter latéralement lorsque la hauteur du remblayage était à 2m de la hauteur finale. Cela s'est produit en dépit du fait que le facteur de sécurité conventionnel d'équilibre limite (EL), calculé à l'aide des hypothèses de conception originelles pour la résistance non drainée, était supérieur à 2. Cet article décrit les analyses par éléments finis (EF) du comportement observé du barrage d'Empingham. Pour ce faire, on a utilisé deux modèles constitutifs différents pour la fondation. Le premier modèle était un modèle « bulle » de durcissement cinématique de type état critique représentant la plasticité avant pic tandis que le deuxième modèle était un modèle plastique Mohr-Coulomb relativement simple et non linéaire représentant l'anti-écrouissage. Cet article montre qu'on peut reproduire le comportement non drainé de la fondation Empingham qui a été observé à l'aide d'analyses EF mais seulement si la résistance au cisaillement non drainé varie d'une section à l'autre.

KEYWORDS: dams; deformation; earthfill; numerical modelling; shear strength; slopes

---

INTRODUCTION

The design of embankments of and on stiff plastic clays that are of sufficiently low permeability to remain undrained during construction has presented problems in the past. Initially, design was empirical (Vaughan *et al.*, 2004). This did not usually lead to stability difficulties unless there were special circumstances, such as sidelong ground and/or pre-sheared plastic clay foundations. However, when higher embankments (e.g. dams) were attempted in the 20th century, instability often occurred. The application of soil mechanics principles led to design in terms of effective stress, with stability analysis by limit equilibrium (LE) methods. However, effective stress analysis could not be applied to

undrained clay, except by using empirical relationships for the generation of pore pressures.

In principle, a design could be undertaken in terms of total stress, but this required the adoption of undrained strength. It was found that this could not be measured reliably; the strength determined by back-analysis of actual failure was typically less than half that measured by conventional tests (Skempton & La Rochelle, 1965). Moreover, the way that stability could be related to measurement by instrumentation of either pore pressure or deformation was limited (Vaughan & Hamza, 1977).

Both design and monitoring could be related to effective stress if a drained design approach was adopted. Gain in strength due to consolidation could then be taken into account. Horizontal drain layers in fills and vertical sand drains in foundations would ensure this could occur. However, the drained approach was often conservative, particularly for plastic clay fills. There are several embankments that have closely spaced drainage blankets where piezometers showed that pore pressures remained negative

Manuscript received 5 May 2006; revised manuscript accepted 31 November 2006.
Discussion on this paper closes on 1 August 2007, for further details see p. ii.
* Geotechnical Consulting Group, London.
† Imperial College, London, UK.

throughout construction and there was nothing to consolidate (Walbancke, 1976). Furthermore, the granular material required for a drained approach was often from off-site, which was both environmentally unpopular and expensive.

Design of large embankments of, and on, stiff plastic clays has potentially changed with the development of advanced numerical methods. These methods can now handle the properties of these clays' coupling mechanical behaviour and flow during and after construction ('coupled' analyses). Such an approach offers considerable advantages in reproducing reality, improving reliability, reducing cost and improving monitoring. It requires realistic soil models and the results from numerical analyses to be checked for internal consistency and against measured behaviour. However, there has been little construction of this type of structure over the last three decades in the UK, upon which such models can be validated.

Empingham Dam is one of the last large embankments to be built on a stiff plastic clay in the UK, using a fill derived from the same clay (Upper Lias Clay (ULC)). It was designed while finite element (FE) analyses were in their infancy and when variants of LE analysis were used for design (Bridle et al., 1985). It was extensively instrumented, allowing the undrained behaviour of the ULC to be studied in more detail. Nevertheless, the undrained behaviour was still considered to be the greatest uncertainty in the design of structures of this type, and improvements in accounting for it were considered the greatest likely source of economy in any new design of embankments of this kind. Recent advances in numerical modelling of stiff plastic clays and fills derived from them (Vaughan et al., 2004) have made further back-analyses of the various aspects of undrained behaviour associated with construction of Empingham Dam viable.

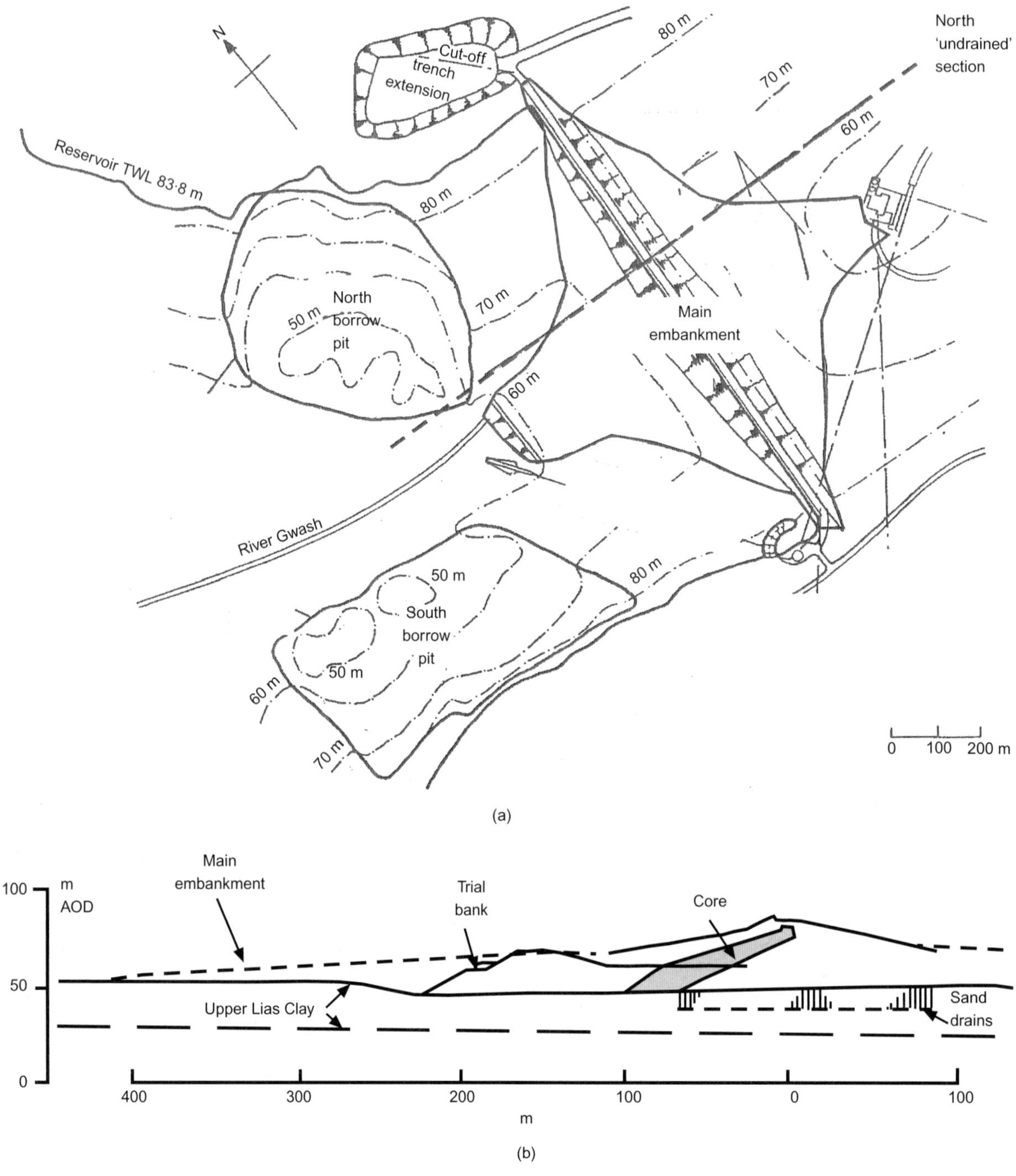

Fig. 1. (a) Plan of Empingham Dam; (b) location of trial bank in main embankment cross-section (after Bridle et al., 1990)

EMPINGHAM DAM

Empingham Dam (Fig. 1(a)) was built for the Rutland Water pumped storage scheme. Site investigations and the preliminary design were undertaken between September 1970 and September 1971 (Bridle *et al.*, 1985). The embankment, nearly 40 m high (Fig. 1(b)), was built between September 1971 and August 1974. The extent of the berms was decided one year after construction had started, when some observations became available. In 1975 the slope protection was placed when impounding started.

The embankment was built on the Upper Lias Clay (ULC), which was overlain by younger deposits of Northampton Sands and Estuarine Series, and by more recent deposits of head and alluvium. In the centre of the valley the ULC was 22 m thick, with thickness increasing towards the valley sides. In the valley slopes the ULC was extensively disturbed by cambering and, in the valley floor, by bulging (Horswill & Horton, 1976), which had involved some 200 m of lateral spreading of the foundation clay (Vaughan, 1976). The clay was then brecciated by periglacial ground freezing.

Brecciated clays, such as the ULC at the site, are characterised by rounded lumps of nearly intact soil closely packed within a rather wetter and softer matrix of disturbed material (Chandler, 1972). This severely disturbed fabric is heterogeneous and therefore causes variation in the undrained shear strength.

The permeability of the brecciated ULC was very low, although below the brecciation the permeability was higher (Bridle *et al.*, 1985). The brecciated material was used as fill, with a core of wetter clay inclined towards the upstream slope to reduce the thrust from the embankment centre (see Fig. 1(b)). The permeability of the fill was even lower than the permeability of the intact clay in the foundation, so the overall dam response during construction was undrained.

The undrained strength of the foundation alone was too low to carry the full height of the embankment. Drainage, to improve strength of the foundation, had been extensively used in the past, and so sand drains were installed below the centre of the embankment, where it was more than 25 m high. However, the extensive toe berms (even in the central portion of the dam) and the ends of the embankment, where it was less than 25 m high, relied solely upon the undrained shearing resistance of the ULC in the foundation. LE analyses, using the original design assumptions, showed that these parts of the dam should be stable without sand drains (Bridle *et al.*, 1985).

As part of the initial construction, an instrumented trial bank over 20 m high and with the upstream slope of 2·5 (horizontal) to 1 (vertical) was built as part of the upstream fill (see Fig. 1(b)) to obtain further information concerning the clay foundation (Bridle *et al.*, 1990). There were two slips in temporary slopes in the borrow pits, which provided further information on the undrained shear strength of the ULC in the foundation. Simple LE analyses of these slips broadly confirmed the assumed undrained strength used for the initial design, and so the trial bank was not taken to failure.

UNDRAINED STRENGTH OF UPPER LIAS CLAY

Figure 2 summarises undrained strength measurements made in the brecciated ULC foundation. Conventional quick undrained triaxial tests on 100 mm samples gave results showing great variability, but quick plate tests gave much less scatter, which, on average, reproduced a similar undrained strength profile.

Maguire (1976) reported the results of the slow triaxial tests with pore pressure measurements and investigated the effect of sample disturbance. As measured for other stiff

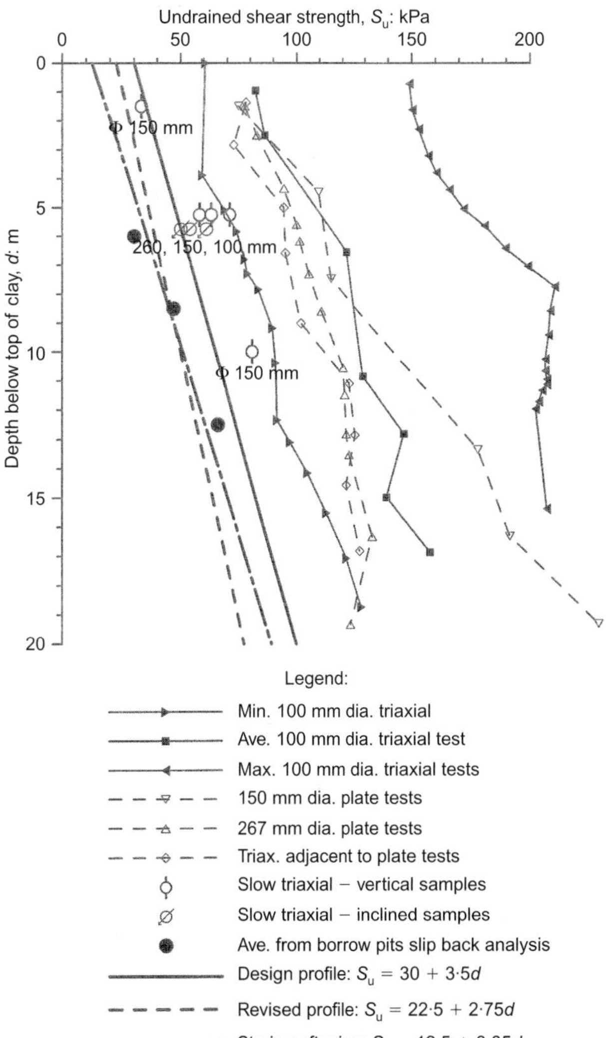

Fig. 2. Undrained strength profiles in Upper Lias Clay foundation at Empingham Dam site

plastic clays (Sandroni, 1977), the undrained strength is influenced by the volume of soil tested, but the effect is largely due to sample disturbance. The undrained strength is dependent on the pore water suction in the sample, which decreases with sample size as the ratio of soil disturbed by the cutting shoe to the undisturbed sample decreases (Vaughan *et al.*, 1992). There was also significant anisotropy; the strength measured in inclined samples was lower than that measured in vertical ones.

It can be seen from Fig. 2 that the quick undrained strength conventionally determined by unconsolidated undrained triaxial tests on 100 mm diameter samples is much higher than the strength at normal engineering loading rates. It had originally been intended to use the results of these tests on 100 mm diameter specimens, obtained by percussion drilling from the grid of holes, to investigate the consistency of the clay across the site. However, these measurements showed considerable scatter. As work proceeded it became apparent that this was a function of variable sample disturbance (Vaughan *et al.*, 1992) and not of the intrinsic properties of the clay. It was pointed out by Vaughan (1994) that the blow counts to take samples during percussion drilling showed a more consistent variation with depth than the undrained strength tests. Thus the blow counts for 100 mm open drive samples offer a useful way of examining clay variability, as will be demonstrated later in the paper.

The slow triaxial tests gave much lower undrained strengths.

These strengths are a function of the mean effective stress after 'perfect sampling' and the effective stress strength (Vaughan *et al.*, 1992; Vaughan, 1994). Assuming that the coefficient of earth pressure at rest is $K_0 = 1$ (a reasonable assumption given the brecciated, frost-shattered and bulged nature of the clay), and that the pore pressure changes as the mean total stress ($\Delta u = \Delta p$), the rate of increase in undrained compression strength (ΔS_u) with depth (Δz) can be expressed as

$$\Delta S_u = \gamma' \tan \phi' \Delta z \qquad (1)$$

where γ' is the clay submerged unit weight and ϕ' is the effective angle of shearing resistance. Substituting $\gamma' = 10$ kN/m^3 and $\phi' = 23°$, the ratio becomes $\Delta S_u = 3{\cdot}54\Delta z$. The original empirical design assumption was $3{\cdot}5$. It should be noted that S_u at the top of the profile depends on the effective stress there and hence on the depth of overburden above the clay. It might be argued that the gradient $\Delta S_u/\Delta z$ should remain constant and that only the intercept should change. However, this would prejudge the way undrained strength changes with depth, and it would imply that K_0 is constant. Thus, in the analyses that follow, both the gradient and the intercept of the undrained shear strength profiles have been varied, as shown in Fig. 2. There is little variation in the strength assumed at depth. The change is mainly at the top of the profile.

OBSERVED RESPONSE AT EMPINGHAM DAM

A small trench was dug to allow the trial bank to be extended to give a slope of constant height (about 20 m) over some 70 m in length. This was thought to be long enough to minimise three-dimensional effects (see Fig. 3(a)). It also gave a sufficient length for representative behaviour, despite the presence of the bulged material (potentially atypical clay) at the valley bottom (Bridle *et al.*, 1985). Two sections were instrumented as shown in Fig. 3(a), one over the bulge. They behaved similarly, the deformations extending 5 m into the ULC below the toe in the 'bulged' section (Fig. 3(b)) and 10 m into the ULC, 5 m below the base of the trench in the section with 'non-bulged' material (Fig. 3(c)). The horizontal strains deduced from the instrumentation were about 2% during trial construction (Bridle *et al.*, 1990), which was probably insufficient to initiate degradation in shear strength owing to strain-softening, and certainly no discontinuous shear surface developed. The instrumentation showed a wave of horizontal strain that developed as the height of the fill increased, and its crest moved downstream during construction (Bridle *et al.*, 1990). A substantial amount of undrained creep occurred after construction until the trial bank was buried over a period of 4 months. During this period strains and displacements increased up to 40%. The total toe movements were approximately 1 m and 0·85 m in the two instrumented sections respectively.

At the time of construction the results were difficult to interpret by conventional approaches because of the high undrained strength of the fill, the likelihood of deep tension cracks, and the uncertainties of LE analysis. It appeared that a high lateral stress induced by compaction played a significant part. A simple non-linear elastic FE analysis performed at that time (Hamza, 1974) indicated that the shear stresses in the foundation were approximately equal to the undrained strength assumed. The slips in the borrow pits were analysed by LE methods, but only an average strength could be deduced. These slips were through a considerable depth of clay, with a strength that increased with depth. Thus the undrained behaviour and the undrained stability were uncertain.

Just before completion of the main dam, the north undrained section (see Figs 1(a) and 16) started to spread (Bridle *et al.*, 1985). Placement of the last 2 m of fill towards the crest of the dam contributed to more than 50 mm of the total 100 mm horizontal movements (see Fig. 18(b)) observed during construction of the whole dam. The deformations observed at the end of construction almost doubled over the next year and a half. They were predominantly due to undrained creep (the permeability of the in situ and compacted ULC is low, and installed piezometers indicated that no significant consolidation took place), although the spreading was accentuated when the rip-rap was placed some 9 months after the embankment had been completed. All of the above suggested potential instability of the north end of the embankment at the end of its construction. It is worth mentioning that the south undrained section spread about half as much as the north section. There was therefore doubt as to whether or not the behaviour of the trial bank and that of the north section of the main embankment were consistent (Bridle *et al.*, 1985).

FINITE ELEMENT ANALYSES

Vaughan *et al.* (2004) summarised work on the analysis of embankments using advanced numerical techniques. They pointed out that there are difficulties in getting agreement between analytical predictions and field observations of movement unless plasticity is incorporated in the analysis pre-peak for both first-time loading and cyclic unloading/reloading. These problems seem to arise when there is principal stress rotation, as a non-linear elastic model relates strain increments to stress increments, whereas field behaviour shows them to be a function of current accumulated stresses as well. Further, elastic models fail to predict the lateral stress in loading (K_0), unless an unrealistic Poisson's ratio is used. Preliminary analyses of the Empingham trial bank were undertaken by Bodas (2002) using relatively simple elasto-plastic constitutive models, but the analyses failed to reproduce the spreading of the embankment.

The analyses reported here are a preliminary attempt to see whether the use of more sophisticated models is able to reproduce the undrained behaviour of the Empingham embankment, whether they can be used to interpret field behaviour, and whether they can allow final stability of undrained embankments to be monitored from interim observations. Although the development of a complete model incorporating both pre-peak plasticity and strain-softening is considered to be feasible, such a model was not available for this study, and so different models that reproduced these effects separately had to be used.

For this study, four cases involving the undrained behaviour of the foundation clay were chosen for back-analysis. They were the trial bank, the slips in the north and south borrow pits, and the north section of the main embankment with the undrained foundation. The locations of the borrow pits and the north 'undrained' section are shown in Fig. 1(a).

MODELS USED IN THE ANALYSES

Two constitutive models were used for the in situ ULC: a 'bubble' model, and a non-linear elastic strain-softening plastic model of the Mohr–Coulomb type. Two sets of parameters for the 'bubble' model were used in the analyses. The model parameters are listed in Tables 1 and 2 for the 'bubble' and Mohr–Coulomb models respectively.

The kinematically hardening 'bubble' model of Al-Tabbaa & Wood (1989), as generalised by Grammatikopoulou (2004), is briefly described in Appendix 1. This is a model of the critical state type, which enables the pre-peak plastic behaviour inside the bounding surface to be accounted for; however, post-peak strain-softening to the residual strength cannot be reproduced. The lateral stresses are defined inde-

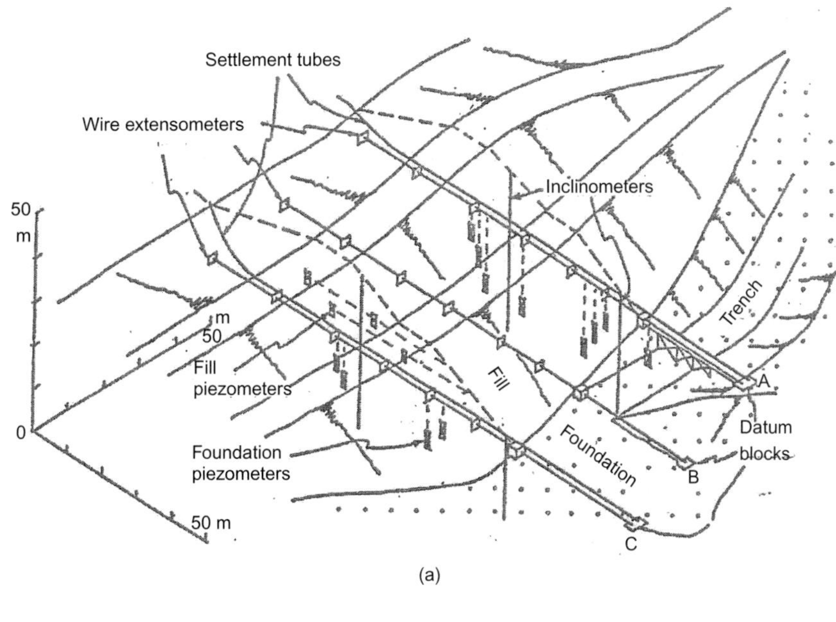

(a)

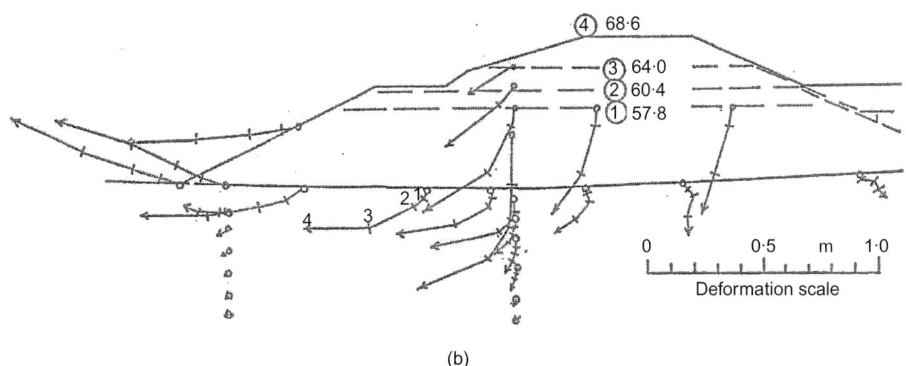

(b)

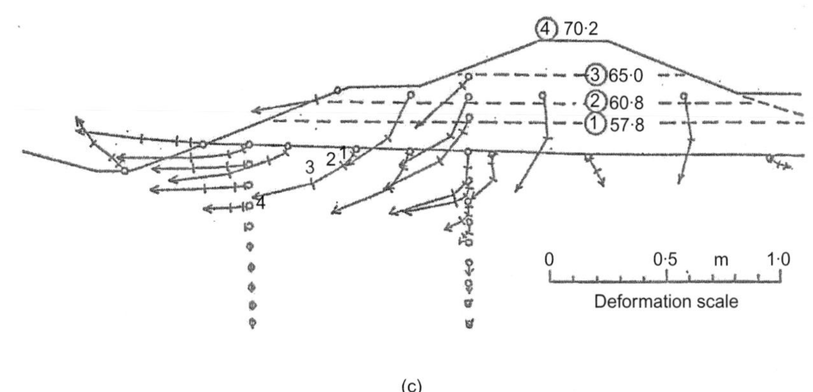

(c)

Fig. 3. (a) Trial bank and its instrumentation; (b) 'bulged' section C; (c) 'non-bulged' section A (after Bridle *et al.*, 1990)

pendently using a K_0 profile. The history of brecciation, valley bulging and cambering means that the horizontal stresses in the foundation are not related to the OCR profile as suggested, for example, by Mayne & Kulhawy's (1982) relationship

$$K_0 = (1 - \sin\phi')OCR^{\sin\phi'} \qquad (2)$$

Based on the limited evidence of sample suctions, the value of K_0 was approximately 1, and this has been adopted in the analyses. The series of undrained triaxial tests performed by Maguire (1976) showed considerable variation of strength in terms of effective stress between samples, but little sample size dependence. Perhaps surprisingly, the tests showed significant anisotropy, with diagonal samples showing a minimum strength. Reassessment of these data suggests a strength of $c' = 0$, $\phi' = 25°$. As there is reason to assume a weighted minimum when there is variation of strength, the original design assumption of $c' = 0$, $\phi' = 23°$ is retained in this work.

The second model was a non-linear elastic strain-softening plastic model of the Mohr–Coulomb type as described in Appendix 2 and used by Potts *et al.* (1997) for the analysis of cut slopes in London Clay. This model did not account for pre-peak plasticity, although in reality clays of this type do exhibit such behaviour. Such clays develop discontinuities

Table 1. 'Bubble' model parameters for in situ Upper Lias Clay and fill derived from it

Parameter	Clay in foundation		Clay fill*
	'First' prediction	'Revised' prediction	
γ: kN/m^3	20·5		20·0
ν_1	2·35		5·18
λ^*	0·08	0·06	0·16
κ^*	0·01		0·03
μ	0·2		0·3
ϕ': degrees	23		20
R	0·3	0·85	0·5
ψ	0·75	0·5	0·75
OCR	2·0–4·8	1·5–3·5	3·5[†]
K_0	1·0		2·5[‡]

λ^* and κ^* are the gradients of the virgin compression and elastic swelling lines respectively in $\ln\nu$–$\ln p'$ space, where ν is the specific volume.
* The compressibility of the pore fluid in the fill was set to be half of the compressibility of the soil skeleton.
[†] Defined in terms of mean effective stress p'.
[‡] As constructed.

Table 2. Mohr–Coulomb model parameters for in situ Upper Lias Clay

Parameter		Value
Bulk unit weight, γ: kN/m^3		20·5
Undrained (peak) shear strength, S_u: kPa		$12\cdot5 + 3\cdot85d$*
Peak strength	Cohesion, c'_p: kPa	0·0
	Angle of shearing resistance, ϕ'_p: degrees	21·0
Residual strength	Cohesion, c'_r: kPa	0·0
	Angle of shearing resistance, ϕ'_r: degrees	13·0
Angle of dilation: degrees		0·0
Plastic shear strain at peak, ε^p_{Dp}:[‡] %		1·0
***Plastic shear strain at residual, ε^p_{Dr}:[‡] %		5·0
Young's modulus, E: kPa		$800p'$[†]
Poisson's ratio, μ		0·2
Coefficient of earth pressure at rest, K_0		1·0

* d is depth below the top of the Upper Lias Clay.
[†] p' is the mean effective stress.
[‡] $\varepsilon^p_D = \{\frac{2}{3}[(\varepsilon^p_1 - \varepsilon^p_2)^2 + (\varepsilon^p_2 - \varepsilon^p_3)^2 + (\varepsilon^p_3 - \varepsilon^p_1)^2]\}^{\frac{1}{2}}$ is the plastic strain invariant.

at peak strength, and post-peak movement occurs as displacement on these discontinuities. The FE code used (ICFEP) is able to simulate a shear zone of approximately half an element thickness. Potts *et al.* (1990) found that, if the rate of strain-softening and the element thickness are adjusted to match the way strength drops to residual post-peak, the field behaviour can be reproduced acceptably.

Behaviour in undrained triaxial compression tests predicted by the 'bubble' model using the 'initial' set of parameters is shown in Fig. 4, where it may be compared with the results of actual laboratory tests. The assumed value of OCR in the model is of crucial importance; the higher OCR value (2·75 instead of 2·30) results in higher undrained shear strength, stiffer pre-peak response and more dilation. The calculated stress–strain responses during undrained triaxial tests in compression for both sets of 'bubble' model parameters ('initial' and 'revised') as well as for the strain-softening Mohr–Coulomb model parameters are shown in Fig. 5. For both models ('bubble' and Mohr–Coulomb), the mean effective stress at the start of undrained shearing was

assumed to be $p'_0 = 100$kPa, whereas the OCR values used in the 'bubble' model were 2·75 and 2·0 for the 'initial' and 'revised' predictions respectively. The various undrained strength profiles simulated by the models are shown in Fig. 2. They will be discussed later in more detail.

The 'bubble' model was also used to characterise the behaviour of the fill. Whereas the ULC in the foundation is a fully saturated material, the fill derived from it is unsaturated. In such media the pore phase is constituted by air and water that are at different pressures. If clays are compacted wet and near the maximum density, their effective stress can be defined as $\sigma' = \sigma - u_w$ as if they are fully saturated. The air and water phase can be regarded as a single pore fluid ('air occluded by water') with a finite compressibility. In the analyses, the development of undrained pore pressure in an effective stress model is simulated by setting the compressibility of the pore fluid as a multiple of the soil skeleton compressibility using a coefficient β. The compressibility of the pore fluid when air is present as occluded bubbles is discussed by Vaughan (2003). It is non-linear and complicated owing to bubble migration. The

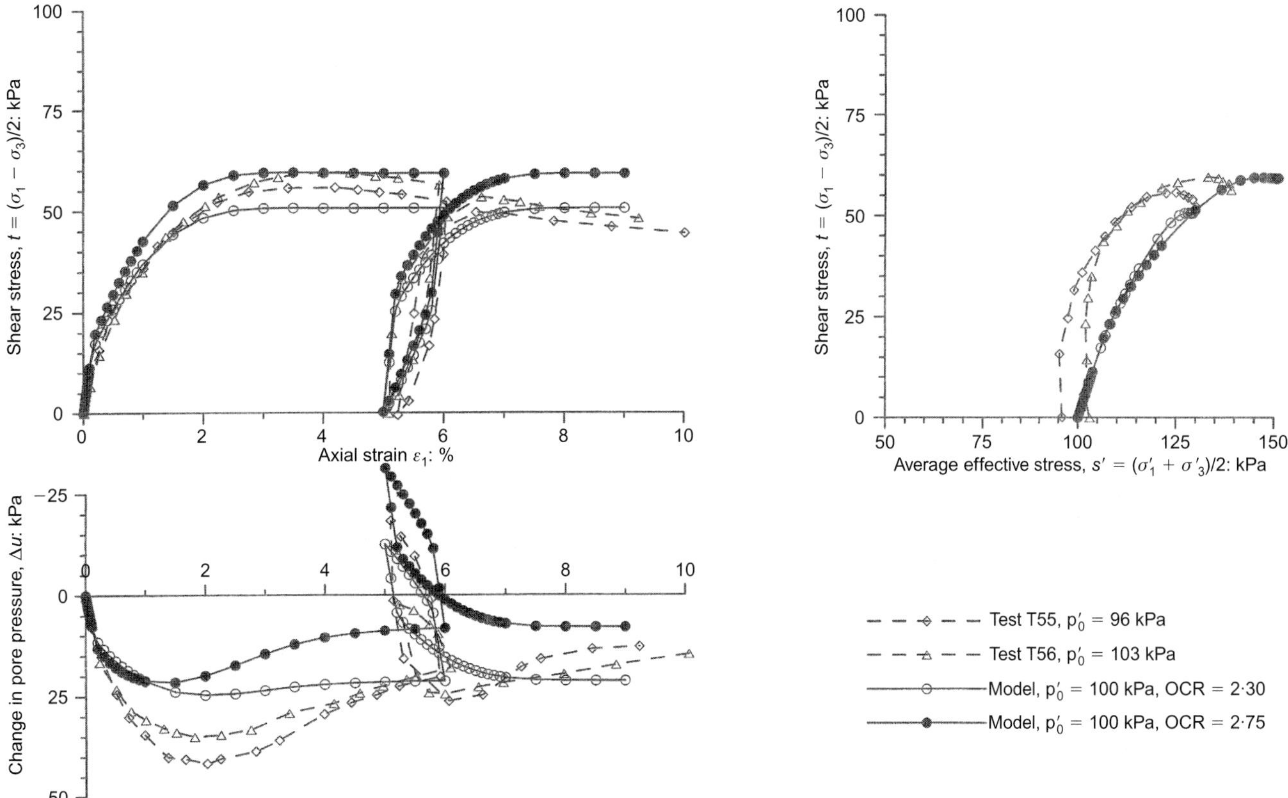

Fig. 4. Observed and predicted (by the 'bubble' model) behaviour of in situ Upper Lias Clay in undrained triaxial compression test

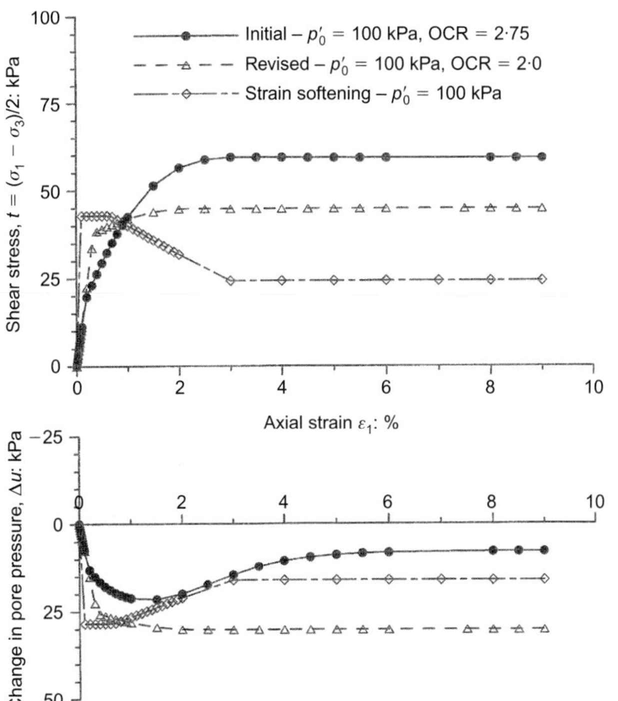

Fig. 5. Predicted behaviour of in situ Upper Lias Clay in undrained triaxial compression test by different models

pore pressure ratio $B_w = \Delta u_w/\Delta\sigma$ becomes unity while the pore fluid is still compressible. However, as an approximation, a linear compressibility was assumed for the analyses reported herein. After adjustment, a compressibility of the pore fluid was chosen ($\beta = 0.5$) that gave the stress–strain relationships as shown in Figs 6(a) (undrained isotropic compression) and 6(b) (undrained triaxial compression), where they can be com-

pared with the available undrained triaxial test results on 100 mm diameter samples of field compacted fill (Sodha, 1974).

The 'strain wave' that developed below the crest of the trial bank as the slope was raised (see Fig. 13) suggested that the driving mechanism for the quite large lateral displacements that developed was the relief of high lateral stresses induced by compaction. What information there is on compaction stresses, and there is little of it, shows them to be high. In the analyses of the trial bank and the north section of the main dam high lateral stresses were introduced in the layer of fill just placed, and were defined by $K_0 = 2.5$. This reduced with lateral strain as further fill was placed according to the 'bubble' model adopted in the analysis.

A more detailed description of modelling techniques used for both the in situ clay and the fill derived from it will be given elsewhere (Kovacevic *et al.*, 2007).

RESULTS OF THE ANALYSES
The trial bank

The 'non-bulged' section (see Fig. 3(c)) was chosen for analysis (see Bridle *et al.*, 1990). Fig. 7 shows the section with instrumentation, and Fig. 8 shows the FE mesh used in the analyses.

The first analysis using the 'bubble' model assumed material properties that gave an undrained compression strength profile the same as the original design assumption: $S_u = 30 + 3.5z$ (in kPa), where z (in m) is the depth below the top of the clay. The analysis underestimated the deformations, and did not produce a significant 'strain wave' as reported by Bridle *et al.* (1990). It did not show any sign of incipient failure as demonstrated by the contours of the undrained shear stress level (a measure of the undrained shear strength mobilised; when the undrained shear strength S_u is fully mobilised, the undrained shear stress level is equal to 1·0) in Fig. 9(a). The predicted shear strains,

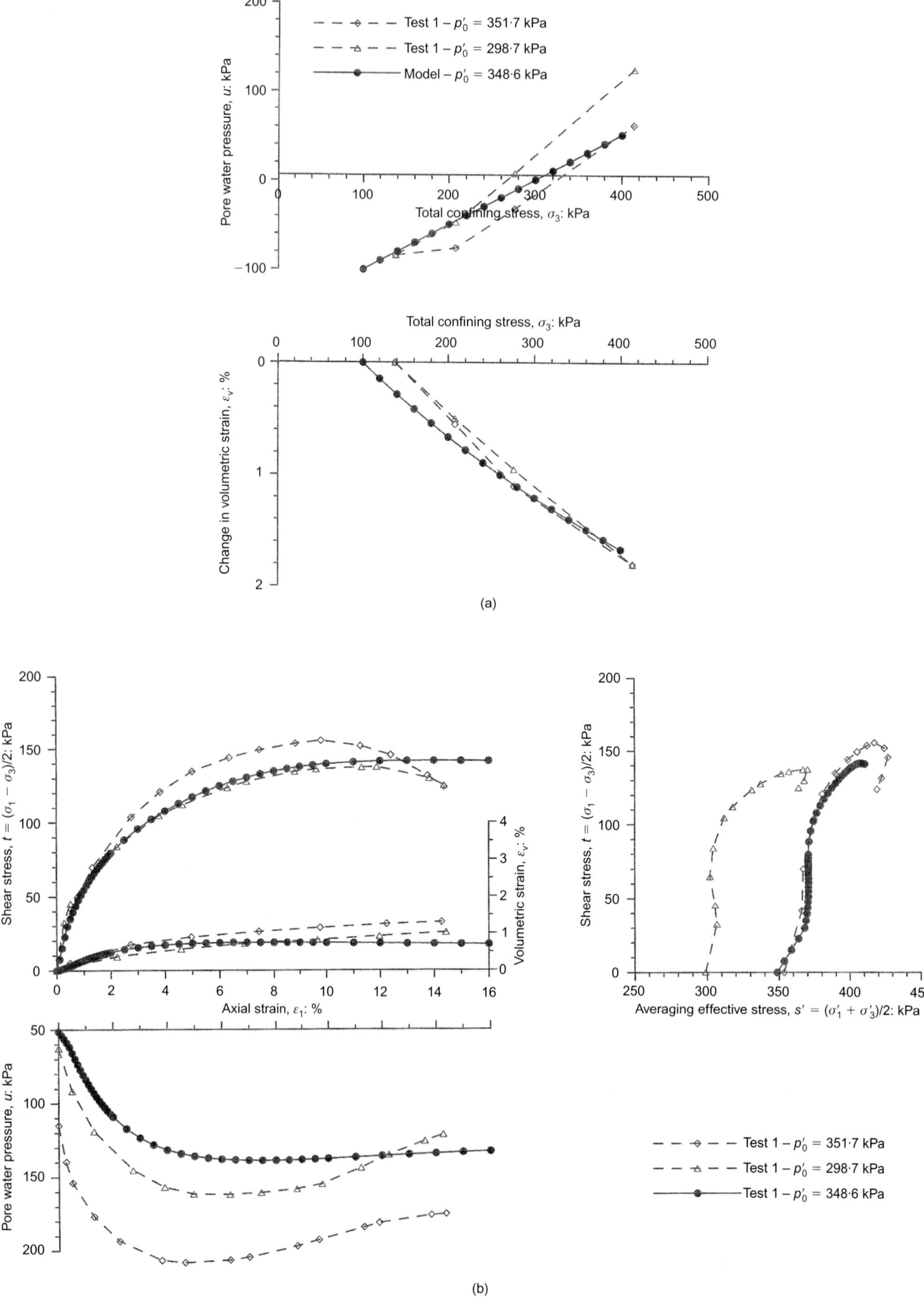

Fig. 6. Observed and predicted (by the 'bubble' model) behaviour of fill derived from Upper Lias Clay in: (a) undrained isotropic test; (b) undrained triaxial compression test

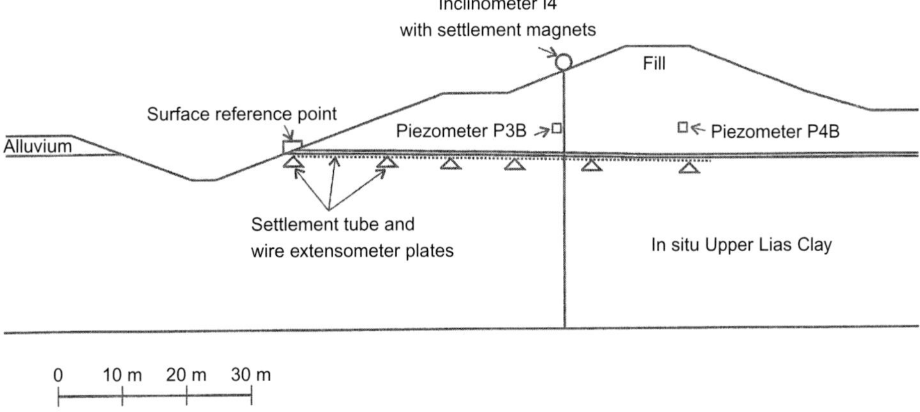

Fig. 7. Section A of trial bank analysed with installed instrumentation

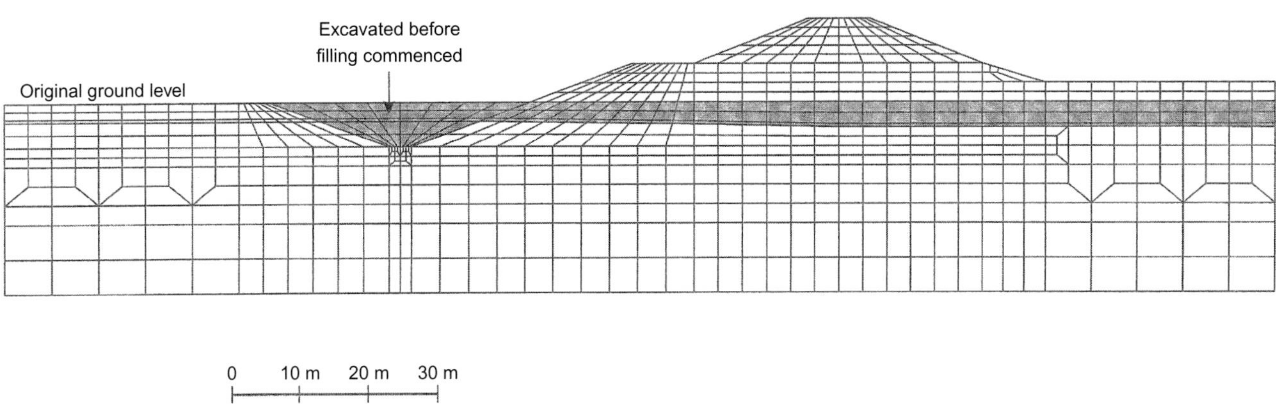

Fig. 8. Finite element mesh used in analyses of trial bank

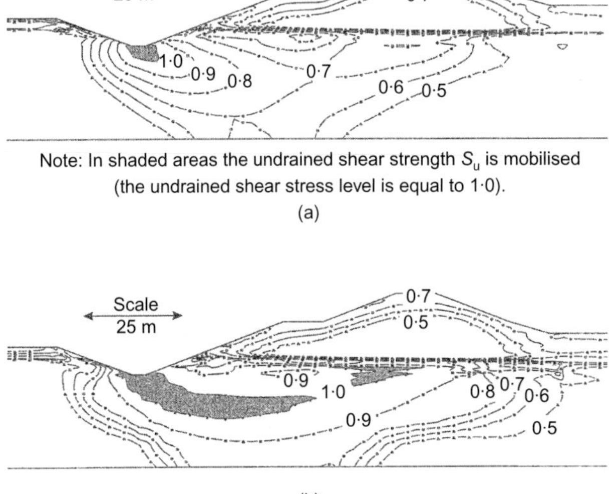

Fig. 9. Contours of undrained shear stress level at end of trial bank construction predicted by 'bubble' model: (a) initial prediction; (b) revised prediction

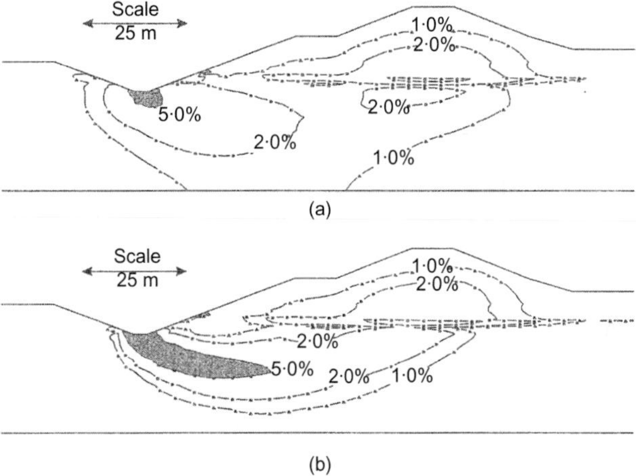

Fig. 10. Contours of shear strain at end of trial bank construction predicted by 'bubble' model: (a) initial prediction; (b) revised prediction

represented by the deviatoric strain invariant (see Appendix 1), were relatively small: 1–2%, as shown in Fig. 10(a).

The analysis was repeated using the 'revised' model parameters (see Table 1). These parameters resulted in a stiffer pre-peak response of the ULC in the foundation; however, the modelled undrained shear strength was lower (see Fig. 5). It should be noted that the 'revised' undrained strength profile given by $S_u = 22{\cdot}5 + 2{\cdot}75z$ was closer to the

undrained strengths back-calculated (by LE method of analysis) from the borrow pit failures, as shown in Fig. 2.

Figure 11 shows the pattern of displacement produced by the 'revised' analysis due solely to placement of the last 1·4 m of fill for the trial bank. A rotational, deep-seated pattern of deformation similar to the observed one (see Fig. 3(c)) can be clearly seen. The predicted contours of the undrained shear stress level (see Fig. 9(b)) show that the undrained shear strength was now fully mobilised on a potential slip surface in the foundation. However, the compacted fill was much stronger than the foundation (owing to the large suctions induced by compaction), and S_u from 100

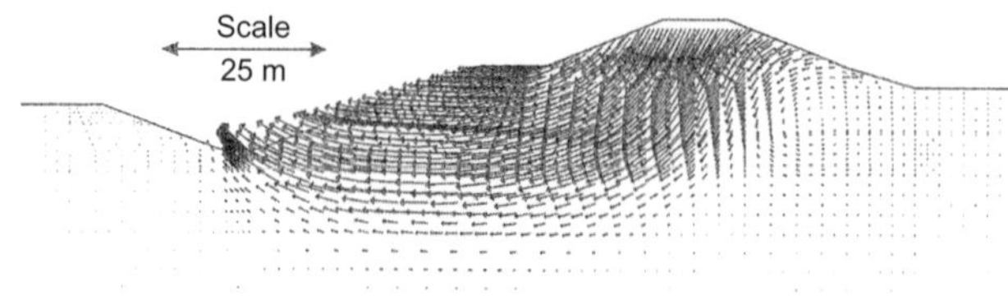

Fig. 11. Incremental displacement vectors at end of trial bank construction predicted by 'bubble' model (revised prediction)

to 120 kPa was predicted by the analysis, values that were lower than the average measured strength ($S_u = 140$ kPa) in quick triaxial tests on 100 mm diameter specimens with a cell pressure of 250 kPa (Bridle *et al.*, 1985). Thus the stronger fill was some way from failure, indicating that the trial bank was still stable. However, the 'revised' analysis predicted relatively large shear strains (in excess of 5%) in a substantial part of the foundation towards the toe of the slope (Fig. 10(b)), increasing the potential for strain-softening and progressive failure. It should be mentioned that a shear surface was observed at the toe of the trial bank during the last stages of construction (Bridle *et al.*, 1990), but it did not extend as far as the first inclinometer at the instrumented section 'A' (see Fig. 3(a)).

Measured and predicted horizontal and vertical deformations of the trial bank at the position of Inclinometer I4 are shown in Figs 12(a) and 12(b) respectively. The predicted and actual 'strain wave' at the top of the ULC foundation is shown in Fig. 13. It is considered that the fit was reasonable and that the method of analysis was satisfactory. Only when the undrained shearing resistance in the clay foundation was reduced significantly below the original design assumption (Fig. 2) was the behaviour of the trial bank reproduced. It is worth noting that the percentage change is greater at the top of the profile.

The slip in the north borrow pit

The slope in the north borrow pit that slipped is shown in Fig. 14(a). Excavation of the slope was simulated with the 'bubble' model and an undrained compression strength profile equivalent to the original design assumption (see Fig. 2). The analysis suggested that the slope collapsed as the last 1·5 m was excavated. Thus the in situ operational undrained strength profile was a little higher than the original design profile. An LE analysis gave a factor of safety of 0·85 for the maximum depth of excavation. The predicted incremental displacement vectors just prior to collapse are shown in Fig. 14(b), and the maximum predicted shear strain at this stage was not more than 5%. This suggests that there would have been little strain-softening or progressive failure, justifying the use of a model without strain-softening. The conclusion is that the operating undrained shearing resistance was greater than that assumed for the original design.

The slip in the south borrow pit

The slope in the south borrow pit that slipped is shown in Fig. 15(a). It was analysed using the 'bubble' model with assumptions that gave the original design undrained strength profile. Excavation to full depth was simulated without failure being predicted by both LE and FE analyses. The analysis was repeated with assumptions that gave an undrained compression strength profile of $S_u = 22·5 + 2·75z$ ('revised' profile). Collapse occurred as excavation of the last 1·7 m of clay was simulated. An LE analysis of the complete slope gave a factor of safety of 0·98, indicating that the operational strength was close to the profile assumed. The maxi-

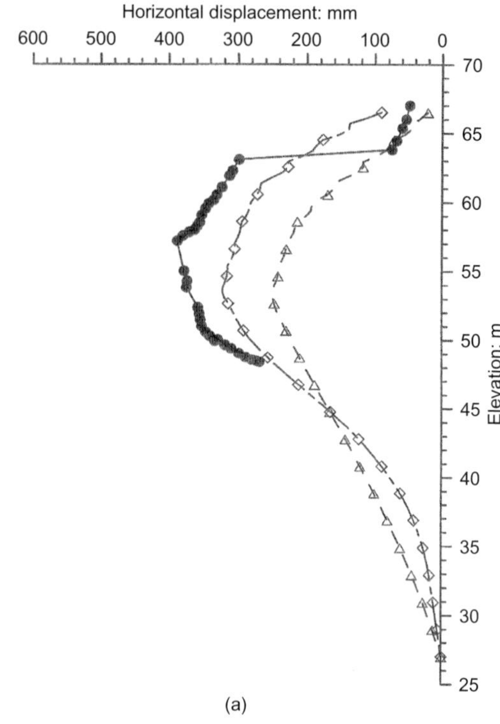

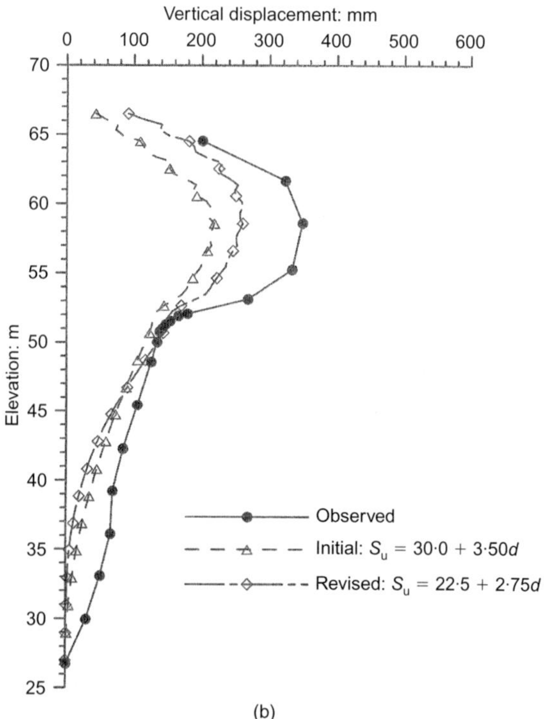

Fig. 12. Measured and predicted (by 'bubble' model) deformations at end of trial bank construction at Inclinometer I4: (a) horizontal; (b) vertical

mum shear strain just prior to collapse was predicted to be in excess of 5%, and so there may have been some strain-softening. The incremental displacement vectors during the last increment of excavation of the south borrow pit are shown in Fig. 15(b).

Undrained behaviour of north section of main dam

The section with the lateral displacement gauges as shown in Fig. 16 was analysed. Its position on the plan of the dam

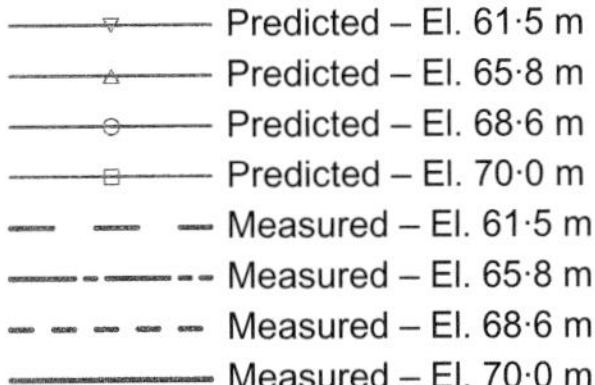

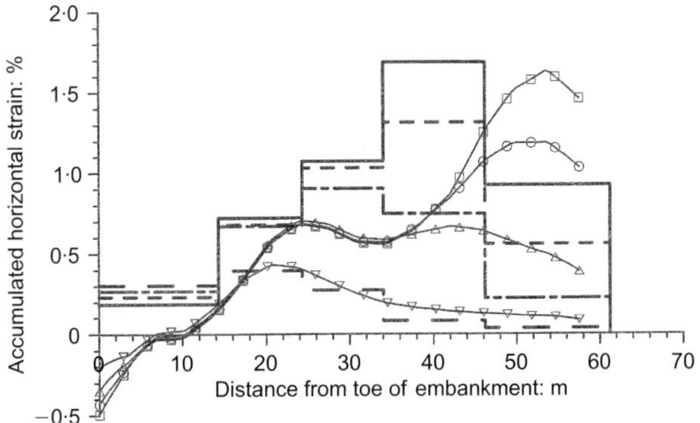

Fig. 13. Measured and predicted (by 'bubble' model) strain waves at top of foundation during trial bank construction (revised prediction)

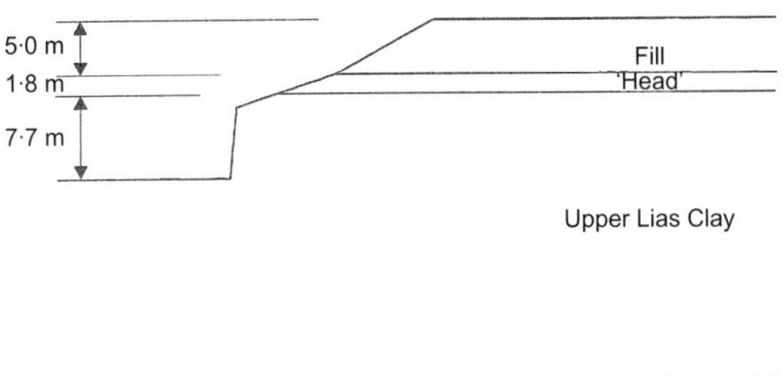

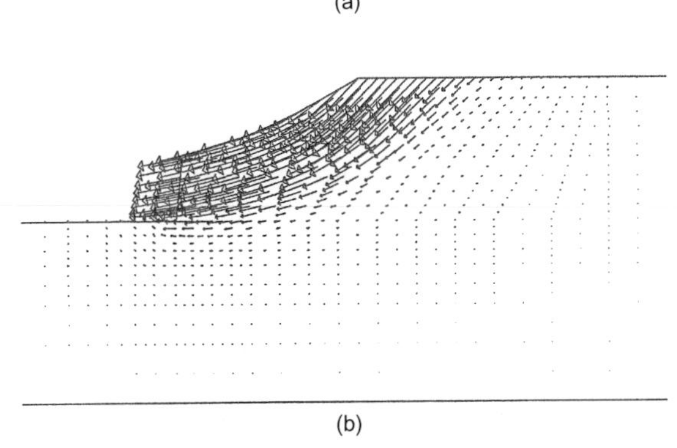

Fig. 14. Predicted slip in temporary slope of north borrow pit by 'bubble' model (initial prediction)

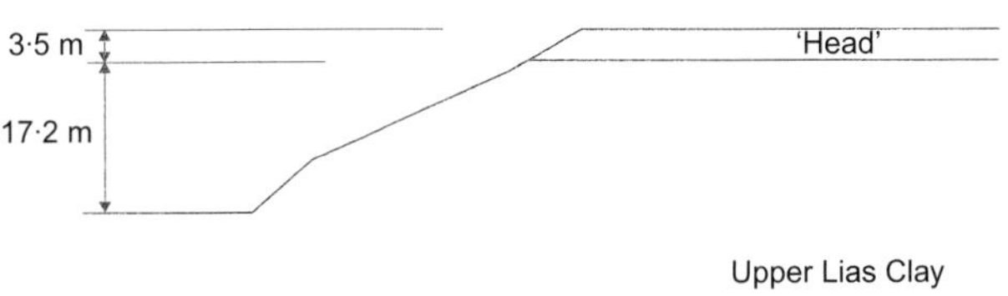

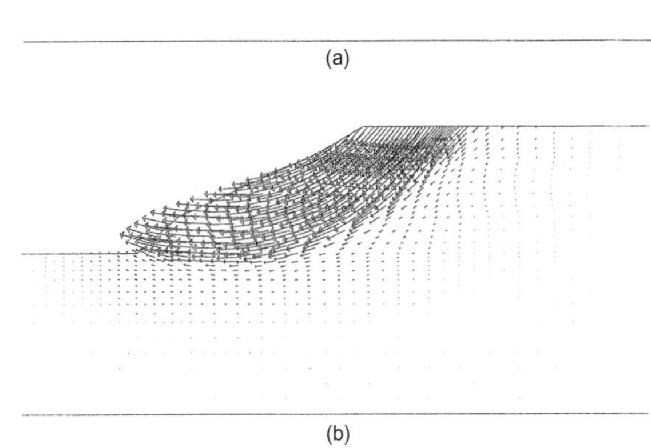

Fig. 15. Predicted slip in temporary slope of south borrow pit by 'bubble' model (revised prediction)

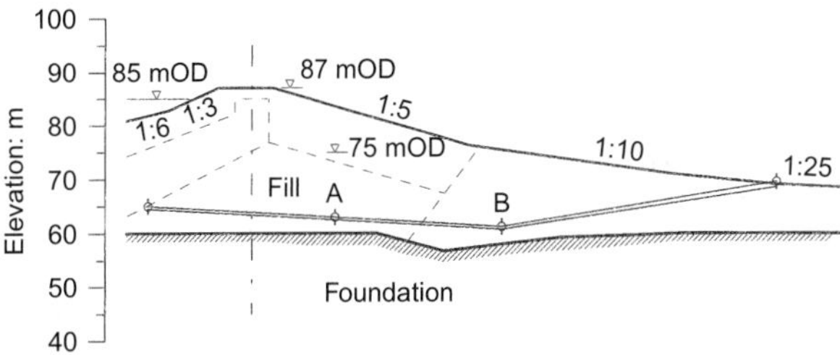

Fig. 16. North section of main dam with position of rod gauge

shear strains (see Appendix 2 and Table 2) were changed until the predicted behaviour matched the observed one more closely (see Fig. 18). An undrained peak strength profile of $S_u = 12·5 + 3·85z$ was simulated in the end, which gave a much reduced shear strength at the top of the foundation (see Fig. 2). The resulting stress–strain curve in undrained triaxial compression is presented in Fig. 5, where it can be compared with the 'bubble' model simulations ('initial' and 'revised'). It can be seen that, to capture the observed behaviour by the FE analysis, it was necessary to model

(*a*) a stiff pre-peak response
(*b*) a reduced (peak) undrained shear strength at the top of the foundation
(*c*) a rather abrupt development of strain-softening.

Figure 19 shows the incremental displacement vectors predicted by the Mohr–Coulomb strain-softening model towards the end of construction. Their absolute magnitude is not important; however, their relative magnitude and direction indicate the current mechanism of behaviour. The predicted contours of plastic shear strains for the same stages of the analysis are shown in Fig. 20.

There was a limited amount of field data available for the north section of the main dam. Nevertheless, the measurement data collected during (and after) dam construction were very consistent (Bridle *et al.*, 1985). The strain-softening analysis showed that the most likely cause of the sudden increase of movement at this section was the low undrained strength at the top of the ULC in combination with strain-softening. The presence of a pre-existing shear surface was virtually precluded by the nature of the brecciation. This

is indicated in Fig. 1(a). The FE mesh used for the analyses is shown in Fig. 17.

The 'bubble' model was used initially in the analyses with assumptions that gave the 'design' and 'revised' undrained compression strength profiles (see Fig. 2). Use of the latter predicted deformations better, as shown in Fig. 18, although they were larger than those measured. This was particularly so at an early stage of loading, and there was no sudden acceleration of movement as the fill approached the crest. It was thought that the abruptness of movement indicated strain-softening. Therefore in the second set of analyses the soil model for the ULC in the foundation was changed, and the non-linear elastic Mohr–Coulomb plastic model with strain-softening was used. The pre-peak elastic stiffness, the peak (and residual) strength and its variation with respect to plastic

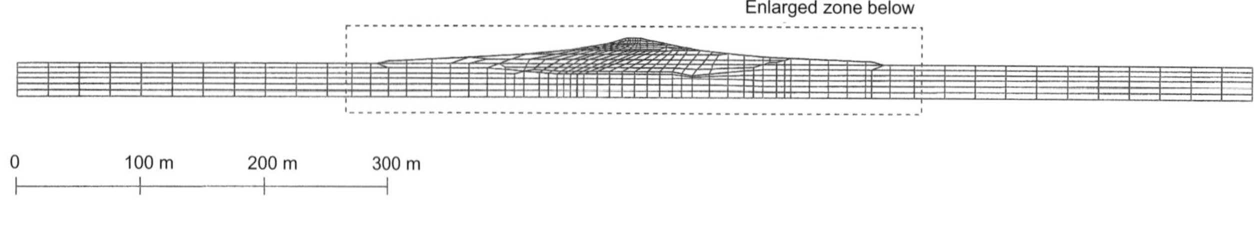

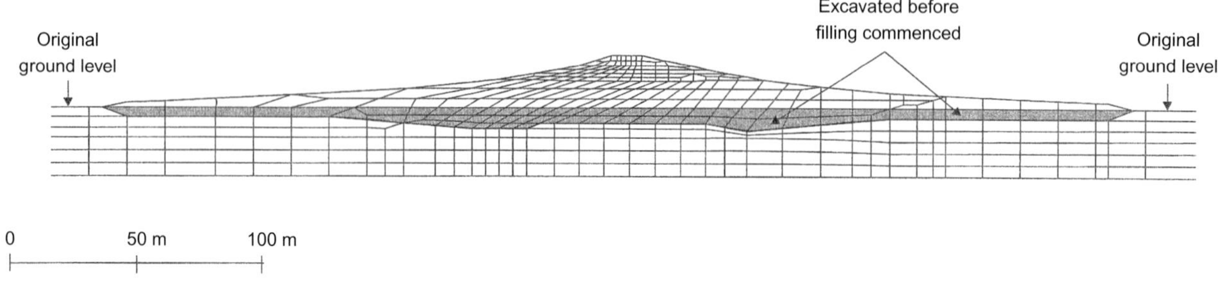

Fig. 17. Finite element mesh used in analyses of north section of main dam

was confirmed by close mapping of the interface between the superficial deposits and the underlying clay (Horswill & Horton, 1976).

CONCLUDING REMARKS

It is understood that a number of schemes are currently being considered that are likely to involve the construction of embankments on stiff plastic clays similar to the Upper Lias Clay (ULC) at Empingham. As demonstrated in this paper, such embankments constructed without incorporating drainage in the foundation could be problematic. The alternative of using drains results in a relatively straightforward design process, but it could involve importing large quantities of drainage materials at high cost and significant environmental impact. It is thus important to utilise the undrained strength of the clay as much as possible. At Empingham, where the embankment was up to 25 m high, it was successfully constructed without relying on drainage to enhance the strength of the foundation. However, the displacements were unexpectedly high, and they were quite close to being unacceptable. This behaviour needs analysis and understanding if the design of large embankments on stiff plastic clays is to be made more efficient.

The undrained deformations penetrate more deeply into the foundation of a steep embankment (the trial bank) than they do into that of a flat one (the north section of the main embankment). The deep penetration beneath a steep embankment was also predicted in a notional analysis of an embankment of London clay fill on a London clay foundation (Vaughan, 1994). Thus this might be an effect that occurs during construction of embankments on stiff plastic clays in general, not just the ULC.

The north and south sections of the main embankment were about the same height, but the undrained movement of the north section was more than twice that of the south section. This implies that the foundation beneath the north section had a lower undrained shearing resistance than on the south section. There are two independent pieces of evidence to support this. The first is the mechanism controlling the slow undrained strength, which makes it a function of mean effective stress. The mean effective stress at the top of the clay depends on the overburden above the ULC. At the south section the overburden was thicker than at the north section. The second piece of

evidence is the blow counts for the undrained 100 mm diameter samples obtained by percussion drilling. Fig. 21 shows them plotted against depth below ground surface for two boreholes at each section. It can be seen that those at the south section are higher. The differences in overburden pressure are reflected in the depth at which the blow count plots start. The profiles of blow count show considerable variability, and the FE analyses of the undrained behaviour at Empingham show that this is consistent with considerable variability of the ULC. This clearly means that, when dealing with construction on soils such as the ULC, the design assumptions have to be carefully considered, and the use of unique relationships may not be appropriate.

It follows from the approach presented herein that the bulk undrained shear strength can be synthesised from the strength in terms of effective stress (with allowance for discontinuities), K_0, and in situ pore pressure. This, together with the pre-peak stiffness and the post-peak strain-softening, can be used to formulate the appropriate model for computation. However, this process is complex, and, as yet, has not been fully checked. It involves uncertainties in the direct measurement of the components of strength, which limit its usefulness in design unless it is correlated with full-scale experience. It is very doubtful whether the undrained shearing resistance of an element of such a clay can be measured directly in a useful manner.

Another way to improve design of embankment dams resting on undrained stiff plastic clay foundations is by observation through instrumentation. Here the extrapolation of observations and calibration of analyses at an early stage of dam construction to predict what will happen at later stages is an approach that is perhaps more promising. If initial stages of construction can be modified to stress the clay more than the complete embankment does, then this could be a way towards an efficient construction process by fully utilising the undrained shearing resistance of the clay in the foundation. However, an integral part of this process is realistic modelling of the construction process to determine future behaviour.

The comparison between observations and the results of analysis presented herein could almost certainly be improved by further analyses. However, the models, and the 'bubble' model in particular, require modification and improvement for optimisation. The results obtained are sufficient to show

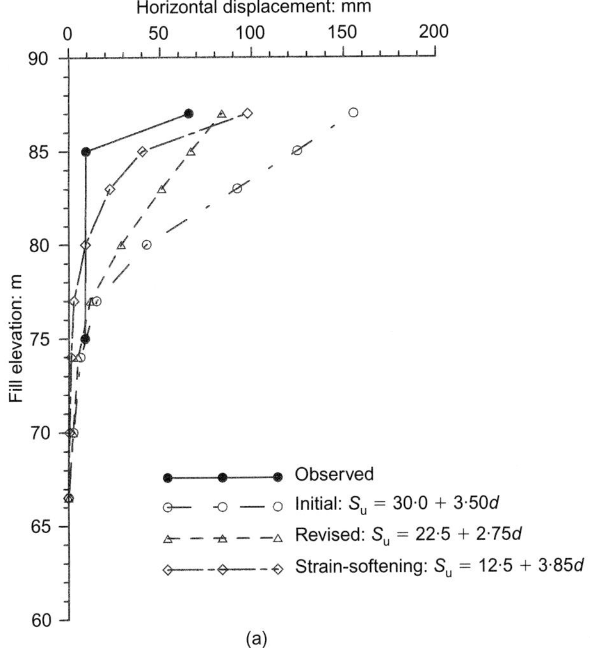

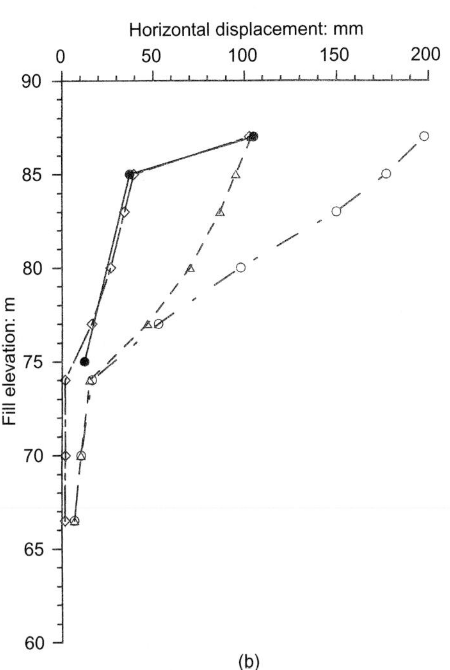

Fig. 18. Observed and predicted horizontal displacements of the main dam using different models at: (a) rod gauge points A; (b) rod gauge point B

that the behaviour of the ULC foundation and fill can be reproduced reasonably well.

Ideally, an analysis should model pre-peak behaviour with plasticity, a non-linear failure envelope in terms of effective stress and strain-softening in such a way that the formation of discontinuities is taken into account even if not directly modelled. It must also model undrained generation of pore pressure in partly saturated fills where necessary. A model simultaneously reproducing all of these features is currently under development.

APPENDIX 1. AL-TABBAA & WOOD'S KINEMATICALLY HARDENING 'BUBBLE' MODEL

The model of Al-Tabbaa & Wood (1989) has been generalised by Grammatikopoulou (2004). It uses the elliptical yield surface of the

modified Cam Clay model (Roscoe & Burland, 1968) to represent the bounding surface defined by

$$\left(p' - \frac{p_0'}{2}\right)^2 + \frac{\mathbf{s} : \mathbf{s}}{2g^2(\theta)} - \frac{p_0''^2}{4} = 0 \tag{3}$$

and employs an inner kinematic yield surface ('bubble') given by

$$(p' - p_a')^2 + \frac{(\mathbf{s} - \mathbf{s}_a) : (\mathbf{s} - \mathbf{s}_a)}{2g^2(\theta)} - R^2\left(\frac{p_0'^2}{4}\right) = 0 \tag{4}$$

assuming it to have the same shape but to be smaller in size than the bounding surface. Here p_a and $\mathbf{s}_a$ are the mean effective stress and the deviatoric stress tensor at the centre of the bubble, and R is the ratio of the bubble size to that of the bounding surface. $g(\theta)$ is defined as

$$g(\theta) = \frac{\sin\phi'}{\cos\theta + \sin\theta \cdot \sin\phi'/\sqrt{3}} \tag{5}$$

where ϕ' is the effective angle of shearing resistance and θ is the Lode angle.

Within the kinematic yield surface the behaviour is assumed to be elastic with a constant Poisson's ratio μ, and a variable bulk stiffness given by

$$K' = \frac{p'}{\kappa^*} \tag{6}$$

where κ^* is the gradient of the elastic swelling line[‡] in $\ln v$–$\ln p'$ space, where v is the specific volume. Otherwise, the behaviour is elasto-plastic with non-associated plasticity assumed and the hardening/softening law, which depends on the experimentally determined parameter ψ. For further details see Grammatikopoulou (2004).

The model requires seven parameters, two more (R and ψ) than the modified Cam Clay model. They are listed in Table 1.

APPENDIX 2. THE GENERALISED MOHR–COULOMB MODEL

A generalised non-linear elastic strain-softening/hardening plastic model incorporating a Mohr–Coulomb yield criterion given by the apparent cohesion c', the angle of shearing resistance ϕ', and the angle of dilation ψ is used.

The Young's modulus E varies according to the mean effective stress p'

$$E = a \cdot p' \tag{7}$$

where a is a dimensionless parameter, and the Poisson's ratio μ is constant.

The yield function is given by

$$F(\sigma') = S - 1 \tag{8}$$

and S is the shear stress level, defined as

$$S = \frac{J}{p' + a}g(\theta) \tag{9}$$

where

$$p' = \frac{\sigma_1' + \sigma_2' + \sigma_3'}{3} \tag{10}$$

$$J^2 = \frac{(\sigma_1' - \sigma_2')^2 + (\sigma_2' - \sigma_3')^2 + (\sigma_3' - \sigma_1')^2}{6} \tag{11}$$

$$g(\theta) = \frac{\sin\phi'}{\cos\theta + \sin\theta \cdot \sin\phi'/\sqrt{3}} \tag{12}$$

$$\theta = \tan^{-1}\left(\frac{2b - 1}{\sqrt{3}}\right) \tag{13}$$

$$b = \frac{\sigma_2' - \sigma_3'}{\sigma_1' - \sigma_3'} \tag{14}$$

a is the intercept of the yield surface on the mean effective stress

---

[‡] λ^* is the gradient of the virgin compression line.

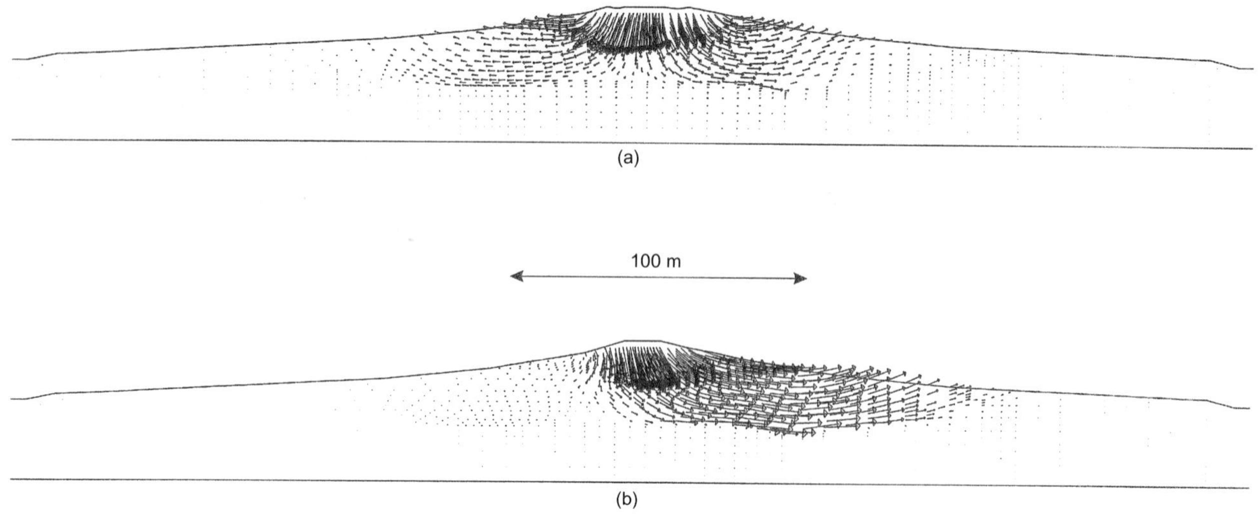

Fig. 19. Predicted incremental displacement vectors at north section of main dam towards end of construction predicted by Mohr–Coulomb strain-softening model, due to fill placing between: (a) El. 83·0 m and El. 85·0 m; (b) El. 85·0 m and El. 87·0 m

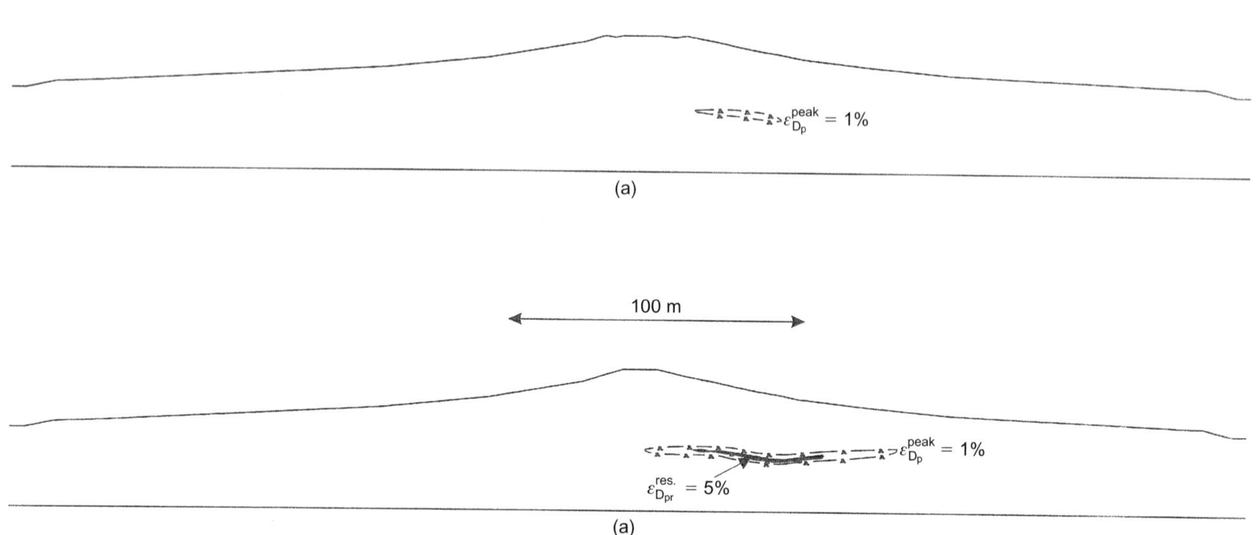

Fig. 20. Predicted contours of plastic shear strain for north section of main dam towards end of construction predicted by Mohr–Coulomb strain-softening model: (a) fill level reached El. 85·0 m; (b) fill level reached El. 87·0 m

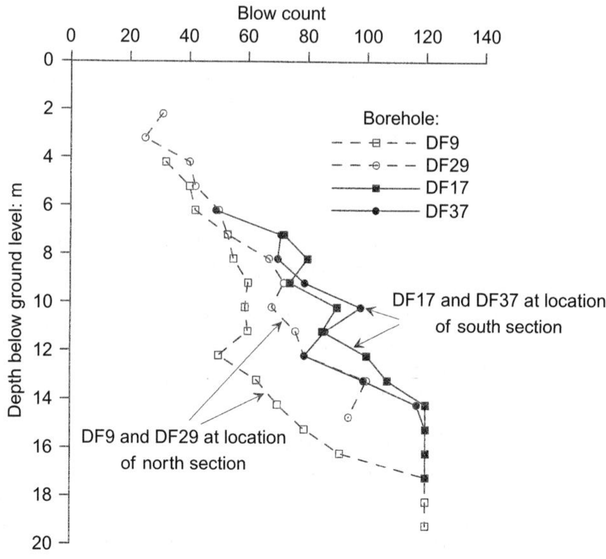

Fig. 21. Comparison of blow counts for north and south sections of main dam. Note different amounts of overburden between north and south sections

(p') axis and is given by

$$\alpha = \frac{c'}{g(\theta = 0)} = \frac{c'}{\sin \phi'} \tag{15}$$

where c' is the cohesion intercept with the Lode angle $\theta = 0$, that is, $\sigma_2' = (\sigma_1' + \sigma_3')/2$.

The plastic potential is also defined in the same way as the yield function, but with the angle of shearing resistance ϕ' replaced by the angle of dilation ψ.

Softening behaviour is introduced by allowing the angle of shearing resistance ϕ', and the cohesion intercept c' to vary with the deviatoric plastic strain invariant ε_D^p, defined as

$$(\varepsilon_D^p)^2 = \frac{2}{3} \left[(\varepsilon_1^p - \varepsilon_2^p)^2 + (\varepsilon_2^p - \varepsilon_3^p)^2 + (\varepsilon_3^p - \varepsilon_1^p)^2 \right] \tag{16}$$

The angle of dilation ψ may also be varied with ε_D^p. The model parameters are listed in Table 2.

NOTATION

a, b dimensionless parameters in Mohr-Coulomb model
B pore pressure ratio
c' effective cohesion intercept
d depth

E Young's modulus
K bulk modulus
K_0 coefficient of earth pressure at rest
OCR   overconsolidation ratio
p' mean effective stress
R parameter defining 'bubble' size
S shear stress level
S_u undrained shear strength
s deviatoric stress tensor
s' average effective stress
t shear stress
u pore pressure
β compressibility parameter
$\varepsilon_1, \varepsilon_2, \varepsilon_3$ principal strains
ε^p_D plastic strain invariant
ϕ' effective angle of shearing resistance
γ, γ' bulk, submerged unit weight
κ^* gradient of elastic swelling line in $\ln v$-$\ln p'$ space
λ^* gradient of virgin compression line in $\ln v$-$\ln p'$ space
μ Poisson's ratio
v specific volume
θ Lode angle
$\sigma_1, \sigma_2, \sigma_3$ principal stresses
ψ angle of dilation, hardening parameter in 'bubble' model

REFERENCES

Al-Tabbaa, A. A. & Wood, D. M. (1989). An experimentally based 'bubble' model for clay. *Proc. 3rd Int. Conf. Numerical Models in Geomechanics, NUMOG III, Niagara Falls*, 91–99.

Bodas, T. M. (2002). *Numerical analysis of the trial bank of Empingham Dam*. MSc dissertation, Imperial College, University of London.

Bridle, R. C., Vaughan, P. R. & Jones, H. N. (1985). Empingham Dam: design, construction and performance. *Proc. Inst. Civ. Engrs, Part 1* **78**, April, 247–289.

Bridle, R. C., Vaughan, P. R. & Wernek, M. L. G. (1990). The instrumented trial slope at Empingham Dam. *Proceedings of the conference on geotechnical instrumentation in civil engineering projects*, London, pp. 415–427.

Chandler, R. J. (1972). Lias clay: weathering processes and their effect on shear strength. *Géotechnique* **22**, No. 3, 403–433.

Grammatikopoulou, A. (2004). *Development, implementation and application of kinematic hardening models for over-consolidated clay*. PhD thesis, University of London.

Hamza, M. M. A. F. (1974). *The analysis of embankment dams by non-linear finite element methods*. PhD thesis, University of London.

Horswill, P. & Horton, A. (1976). The valley of the River Gwash with special reference to cambering and valley bulging. *Phil Trans. R. Soc. London Ser. A*, **283**, December, 451–461.

Kovacevic, N., Higgins, K. G., Vaughan, P. R. & Potts, D. M. (2007). Back analysis of the trial bank at Empingham Dam site. *10th Int. Conf. Numerical Models in Geomechanics, NUMOG X, Rhodes* (submitted).

Maguire, W. M. (1976). *The undrained strength and stress–strain behaviour of brecciated Upper Lias clay*. PhD thesis, University of London.

Mayne, P. W. & Kulhawy, F. H. (1982). K_0–OCR relationship in soils. *J. Geotech. Engng Div. ASCE* **108**, No. GT6, 851–872.

Potts, D. M., Dounias, G. T. & Vaughan, P. R. (1990). Finite element analysis of progressive failure of Carsington embankment. *Géotechnique* **40**, No. 1, 79–101.

Potts, D. M., Kovacevic, N. & Vaughan, P. R. (1997). Delayed collapse of cut slopes in stiff clay. *Géotechnique* **47**, No. 5, 953–982.

Roscoe, K. H. & Burland, J. B. (1968). On the generalised stress–strain behaviour of 'wet' clay. In *Engineering plasticity* (eds J. Heymann and F. A. Leckie), pp. 535–609. Cambridge: Cambridge University Press.

Sandroni, S. S. (1977). *The strength of London Clay in total and effective stress terms*. PhD thesis, University of London.

Skempton, A. W. & La Rochelle, P. (1965). The Bradwell slip, a short-term failure in London Clay. *Géotechnique* **15**, No. 3, 221–242.

Sodha, V. G. (1974). *The stability of embankment dam fills of plastic clays*. MPhil thesis, University of London.

Vaughan, P. R. (1976). The deformations of the Empingham valley slope. In Horswill, P. 'Cambering and Valley Bulging in the Gwash Valley at Empingham, Rutland', Horswill, P. & Horton, A. (eds). Appendix, *Phil. Trans. Royal Soc. London, Ser. A*, **283**, 427–462.

Vaughan, P. R. (1994). Assumption, prediction and reality in geotechnical engineering. 34th Rankine Lecture. *Géotechnique* **44**, No. 4, 573–609.

Vaughan, P. R. (2003). Observations on the behaviour of clay fill containing occluded air bubbles. *Géotechnique* **53**, No. 2, 265–272.

Vaughan, P. R. & Hamza, M. M. A. F. (1977). Clay embankments and foundations: Monitoring stability by measuring deformations. *Proc. 9th Int. Conf. Soil Mech. Found. Engng, Tokyo, Speciality Session No. 8, Deformations of Earth–Rock Dams*, 37–48.

Vaughan, P. R., Chandler, R. J., Apted, J. P., Maguire, J. P. & Sandroni, S. S. (1992). Sample disturbance: with particular reference to its effect on stiff clays. *Predictive Soil Mechanics, Proc. Wroth Memorial Symp.*, pp. 685–708. London: Thomas Telford.

Vaughan, P. R., Kovacevic, N. & Potts, D. M. (2004). Then and now: some comments on the design and analysis of slopes and embankments. *Proc. Skempton Conference*, Vol. 1, pp. 241–290. London: Thomas Telford.

Walbancke, H. J. (1976). *Pore pressure in clay embankments and cuttings*. PhD thesis, University of London.

Richards, D. J., Powrie, W., Roscoe, H. & Clark, J. (2007). *Géotechnique* **57**, No. 2, 197–205

Pore water pressure and horizontal stress changes measured during construction of a contiguous bored pile multi-propped retaining wall in Lower Cretaceous clays

D. J. RICHARDS*, W. POWRIE*, H. ROSCOE† and J. CLARK‡

The design and construction of a contiguous bored pile propped retained cut on the approaches to Ashford International Station raised some challenging design issues. Design problems associated with under-drainage and groundwater control (including the performance of semi-permeable propped retaining walls) and the effects of wall installation and movement on in situ stresses were compounded by the relatively limited geotechnical engineering experience of the Atherfield and Weald Clay formations. Differences in the permeabilities of these two strata were expected, but the impact of these on the design was only fully established at the start of construction. Some aspects of the engineering behaviour of the Atherfield and Weald Clays are discussed with reference to an instrumented section of the propped retained cut. The changes in total horizontal stress and pore water pressures measured during and over a six-year period following construction of the retained cut are presented, and their significance is discussed.

KEYWORDS: case history; clays; field instrumentation; retaining walls

La conception et la construction d'une coupe retenue consolidée à pieu foré contigu sur les voies d'accès d'Ashford International Station a soulevé des problèmes de conception ambitieux. Aux problèmes de conception associés à l'infradrainage et au contrôle de l'eau sous-terraine (y compris la performance des murs de retenue consolidés semi-perméables) et aux effets de l'installation du mur et du mouvement des contraintes in situ, s'est ajouté le fait qu'on avait une expérience d'ingénierie géotechnique relativement limitée des formations argileuses de Atherfield et de Weald. On s'attendait à des différences de perméabilité entre ces deux couches mais ce n'est qu'au début de la construction qu'on a établi l'impact de ces différences sur la conception. L'article décrit certains aspects du comportement d'ingénierie des formations argileuses d'Atherfield et Weald, en se référant à une section munie d'instruments de la coupe retenue consolidée. L'article décrit aussi les changements dans la contrainte horizontale totale et les pressions de l'eau interstitielle mesurées pendant une période de six ans, suite à la construction de la coupe retenue. Il discute ensuite de l'importance de ces changements.

INTRODUCTION

The Channel Tunnel Rail Link (CTRL) is the first new railway line to be constructed in the UK for over a century. It consists of 109 km of high speed track and associated infrastructure linking the Channel Tunnel at Folkestone, Kent, to the London terminus at St Pancras. It was constructed in two sections: Section 1 comprises 74 km of trackway from the Channel Tunnel and passes through Ashford International Station, Kent.

Various measures were adopted to minimise the impact of the CTRL in terms of noise and visual intrusion, particularly in urban areas. At Ashford, the railway runs through approximately 1·8 km of cut and cover tunnels and associated retained cuttings to minimise noise and to avoid crossing existing road and rail routes at grade. The tunnels and propped retained cuts were constructed between contiguous bored pile retaining walls, using piles varying in diameter from 900 mm to 1350 mm (Roscoe & Twine, 2001).

The side walls of the Ashford tunnels and retained cuts for the most part retain and/or are embedded into the Atherfield and Weald Clays. At the time the CTRL was designed, there was little direct experience of the engineering properties of these strata and how they would affect the performance of the retaining walls. Preliminary designs had been made in the expectation that a more detailed understanding could be developed during a 6-month value engineering period at the start of construction, but strategic changes to the overall phasing of the project meant that this was not possible. The key issues that the designers faced at the start of construction included

(*a*) the in situ lateral earth pressures in the Atherfield and Weald Clays, the effect of wall installation and subsequent construction, and the long-term lateral stresses that would act on the wall

(*b*) the operational field permeabilities of the Atherfield and Weald Clays, and their influence on pile construction and groundwater control

(*c*) the effect on pore water pressures behind and around the walls following the decision to use contiguous bored piled walls rather than diaphragm wall panels, as originally envisaged

(*d*) the possibility of a potentially independent lower groundwater level in the Weald Clay, which was addressed by the construction of sand drains in front of the wall to provide possible pore pressure relief to base slab level, if required.

This paper presents measurements of pore water pressures and horizontal stresses in the soil around the bored pile retaining walls, and discusses them in the context of the structure and some of the engineering properties of the Atherfield and Weald Clays.

INSTRUMENTATION

Construction monitoring

About 30 cross-sections on the CTRL in the Ashford area were instrumented to provide a basis for modifying construc-

Manuscript received 11 May 2006; revised manuscript accepted 17 January 2007.
Discussion on this paper closes on 1 August 2007, for further details see p. ii.
* University of Southampton, UK.
† Atkins Design Environment & Engineering, UK (formerly with Skanska Cementation Foundations).
‡ Gifford, UK (formerly University of Southampton).

tion and to accelerate the works. More than 700 instruments (mainly piezometers, inclinometers within individual piles, and surface and embedment strain gauges to monitor temporary and permanent loads) were installed for this purpose (Holmes *et al.*, 2005).

Research monitoring

A 12·15 m wide propped retained cut was constructed from contiguous bored piled walls formed of 1·05 m diameter piles approximately 20 m long and spaced at 1·35 m centres. An 11 m-long section of this structure was monitored more intensively as part of a detailed study of in situ retaining wall behaviour in a stiff overconsolidated clay. In addition to the general construction monitoring instrumenta-

tion described by Holmes *et al.* (2005), push-in pressure cells (spade cells) incorporating piezometers were used to measure horizontal stresses and pore water pressures at different depths and distances from the wall (Fig. 1) before, during and after wall installation, during construction, and into the long term. To determine in situ horizontal stresses, all spade cell readings were corrected by $0.35C_u$ as described by Richards *et al.* (2007).

GEOTECHNICAL CHARACTERISATION
Site geology and ground conditions

The Ashford tunnels and tunnel approaches are constructed in clays of the Lower Cretaceous period of the Mesozoic Era (Smart *et al.*, 1966; Humpage & Booth,

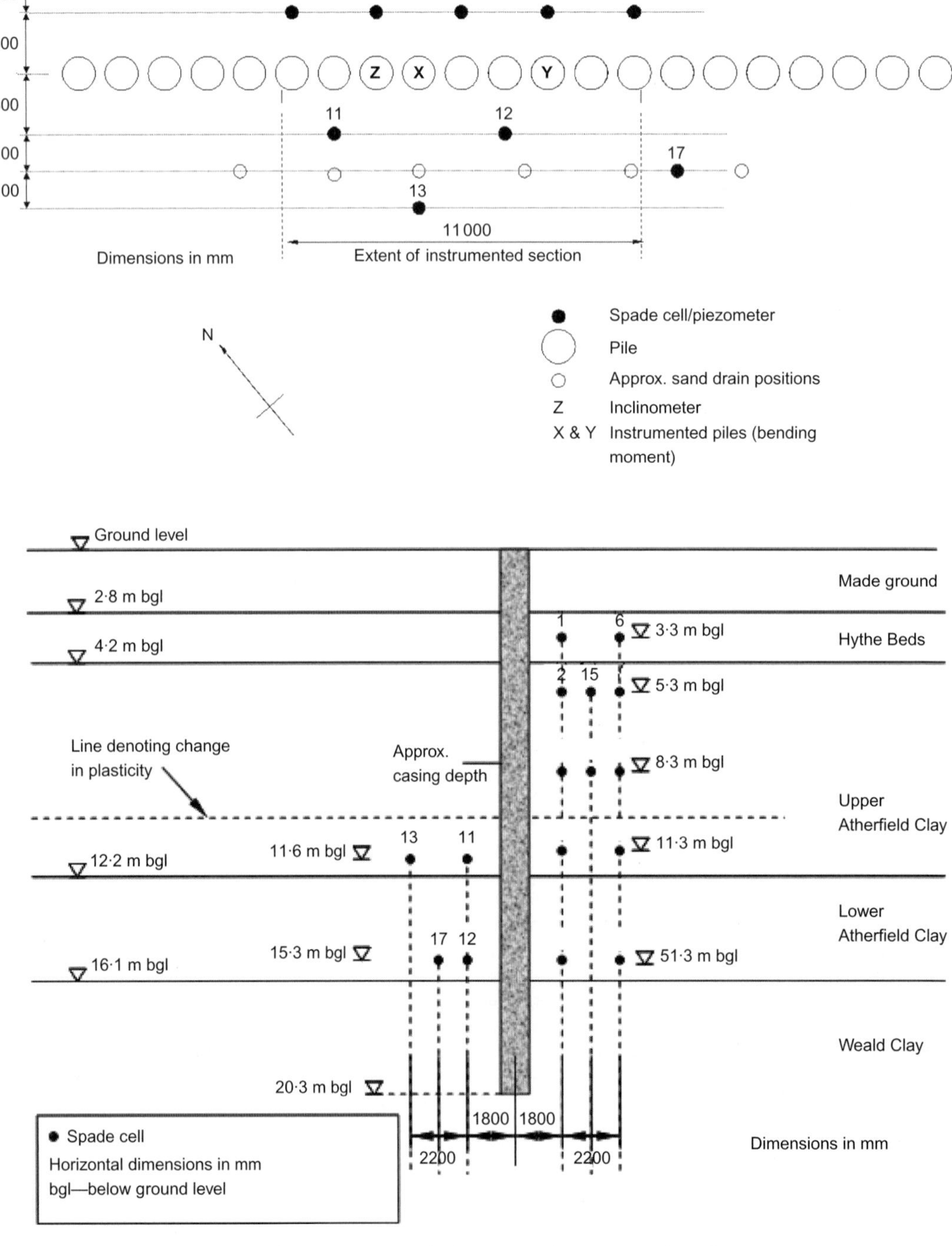

Fig. 1. Spade cell/piezometer positions and geological strata at instrumented section

2000). The regional dip in the Ashford area is approximately 3° to the north-east. The general succession of strata is given in Table 1.

At the instrumented section, only the Hythe Beds, Atherfield Clay and Weald Clay are present, and the railway is aligned with the escarpment formed by the outcrop of the Hythe Beds. The geology at this location is described in detail by Richards *et al.* (2006): for the purposes of the present paper, the key points are the disturbance caused by solifluction and other mass movements (cf. Skempton & Weeks, 1976) and the laminated/fissured structure of the clays.

In situ lateral stresses and pore water pressures

In situ lateral earth pressure coefficients (K_o) were calculated using (Mayne & Kulhawy, 1982)

$$K_0 = (1 - \sin\phi')\mathrm{OCR}^{\sin\phi'} \tag{1}$$

with ϕ' measured from effective stress tests, and overconsolidation ratios (OCR) estimated by comparing measured undrained strengths with those of normally consolidated clays of similar plasticity following Schmertmann (1978) (Table 2). Horizontal in situ stresses measured in self-boring pressuremeter tests during the site investigation and later using the research spade cells were generally less than the overburden pressure (Fig. 2). On the basis of the pressuremeter data, a value of $K_0 = 1$ was adopted for the design of the retaining structures.

In the upper Atherfield Clay the initial in situ pore water pressures were slightly less than hydrostatic, with a water table 1·2 m below the original ground surface. In the lower part of this stratum, the deviation of the pore water pressures from hydrostatic was more pronounced (Fig. 3). This is consistent with underdrainage into the Weald Clay, but may also have been influenced by an ejector well pore pressure control system in operation approximately 100 m from the instrumented section. Measured and idealised in situ horizontal stresses and pore water pressures at the Gasworks Lane site are shown in Fig. 3.

Permeabilities

In situ tests carried out in boreholes during the initial site investigation indicated average horizontal permeabilities of 3 $\times$ 10^{-8} m/s (range 2 $\times$ 10^{-9} to 9 $\times$ 10^{-8} m/s) for the Atherfield Clay, and 8 $\times$ 10^{-8} m/s (range 1 $\times$ 10^{-9} to 3 $\times$ 10^{-7} m/s) for the Weald Clay. Water strikes during drilling suggest that, in the Weald Clay, the true horizontal permeability is probably at least an order of magnitude greater (Roscoe *et al.*, 2002).

The difference in permeability and consolidation behaviour between the Weald and Atherfield Clays was further demonstrated by the trial bores that were made at the start of the piling contract. Open bores within the Atherfield Clay remained stable for >24 h, but within the Weald Clay deteriorated rapidly owing to water seepages and inflows (Roscoe *et al.*, 2002). Subsequent field pumping tests confirmed that groundwater flow in the Weald Clay is dominated by the horizontal silt partings and laminations, and that pore water pressures would respond very rapidly to construction activities.

Stiffness

Figure 4 shows the variation in undrained secant modulus with $\log_{10}$(axial strain) for an undisturbed sample of Atherfield Clay obtained by continuous rotary wire line coring in the vicinity of the instrumented wall; full details of the sample preparation and testing procedures are given in Clayton *et al.* (2006).

DESIGN CONSIDERATIONS
Retaining wall installation and construction sequence

The retaining wall installation and construction sequence is summarised in Table 3.

Table 1. General geological succession in the central Ashford area

Stratum	Description	Typical thickness
Gault Clay	Very stiff fissured clay	
Folkestone Beds	Very dense sand	0–6
Sandgate Beds	Silty sand and silty clay	
Hythe Beds	Orange brown silty or clayey fine sand with one or two poorly developed layers of sandstone (more usually interbedded Limestone (Rag) and very dense sand or weak sandstone (Hassock))	0–8
Atherfield Clay	Stiff to very stiff plastic clay, frequently closely fissured. Zones containing thin silt partings were present at some locations, occurring more frequently with depth	0–13
Weald Clay	Stiff to very stiff clay with numerous fissures, silt partings and laminations and bands of siltstone typically 100 – 200 mm thick	Up to 120

Table 2. Soil characterisation and design parameters and estimated in situ earth pressure coefficients

Stratum	Unit weight: Mg/m³	I_P w_{LL}-w_{PL}	ϕ': degrees	Estimated OCR	K_0 (using equation (1))
Hythe Beds	1·95	45%	27	7·3–9·3	1·35–1. 5
Atherfield Clay	2·1	(Upper) 50% (Lower) 20–30%	22	3·5–10·3	1·0–1·5
Weald Clay	2·1	10–30%	21	4·5–10·7	1·1–1·5

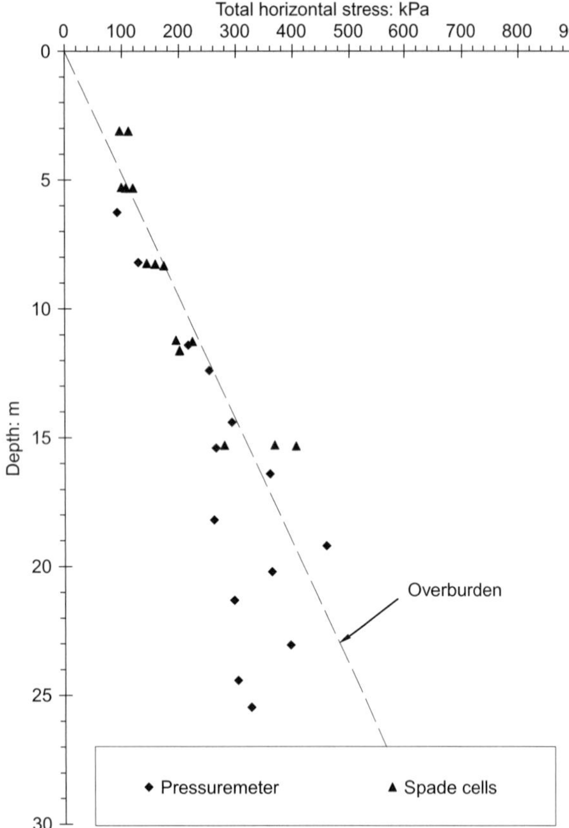

Fig. 2. Comparison between spade cell readings (corrected by $0.35\,c_u$) and self-boring pressuremeter test results

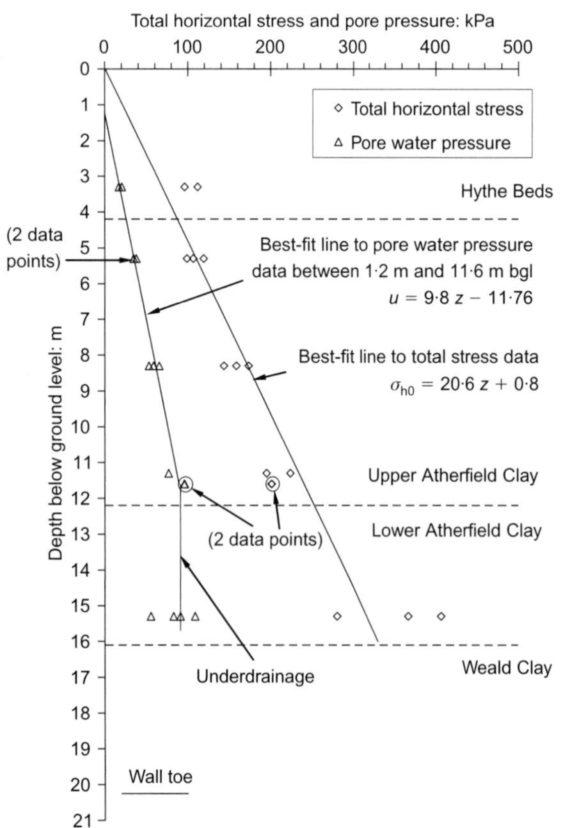

Fig. 3. In situ total horizontal stress and pore water pressure profile prior to wall installation

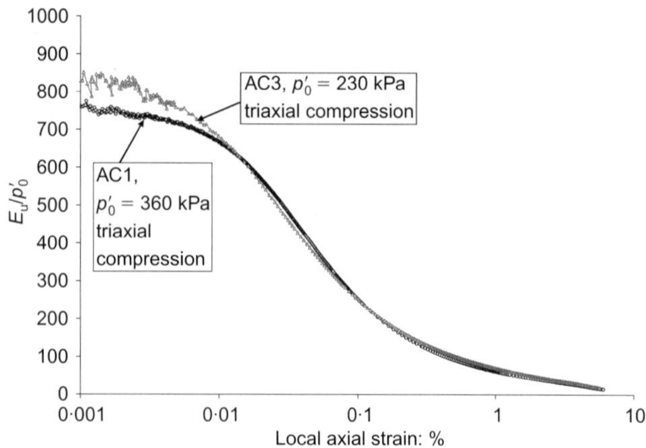

Fig. 4. Normalised undrained Young's modulus E'_u/p'_0 against $\log_{10}$(axial strain) for Atherfield Clay (Clayton *et al.*, 2006)

Lateral stresses

Assessment of the long-term lateral stresses on a retaining wall in an overconsolidated deposit is a subject of some uncertainty, but is an important factor for safe and economical design. BD42/00, the Highways Agency (2000) standard relevant to the design of embedded retaining walls and bridge abutments, suggests that embedded retaining walls propped at the crest should be designed to withstand in-service earth pressure coefficients in the retained ground of up to 1·5. This could lead to the over-design of the wall if in reality the lateral stresses are reduced during wall installation and excavation and the in situ stresses do not become re-established adjacent to the wall in the long term.

Pore pressures

The initial conceptual design assumed flow beneath an impermeable diaphragm wall to a drainage blanket below the base slab. The contractor's alternative design substituted a contiguous bored piled wall for the diaphragm wall, and the drainage blanket was dispensed with during a value engineering exercise. Shortly after completion of the base slab the exposed faces of the piles were shotcreted to retain debris, but this had no effect on the semi-permeable nature of the retaining wall below excavation level.

Preliminary designs at tender stage were made on the basis that the Weald Clay would act as a drained material. The possibility of treating the Atherfield Clay as an undrained material during construction was considered but rejected because of the relatively short drainage path length—about 6 m vertically to the upper and lower boundaries of the stratum, as discussed below.

To control pore pressures during excavation and construction, an ejector well dewatering system was installed some 100 m up-line of the instrumented section. At the instrumented section itself, passive pore pressure control measures consisting of lines of vertical drains parallel to and at a distance of 3 m from each wall on the excavated side were installed. The vertical drains comprised 150 mm diameter bores at 3 m centres, approximately 30 m deep and filled with 10 mm gravel, and were installed to relieve any excess pore water pressures in the passive zone that might result from flow through the contiguous piled walls.

The control of groundwater pressures through the use of the temporary pumped dewatering and passive drainage systems allowed significant changes to be made to the original construction sequence at the propped cut. An intermediate level of temporary props was eliminated. The excavation was carried out top down using the permanent reinforced concrete props at ground level, instead of bottom

Table 3. Main construction events

Stage	Name	Schematic	Day	Date
1	Spade cell installation		1–13	8–20 October 1999
2	Pile installation		47–71	23 November to 17 December 1999
3	Sand drain installation		349–352	20–23 September 2000
4	Capping beam construction		440–442	approximately 20–22 December 2000
5	RC (reinforced concrete) prop construction		465–467	Props 1 and 2: 14 January 2001 Prop 3: 16 January 2001
6	Excavation phase 1	5·4 m	483–509	1–27 February 2001 (no work 7–21 February inclusive)
7	Temporary prop installation		512–522	Prop 1: 2 March 2001 Props 2 and 3: 12 March 2001
8	Excavation phase 2	3·7 m	530–537	20–27 March 2001
9	Base slab construction		579	8 May 2001
10	Temporary prop removal		581	Prop 1: 10 May 2001 Props 2 and 3: 24 May 2001
			595	

piezometers behind the wall. This may be a result of the horizontal drainage of groundwater through the semi-permeable contiguous pile wall and into the passive vertical drains.

Figures 5 and 6 show that the lateral total stress behind the wall reduced substantially during excavation, at distances of both 1·275 m and 3·475 m behind the wall. Again as would be expected, the stress reductions were smaller at a distance of 3·475 m than at 1·275 m. Some slight increases in lateral total stress were recorded at a distance of 1·275 m from the wall during the period after the end of excavation, in addition to minor changes associated with the construction of the base slab and the removal of the temporary props. However, these increases were not seen at the spade cells 3·475 m from the wall, and may therefore be assumed to be due to stress redistribution in the vicinity of the wall rather than any tendency for the re-establishment of in situ values.

In front of the wall, reductions in horizontal total stress of up to about 77% of the removal of overburden (P/S17) occurred as a result of the combined effects of the removal of overburden (which on its own would be expected to result in a reduction in lateral stress) and the movement of the wall into the soil (which on its own would be expected to cause an increase). Horizontal total stresses increased (with the pore pressures) following casting of the base slab and removal of the temporary props. After this, no clear trend of any change in total horizontal stress in front of the wall after construction is apparent.

Wall installation and construction has resulted in substantial reductions in lateral stress in the retained soil. The earth pressure coefficients (ratios σ_h'/σ_v') reached after construction at spade cells S2 and S3 are below the normal active limit, even allowing for the effects of full wall friction. However, it is shown by Clark (2006) that they are consistent with the bending moments measured in the wall. These low earth pressure coefficients occur in the zone where the bending deflection (Fig. 10) of the wall is greatest, and it is possible that they are a result of stress redistribution due to wall bending.

CONCLUSIONS
(a) During excavation in front of an in situ bored pile retaining wall in the Atherfield Clay, the pore water pressures and horizontal stresses in the soil on both sides of the wall reduced markedly. As would be expected, the reductions in lateral stress and pore pressure in the retained soil became smaller with increasing distance from the wall. In front of the wall, the reduction in lateral stress due to the removal of overburden was a more dominant effect than the increase in lateral stress due to wall movement.

(b) Pore pressures both behind and in front of the wall recovered quite rapidly to their long-term equilibrium values, over periods of 4 to 15 weeks. The timescales and patterns of negative excess pore pressure dissipation behind the wall were consistent with a consolidation-type response, with independently measured values of permeability and one-dimensional modulus, and assuming vertical drainage. This confirmed that the Atherfield Clay would not behave as a substantially undrained material for periods longer than a few days to a week.

(c) The long-term pore water pressures in the retained ground are consistent with the contiguous pile wall acting as a drain, rather than an impermeable boundary, and also reflect the influence of the passive permanent pore pressure control measures. This demonstrates that

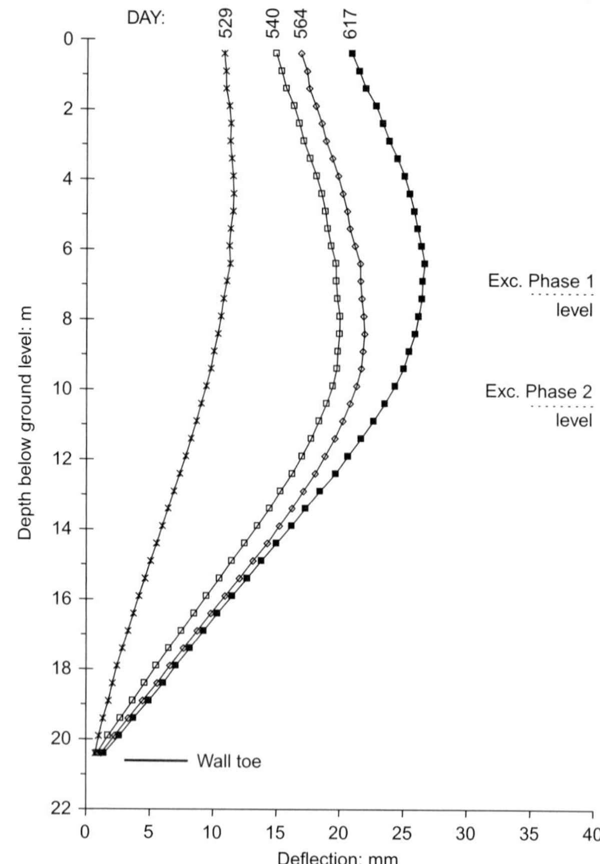

Fig. 10. Measured wall deflections (pile Z) obtained during excavation Phase 2 and following temporary prop removal (day 617)

it is possible and feasible to install passive drainage measures to reduce the long-term pore water pressures behind an embedded retaining wall in an overconsolidated deposit of this nature.

(d) Changes in measured lateral stress over the period of 2200 days following construction have been minimal. Certainly, there has been no consistent increase in lateral stresses that might indicate the possible long-term re-establishment of the in situ conditions.

ACKNOWLEDGEMENTS
The authors are grateful for the financial support provided by the Engineering and Physical Sciences Research Council (Grant reference GR/M95011), and to Rail Link Engineering and Skanska Cementation Foundations for facilitating the research and their substantial practical support on site. We would also like to thank the design and construction personnel at CTRL Contract 430, Ashford, for their help and assistance in establishing the instrumented wall section; and Dr M. Xu for his assistance during the preparation of this paper.

NOTATION

c_u undrained shear strength
c_v consolidation coefficient
d drainage path length
E' Young's modulus
E'_0 one-dimensional stiffness modulus
E_u undrained stiffness modulus-
G shear modulus
I_p plasticity index
K_0 in situ ratio of horizontal to vertical effective stress

k	hydraulic conductivity
OCR	overconsolidation ratio
p'_0	constant average in situ effective stress
T	dimensionless time factor in consolidation problems
t	time
u	pore water pressure
w_{LL}	liquid limit
w_{PL}	plastic limit
z	depth below ground level (m)
γ	unit weight of soil
γ_w	unit weight of water
Δu	change in pore water pressure
$\Delta\sigma_h$	change in total horizontal stress
$\Delta\sigma_v$	change in total vertical stress
v'	Poisson's ratio
v_u	undrained Poisson's ratio
σ'_h	effective horizontal stress
σ'_v	effective vertical stress
σ'_{ho}	in situ effective horizontal stress
σ'_{vo}	in situ effective vertical stress
Φ'	effective angle of friction

REFERENCES

Clark, J. (2006). *Performance of a propped retaining wall at the Channel Tunnel Rail Link, Ashford.* PhD thesis, University of Southampton.

Clayton, C. R. I., Xu, M. & Bloodworth, A. G. (2006). Intrinsic soil behaviour and the development of earth pressure behind integral bridge abutments. *Géotechnique* **56**, No. 8, 561–571.

Highways Agency (2000). *Design Manual for Roads and Bridges. Part 2: Design of Embedded Retaining Walls and Bridge Abutments*, BD42/00 London: HMSO.

Holmes, G., Roscoe, H. & Chodorowski, A. (2005). Construction monitoring of cut and cover tunnels. *Proc. Instn Civ. Engrs Geotech. Engng* **158**, No. 4, 187–196.

Humpage, A. J. & Booth, K. A. (2000). *The Lower Greensand in the Mid-Kent and Ashford contract sectors of the Channel Tunnel Rail Link construction works*, British Geological Survey Commercial Report CR/00/19. Keyworth: British Geological Survey.

Mayne, P. W. & Kulhawy, F. H. (1982). K_0–OCR relationships in soil. *J. Geotech. Engng Div. ASCE* **108**, GT6, 851–872.

Richards, D. J., Clark, J. & Powrie, W. (2006). Installation effects of a bored pile wall in overconsolidated clay. *Géotechnique* **56**, No. 6, 411–425.

Richards, D. J., Clark, J., Powrie, W. & Heymann, G. (2007). Performance of push-in pressure cells in overconsolidated clay. *Proc. Instn Civ. Engrs Geotech. Engng* **160**, No. 1, 31–41.

Roscoe, H., Chodorowski, A. & Wilcher, P. (2002). Piling the Gateway to Europe. *Proc. 9th Int. Conf. on Piling and Deep Foundations, Nice*, 681–688.

Roscoe, H. & Twine, D. (2001). Design collaboration speeds Ashford Tunnels. *World Tunnelling* **14**, No. 5, 32–69.

Schmertmann, J. H. (1978). *Guidelines for cone penetration test: Performance and design*, Report FHWA-TS-78–209. Washington, DC: US Department of Transport, Federal Highways Administration, Offices of Research and Development.

Skempton, A. W. & Weeks, A. G. (1976). The Quaternary history of the Lower Greensand Escarpment and Weald Clay Vale near Sevenoaks, Kent. *Phil. Trans. R. Soc. Ser. A.* **283**, 493–526.

Smart, J. G. O., Bisson, G. & Worssam, B. C. (1966). *Geology of the country around Canterbury and Folkestone*. London: HMSO.

INFORMAL DISCUSSION

Session 4
Engineering in Stiff Clays

CHAIR: MR DUNCAN NICHOLSON & DR JOHN POWELL

Dr John Powell *Chairman*

We would like to start the discussion with some questions arising from the paper by Kovacevic, Higgins, Potts and Vaughan 'Undrained behaviour of brecciated Upper Lias Clay at Empingham Dam'. Our first question is, in the light of back-analysis of the behaviour of the Upper Lias Clay at the site, what site investigation techniques would the authors recommend in similar situations?

Dr Nesha Kovacevic, *Geotechnical Consulting Group*

It is important to recognise that the Empingham Dam case history was a subject of extensive research, mainly carried out at Imperial College in the 1970s. No less than four PhD theses focussed on the overall dam behaviour and particularly the behaviour of the in-situ Upper Lias Clay in the dam foundation and the fill derived from it and used in dam construction. The research involved extensive laboratory and field testing, monitoring, and numerical analysis. However, even with the benefit of all this work it was still very difficult to characterise the behaviour of these materials for numerical analyses. The most important parameter when assessing the stability of an earth-fill dam on such stiff and disturbed clays is their undrained shear strength. We followed an effective stress approach in deducing the undrained shear strength profiles but this approach proved problematic due to a wide range of strengths displayed by the Upper Lias Clay on the site. The only satisfactory method to measure the undrained shear strength of stiff plastic clays in the laboratory is to use large diameter triaxial tests. For the Empingham Dam study undisturbed samples of various diameters, up to 260 mm, were tested and an influence of sample size on the undrained shear strength measurements was clearly observed. A slow rate of shearing after consolidation of the samples also contributed to significantly lower strenghts. We discussed with Professor Chandler the possible causes of the variation in undrained shear strength on account of soil fabric and in situ shear planes. Although it was known from the geology that there had been very large translational movements due to valley bulging and cambering amounting to some 200 metres, surprisingly there were no signs of distinct shear planes at the site. The top of the Upper Lias Clay was investigated in considerable detail and continuous shear surfaces were not observed, suggesting that the process of brecciation may have completely destroyed the clay structure due to the previous history. This observation also relates to the assessment of the in situ K_0. The disturbance to the structure of the clay meant that conventional empirical relationships between K_0 and OCR were not applicable.

Dr John Powell *Chairman*

I note from your paper that the original SI data on undrained shear strengths gave values which were much higher than the values which were determined from back-analysis of the borrow pit slips. Your large diameter slow triaxial tests gave values which were closer to the observed field strengths but were still too high. You explained the observed behaviour using a material model with peak and post-peak strengths depending on the shear strains mobilised. If we were to use the sort of model which you used for the back-analysis of Empingham Dam for new design work, how would we derive the parameters and what developments of the models do you see in the future for such analyses?

Dr Nesha Kovacevic, *Geotechnical Consulting Group*

I will first start with the models. Our previous experience, particularly from work on modelling the behaviour of various rock-fill embankments, has clearly shown the importance of modelling the plastic behaviour of soils before the peak strength is mobilised. On the trial embankment at Empingham there was no sign of progressive failure and we consequently decided to use the kinematically hardening elasto-plastic soil model of a 'bubble' type which accounted for the plastic behaviour pre-peak but did not include the post-peak strain-softening. This model worked well for the trial embankment. However, when we used the same model and model parameters for the analysis of the main dam we were not able to match the observed behaviour, particularly the sudden spreading of the northern 'undrained' section of the dam towards the end of its construction. To capture to the observed spreading we introduced the post-peak strain-softening within the in-situ Upper Lias Clay by means of the general Mohr-Coulomb elasto-plastic model. So effectively we used two constitutive soil models to characterise two different aspects of soil behaviour. A logical future development would be to incorporate both features into one model. This work is currently under development. As far as the model parameters are concerned, their derivation requires a substantial amount of effort and expertise but 'standard' laboratory testing on large samples such as triaxial, direct and/or ring shear box test are usually adequate.

Dr John Powell *Chairman*

I would like to move on to discuss some points relating to excavations in stiff clays as described in the paper by Kovacevic, Hight and Potts, 'Predicting the stand-up time of temporary London Clay slopes at Terminal 5, Heathrow Airport'. An interesting issue raised in this paper and in some other papers in the symposium is the permeability of stiff clays and the importance of the anisotropy of permeability. I would like to ask Dr Kovacevic how he modelled that and how he would recommend that people incorporate anisotropy of permeability into the models, for analyses of short and long term behaviour.

Dr Nesha Kovacevic, *Geotechnical Consulting Group*

For the excavation slope stability analyses at T5 initially we used an anisotropic permeability model because there were quite strong indications that the permeability of the London Clay at the site was anisotropic. The permeability in the horizontal direction was typically assumed to be 3 to 5 times higher than that of the vertical permeability. With some more permeable zones in which even higher horizontal permeabilities were assumed due to the presence of silt seams. It became apparent early in the study that these anisotropic permeability profiles which did not change during the analyses governed the global slope stability. However, the local instability of the excavation berms was likely to occur before deep seated failures of the entire cut slope. For these shallow failures, we decided to change the anisotropic permeability model to an isotropic, but non-linear, model which captured the important linkage between permeability and effective stress. We found out that changes of permeability arising from large changes of effective stress close to the face of the excavation as it swells had a dominant effect on predicted times to failure.

Dr John Powell *Chairman*

Your paper describes the slope stability assessment at T5 as a class A prediction. Have you now had the opportunity to compare the prediction with the field behaviour?

Dr Nesha Kovacevic, *Geotechnical Consulting Group*

No, we have not yet received the field data which would enable us to make that comparison, but we do hope to compare the field performance with our predictions in due course. One of the variables which played an important role in our analyses was imposed soil suction values at the exposed London Clay surface and it would be very useful to obtain measured suction data for comparison with our assumptions and predictions.

Dr John Powell *Chairman*

So was the construction controlled by an observational method based on your analytical predictions?

Dr Nesha Kovacevic, *Geotechnical Consulting Group*

Yes, I presume so. We focussed on two parameters which could be relatively easily measured in the field. The first of these was soil suction as this controlled the effective stresses and permeability at shallow depth. The second parameter was the horizontal displacements monitored by inclinometers conveniently installed from the berms formed in the slopes.

Dr John Powell *Chairman*

I would like now to move onto the paper by Richards, Powrie, Roscoe and Clark concerning the major bored pile retaining wall for the CTRL at Ashford and invite questions from the floor.

[Unnamed contribution]

I noted in the paper by Richards and his colleagues that some of the piezometers have shown a negative pore pressure response even in the long term. Would the authors like to comment on these readings?

Mr Howard Roscoe, *Atkins*

Before we discuss the piezometer readings at Ashford, I would like to pick up on a couple of points raised by Mr Nicholson in his introduction to this session. He mentioned in particular the value of case histories for engineering in stiff clays. Unfortunately, there don't happen to be any large cities built on either Atherfield Clay or Gault Clay, so we did not have a body of case histories to consult for the Ashford walls. The introduction also mentioned back analysis—I would make the point that this paper gives the results of the long term monitoring carried out by Southampton University's research project. There are other papers in publication dealing with some of the other aspects of the work at Ashford and one of these will deal with the thorough back analysis we did as part of the works in order to deliver some savings in time and cost using the observational method. I would also like to deal with the question that you raised of different pore water pressures shown on two of the figures. The reason for this is that Figure 3 of our paper shows the pore water pressures prior to wall installation whereas Figure 9 shows them at the end of construction. As you will all have realised, the construction included the installation of permanent sand drains through the bottom of the excavation which of course would affect the pore water pressure regime. We have not commented in the paper on that particular detail but that does highlight one of the difficulties in reporting case histories of what you include and what you leave out. Perhaps on reflection there is some discussion we should add to that. Now David Richards is going to make some observations about K_0 values and the piezometer readings.

Dr David Richards, *Southampton University*

Yes, I would like to deal with the issue of the large measured suctions. I think that the piezometers recorded the undrained response and correctly showed the negative pressures that occurred following construction. With any of these types of integral piezometer instruments once significant negative pore pressures are generated, the saturation of the filter stone and the response of the transducer then becomes a bit problematic. I think if we had been able to flush the piezometers we would have seen a more rapid recovery of the pore pressures rather than almost permanent negative pore pressures after construction. Nonetheless we have seen an increase in the pore pressures post-excavation showing that the pore pressures are starting to recover and we used that as the basis of the back analysis later on in the paper.

As for the K_0 issue, we calibrated the spade cells independently which, as Mr Roscoe has indicated, is covered in a previous paper. We corrected the spade cell readings by 0·35 s_u and correlated them to the pressuremeter test data. The instruments showed there was an initial in situ K_0 value of approximately 1 and there was a reduction in K to about 0·7 following the installation of the piles. We consider that the low measured K_0 following pile installation is a real effect which is seen in many case studies. These values of K_0 are considerably lower than design recommendations such as those of the HA. The back analysis of early TRL case studies seems to indicate that following installation of a bored pile wall a K of 1 is a reasonable basis for design and there is a lot of additional evidence to back that proposition.

Regarding the performance of the instrumentation we installed: the spade cells performed very well while the piezometers were somewhat less satisfactory. I think we could have improved the pore pressure readings if we had used flushable piezometers.

Dr John Powell, *Chairman*

I was particularly interested to note that the Ashford wall was designed as a leaky wall. Can you indicate why this form of design was chosen and whether it has been effective?

Dr David Richards, *Southampton University*

The leaky wall has proved very effective. The piezometric head on the retained side of the wall in the long term has reduced from the pre-construction level which was close to ground level to a new level of about half the excavation depth. We go down to site fairly regularly to service the data loggers which are still recording prop loads, horizontal stresses and pore pressures more than 7 years after construction began. There is evidence that there is leakage through the wall which remains semi-permeable and this is having a considerable impact on reducing the pressure acting on the back of the wall. From a design perspective it is very interesting observation and suggests that the technique could be used more widely in situations where small amounts of leakage are acceptable.

Dr Kenichi Soga, *Cambridge University*

On the St James's Park tunnels I would first like to give credit to Dr Jamie Standing for the field data used in our analysis. His data enabled us to gain a real insight into the long term behaviour of the stiff clay soil around these particular tunnels. Understanding the permeability is very important for modelling the long term behaviour of the tunnel and the paper only describes some of the work undertaken. Dr Wongsaroj undertook a series of trials to find out which model would best represent the observed behaviour. The full details are given in his PhD thesis. The other point which came out of the study was the effect of leakage around the tunnel lining. The behaviour is complex because the field data show that the soil around the tunnel crown was swelling while around the springing it was consolidating. The question was how could we reproduce that behaviour in our numerical analysis? We concluded that maybe each lining segment was part of the drainage system. However, there is still more work to be done on modelling the tunnel linings in order to understand the long term behaviour of the ground adjacent to the tunnel.

Dr John Powell *Chairman*

I would recommend this paper by Wongsaroj, Soga and Mair as a most interesting paper with very good observational data.

And finally of the papers in this section there are reports on very deep tunnelling in the Boom Clay from Mol in Belgium. I would like to ask the authors about their observations concerning anisotropy of permeability.

Li Xiang Ling, *Euridice, Belgium*

Concerning the first question, yes, there is clearly some anisotropy of permeability in the Boom Clay around our tunnels.

Dr John Powell *Chairman*

Please could you comment on the role of soil suctions generated during tunnelling in the Boom Clay, as measured by your instrumentation.

Li Xiang Ling, *Euridice, Belgium*

During the tunnel excavation we measured the development of soil suction in different in situ tests. There was some variation in the development of suction but we were not able to measure suctions lower than minus one atmosphere with our field instrumentation. In the recent laboratory tests reported in our Technical Note we describe our investigation of the development of suctions due to total stress release in more detail. These results suggest that soil suctions ranging up to about 2·5 MPa may be generated by stress relief adjacent to our deep tunnels.

Dr John Powell *Chairman*

Would you also like to comment on your experiments concerning cracking and healing of cracks in the Boom Clay?

Li Xiang Ling, *Euridice, Belgium*

Our field work suggests that the tunnelling process generates a pattern of cracks around the advancing tunnel. We have studied the healing of such cracks in laboratory experiments. The very fast sealing of cracks which we observed in the laboratory appears to be related to the swelling and consolidation properties of the clay.

Professor David Potts, *Imperial College*

I would like to raise a query with Dr Soga about the modelling of the St. James's Park tunnel. In the paper a 6 metre length of unsupported tunnel is described between the face and the tail of the shield. My 3-D finite element analyses on the same tunnels suggests that the volume losses would be excessive if a 6 m unsupported length is modelled, and the values would be in excess of 5%. To get more realistic volume losses my analyses indicated a maximum unsupported excavation length of only 2 to 2·5 metres at the most. I suspect that Dr Soga's modelling has generated very large volume losses and that as a consequence the effective stresses around the tunnel at the end of construction may not be very realistic. The model for subsequent pore pressure changes is dependent on the accuracy of the post-construction effective stresses so what happens in the long term is going to be affected but the question is by how much.

Dr Kenichi Soga, *Cambridge University*

In our model, the 6 metre unsupported length comes from the length of the shield plus the backhoe reach extending in front of the shield. We found out with our model that the 6

Lings, M. L., Pennington, D. S. & Nash, D. F. T. (2000). *Géotechnique* **50**, No. 2, 109–125

Anisotropic stiffness parameters and their measurement in a stiff natural clay

M. L. LINGS,* D. S. PENNINGTON† and D. F. T. NASH*

The framework of anisotropic elasticity has been used to develop relationships between various anisotropic stiffness parameters found in the literature. It is shown that they are functions of the two Young's moduli and two Poisson's ratios that describe a cross-anisotropic soil, but not of the independent shear modulus. Multiple drained triaxial stress path excursions have been performed on 100 mm dia. samples of natural Gault Clay from Madingley in a stress path cell. Anisotropic parameters at small strains along different stress paths are reported and compared. It is shown that there are some advantages in performing tests at constant vertical and constant horizontal effective stress. Horizontally mounted bender elements on the same trixial samples have enabled the two anisotropic elastic shear moduli to be measured. Combining results from both sets of tests has enabled all five independent cross-anisotropic elastic parameters to be estimated. The Gault Clay at Madingley is highly anisotropic, although the degree of anisotropy depends on the way it is defined.

KEYWORDS: anisotropy; clays; elasticity; laboratory tests; stiffness.

Nous avons utilisé l'élasticité anisotrope comme cadre de travail pour développer des relations entre divers paramètres de rigidité anisotrope trouvés dans les textes. Nous montrons que ces paramètres sont fonction des deux modules de Young et deux coefficients de Poisson qui décrivent un sol à anisotropie croisée mais que ces paramètres ne sont pas fonction du module de cisaillement indépendant. Nous avons fait des excursions multiples du parcours de contrainte triaxiale drainée sur des échantillons de 100 mm de diamètre dans de l'argile de Gault venant de Madingley, dans une cellule de parcours de contrainte. Nous avons noté et comparé les paramètres anisotropes pour les petites déformations le long des divers parcours de contrainte. Nous montrons qu'il existe certains avantages à faire les tests en utilisant une contrainte effective verticale et horizontale constantes. Les éléments de flexion montés horizontalement sur les mêmes échantillons triaxiaux ont permis de mesurer les deux modules de cisaillement élastiques anisotropes. En combinant les résultats des deux lots d'essais, nous avons pu évaluer les cinq paramètres indépendants d'élasticité et d'anisotropie croisée. L'argile de Gault venant de Madingley est extrêmement anisotrope mais le degré d'anisotropie dépend de la façon dont celui-ci est défini.

INTRODUCTION

Anisotropy is now recognized as an important feature when describing the behaviour of many types of soil. As far as stiff, heavily overconsolidated clays are concerned, Simpson *et al.* (1996) have shown for tunnels in London Clay that, by using numerical analyses which incorporate significant anisotropy in the shear moduli, calculated surface settlement troughs match well with field measurements.

That moduli are non-linear with strain, being greatest at very small strains, has been well established for at least the past decade (Atkinson & Sällfors, 1991). The flattening of the curve of modulus against logarithmic strain at strains less than about 0·001% has been interpreted as evidence of a linear elastic region at very small strain. Strictly speaking, the existence of a plateau would of itself only show linear behaviour, as elastic behaviour would also require full reversibility on unloading, which may or may not be true.

It is also well established that moduli are strongly affected by recent stress history (Atkinson *et al.*, 1990). The two main influences are an extended holding period at the current stress state and a sudden change in the direction of the stress path. Moduli generally increase on resting, and the highest values of moduli are found after a full 180° reversal in the stress path direction.

Dynamic testing of soils, for example in the resonant-column apparatus, shows clear reductions in shear modulus from a plateau with increasing strain amplitude. Although there is often a clear threshold strain beyond which the shear modulus starts to decrease, there is some doubt about whether this marks the limit of genuine elastic behaviour, on account of possible pre-straining and rate effects (Jardine, 1992).

Improvements in local strain measurements have enabled static moduli to be determined at smaller and smaller strain levels. However, even resolving strains down to 0·0001%, the identification of a plateau has been elusive in many soils, although it is clearly evident in soft rocks (Cuccovillo & Coop, 1997).

The general framework of soil behaviour described by Jardine (1992) probably represents the currently accepted view. At very small strains there is thought to be a linear elastic zone within which behaviour is truly linear elastic; it is also likely to be anisotropic. This zone (Zone I) is very small, although its size is difficult to gauge. At slightly larger strains there is a recoverable zone (Zone II), where the stress–strain behaviour is non-linear and hysteretic, but strains are still recoverable. The limiting distortional strain defining the boundary of this second zone is perhaps 0·01 to 0·04% for clays (Jardine, 1992). Within this zone there is a marked dependence of modulus on recent stress history. Beyond this is the plastic zone, characterized by the development of irrecoverable strains.

Linear anisotropic elasticity is the appropriate soil model to use within Zone I. In Zone II, the incremental stress–strain behaviour can still be described by the same anisotropic equations as in Zone I, but now the parameters represent tangent slopes of the appropriate stress–strain curves, and are not elastic moduli. A non-linear elastic representation of behaviour would only be valid for monotonic loading, because stress path reversal (or a long rest period) leads to high initial modulus values.

The various aspects of behaviour summarized here have been incorporated into numerical soil models, often employing kinematic-hardening plasticity, for example, Stallebrass & Taylor (1997) and Puzrin & Burland (1998). A different approach has been taken by Simpson (1992), but similar aspects of behaviour are accounted for. In contrast, Hird & Pierpoint (1997) used a non-linear anisotropic elastic formulation to model the behaviour of a stiff natural clay. In all cases, the elastic moduli are the anchor points for the subsequent degradation in stiffness.

In the first part of this paper, the anisotropic formulations developed by various authors are explored, and the relationships between different anisotropic parameters are derived. As indicated above, the equations are interpreted as linear elastic with-

Manuscript received 26 June 1999; revised manuscript accepted 26 August 1999. Discussion on this paper closes 4 August 2000; for further details see p. ii.
* University of Bristol.
† Arup Geotechnics; formerly University of Bristol.

in Zone I, and the term 'elastic' will be reserved for this case. In Zone II, the equations are still relevant, even though the behaviour is no longer linear elastic.

In the second part of the paper, results from a range of tests carried out on 100 mm samples of natural Gault Clay in a triaxial stress path cell are reported. Multiple drained stress path excursions have been performed on single samples, enabling a range of anisotropic parameters to be measured and compared. Bender elements mounted horizontally on the same specimens (Pennington et al., 1997) have enabled both anisotropic elastic shear moduli to be measured. Results from both static and dynamic tests have then been combined to estimate values for all five independent cross-anisotropic elastic parameters. One objective in deriving anisotropic parameters for the Gault Clay is to enable further back analyses of the Lion Yard deep excavation (Lings et al., 1991) to be carried out in the future.

The magnitude of shear strains developed during bender element tests has been estimated as 0·0001% by Pennington (1999) and it is considered reasonable to treat the measured moduli as elastic. Static moduli have been resolved down to strains of around 0·0005%, which represents the limit of the equipment used. To obtain the highest values of moduli, 180° reversals of stress path direction have been employed, thereby encountering the high stiffness of Zone I. In the absence of a clearly defined plateau, the assumption has been made that these maximum moduli are close to being elastic values, and are treated as such. As explained later in the paper, no direct comparison between dynamic and static measurements is possible, because different anisotropic moduli are being measured in each case.

ANISOTROPIC ELASTICITY

Many soils have been deposited over areas of large lateral extent, and the deformations they have experienced subsequent to deposition have been essentially one-dimensional. The consequence is that material properties are often different horizontally and vertically, departing significantly from the assumptions of isotropy. With a single vertical axis of symmetry, the most common type of anisotropy considered is usually known as 'cross anisotropy' in the soil mechanics literature. It is also referred to as 'transverse isotropy' (Graham & Houlsby, 1983), and in the geophysics literature as 'hexagonal anisotropy' (Crampin, 1981).

The relationship between effective-stress increments and strain increments for a material with a single vertical axis of symmetry is described by equation (1):

$$
\begin{bmatrix} \delta\varepsilon_{xx} \\ \delta\varepsilon_{yy} \\ \delta\varepsilon_{zz} \\ \delta\gamma_{yz} \\ \delta\gamma_{zx} \\ \delta\gamma_{xy} \end{bmatrix} =
\begin{bmatrix}
\dfrac{1}{E_h} & \dfrac{-\nu_{hh}}{E_h} & \dfrac{-\nu_{vh}}{E_v} & \cdot & \cdot & \cdot \\[1mm]
\dfrac{-\nu_{hh}}{E_h} & \dfrac{1}{E_h} & \dfrac{-\nu_{vh}}{E_v} & \cdot & \cdot & \cdot \\[1mm]
\dfrac{-\nu_{hv}}{E_h} & \dfrac{-\nu_{hv}}{E_h} & \dfrac{1}{E_v} & \cdot & \cdot & \cdot \\[1mm]
\cdot & \cdot & \cdot & \dfrac{1}{G_{hv}} & \cdot & \cdot \\[1mm]
\cdot & \cdot & \cdot & \cdot & \dfrac{1}{G_{hv}} & \cdot \\[1mm]
\cdot & \cdot & \cdot & \cdot & \cdot & \dfrac{1}{G_{hh}}
\end{bmatrix}
\times
\begin{bmatrix} \delta\sigma'_{xx} \\ \delta\sigma'_{yy} \\ \delta\sigma'_{zz} \\ \delta\tau_{yz} \\ \delta\tau_{zx} \\ \delta\tau_{xy} \end{bmatrix}
\tag{1}
$$

where the stress increments and strain increments are referred to rectangular Cartesian axes, with the z axis vertical. This enables two distinct horizontal directions to be identified, but shows the material properties to be the same in both horizontal directions. The seven elastic parameters shown in equation (1) are defined as follows: E_v, Young's modulus in the vertical direction; E_h, Young's modulus in the horizontal direction; ν_{vh}, Poisson's ratio for horizontal strain due to vertical strain; ν_{hv}, Poisson's ratio for vertical strain due to horizontal strain; ν_{hh}, Poisson's ratio for horizontal strain due to horizontal strain at right angles; G_{hv}, shear modulus in the vertical plane (also written as G_{vh}); and G_{hh}, shear modulus in the horizontal plane. All parameters relate to effective stresses, and the use of subscripts follows that adopted by Pickering (1970).

However, not all of these seven parameters are independent. Because the horizontal plane is a plane of isotropy, the term G_{hh} is a dependent parameter related to E_h and ν_{hh}:

$$
G_{hh} = \frac{E_h}{2(1 + \nu_{hh})}
\tag{2}
$$

For an elastic material, there is a thermodynamic requirement that the compliance matrix must be symmetric (Love, 1927). Therefore parameters in the third row and third column can be equated, giving

$$
\frac{\nu_{hv}}{E_h} = \frac{\nu_{vh}}{E_v}
\tag{3}
$$

Thus a full description of a cross-anisotropic elastic material requires five independent parameters, which are normally chosen as E_v, E_h, ν_{vh}, ν_{hh} and G_{hv} as shown in equation (4):

$$
\begin{bmatrix} \delta\varepsilon_{xx} \\ \delta\varepsilon_{yy} \\ \delta\varepsilon_{zz} \\ \delta\gamma_{yz} \\ \delta\gamma_{zx} \\ \delta\gamma_{xy} \end{bmatrix} =
\begin{bmatrix}
\dfrac{1}{E_h} & \dfrac{-\nu_{hh}}{E_h} & \dfrac{-\nu_{vh}}{E_v} & \cdot & \cdot & \cdot \\[1mm]
\dfrac{-\nu_{hh}}{E_h} & \dfrac{1}{E_h} & \dfrac{-\nu_{vh}}{E_v} & \cdot & \cdot & \cdot \\[1mm]
\dfrac{-\nu_{vh}}{E_v} & \dfrac{-\nu_{vh}}{E_v} & \dfrac{1}{E_v} & \cdot & \cdot & \cdot \\[1mm]
\cdot & \cdot & \cdot & \dfrac{1}{G_{hv}} & \cdot & \cdot \\[1mm]
\cdot & \cdot & \cdot & \cdot & \dfrac{1}{G_{hv}} & \cdot \\[1mm]
\cdot & \cdot & \cdot & \cdot & \cdot & \dfrac{2(1 + \nu_{hh})}{E_h}
\end{bmatrix}
\times
\begin{bmatrix} \delta\sigma'_{xx} \\ \delta\sigma'_{yy} \\ \delta\sigma'_{zz} \\ \delta\tau_{yz} \\ \delta\tau_{zx} \\ \delta\tau_{xy} \end{bmatrix}
\tag{4}
$$

Although all five parameters are independent, there are bounds on the values that they can take, because of the thermodynamic requirement that strain energy be positive in an elastic material. Pickering (1970) has shown that E_v, E_h and G_{hv} must all be positive, and that $-1 < \nu_{hh} < 1$. He also showed that E_v, E_h, ν_{vh} and ν_{hh} must satisfy an inequality, which is equivalent to an expression given by Raymond (1970). Both may be expressed more conveniently as

$$
\frac{E_v}{E_h}(1 - \nu_{hh}) - 2\nu_{vh}^2 \geqslant 0
\tag{5}
$$

Raymond (1970) also showed that G_{hv} is bounded by the expression

$$
G_{hv} \leqslant \frac{E_v}{2\nu_{vh}(1 + \nu_{hh}) + 2\sqrt{(E_v/E_h)(1 - \nu_{hh}^2)[1 - (E_h/E_v)\nu_{vh}^2]}}
\tag{6}
$$

Crampin (1981) describes many different types of axisymmetric anisotropy, ranging from cubic (three parameters), through hexagonal (five parameters) and tetragonal (six parameters), to orthorhombic (nine parameters). Gault Clay does not possess cubic anisotropy, with $G_{hv} = G_{hh}$, because the measured elastic moduli are so very different (Pennington et al., 1997). Nor does it possess orthorhombic anisotropy, with all three shear moduli independent, since measurements of G_{vh} in two orthogonal planes have consistently shown them to be the same. Tetragonal anisotropy has G_{hh} as an independent parameter, i.e. equation (2) would not be valid, and there is no direct evidence either for or against this as a more correct model. Thus adoption of five-parameter cross anisotropy (hexagonal anisotropy) to explore laboratory behaviour is, strictly speaking, an assumption, although one commonly made.

The assumption of axisymmetry may itself be a simplification, particularly if the *in situ* horizontal stresses show marked variations with direction in plan. Also, significant layering of the soil may complicate the picture, leading to $G_{vh} \neq G_{hv}$ in the field (Simpson *et al.*, 1996). Such behaviour is probably due to non-homogeneity rather than anisotropy, and bench top tests on both natural and reconstituted Gault Clay samples in the laboratory have consistently shown $G_{vh} = G_{hv}$ (Pennington, 1999).

INVESTIGATING ANISOTROPY USING TRIAXIAL TESTING

The stress variables used in the triaxial plane are the mean effective stress p' and the deviator stress q. They are related to the vertical and horizontal effective stresses σ'_v and σ'_h by

$$\begin{bmatrix} p' \\ q \end{bmatrix} = \begin{bmatrix} 1/3 & 2/3 \\ 1 & -1 \end{bmatrix} \begin{bmatrix} \sigma'_v \\ \sigma'_h \end{bmatrix} \tag{7}$$

The corresponding strain variables are the volumetric strain ε_p and the distortional strain ε_q. They are related to the vertical and horizontal strains ε_v and ε_h by

$$\begin{bmatrix} \varepsilon_p \\ \varepsilon_q \end{bmatrix} = \begin{bmatrix} 1 & 2 \\ 2/3 & -2/3 \end{bmatrix} \begin{bmatrix} \varepsilon_v \\ \varepsilon_h \end{bmatrix} \tag{8}$$

$G'K'J'$ and G^*K^*J formulations

When small increments of stress and strain are considered, a constitutive equation can be written (Atkinson *et al.*, 1990) as

$$\begin{bmatrix} \delta\varepsilon_p \\ \delta\varepsilon_q \end{bmatrix} = \begin{bmatrix} \dfrac{1}{K'} & \dfrac{1}{J'_{qp}} \\ \dfrac{1}{J'_{pq}} & \dfrac{1}{3G'} \end{bmatrix} \begin{bmatrix} \delta p' \\ \delta q \end{bmatrix} \tag{9}$$

where G' is a shear modulus, K' is a bulk modulus, J'_{pq} is a coupling modulus linking changes in mean effective stress and changes in distortional strain and J'_{qp} is a coupling modulus linking changes in deviator stress and changes in volumetric strain. (The subscripts qp and pq have been used, following Hird & Pierpoint (1994), rather than the 1 and 2 used by Atkinson *et al.* (1990).) Each parameter in equation (9) can be evaluated separately by conducting drained probing tests either at constant p' or at constant q. For an elastic material the compliance matrix must be symmetric, thus $J'_{qp} = J'_{pq} = J'$. For a material that is also isotropic, there is no coupling between volumetric and distortional behaviour, and the $1/J'$ terms are zero ($J' = \pm\infty$).

As an alternative to a compliance matrix, a constitutive equation can be written in terms of a stiffness matrix (Graham & Houlsby, 1983):

$$\begin{bmatrix} \delta p' \\ \delta q \end{bmatrix} = \begin{bmatrix} K^* & J \\ J & 3G^* \end{bmatrix} \begin{bmatrix} \delta\varepsilon_p \\ \delta\varepsilon_q \end{bmatrix} \tag{10}$$

where G^* is a shear modulus, K^* is a bulk modulus and J is a coupling modulus. For a material that is isotropic, here too there is no coupling between volumetric and distortional behaviour, and the J terms are zero. It should be emphasized that there is no equivalence between the G', K' and G^*, K^* parameters *unless* the material behaviour is uncoupled (implying isotropy), when $G' = G^*$ and $K' = K^*$. The G^* parameter can be directly determined from undrained tests, for which $\delta\varepsilon_p = 0$.

Conversion between the two sets of parameters is readily undertaken by inverting the matrices. This leads to the equations

$$G^* = \frac{G'J'^2}{J'^2 - 3K'G'}$$

$$K^* = \frac{K'J'^2}{J'^2 - 3K'G'}$$

$$J = \frac{-3G'K'J'}{J'^2 - 3K'G'} \tag{11}$$

and

$$G' = \frac{3G^*K^* - J^2}{3K^*}$$

$$K' = \frac{3G^*K^* - J^2}{3G^*}$$

$$J' = \frac{-(3G^*K^* - J^2)}{J} \tag{12}$$

THREE- AND FOUR-PARAMETER DESCRIPTIONS OF ANISOTROPY

If a triaxial test is carried out on a cross-anisotropic soil, no shear stresses (τ_{yz}, τ_{zx}, τ_{xy}) can be applied and no shear strains (γ_{yz}, γ_{zx}, γ_{xy}) can be measured. Hence only the top left-hand corner of the compliance matrix in equation (4) can be investigated. For the conditions in the triaxial cell, $\delta\varepsilon_{xx} = \delta\varepsilon_{yy} = \delta\varepsilon_h$ and $\delta\sigma'_{xx} = \delta\sigma'_{yy} = \delta\sigma'_h$; hence equation (4) can be simplified and rewritten as

$$\begin{bmatrix} \delta\varepsilon_v \\ \delta\varepsilon_h \end{bmatrix} = \begin{bmatrix} \dfrac{1}{E_v} & \dfrac{-2\nu_{vh}}{E_v} \\ \dfrac{-\nu_{vh}}{E_v} & \dfrac{1 - \nu_{hh}}{E_h} \end{bmatrix} \begin{bmatrix} \delta\sigma'_v \\ \delta\sigma'_h \end{bmatrix} \tag{13}$$

$G'K'J'$ formulation

By making use of equations (8) and (13), it can be shown (Appendix 1) that the G', K', J' parameters from equation (9) can be expressed as

$$G' = \frac{3}{4[(1 + 2\nu_{vh})/E_v + (1 - \nu_{hh})/2E_h]} \tag{14}$$

$$K' = \frac{1}{[(1 - 4\nu_{vh})/E_v + 2(1 - \nu_{hh})/E_h]} \tag{15}$$

$$J' = \frac{3}{2[(1 - \nu_{vh})/E_v - (1 - \nu_{hh})/E_h]} \tag{16}$$

It can be seen that G', K', J' are each functions of all four independent elastic parameters E_v, E_h, ν_{vh} and ν_{hh} that appear in equation (13).

G^*K^*J formulation

Again making use of equations (8) and (13), it can be shown (Appendix 2) that the G^*, K^*, J parameters from equation (10) can be expressed as

$$G^* = \frac{E_v}{6} \left[\frac{2(1 - \nu_{hh})E_v + (1 - 4\nu_{vh})E_h}{(1 - \nu_{hh})E_v - 2\nu_{vh}^2 E_h} \right] \tag{17}$$

$$K^* = \frac{E_v}{9} \left[\frac{(1 - \nu_{hh})E_v + 2(1 + 2\nu_{vh})E_h}{(1 - \nu_{hh})E_v - 2\nu_{vh}^2 E_h} \right] \tag{18}$$

$$J = \frac{E_v}{3} \left[\frac{(1 - \nu_{hh})E_v - (1 - \nu_{vh})E_h}{(1 - \nu_{hh})E_v - 2\nu_{vh}^2 E_h} \right] \tag{19}$$

It is clear that G^*, K^*, J are also each functions of all four independent elastic parameters E_v, E_h, ν_{vh} and ν_{hh}.

304 LINGS, PENNINGTON AND NASH

$E^* v^* \alpha$ formulation

Graham & Houlsby (1983) also proposed a three-parameter $E^* v^* \alpha$ formulation, where E^* and v^* are the Young's modulus and Poisson's ratio terms, respectively, and α is an anisotropy factor. By making certain assumptions, their model allows values to be computed for all the elastic parameters:

$$E_v = E^*$$

$$E_h = \alpha^2 E^*$$

$$v_{vh} = v^*/\alpha$$

$$v_{hh} = v^*$$

$$G_{hv} = \alpha E^*/2(1 + v^*)$$

$$G_{hh} = \alpha^2 E^*/2(1 + v^*) \tag{20}$$

Within the restrictions of their model, the anisotropy factor α controls the ratios of the Young's modulus, Poisson's ratio and shear modulus terms, requiring them all to be the same:

$$\alpha = \sqrt{\frac{E_h}{E_v}}$$

$$\alpha = \frac{v_{hh}}{v_{vh}}$$

$$\alpha = \frac{G_{hh}}{G_{hv}} \tag{21}$$

Explicit equations giving G^*, K^*, J in terms of E^*, v^*, α are given by Muir Wood (1990). If these equations are combined with equations (17)–(19), manipulation leads to

$$E^* = E_v \tag{22}$$

$$v^* = \frac{v_{vh} E_h [-v_{vh} + \sqrt{v_{vh}^2 + 4(E_v/E_h)(1 - v_{hh})}]}{2 E_v (1 - v_{hh})} \tag{23}$$

$$\alpha = \frac{E_h [-v_{vh} + \sqrt{v_{vh}^2 + 4(E_v/E_h)(1 - v_{hh})}]}{2 E_v (1 - v_{hh})} \tag{24}$$

From equations (23) and equation (24) it is clear that $v^* = v_{vh}\alpha$. Comparing this and equation (22) with equation (20) reveals that E_v and v_{vh} are *always* correctly computed with the Graham & Houlsby (1983) model.

Equations linking G', K', J' and E^*, v^*, α are given by Hird & Pierpoint (1994).

An alternative three-parameter formulation

Thus far, formulations involving triaxial stress and strain variables have been used to obtain parameters describing anisotropy. However, if tests are performed at constant σ_v' and at constant σ_h', equation (13) yields explicit relationships between the anisotropic parameters and the vertical and horizontal stresses and strains measured in triaxial tests:

$$E_v = \left(\frac{\delta\sigma_v'}{\delta\varepsilon_v}\right)_{\delta\sigma_h'=0} \tag{25}$$

$$v_{vh} = -\left(\frac{\delta\varepsilon_h}{\delta\varepsilon_v}\right)_{\delta\sigma_h'=0} \tag{26}$$

$$\frac{E_h}{(1 - v_{hh})} = \left(\frac{\delta\sigma_h'}{\delta\varepsilon_h}\right)_{\delta\sigma_v'=0} \tag{27}$$

$$\frac{2v_{hv}}{(1 - v_{hh})} = -\left(\frac{\delta\varepsilon_v}{\delta\varepsilon_h}\right)_{\delta\sigma_v'=0} \tag{28}$$

In obtaining equation (28), E_h and v_{hv} have been reintroduced into the top right-hand term of the compliance matrix in equation (13) (i.e. equation (3) has not been invoked) so that the symmetry of the compliance matrix in equation (1) can be investigated.

Inspection of equations (14)–(16) for G', K', J' in terms of E_v, E_h, v_{vh} and v_{hh} shows that the parameters E_h and v_{hh} always appear together in the combination found in equation (27). If we write

$$F_h = \frac{E_h}{1 - v_{hh}} \tag{29}$$

then the parameters E_v, v_{vh}, F_h can provide as complete a description of the behaviour in a triaxial test as the parameters G', K', J'. This is demonstrated by rewriting equations (14)–(16) as

$$G' = \frac{3 E_v F_h}{4 F_h + 8 v_{vh} F_h + 2 E_v} \tag{30}$$

$$K' = \frac{E_v F_h}{F_h - 4 v_{vh} F_h + 2 E_v} \tag{31}$$

$$J' = \frac{3 E_v F_h}{2 F_h - 2 v_{vh} F_h - 2 E_v} \tag{32}$$

Manipulation of these equations leads to expressions for E_v, v_{vh}, F_h in terms of G', K', J':

$$E_v = \frac{9 G' K' J'}{6 G' K' + G' J' + 3 K' J'} \tag{33}$$

$$v_{vh} = -\frac{1}{2} \frac{3 G' K' + 2 G' J' - 3 K' J'}{6 G' K' + G' J' + 3 K' J'} \tag{34}$$

$$F_h = \frac{-18 G' K' J'}{12 G' K' - 4 G' J' - 3 K' J'} \tag{35}$$

In all cases, the parameters of one formulation can be expressed as functions of all three parameters of the other formulations.

In summary, if symmetry of the compliance matrix is assumed, then only three independent parameters can be measured in a triaxial test. These can be done in a number of ways, and four different three-parameter formulations have been described. However, if the soil is cross-anisotropic, there are four independent parameters that control the behaviour in triaxial tests, and relationships between the three- and four-parameter formulations have been derived.

INVESTIGATING FIVE-PARAMETER ANISOTROPY

Pennington *et al.* (1997) described a modified 100 mm stress path triaxial apparatus which can propagate orthogonally polarized shear waves both vertically and horizontally through the same sample using bender elements. This apparatus enables the measurement of three separate shear moduli, G_{0vh}, G_{0hv} and G_{0hh}, where the subscript 0 indicates the very small strain level of approximately 0·0001% of the tests, and the two other subscripts show the directions of wave propagation and polarization, respectively.

In tests on natural and reconstituted Gault Clay, the measured values of G_{0hv} were consistently higher than G_{0vh} (see for example Fig. 5(b) of Pennington *et al.*, 1997). Yet in isotropic tests on the bench, no significant differences could be found. This has led to the conclusion that end effects associated with rigid triaxial platens have affected the G_{0vh} measurements, with underestimates of up to 30% (Pennington, 1999). Consequently, G_{0vh} measurements are regarded as less reliable than those of G_{0hv}, and only G_{0hv} and G_{0hh} values will be considered here. (This has also led to the adoption of G_{hv} in preference to G_{vh} throughout the paper.)

Three parameters can be measured in the triaxial apparatus, and two parameters can be measured using orthogonal horizontal bender elements. Because there is some overlap in these two groups of parameters, further manipulation is possible. Bender element tests yield values of G_{0hv} and $G_{0hh} = E_{0h}/2(1 + v_{0hh})$ (see equation (2)), and very-small-strain data from triaxial tests can yield values of E_{0v}, v_{0vh} and $F_{0h} = E_{0h}/(1 - v_{0hh})$ (see equation (29)). The equations for F_{0h} and G_{0hh} can be combined to give E_{0h} and v_{0hh}:

$$E_{0h} = \frac{4 F_{0h} G_{0hh}}{F_{0h} + 2 G_{0hh}} \tag{36}$$

$$\nu_{0hh} = \frac{F_{0h} - 2G_{0hh}}{F_{0h} + 2G_{0hh}} \tag{37}$$

Thus in the very-small-strain region, where the soil behaviour may be assumed elastic, it is in principle possible to determine all five independent cross-anisotropic elastic parameters.

TESTING PROGRAMME

High-quality undisturbed samples of Gault Clay from Madingley, near Cambridge, were obtained and prepared for testing as described by Pennington *et al.* (1997). The *in situ* stress state has been estimated as $K_0 \approx 2$ between 6 and 8 m below ground level (Butcher & Lord, 1993). The stress path to reach the *in situ* stress state involved isotropic reconsolidation to a mean effective stress equal to the *in situ* vertical effective stress, followed by a horizontal effective-stress increase (at constant vertical effective stress) until an estimated K_0 of 2 was reached.

This procedure was adopted after finding sample suctions measured in the sampling tubes to be less than the vertical effective stress. This stress path offered the shortest route to the *in situ* stress state with the minimum amount of sample strain and hence possible sample damage. As the deformation characteristics being investigated involved 180° changes in stress path direction, it was considered unnecessary to attempt a more elaborate reconsolidation and swelling strategy to model the stress history of each sample.

Drained stress path excursions were carried out along paths where one stress was held constant for a small change in strain (approximately 0·1%). Further stress path excursions in different directions were carried out on the same sample to derive the maximum benefit. The use of multiple mini-stress-path excursions follows its successful adoption by Hird & Pierpoint (1994, 1997) for stiff Oxford Clay.

Local measurements of axial and radial strain were made using Hall effect gauges similar to those described by Clayton & Khatrush (1986). Testing was carried out in a temperature-controlled laboratory, where continuous monitoring of the cell water temperature confirmed gross temperature changes during any stress path excursion to be less than 0·3°C. The stability and resolution of the Hall effect gauges under these conditions led to accuracies of individual data points of approximately ±0·001% for both radial and axial strain.

The results from tests R14 and R23 are reported here, and the stress paths actually followed during these tests are shown in Figs 1 and 2. Rates of loading were typically 1 kPa per hour, and tests were halted at each of the symbol markers shown in the figures for bender element tests to be performed. Before these tests were done, a pause ranging between 18 and 30 h was allowed, after which vertical strain rates were negligible, typically less than 0·003% per hour.

Test R14 (Fig. 1) consisted of a number of constant-σ'_v sectors (Oa, ab, bc) to the estimated *in situ* stress state at c, followed by a 180° reversal to d, and a further 180° reversal back to c. After a constant-σ'_h sector to e, the test followed a constant-σ'_v sector to f. Further excursions were planned; however, a leak developed and the test was stopped.

Test R23 (Fig. 2) provided a comprehensive range of stress path excursions, not only at constant σ'_v and σ'_h but also at constant p' and constant q. These were concentrated in the region of the estimated *in situ* stress state (paths ab, bc, cb, bd, db, be). A large multistage drained triaxial shearing sector was then carried out at constant cell pressure from the *in situ* stress state to a point with $\sigma'_h/\sigma'_v = 0·5$ (paths be, ef, fg and gh). The stresses were then returned to isotropic at f, and thence to O. There was no evidence of leakage throughout the 55 day duration of this test.

An important finding from test R23 was that values of G_{0hv} and G_{0hh} measured on returning to O were no different from the initial values measured at the start of the test (Pennington, 1999). This was in spite of the large triaxial shearing sector to h, and is evidence that no significant destructuration occurred during the test. This confirms the observation of Jovičić & Coop

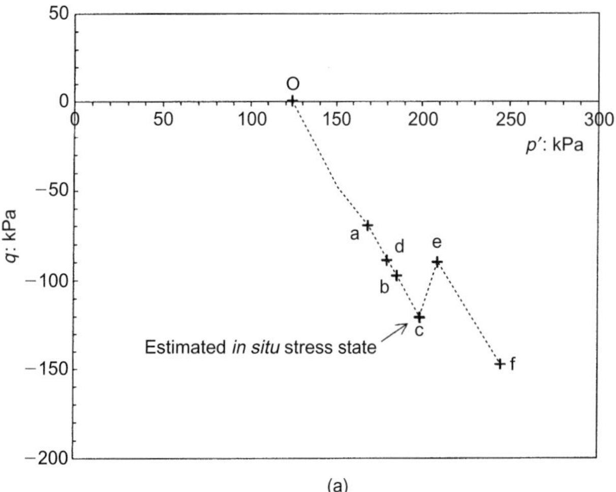

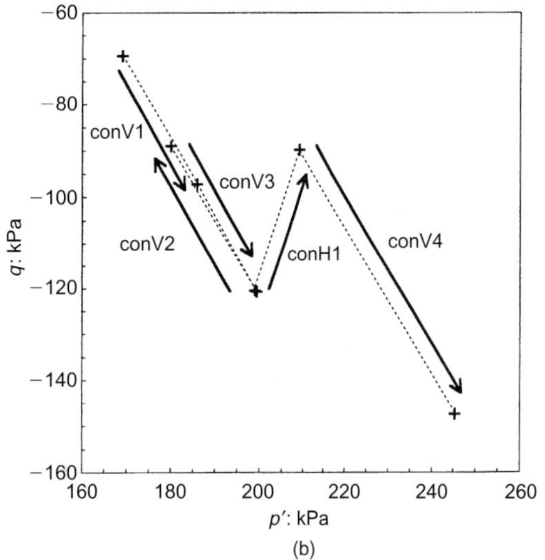

Fig. 1. Stress paths followed in test R14: (a) reconsolidation and stress excursions; (b) details of stress excursions

(1998) that very large plastic strains are required to affect the inherent anisotropy of stiff clays, and contrasts with the behaviour of soft structured clays. In hindsight, the care with which sample preparation and reconsolidation were carried out may not have been entirely necessary.

At the time the tests were carried out, a number of different aspects of behaviour were being investigated. If the tests had been done solely to explore cross anisotropy as described in the first part of the paper, a somewhat different sequence of tests would have been performed. An ideal test programme would have used a single natural sample, and started all stress probes from the same *in situ* stress state, with each probe following a 180° stress path reversal. This ideal programme was not always followed, and hence there are additional uncertainties in interpreting the results.

EXPERIMENTAL RESULTS

All results, for both stress increments and moduli, are presented as normalized values. This has been done because some stress excursions went a long way from the assumed *in situ* values, yet the derived moduli need to be combined on a common basis. Also, tests were done on two samples from different depths, but results from both needed to be used. Normalization has been carried out by dividing by the value of p' at the start of the relevant stress probe, referred to as p'_0.

There clearly are arguments for using different methods of normalization. Viggiani & Atkinson (1995) found that the shear

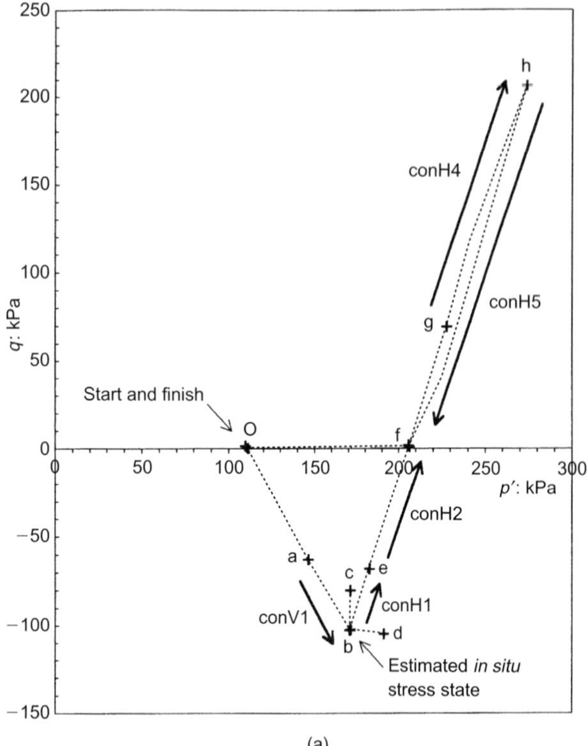

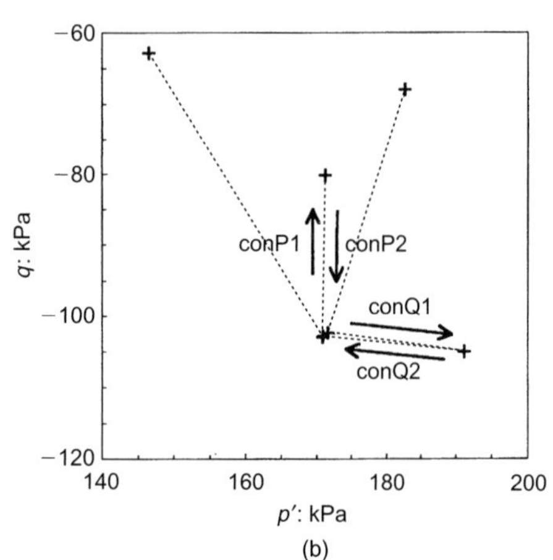

Fig. 2. Stress paths followed in test R23: (a) reconsolidation and stress excursions; (b) details of constant-p' and constant-q stress excursions

moduli varied with p' raised to the power n, with values of n in the range of 0·6 to 0·8 when considering G_{0vh}. At large strains, the value of n was found to be approximately 1 when considering G'. Hird & Pierpoint (1997) found a similar dependence on p' in tests on natural Oxford Clay, but found $n = 0·67$ when considering G', with no clear dependence on strain level. Normalization by p' (i.e. $n = 1$) has been widely adopted by many others in the past, and has been done here largely for reasons of simplicity. A rigorous normalization of the modulus over the full strain range would require a detailed knowledge of the variation of n with strain level, which at the present time is not possible. It is also not yet known how such a strain level dependence may vary between the different anisotropic parameters. Full normalization would also take into account the changing values of p' during the course of a test, but this has not been attempted here either.

The absolute values of both the normalized stress increments and the resulting strain increments are plotted in this paper to enable comparison between data regardless of stress path direc-

tion. Following the work of Atkinson *et al.* (1990), the recent stress history at the start of each probing test is indicated. Each set of data is annotated with the change in stress path direction, measured in degrees from the previous path direction in the p'–q plane. No distinction has been made between clockwise and anticlockwise rotation angles, although this information is available in Figs 1 and 2.

All stress-probing tests were done drained. This meant that the testing rates were orders of magnitude slower than, for example, the undrained stress probes of Coop *et al.* (1997), where strain resolutions down to 0·0001% were achieved. In order to evaluate moduli, polynomial curves were fitted to the raw stress–strain data. Tangent moduli were then obtained by differentiation of the polynomial, allowing evaluation at any desired strain value. This procedure was performed in semiautomated Excel workbooks, where the data were plotted for a number of strain ranges, so that the quality of fit could be assessed, and the strain range or order of the polynomial adjusted to give the best fit for each excursion.

An example of such a curve is given in Fig. 3. The minimum resolution of each local vertical-strain measurement is $\approx 0·0015\%$ (Pennington, 1999). Since $\delta\varepsilon_q = (2/3)\delta\varepsilon_v - (2/3)\delta\varepsilon_h$, and the horizontal strain change was less than the resolution of the horizontal belt, i.e. $\delta\varepsilon_h \approx 0$, it follows that $\delta\varepsilon_q \approx (2/3)\delta\varepsilon_v$ and the minimum resolution per local vertical gauge is $\approx 0·001\%$. With two such gauges the effective resolution becomes $\pm 0·0005\%$. This has been represented by the error bars included in the figure, and the best-fit curve fits within these extremes. In general, curve fitting was taken to strain levels of 0·001%, but sometimes as low as 0·0002% when justified by the quality of the fit.

One consequence of curve fitting using a polynomial is that a plateau in the modulus–strain curves cannot be achieved. This is because the second derivative of the stress–strain curve can never be zero, which is needed to define a truly linear portion. Higher-resolution instrumentation would be required to obtain direct evidence of linear behaviour.

Tests at constant σ_v' and constant σ_h'

The results from stress-probing tests conducted at constant σ_v' and constant σ_h' are presented first. Fig. 4 shows vertical stress–strain curves; Fig. 5 shows horizontal stress–strain curves. Both sets of data show increasing gradients, and thus increasing moduli, as the angle of stress path rotation increases.

Figures 6 and 7 show strain paths for tests R23 and R14, respectively. The slopes of these plots (or their inverse) represent Poisson's ratio terms, and it can be seen that, apart from curves R23conH4 and R14conV1, these values are initially zero.

Figure 8 shows how E_v and ν_{vh} vary with vertical strain on a logarithmic scale. Fig. 9 shows how $E_h/(1 - \nu_{hh})$ and $\nu_{hv}/(1 - \nu_{hh})$ vary with horizontal strain. In this and similar figures, tangent moduli are presented.

The consequences of recent stress history can be seen in Fig. 8(a), not only in terms of the strong influence of stress path rotation angle, but also in the effect of holding stresses constant. Excursions R23conH1, H2 and H4 were all in the same direction and show decreasing modulus with increasing strain. When stress paths continue after a pause, but with no change in direction, initial increases in modulus are observed in each case. In the same tests, non-zero initial values of Poisson's ratio are observed in Fig. 8(b), and pauses seem to have almost no effect (compare the start of R23conH4 with the end of R23conH2). The values clearly return to zero following full stress path reversal (e.g. R23conH5).

At very small strain, the normalized values chosen after 180° stress path reversal were $E_{0v} = 550$ from test R23conH5 and $E_{0h}/(1 - \nu_{0hh}) = 2100$ from test R14conV2. This test was chosen in preference to R14conV3 (with a value of 1950) as it shows a higher value. It also starts from *in situ* stresses. The values of ν_{0vh} and ν_{0hv} are both zero.

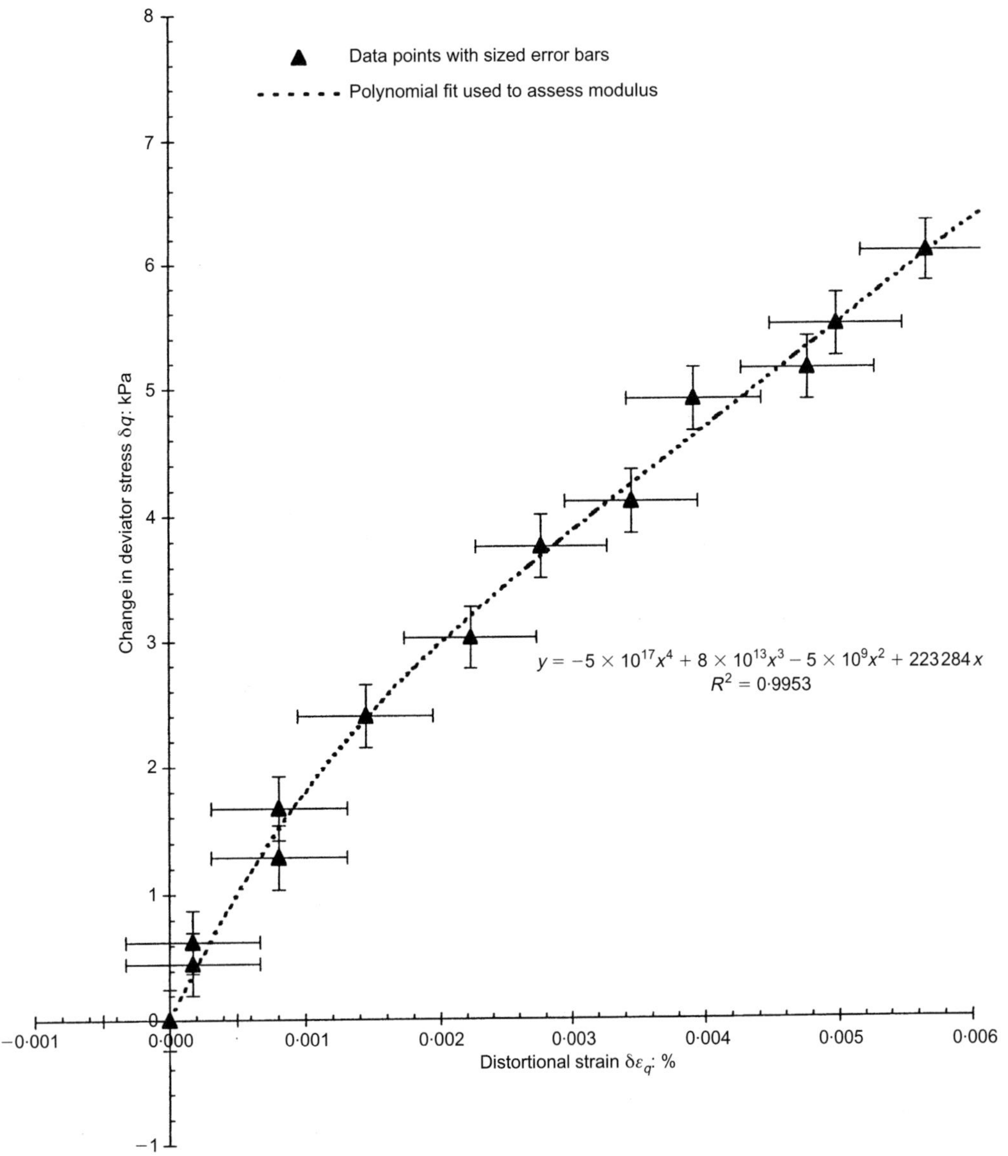

Fig. 3. Example of stress–strain curve fitting for test R23conP1

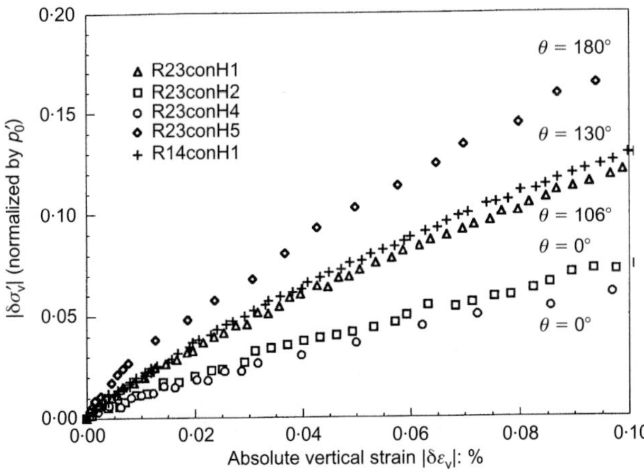

Fig. 4. Normalized vertical stress–strain curves at constant σ_h'

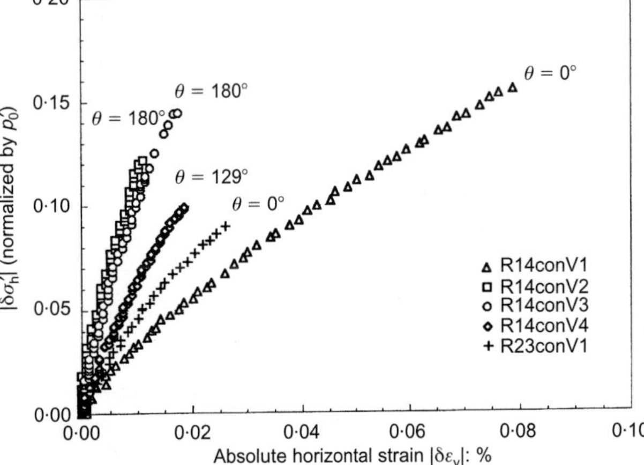

Fig. 5. Normalized horizontal stress–strain curves at constant σ_v'

Bender element tests

Shear wave velocity measurements were made at the start of each stress probe. Normalization by p_0' has been carried out with these data too, despite the evidence that shear moduli are more dependent on components of stress in the plane of the shear wave (Roesler, 1979; Jamiolkowski *et al.*, 1995; Nash

et al., 1999). The pragmatic reason for this apparent inconsistency is the need to combine bender element data with small-strain data, for which there is a long history of normalization by p_0'. Normalized G_{0hh} and G_{0hv} values for tests R23 and R14 are plotted against p' in Fig. 10. The normalized moduli clearly vary with confining pressure, showing the deficiencies of this

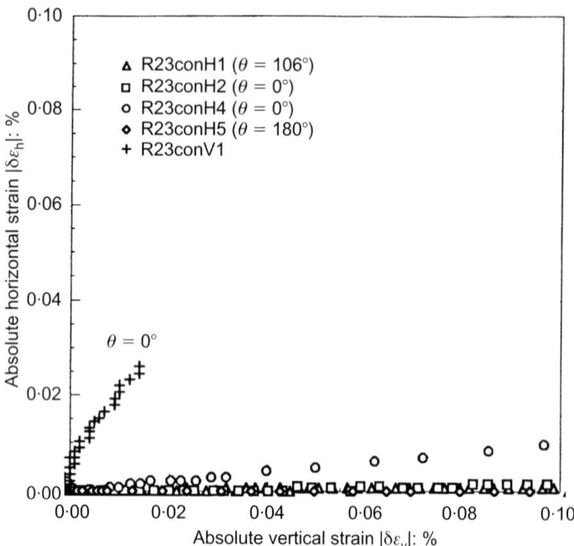

Fig. 6. Absolute strain paths in test R23

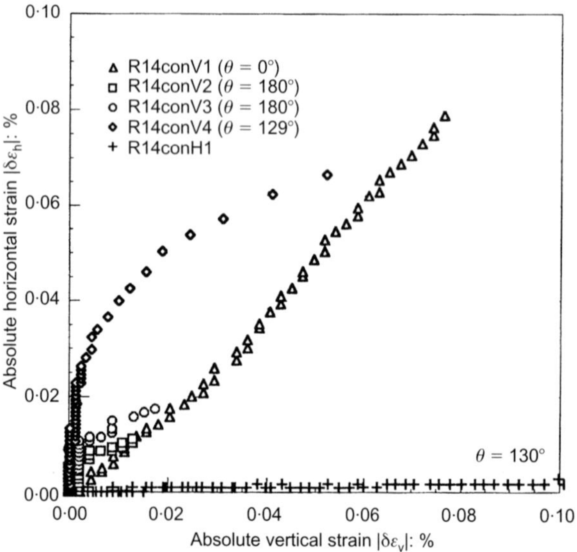

Fig. 7. Absolute strain paths in test R14

normalization. But the important points to note are that, at the relevant *in situ* stresses, the values of G_{0hh} and G_{0hv} are extremely well defined, and the values are virtually indistinguishable between tests R23 and R14. A more comprehensive assessment of the effects of stress level and void ratio is given by Nash *et al.* (1999). The averaged values at the *in situ* stress state for test R23 have been adopted, giving $G_{0hv} = 507$ and $G_{0hh} = 1140$.

Five independent elastic parameters

As described earlier, all five independent elastic parameters for a cross-anisotropic soil can be found from E_{0v}, ν_{0vh} and F_{0h}, obtained from local small-strain data after 180° stress path reversal, and G_{0hv} and G_{0hh}, measured using bender elements. The normalized values of each parameter are presented in Table 1, together with the resulting normalized values of E_{0v}, E_{0h}, ν_{0vh}, ν_{0hh} and G_{0hv} obtained after using equations (36) and (37). The likely accuracies are discussed later.

Tests at constant p' and constant q

Test R23 included two stress probes at constant p' and two at constant q. In Fig. 11 $\delta p'$ and δq are plotted against the distortional strain $\delta\varepsilon_q$; in Fig. 12 they are plotted against the volumetric strain $\delta\varepsilon_p$.

Figure 13 shows how G' and J'_{pq} vary with distortional strain, and Fig. 14 shows how K' and J'_{qp} vary with volumetric strain, both on a logarithmic scale. The two plots for G' show excellent agreement, whereas those for K' do not, at least at very small strains. The reason becomes clear once the angle of stress path rotation is taken into account. The plots of J'_{pq} and J'_{qp} show quite a range of values at very small strain, with the largest value occurring with a stress path rotation of less than 90°. A similar variability in the J' parameters was found by Hird & Pierpoint (1997), although the reasons for this are unclear.

Difficulties were experienced in controlling q precisely during the constant-q stress path excursions (see Fig. 2), where changes in deviator stress of some 2 to 3 kPa occurred. The effect of this change was small, but it has nevertheless been accounted for by using the coupling implied by equation (9) and results from the constant-p' tests.

Incremental strain energy

To make comparisons between the various measured parameters, some means of transforming vertical, horizontal, volumetric and distortional strains to a unified strain representation is required. Various definitions of a unified strain have been proposed and used to enable moduli determined with respect to different strains to be compared. Burland (1989) expressed the opinion that there were particular advantages in the use of incremental strain energy. Equations can be written in terms of the variables used here as follows:

$$\Delta\varepsilon_{\text{unified}} \equiv U = \sum_0^{\varepsilon_p}(p' - p'_0)\delta\varepsilon_p + \sum_0^{\varepsilon_q}(q - q_0)\delta\varepsilon_q \tag{38}$$

where the initial stress state is defined by (p'_0, q_0), and the incremented state is defined by (p', q), and

$$\Delta\varepsilon_{\text{unified}} \equiv U = \sum_0^{\varepsilon_v}(\sigma'_v - \sigma'_{v0})\delta\varepsilon_v + 2\sum_0^{\varepsilon_h}(\sigma'_h - \sigma'_{h0})\delta\varepsilon_h \tag{39}$$

where the initial stress state is defined by $(\sigma'_{v0}, \sigma'_{h0})$, and the incremented state is defined by (σ'_v, σ'_h). It is a simple procedure to show the equivalence of equations (38) and (39).

Measured values of G', K' and J' are plotted against strain energy on a logarithmic scale in Fig. 15. Measured values of E_v, ν_{vh} and F_h are plotted against strain energy in Fig. 16; in this case only tests R23conH5 and R14conV2 are shown, being those with full stress path reversal.

Comparison of three-parameter formulations

It is now possible to compare measured values of E_v, ν_{vh}, F_h and G', K', J' at the same strain energy using equations (30)–(35). The measured values of E_v, ν_{vh}, F_h in Fig. 16 have been used to calculate values of G', K', J', which are shown in Fig. 15. The measured values of G', K', J' in Fig. 15 have been used to calculate E_v, ν_{vh}, F_h. When values of J' from R23conP1 and R23conQ2 were used, the calculated values of F_h were negative at certain strain energies, and generally showed wide fluctuations. The calculated values of E_v and F_h shown in Fig. 16 are therefore only those calculated using J' values from tests R23conP2 and R23conQ1. The calculated values of ν_{vh} show significant variation depending on the J' value.

At very small strains, the values of E_{0v}, ν_{0vh}, F_{0h} and G'_0, K'_0, J'_0 can be compared. Table 2 shows both measured and calculated values, and the difference between them, at a common strain energy of 10^{-6} kJ/m³. The largest value of J'_0, measured in test R23conQ1, has been used in this calculation. It appears that, apart from F_{0h}, the agreement is quite good, with the difference in G'_0 being only 5%. The calculated value of F_{0h} is sensitive to the measured J'_0 value, and the difference here is 24%. Given the uncertainties over normalization, any

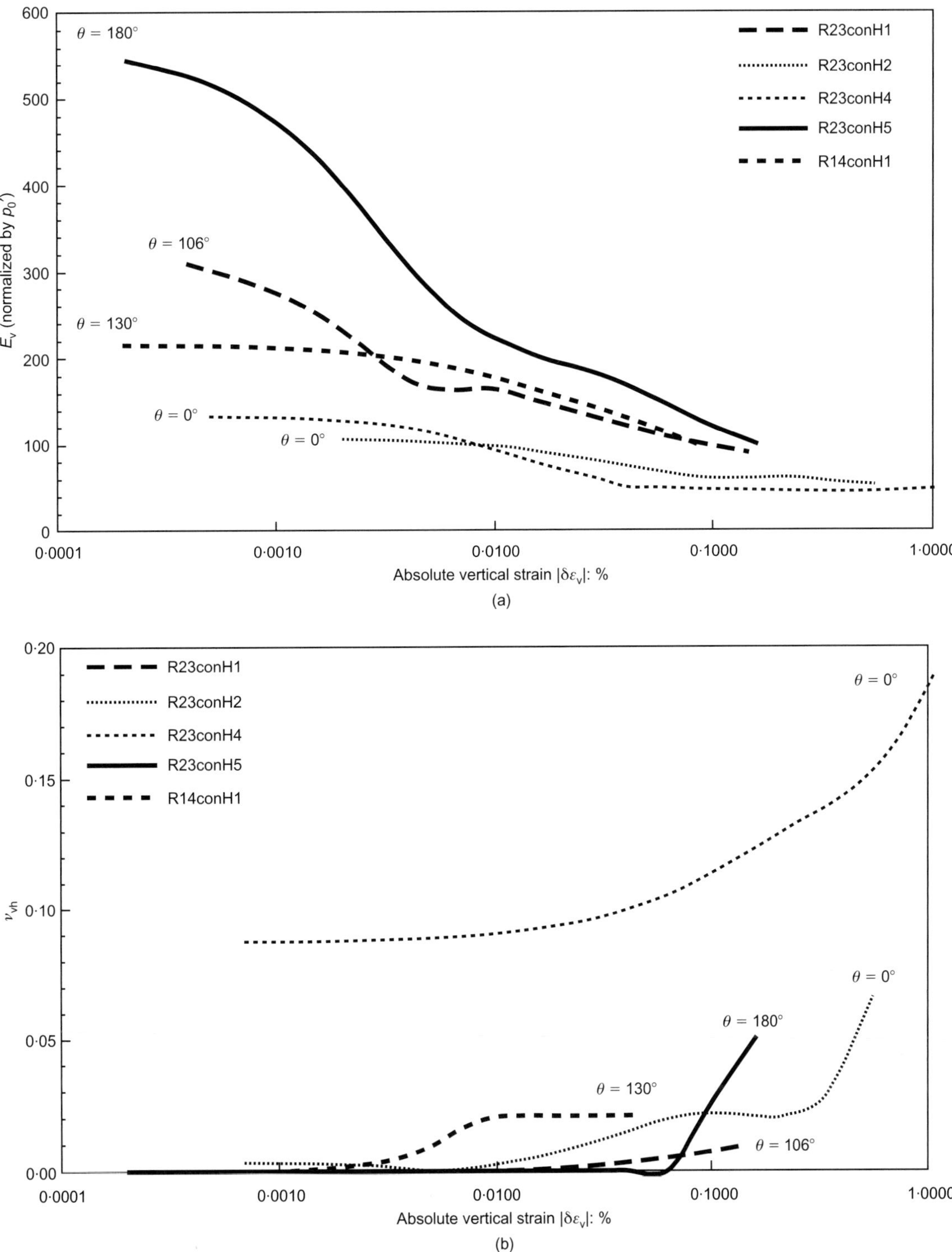

Fig. 8. (a) Normalized E_v plotted against logarithm of absolute vertical strain; (b) ν_{vh} plotted against logarithm of absolute vertical strain

agreement between these two sets of parameters may be fortuitous.

Comparison of five-parameter formulations

At very small strains it is also possible to compare the best estimated measured values of the five independent elastic parameters E_{0v}, E_{0h}, ν_{0vh}, ν_{0hh} and G_{0hv} with those computed from the Graham & Houlsby (1983) model. The measured values of E_{0v}, ν_{0vh} and $E_{0h}/(1 - \nu_{0hh})$ have been converted to E^*, ν^* and α using equations (22)–(24), which have been converted to five elastic parameters using equation (20). The results are shown in Table 3, together with the percentage differences between the two sets of parameters; values of G_{0hh} are also shown.

First, it should be noted that the parameters E_{0v} and ν_{0vh} are unchanged in passing through this transformation. This will always be the case, as comparison of equation (20) and equations (22)–(24) has shown. All the other parameters show some deviation from the measured values.

Second, it is clear that the agreement for Gault Clay is generally good. Differences in ν_{0hh} have been expressed in terms of $(1 - \nu_{0hh})$, and because E_{0h} is underestimated and ν_{0hh} is overestimated, G_{0hh} is the parameter with the largest underestimation. The value of G_{0hv} is overestimated. In the Graham & Houlsby (1983) model, the parameter α expresses a fixed ratio between $\sqrt{E_{0h}/E_{0v}}$ and G_{0hh}/G_{0hv} (see equation (21)), and for Gault Clay α is 1·95. In contrast, the measured ratios are $\sqrt{E_{0h}/E_{0v}} = 2·0$ and $G_{0hh}/G_{0hv} = 2·25$.

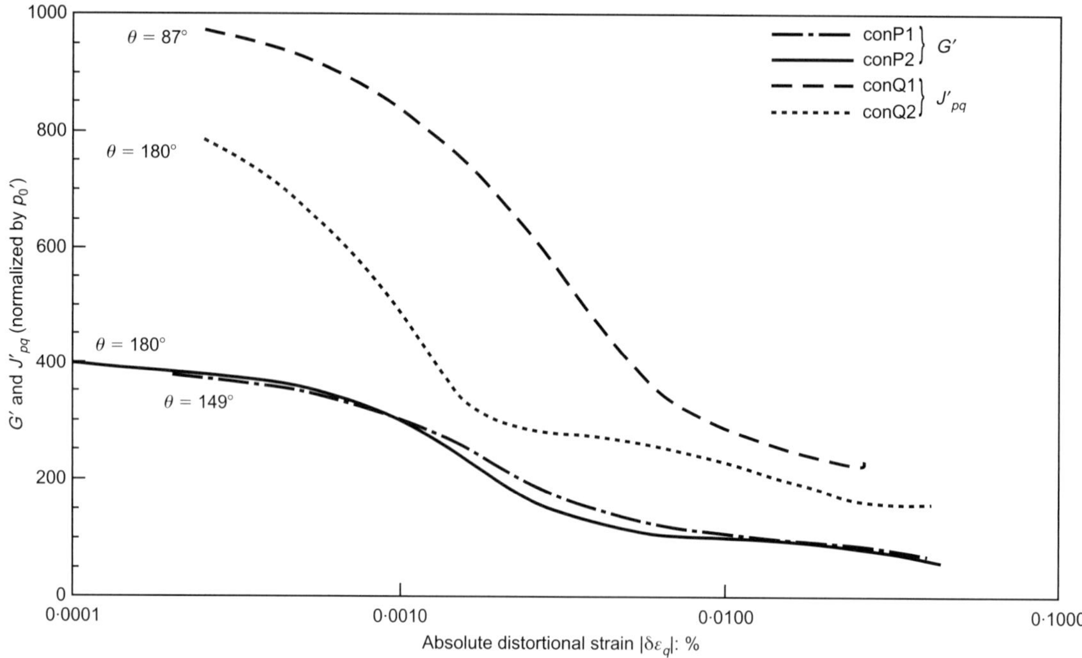

Fig. 13. Normalized G' and J'_{pq} plotted against logarithm of absolute distortional strain

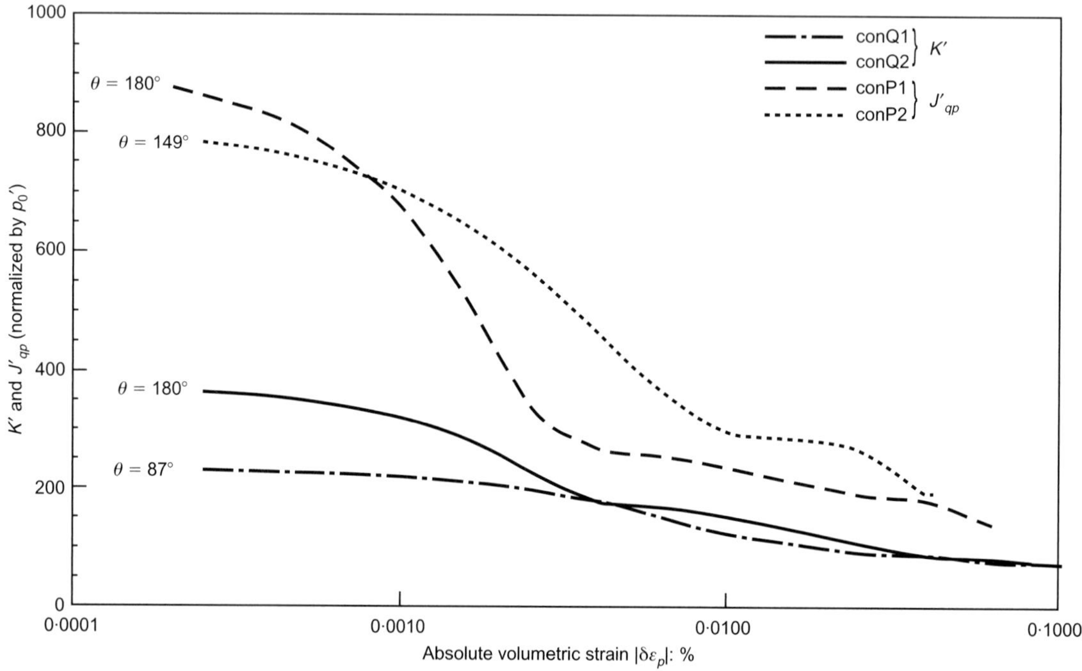

Fig. 14. Normalized K' and J'_{qp} plotted against logarithm of absolute volumetric strain

larger strains, and, strictly speaking, there is none on the E_h/E_v ratio either. However, if it is assumed that the value of ν_{hh} remains small, then an approximate value of the ratio can be obtained from F_h/E_v. The variation of this ratio is plotted in Fig. 16(a), and it can be seen that it remains very close to 4 until a strain energy of about $10^{-4}\,\mathrm{kJ/m^3}$, after which it reduces steadily towards 2.

Triaxial testing

It has been shown that *all* the three-parameter formulations describing the soil response in the triaxial apparatus ($G'K'J'$, G^*K^*J, $E^*\nu^*\alpha$ and $E_v\nu_{vh}F_h$) can be expressed as unique functions of the four independent parameters E_v, E_h, ν_{vh} and ν_{hh}. Almost all of them depend on all four; the exceptions are

E_v and ν_{vh} (and $E^* \equiv E_v$), which are single parameters, and F_h, which is a function of just two parameters (E_h and ν_{hh}).

If stress path triaxial testing is to be carried out to obtain soil parameters, which when used in subsequent modelling will probably be assumed to be stress-path-independent, there appear to be arguments in favour of using an $E_v\nu_{vh}F_h$ formulation. The first is the directness of measurement: tests at constant σ'_h will produce two of the required parameters directly (E_v and ν_{vh}), and tests at constant σ'_v will obtain the third parameter (F_h) as a function of only the remaining two. This contrasts with measurements at constant p' and constant q, when the resulting parameters are each functions of all four independent parameters.

The second is a practical one: measurements of J' seem to show significant scatter, which will affect any subsequently

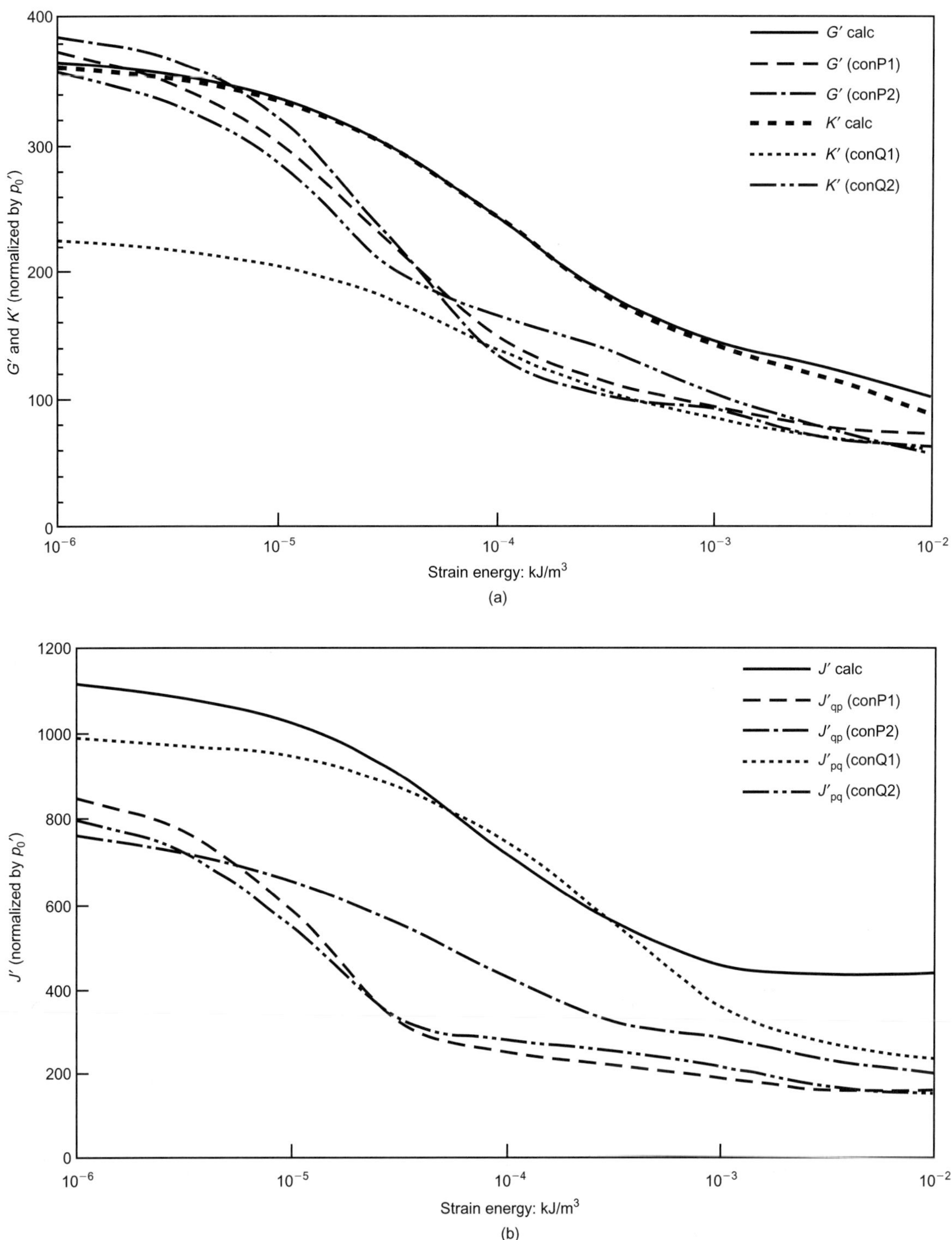

Fig. 15. (a) Measured and calculated normalized G' and K' plotted against logarithm of strain energy; (b) measured and calculated normalized J' plotted against logarithm of strain energy

derived parameters. Inspection of Fig. 16(b) shows how four different tests at constant p' or q have been unable to produce the clarity of the direct measurement (at constant σ'_h) that shows $\nu_{vh} = 0$, although this may also reflect some inherent stress path dependence in the parameters.

The third is simply one of clarity: measurements of shear, bulk and coupling moduli (e.g. G', K', J') can lead to the feeling that all aspects of behaviour have been covered. With models such as that of Graham & Houlsby (1983) enabling easy conversion from three to five anisotropic parameters, the missing two measurements can be overlooked. With an $E_v \nu_{vh} F_h$ formulation, it is made clear that *no* shear modulus information has been obtained.

Independent shear modulus

The independent shear modulus G_{hv} cannot be measured in a stress path triaxial test, and has no influence on the measured soil behaviour. Yet it clearly affects the soil response in any plane strain application, as shown by Simpson *et al.* (1996). For such an analysis, a full five-parameter description of anisotropy is required, and no three-parameter description can by itself be sufficient.

The 'shear modulus' is widely referred to in the literature as a fundamental parameter, usually within an assumed isotropic framework. If an anisotropic soil were (wrongly) assumed to be isotropic, then a pseudo-isotropic shear modulus could be found from

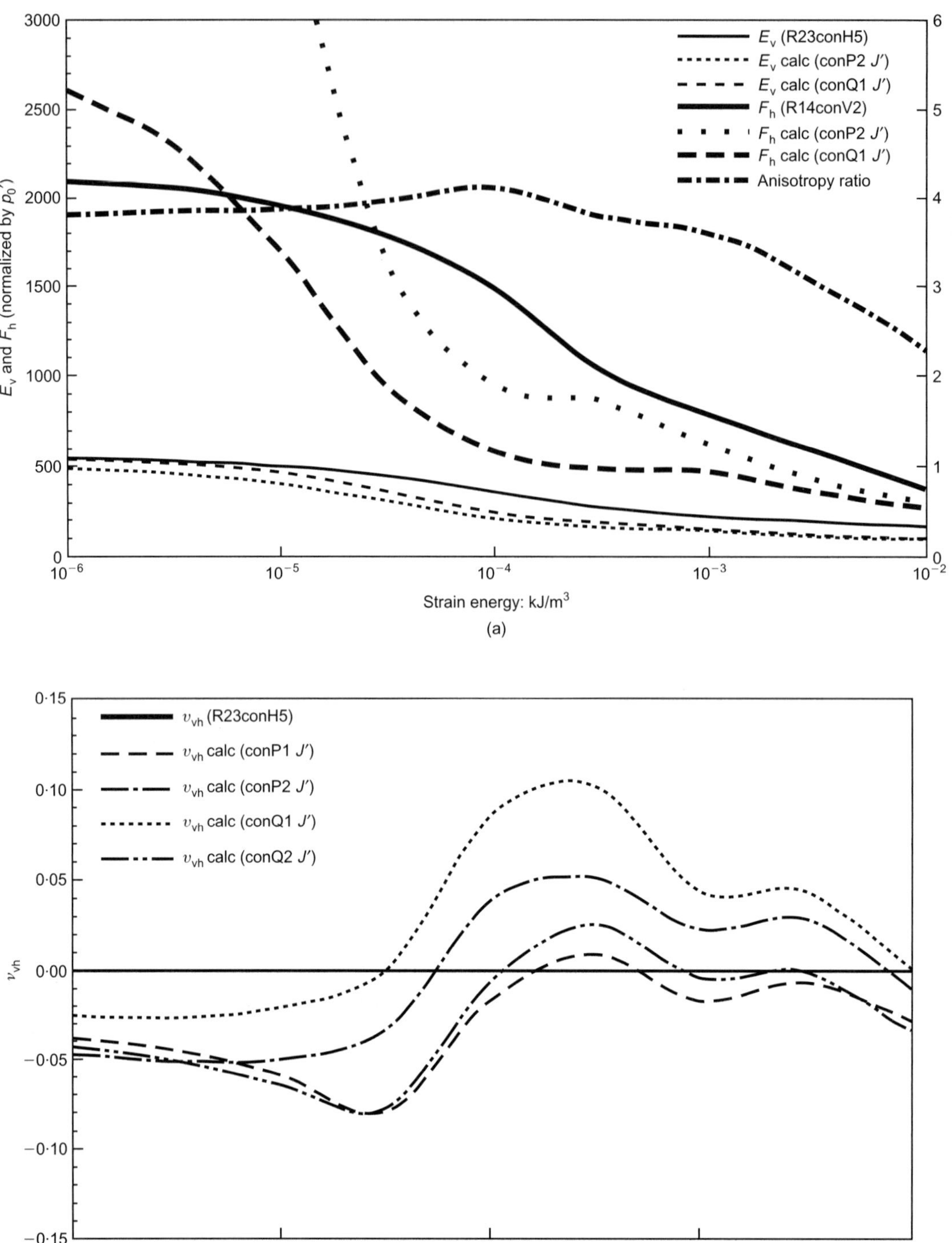

Fig. 16. (a) Measured and calculated normalized E_v and F_h and anisotropy ratio, plotted against logarithm of strain energy; (b) measured and calculated ν_{vh} plotted against logarithm of strain energy

Table 2. Comparison of different normalized parameters

	E_{0v}	ν_{0vh}	F_{0h}
Measured	550	0	2100
Calculated (from G_0', K_0', J_0')	540	−0·025	2607
Difference	2%	–	24%

	G_0'	K_0'	J_0'
Measured	385	356	990
Calculated (from E_{0v}, ν_{0vh}, F_{0h})	365	361	1118
Difference	5%	1%	11%

$$G_{iso} = \frac{E_v}{2(1 + \nu_{vh})} \tag{40}$$

because such an equation would be correct for an isotropic soil.

On the basis of the measured normalized values of E_{0v}, E_{0h}, ν_{0vh} and ν_{0hh} for Gault Clay, it is instructive to calculate a range of possible normalized values for the 'shear modulus'. Use of equation (14) yields a value of $G_0' = 365$; use of equation (17) yields a value of $G_0^* = 533$; and use of equation (40) yields a value of $G_{0iso} = 275$. The measured value of G_{0hv} is 507. This illustrates not only its independence, but also the

Table 3. Comparison of measured and Graham & Houlsby (1983) normalized parameters

Parameter	Measured (small strain/benders)	Calculated (equations (22)–(24))	Calculated (equation (20))	Difference
E_{0v}	550		550	–
E_{0h}	2186		2100	4%
ν_{0vh}	0		0	–
ν_{0hh}	$-0\cdot04$		0	4%†
G_{0hv}	507		537	6%
G_{0hh}	1140		1050	9%
E^{*}		550		
ν^{*}		0		
α		$1\cdot95$		

† Based on $(1 - \nu_{hh})$.

importance of being very clear about which parameter is being referred to when describing the 'shear modulus'.

Earlier work has tried to show the continuity between dynamic and static measurements of stiffness, but has compared dynamic G_{hv} measurements with G^{*} measurements in an undrained triaxial test (e.g. Georgiannou *et al.*, 1991; Coop *et al.*, 1997). If the soil is anisotropic, there is no reason to *expect* agreement between these two parameters, as they are independent (see equation (17)). Equivalence between dynamic and static measurements has not been shown here either, with no comparison possible in a strongly anisotropic material. However, measurement of G^{*} in an undrained excursion might still be a good method of checking the credibility of E_v, ν_{vh}, F_h values obtained in drained stress path excursions.

CONCLUSIONS

Various three-parameter formulations used to describe the cross-anisotropic behaviour of soils have been investigated. It has been shown that all of them are functions of the four independent parameters E_v, E_h, ν_{vh} and ν_{hh}. An alternative three-parameter formulation comprising E_v, ν_{vh} and F_h $(= E_h/(1 - \nu_{hh}))$ has been presented, and shown to provide as complete a description of soil behaviour as other three-parameter formulations and to possess certain advantages. The Graham & Houlsby (1983) model has been investigated and shown always to produce correct values of E_v and ν_{vh}. Conversion equations between the various formulations have been developed.

High-quality triaxial testing on samples of natural Gault Clay from Madingley has been carried out, involving multiple drained stress path excursions and orthogonal determinations of horizontal shear wave velocity using bender elements. At very small strains, data from both engineering strain and shear wave velocity measurements have been combined to derive all the five independent parameters needed to describe a cross-anisotropic elastic material. The vertical Poisson's ratio is zero, and the horizontal Poisson's ratio is small but negative. All moduli have been expressed as normalized values by dividing by p'_0. The measured value of G_{hv}, the independent shear modulus, was found to be below the upper limit defined by Raymond (1970) as a function of E_v, E_h, ν_{vh} and ν_{hh}.

Atkinson *et al.* (1990) have shown the dependence of moduli on holding periods and angle of stress path rotation for reconstituted soils for strains generally in excess of 0·01%. This has been extended here to a natural soil over a wider strain range (down to 0·001%), and similar effects of recent stress history have been observed. The moduli used for the assessment of elastic parameters have been measured after a full reversal in stress path direction.

When the measured normalized values of the five independent cross-anisotropic elastic parameters were compared with those computed using the Graham & Houlsby (1983) model, broad agreement was found for natural Gault Clay, the maximum difference being 9%. The agreement is probably strongly influenced by the clear experimental finding that $\nu_{0vh} = 0$.

The anisotropy of natural Gault Clay has been investigated under anisotropic *in situ* stresses. Anisotropy ratios of $E_{0h}/E_{0v} = 4$ and $G_{0hh}/G_{0hv} = 2\cdot25$ have been measured, showing the natural Gault Clay to be highly anisotropic. The anisotropy ratio depends on how it is defined. Both G_{0hh} and G_{0hv} were found, on return to the same stress state, to be extremely insensitive to large plastic strains. Natural Gault Clay appears to behave as a cross-anisotropic elastic material at very small strains.

The independent shear modulus cannot be measured in a conventional triaxial test, yet it is important to measure G_{hv} because of its influence in practical applications.

ACKNOWLEDGEMENTS

The authors are grateful to the Building Research Establishment, who recovered intact samples of Gault Clay from Madingley. They also gratefully acknowledge financial support for this work from the Engineering and Physical Sciences Research Council.

APPENDIX 1. DERIVATION OF G', G', J' IN TERMS OF E_v, E_h, ν_{vh} and ν_{hh}

In constant-p' stress path excursions, G' and J'_{qp} can be evaluated from

$$G' = \frac{1}{3}\left(\frac{\delta q}{\delta \varepsilon_q}\right)_{\delta p'=0} \tag{41}$$

$$J'_{qp} = \frac{1}{3}\left(\frac{\delta q}{\delta \varepsilon_p}\right)_{\delta p'=0} \tag{42}$$

Since p' is held constant, $\delta\sigma'_h = -\delta\sigma'_v/2$, and substitution into equation (13) leads to

$$\delta\varepsilon_v = \frac{1}{E_v}\delta\sigma'_v + \frac{\nu_{vh}}{E_v}\delta\sigma'_v = \frac{1+\nu_{vh}}{E_v}\delta\sigma'_v \tag{43}$$

and

$$\delta\varepsilon_h = \frac{-\nu_{vh}}{E_v}\delta\sigma'_v - \frac{1-\nu_{hh}}{2E_h}\delta\sigma'_v = -\left(\frac{\nu_{vh}}{E_v} + \frac{1-\nu_{hh}}{2E_h}\right)\delta\sigma'_v \tag{44}$$

Combining these into one equation using the relationship for deviatoric strain ε_q given in equation (8), and noting that since $\delta\sigma'_h = -\delta\sigma'_v/2$, $\delta q = 3\delta\sigma'_v/2$,

$$\delta\varepsilon_q = \frac{2}{3}(\delta\varepsilon_v - \delta\varepsilon_h) = \frac{2}{3}\left(\frac{1+2\nu_{vh}}{E_v} + \frac{1-\nu_{hh}}{2E_h}\right)\frac{2}{3}\delta q \tag{45}$$

This can be rearranged to give an expression for G':

$$\frac{1}{3}\left(\frac{\delta q}{\delta\varepsilon_q}\right)_{\delta p'=0} = G' = \frac{3}{4[(1+2\nu_{vh})/E_v + (1-\nu_{hh})/2E_h]} \tag{46}$$

Similarly, using the relationship for volumetric strain ε_p given in equation (8),

$$\delta\varepsilon_p = \delta\varepsilon_v + 2\delta\varepsilon_h = \left(\frac{1-\nu_{vh}}{E_v} - \frac{1-\nu_{hh}}{E_h}\right)\frac{2}{3}\delta q \tag{47}$$

This can be rearranged to give an expression for J'_{qp}:

$$\left(\frac{\delta q}{\delta \varepsilon_p}\right)_{\delta p'=0} = J'_{qp} = \frac{3}{2[(1-\nu_{vh})/E_v - (1-\nu_{hh})/E_h]} \quad (48)$$

In constant-q stress path excursions, K' and J'_{pq} can be evaluated from

$$K' = \left(\frac{\delta p'}{\delta \varepsilon_p}\right)_{\delta q=0} \quad (49)$$

$$J'_{pq} = \left(\frac{\delta p'}{\delta \varepsilon_q}\right)_{\delta q=0} \quad (50)$$

Since q is held constant, $\delta\sigma'_h = \delta\sigma'_v$ and substitution into equation (13) leads to

$$\delta\varepsilon_v = \frac{1}{E_v}\delta\sigma'_v - \frac{2\nu_{vh}}{E_v}\delta\sigma'_v = \frac{1-2\nu_{vh}}{E_v}\delta\sigma'_v \quad (51)$$

and

$$\delta\varepsilon_h = \frac{-\nu_{vh}}{E_v}\delta\sigma'_v + \frac{1-\nu_{hh}}{E_h}\delta\sigma'_v = \left(\frac{-\nu_{vh}}{E_v} + \frac{1-\nu_{hh}}{E_h}\right)\delta\sigma'_v \quad (52)$$

Combining these into one equation using the relationship for volumetric strain ε_p given in equation (8), and noting that since $\delta\sigma'_h = \delta\sigma'_v$, $\delta p' = \delta\sigma'_v$,

$$\delta\varepsilon_p = \delta\varepsilon_v + 2\delta\varepsilon_h = \left(\frac{1-4\nu_{vh}}{E_v} + \frac{2(1-\nu_{hh})}{E_h}\right)\delta p' \quad (53)$$

This can be rearranged to give an expression for K':

$$\left(\frac{\delta p'}{\delta \varepsilon_p}\right)_{\delta q'=0} = K' = \frac{1}{(1-4\nu_{vh})/E_v + 2(1-\nu_{hh})/E_h} \quad (54)$$

Similarly, using the relationship for deviatoric strain ε_q given in equation (8),

$$\delta\varepsilon_q = \frac{2}{3}(\delta\varepsilon_v - \delta\varepsilon_h) = \frac{2}{3}\left(\frac{1-\nu_{vh}}{E_v} - \frac{1-\nu_{hh}}{E_h}\right)\delta p' \quad (55)$$

This can be rearranged to give an expression for J'_{pq} which is the same as that given for J'_{qp} in equation (48).

APPENDIX 2. DERIVATION OF G^*, K^*, J IN TERMS OF E_v, E_h, ν_{vh} and ν_{hh}

G^*, K^*, J can be evaluated under constant-strain conditions, giving

$$G^* = \frac{1}{3}\left(\frac{\delta q}{\delta \varepsilon_q}\right)_{\delta\varepsilon_p=0}$$

$$K^* = \left(\frac{\delta p}{\delta \varepsilon_p}\right)_{\delta\varepsilon_q=0}$$

$$J = \left(\frac{\delta q}{\delta \varepsilon_p}\right)_{\delta\varepsilon_q=0} = \left(\frac{\delta p}{\delta \varepsilon_q}\right)_{\delta\varepsilon_p=0} \quad (56)$$

If $\delta\varepsilon_p = 0$ then $\delta\varepsilon_v + 2\delta\varepsilon_h = 0$. Requiring this condition in equation (13) yields

$$\frac{\delta\sigma'_h}{\delta\sigma'_v} = -\frac{1}{2}\left(\frac{(1-2\nu_{vh})E_h}{(1-\nu_{hh})E_v - \nu_{vh}E_h}\right) \quad (57)$$

Hence

$$\delta q = \delta\sigma'_v - \delta\sigma'_h = \delta\sigma'_v\left(1 - \frac{\delta\sigma'_h}{\delta\sigma'_v}\right)$$

$$= \frac{\delta\sigma'_v}{2}\left(\frac{2(1-\nu_{hh})E_v + (1-4\nu_{vh})E_h}{(1-\nu_{hh})E_v - \nu_{vh}E_h}\right) \quad (58)$$

Substituting equation (57) in equation (13) for $\delta\varepsilon_v$ gives

$$\delta\varepsilon_v = \delta\sigma'_v\left(\frac{1}{E_v} + \frac{\nu_{vh}}{E_v}\frac{(1-2\nu_{vh})E_h}{(1-\nu_{hh})E_v - \nu_{vh}E_h}\right) \quad (59)$$

Because $\delta\varepsilon_v + 2\delta\varepsilon_h = 0$, it follows that $\delta\varepsilon_q = \delta\varepsilon_v$. Combining equations (58) and equation (59) gives

$$\frac{1}{3}\left(\frac{\delta q}{\delta\varepsilon_q}\right)_{\delta\varepsilon_p=0} = G^* = \frac{E_v}{6}\left(\frac{2(1-\nu_{hh})E_v + (1-4\nu_{vh})E_h}{(1-\nu_{hh})E_v - 2\nu_{vh}^2 E_h}\right) \quad (60)$$

Similar manipulations with $\delta\varepsilon_q = 0$ yield expressions for K^* and J:

$$K^* = \frac{E_v}{9}\left(\frac{(1-\nu_{hh})E_v + 2(1+2\nu_{vh})E_h}{(1-\nu_{hh})E_v - 2\nu_{vh}^2 E_h}\right) \quad (61)$$

$$J = \frac{E_v}{3}\left(\frac{(1-\nu_{hh})E_v - (1-\nu_{vh})E_h}{(1-\nu_{hh})E_v - 2\nu_{vh}^2 E_h}\right) \quad (62)$$

NOTATION

E^*	Young's modulus parameter (Graham & Houlsby, 1983)
E_h	horizontal Young's modulus
E_v	vertical Young's modulus
F_h	horizontal modulus $(= E_h/(1-\nu_{hh}))$
G^*	shear modulus parameter (Graham & Houlsby, 1983)
G'	shear modulus parameter (Atkinson et al., 1990)
G_{hh}	horizontal shear modulus
G_{hv}	vertical shear modulus $(\equiv G_{vh})$
G_{iso}	pseudo-isotropic shear modulus
J	coupling modulus parameter (Graham & Houlsby, 1983)
J'	coupling modulus parameter (Atkinson et al., 1990)
J'_{pq}, J'_{qp}	coupling modulus parameter (Hird & Pierpoint, 1994)
K^*	bulk modulus parameter (Graham & Houlsby, 1983)
K'	bulk modulus parameter (Atkinson et al., 1990)
p'	mean effective stress
p'_0	normalizing mean effective stress
q	deviator stress
x	horizontal axis
y	horizontal axis
z	vertical axis
α	anisotropy parameter (Graham & Houlsby, 1983)
γ_{xy}	shear strain in horizontal plane
γ_{yz}, γ_{zx}	shear strain in vertical plane
ε_h	horizontal direct strain
ε_p	volumetric strain
ε_q	distortional strain
ε_v	vertical direct strain
$\varepsilon_{xx}, \varepsilon_{yy}$	horizontal direct strain
ε_{zz}	vertical direct strain
ν^*	Poisson's ratio parameter (Graham & Houlsby, 1983)
ν_{hh}	Poisson's ratio (horizontal to horizontal)
ν_{hv}	Poisson's ratio (horizontal to vertical)
ν_{vh}	Poisson's ratio (vertical to horizontal)
σ'_h	effective horizontal stress
σ'_v	effective vertical stress
$\sigma'_{xx}, \sigma'_{yy}$	effective horizontal stress
σ'_{zz}	effective vertical stress
τ_{xy}	shear stress in horizontal plane
τ_{yz}, τ_{zx}	shear stress in vertical plane
Prefix δ	indicates small increment
Subscript 0	indicates value at very small strain

REFERENCES

Atkinson, J. H. & Sällfors, G. (1991). Experimental determination of stress–strain–time characteristics in laboratory and in-situ tests. General report. Proc. 10th Eur. Conf. Soil Mech. Found. Engng, Florence 3, 915–956.

Atkinson, J. H., Richardson, D. & Stallebrass, S. E. (1990). Effect of recent stress history on the stiffness of overconsolidated soil. Géotechnique 40, No. 4, 531–540.

Burland, J. B. (1989). Ninth Laurits Bjerrum Memorial Lecture: 'Small is beautiful'—the stiffness of soils at small strains. Can. Geotech. J. 26, 499–516.

Butcher, A. P. & Lord, J. A. (1993). Engineering properties of Gault Clay in and around Cambridge. Geotechnical Engineering of Hard Soils–Soft Rocks. (eds Anagnostopolos, A., Schlosser, F., Kalteziotis, N. and Frank, R.), 1, 1, pp. 405–416. Rotterdam: Balkema.

Clayton, C. R. I. & Khatrush, S. A. (1986). A new device for measuring local axial strains on triaxial specimens. Géotechnique 36, No. 4, 593–597.

Coop, M. R., Jovičić, V. & Atkinson, J. H. (1997). Comparisons between soil stiffnesses in laboratory tests using dynamic and continuous loading. Proc. 14th Int. Conf. Soil Mech. Found. Engng, Hamburg 1, 267–270.

Crampin, S. (1981). A review of wave motion in anisotropic and cracked elastic media. Wave Motion 3, 343–391.

Cuccovillo, T. & Coop, M. R. (1997). The measurement of local axial strains in triaxial tests using LVDTs. Géotechnique 47, No. 1, 167–171.

Georgiannou, V. N., Rampello, S. & Silvestri, F. (1991). Static and dynamic measurements of undrained stiffness on natural overconsolidated clays. *Proc. 10th Eur. Conf. Soil Mech. Found. Engng, Florence* **1**, 91–95.

Graham, J. & Houlsby, G. T. (1983). Anisotropic elasticity of a natural clay. *Géotechnique* **33**, No. 2, 165–180.

Hird, C. C. & Pierpoint, N. D. (1994). A non-linear anisotropic elastic model for overconsolidated clay based on strain energy. *Proc. 3rd Eur. Conf. Numer. Methods Geotech. Engng, Manchester*, 67–74.

Hird, C. C. & Pierpoint, N. D. (1997). Stiffness determination and deformation analysis for a trial excavation in Oxford Clay. *Géotechnique* **47**, No. 3, 665–691.

Jamiolkowski, M., Lancellotta, R. & Lo Presti, D. C. F. (1995). Remarks on the stiffness at small strains of six Italian clays. In *Developments in Deep Foundations and Ground Improvement Schemes* (eds A. S. Balasubramaniam *et al.*), pp. 197–216. Rotterdam: Balkema.

Jardine, R. J. (1992). Some observations on the kinematic nature of soil stiffness. *Soils Found.* **32**, No. 2, 111–124.

Jovičić, V. & Coop, M. R. (1998). The measurement of stiffness anisotropy in clays with bender element tests in the triaxial apparatus. *Geotech. Test. J.* **21**, No. 1, 3–10.

Lings, M. L., Nash, D. F. T., Ng, C. W. W. & Boyce, M. D. (1991). Observed behaviour of a deep excavation in Gault clay: a preliminary appraisal. *Proc. 10th Eur. Conf. Soil Mech. Found. Engng, Florence* **2**, 467–470.

Love, A. E. H. (1927). *A treatise on the mathematical theory of elasticity*. 4th edn. Cambridge: Cambridge University Press.

Muir Wood, D. (1990). *Soil behaviour and critical state soil mechanics*. Cambridge: Cambridge University Press.

Nash, D. F. T., Pennington, D. S. & Lings, M. L. (1999). The dependence of anisotropic G_0 shear moduli on void ratio and stress state for reconstituted Gault clay. *2nd Int. Symp. Pre-failure Deformation Characteristics of Geomaterials, Turin*, 229–238.

Pennington, D. S. (1999). *The anisotropic small strain stiffness of Cambridge Gault clay*. PhD thesis, University of Bristol.

Pennington, D. S., Nash, D. F. T. & Lings, M. L. (1997). Anisotropy of G_0 shear stiffness in Gault Clay. *Géotechnique* **47**, No. 3, 391–398.

Pickering, D. J. (1970). Anisotropic elastic parameters for soil. *Géotechnique* **20**, No. 3, 271–276.

Puzrin, A. M. & Burland, J. B. (1998). Non-linear model of small-strain behaviour of soils. *Géotechnique* **48**, No. 2, 217–233.

Raymond, G. P. (1970). Discussion: Stresses and displacements in a cross-anisotropic soil, by L. Barden. *Géotechnique* **20**, No. 4, 456–458.

Roesler, S. K. (1979). Anistropic shear modulus due to stress anisotropy. *J. Geotech. Eng. Div.* A.S.C.E. **105**, GT7, July, 871–880.

Simpson, B. (1992). Thirty-second Rankine Lecture: Retaining structures: displacement and design. *Géotechnique* **42**, No. 4, 541–576.

Simpson, B., Atkinson, J. H. & Jovičić, V. (1996). The influence of anisotropy on calculations of ground settlements above tunnels. *Proceedings of the international symposium on geotechnical aspects of underground construction in soft ground*, pp. 591–594. Rotterdam: Balkema.

Stallebrass, S. E. & Taylor, R. N. (1997). The development and evaluation of a constitutive model for the prediction of ground movements in overconsolidated clay. *Géotechnique* **47**, No. 2, 235–253.

Viggiani, G. & Atkinson, J. H. (1995). Stiffness of fine-grained soil at very small strains. *Geotechnique* **45**, No. 2, 249–265.

Zdravkovic, L., Potts, D. M. & St John, H. D. (2005). *Géotechnique* **55**, No. 7, 497–513

Modelling of a 3D excavation in finite element analysis

L. ZDRAVKOVIC*, D. M. POTTS* and H. D. ST JOHN†

This paper investigates a number of issues related to the modelling of a retaining structure used to support an excavation in 3D finite element analyses. In particular, the effects of wall stiffness in different coordinate directions and the rotational fixity in the corner of the excavation are examined. Both square and rectangular excavations are analysed and compared with the equivalent axisymmetric and plane strain analyses, normally used as approximations for modelling purposes. The chosen geometry, construction sequence and soil conditions are based on a proposed deep excavation at Moorgate in London (next to the Moor House development), which will form part of an underground station for the Crossrail project. The objective of the study is to provide a detailed assessment of wall and ground movements and structural forces in the wall in the light of different modelling assumptions. The study has wider application to a variety of projects that include the development of infrastructure, the construction of deep basement car parks and buried structures, and the effect that these have on the surrounding areas.

KEYWORDS: deep excavation; numerical modelling and analysis; retaining walls

Cet exposé étudie plusieurs questions liées à la modélisation d'une structure de soutènement utilisée pour soutenir une excavation dans les analyses d'éléments finis en 3D. Nous examinons plus particulièrement les effets de la rigidité du mur dans diverses directions coordonnées et la fixité rotationnelle dans l'angle de l'excavation. Des excavations carrées et rectangulaires sont analysées et comparées avec les analyses de déformations axisymétriques et planes équivalentes normalement utilisées comme approximations à des fins de modélisation. La géométrie, la séquence de construction et les conditions de sol choisies sont basées sur une excavation profonde proposée à Moorgate à Londres (près du développement de Moor House), qui fera partie d'une station souterraine du projet Crossrail. L'objectif de cette étude est de donner une évaluation détaillée des mouvements du mur et du sol ainsi que des forces structurales dans le mur à la lumière de différentes hypothèses de modélisation. Cette étude peut s'appliquer plus largement à une variété de projets qui incluent le développement infrastructural et la construction de parkings en sous-sol et de structures enterrées ; elle pourra s'appliquer aussi à l'effet que ces structures ont sur les zones environnantes.

INTRODUCTION

The construction of tunnels and station boxes in urban areas, such as London, requires a detailed assessment of the effects that such construction might have on existing structures. Sometimes, if there is enough information about previous similar undertakings, it is possible to make this assessment on the basis of experience. However, if this is not the case, then it is necessary to use numerical techniques to make the necessary predictions.

Current design practice suggests that, in a general rectangular excavation, plane strain two-dimensional (2D) analysis should be applied to assess the wall and ground movements in the centre of the excavation (along its longer side), whereas an axisymmetric analysis should be applied to assess conditions in the corner and the shorter side of the excavation (see Fig. 1). To date, full three-dimensional (3D) analyses have rarely been carried out because of time and cost constraints.

St John (1975) compared the predictions of ground movements for plane strain, axisymmetric and square excavations modelled assuming a uniform linear-elastic soil and no wall, in an attempt to explain the variation of surface ground movements measured at the Houses of Parliament in London. A number of recent publications describe the 3D modelling of deep strutted excavations in a variety of soil conditions and compare the results with those from 2D

analyses (Ou *et al.*, 1996; Ou & Shiau, 1998; Moormann & Katzenbach, 2002). These analyses have gone further in modelling the soil as an elasto-plastic material, but the retaining walls are still assumed to be isotropic elastic.

In reality, however, a concrete retaining wall, for example, is not an isotropic solid. Whether it is a diaphragm wall, a contiguous wall, a secant pile wall, or even a sheet pile wall, it has continuous vertical elements (e.g. diaphragm panels, piles), but is discontinuous in the horizontal direction, along the sides of the excavation; see Fig. 2. Consequently, it cannot sustain any significant out-of-plane bending, and also the horizontal axial stiffness of the wall is much smaller than the stiffness of the solid concrete, as a consequence of joints between the vertical elements. The assumption of isotropic stiffness (i.e. the same stiffness in all coordinate directions) therefore introduces a significant limitation to any analysis. An axisymmetric analysis with isotropic wall stiffness predicts small wall and ground movements and shows that the support is provided by hoop stresses, rather than bending resistance along the vertical direction as would be the case for a circular shaft sunk using traditional techniques (Cabarkapa *et al.*, 2003). Even for a truly circular shaft constructed with an *in situ* retaining wall this assumption is unrealistic, as the behaviour of the wall will be dominated by the compression of the joints between the elements of the wall (e.g. panels or piles). Consequently, if realistic predictions of wall and ground movements and structural forces are to be achieved when modelling either axisymmetric or full 3D excavations, it is necessary to reduce the out-of-plane wall stiffness (both axial and bending) to an appropriate value.

This paper first investigates the effects of wall stiffness in a square excavation based on the geometry and ground conditions for a proposed deep excavation on the Crossrail

Manuscript received 3 September 2004; revised manuscript accepted 3 June 2005.
Discussion on this paper closes on 1 March 2006, for further details see p. ii.
* Department of Civil and Environmental Engineering, Imperial College, London, UK.
† Geotechnical Consulting Group, London, UK.

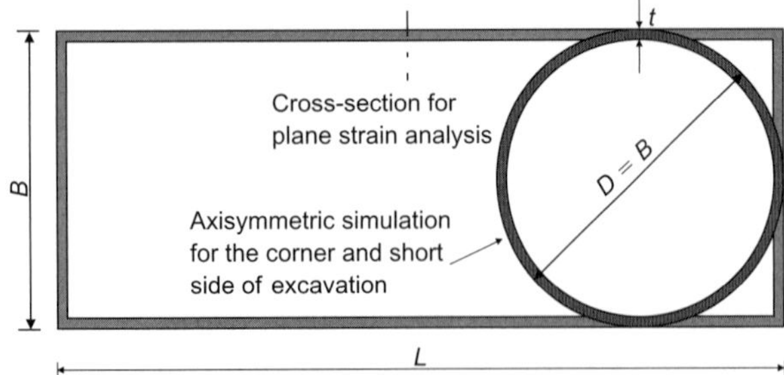

Fig. 1. Schematic approximation for appropriate 2D analyses

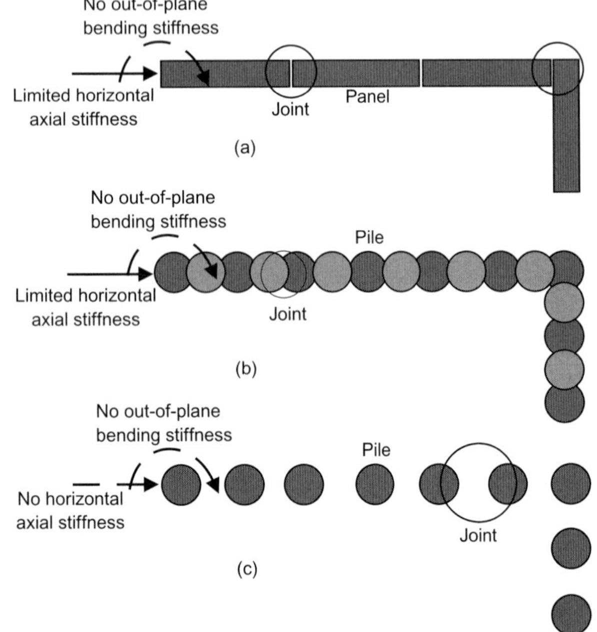

Fig. 2. Schematic view of different wall types: (a) diaphragm wall; (b) secant pile wall; (c) contiguous pile wall

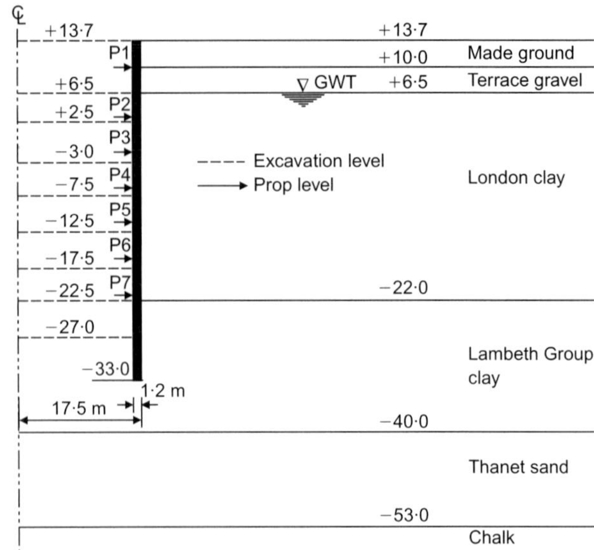

Fig. 3. Ground profile and construction sequence

constructed and excavation carried out to a level of + 2·5m. This sequence of propping and excavation then continues until the final excavation level of −27·0 mOD is reached, making the total excavation depth 40·7 m. Such a large depth of excavation is required because of the necessity for the new tunnels to run below the existing London Underground tunnels, and at this location they must be at 30 m depth. Although this is an exceptionally deep excavation compared with usual excavation depths for developments in urban areas, it will be shown in this paper that the results from the analyses presented here can be used to assess the behaviour at shallower excavation depths.

The pore water pressure and K_0 profiles adopted in the analyses are shown in Fig. 4. The clay layers are modelled as undrained, but the remaining layers are drained.

Soil constitutive models

The non-linear elasto-plastic Mohr–Coulomb model (Potts & Zdravkovic, 1999) is used to model all soil units, apart from the made ground, which is modelled with a linear elastic Mohr–Coulomb model. The non-linearity below yield is simulated with the Jardine *et al.* (1986) small-strain stiffness model. Model parameters for all soil units are summarised in Tables 1 and 2, and the variation of normalised shear $(3G/p')$ and bulk (K/p') stiffness with deviatoric (E_d) and volumetric (ε_v) strain respectively is shown in Fig. 5.

Geometry

Both 2D and 3D finite element analyses are performed in this study, using the Imperial College Finite Element Program (ICFEP; Potts and Zdravkovic, 1999). The results from axisymmetric and plane strain analyses are used as a reference for comparison with those from 3D analyses. The Moorgate excavation geometry is square in plan (35 m × 35 m outer dimensions), and therefore only half of the central cross-section is modelled in 2D analyses. The mesh used for the 2D analyses is shown in Fig. 6; it consists of 800 eight-noded quadrilateral isoparametric elements. The props are modelled as two-noded bar elements that can transmit only axial force, and the wall is modelled with either solid or beam elements (Potts & Zdravkovic, 1999). In the 3D analyses advantage is taken of a fourfold symmetry, and this mesh is shown in Fig. 7. The soil here is

FINITE ELEMENT ANALYSES

Soil conditions and construction sequence

The ground conditions adopted in the analyses reflect a typical soil profile in central London (see Fig. 3), with the groundwater table at the top of the London clay. Also shown in the figure is the construction sequence that is envisaged for the site at Moorgate station. The dashed lines represent ground level at different stages of excavation, and the arrows represent props. The wall behaves as an embedded cantilever up to the excavation level of +6·5 mOD. Prop P1 is then

route (Moorgate station), the latest tunnelling project that aims to connect the east and west ends of London via 19 km long tunnels running beneath central London. This particular excavation is to serve as a launching platform for the tunnel boring machine for one of the Crossrail tunnels and also as a part of a future station with escalators. Some preliminary results from this study are described in Potts (2003) and Torp-Petersen *et al.* (2003). The effects of different moment connections in the corner of such an excavation are also examined. Having in this way established the most appropriate approach for modelling a 3D excavation, the study is then extended to the analyses of rectangular excavations, with length, L, to width, B, ratios of 2:1 and 4:1. All the results are compared with the appropriate axisymmetric and plane strain predictions.

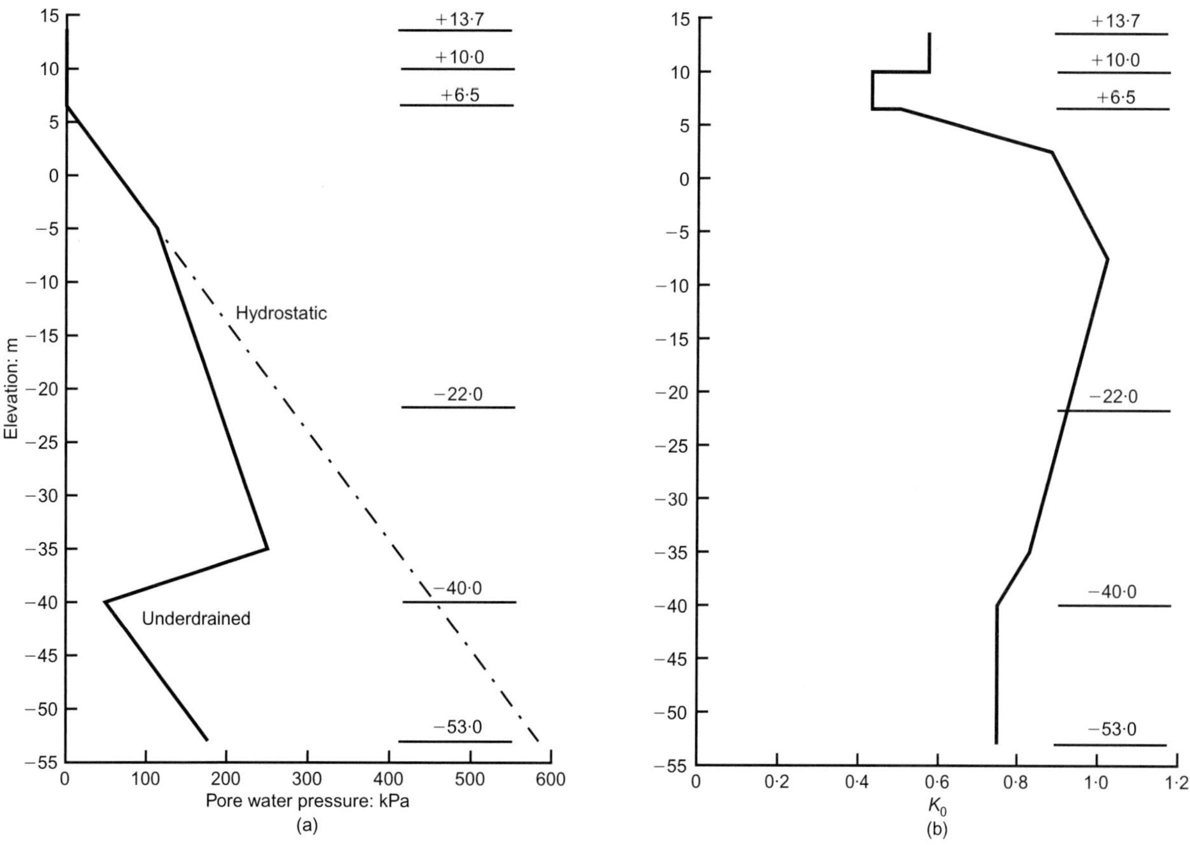

Fig. 4. Soil conditions: (a) pore water pressure: (b) K_0 profiles

Table 1. Soil properties

Layer	Angle of shearing resistance, ϕ': deg	Cohesion, c': kPa	Angle of dilation, v: deg	Young's modulus, E: kPa	Poisson's ratio, μ
Made ground	25	0	12·5	10 000	0·2
Terrace gravel	35	0	17·5	Small strains (see Table 2)	0·2
London clay	22	0	11	Small strains (see Table 2)	0·3
Lambeth clay	22	0	11	Small strains (see Table 2)	0·3
Thanet sand	32	0	16	Small strains (see Table 2)	0·2

discretised with 4500 20-noded hexahedral isoparametric elements, whereas the props are modelled using eight-noded membrane elements that can transmit only in-plane axial forces. The wall is modelled with either 20-noded solid or eight-noded shell elements (Schroeder, 2002). In all the analyses, structural elements are modelled as elastic, and the adopted wall thickness is equivalent to a 1·2 m thick diaphragm wall. The Young's modulus of 3×10^6 kPa for the props was estimated from the equivalent stiffness of tubular steel pipes that would normally be used in such an excavation. The elastic properties of the walls are summarised in Table 3. It should be noted that full interface friction was assumed between the soil and the wall: consequently no interface elements were used in the analyses.

REFERENCE ANALYSES

The plane strain and axisymmetric analyses that serve as a reference for comparison with the 3D results are performed by modelling the retaining wall with either solid or beam elements. A circular shape is inscribed in a square for the axisymmetric analysis, similar to the sketch in Fig. 1. In

the plane strain analysis the wall stiffness (in the vertical z-direction) is specified as $E_z = 28 \times 10^6$ kPa, to simulate properties of concrete. In the axisymmetric analysis the same value is specified for the axial wall stiffness (E_z), but zero stiffness is prescribed in the circumferential direction (E_θ), to account for a discontinuous wall in this direction.

The analyses with the wall modelled with solid elements are performed first, and the horizontal wall movements after the complete construction sequence (i.e. excavation to a depth of 40·7 m) are shown in Fig. 8. As expected, the axisymmetric analysis predicts smaller movements, and for this case the maximum value (at −21·0 mOD) is about 70% of that predicted in the plane strain analysis. The two analyses are also repeated with the wall modelled with beam elements, placed on the excavation side of the solid elements (see Fig. 6). The relative difference between the two wall deflections is similar to the analyses with solid elements; however, in each of the analyses the wall deflection is larger than when the wall is modelled with solid elements. This is a consequence of the lack of the beneficial action of shear stresses mobilised on the back of the wall. In the case of solid elements this shear stress acts downward at a certain

Table 2(a). Small-strain soil properties: coefficients for elastic shear modulus

Layer	A	B	$C \times 10^{-4}$: %	α	γ	$E_{d,min} \times 10^{-4}$: %	$E_{d,max}$: %	G_{min}: kPa
Terrace gravel	1104	1035	5	0·974	0·940	8·83346	0·3464	2000
London clay	1400	1270	1	1·335	0·617	8·66025	0·6928	2667
Lambeth clay	1400	1270	1	1·335	0·617	8·66025	0·6928	2667
Thanet sand	930	1120	2	1·100	0·700	3·63731	0·1645	2000

Table 2(b). Small-strain soil properties: coefficients for elastic bulk modulus

Layer	R	S	$T \times 10^{-3}$: %	δ	η	$\varepsilon_{v,min} \times 10^{-3}$: %	$\varepsilon_{v,max}$: %	K_{min}: kPa
Terrace gravel	275	225	2	0·998	1·044	2·1	0·20	5000
London clay	686	633	1	2·069	0·420	5·0	0·15	5000
Lambeth clay	686	633	1	2·069	0·420	5·0	0·15	5000
Thanet sand	190	110	1	0·975	1·010	1·1	0·20	5000

Coefficients in Tables 2(a) and 2(b) are material constants used in the following equations to give a variation of tangent shear and bulk stiffness with both stress and strain level:

$$\frac{3G}{p'} = A + B \cos\left\{\alpha\left[\log_{10}\left(\frac{E_d}{C\sqrt{3}}\right)\right]^{\gamma}\right\} - \frac{B\alpha\gamma\left[\log_{10}\left(\frac{E_d}{C\sqrt{3}}\right)\right]^{\gamma-1}}{2\cdot303}\sin\left\{\alpha\left[\log_{10}\left(\frac{E_d}{C\sqrt{3}}\right)\right]^{\gamma}\right\}$$

$$\frac{K}{p'} = R + S \cos\left\{\delta\left[\log_{10}\left(\frac{|\varepsilon_v|}{T}\right)\right]^{\eta}\right\} - \frac{S\delta\eta\left[\log_{10}\left(\frac{|\varepsilon_v|}{T}\right)\right]^{\eta-1}}{2\cdot303}\sin\left\{\delta\left[\log_{10}\left(\frac{|\varepsilon_v|}{T}\right)\right]^{\eta}\right\}$$

where E_d and ε_v are the deviatoric and volumetric strains respectively.

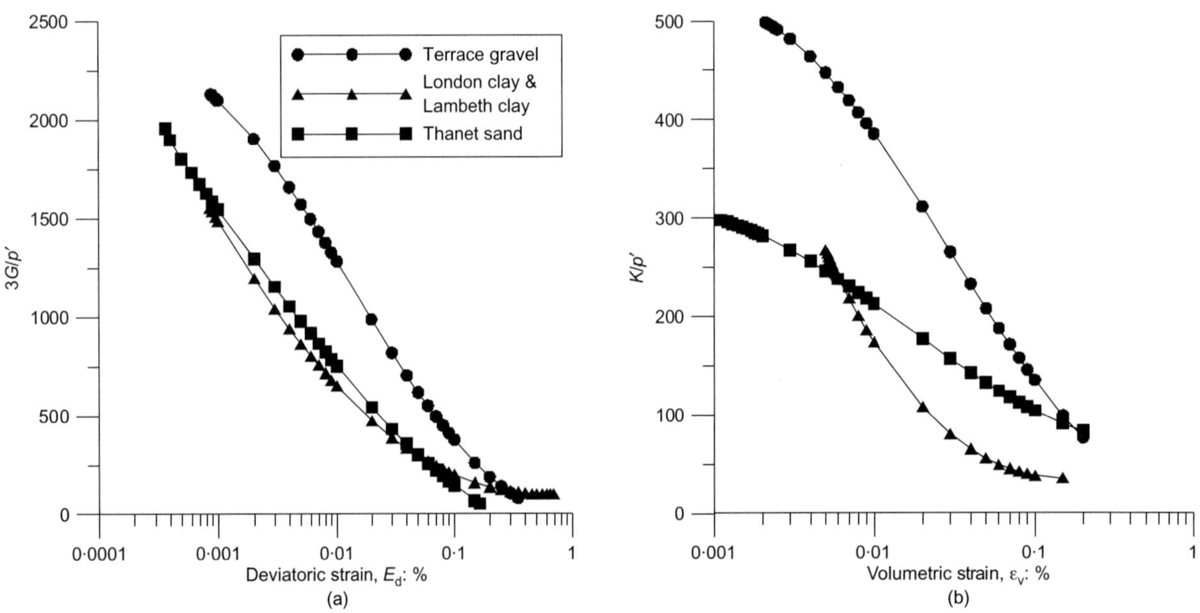

Fig. 5. Non-linear stiffness used in the analyses: (a) shear stiffness; (b) bulk stiffness

distance from the neutral axis of the wall, and therefore produces a clockwise moment about this axis that reduces the anticlockwise moment generated by the horizontal stresses acting on the back of the wall. When the wall is modelled with beam elements, although its properties take account of the wall thickness, the actual beam elements do not have thickness in the finite element mesh, and therefore there is no clockwise moment from the shear stresses on the back of the wall, thus resulting in larger horizontal movements.

A parametric study in which the ratio E_θ/E_z in the axisymmetric analysis was varied between 0 and 1 by an order of magnitude (i.e. 1·0, 0·1, 0·01, etc.) was also performed and showed that if $E_\theta/E_z \leqslant 0\cdot001$ there is no difference in predicted wall deformation between the analyses performed with different E_θ/E_z values. Fig. 8 also shows the wall deflection for the case of isotropic wall stiffness (i.e. $E_\theta/E_z = 1\cdot0$, termed 'stiff wall' on the figure), which demonstrates that such a simulation is clearly unrealistic as it predicts negligible wall movements. A similar relationship between the analyses' predictions is observed for the surface settlement behind the wall. This is an

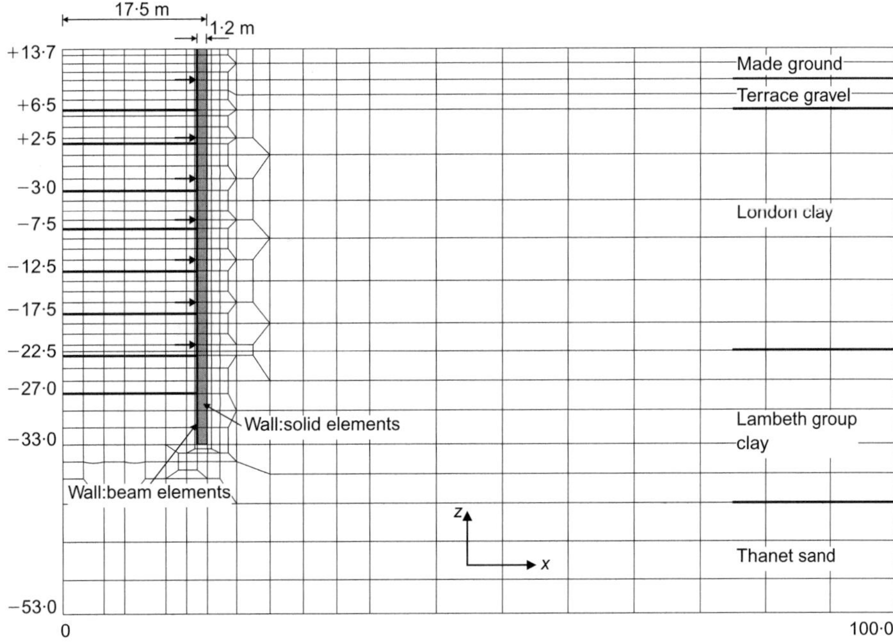

Fig. 6. 2D finite element mesh

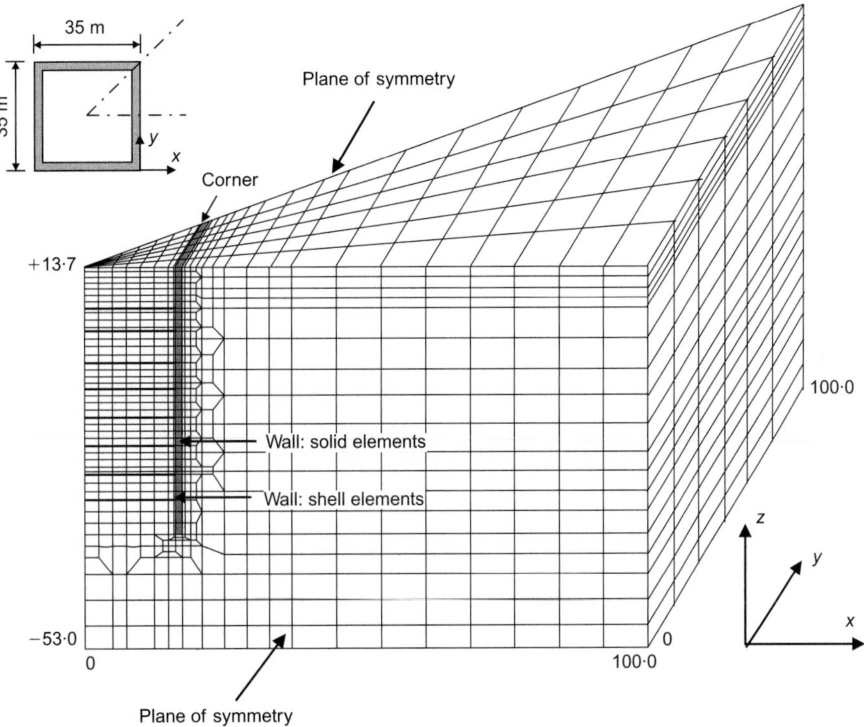

Fig. 7. 3D finite element mesh for square excavation

important aspect of any axisymmetric analysis of a retaining wall, because the inclusion of any significant circumferential stiffness, E_θ, results in the resistance to soil pressure on the back of the wall being provided by the hoop (i.e. circumferential) stresses, rather than by the bending of the wall in the vertical plane, which is unrealistic.

3D ANALYSES OF SQUARE EXCAVATION: SOLID ELEMENT WALL

In the first set of 3D analyses the wall is modelled using 20-noded hexahedral solid elements. Two analyses are per-formed, one with an isotropic wall stiffness (i.e. $E_x = E_y = E_z = 28 \times 10^6$ kPa) and the other with an anisotropic wall stiffness ($E_x = E_z = 28 \times 10^6$ kPa, $E_y/E_z = 10^{-5}$; see Fig. 7 for coordinate directions). As the chosen ratio of E_y/E_z is smaller than the minimum threshold of 10^{-3} established in the axisymmetric analyses, the movements of the wall will be the maximum possible. Because of the very low stiffness in the y-direction, the latter analysis broadly simulates the conditions in a contiguous pile wall. The results are pre-sented in comparison with the equivalent plane strain and axisymmetric analyses from Fig. 8 (i.e. solid wall simula-tions).

Table 3(a). Elastic wall properties: solid elements

Wall type	E_x: kPa	E_y: kPa	E_z: kPa	Poisson's ratio: μ	Wall thickness: m
Plane strain	28×10^6	N/A	28×10^6	0·2	1·2
Axisymmetric	28×10^6	28×10^1	28×10^6	0·2	1·2
3D isotropic	28×10^6	28×10^6	28×10^6	0·2	1·2
3D anisotropic	28×10^6	28×10^1	28×10^6	0·2	1·2

Table 3(b). Elastic wall properties: shell elements

Wall type	E: kPa	μ	t: m	Vertical axial stiffness: $\%(EA)$	Vertical bending stiffness: $\%(EI)$	Horizontal axial stiffness: $\%(EA)$	Horizontal bending stiffness: $\%(EI)$
Plane strain	28×10^6	0·2	1·2	100	100	N/A	N/A
Axisymmetric	28×10^6	0·2	1·2	100	100	0·01	0·01
3D isotropic	28×10^6	0·2	1·2	100	100	100	100
3D anisotropic	28×10^6	0·2	1·2	100	100	20	1

A is the cross-sectional area of the wall per metre length of wall; I is the second moment of inertia of the wall per metre length of wall.

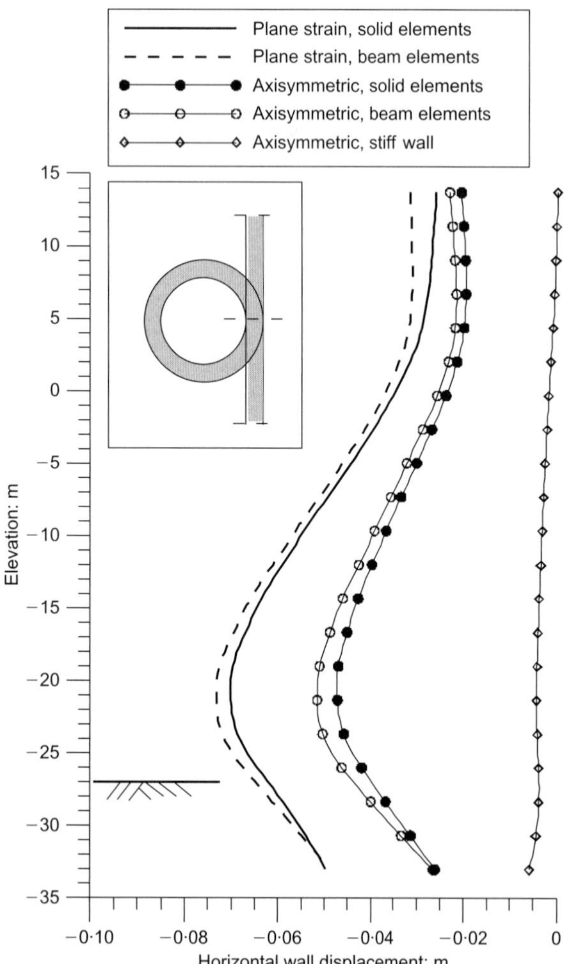

Fig. 8. Comparison of wall deflections in plane strain and axisymmetric analyses for different wall models

Wall deflections in the centre and corner of the excavation at the end of the complete construction sequence are shown in Fig. 9. At the centre, modelling the wall as an isotropic solid predicts about 20% smaller maximum horizontal wall movement, whereas the movement of the top of the wall is nearly three times smaller, when compared with the anisotropic wall analysis. At the corner, the wall movement from the isotropic analysis is seen to be negligible, but significant movement is seen in the anisotropic analysis. As explained earlier, this is a consequence of essentially modelling the wall as a continuous stiff membrane in the ground.

Surface settlement troughs behind the wall, in the centre and corner, are shown in Fig. 10. They follow a relationship similar to that of the wall deflections, with the isotropic wall having the smallest settlement in both cross-sections. It is interesting to note that, although the maximum horizontal wall movement in the centre of the excavation of an anisotropic wall (Fig. 9(a)) is only about 13% smaller than that of the plane strain analysis (thus suggesting that plane strain may not be an unreasonable simplification), the maximum surface settlement in the same central cross-section is significantly overpredicted by the plane strain analysis, being 1·6 times larger than that of the anisotropic wall analysis. This clearly demonstrates the effects of a 3D geometry on ground movements.

The vertical wall bending moments M_1, corresponding to rotation about the y-axis of the wall, at the centre of the excavation (Fig. 11(a)) are broadly similar for all analyses, because of the similarity of the curvatures of the deformed wall. In the corner of the excavation (Fig. 11(b)) the anisotropic wall gives bending moments M_1 that are generally half of those predicted at the centre. This indicates, for uniform walls, that the corners of the excavation are safe, as the reinforcement necessary for the centre is sufficient to cover the bending moments in the corner. However, the isotropic wall, which essentially simulates a full moment connection, shows the opposite trend to the other analyses. This implies that, for a diaphragm wall for example, the corner panels would have to be reinforced differently from the central panels, which is not normally done in practice.

A further drawback in modelling the wall with an isotropic stiffness is shown in Fig. 12. Fig. 12(a) shows the distribution of the out-of-plane horizontal bending moment M_2, corresponding to rotation about the z-axis of the wall, at a level of $-24·0$ mOD (which is the level of the maximum vertical bending moment, M_1, in Fig. 11). The anisotropic wall cannot transmit any moment in this direction, but the magnitude of this moment in the isotropic wall is similar to the magnitude of the moment M_1, and it also changes sign towards the corner of the excavation. In a similar way the horizontal axial force in an isotropic wall is more than five times larger than that in an anisotropic wall (Fig. 12(b)). As a result of the jointed nature of any wall type (as sketched in Fig. 2), such high structural forces in this direction are considered unlikely.

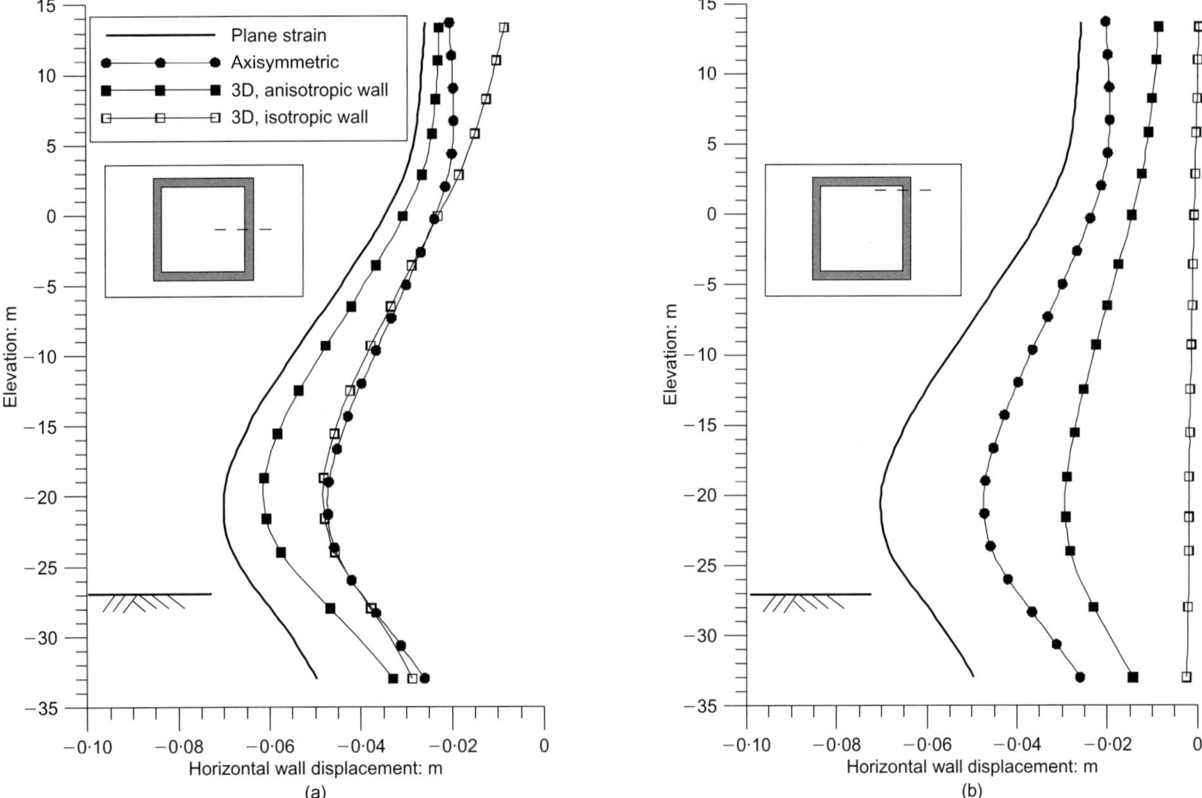

Fig. 9. Horizontal wall movements in square excavation (solid element wall): (a) in centre; (b) in corner

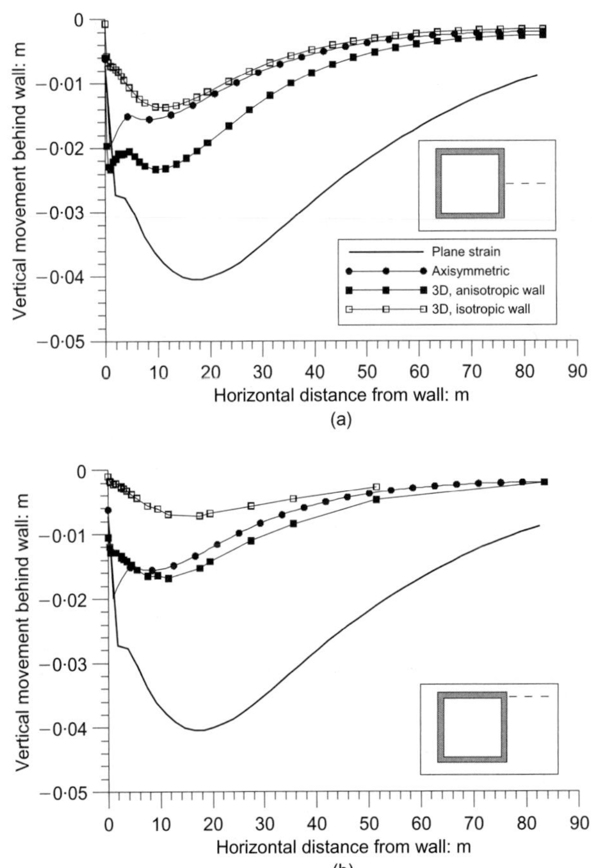

Fig. 10. Surface settlements behind the wall in square excavation (solid element wall): (a) in centre; (b) in corner

3D ANALYSES OF SQUARE EXCAVATION: SHELL ELEMENT WALL

General

In the following study the same square excavation is analysed, but this time with the wall modelled using shell elements (Schroeder, 2002). One advantage of using shell elements is their formulation in terms of structural forces, rather than stresses, so that the magnitudes of these come as a direct result from the analyses. In the case of solid elements in the previous section, structural forces have to be calculated from the stresses at element integration points, which makes the whole process slightly cumbersome.

In addition to this, apart from displacement degrees of freedom, shell elements also have rotational degrees of freedom, which gives greater choice for modelling the moment conditions in the corner of the excavation. In this study the shell elements are modelled as elastic, but with the freedom of having different axial and bending stiffness in the vertical and horizontal directions.

Five analyses are performed that, because of the properties assigned to the shell elements, are considered to simulate the conditions in a diaphragm wall. In analysis 1 (a1), the shell wall is modelled as isotropic, with $E_z = E_y = 28 \times 10^6$ kPa, and the rotational degrees of freedom in the corner are fixed (i.e. full moment connection). This scenario is similar to that of the isotropic solid element wall in the previous section. Analysis 2 (a2) also models the wall as isotropic, but releases the rotational degrees of freedom in the corner (i.e. moment-free connection). The purpose of this analysis is to investigate whether just this change in modelling is sufficient to provide more realistic results. Analysis 3 (a3) introduces an anisotropic shell wall (i.e. smaller axial and bending stiffness in the horizontal y-direction), with fixed rotational degrees of freedom in the

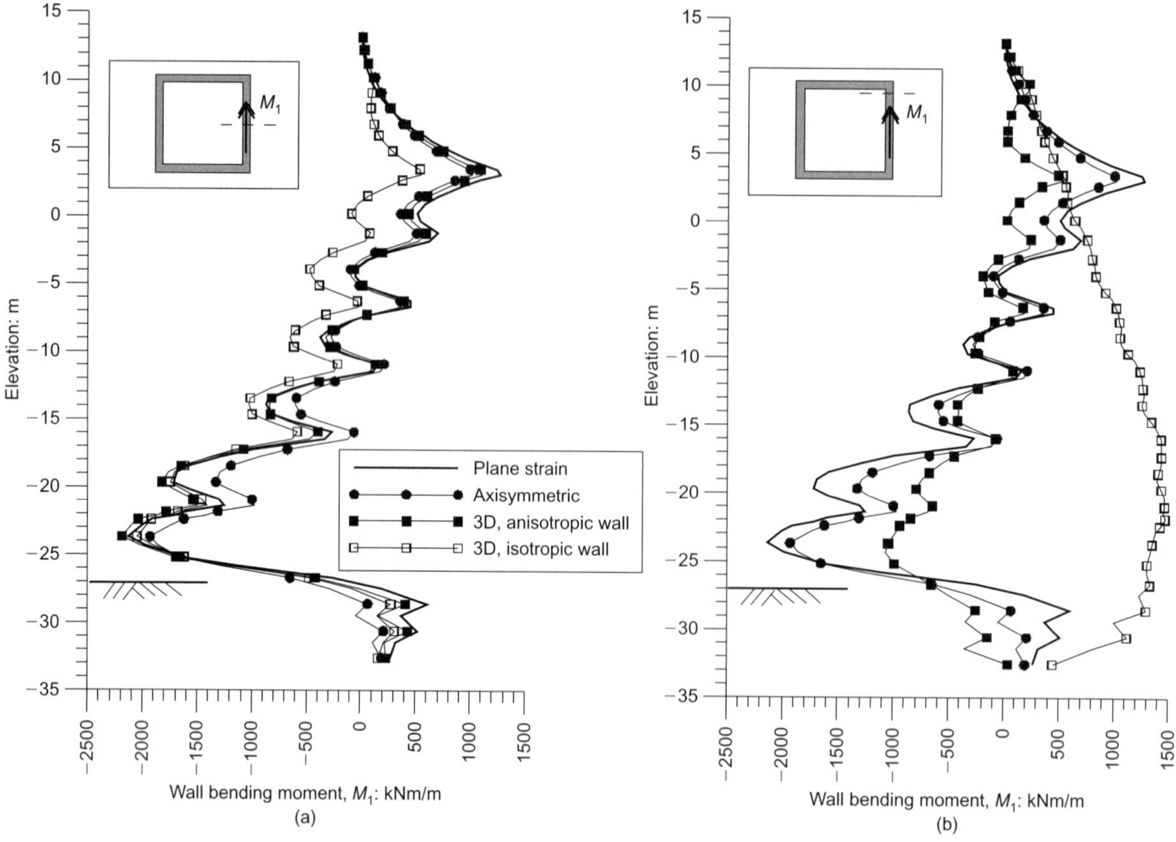

Fig. 11. Wall bending moments in square excavation (solid element wall): (a) in centre; (b) in corner

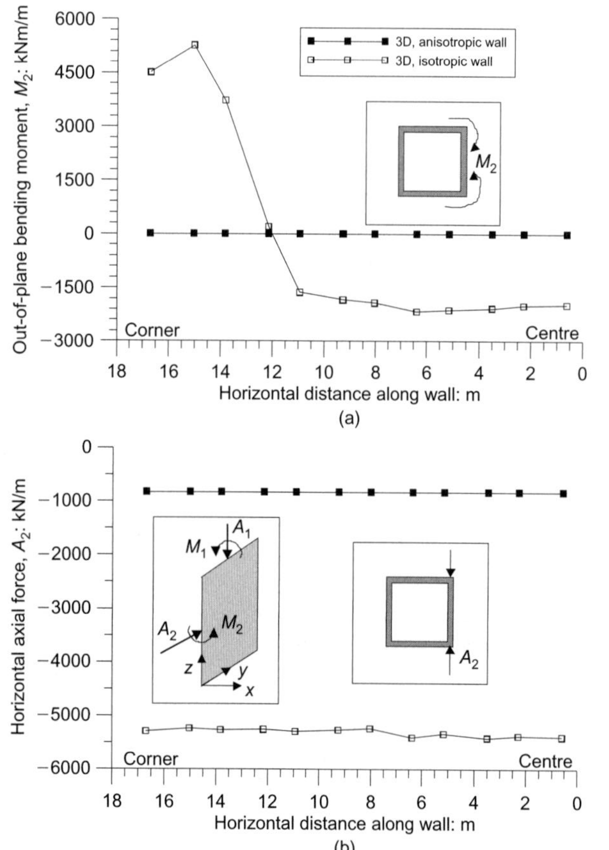

Fig. 12. Horizontal axis of wall in square excavation (solid element wall): (a) bending moment; (b) axial force

corner. The purpose of this is to investigate whether the introduction of anisotropy, but still with full moment connection in the corner, provides more realistic results than analysis (a1). Finally, analysis 4 (a4) introduces anisotropy in both the axial and bending stiffness of the wall (the same as a3), and releases the rotational degrees of freedom in the corner. This is thought to represent the most realistic model of a diaphragm wall in a 3D excavation. Additional analysis 5 (a5) investigates the effect of a capping beam that is normally constructed on the top of the wall to connect all the structural elements. This is achieved by modelling the top $1 \cdot 7$ m of shell elements as isotropic and with full moment connection in the corner, whereas the rest of the wall is anisotropic and with a moment-free connection in the corner, as in a4.

In the analyses where the shell wall is anisotropic, this is achieved by assigning the shell elements a full axial and bending stiffness in the vertical z-direction ($E_z = 28 \times 10^6$ kPa), whereas in the horizontal y-direction the axial stiffness is 20% of the vertical value, and the bending stiffness is only a nominal 1% of the vertical value. The horizontal axial stiffness is estimated on the basis that the joints between the panels of a typical diaphragm wall may close by an assumed 1 mm, taking the axial shortening of each panel at maximum excavation depth from an isotropic analysis, and reducing the stiffness so that the total shortening (panel shortening plus gap closure) would be produced under the same load conditions. Although this clearly underestimates the movements where the axial loads are lower, short of modelling each panel individually, it is a reasonable approximation.

Figures 13–16 compare the predictions of wall and ground movements, as well as the structural forces in the wall, for the five analyses described above.

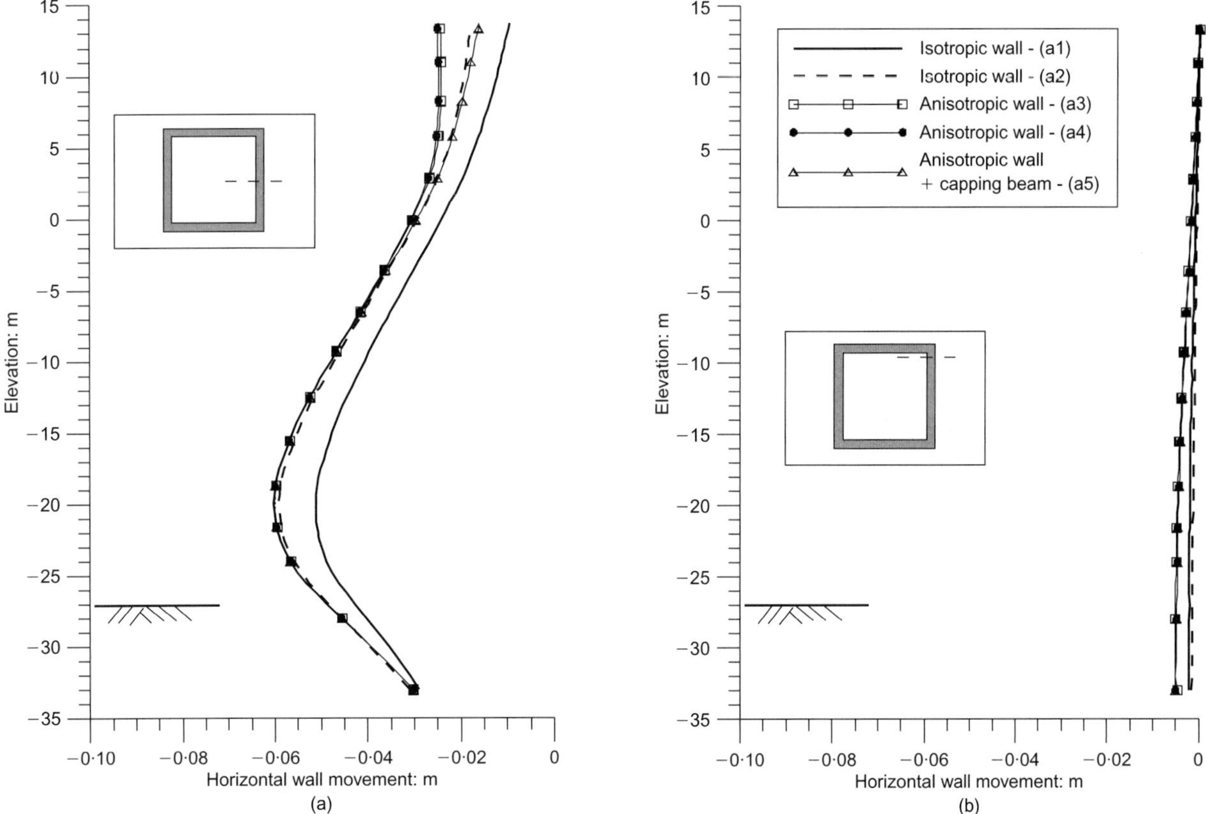

Fig. 13. Effect of modelling assumptions in square excavation on horizontal wall movements (shell element wall): (a) in centre; (b) in corner

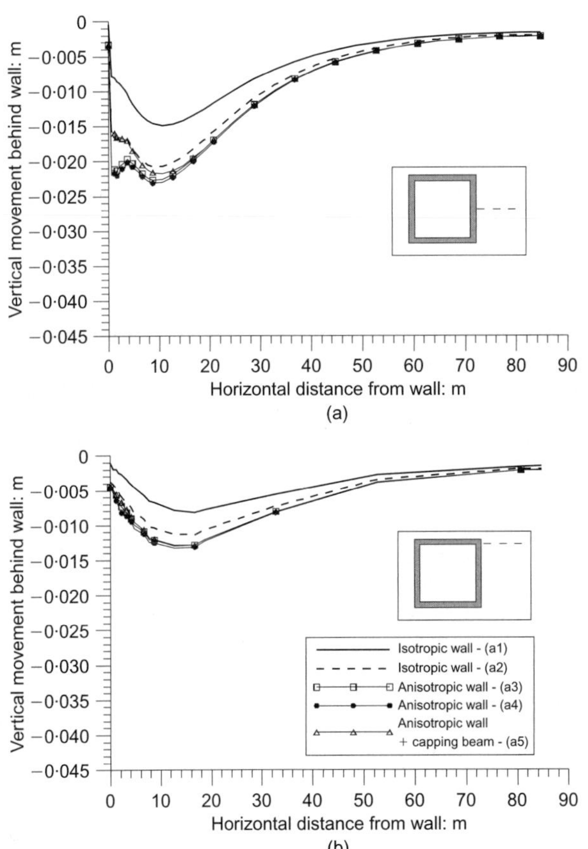

Fig. 14. Effect of modelling assumptions in square excavation on surface settlements behind wall (shell element wall): (a) in centre; (b) in corner

Movements

Figure 13 shows the horizontal wall movements in the centre and corner of the excavation. In the centre, the isotropic shell wall with the full moment connection (a1) results in the smallest deflection, as expected from the similar analyses with the solid element wall in Fig. 9(a). Comparison with this figure also shows that the shell wall predicts slightly larger horizontal movements than the solid element wall, which was explained earlier as a consequence of the zero thickness of the shell wall in the finite element mesh. A similar result will be seen when comparing surface settlement behind the wall for these two analyses (Figs 10 and 14).

The remaining four analyses predict almost identical maximum horizontal wall displacement. It appears that the release of the full moment connection in the corner of the isotropic wall (a2) is sufficient to give a reasonable prediction of wall deflection in the centre of the excavation, and in particular the maximum value. The addition of the capping beam (a5) only restricts the movement of the top part of the wall; it doesn't affect the rest of it. All analyses, apart from a1, also predict almost identical maximum horizontal wall displacement to that of the anisotropic solid element wall in Fig. 9(a). Although the conditions in the corner for this analysis are similar to those of no moment connection, this wall also has negligible horizontal axial stiffness (compared with the shell element wall for which this stiffness is 20% of the vertical axial stiffness). This difference in the magnitude of the horizontal axial stiffness in shell and solid element wall analyses does not appear to affect the maximum wall deformation in the centre. However, wall movements in the corner of the excavation (Fig. 13(b)) are all negligibly small compared with that of the anisotropic solid element wall in Fig. 9(b). This appears to be the conse-

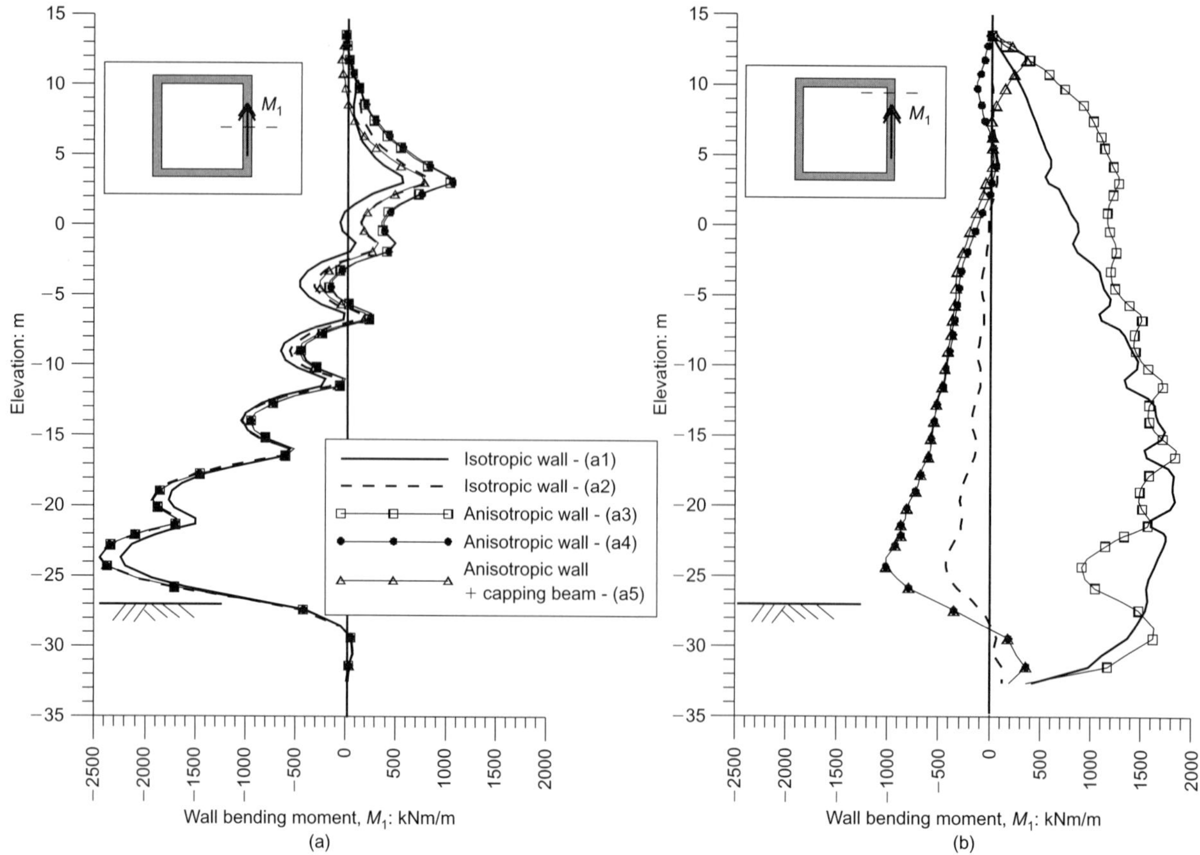

Fig. 15. Effect of modelling assumptions in square excavation on wall bending moments (shell element wall): (a) in centre; (b) in corner

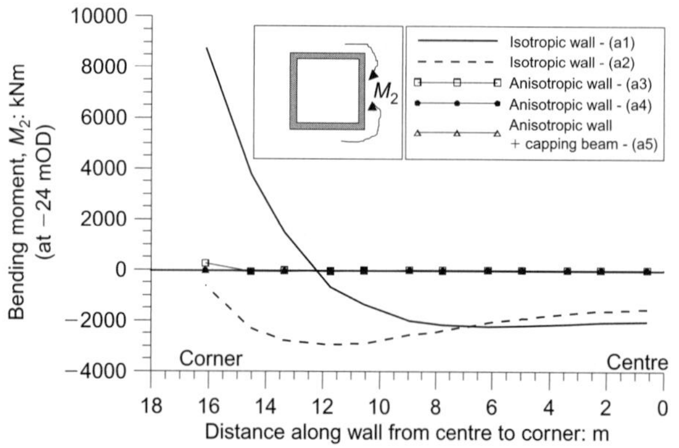

Fig. 16. Effect of modelling assumptions in square excavation on out-of-plane bending moment; shell element wall

quence of both isotropic and anisotropic shell walls having a larger axial force, due to larger stiffness, in the horizontal y-direction.

Similar conclusions can be drawn for the surface settlements behind the wall in Fig. 14, in the centre and corner of the excavation. Again, it is of interest to note that the maximum surface settlement behind the isotropic wall (a2) is on average only about 12% smaller than that of the anisotropic wall (a4), which is considered as the most appropriate model of the wall. For practical purposes this would normally be considered acceptable. The inclusion of the capping beam (a5) reduces significantly the settlement immediately behind the wall in the central section, because it is modelled as an isotropic element, with the same vertical and horizontal stiffness and full moment connection in the

corner. However, the maximum value is only about 6% smaller than in the (a4) wall. Consequently, not taking the capping beam into account gives a slightly conservative prediction of wall and ground movements, which justifies its omission in other analyses. The presence of the capping beam does not appear to influence the surface settlement in the corner of the excavation.

Structural forces

Figure 15 presents the bending moments M_1 in the centre and corner of the excavation. Whereas all five analyses give similar predictions in the centre of the wall (Fig. 15(a)), the picture is quite different in the corner (Fig. 15(b)). Analyses (a4) and (a5), apart from the top part of the wall, give almost identical bending moment diagrams, whose magnitude is almost half of that in the centre. This result is also similar to that of the anisotropic solid element wall in Fig. 11(b). Analysis (a2) gives smaller bending moments, but of the same sign as the previous two analyses. This again demonstrates that, although the wall is isotropic, the moment-free connection in the corner is sufficient to give a more realistic prediction of the bending moment M_1. In analysis (a3), although the wall has appropriate anisotropic properties (the same as in the (a4) analysis), the full moment connection in the corner causes a change of sign of the bending moment M_1, similar to that in (a1). These are comparable to the isotropic solid element wall analysis in Fig. 11(b).

Figure 16 shows the distribution of the out-of-plane moment M_2 along the horizontal y-axis, at -24 mOD, which is the same level as in the solid element wall analyses. All three of the anisotropic shell walls (a3, a4 and a5) show that a negligible M_2 moment is transmitted in this direction. Both of the isotropic shell walls (a1 and a2) transmit a significant

M_2 moment (with a magnitude similar to that of the M_1 bending moment), which is unrealistic for any wall that is discontinuous in the y-direction. The only difference between the two is that, whereas in wall (a1) the bending moment M_2 switches sign in the corner owing to the full moment connection (similar to the isotropic solid element wall in Fig. 12(a)), the corner moment in the wall (a2) goes to zero because of the moment-free connection.

3D ANALYSES OF RECTANGULAR EXCAVATIONS
General

In the remainder of this study further analyses are performed to investigate the behaviour of rectangular excavations. Two geometries are considered, one with a width, B, to length, L, ratio of 1:2, and the other with $B{:}L = 1{:}4$. The width B is kept the same as in the square excavation, whereas the length L is changed accordingly. Because of symmetry only a quarter of the geometry is analysed; see Fig. 17. The depth of the excavation and the construction sequence, as well as the soil profile and material properties, are the same as in the previous analyses. The wall is represented with shell elements, with the most appropriate wall model that resulted from the square analyses. This was considered to be analysis (a4), and consequently in the rectangular analyses the same anisotropic properties are assigned to the wall. However, whereas in (a4) it was possible to release the rotational degrees of freedom of the shell elements in the corner (because it was on the boundary of the mesh; see Fig. 7), this is not possible in the analysis of rectangular excavations, and consequently this corner has a full moment connection. This is not considered to be a serious drawback in the rectangular analyses, as the results from (a3) for the square excavation, which has the same wall model, showed that the effect of the full moment connection is only in predicting a high, and of opposite sign, bending moment M_1 in the corner of the excavation (see Fig. 15(b)). Consequently, this bending moment will not be presented for these analyses.

In the following, the results from the two rectangular analyses are compared with the square and appropriate plane strain and axisymmetric analyses (i.e. in which the wall was modelled using beam elements; see Fig. 8), in order to assess at which B/L ratio the plane strain conditions are met along the longer side of the wall. Also, the results are compared at three different stages of excavation, in order to assess whether these conditions are met at earlier stages of excavation. The stages considered are the following (see Fig. 3):

(a) stage 1: excavation to +6·5 mOD (i.e. top of London clay), at which the wall acts as an embedded cantilever
(b) stage 2: excavation to −7·5 mOD (i.e. depth of excavation 21·2 m, which is a more usual depth for developments in London), at which stage the wall is propped at three levels (props P1 to P3)
(c) stage 3: full excavation to −27 mOD (i.e. 40·7 m excavation depth), with all seven propping levels.

Movements

The horizontal wall movements in the centre of the long and short wall sides, together with those from the square, axisymmetric and plane strain analyses, are shown in Fig. 18. The movements in all three stages are bounded by the axisymmetric prediction on the lower side, and the plane strain prediction on the upper side. In this, the maximum movements of the short side of the wall and that of the square excavation are grouped towards the axisymmetric value, whereas those of the long side are grouped towards the plane strain value. The maximum deflection of the long side of the wall, for the first two stages of excavation, is smaller than the deflection of the wall in plane strain conditions by 8% and 3% for $L/B = 2$ and 4 respectively. This difference increases with depth of excavation, but at stage 3 it is still only 12% and 5% for $L/B = 2$ and 4 respectively. In all stages the maximum movement of the long side of the wall is for $L/B = 4$, followed by $L/B = 2$ and then $L/B = 1$ (i.e. square excavation). However, the excavation depth appears to have a greater effect on the maximum movement of the short side of the wall, which does not show a clear pattern of deformation dependence on plan geometry.

Comparing the maximum horizontal movements from each analysis at the three stages it can be seen that, in the first 20 m of excavation (stages 1 and 2), although the position of the maximum deflection moves down with excavation and propping, its magnitude increases only marginally (by less than 10%). However, with further excavation the magnitude of the maximum deflection increases dramatically, such that after another 20 m of excavation (stage 3) it is 70% higher than in stage 2.

The horizontal wall movements in the corner of the excavation are very small, and similar to those presented in Fig. 13(b): consequently they are not shown here.

The settlement troughs in the central sections behind the short and long sides of the walls in the rectangular excavations, together with the plane strain, axisymmetric and square predictions, are shown in Fig. 19 for all three stages

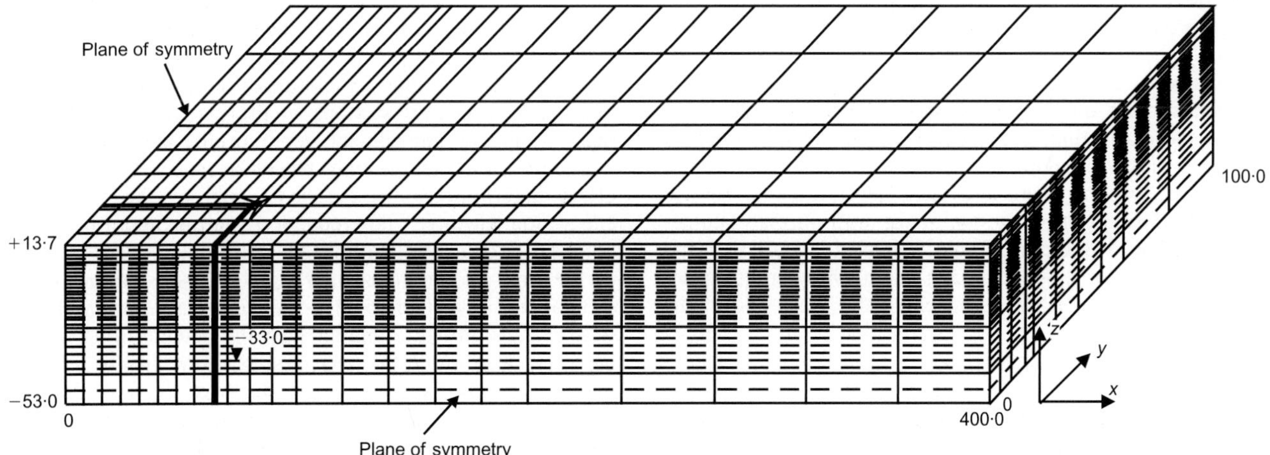

Fig. 17. 3D finite element mesh for rectangular excavation

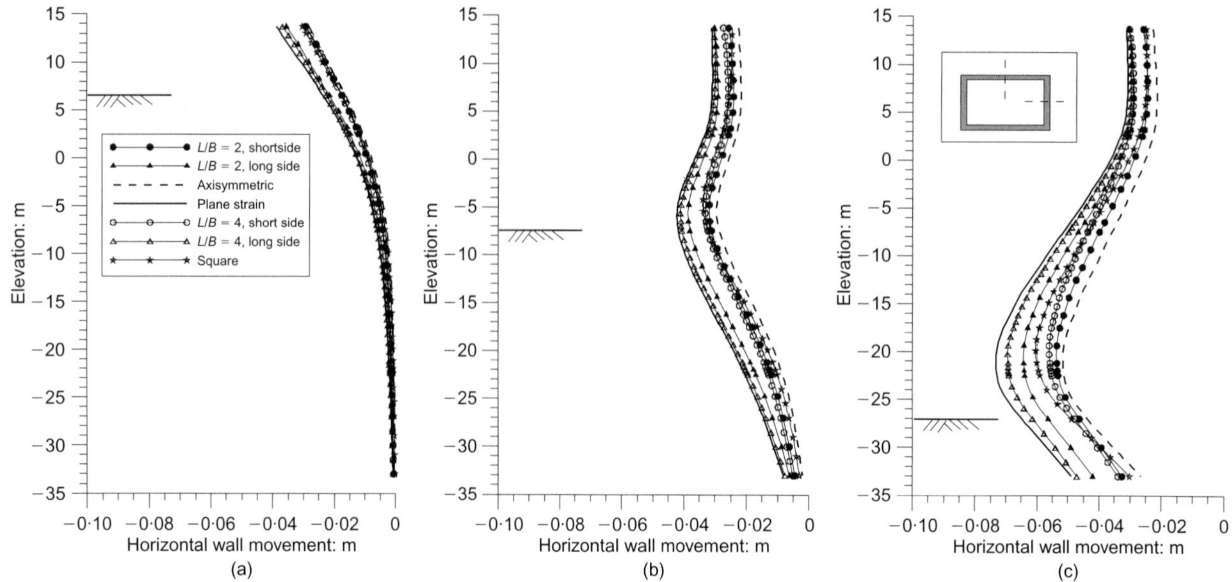

Fig. 18. Comparison of wall movements at different stages of excavation (shell element wall): (a) stage 1; (b) stage 2; (c) stage 3

of excavation. Similar to the wall deflections, the axisymmetric prediction provides a lower bound, and the plane strain prediction an upper bound to the results. Even for $L/B = 4$, the maximum surface settlement behind the long side of the wall at the end of excavation (stage 3) is about 10% smaller than that of the plane strain prediction, whereas for $L/B = 2$ it is some 30% smaller. The $L/B = 4$ prediction on the long side appears to be closer to the plane strain prediction at shallower depths of excavation.

Contrary to the wall deflections in Fig. 18, the changes in the maximum surface settlement with depth of excavation are more pronounced. For each analysis the maximum settlement at stage 2 is about 35% larger than that in stage 1, whereas in stage 3 it is about 70% larger than in stage 2.

Surface settlements in the corner of the excavation are shown in Fig. 20, for all three stages. Note that for clarity the plane strain prediction for stage 3 is not presented, as its magnitude is higher than the adopted scale. These settlements also increase with depth of excavation, but for shallow depths (stage 2) the maximum settlements are close to the axisymmetric prediction. However, the shapes of settlement troughs, especially in the first 10 m from the wall, are different from the axisymmetric prediction, as the corner of the excavation does not appear to be affected by the propping system in the same way as the centre of the excavation (Fig. 19), or plane strain and axisymmetric geometries.

Figure 21 shows the contours of ground surface settlements at the end of excavation (i.e. stage 3) for the 3D analyses with $L/B = 1$, 2 and 4. The L/B ratio has a significant effect on the displacements adjacent to the long side of the excavation, but it has a much smaller effect on the short side. In addition, as noted above for the wall movements, whereas there is a clear dependence of surface settlements behind the long side of the wall on plan geometry (i.e. $L/B = 4$ is the maximum, followed by $L/B = 2$ and then $L/B = 1$) for any excavation depth, this pattern is not so clear for surface settlements behind the short side of the wall.

Bending moments
Bending moments M_1 in the centre of both short and long sides of the wall are shown in Fig. 22, for all three stages of excavation. Again, because of the similar curvatures of the

deflected walls in all the analyses, the bending moments are very similar at all three stages of excavation, with the plane strain prediction being an upper bound for all results.

The out-of-plane bending moment M_2 along the horizontal axis of either the short or long side of the wall is always negligible (similar to that shown in Fig. 16 for anisotropic walls) and, for brevity, it is not shown here.

THE EFFECT OF WALL DEPTH
At the beginning of this study it was recognised that the excavation at Moorgate station was exceptionally deep, with a significant number of props. This poses the question as to whether the results presented so far in this paper can be used to assess the behaviour of a wall and the surrounding soil at smaller excavation depths (and therefore shallower walls), which are more common in building construction in urban areas. For this purpose, an additional analysis is performed with the $L/B = 2$ rectangular geometry, in which the maximum excavation depth is 21·2 m (−7·5 mOD), and the depth of the wall is only another 7 m below the maximum excavation depth (−14·5 mOD). The excavation stages up to this level are the same as in the previous analyses, and the wall is propped by props P1 to P3 (see Fig. 3). The embedded depth of this wall is similar to that of the wall in the deep excavation.

The results from this analysis are compared with those of stage 2 for the analysis with the same $L/B = 2$ ratio but with the deep wall, as this stage has the same excavation depth of 21·2 m. Fig. 23 compares horizontal wall movements in the central sections of the excavation, and shows that the longer embedment depth of the wall reduces the horizontal movement mainly below the excavation level, the effect being larger on the short side of the rectangular excavation. However, this reduction is a maximum of 10%.

Surface settlements in both the central and corner sections of the wall are compared in Fig. 24. These show negligible differences (less than 5%) between the shallow and deep wall excavations. In a similar way, the bending moments M_1 are very close in the two analyses, as shown in Fig. 25.

Consequently, the embedment depth of the wall does not appear to have a significant effect on the behaviour of the wall and the surrounding soil, and the results from the analysis of a deep wall can be used to assess the behaviour of shallow excavations retained by shallower walls.

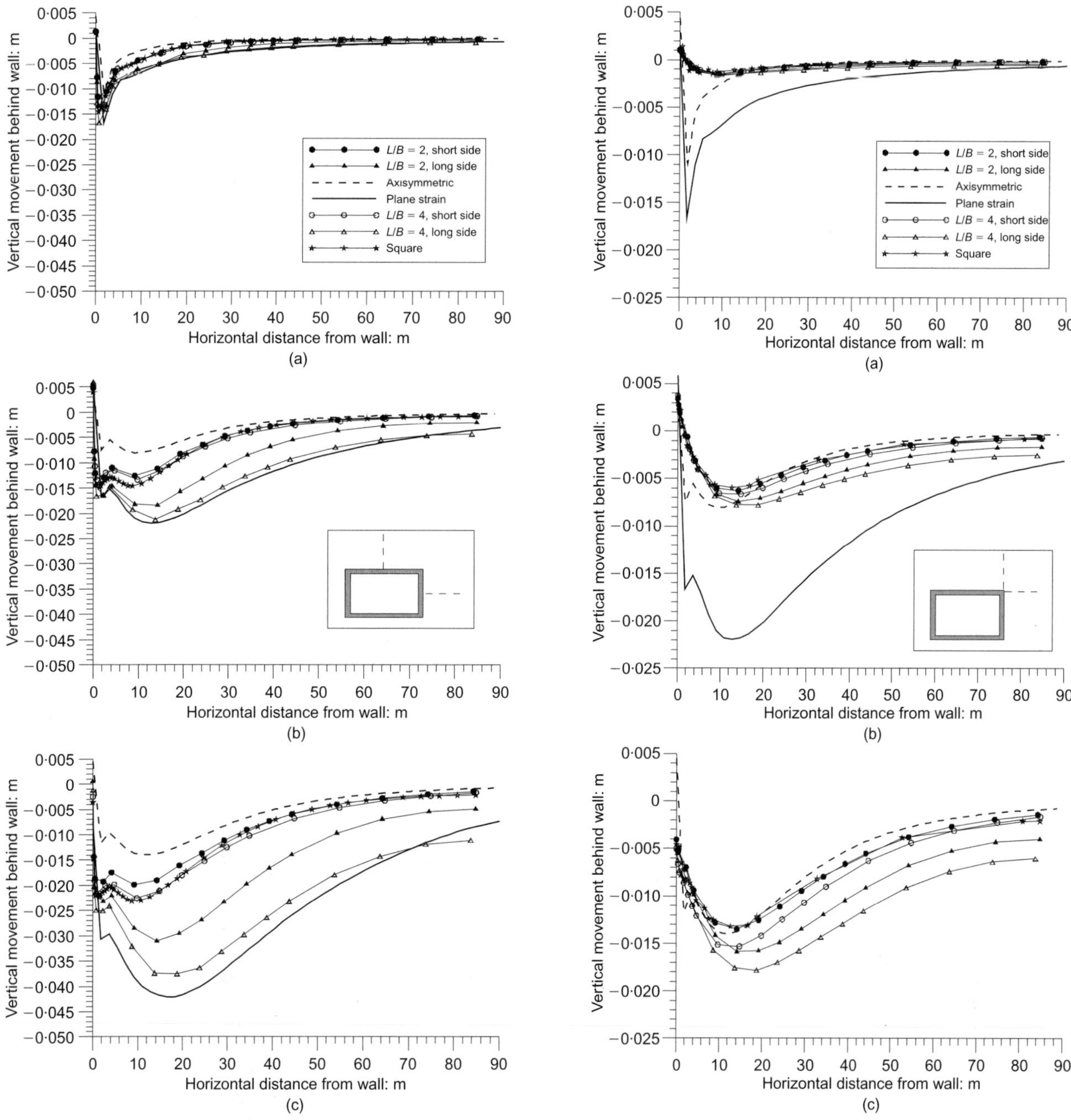

Fig. 19. Comparison of surface settlements in the centre at different stages of excavation (shell element wall): (a) stage 1; (b) stage 2; (c) stage 3

Fig. 20. Comparison of surface settlements in corner at different stages of excavation (shell element wall): (a) stage 1; (b) stage 2; (c) stage 3

CONCLUSIONS

The objective of this paper is to investigate possible ways of modelling a retaining wall in square and rectangular excavations, and provide guidance for the most appropriate approach to be used in any 3D finite element analysis. The paper also shows how the 3D predictions compare with those obtained from equivalent plane strain and axisymmetric analyses, and gives guidance for practical use of these results.

The following main conclusions can be drawn from the study.

Modelling of the wall
(a) The retaining wall can be modelled in finite element analysis using either solid or beam/shell elements. The latter type of element predicts larger wall and ground movements because they do not have a thickness in the finite element mesh and therefore cannot develop the beneficial reducing moment generated by the downward-acting shear stresses on the back of the wall (Fig. 8).

(b) Any retaining wall is unlikely to be a continuous membrane along its perimeter, as it is made from a number of vertical elements (diaphragms or piles) that are not fully connected in this direction. Therefore, to obtain realistic results in axisymmetric and 3D analyses, the axial and bending stiffness of the wall along its perimeter must be reduced.

(c) In practice, the realistic conditions in the corner of the excavation are such that the full moment is not transmitted. In the analysis, this can be achieved either by modelling a wall with anisotropic solid elements, or with anisotropic shell elements that have released rotational degrees of freedom in the corner of the excavation.

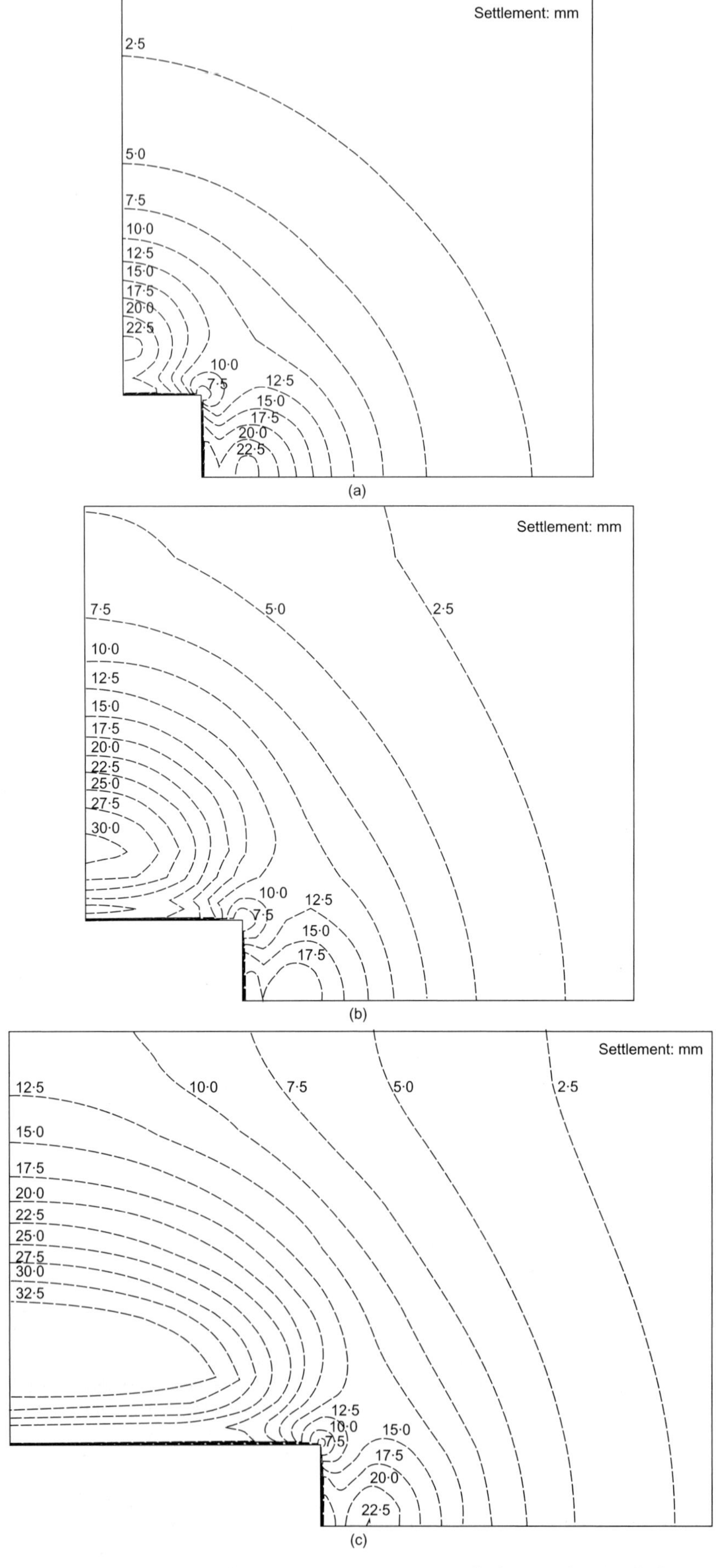

Fig. 21. Comparison of surface settlement contours at end of excavation (shell element wall): (a) $L/B = 1$; (b) $L/B = 2$; (c) $L/B = 4$

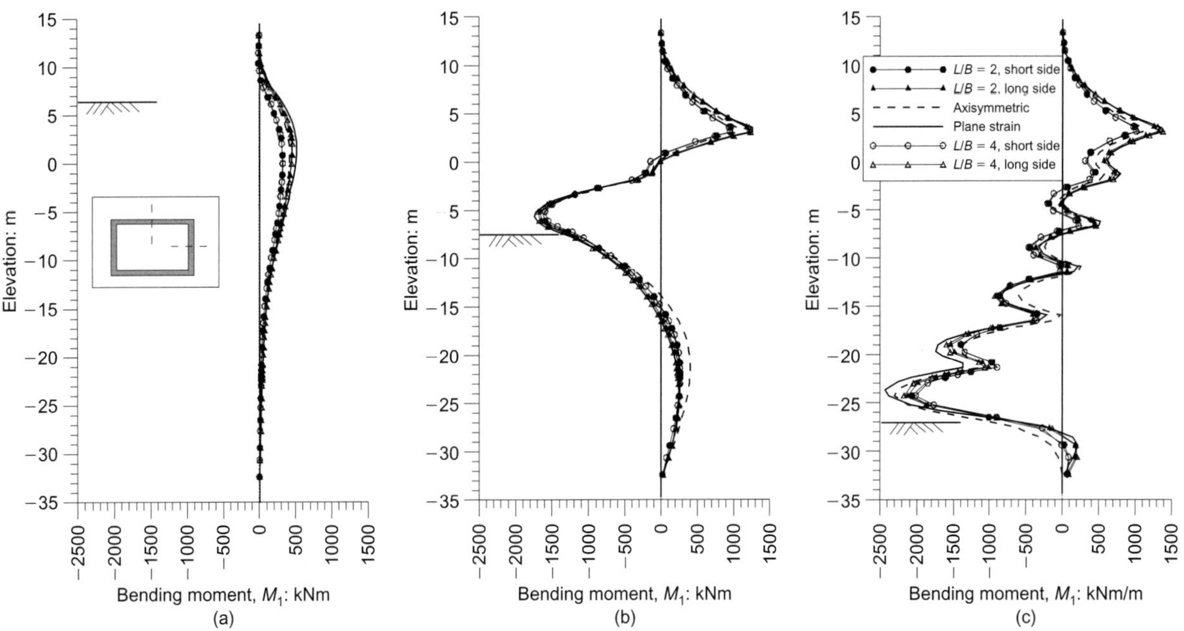

Fig. 22. Comparison of wall bending moments in centre at different stages of excavation (shell element wall): (a) stage 1; (b) stage 2; (c) stage 3

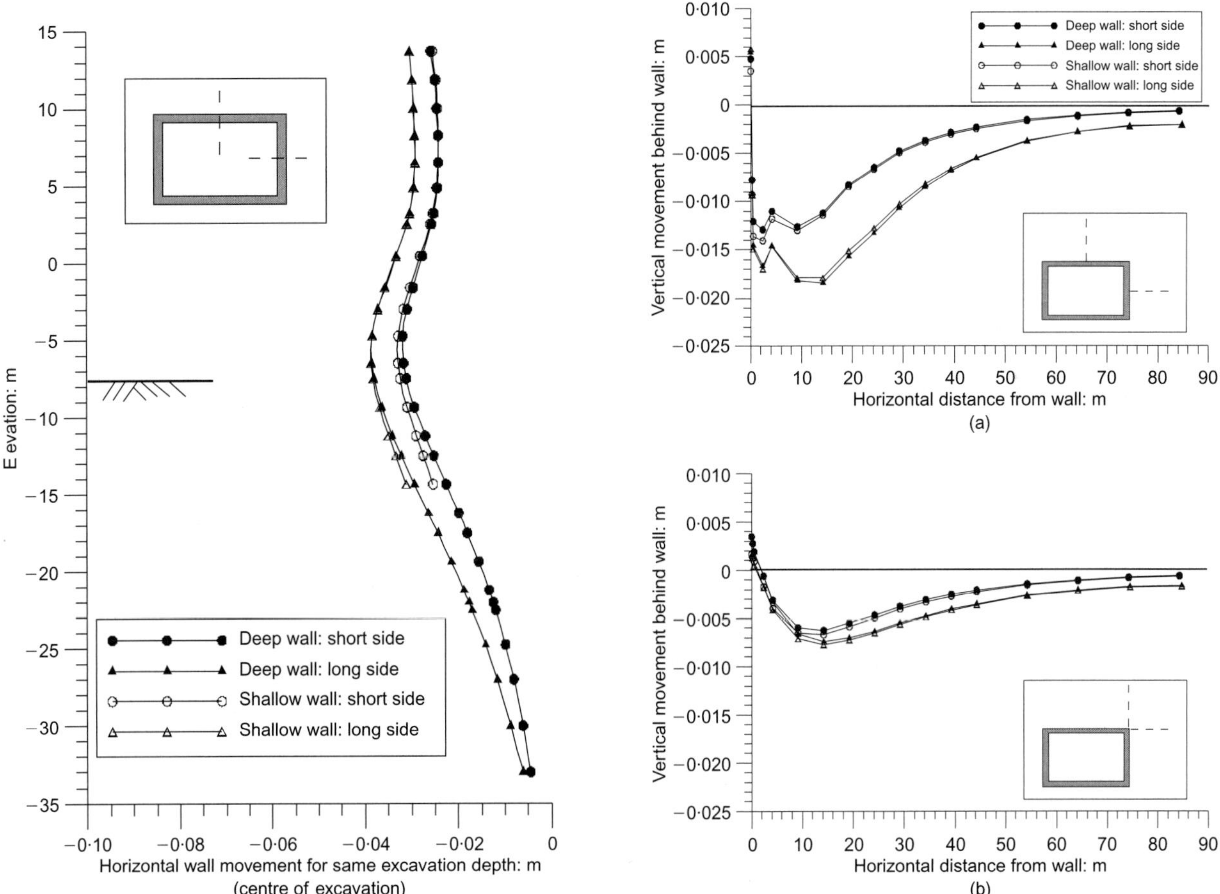

Fig. 23. Effect of wall embedment depth on horizontal wall movement (shell element wall)

Fig. 24. Effect of wall embedment depth on surface settlement (shell element wall): (a) in centre; (b) in corner

(*d*) If there is no ability in the software to account for anisotropic wall properties (i.e. it has to be modelled as isotropic), but if the corner of the excavation can be modelled as a moment-free connection, then predictions of wall deflections and surface settlements (Figs 13 and 14), as well as bending moment M_1, that are reasonable, although on the lower side, can be obtained.

However, the out-of-plane bending moment M_2 will be unrealistically high (Fig. 16).

(*e*) On the other hand, if the wall can be modelled as anisotropic, but the condition in the corner has to be that of a full moment connection, then the only unrealistic prediction will be that of the bending moment M_1 in the corner of the excavation (Fig. 15(b)).

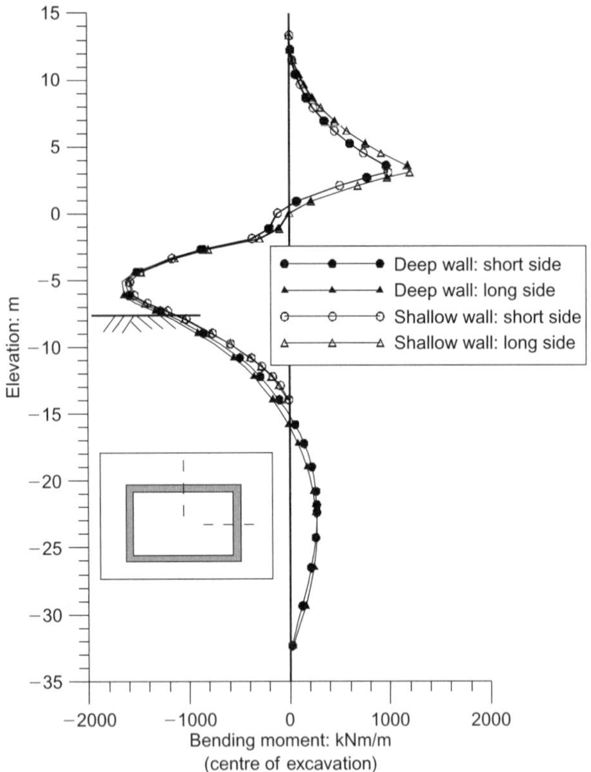

Fig. 25. Effect of wall embedment depth on wall bending moment (shell element wall)

(*f*) Taking account of the capping beam at the top of the wall has negligible (reducing) effects on movements and structural forces, and it is therefore reasonable to ignore it (Figs 13 to 16).

Movements and structural forces

(*g*) Wall movements and surface settlements behind the wall in the centre of a square excavation are closer to the results of axisymmetric than plane strain analysis (Figs 18 and 19).

(*h*) In rectangular excavations, even for a length-to-width ratio L/B of 4, the conditions in the centre of the longer side of the excavation are not fully plane strain in terms of wall movements and surface settlements behind the wall, although they are at most 10% smaller than plane strain predictions, even for the full depth of excavation. For shallower depths of excavation the difference is negligible (Figs 18 and 19). For L/B of 2 the influence of depth of excavation is more significant.

(*i*) The maximum surface settlements behind the wall in the corner of square/rectangular excavations are on average about 30–50% smaller than the maximum values in the central sections of the excavation. At shallower depths (around 20 m) they appear to be well represented by the predictions of an equivalent axisymmetric analysis.

(*j*) The effect of the embedment depth of a wall on movements and structural forces in the excavations analysed is negligible. Therefore the results from the intermediate stages of an analysis of a deep excavation can be used to assess the wall and ground movements and structural forces at shallower depths, without having to repeat the analysis with a shallower wall.

Although the presented study is based on a particular soil profile and a particular excavation geometry and construction sequence, the conclusions are general in a sense that they result from analyses in which only the boundary conditions on the wall and the wall properties are varied, while the soil properties, the remaining boundary conditions in the mesh and the construction sequence are always the same. In this respect the ground surface settlements predicted in the analyses can be compared with the observations of well-monitored excavations and wall systems in stiff clay, summarised in Gaba *et al.* (2003) and reproduced in Fig. 26. Also shown in this figure are the normalised maximum settlements and settlements at a distance equal to two excavation depths away from the wall (the maximum excava-

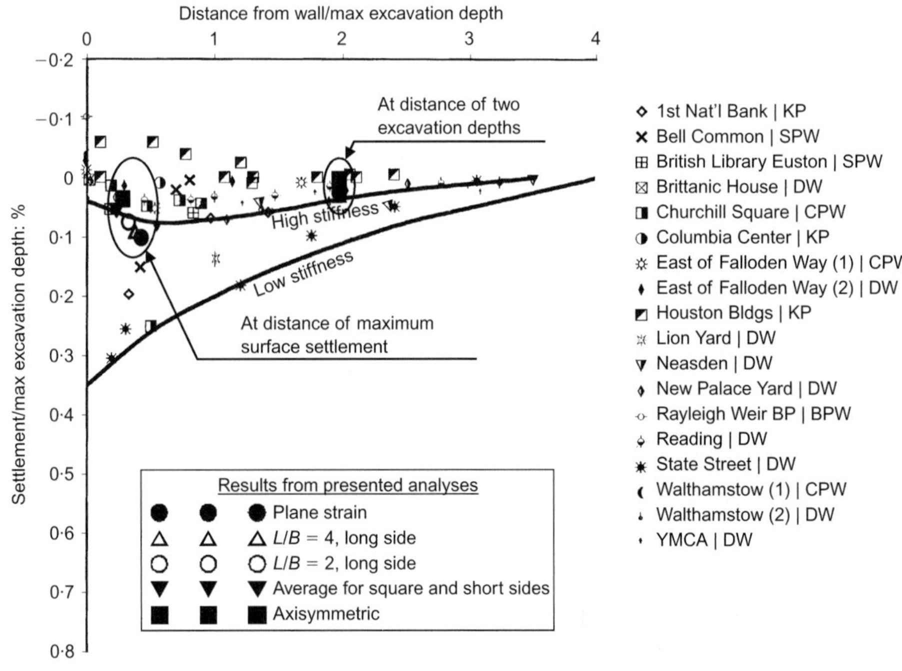

Fig. 26. Ground surface settlements due to excavation in front of wall in stiff clay (from Gaba *et al.*, 2003) (BP: bored piles; BPW: bored pile wall; CPW: contiguous bored pile wall; DW: diaphragm wall; KP: king post wall; SPW: secant bored pile wall)

tion depth being 40·7 m, i.e. settlement troughs in Fig. 19(c)), for the analyses with little or no out-of-plane flexural stiffness. The results are clearly in agreement with the empirical data for walls with high in-plane stiffness.

The analyses also indicate that the current design practice for rectangular excavations (of using plane strain analysis to assess the long side and axisymmetric analysis to assess corners and the short side of the excavation) is broadly appropriate.

NOTATION

A_1	axial force in vertical direction (along z axis)
A_2	axial force in horizontal direction (along y axis)
B	excavation width
c'	cohesion
D	excavation diameter
E_d	deviatoric strain
E_x, E_y, E_z	Young's modulus in x, y and z directions respectively
E_θ	Young's modulus in circumferential direction
G	shear modulus
K	bulk modulus
K_0	coefficient of earth pressure at rest
L	excavation length
M_1	bending moment in vertical plane (rotation about y axis)
M_2	out of plane bending moment (rotation about z axis)
p'	mean effective stress
t	wall thickness
ε_v	volumetric strain
μ	Poisson's ratio
ν	angle of dilation
φ'	angle of shearing resistance

REFERENCES

Cabarkapa, Z., Milligan, G. W. B., Menkiti, C. O., Murphy, J. & Potts, D. M. (2003). Design and performance of a large diameter shaft in Dublin Boulder Clay. Foundations: innovations, observations, design and practice: *Proc. BGA Int. Conf.* (ed. T. A. Newson), pp. 176–185. London: Thomas Telford

Gaba, A. R., Simpson, B., Powrie, W. & Beadman, D. R. (2003). *Embedded retaining walls: guidance for economic design*, CIRIA C580 Report. London: CIRIA.

Jardine, R. J., Potts, D. M., Fourie, A. B. & Burland, J. B. (1986). Studies of the influence of nonlinear stress–strain characteristics in soil–structure interaction. *Géotechnique* **36**, No. 3, 377–396.

Moormann, C. & Katzenbach, R. (2002). Three-dimensional effects of deep excavations with rectangular shape. *Proc. 2nd Int. Conf. Soil–Structure Interaction, Zurich* **1**, 135–142.

Ou, C. Y. & Shiau, B. Y. (1998). Analyses of the corner effect on excavation behaviour. *Can. Geotech. J.* **35**, 532–540.

Ou, C. Y., Chiou, D. C. & Wu, T. S. (1996). Three-dimensional finite element analysis of deep excavations. *ASCE J. Geotech. Engng* **122**, No. 5, 337–345.

Potts, D. M. (2003). Numerical analysis: a virtual dream or practical reality? 42nd Rankine Lecture. *Géotechnique* **53**, No. 6, 535–573.

Potts, D. M. & Zdravkovic, L. (1999). *Finite element analysis in geotechnical engineering: Theory*. London: Thomas Telford.

Schroeder, F. C. (2002). *The influence of bored piles on existing tunnels*. PhD thesis, Imperial College, University of London.

St John, H. D. (1975). *Field and theoretical studies of the behaviour of ground around deep excavations in London Clay*. PhD thesis, Cambridge University.

Torp-Petersen, G. E., Zdravkovic, L., Potts, D. M. & St John, H. D. (2003). The prediction of ground movements associated with the construction of deep station boxes. *(Re)Claiming the underground space: Proc. ITA World Tunnelling Cong.* (ed. J. Saveur), pp 1051–1058. Lisse: Balkema.

Smethurst, J. A., Clarke, D. & Powrie, W. (2006). *Géotechnique* **56**, No. 8, 523–537

Seasonal changes in pore water pressure in a grass-covered cut slope in London Clay

J. A. SMETHURST*, D. CLARKE* and W. POWRIE*

In temperate European climates, the season of peak water demand by vegetation (summer) is out of phase with the season of greatest rainfall (winter). This results in seasonal fluctuations in soil water content and, in clay soils, associated problems of shrinking and swelling that can in turn contribute to strain-softening and progressive slope failure. This paper presents field measurements of seasonal moisture content and pore water pressure changes within the surface drying zone of a cut slope in the London Clay at Newbury, Berkshire, UK. A climate station was installed at the site to measure the parameters needed to determine specific plant evapotranspiration. This information was used to carry out a water balance calculation to estimate the year-round soil moisture deficit caused by the vegetation. The calculated soil moisture deficit matches reasonably closely the field measurements of soil drying. The field measurements of seasonal changes in pore water pressure and suction are linked quantitatively to the measured changes in water content using the soil water characteristic curve for the London Clay. The suctions generated by the light vegetation cover at Newbury were found not to persist into the winter and early spring.

Dans les zones européennes de climat tempéré, la période de demande d'approvisionnement en eau la plus importante pour la végétation ne correspond pas à la saison de pluviosité la plus forte (hiver). Il en résulte des fluctuations du taux d'humidité dans le sol, et, pour les terrains argileux, des problèmes associés de gonflement et rétrécissement qui peuvent, à leur tour, contribuer à un écrouissage négatif et une fracture progressive du versant. Cet article présente des expérimentations réalisées in situ pour mesurer le taux d'humidité et les variations de pression hydrique interstitielle saisonniers à l'intérieur de la zone d'assèchement de la surface d'un versant découpé dans le London Clay à Newbury, dans le Berkshire, au Royaume-Uni. Une station climatologique a été installée sur le site afin de relever les paramètres nécessaires pour déterminer l'évapotranspiration végétale spécifique. Ces informations ont été utilisées pour calculer le bilan hydrique et estimer ainsi, sur une année, le déficit hydrique que la végétation provoque au niveau du sol. Les résultats de ces calculs en matière de déficit hydrique montrent une corrélation raisonnablement étroite avec les mesures de l'assèchement du sol in situ. Pour la succion et la pression interstitielle de l'eau, les mesures des variations saisonnières, relevées in situ, sont qualitativement liées aux variations de teneur en eau obtenues à l'aide de la courbe caractéristique d'humidité du sol pour le London Clay. Nous avons pu observer que les succions produites par la légère couverture végétale à Newbury ne persistent pas en hiver et au début du printemps.

KEYWORDS: clays; field instrumentation; monitoring; pore pressures; slopes; suction

---

BACKGROUND

It is well known that plants can influence the engineering performance of slopes. Beneficial effects include root reinforcement, generation of pore water suctions, buttressing and arching. Detrimental effects include loading of the upper part of the slope and loading associated with uprooting or overturning (Barker, 1986; Coppin & Richards, 1990). In temperate European climates, the season of peak water demand by vegetation (summer) is out of phase with the season of greatest rainfall (winter). This results in seasonal fluctuations in soil moisture content and associated problems of shrinking and swelling in clay soils that can in turn lead to major serviceability problems. For example, large volume losses due to soil shrinkage in vegetated railway embankments can cause significant distortion of railway tracks (Andrei, 2000).

In clays, the cyclic changes in water content will result in corresponding cyclic changes in pore water pressure and hence effective stress. In a conventional static slope stability

analysis in which worst-case pore pressure conditions are considered, suctions generated by vegetation will have a beneficial effect only if they are carried through into winter and early spring. Mature trees growing in low-permeability clays have been demonstrated to generate sufficiently large suctions in the summer to prevent full re-wetting of the soil in winter and spring (Biddle, 1983, 1998; Driscoll, 1983); however, the suctions generated by light shrub and/or grass cover are rarely sustained through the winter period (Croney, 1977; Greenwood *et al.*, 2001, 2004). Furthermore, there is recent evidence from both centrifuge model tests and numerical analyses of clay slopes that cyclic stresses thought to be representative of those induced by vegetation can cause strain-softening to occur, starting from the toe of the slope (Nyambayo *et al.*, 2004; Take & Bolton, 2004; O'Brien *et al.*, 2004; Vaughan *et al.*, 2004). Over a period of several years, these cycles of pore pressure and effective stress can result in progressive failure.

Knowledge of the effects of climate and vegetation on pore water pressures is important for the validation of numerical models of slope stability that attempt to take into account the complex interactions between climate, vegetation and the ground. Such models are essential if the cyclic changes in effective stress that lead potentially to progressive failure and the influence of climate change on long-term slope stability are to be understood. There have been

Manuscript received 19 May 2005; revised manuscript accepted 18 July 2006.
Discussion on this paper closes on 2 April 2007, for further details see p.ii.
* School of Civil Engineering and the Environment, University of Southampton, UK.

several studies of pore water pressures in vegetated clay infrastructure slopes (Walbancke, 1976; Anderson & Kneale, 1980; Crabb *et al.*, 1987; Ridley *et al.*, 2004a, 2004b), including recent field investigations to assess the impact of different types of vegetation (Greenwood *et al.*, 2001; MacNeil *et al.*, 2001). However, few of these have tried to link closely the observed changes in pore water pressure to a full soil moisture deficit calculation based on vegetation properties and climatic parameters, as advocated by Blight (1997).

This paper describes and analyses the results of a field study carried out to quantify the hydrological environment and pore water pressures in a cut slope in London Clay. Data were collected over a complete winter/summer cycle, including a wet winter (2002) followed by an exceptionally dry summer (2003). A full soil water balance is developed, and linked to the observed pore water pressures.

SITE DESCRIPTION AND SOIL CHARACTERISTICS

A cutting on the A34 Newbury bypass in southern England (Fig. 1; OS grid reference SU455652) was chosen for the study owing to its accessibility and relatively uniform soil conditions and vegetation characteristics (mainly grass). The instrumented slope is east facing, 8 m high and 28 m long (Fig. 2). The cutting was constructed in 1997, and is entirely within the London Clay. The London Clay at the site is about 20 m thick, highly weathered to a depth of about 2·5 m below original ground level, and underlain by Lambeth Group deposits and the Upper Chalk. After the cutting was excavated, up to 0·4 m of topsoil was placed over the cut London Clay surface to facilitate the planting of vegetation on the slope. A gravel fin drain approximately 600 mm deep, 4 m from the toe of the slope, was installed to drain the road sub-base. The fin drain connects at intervals into a sealed carrier drain that outfalls to the south of the cutting, and to which the road gullies are also connected. A cross-section through the slope is shown in Fig. 3.

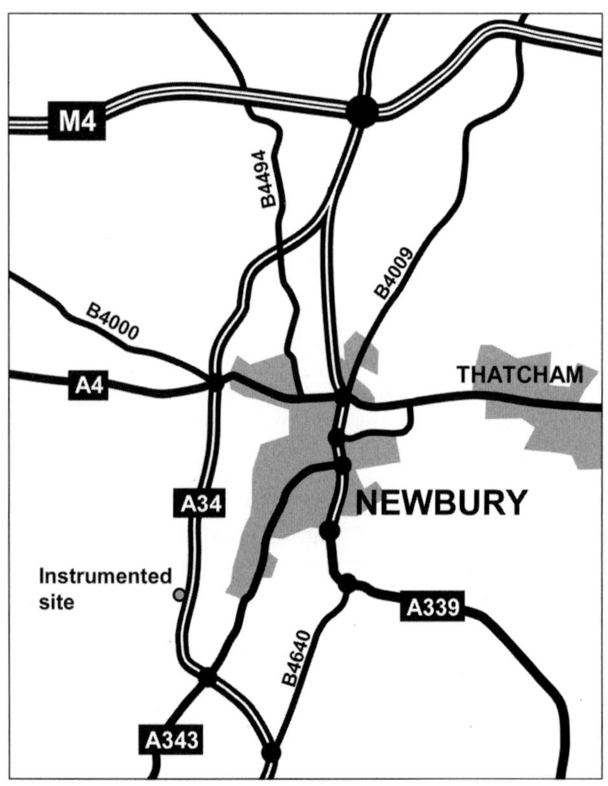

Fig. 1. Location map

Fig. 2. Photograph of site

The current groundwater regime in the slope is effectively hydrostatic below the groundwater level. This indicates that, within the near-surface zone being monitored, any negative excess pore pressures induced by unloading as a result of excavating the cutting have substantially dissipated.

The vegetation is primarily rough grass and herbs with a few small shrubs less than 0·5 m high. The grass and herbs were initially mowed periodically to help the development of shrubs planted on the slope, but have remained uncut since October 2002. Mature beech, oak and silver birch trees fringe the top of the slope.

The London Clay is predominantly a stiff grey clay, but contains several bands of silty clay up to 50 mm thick and bands of large flints. The weathered London Clay is spatially very variable, changing from a stiff orange brown clay to a clayey silt over small distances and depths, similar to the description given by Perry *et al.* (2000). The permeability, unit weight and plasticity index of the London Clay at the site are summarised in Table 1. The saturated vertical permeability was obtained from tests on undisturbed samples from depths of 0·5, 1·0, 1·5, 2·0 and 3·0 m carried out at effective confining pressures of 10, 15, 20, 25 and 35 kPa in the triaxial apparatus. Estimates of the in situ permeability, obtained from bailing out tests carried out in April 2003 in hand-augered boreholes 3 m deep, were typically one to two orders of magnitude larger owing to the effects of anisotropy and fabric (including silt partings and fissures) not fully captured in the triaxial samples. The dry unit weight was measured from undisturbed samples obtained from 0·5 m depth for both the weathered and unweathered clays.

Relationships between soil water content and suction for both the drying and wetting of (undisturbed) samples of London Clay have been published by Croney (1977), and are reproduced in Fig. 4. Fig. 4(a) shows the full set of data obtained by Croney (1977) plotted on a log scale for

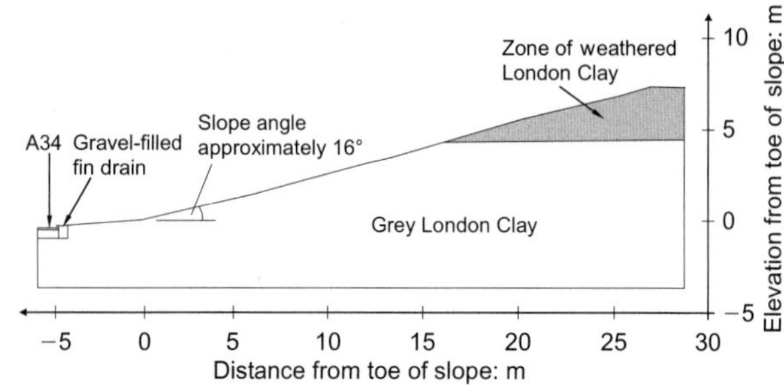

Fig. 3. Cross-section

Table 1. Permeability, unit weight and plasticity index of grey and weathered London Clay at the Newbury test site

Property	Grey London Clay		Weathered London Clay	
	Range	Average	Range	Average
Saturated vertical permeability from triaxial tests: m/s	$3{\cdot}9 \times 10^{-11}$ to $6{\cdot}6 \times 10^{-10}$	$2{\cdot}3 \times 10^{-10}$	$5{\cdot}0 \times 10^{-10}$ to $1{\cdot}6 \times 10^{-9}$	$8{\cdot}7 \times 10^{-10}$
Saturated permeability from borehole bail-out tests: m/s	$2{\cdot}3 \times 10^{-9}$ to $4{\cdot}4 \times 10^{-9}$	$3{\cdot}7 \times 10^{-9}$	$3{\cdot}6 \times 10^{-8}$ to $5{\cdot}0 \times 10^{-8}$	$4{\cdot}3 \times 10^{-8}$
Dry unit weight, γ_d: kN/m^3	13·2 to 15·2	14·6	13·2 to 16·2	16·0
Plasticity index, I_D: %	32·5 to 36·4	34·8	31·7*	31·7*

*Only one of five samples tested for the weathered London Clay exhibited plasticity, the remainder were a silt.

suction; Fig. 4(b) shows the same data plotted up to a typical plant wilting point of pF 4·2 or 1500 kPa (Kabat & Beekma, 1994) on a linear scale that more clearly shows the rapid increase in suction that occurs as the soil dries.

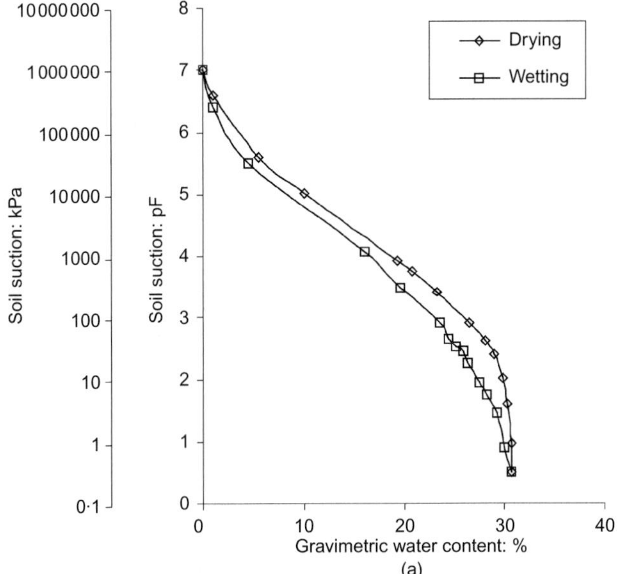

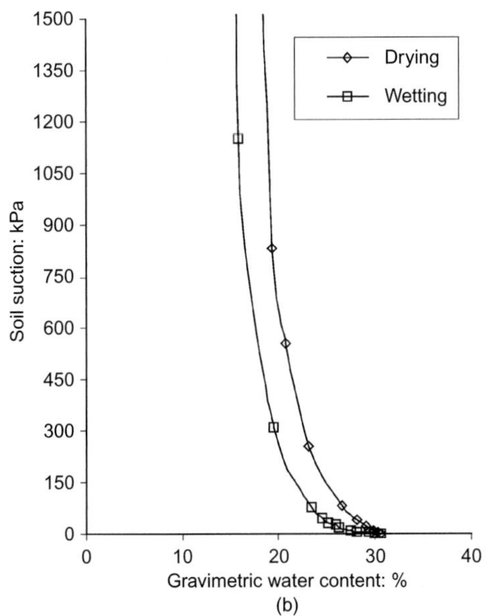

Fig. 4. Graphs of gravimetric water content against suction for undisturbed samples of London Clay (redrawn from Croney, 1977): (a) plotted on a log scale of suction; (b) plotted onto a linear scale of suction up to a typical plant wilting point of 1500 kPa

Vegetation will remove water from the soil profile when evapotranspiration is greater than rainfall, and during the summer the soil near the surface will begin to dry out. Intact clay peds will remain saturated even at quite high suctions owing to their small pore size. However, larger pores and fissures—especially near the surface—will desaturate at lower suctions.

INSTRUMENTATION

The site was instrumented to monitor the soil water content, pore water pressure, soil temperature, the free water surface, rainfall, runoff and climatic data required to estimate evapotranspiration.

Figure 5 shows the layout of the instrumentation. Arrays of time domain reflectometry (TDR) probes for measuring soil water content, flushable vibrating wire piezometers, water filled tensiometers and equitensiometers were installed in four groups spaced 6 m apart down the slope. The sensors were installed at depths between 0·3 m and 3·5 m, at intervals of 0·3 m or 0·5 m. Table 2 summarises the sensor types and the depths at which they were installed.

Each sensor was installed into its own hole, which was hand-augered to a diameter slightly larger than the body of the sensor. The piezometer, equitensiometer and tensiometer ceramics were pushed into a stiff speswhite kaolin paste that had been placed at the base of the hole, to transmit suction effectively from the soil to ceramic of the sensor. A 20 cm depth of dry bentonite powder was used to form a plug above the instrument, before the hole was backfilled either with a stiff slurry mixed from bentonite and the extracted London Clay (shallow installations) or cement bentonite grout (deep installations). Only one sensor was installed in each hole, to minimise the possibility of errors due to an imperfect seal.

Rainfall at the site was recorded using two gauges in case of instrument failure. A climate station was placed on the slope to record air temperature, humidity, wind speed, solar radiation and soil temperature at 30 cm depth. Except for the soil temperature, these parameters were used to estimate potential evapotranspiration using the Penman–Monteith method (Allen *et al.*, 1998).

Surface runoff together with interflow (i.e. flow of water through the topsoil) was measured using an interceptor drain cut across the face of the slope. The interceptor drain was 6 m long and 35 cm deep, dug through the topsoil into the top of the undisturbed London Clay. A slotted drainpipe with gravel packing channelled water into a stilling tank to allow silt to settle out, and thence into a tipping bucket flow gauge (Unidata Ltd, Sheffield) capable of measuring up to 15 l/min.

All of the above sensors were connected to a Campbell Scientific CR10X data logger with a GSM modem connection, powered by a 12 V car battery recharged by a 10 W

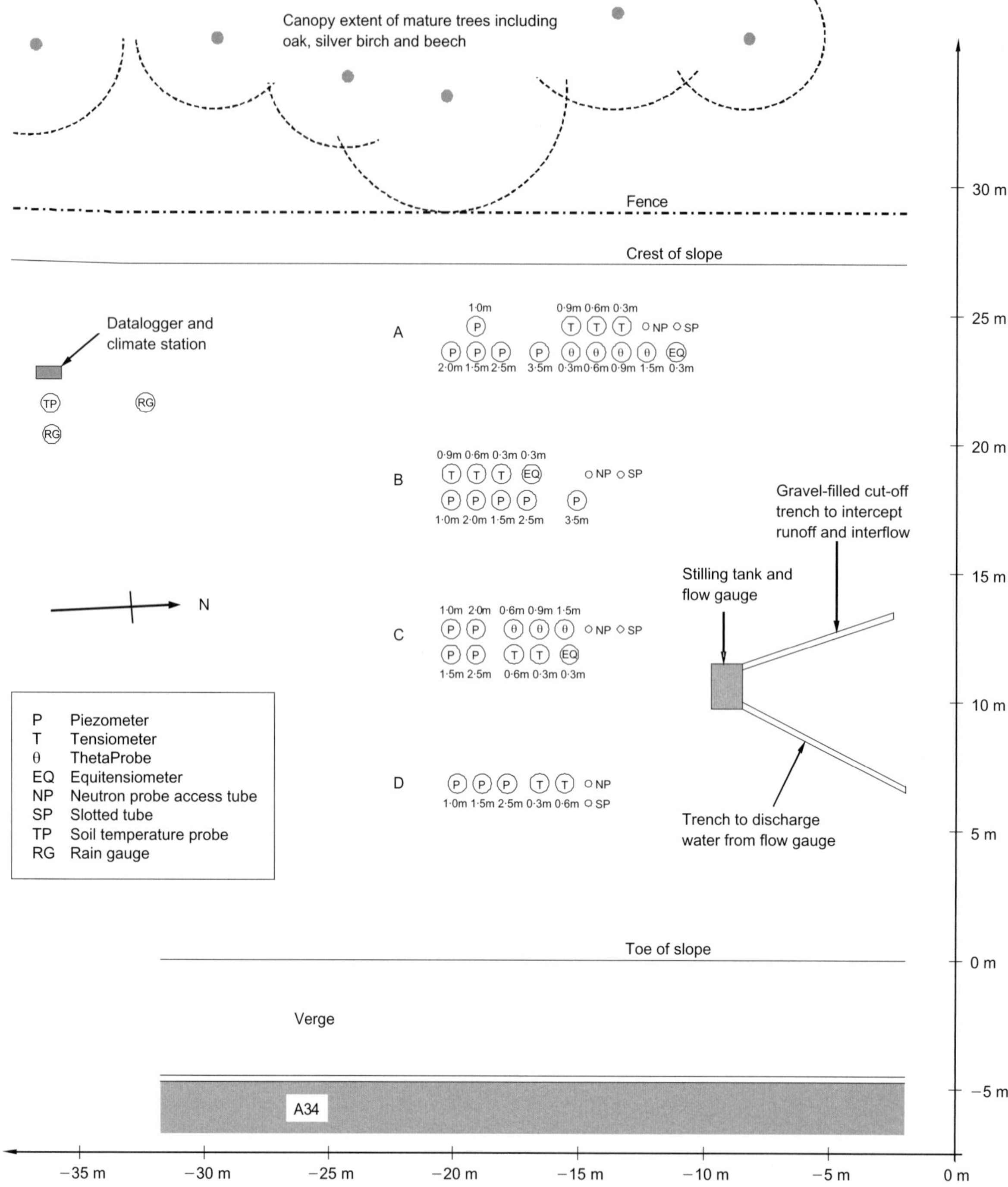

Fig. 5. Plan of site showing layout of instrumentation

Table 2. Summary of soil suction and water content sensors installed

Measurement	Type of instrument	Quantity and depths	Measuring range/accuracy	Source/references
Soil suction	Tensiometer	10 no., depths of 0·3, 0·6, 0·9 m	Matric suction up to 90 kPa	Delta-T Devices Ltd, Cambridge, UK
Soil suction/pore water pressure	Flushable piezometer	16 no., depths of 1·0, 1·5, 2·0, 2·5, 3·0, 3·5 m	Pore pressure between 300 and −90 kPa	Soil Instruments Ltd, Uckfield, UK
Soil suction	Equitensiometer	3 no., all at 0·3 m depth	Matric suction up to 1500 kPa, over 100 kPa accuracy ±5% of reading	Delta-T Devices Ltd, Cambridge, UK
Soil water content	TDR 'ThetaProbe'	8 no., depths of 0·3, 0·6, 0·9, 1·5 m	Volumetric water content, 0–50%	Delta-T Devices Ltd, Cambridge, UK

solar panel. The data were logged at 10 min intervals and recorded as hourly averages, commencing in October 2002.

Aluminium access tubes for a neutron probe were installed to enable point measurements of water content adjacent to the logger sensors. A Wallingford neutron probe (Bell, 1987), calibrated against samples for which the water content was determined gravimetrically, was used to measure soil water profiles at approximately monthly intervals. Finally, four 50 mm diameter slotted plastic tubes were installed to act as observation wells at intervals down the slope to locate the position of the free water surface. The auger holes for the instruments and observation tubes provided a large number of soil samples from which gravimetric water contents were obtained.

WATER BALANCE

In the absence of artificial recharge or irrigation, water may enter the soil through the ground surface from rainfall and leave as a result of evapotranspiration by plants. For the zone of major seasonal wetting and drying at the slope surface, the full water balance may be written (after Blight, 2003) as

$$\sum (R - \mathrm{RO}) - \sum \mathrm{ET} + S - \mathrm{RE} \approx 0 \qquad (1)$$

where R is the rainfall, RO is the runoff, ET is the actual evapotranspiration, S is the change in stored water within the soil, and RE is the net recharge from the surrounding soil. Rainfall is simple to measure, and although it can be very site specific, long records are available for the UK. Runoff is likewise site specific but measurable. Evapotranspiration is a function of the interactions between the elements of the plant–soil–atmosphere system. It depends on plant type, climate, soil characteristics and soil moisture conditions, and is more difficult to quantify owing to the variability of climate, soil and plant types. A simplified approach is to define a standard crop and soil condition so that evapotranspiration is then a function only of climate. This is known as the *reference crop* or *potential* evapotranspiration, typical of that from a well-watered short green crop, such as 10–15 cm long healthy grass.

Potential evapotranspiration can be estimated with reasonable confidence using the Penman–Monteith equation (Allen *et al.,* 1994), which is based on the energy available to change liquid water into water vapour and the vapour gradient from the evaporating surface to the atmosphere. Daily totals and monthly averages of potential evapotranspiration (PET) calculated using the climate data from the Newbury site are shown, and the latter compared with data for Southampton averaged over the period 1961–1990, in Fig. 6. The potential evapotranspiration at Newbury varied from almost zero in winter to a peak of 4·3 mm/day in July 2003, with significant daily fluctuations. The monthly averages were generally some 20–30% lower than the long-term data for Southampton because the slope at Newbury faces east and is in shade in the mid/late afternoon.

In the context of an investigation into the effect of vegetation on slope stability and behaviour, the purpose of a water balance calculation is to estimate changes in soil water content, which are related to the change in the volume of water stored within the soil, S. In agricultural science, S is usually expressed as the soil moisture deficit (SMD), calculated in mm as a volume of water per unit area. A soil with zero SMD is at 'field capacity', that is, the equilibrium water content within a soil free to drain downward under gravity. Water is held in the soil by capillary action: hence field capacity is a function of pore size. For many soils, SMD = 0 mm usually occurs 1–2 days after rainfall and corresponds to a suction of about pF 2 or 10 kPa (Kabat & Beekma, 1994). An intact clay soil would probably still be saturated at field capacity, but in a more structured material there is likely to be air present in the larger fissures and voids.

The soil moisture deficit changes dynamically in response to the inflows and outflows of water in the field, and if the water balance calculation covers too long a period of time, the results will be meaningless. For example, the climate of southern England generally has an annual net water surplus: that is, the total annual rainfall exceeds the total annual potential evapotranspiration (Table 3). However, more rain

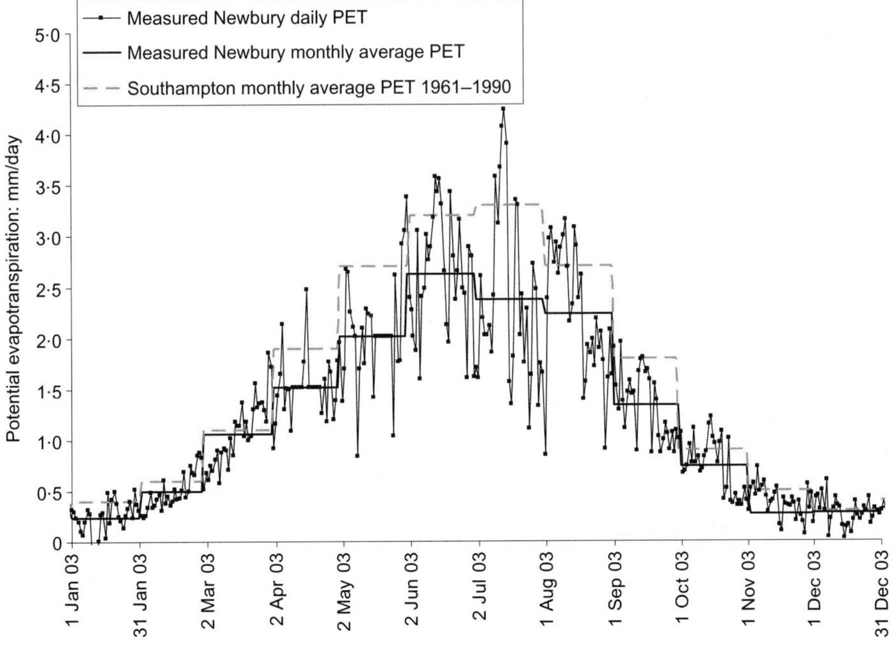

Fig. 6. Potential evapotranspiration for Newbury site calculated using measured weather parameters

Table 3. Rainfall and potential evapotranspiration for the Newbury site, 2003

	Jan	Feb	Mar	Apr	May	June	July	Aug	Sept	Oct	Nov	Dec	Total
Rainfall: mm	112·0	49·0	30·1	52·1	65·8	54·9	56·8	20·2	10·1	50·1	135·8	86·9	723·8
PET: mm	7·4	13·7	32·9	45·3	62·6	78·9	73·8	69·4	39·9	22·9	11·7	8·7	467·2

falls in the winter months than in the summer, but potential evapotranspiration is greatest in the summer. This leads to a significant water surplus in the winter months, resulting in runoff, and a water deficit during the summer, which may lead to the vegetation becoming stressed. In the latter case, the actual rate of evapotranspiration is likely to fall below the potential evapotranspiration calculated using the Penman–Monteith equation, because a condition is reached whereby the residual soil moisture is not readily accessible by plants.

The low permeability of the clay means that, despite the likely establishment of a hydraulic potential gradient, the ability of the plants to remove water from the soil below the rooting zone (typically 0·6 to 0·9 m in clays: Allen *et al.*, 1998; Greenwood *et al.*, 2001) is limited. This is illustrated by data from the Newbury site shown in Figs 10, 12 and 18 and discussed in detail later: during early autumn 2003, the top 0·5–1·0 m of soil experienced significant drying, and despite a suction gradient extending to a depth of several metres there was little change in the water content below 1·0 m depth. Taking the largest measured suction gradient (from 50 kPa at 1 m depth to 0 kPa at 4·5 m depth, in mid September 2003) and a high permeability for the London Clay (1×10^{-9} m/s; Table 1) gives an upward recharge rate according to Darcy's law of 0·04 mm/day. This

is negligible in comparison with the potential evapotranspiration of 1·5 mm/day for the same period. Hence it is reasonable to ignore recharge from below the rooting zone in the water balance calculation (equation (1)) in comparison with the other inputs and outputs, giving

$$\sum(R - \mathrm{RO}) - \sum \mathrm{ET} + S \approx 0 \qquad (2)$$

For the purposes of calculation, it might reasonably be further assumed that all of the rainfall R infiltrates until the ground is at field capacity (i.e. the runoff RO is zero), after which all of the rainfall runs off (i.e. RO = R).

Assuming that the vegetation is not stressed, the actual evapotranspiration (ET) is calculated by scaling the potential evapotranspiration (PET) calculated using the Penman–Monteith equation by a crop factor K_c specific to the vegetation type.

In calculating the actual evapotranspiration it is necessary to consider not only the total available water in the active root zone, TAW (i.e. that not bonded to the surface of the clay particles), but also the remaining readily available water, RAW (i.e. that which the plants can access without stress). Both RAW and TAW are expressed as volumes of water per unit area within the zone of drying, and therefore have units of mm, the same as the soil moisture deficit (SMD).

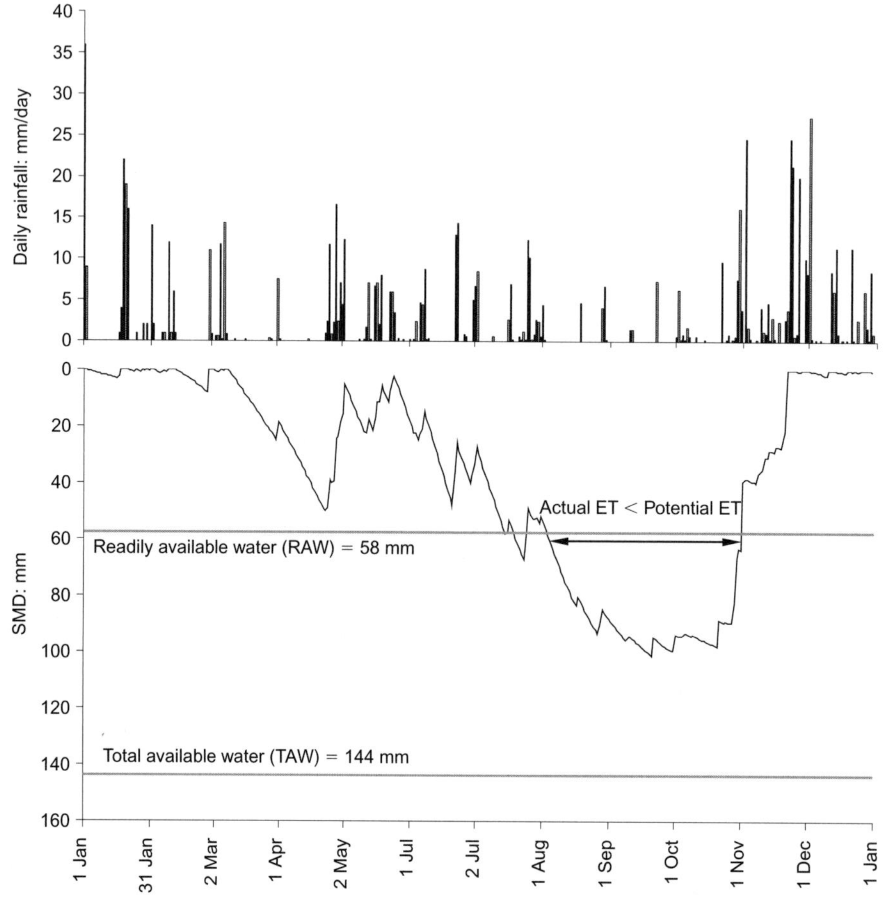

Fig. 7. Daily rainfall measured and soil moisture deficit calculated for Newbury site over year 2003

While the SMD is less than the RAW, evapotranspiration can be assumed to occur at the potential rate for the actual crop (i.e. PET $\times K_c$). When the SMD exceeds the RAW, evapotranspiration is assumed to fall below the potential rate in proportion to the ratio of non-readily available water (TAW − RAW) extracted: that is,

$$ET = PET \times K_c \qquad \text{for } 0 \leqslant SMD \leqslant RAW \qquad (3a)$$

and

$$ET = PET \times K_c \times \frac{TAW - SMD}{TAW - RAW} \qquad \text{for } SMD \geqslant RAW \qquad (3b)$$

This is the basis of the program CROPWAT (Clarke et al., 1998), which was used to calculate the water balance for the site at Newbury on the following basis.

(a) Recharge from the ground (RE = 0) and runoff occur as already stated.

(b) The plant rooting depth for grass and herbs was taken as 800 mm, and the water balance calculation was carried out for this depth of soil.

(c) At the start of the calculation period (1 January 2003), the 800 mm depth of soil was assumed to be at or above field capacity (i.e. SMD = 0). This is reasonable given high rainfall during the preceding three months (October–December 2002).

(d) The crop factor for the long grass and herb vegetation on the slope at Newbury was taken as $K_c = 1.0$. (Typical values of K_c vary from 0·85 for short lawn grass to 1·2 for field crops such as wheat; Allen et al., 1998).

(e) Although a saturated clay may be 50% water (by volume), much of the water is chemically attached to the clay particles or held in small capillaries from which it is difficult for plants to extract. The total volume of water potentially available to the plants (TAW) was taken as 18% of the total soil volume: this is consistent with the values measured for a range of soils by Hall et al. (1977) and Jarvis & Mackney (1979), and for a soil depth of 800 mm gives a TAW of 144 mm. The TAW corresponds to the change in water up to the wilting point, typically a soil suction of about pF 4·2 or 1500 kPa (Kabat & Beekma, 1994).

(f) The readily available water (RAW) is taken as the change in moisture up to the point where the vegetation

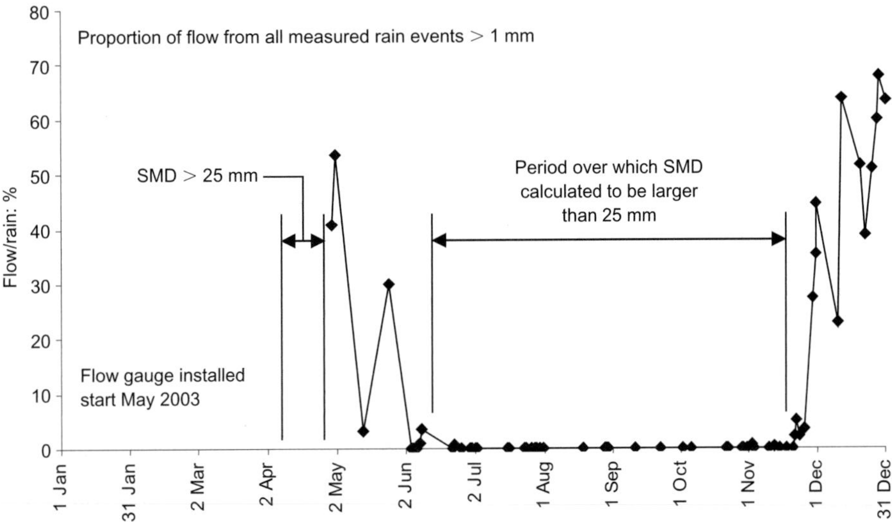

Fig. 8. Proportion of each rainfall event measured by flowgauge (graph is for 2003)

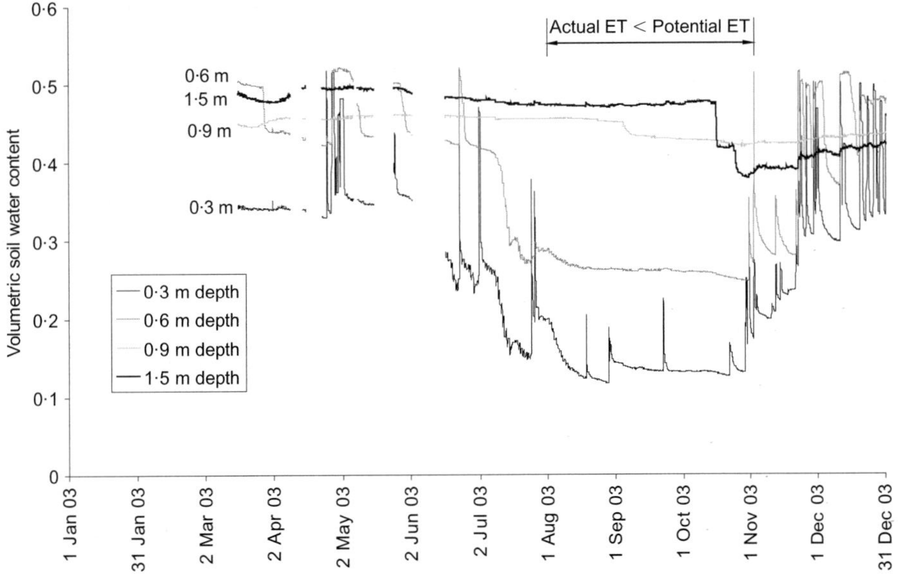

Fig. 9. Readings from TDR probes installed in weathered London Clay at instrument group A

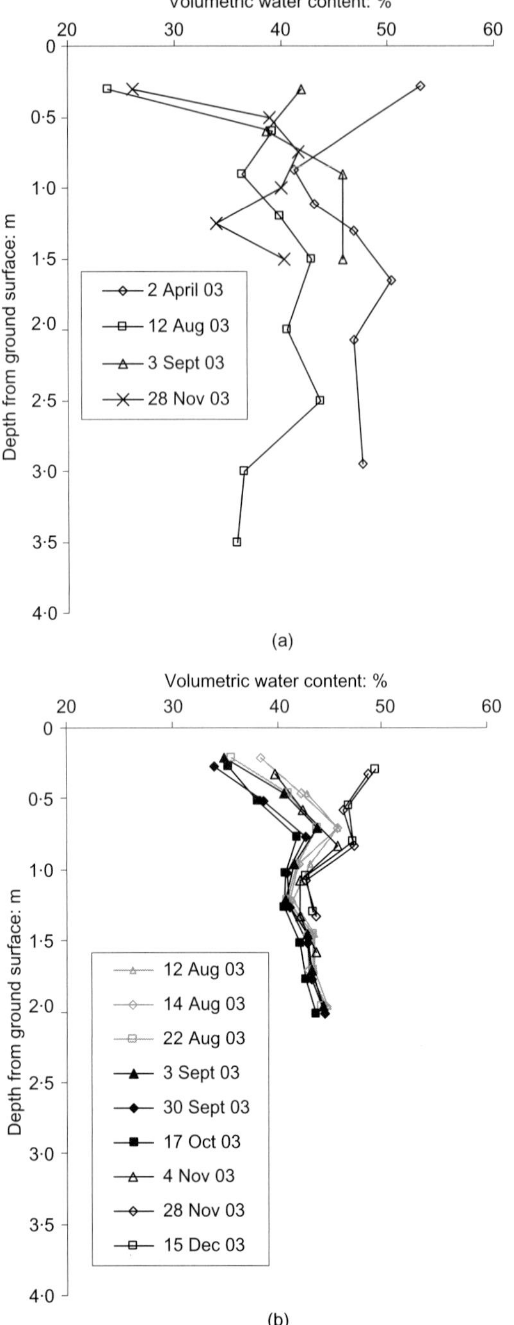

Fig. 10. Volumetric water content against depth for instrument group A: (a) soil samples taken to laboratory and oven-dried (converted from gravimetric to volumetric water content using average dry bulk density shown in Table 1); (b) neutron probe readings

starts to become stressed; this typically occurs at a soil suction of pF 2·6–3·0 or 40–100 kPa depending on the soil properties and the vegetation type (Kabat & Beekma, 1994). Measurements made in the London Clay at Newbury indicate that the grass and herbs start to experience difficulty extracting water from the soil at about 45 kPa (shown by the onset of large diurnal variations in observed suction, Fig. 16). Considering the suction/water content characteristic curve in Fig. 4(b), 45 kPa suction occurs when the water content is reduced to 40% of the TAW (where the TAW corresponds to 10–1500 kPa on the drying curve). Therefore the RAW was taken as 40% of the TAW, giving RAW = 0·4 × 144 mm = 58 mm.

The soil moisture deficit was calculated using the above rules for the Newbury site for each day of 2003. This included an exceptionally hot dry summer period. The soil water balance (Fig. 7) suggests the development of a soil moisture deficit from mid February to late November. The calculated SMD exceeded the RAW from early August to late October, and reached a maximum of 101 mm in late September. This is consistent with the behaviour of the vegetation on the slope, which by early August had begun to turn brown owing to a lack of water.

In the winter months, low evapotranspiration and higher rainfall resulted in a calculated SMD close to zero. Rainfall during these periods could therefore not be stored in the soil profile and ran off. Fig. 8 shows the runoff and interflow measured by the flow gauge for individual rain events, starting in May 2003 when the flow gauge was installed at the site. The runoff events measured in April–May and December 2003 correspond with periods when CROPWAT calculated the SMD to be close, or equal, to zero. In addition, Fig. 8 shows that, for the period of the year in which the calculated SMD exceeded 25 mm, no runoff was measured on the site (the largest daily rainfall event was 25 mm; Fig. 7).

MEASUREMENT OF SOIL WATER CONTENT AND SUCTION

The water balance model for the soil–slope system was tested using the extensive measurements of soil water content and suction made at Newbury during 2003.

Soil water content

Volumetric soil water contents (w_{vol} = volume of water ÷ total volume) measured using the TDRs installed at the top of the slope in the 2·5 m thick layer of weathered London Clay at location A (Fig. 5) are shown in Fig. 9. Significant reductions in water content were measured at 0·3 m and 0·6 m depth between mid July to the end of October, reflecting the period when the calculated soil moisture deficit exceeded the limit of readily available water. Water content changes at 0·9 m and 1·5 m depth were smaller. Transient effects of rainfall events were noticeable at 0·3 and 0·6 m, reflecting the saturation of the surface layers during rainfall and the subsequent redistribution of water into the ground.

Vertical profiles of volumetric water content based on oven-drying of samples taken at instrument group A during 2003 are shown in Fig. 10(a). These data were converted from gravimetric water contents (w = mass of water ÷ mass of soil solids) using the average value of dry unit weight given in Table 1 ($w_{vol} = w.\rho_{dry}/\rho_w$. Note that the same dry density has been used to convert all four profiles; in reality the value may vary as the soil dries). Fig. 10(b) shows corresponding water content profiles measured at the same location using the neutron probe. In both cases, the most significant changes are over the top 1·0 m. Below 1·0 m depth, the neutron probe suggests a fairly consistent water content of about 42% by volume, whereas the profiles obtained by oven-drying samples were more variable, probably because of the small distances (of up to about 2·5 m) between the sampling locations and the spatially variable nature of the weathered London Clay.

Further down the slope at location C, the topsoil and weathered clay is only 40 cm thick and the sensors are mainly in the unweathered grey London Clay. Fig. 11 shows the volumetric soil water contents measured by the TDR sensors, and Fig. 12 shows the volumetric water content profiles with depth from borehole sampling and neutron probe surveys. The changes in water content here are

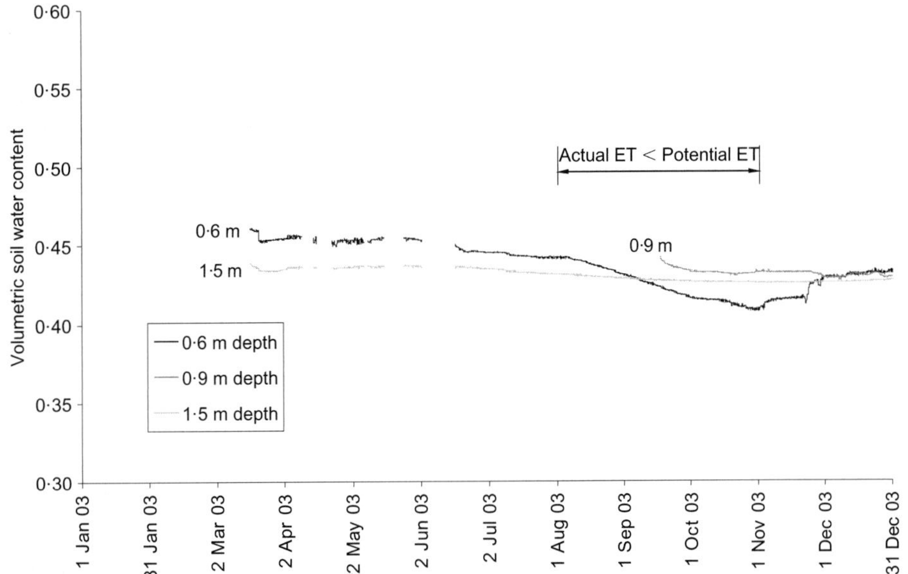

Fig. 11. Readings from TDR probes installed into grey London Clay at instrument group C

smaller than those measured at location A in the weathered London Clay (Figs 9 and 10). For example, at 0·6 m depth the reduction in volumetric water content over the summer measured by the TDRs was only 4% at C compared with 20% at A. Neutron probe data show a smaller difference at 0·6 m, with 5% drying at C compared with 8% at A. Comparison of Fig. 12 with Fig. 10 confirms that the depth of significant summer drying is slightly less ($\sim$800 mm compared with $\sim$1000 mm according to the neutron probe data) in the grey London Clay than in the weathered material. The generally smaller amount of drying at C reflects the difficulty that the plant roots had in penetrating or removing water from the stiff intact clay peds and/or the likely groundwater flow regime of seepage out of the toe of the slope.

The measured water contents in the upper soil profile may be compared with the SMD calculated using CROPWAT, by noting that

$$\Delta w_{vol} = \frac{\Delta V_w}{V_t} = -\frac{\Delta SMD \cdot A}{h \cdot A} = \frac{-\Delta SMD}{h} \qquad (4)$$

where h is the depth of the drying zone. SMDs determined from the neutron probe data (Figs 10(b) and 12(b)) and the TDR data at instrument group A (Fig. 9) on this basis, with $h = 800$ mm and Δw_{vol} taken as the average over this depth and the wettest profile taken to correspond to SMD = 0, are compared in Fig. 13 with the SMDs calculated using CROP-WAT. (The absence of a TDR probe at 0·3 m depth at instrument group C makes it difficult to calculate the drying from the TDRs at this location.)

In general, the match between the CROPWAT calculation and the neutron probe data is close. While the data from the TDR probes give a reasonable match to the CROPWAT calculation in the early part of the year (March–July 2003), during the summer the TDRs show a much greater extent of drying. Although located fairly close together on the slope, the samples taken from the auger holes for water content determination show that the neutron probe tube is installed predominantly in clay whereas the TDR probes are in clayey sandy silt. Both the total and readily available water would be expected to be slightly larger for a silt than for a clay, allowing a greater extent of soil drying by the vegetation. The water-holding capacity used in the CROPWAT calcula-tion can be varied to account for this, but it does not fully

explain the apparent discrepancy between the TDR and neutron probe data. The difference might be due to addi-tional drainage processes in the silt not taken into account in the CROPWAT calculation, or the fact that the TDR probes respond to and represent a rather smaller volume of soil.

Pore pressures and suctions

Figure 14 shows data from vibrating-wire peizometers between 1·0 and 3·75 m deep at Group A at the top of the slope in the weathered clay. The winter of 2002–2003 was very wet, with 300 mm of rainfall in November and Decem-ber 2002 resulting in positive pore water pressures being recorded at all instruments between December 2002 and March 2003. The summer of 2003 was unusually hot and dry. From May to September 2003 evapotranspiration was greater than rainfall, resulting in the development of a large soil moisture deficit and associated negative pore pressures. The highest suction recorded was 25 kPa at 1·0 m depth in mid October 2003. Heavy rainfall during November and December 2003 infiltrated into the soil profile, increasing the water content and reducing the soil suction. However, at the end of December 2003, most of the group A piezometers still measured a suction and had not recovered to their January 2003 values.

Water-filled tensiometers at 0·3 and 0·6 m depth within the root zone at Group A recorded suctions that increased rapidly in June 2003 (Fig. 15). Tensiometers can only measure suctions of up to 90 kPa before the water in the device is drawn into the surrounding soil and air enters the instrument. On refilling, the soil again removes the water, as indicated by the repeated vertical lines in Fig. 15. Suctions at 0·9 m took longer to increase and reached a maximum of only 65 kPa in October 2003. Diurnal variations in pore pressure of up to ±5 kPa were recorded during the build-up of suction in the tensiometers, caused by the relaxation of vegetation water demand during the night. As the suction increases, further increases in suction yield less and less water, as indicated by the steepening of the SWCC curve (Fig. 4). Relaxation of water demand at night results in bigger diurnal cycles of suction, especially after about 45 kPa (Fig. 16).

Data from the piezometers installed in Group C are shown in Fig. 16. The late-summer suctions are larger than at A,

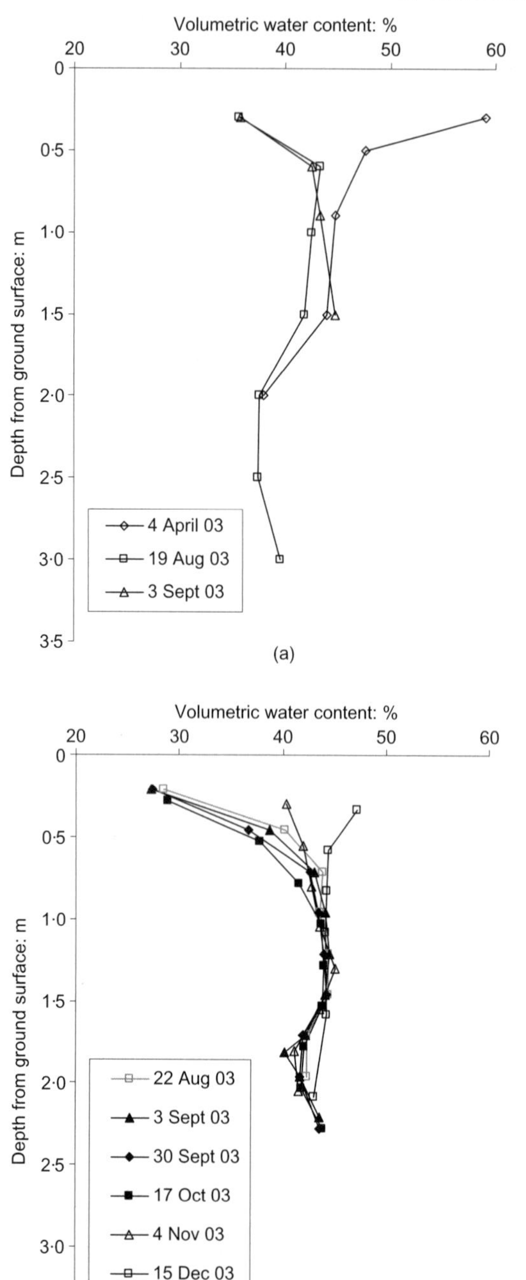

Fig. 12. Volumetric water content against depth for instrument group C: (a) soil samples taken to laboratory and oven-dried (converted from gravimetric to volumetric water content using average dry bulk density shown in Table 1); (b) neutron probe readings

exceeding 65 kPa at 1·0 m depth between August and October. Suctions in excess of 25 kPa were recorded for several weeks at 1·5 m and 2·0 m depth. Heavy rainfall during November and December caused an increase in water content and a corresponding decrease in soil suction. The sudden large changes in pore pressure at 1·0 m and 1·5 m depth are the result of rain filling tension cracks close to the instrument. Pore pressures at the end of December 2003 had not recovered to their January 2003 values owing to the particularly dry summer and the fact that the previous winter (to January 2003) had been exceptionally wet.

Observations of suctions near the surface (to 0·6 m depth) at C were made using water-filled tensiometers (Fig. 17). Suctions to the maximum recordable by the tensiometers

(about 90 kPa) developed rapidly in June 2003 at 0·3 m depth and by the end of July at 0·6 m depth. Deeper piezometers recorded increasing suctions until November 2003. Suctions greater than 90 kPa were recorded using an equitensiometer, which consists of a TDR sensor encapsulated in a fine ceramic with known moisture content–suction characteristics. This was installed at 0·3 m depth. Suctions greater than 70 kPa were recorded during the period over which the calculated SMD was greater than the RAW (Fig. 7), with a maximum suction of 440 kPa being measured in September 2003 (Fig. 17). However, the equitensiometer data are very sensitive to the water content–suction relationship of the ceramic, so these data should be treated as indicative only rather than quantitative.

The field measurements of pore water pressure and water content highlight a deficiency in the detail of the CROPWAT water balance model in the context of the current application. The gradual drying process in the summer is described well by CROPWAT, but during rapid re-wetting it assumes full redistribution of soil water at the end of each day. This does not allow for the low permeability of the clay, which prolongs the re-wetting process, especially for intact peds. CROPWAT assumes that any water in excess of zero SMD is redistributed or drains off the slope during the one-day time step calculation, which may not occur in reality. This is illustrated in Fig. 9, where the shallow TDR soil water content sensors show spikes following rainfall. These spikes reflect rapid increases in water content to saturation, which then returns to field capacity about 2 days after the rainfall.

Figure 18 shows the profiles with depth of maximum and minimum pore pressure measured by the piezometers and tensiometers in each of the four instrument groups. The (maximum) pore water pressures in early January 2003 were generally hydrostatic below a water table at most 0·5 m below the slope surface (note that the tensiometers had not yet been installed in January 2003). By late September the (minimum) pore water pressures were negative to at least 4 m depth, with the overall change in pore water pressure being greatest at the surface and decreasing non-linearly with depth. The high suctions ($>$ 90 kPa) close to the surface correspond with the significant soil drying measured from 0 to 800–1000 mm depth (Figs 10 and 12). The extensive drying in the clay in this upper zone is caused by the direct removal of water by the plant roots, and the depth of the drying zone is therefore dependent on the ability of the roots to penetrate the clay.

Beneath this major drying zone, the suctions are smaller (Fig. 18), and the gradient denotes some upward movement of water (although for the purposes of the water balance calculation this was shown to be small, and consequently ignored). The reason why the high suctions close to the surface are not transmitted to the soil below the major drying zone is likely to be the low permeability of the clay, which is exacerbated by the high degree of drying of the soil that occurs at the surface. In a stiff clay, the vegetation therefore has limited ability to draw water from below its root zone.

The seasonal range of pore water pressures shown in Fig. 18 is similar to that determined by Walbancke (1976; replotted by Vaughan *et al.*, 2004) from measurements in a number of grassed embankment and cutting slopes, where the largest suctions occurred over the top 2·0 m of soil and the maximum winter pore pressures in the same 2·0 m zone are hydrostatic from the slope surface.

Relating the observed pore water pressures and soil moisture contents

The envelope of maximum soil suction shown in Fig. 18 has been converted into a profile of water content with depth

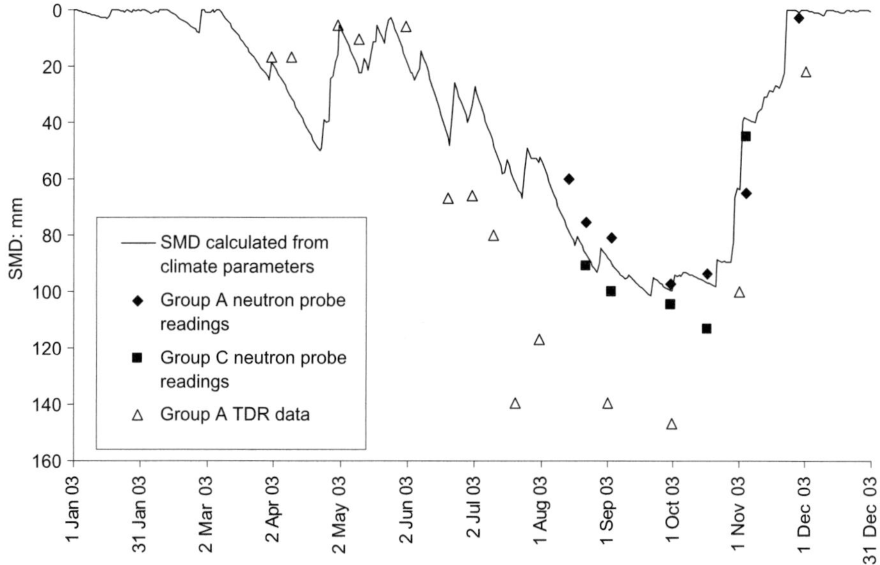

Fig. 13. Profile of daily soil moisture deficit obtained from CROPWAT plotted with measured drying taken from neutron probe readings and TDR data

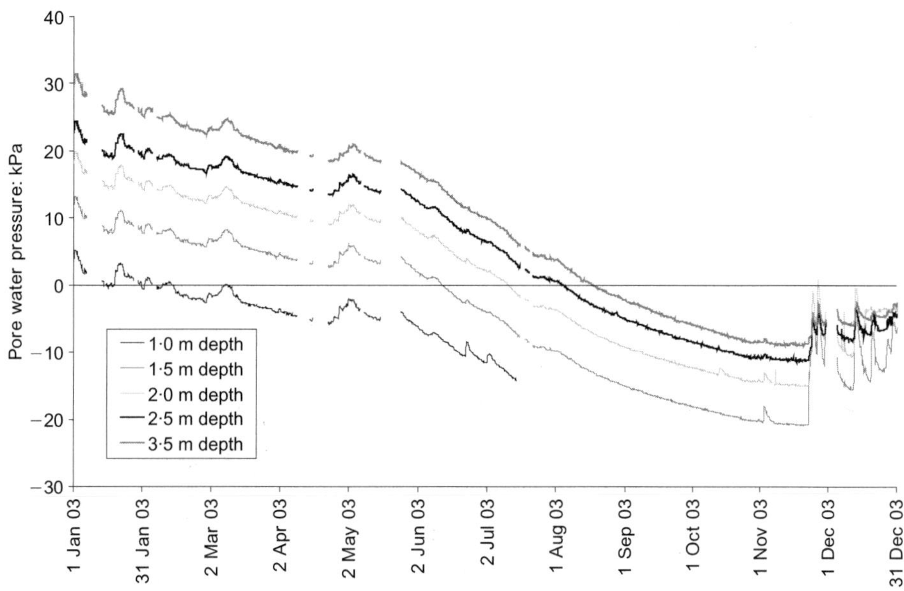

Fig. 14. Piezometer readings from instrument group A

using the soil water characteristic curve (SWCC) for the London Clay given in Fig. 4. This is plotted with the measured neutron probe water content profiles in Fig. 19. In making the calculation from the maximum soil suction to the corresponding volumetric water content profile, the following assumptions were made.

(a) The suction at 0·3 m depth was that measured by the equitensiometer at Group C of 440 kPa. In the absence of any measurements above 0·3 m depth, the maximum suction is assumed to be 440 kPa.

(b) The drying curve shown in Fig. 4 was used, as the soil-drying processes in July and August 2003 that give the maximum suction are rapid and reasonably continuous.

(c) The drying curve of the SWCC, which has been obtained in terms of gravimetric water content, is assumed to correspond to a total change of 18% in volumetric water content between field capacity (-10 kPa) and wilting point (-1500 kPa), the full width of the curve shown in Fig. 4(b). The 18% change in volumetric water content is

consistent with the total available water (TAW) used in the CROPWAT calculation.

Saturated conditions are assumed to be represented by a uniform-with-depth water content of 46%: this is consistent with the wettest profile measured by the neutron probe.

The envelope of water content change given by the maximum suction profile correlates reasonably well with measured neutron probe data, particularly over the major drying zone.

IMPLICATIONS FOR ASSESSMENT AND MODELLING OF SLOPE STABILITY

In terms of the stability of the slope, it is clear that rough grass and herb cover does not necessarily generate sufficient soil drying in the summer months for soil suctions to be be retained into the winter and early spring when the low evapotranspiration and higher rainfall make slope stability most critical. However, the drying caused by the rough

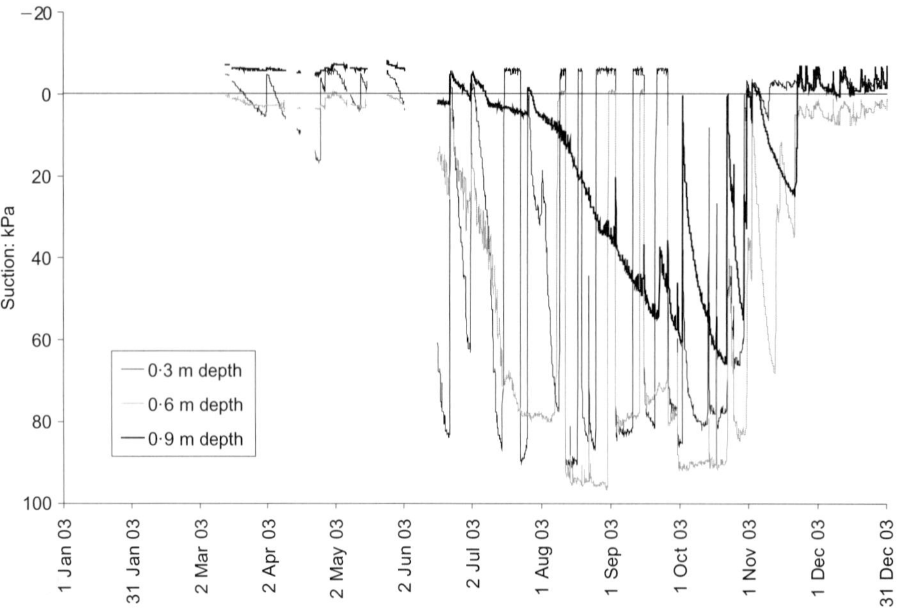

Fig. 15. Tensiometer readings from instrument group A

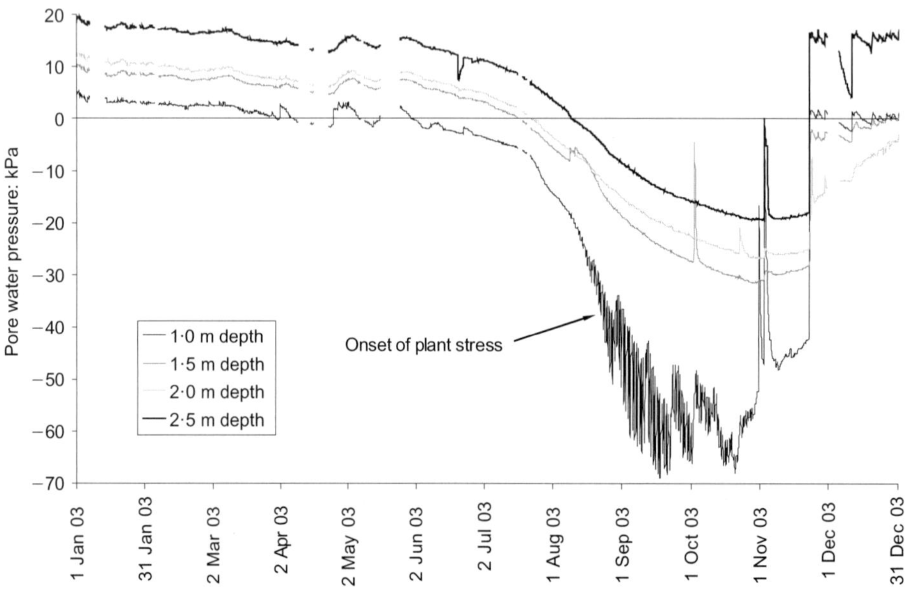

Fig. 16. Piezometer readings from instrument group C

grass/herbs (and the prolonged time taken to re-wet the peds of the soil) does prevent critical winter and spring pore pressures being reached as often or for as long a period of time as when vegetation is absent from the slope. There are also other biomechanical benefits of vegetation, such as reinforcement of the soil by the plant roots, that will act to aid slope stability during the winter and spring months, and help to prevent erosion by surface runoff.

It is clear that the vegetation causes a large cyclic change in effective stress within the major drying zone (top 1·0 m depth of the profile) through the winter–summer–winter cycle for which data were presented. Recent work by Nyambayo *et al.* (2004), Take & Bolton (2004), O'Brien *et al.* (2004) and Vaughan *et al.* (2004) has shown that the cyclic change in stress could promote progressive failure of the slope. The significant displacements caused by the cycles of shrink and swell cause damage to infrastructure and the need for regular maintenance and repair. With the possibility of drier summers and wetter winters as a result of climate

change, it is likely that the problems of deep-seated progressive failure and serviceability problems within infrastructure slopes and embankments will worsen.

The work here has shown that a simple model of the plant–soil–atmosphere system based on the CROPWAT method can give a close estimate of the soil moisture deficit beneath a vegetated clay slope surface. The measured soil moisture deficit may then be linked to the pore water pressures using the SWCC curve for the clay. These results demonstrate the feasibility of creating a numerical model for a clay soil based on a water balance calculation in which the water inputs and outputs (rainfall infiltration and evapotranspiration) control the degree of saturation in the model soil, and the stress changes that take place are then calculated from the soil water content–suction relationship. Such an approach would enable the changes in water content and suction that occur in a clay soil with variability in the climate and water uptake by different types of vegetation to be calculated. The correct modelling of soil permeability

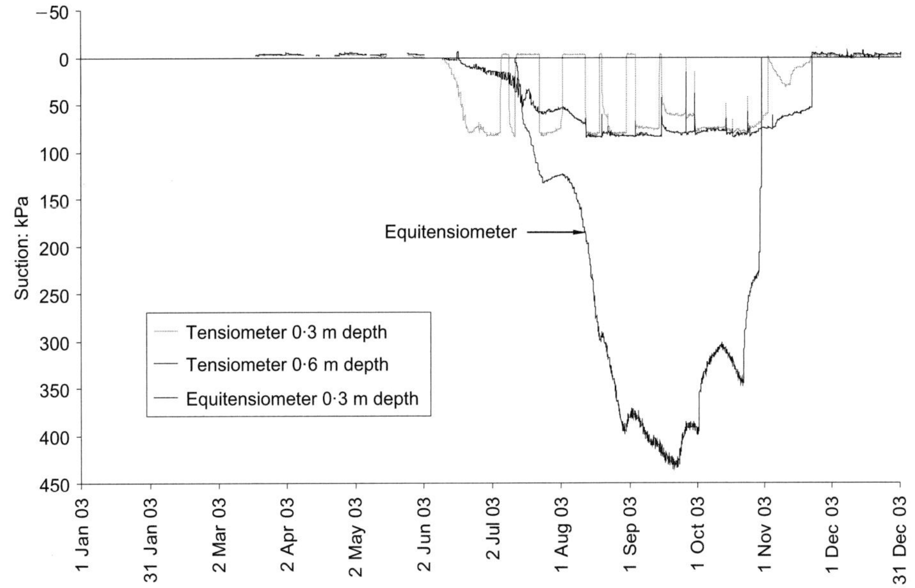

Fig. 17. Tensiometer and equitensiometer readings from instrument group C

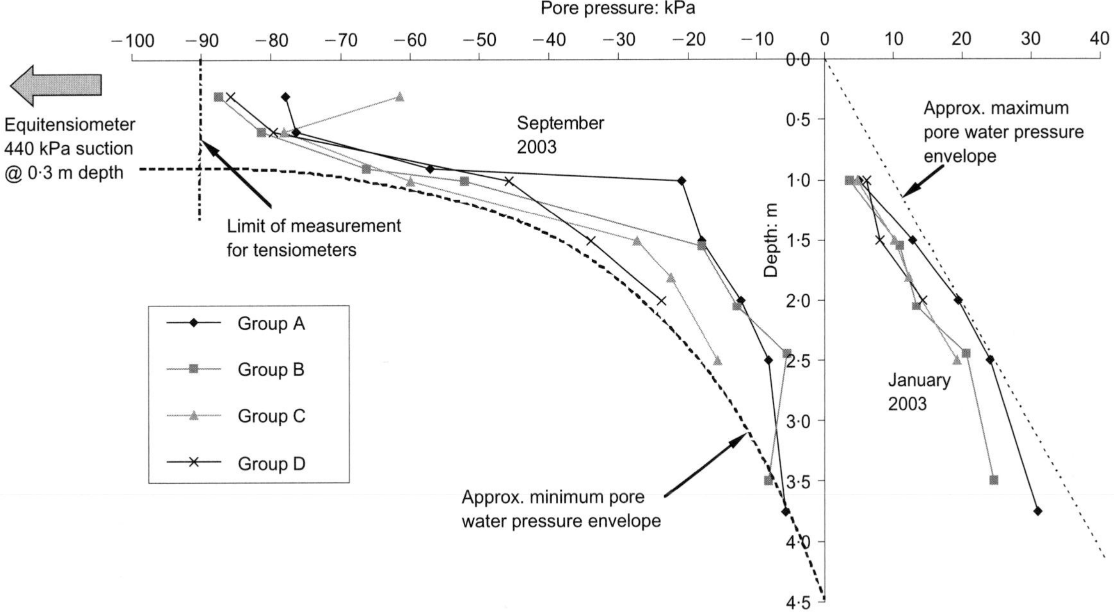

Fig. 18. Distribution of pore water pressure with depth

and flow should also improve the water balance (when compared with CROPWAT), with the delayed wetting up due to the low permeability of the intact lumps (peds) in a structured soil more accurately represented.

CONCLUSIONS

(a) The soil moisture deficit calculated by the water balance model CROPWAT was found to correlate well with the changes in soil water content measured in the slope throughout 2003. However, the early winter re-wetting process is not so well modelled because CROPWAT assumes full redistribution of water each day. Measurement of site-specific climate parameters, incorporating the east-facing nature of the slope, was found to be important in the calculation of potential evapotranspiration.

(b) The water content measurements show that the vegeta-

tion had caused a significant soil moisture deficit in the top 0·8–1·0 m depth of the clay by the late summer. Instruments installed within this zone measured a summer maximum profile of suction of about 450 kPa at 0·3 m depth, decreasing to about 50 kPa at 1·0 m depth. The extent and depth of soil drying were smaller in the grey London Clay at the toe of the slope, probably as a result of the natural seepage regime and/or the difficulty of plant roots penetrating into the stiff clay.

(c) The seasonal variation in water content in the London Clay below 1·0 m depth is small, typically < 4% by volume. In late summer, this small reduction in water content gave a suction varying between about 50 kPa at 1·0 m depth and 0 kPa at 4·5 m depth. The upward flow of water resulting from this suction gradient into the drying (root) zone at the slope surface is negligible in comparison with rainfall and evapotranspiration, and may be ignored in water balance calculations. In a clay

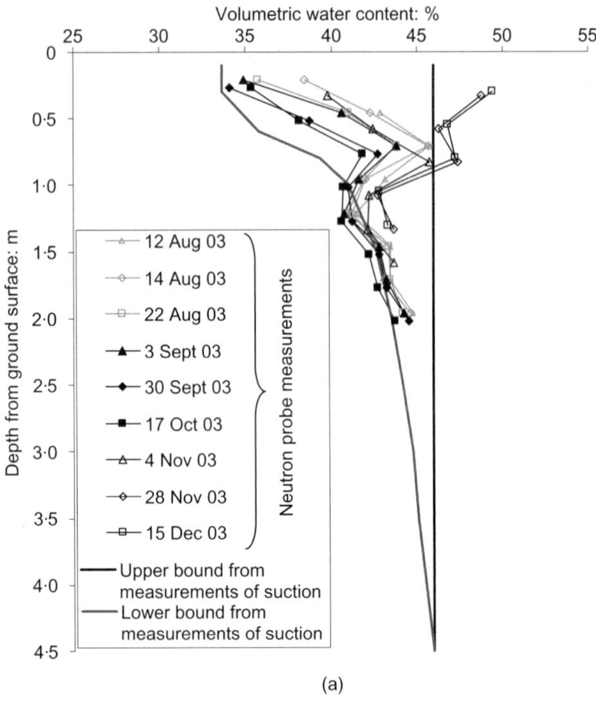

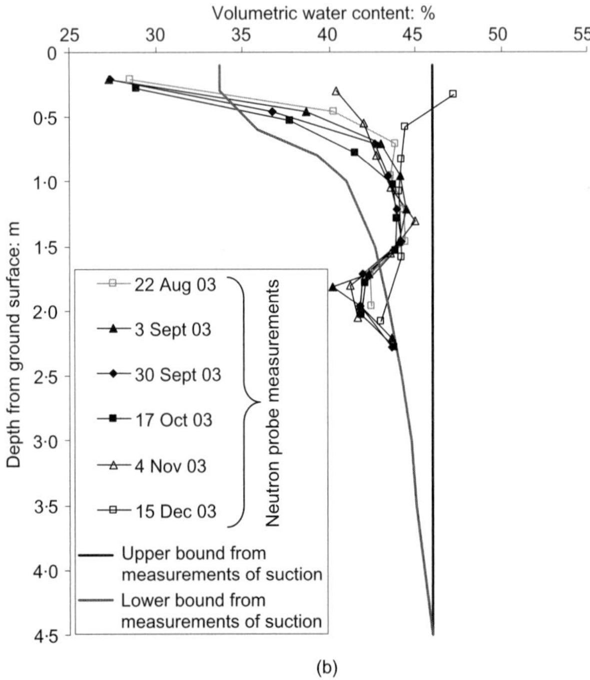

Fig. 19. Profiles of measured water content, with envelope of change in water content determined from measured suction profile: (a) instrument group A; (b) instrument group C

soil, the plants cause significant drying only within the rooting zone.

(*d*) The soil moisture deficit returned to zero through the winter months, meaning that the suctions generated by the rough grass and herb vegetation are unlikely to be able to improve slope stability in the winter and early spring.

(*e*) The measured water contents are reasonably consistent with measured suctions according to the laboratory-derived SWCC. This, together with the ability to calculate SMD with CROPWAT, demonstrates the feasibility of using a numerical model based on a surface flow boundary (representing infiltration and evapotranspiration) to predict the seasonal changes in soil water content and suction.

ACKNOWLEDGEMENTS

The work described in this paper was funded by EPSRC grant number GR/R72341/01. The authors are grateful to the Highways Agency for permission to use the slope at Newbury, and to John Perry and Martin Field of Mott MacDonald, who helped to make the arrangements with the Highways Agency and provided information on the construction and maintenance of the Newbury bypass. Thanks also go to Harvey Skinner, who assembled and maintained the datalogger systems.

NOTATION

A	unit area
ET	actual evapotranspiration
h	height of the drying (rooting) zone
I_D	plasticity index
K_c	crop coefficient
PET	potential evapotranspiration
pF	base 10 log of the head of water in cm
R	rainfall
RAW	readily available water
RE	recharge of water from surrounding soil
RO	runoff
S	change in stored soil water
SMD	soil moisture deficit
SWCC	soil water characteristic curve
TAW	total available water
V_w	volume of water
V_t	total volume
w	gravimetric water content
w_{vol}	volumetric water content
γ_d	soil dry unit weight
ρ_{dry}	soil dry density
ρ_w	density of water

REFERENCES

Allen, R. K., Smith, M., Perrier, A., & Pereira, L. S. (1994). An update for the calculation of Reference Evapotranspiration. *ICID Bulletin* **43**, No. 2, 35–92.

Allen, R. K., Pereira, L. S., Raes, D., & Smith, M. (1998). *Crop evapotranspiration: Guidelines for computing crop water requirements.* Food and Agricultural Organization's Irrigation and Drainage Paper, No. 56.

Anderson, M. G., & Kneale, P. E. (1980). Pore water pressure and stability conditions on a motorway embankment. *Earth Surface Processes* **5**, No. 1, 37–46.

Andrei, A. (2000). Embankment stabilisation works between Rayners Lane and South Harrow Underground stations. *Ground Engng*, **33**, No. 1, 24–26.

Barker, D. H. (1986). Enhancement of slope stability by vegetation. *Ground Engng*, **19**, No. 3, 11–15.

Bell, J. P. (1987). *Neutron probe practice*, Institute of Hydrology report 19, 3rd edition. http://www.ceh.ac.uk/products/publications/documents/IH19NEUTRONPROBEPRACTICE.pdf

Blight, G. E. (2003). The vadose zone soil water balance and transpiration rates of vegetation. *Géotechnique* **53**, No. 1, 55–64.

Blight, G. E. (1997). Interactions between the atmosphere and the earth. *Géotechnique* **47**, No. 4, 715–767.

Biddle, P. G. (1983). Patterns of soil drying and moisture deficit in the vicinity of trees on clay soils. *Géotechnique* **33**, No. 2, 107–126.

Biddle, P. G. (1998). *Tree root damage to buildings*. Wantage: Willowmead Publishing.

Coppin, N. J. & Richards, I. G. (1990). *Use of vegetation in civil engineering*. London: Butterworth.

Crabb, G. I., West, G. & O'Reilly, M. P. (1987). Groundwater conditions in three highway embankment slopes. *Proc. 9th Eur. Conf. Soil Mech. Found. Engng, Dublin*, 401–406.

Croney, D. (1977). *The design and performance of road pavements*. London: Her Majesty's Stationery Office.

Clarke, D., Smith, M. & El-Askari, K. (1998). New software for crop water requirements and irrigation scheduling. *ICID Bulletin* **47**, No 2.

Driscoll, R. (1983). The influence of vegetation on the swelling and shrinking of clay soils in Britain. *Géotechnique* **33**, No. 2, 93–105.

Greenwood, J. R., Vickers, A. W., Morgan, R. P. C., Coppin, N. J. & Norris, J. E. (2001). *Bioengineering: the Longham Wood Cutting field trial*, CIRIA Project Report 81. London: Construction Industry Research and Information Association.

Greenwood, J. R., Norris, J. E. & Wint, J. (2004). Assessing the contribution of vegetation to slope stability. *Proc. Inst. Civ. Engrs Geotech. Engng* **157**, No. 4, 199–207.

Hall, D. G. M., Reeve, M. J., Thomasson, A. J. & Wright, V. F. (1977). *Water retention, porosity and density of field soils*, Soil Survey Technical Monograph No. 9. Harpenden: Soil Survey of England and Wales.

Jarvis, M. G. & Mackney, D. (1979). *Soil survey applications*, Soil Survey Technical Monograph No. 13. Harpenden: Soil Survey of England and Wales.

Kabat, P. & Beekma, J. (1994). Water in the unsaturated zone, in *Drainage principles and applications* (ed. H. P. Ritzema), ILRI publication 16. Wageningen: International Institute for Land Reclaimation and Improvement, 383–434.

MacNeil, D. J., Steele, D. P., McMahon, W. & Carder, D. R. (2001). *Vegetation for slope stability*, TRL Report 515. Crowthorne: Transport Research Laboratory.

Nyambayo, V. P., Potts, D. M. & Addenbrooke, T. I. (2004). The influence of permeability on the stability of embankments experiencing seasonal cyclic pore water pressure changes. *Advances in Geotechnical Engineering: Proceedings of the Skempton Conference*, Imperial College, London, Vol. 2, pp. 898–910.

O'Brien, A. S., Ellis, E. A., & Russell, D. (2004). Old railway embankment clay fill: laboratory experiments, numerical modelling and field behaviour. *Advances in Geotechnical Engineering: Proceedings of the Skempton Conference*, Imperial College, London, Vol. 2, pp. 911–921.

Perry, J., Field, M., Davidson, W. & Thompson, D. (2000). The benefits from geotechnics in construction of the A34 Newbury Bypass. *Proc. Inst. Civ. Engrs Geotech. Engng* **143**, No. 2, 83–92.

Ridley, A., McGinnity, B., & Vaughan, P. (2004a). Role of pore water pressures in embankment stability. *Proc. Inst. Civ. Engrs Geotech. Engng* **157**, No. 4, pp. 193–198.

Ridley, A. M., Vaughan, P. R., McGinnity, B. & Brady, K. (2004b). Pore pressure measurements in infrastructure embankments. *Advances in Geotechnical Engineering: Proceedings of the Skempton Conference*, Imperial College, London, Vol. 2, pp. 922–932.

Take, W. A. & Bolton, M. D. (2004). Identification of seasonal slope behaviour mechanisms from centrifuge case studies. *Advances in Geotechnical Engineering: Proceedings of the Skempton Conference*, Imperial College, London, Vol. 2, pp. 992–1004.

Vaughan, P. R., Kovacevic, N. & Potts, D. M. (2004). Then and now: some comments on the design and analysis of slopes and embankments. *Advances in Geotechnical Engineering: Proceedings of the Skempton Conference*, Imperial College, London, Vol. 1, pp. 241–290.

Walbancke, H. J. (1976). *Pore pressures in clay embankments and cuttings*. PhD thesis, Imperial College, London.

Richards, D. J., Clark, J. & Powrie, W. (2006). *Géotechnique* **56**, No. 6, 411–425

Installation effects of a bored pile wall in overconsolidated clay

D. J. RICHARDS*, J. CLARK† and W. POWRIE*

In situ embedded retaining walls are used to form basements and cuttings in urban areas throughout the world, particularly where space is restricted. It is generally accepted that the process of installing such a wall may influence the stress state of the soil and hence the post-construction displacements, wall bending moments and prop loads. However, wall installation is a complex three-dimensional problem, and its effects are not well understood. This paper describes a study carried out on a section of contiguous bored pile retaining wall in an overconsolidated clay in Kent, England, using instrumentation to monitor total horizontal stresses and pore water pressures before, during and after wall installation. Reductions in horizontal stress were recorded during wall installation, bringing the ratio of horizontal to vertical effective stress, K, from an initial value of about 1 to about 0·8. Following wall installation there was no further change in horizontal stress over an additional period of 10 months during which no further construction work took place.

KEYWORDS: diaphragm and in situ walls; field instrumentation

Dans le monde entier, des murs de soutènement enfouis in situ sont utilisés pour former les fondations et les tranchées dans les zones urbaines, particulièrement là où la place manque. On admet généralement que le processus d'installation de ces murs peut avoir une influence sur l'état de contrainte du sol et donc sur les déplacements post-construction, le moment de flexion du mur et les charges d'étayement. Cependant, l'installation d'un mur est un problème tridimensionnel complexe dont les effets ne sont pas bien compris. Cet exposé décrit une étude effectuée sur une section de mur de soutènement à pieu foré contigu dans une argile surconsolidée du Kent en Angleterre, utilisant les instruments pour suivre les contraintes horizontales totales et les pressions d'eau interstitielle avant, pendant et après l'installation du mur. Des réduction des contraintes horizontales ont été enregistrées pendant l'installation du mur, faisant baisser le taux de contrainte effective horizontale/verticale, K, d'une valeur initiale d'environ 1 à environ 0,8. Après l'installation du mur, il n'y a plus eu de changement de la contrainte horizontale sur une période supplémentaire de 10 mois, pendant laquelle aucun autre travail de construction n'a eu lieu.

INTRODUCTION

It is well known that the stress state of the soil adjacent to an in situ retaining wall following installation may have a significant effect on the behaviour as calculated in a numerical analysis (Potts & Fourie, 1984; Fourie & Potts, 1989; Powrie & Li, 1991), and that the stress state in the surrounding soil is affected by the process of wall installation (Gunn *et al.*, 1993; Symons & Carder, 1993; Gourvenec & Powrie, 1999). Wall installation will affect the recent stress history of the soil and hence its subsequent stress–strain response (Powrie *et al.*, 1998), and in soft soils may cause significant ground movements in its own right (Stroud & Sweeney, 1977; Powrie & Kantartzi, 1996).

Numerical analyses of in situ walls in which the effects of wall installation are neglected generally give much higher bending moments and prop loads than both those obtained using conventional limit equilibrium methods based on fully active pressures behind the wall and those observed directly in the field (e.g. Potts & Fourie, 1984, 1985; Tedd *et al.*, 1984). A better understanding of the changes in stress that occur during installation of a retaining wall could therefore result in more realistic estimates of prop loads, wall bending moments and ground movements than have been possible in the past. Several finite element analyses of in situ retaining wall installation have been reported in the literature (e.g. Ng *et al.*, 1995; Ng and Yan, 1999; Gourvenec and Powrie,

1999), but they generally all model slightly different ground conditions and wall geometries, and all present their results in different ways. It is therefore difficult to draw any general conclusions.

To obtain comprehensive field data on wall installation effects, a section of an in situ retaining wall on part of the Channel Tunnel Rail Link in Kent formed using contiguous bored piles has been instrumented and monitored. Push-in pressure cells (spade cells) were used to measure horizontal stresses and pore water pressures at different depths and at different distances from the wall before, during and after wall installation. Some preliminary data were presented by Clark *et al.* (2004). In this paper the site geology and wall installation sequence are summarised, and further data are presented and interpreted with reference to classical analyses to gain additional insights into the fundamental mechanisms of behaviour involved. Finally, some conclusions are made concerning the effects of wall installation on horizontal total stress distributions.

SITE GEOLOGY

The study was carried out on a section of retaining wall forming part of a propped cutting on part of the Channel Tunnel Rail Link (CTRL) at Ashford, Kent. The main geological deposits at the site are from the Lower Cretaceous Hythe Beds and Atherfield Clay from the Lower Greensand, and Weald Clay from the Wealden formation.

The Weald Clay underlies the site and is up to 120 m thick in the vicinity of the instrumented retaining wall. It is generally a stiff to very stiff clay, in places thinly interbedded with silt. The upper part of the deposit contains many silt laminations and occasional bands of siltstone 100–200 mm thick. Because of the variable silt content, the plasticity index of the Weald Clay ranges between 10% and 30%.

Manuscript received 15 March 2005; revised manuscript accepted 26 April 2006.
Discussion on this paper closes on 1 February 2007, for further details see p. ii.
* School of Civil Engineering and the Environment, University of Southampton, UK.
† Gifford Consulting Engineers; formerly School of Civil Engineering and the Environment, University of Southampton, UK.

An unconformity marks the boundary between the Weald Clay and the overlying Atherfield Clay (Smart *et al.*, 1966). The lower Atherfield Clay is a very stiff, brown clay, about 4·5 m thick with a plasticity index of 20–30%. It is delineated by a distinctive band of light brown material, 400–500 mm thick, at the top. The upper Atherfield Clay is very stiff to stiff, fissured, closely to extremely closely bedded, grey, sandy, and about 8 m thick with a plasticity index of approximately 50%. It contains zones with thin silt partings, which become more frequent towards the bottom of the layer. During sampling, the upper Atherfield Clay fell apart along the fissures into hard peds (i.e. individual natural soil aggregates), which were in places surrounded by a softer matrix of clayey material. The size of the peds reduces with depth as the soil becomes less structured. This observation is consistent with the description of the Atherfield Clay near Ashford given by Smart *et al.* (1966) as a 'grey clay that crumbles into cubical lumps'.

The Hythe Beds overlying the Atherfield Clay outcrop over much of the CTRL Ashford site and are approximately 1·5 m thick at the main instrumented section. They consist of firm to stiff yellow/brown/light grey mottled sandy clay with alternating layers of Ragstone (a hard, sandy limestone) and Hassock (a grey to brownish grey, glauconitic, calcareous sand or soft sandstone). The junction between the Hythe Beds and the Atherfield Clay is not distinct, with the base of the Hythe Beds being generally clayey sand and the top of the Atherfield Clay being sandy and glauconitic (Gallois, 1965; Skempton & Weeks, 1976). The Hythe Beds are overlain by about 2 m of made ground, which comprises firm mottled dark brown/grey organic sandy clay with rootlets, pottery tile and rotten wood fragments.

There is evidence to suggest that there has been historic movement of the upper regions of the Atherfield and possibly the Weald Clay in the area of the instrumented section. Shear planes were observed within the Atherfield Clay and Weald Clay formations during site investigation and construction of the CTRL. A shear plane was observed in the excavation 50 m to the west of the main instrumented section. A discontinuity (possibly a shear plane) was also observed in the base of the excavation at the instrumented section before blinding concrete was laid for the base slab. Movement of the soil by faulting or a landslide may reduce the in situ horizontal earth pressure coefficients to below those that might otherwise be expected for a stiff over-consolidated clay on the basis of its stress history.

A soil profile from a continuous wire-line rotary core sample obtained approximately 30 m to the north-west of the instrumented section, including representative data of liquid limits, plastic limits and moisture contents, is shown in Fig. 1. A slight variance in the base levels of the geological strata from those encountered at the instrumented section was noted. The moisture content is generally at or near the plastic limit, and there is a clear change in plasticity within the upper Atherfield Clay. The bulk density of the Atherfield and Weald Clays was determined, from undisturbed block samples taken during the main site investigation for the CTRL and from samples taken from the excavation, as 21 kN/m^3.

Figure 2 shows undrained shear strengths (c_u) as a function of depth z in metres below original ground level, determined from unconsolidated undrained triaxial tests on 100 mm diameter samples and estimated from standard penetration test (SPT) results. The 'undisturbed' 100 mm diameter samples were obtained using light percussion boring techniques and open drive sampling equipment. The SPT blowcount N multiplied by 4·5 kPa (Stroud, 1974) correlates closely with the laboratory data for the Atherfield Clay, but gives values generally in excess of the laboratory tests for

the Weald Clay. The best-fit line to all the data in the Atherfield Clay is indicated, and is given by the equation

$$c_u \text{ (kPa)} = 22 + 7z \tag{1}$$

WALL INSTALLATION SEQUENCE

The contiguous bored pile retaining wall was formed from piles approximately 20 m long and 1·05 m in diameter, spaced at 1·35 m centres, and was installed through the instrumented section in November and December 1999. In most cases the uppermost 8 m of the pile bore was excavated during one working day, following which a casing was inserted to support the sides. The remainder of the bore was excavated the next working day without any further form of support, although on reaching the required depth bentonite slurry was introduced up to original ground level. The reinforcement cage was then lowered into the hole, and concrete tremied in from the base at the same rate that bentonite was removed. (This sequence was implemented after observation of trial boreholes left open for several days showed that those in the Atherfield Clay would remain stable for at least 24 h whereas deterioration occurred more quickly in the Weald Clay; Roscoe & Twine, 2001.) The exception to this installation sequence was a single pile in stage D1 for which excavation and concreting was carried out within one afternoon, immediately after stage D. Fig. 3 shows the layout on plan of the instrumentation and the nearby piles. The lettering on the piles indicates the sequence in which the piles were installed, and is detailed in Table 1. The pile installation dates are identified in days, where day 1 was 8 October 1999, the day the first spade cell was installed.

INSTRUMENTATION

Sixteen Soil Instruments vibrating-wire spade cells with integral vibrating-wire piezometers were installed to measure the total horizontal stresses and pore water pressures near the retaining wall, as indicated in Figs 3(a) and 3(b). Each spade cell consists of a pointed rectangular spade-shaped flat cell about 7 mm thick and 100 mm wide, formed from two sheets of steel welded around the edge (Fig. 4). The narrow gap between the plates is filled with oil and connected to a vibrating-wire pressure transducer, forming a closed hydraulic system. This results in a high cell stiffness—5·44 MPa for the instrument used in this study (Richards *et al.*, forthcoming)—so that any under-read due to cell over-compliance (cell action factor effects) is likely to be small (Peattie & Sparrow, 1954; Clayton & Bica, 1993). Each spade cell was installed by drilling a vertical borehole to approximately 0·5 m above its final installation depth and then pushing it into the ground, taking great care to ensure that it was aligned to measure the horizontal stress perpendicular to the wall. It may be demonstrated with reference to Fig. 11 (see later) that, provided the spade cells are not misaligned by more than 9°, the error in horizontal stress measurement will be less than 5%. Readings were recorded every 2 h throughout the period of wall installation, using a Campbell Scientific CR10X datalogger.

Spade cells installed in stiff overconsolidated clays over-estimate the magnitude of the in situ horizontal stress because of the complex local stress field created in the ground during installation (Tedd & Charles, 1981). By comparing the readings of spade cells installed to measure vertical stress with the over-burden and measurements of horizontal stress from spade cells and self-boring pressuremeter tests, Tedd & Charles (1983), Tedd *et al.* (1984) and Ryley & Carder (1995) determined empirical corrections of generally between 0·5 and 0·8 times the undrained shear

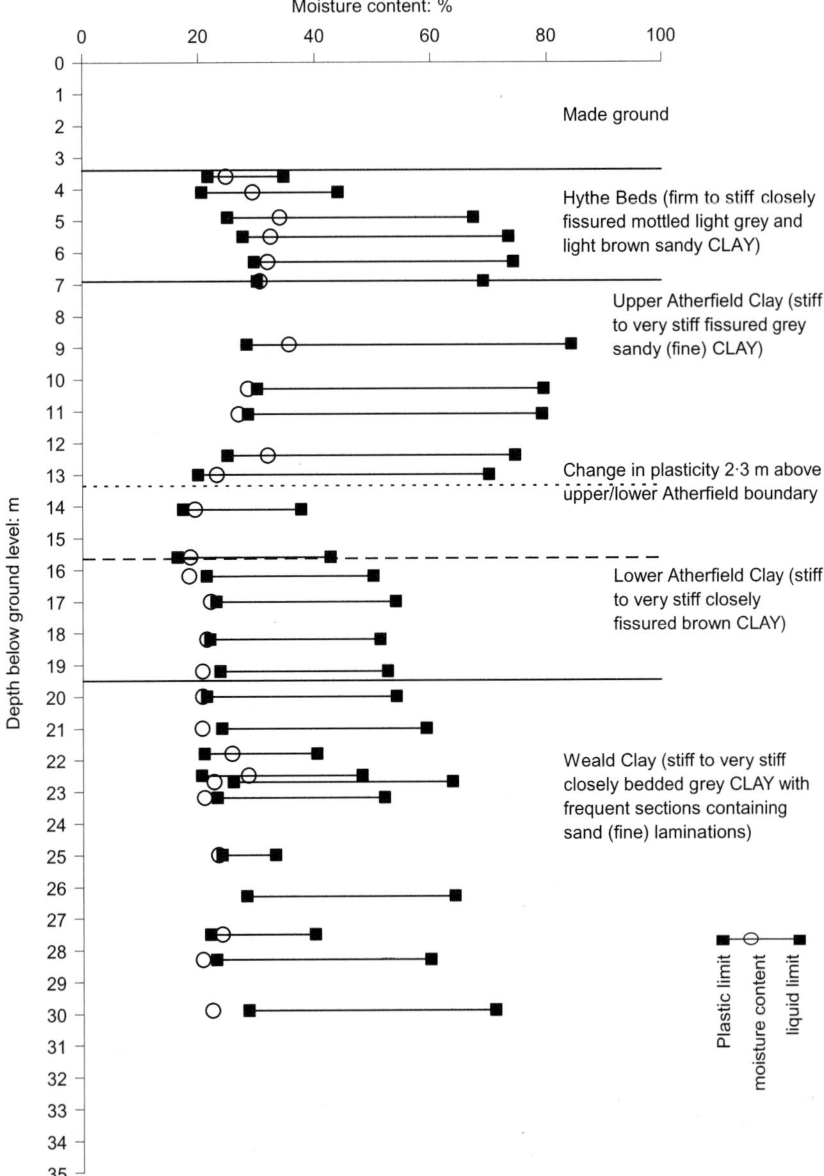

Fig. 1. Geotechnical profile including liquid and plastic limit data

strength c_u (increasing with c_u) for the London Clay. In this study, the raw spade cell data have been corrected for over-read due to installation effects by subtracting $0.35c_u$, as determined in a calibration exercise in the Atherfield Clay at the CTRL Ashford site (Richards *et al.*, forthcoming).

Following wall installation in November and December 1999, no major construction activities were carried out in the instrumented area until 20 September 2000. In general the spade cells were very reliable, with the exception of piezometer P9 and total stress cell S10, which read only intermittently. Unfortunately, because of the congested linear nature of the site, it was not possible to install inclinometers or other instruments to measure horizontal movements of the ground during wall installation. However, inclinometers were installed in selected piles to measure post-excavation wall movements.

RESULTS

Stabilisation of spade cells following insertion

Figure 5 shows the corrected readings of total horizontal stress and pore water pressure during spade cell insertion and for a period of one month afterwards. (Readings from spade cell or piezometer number n are referred to as Sn and Pn respectively. The graph is plotted with the time in days as described above.) Because of construction activities it was necessary to re-site the data-logger between day 13 and day 33, corresponding to the period of missing data highlighted in Fig. 5. The excess pore water pressures induced during spade cell insertion had dissipated and readings of total horizontal stress had stabilised prior to the installation of the bored pile wall.

In situ total horizontal stresses and pore water pressures

Figure 6 shows the corrected, stabilised readings of total horizontal stress and pore water pressure from all spade cells and piezometers, plotted against depth below original ground level. The initial in situ pore water pressures were slightly less than hydrostatic beneath a water table 1·2 m below the original ground surface in the upper Atherfield Clay. In the lower part of the stratum, the deviation of the pore water pressures from hydrostatic was more pronounced. This distribution of pore water pressure with depth is consistent with

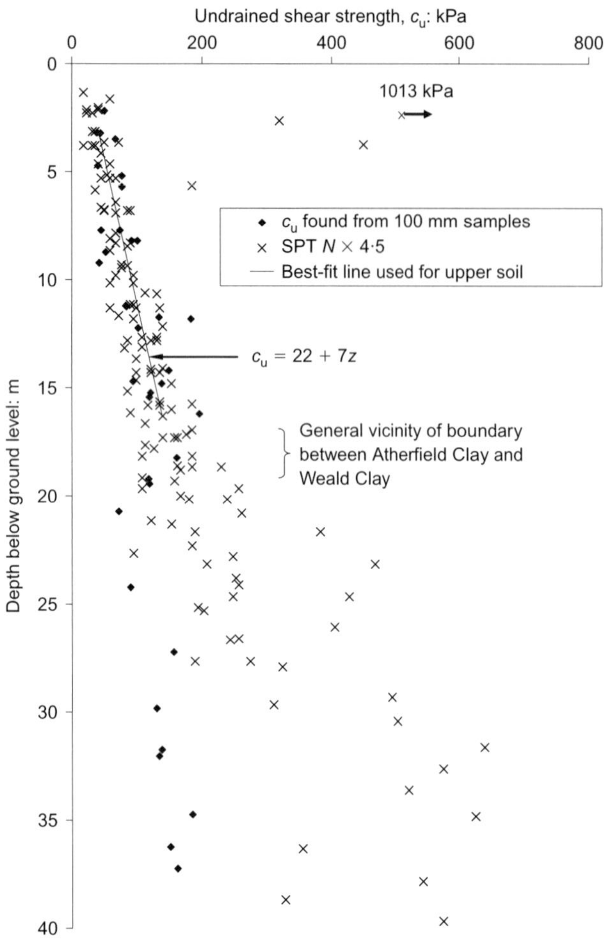

Fig. 2. Profile of undrained shear strength (SPTs and laboratory tests)

underdrainage of the Atherfield Clay into the underlying Weald Clay, within which an ejector well pore pressure control system was in operation about 100 m from the instrumented section (Roscoe, 2003).

The variation of in situ total horizontal stress σ_{h0} with depth z (below original ground level) approximated as a linear function was

$$\sigma_{h0} = 0.8 + 20.6z \tag{2}$$

Effective stress profile

The profile of effective horizontal stress σ'_{h0} with depth is shown in Fig. 7. The implied in situ horizontal earth pressure coefficient K_0 ($= \sigma'_{h0}/\sigma'_{v0}$) is generally within the range 0.7 to 1.5, with an average of 1.04. This is rather less than would be expected on the basis of the one-dimensional stress history of the deposit, possibly as a result of large-scale movements due to faulting or landslip. However, it is consistent with other measurements at the site, made using a self-boring pressuremeter (Clark *et al.*, 2004).

The best-fit linear approximation to the in situ effective horizontal stress below the groundwater level of 1.2 m below ground level is

$$\sigma'_{h0} = 15.3 \, (z - 1.2) + 10.9 \quad \text{(for } z \geqslant 1.2\text{m)} \tag{3}$$

Pore water pressure changes during and after wall installation

Figure 8 shows the variations in pore water pressure measured during the period of wall installation, together with the

periods of installation of each pile group. Detailed interpretation of the data is complicated by the sequential nature of pile installation, together with the fact that the spade cells could not all be installed at one cross-section but had to be spread along the length of the wall. However, the general response of each transducer comprises a reduction in pore water pressure on excavation followed by an overcompensating increase in pore water pressure on concreting. The pore water pressure then gradually falls to the original value over a relatively short period of time (in this case less than a day) as the concrete begins to set. This response is similar to that measured in the field by Symons & Carder (1993) and in centrifuge model tests by Powrie & Kantartzi (1996). The magnitudes of the measured changes vary, but generally reduce with distance from the pile being installed. The changes in pore pressure were consistently most pronounced at piezometer P4: the reason for this is unknown.

Pile installation may be idealised as an expanding or contracting circular cylindrical cavity. Analysis following Gibson & Anderson (1961) shows that, if it remains elastic, the soil surrounding the cavity deforms at constant volume and constant average total stress $p = (\sigma_r + \sigma_\theta + \sigma_v)/3$. In an elastic medium, deformation at constant volume occurs at constant average effective stress p': hence while p remains constant there should be no change in pore water pressure. However, a reduction or increase in cavity pressure (from the in situ condition) equal to the undrained shear strength of the soil, c_u, is sufficient to cause yield and hence potential changes in pore water pressure. In the present case, the in situ total horizontal stress (given by equation (2), $\sigma_{h0} = 0.8 + 20.6z$) is greater than the undrained shear strength (given by equation (1), $c_u = 22 + 7z$) for depths z greater than 1.6 m. Unloading the pile bore from an initial in situ total horizontal stress σ_{h0} to a cavity support pressure P [where $(\sigma_{h0} - P) \geqslant c_u$] will cause yielding of the surrounding soil within a plastic radius r_p given by

$$r_p = R \exp\left(\frac{\sigma_{h0} - c_u - P}{2c_u}\right) \tag{4}$$

where R is the cavity radius. In the present case, taking the cavity support pressure $P = 0$, the cavity radius $R = 0.525$ m and the profiles of in situ total horizontal stress and undrained shear strength with depth given by equations (1) and (2) respectively, the width of the plastic zone ($r_p - R$) varies from zero at a depth of 1.6 m to 0.6 m at the bottom of the pile bore.

Even the closest pore water pressure transducers were outside the theoretical plastic zone, and might therefore have been expected to show no change in pore water pressure. However, it is well known that soil does not behave as an ideal elastic material. Also, the neglect in this simple analysis of both the potential for vertical load transfer to the soil below the pile toe and the effect of any piles already installed introduces further degrees of approximation. Nonetheless, the measured pore water changes are generally not large, and decrease rapidly with distance from the wall as the effects of the approximations become less significant.

For a wall constructed from diaphragm wall panels rather than individual piles, a greater pore water pressure response would be expected. Excavation in plane strain results in a reduction in average total stress p of half the reduction in horizontal pressure within the trench, in the soil close to the trench (Powrie & Kantartzi, 1996). If the soil is again assumed to deform elastically with $p' = $ constant in undrained conditions, the change in pore water pressure would be expected to be the same as the change in p: that is, half the reduction in horizontal stress within the trench. Powrie & Kantartzi (1996) report centrifuge tests modelling dia-

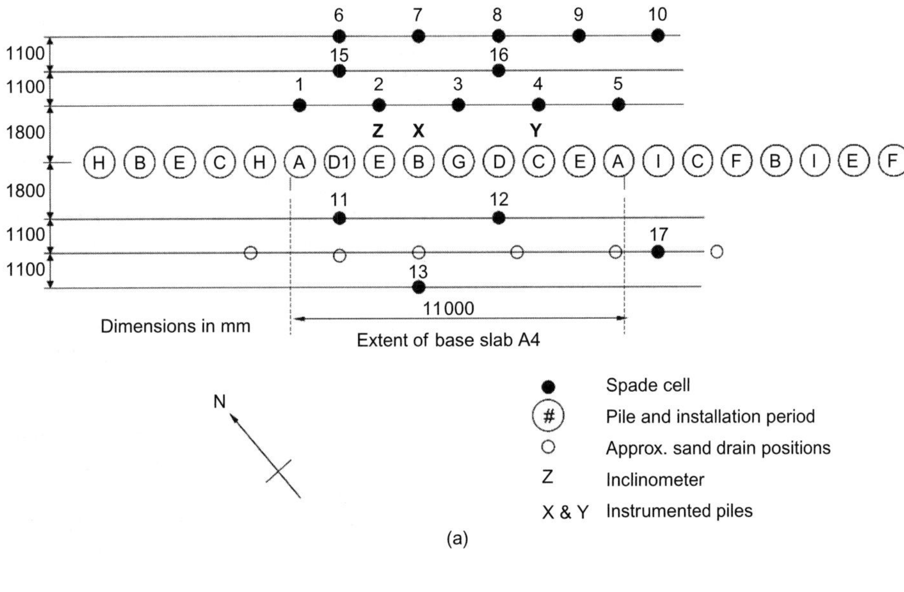

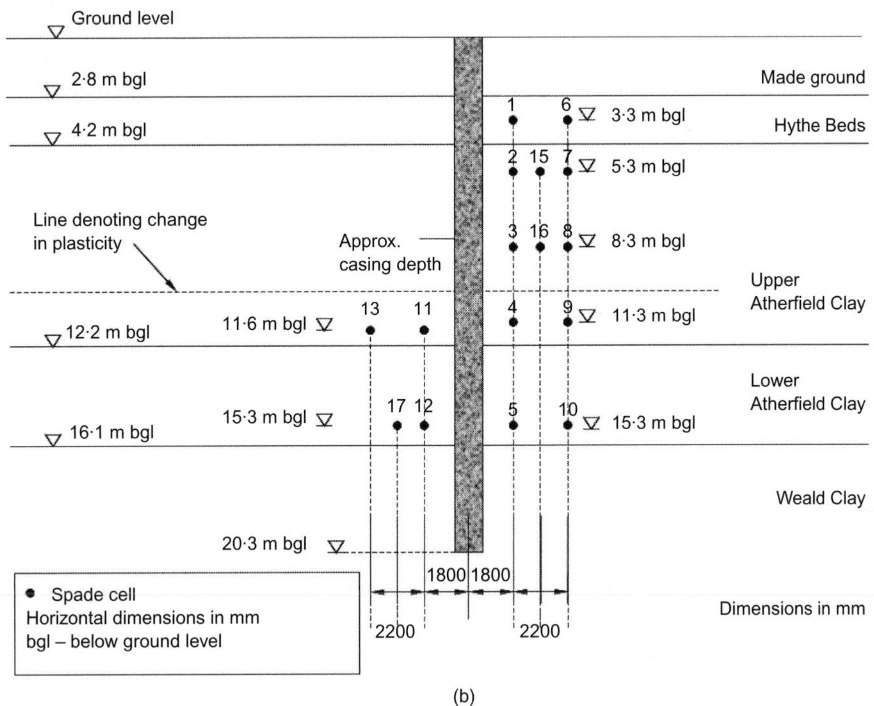

Fig. 3. Spade cell locations: (a) plan; (b) elevation

phragm wall panel installation in which the measured changes in pore water pressure clearly reduced with reducing panel length. However, the difference in pore water pressure response between bored pile and diaphragm walls is not so clear from the field observations described, but not presented in any great detail, by Symons & Carder (1993).

Figure 9 shows the measured pore water pressures plotted against time for periods before, during and 10 months after wall installation. When viewed in this way, the fluctuations in pore water pressure due to pile excavation and concreting are little more than noise, and the data show clearly that the overall net effect of wall installation on the in situ pore water pressures was negligible. This is consistent with the observations of Symons & Carder (1993) and Powrie & Kantartzi (1996), and confirms that, in an overconsolidated clay, the overall effect of bored pile or diaphragm wall installation on the in situ pore water pressures may reasonably be neglected. In the 10 months following wall installa-

tion the pore water pressure slowly reduced at the positions of many of the spade cells. The pore water pressure began to fall in March (approximately day 150), probably because of seasonal variations and/or construction dewatering being carried out about 500 m from the instrumented section.

Horizontal total stress changes due to wall installation

Figure 10 shows the total horizontal stresses measured during installation of the wall. As with the pore water pressures, interpretation is complicated by the sequential installation of the piles and the staggered pattern of the spade cells along the length of the wall. Nonetheless, the influence of nearby pile construction on the individual traces is clear: generally, a reduction in total horizontal stress on excavating at least the lower portion of the pile, and an increase in total horizontal stress on concreting. Overall, pile installation resulted in a reduction in total horizontal stress

Table 1. Sequence of pile installation

Installation period	Boring started		Casing depth reached, casing inserted	Boring restarted		Pile completed, bentonite added	Concreting started	Concreting ended
	Day	Time	Time	Day	Time	Time	Time	Time
A	47	1510–1550	1540–1620	48	1155–1255	1235–1330	1420–1550	1453–1638
B	49	1440–1720	1600–1740	50	0805–0916	0845–1001	0916–1225	1002–1311
C	53	1459–1625	1520–1646	54	0800–0934	0835–1000	0955–1227	1032–1315
D	55	1518	1542	56	0910	1000	1109	1150
D1	56	1220	1255	56	1255	1350	1626	1724
E	57	1150–1435	1218–1500	60	0906–1310	0941–1335	1112–1540	1156–1627
F	64	1040–1108	1130–1200	67	0810–0842	0842–0905	0950–1114	1036–1201
G	67	1715	1746	68	0848	0908	1058	1145
H	69	1631–1659	1657–1721	70	0810–0841	0840–0905	0933–1040	1006–1117
I	71	1135–1203	1200–1225	74	1120–1148	1146–1210	1430–1605	1545–1705

In each installation period, between one and four piles were installed. Where more than one pile was constructed in a period the range indicates the entire time over which pile installation processes were being undertaken.

to below the in situ value, the magnitude of the reduction generally decreasing with distance from and along the wall.

The changes in stress measured due to the installation of a single pile can be compared with values calculated using an elastic analysis, assuming as an upper limit that the change in stress at the pile bore is from the in situ stress to zero (as the pile was initially bored with no support). The change in radial stress $\Delta\sigma_r$, at a distance r from the centre of the pile, is related to the change in stress at the pile bore, $\Delta\sigma_{PB}$, and the radius R of the pile bore by

$$\Delta\sigma_r = \frac{R^2}{r^2}\,\Delta\sigma_{PB} \text{ for } r > R \tag{5}$$

(See for example Fjaer et al., 1992.)

The installation of any pile affects the stress state of the ground at the positions where further piles are to be installed. At the centres of piles yet to be installed, the stress in the direction perpendicular to the wall increases (due to arching) and the stress in the direction along the wall decreases. Calculation of the changes in stress seen by a spade cell due to the installation of piles after the first pile is therefore more complicated. A lower limit to the change in stress may be estimated by assuming that the in situ stress at the position of the piles not yet installed is reduced (as happens to the stress in the direction along the wall). An upper limit may be estimated by assuming that the in situ stress increases (as happens to the stress in the direction perpendicular to the wall).

Consideration of the Mohr circle of stress shows that the change in horizontal stress acting on the plane of a spade cell, $\Delta\sigma_{SC}$, due to the installation of any pile is given by

$$\Delta\sigma_{SC} = \Delta\sigma_r \times \cos 2\theta \tag{6}$$

where θ is the angle between the normal to the wall and the line joining the pile to the spade cell (Fig. 11) and $\Delta\sigma_r$ is the change in radial stress at the spade cell (as given by equation (5)).

Figure 12 shows the measured and calculated changes in stress, normalised with respect to the in situ stress, plotted as a function of the distance along the wall (in pile spacings) between the spade cell and the nearest pile being installed in a particular pile installation period. Because of the complexity of the installation sequence, the following rules have been adopted.

(a) Where two piles were installed simultaneously at an equal number of pile spacings from a spade cell, the change in stress measured at the spade cell attributed to the installation of a single pile was assumed to be half the total. For example, two piles, each four pile spacings from spade cell 3, were installed during Period A, so half the measured total stress change has been attributed to each.

(b) Where two piles were installed simultaneously at distances from a spade cell differing by one or two pile spacings, the change in stress has been halved and assigned to a pile at the average distance from a spade cell. For example, piles were installed two and three pile spacings from spade cell 12 in Period E, so half the measured total stress change has been plotted as a change due to pile installation 2·5 pile spacings away.

(c) Where in any installation period the difference in the distance between the spade cell and the nearest pile installed and the second nearest pile was more than two pile spacings, the entire change in stress has been assigned to the nearest pile. For example, the two piles installed in period A were two and six pile spacings from spade cell 2, so the change measured at spade cell

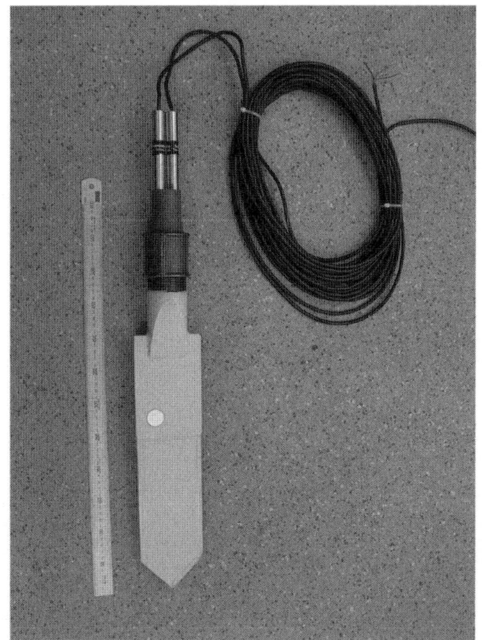
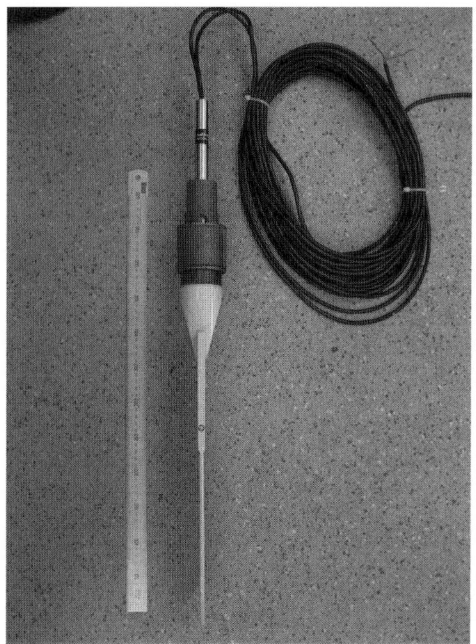

Fig. 4. Spade cell

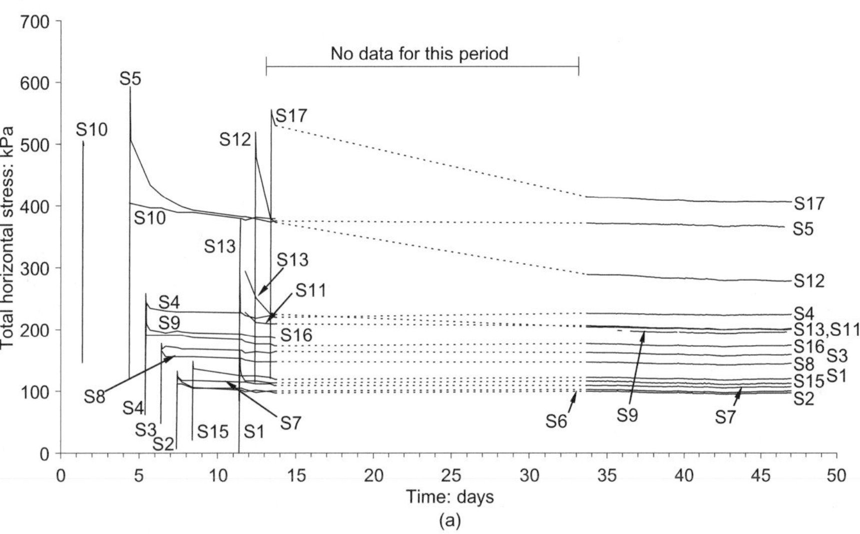

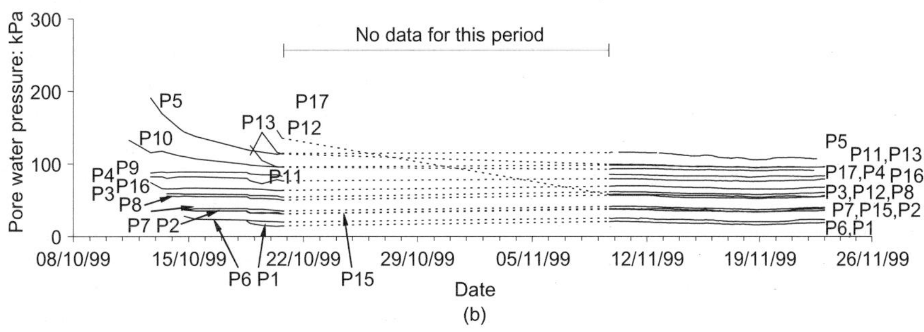

Fig. 5. Spade cell measurements taken between their installation and wall installation: (a) total horizontal stresses; (b) pore water pressures

2 during period A has been plotted as a change due to a pile installation two pile spacings away.

The change in stress was calculated over the period from 2 h before the pile installation process began to 2 h after it finished. Fig. 12 shows that the reduction in total horizontal stress measured on installation of the pile in line with a spade cell was approximately 0–20% of the in situ value 1·275 m from the edge of the wall, 3–9% 2·375 m from the edge of the wall, and 1 to 5% 3·475 m from the edge of the wall.

Figure 12 shows that the measured reduction in stress due

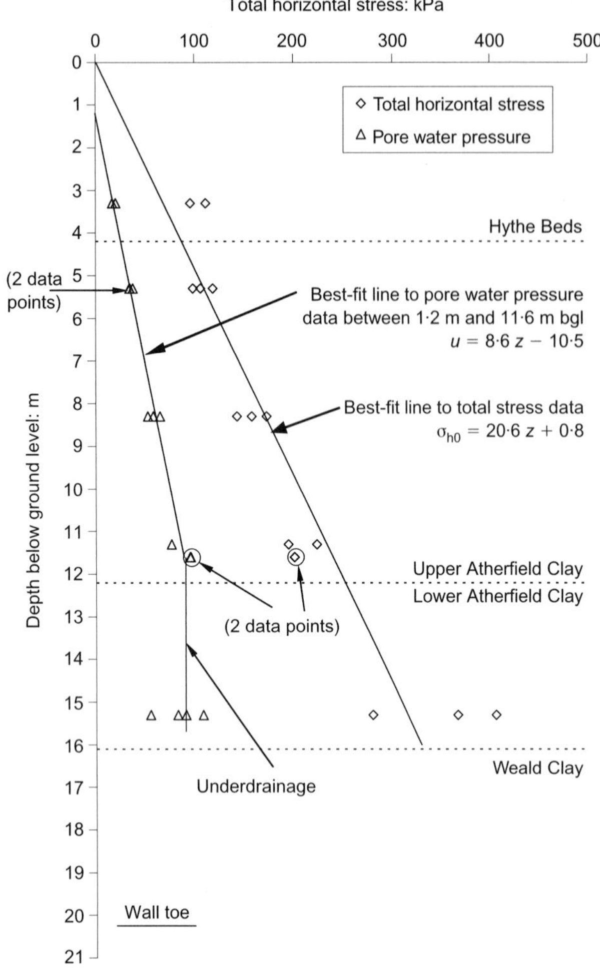

Fig. 6. Stabilised readings of total horizontal stress and pore water pressure from all spade cells and piezometers, plotted against depth below original ground level (before wall installation)

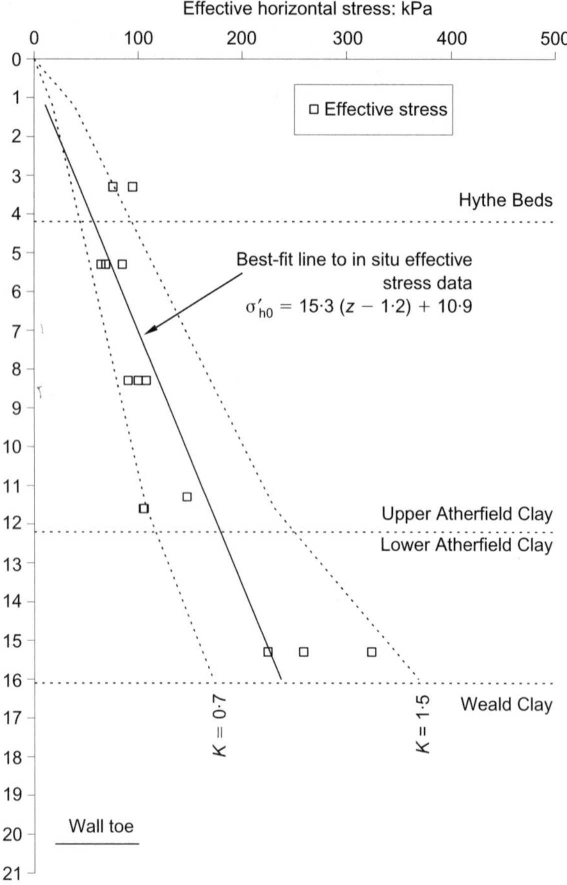

Fig. 7. Original in situ effective horizontal stress σ'_{h0} against depth (obtained by subtracting measured pore pressure from total horizontal stress at each location)

to pile installation is generally larger than that calculated using the simple elastic analysis—even assuming zero pressure (rather than the pressure of bentonite or wet concrete) in the pile bore and an in situ stress corresponding to the upper limit value resulting from the installation of previous piles. This could be a result of shrinkage of the concrete (possibly due to thermal effects) as it sets, and/or vertical arching and stress transfer below the bottom of the bore, which is not taken into account in the simple analysis. The underestimation of the measured reductions in horizontal stress is confirmed by Fig. 13, which shows the calculated and average measured reductions in stress (expressed as a percentage of the in situ value), for piles directly opposite the spade cell (i.e. in Fig. 12 a distance of zero pile spacings along the wall).

Figure 14 shows the measured in situ and post-installation total horizontal stresses at a distance of 1·275 m from the wall. The best-fit linear approximation to the in situ total horizontal stress is given by equation (2). The best-fit linear approximation to the post-installation total horizontal stress measured by the instruments 1·275 m from the wall is

$$\sigma_h = 7·4 + 16·7z \tag{7}$$

On the basis of the best-fit linear approximations, the reduction in total horizontal stress ranges from 10% at the top of the Atherfield Clay to 17% 15·3 m below ground

level. However, this masks the true nature of the variation with depth of the change in horizontal effective stress, which is much greater (about 30%) at the mid-depth of the wall than at ground level or at the toe (approximately 4%), as shown in Fig. 15. This is consistent with the transfer of lateral stress to the soil below the wall (sometimes referred to as vertical arching), which results in an increase in horizontal stress below the toe, as shown by Ng *et al.* (1995). It also suggests that an attempt to quantify the effects of wall installation as a uniform with depth percentage reduction may be too much of an over-simplification.

Effective stress changes due to wall installation

Figure 16 shows all the in situ effective horizontal stresses and the post-installation effective horizontal stresses measured 1·275 m from the wall.

The best-fit linear approximation to the post-installation effective horizontal stress below groundwater level is

$$\sigma'_h = 11·6\,(z - 1·2) + 5·5 \quad \text{(for } z \geqslant 1·2\text{m)} \tag{8}$$

The change in effective horizontal stress calculated from the linear best-fit approximations to the measurements is 30% 3·3 m below ground level, falling to 25% 15·3 m below ground level (Fig. 17). However, the linear approximations again mask the true pattern, and the actual measurements show a similar trend to the total stress changes (Fig. 15) with the greatest reduction occurring at 8·3 m below ground level. The percentage changes in effective stress are greater

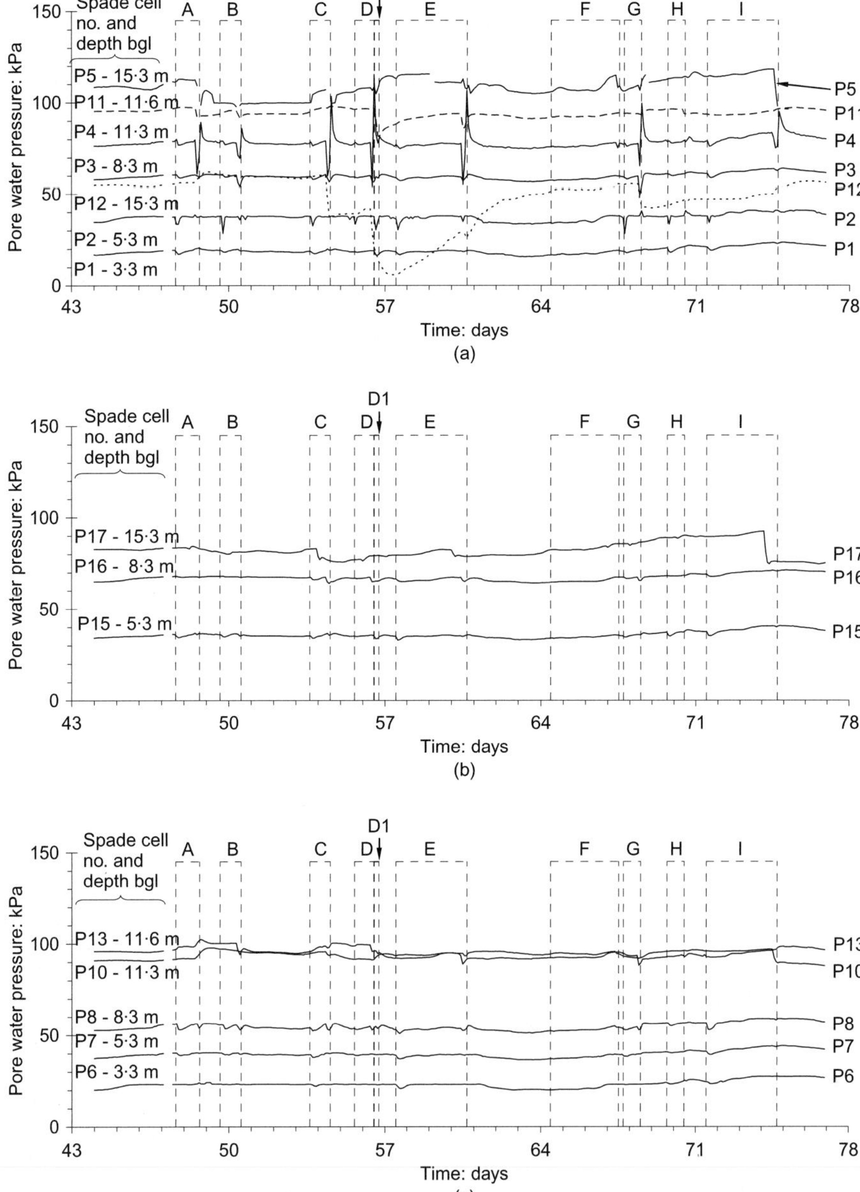

Fig. 8. Piezometer readings over the period of wall installation: (a) 1·275 m from wall; (b) 2·375 m from wall; (c) 3·475 m from wall

than the percentage changes in total stress because there was no overall change in pore water pressures.

Current guidance given in CIRIA Report C580 (Gaba *et al.*, 2003) is that installation of a bored pile wall might be expected to reduce the in situ earth pressure coefficient by about 10%. This is based largely on work by Symons & Carder (1993), who instrumented a contiguous bored pile wall comprising piles 1·5 m diameter and 24 m long, installed in London Clay, and measured a reduction in lateral earth pressure coefficient K from 2·1 to 1·9 ($\sim$10%). At Ashford the lateral earth pressure coefficient reduced from 1·04 to approximately 0·8 ($\sim$25%), on the basis of the linear idealisations to the data indicated in Fig. 16. Although there appears to be a large discrepancy between these changes when expressed in percentage terms, they are consistent with a similar absolute reduction in horizontal total or effective stress of about $0{\cdot}25\gamma z$ in each case (where γ is the bulk density of the soil). This suggests that the likely reduction in earth pressure due to in situ wall installation could be estimated in absolute terms. Given that the process might be expected to be controlled by the boundary stresses at the pile bore this is perhaps surprising, but may be more credible than an approach based on a reduction in K_0 in percentage terms. However, it has already been noted that idealisation of the percentage stress change as uniform with depth may be inappropriate.

Horizontal total and effective stress changes after wall installation

Concern is sometimes expressed that the in situ horizontal stresses become re-established in the long term. In this case, measured total horizontal stresses showed no tendency to change for a period of 10 months after wall installation (Fig. 18), suggesting that the reductions in horizontal stress due to wall installation are permanent. This is consistent with long-term observations of horizontal pressures behind several in-service retaining walls, which show no significant change in

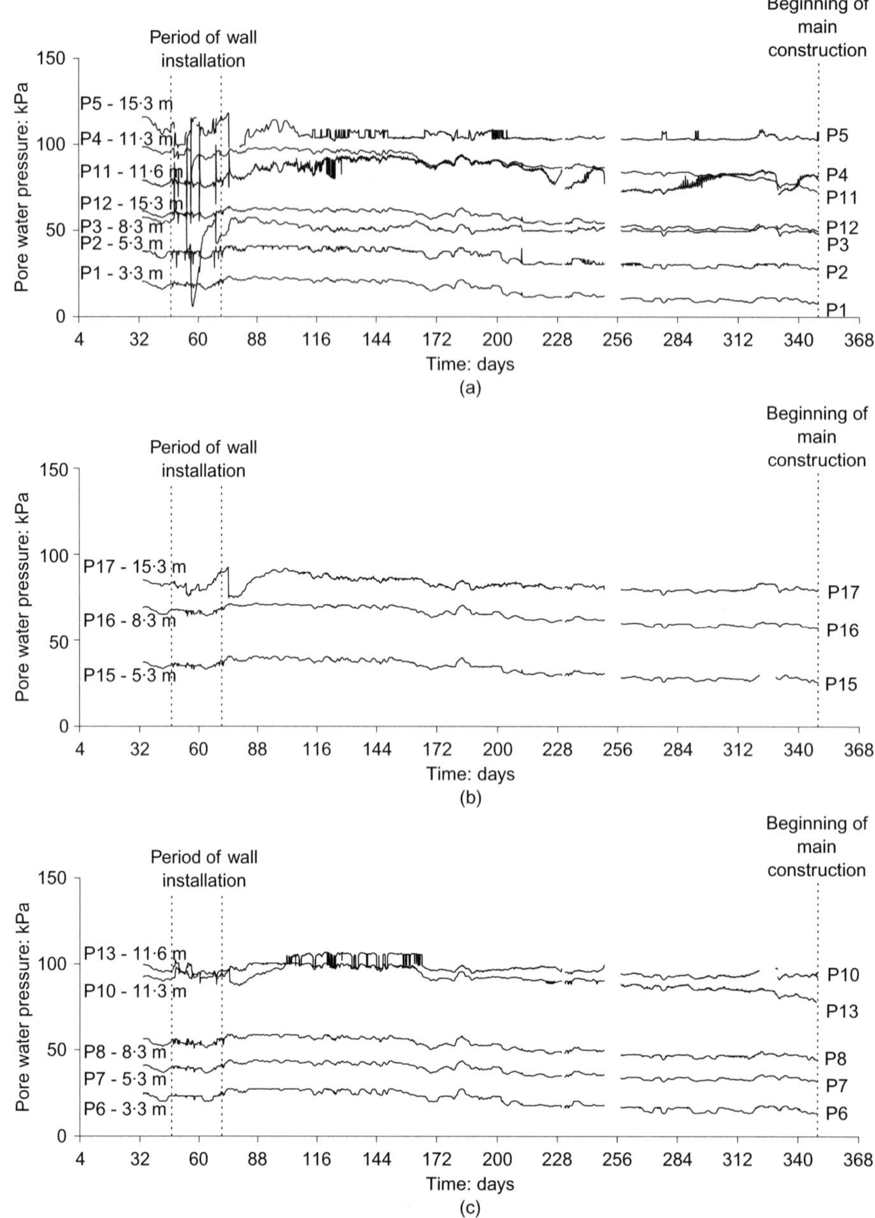

Fig. 9. Pore water pressures measured before, during and 10 months after wall installation: (a) 1·275 m from wall; (b) 2·375 m from wall; (c) 3·475 m from wall

total horizontal stress over periods of between 5 and 24 years after construction (Carder & Darley, 1998).

Although the in situ earth pressure coefficient of the Atherfield and Weald Clay at this site is low relative to other overconsolidated clay deposits, the reduction in total horizontal stress was similar to that for a clay with $K_0 = 2$ excavated under bentonite, assuming a water table at ground level.

CONCLUSIONS

Measurement of the changes in total horizontal stress and pore water pressure during installation of a contiguous bored pile retaining wall forming part of the Channel Tunnel Rail Link at Ashford has shown that:

(*a*) Pore water pressures fell during pile excavation and then increased during concreting. The magnitude of the changes decreased with increasing distance from the pile. In theory, the change in pore water pressure

should be zero for a circular cylindrical cavity if the surrounding soil behaves elastically. However, different geometries would produce different theoretical results: for example, in plane strain $\Delta u = 0.5\Delta\sigma_h$ close to the trench.

(*b*) Overall, wall installation had no effect on pore water pressures, which rapidly returned to their in situ values on completion of the process. This is in agreement with previous observations for both diaphragm and bored pile walls. In contrast, there was a very clear reduction in the horizontal effective stresses and lateral earth pressure coefficients.

(*c*) The in situ total horizontal stresses did not re-establish during the 10 months that elapsed between wall installation and excavation in front of the wall.

(*d*) Although the soil almost certainly yielded during installation, the size of the plastic zone was small. However, the measured changes in horizontal stress due to individual pile installation were generally greater than calculated in a simple elastic analysis, even

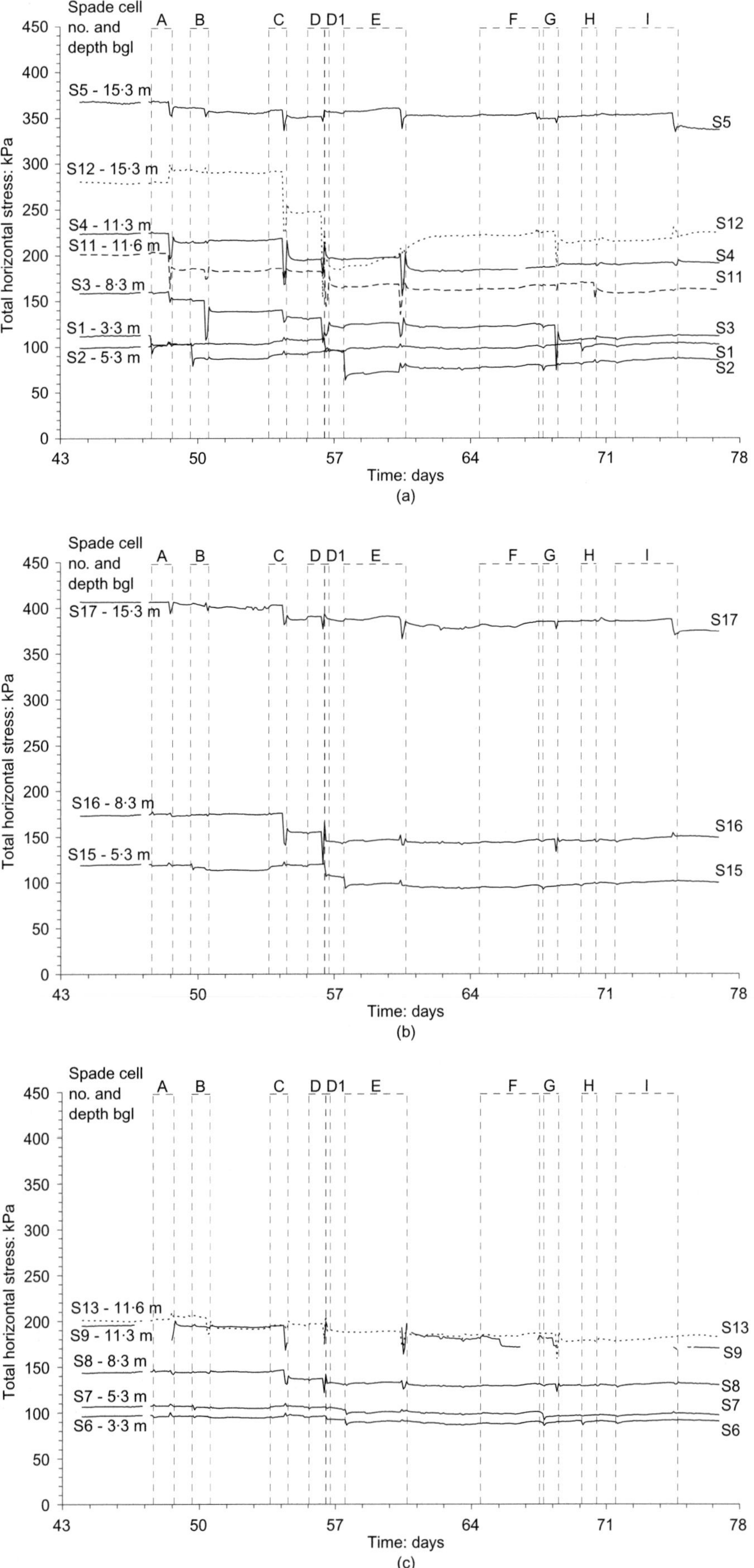

Fig. 10. Total horizontal stress measurements during the period of wall installation: (a) 1·275 m from wall; (b) 2·375 m from wall; (c) 3·475 m from wall

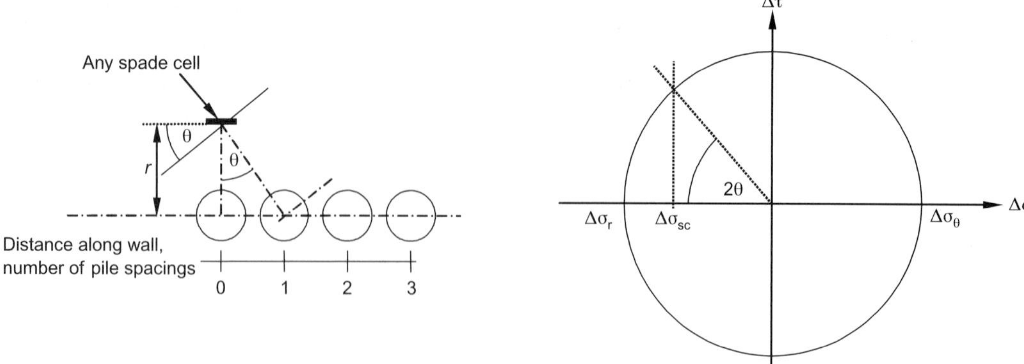

Fig. 11. Pile installation sequence for elastic analysis, and Mohr circle showing calculation of correction for stress change measured on spade cell

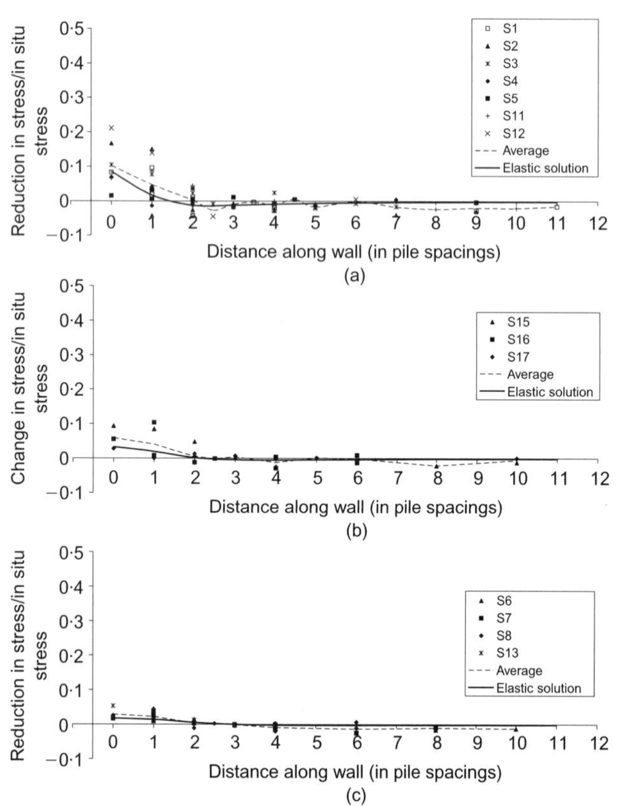

Fig. 12. Measured reduction in total horizontal stress normalised with respect to in situ total horizontal stress (showing prediction from elastic analysis): (a) 1·275 m; (b) 2·375 m; (c) 3·475 m from wall (see Figs 9 and 10)

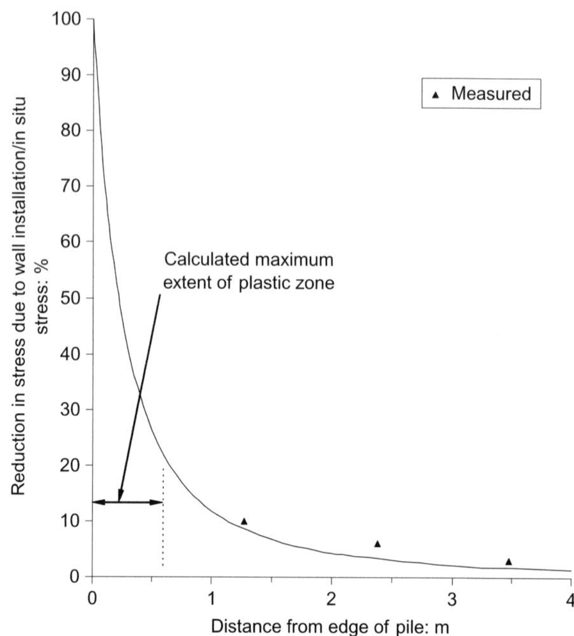

Fig. 13. Reduction in total horizontal stress due to installation of pile nearest to spade cell only with distance from wall compared with elastic prediction

total horizontal stress that is uniform with depth may be an over-simplification.

assuming zero pressure at the pile bore rather than the pressure of bentonite or wet concrete. This may have been due to shrinkage of the concrete due to thermal effects, and/or vertical arching not taken into account in the analysis.

(e) The profiles of reduction in total and effective horizontal stress were not linear with depth, but greatest at the mid-depth of the pile. This is consistent with the transfer of lateral stress to the soil below the wall (sometimes referred to as vertical arching). In this case study, wall installation caused a reduction in total horizontal stress varying from about 4% at the top and the bottom of the wall to approximately 30% at mid-depth. Attempting to quantify the effects of wall installation as a percentage reduction in

ACKNOWLEDGEMENTS
The work described in this paper was carried out with the support of the Engineering and Physical Sciences Research Council (Grant GR/M95011), Rail Link Engineering, Union Railways and Skanska. The authors are particularly grateful to David Twine, Howard Roscoe, Dr Gary Holmes and the site staff at Contract 430 for their support, encouragement and help at various stages of the project.

NOTATION

K — ratio of horizontal to vertical effective stress
K_0 — in situ ratio of horizontal to vertical effective stress
c_u — undrained shear strength
p — constant average total stress
p' — constant average effective stress

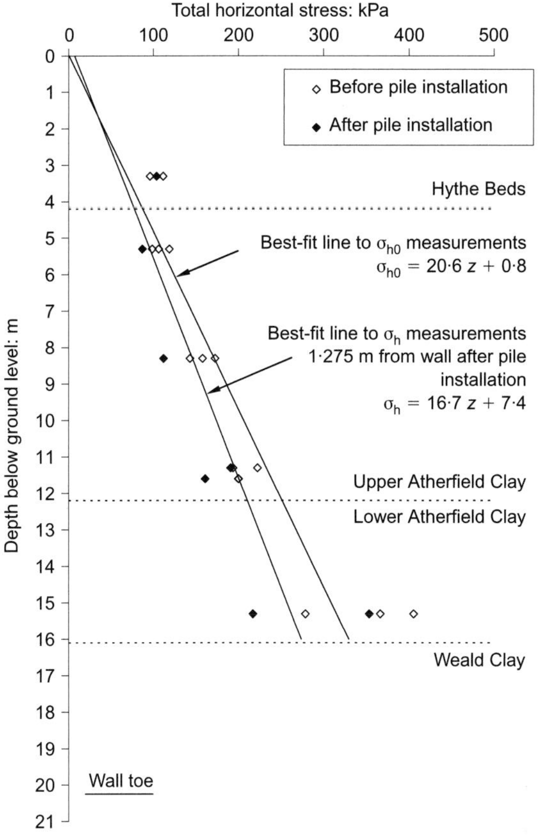

Fig. 14. Total horizontal stresses measured by all spade cells before wall installation and by those 1·275 m from edge of wall after wall installation

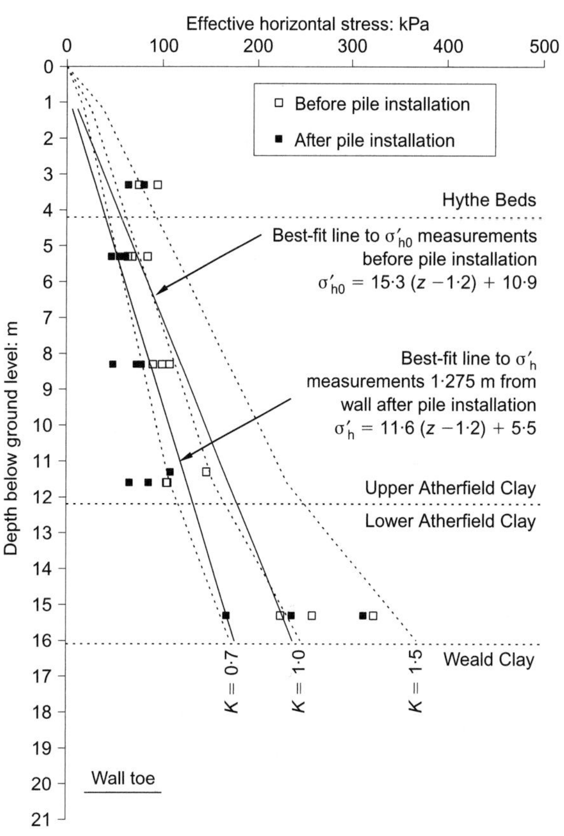

Fig. 16. Effective horizontal stresses measured by all spade cells before wall installation and by those 1·275 m from edge of wall after wall installation

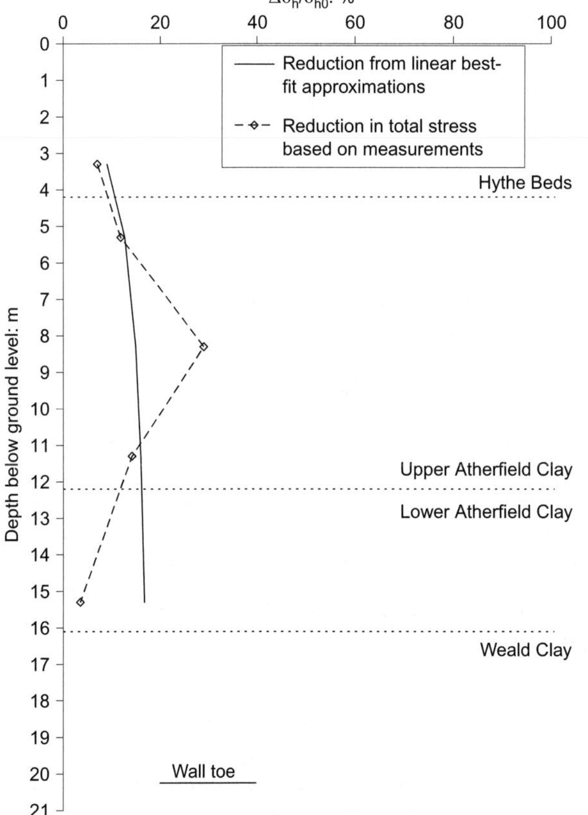

Fig. 15. Change in total horizontal stress

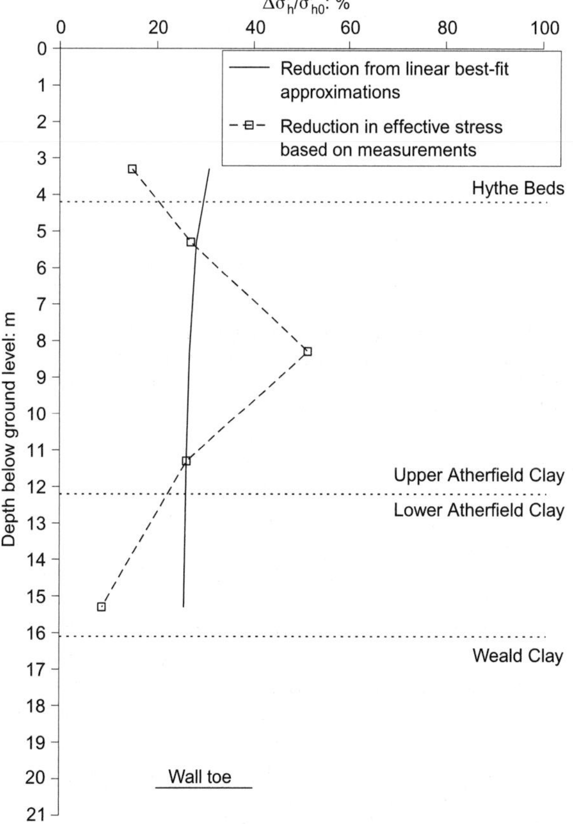

Fig. 17. Change in effective horizontal stress

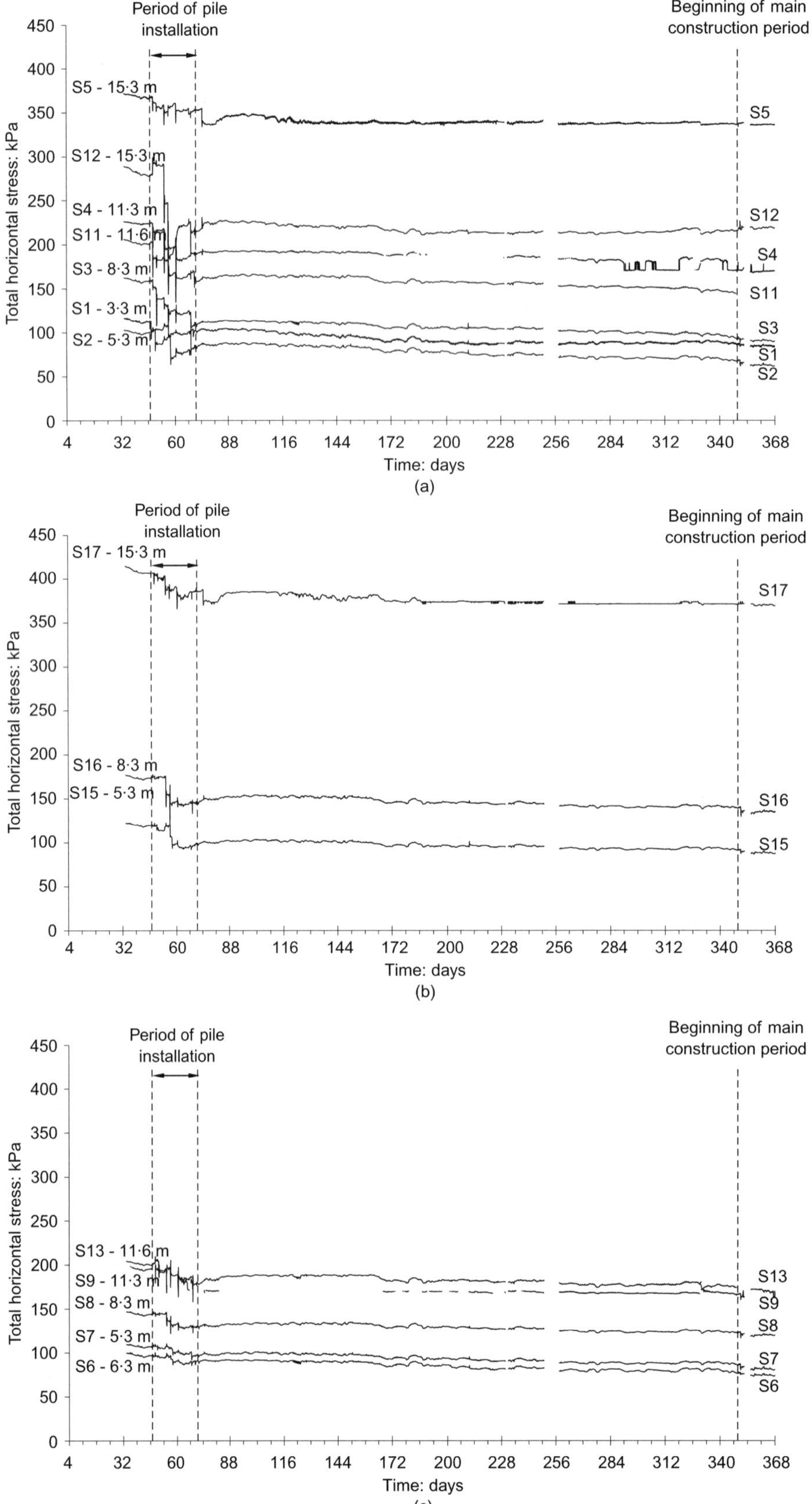

Fig. 18. Total horizontal stress measured before, during and 10 months after wall installation: (a) 1·275 m from wall; (b) 2·375 m from wall; (c) 3·475 m from wall

r	distance measured from pile centre	Δu	change in pore water pressure
r_p	plastic radius	$\Delta\sigma_h$	change in total horizontal stress
P	cavity support pressure	$\Delta\sigma_r$	change in radial stress
R	radius of pile	$\Delta\sigma_{PB}$	change in stress at pile bore
z	depth below ground level	$\Delta\sigma_{SC}$	change in stress acting on plane of spade cell

γ bulk density
σ_{h0} in situ total horizontal stress
σ'_{h0} in situ effective horizontal stress
σ_r total radial stress
σ_v total vertical stress
σ'_{v0} in situ effective vertical stress
σ_θ total hoop stress

REFERENCES

Carder, D. R. & Darley, P. (1998). *The long term performance of embedded retaining walls,* TRL Report 381. Crowthorne: Transport Research Laboratory.

Clark, J., Richards, D. J. & Powrie, W. (2004). Wall installation effects: preliminary findings from a field study at the CTRL, Ashford. *Proceedings of the Skempton Memorial Conference,* **2**, 691–699. London: Institution of Civil Engineers.

Clayton, C. R. I. & Bica, A. V. D. (1993). The design of diaphragm-type boundary total stress cells. *Géotechnique* **43**, No. 4, 523–535.

Fjaer, E., Holt, R. M., Horsrud, P., Raaer, R. M. & Risnes, R. (1992). *Petroleum related rock mechanics.* Amsterdam: Elsevier.

Fourie, A. B. & Potts, D. M. (1989). Comparison of finite element and limiting equilibrium analyses for an embedded cantilever retaining wall. *Géotechnique* **39**, No. 2, 175–188.

Gaba, A. R., Simpson, B., Powrie, W. & Beadman, D.R. (2003). *Embedded retaining walls, guidance for economic design,* CIRIA Report C580. London: Construction Industry Research and Information Association.

Gallois, R. W. (1965). *British regional geology: The Wealden District,* 4th edn. London: Her Majesty's Stationery Office.

Gibson, R. E. & Anderson, W. F. (1961). *In situ* measurement of soil properties with the pressuremeter. *Civ. Engng Pub. Works Rev.* **56**, No. 658, 615–618.

Gourvenec, S. M. & Powrie, W. (1999). Three-dimensional finite-element analysis of diaphragm wall installation *Géotechnique* **49**, No. 6, 801–823.

Gunn, M. J., Satkunananthan, A. & Clayton, C. R. I. (1993). Finite element modelling of installation effects. In *Retaining structures* (ed. C. R. I. Clayton). London: Thomas Telford, 16–55.

Ng, C. W. W. & Yan, R. W. M. (1999). Three-dimensional modelling of a diaphragm wall construction sequence. *Géotechnique* **49**, No. 6, 825–834.

Ng, C. W. W., Lings, M. L., Simpson, B. & Nash, D. F. T. (1995). An approximate analysis of the three-dimensional effects of diaphragm wall installation. *Géotechnique* **45**, No. 3, 497–507.

Peattie, K. R. & Sparrow, R. W. (1954). The fundamental action of earth pressure cells *J. Mech. Phys. Solids* **2**, 141–155.

Potts, D. M. & Fourie, A. B. (1984). The behaviour of a propped retaining wall: results of a numerical experiment. *Géotechnique* **34**, No. 3, 383–404.

Potts, D. M. & Fourie, A. B. (1985). The effect of wall stiffness on the behaviour of a propped retaining wall. *Géotechnique* **35**, No. 3, 347–352.

Powrie, W. & Kantartzi, C. (1996). Ground response during diaphragm wall installation in clay: centrifuge model tests. *Géotechnique* **46**, No. 4, 725–739.

Powrie, W. & Li, E. S. (1991). Finite element analyses of an *in situ* wall propped at formation level. *Géotechnique* **41**, No. 4, 499–514.

Powrie, W., Pantelidou, H. & Stallebrass, S. E. (1998). Soil stiffness in stress paths relevant to diaphragm walls in clay. *Géotechnique* **48**, No. 4, 483–494.

Richards, D. J., Clark, J., Powrie, W. & Heymann, G. (2006). An evaluation of total horizontal stress measurements using push-in pressure cells in an overconsolidated clay deposit. *Proc. Instn Civ. Engrs Geotech. Engng,* forthcoming.

Roscoe, H. (2003). Retaining wall movements, CTRL Contract 430: Ashford Tunnels. *Proceedings of the BGA international conference on foundations,* Dundee, 757–766.

Roscoe, H. & Twine, D. (2001). Design collaboration speeds Ashford tunnels. *World Tunnelling,* **14**, No. 5. 237–241.

Ryley, M. D. & Carder, D. R. (1995). The performance of push-in spade cells installed in stiff clay. *Géotechnique* **45**, No. 3, 533–539.

Skempton, A. W. & Weeks, A. G. (1976). The Quaternary history of the Lower Greensand escarpment and Weald Clay vale near Sevenoaks, Kent. *Phil. Trans. R. Soc. London Ser. A* **283**, 493–526.

Smart, J. G. O., Bisson, G. & Worssam, B. C. (1966). *Geology of the country around Canterbury and Folkestone.* London: Her Majesty's Stationery Office.

Stroud, M. A. (1974). The standard penetration test in insensitive clays and soft rocks. *Proc. Eur. Symp. on Penetration Testing (ESOPTI), Stockholm* **2:2**, 367–375.

Stroud, M. A. & Sweeney, D. J. (1977). Discussion appendix. In *A review of diaphragm walls,* pp. 142–148. London: Institution of Civil Engineers.

Symons, I. F. & Carder, D. R. (1993). Stress changes in stiff clay caused by the installation of embedded retaining walls. In *Retaining structures* (ed. C. R. I. Clayton). London: Thomas Telford, 227–236.

Tedd, P. & Charles, J. A. (1981). *In situ* measurement of horizontal stress in overconsolidated clay using push-in spade-shaped pressure cells. *Géotechnique* **31**, No. 4, 534–538.

Tedd, P. & Charles, J. A. (1983). Evaluation of push-in pressure cell results in stiff clay. *Proceedings of the international symposium on soil and rock investigation by in situ testing, Paris,* Vol. 2, pp. 579–584.

Tedd, P., Chard, B. M., Charles, J. A. & Symons, I. F. (1984). Behaviour of a propped embedded retaining wall in stiff clay at Bell Common. *Géotechnique* **34**, No. 4, 513–532.

Gasparre, A. & Coop, M. (2006). *Géotechnique* **56**, No. 7, 491–495

TECHNICAL NOTE

Techniques for performing small-strain probes in the triaxial apparatus

A. GASPARRE* and M. COOP†

KEYWORDS: clays; elasticity; laboratory equipment; laboratory tests; stiffness

INTRODUCTION

Field measurements of the behaviour of geotechnical structures and the need for improved geotechnical design have highlighted the importance of accurately simulating the behaviour of soils at very small strains (e.g. Simpson *et al.*, 1979, 1981; Puzrin & Burland, 1998). In recent years, the study of soil behaviour at small strains has been improved by the development of high-resolution instrumentation, which allows strain measurements in the laboratory to be resolved to magnitudes as small as 0·0001% (e.g. Cuccovillo & Coop, 1997; Tatsuoka *et al.*, 2000). As highlighted in this paper, at this order of magnitude the effects of other factors, such as creep and temperature, can became significant and may interact with the measurements of the displacements induced by the changes of load. Examples of temperature and creep effects at small strains will be shown, together with some suggestions for some simple techniques that allow their effects to be minimised.

APPARATUS AND SOIL TESTED

Small-strain probes were conducted on 100 mm diameter samples of London Clay taken from rotary cores and from blocks retrieved from Heathrow Terminal 5. The tests were performed in a modified version of the hydraulic stress path apparatus described by Bishop & Wesley (1975). The cell pressure and the pore or back-pressure, supplied through air–water interface systems, were measured by Druck pressure transducers of 1700 kPa capacity, with a resolution of 0·01 kPa. The deviatoric force F_a was measured by a load cell of Imperial College type, with a resolution of load of about 0·2 N. At the very small strains investigated here, it was found that instead of using the usual method of controlling the oil pressure in the base piston through an air pressure controller and air–oil interface, more accurate control could be achieved using the constant rate of strain pump to control the axial stress as well as the axial strain. The apparatus was also equipped with a standard displacement transducer with a maximum travel length of 25 mm to measure the external displacements, and a volume gauge of Imperial College type to measure the volumetric changes of the samples.

In order to resolve the small-strain region, much effort was put into using sophisticated local instrumentation and into improving its accuracy. The apparatus was therefore equipped

Manuscript received 4 November 2005; revised manuscript accepted 13 April 2006.
Discussion on this paper closes on 1 March 2007, for further details see p. ii.
* Geotechnical Consulting Group, formerly Imperial College, London, UK.
† Soil Mechanics Section, Department of Civil and Environmental Engineering, Imperial College London, UK.

with an internal thermometer probe, a piezometer probe (Druck model PDCR-81) to measure the pore pressure at the sample mid-height (Hight, 1983), and bender elements mounted on the side of the sample to measure the elastic stiffness of the specimen (Pennington *et al.*, 1997). The local displacements were measured by using axial linear variable displacement transducers (LVDTs), as described by Cuccovillo & Coop (1997), and a radial strain belt, also fitted with an LVDT. At the start of the shearing probes the LVDTs were set at their electrical zero by adjusting the zero potentiometer in the transducer amplifier, which allowed the data logger to work in its most sensitive range, so maximising its resolution. The noise of the LVDTs was then reduced to about 2×10^{-5} mm by programming the data logger to integrate the set of readings (i.e. all local transducers) over a relatively long period of about 3 s. This technique was also applied to the load cell readings to improve its resolution. The control program used was the Durham University Control System (Toll, 1993), which monitored the pressures and displacements, controlled the stresses and strains, and allowed the user to define the triaxial testing stages with automatic changes of the stress paths.

SMALL-STRAIN PROBES

The study of the behaviour of London Clay at small strains carried out by Gasparre (2005) allowed the evaluation of the elastic parameters of this material, the investigation of the nature of the kinematic yield surfaces, and the definition of their relationships with the lithology of the clay. Static probes were conducted using the high-resolution locally mounted LVDTs, which allowed the identification of an elastic behaviour for the London Clay over the range of axial strains lower than about 0·002%. The probes were conducted on samples that had been consolidated in the triaxial apparatus to their in situ stresses, and each probe was drained and carried out with a control of the applied stresses rather than the strains. The analysis of the probes assumed that the London Clay was cross-anisotropic. The vertical Young's modulus E_v and the Poisson's ratio ν_{vh} were measured from axial compression probes, and the parameters E_h, ν_{hh} and ν_{hv} were derived from radial compression probes. The shear moduli G_{hh} and G_{hv} were measured with the laterally mounted bender elements. The combination of the independent parameters measured also allowed the calculation of the equivalent shear modulus G_{eq}, the bulk modulus K, and the coupling moduli J_{qp} and J_{pq}, which were also measured for comparison from constant p' and constant q probes. Good agreement between the measured G_{eq} and K values and those predicted from the other independent elastic parameters confirmed that the assumption of cross-anisotropy was valid.

The strains developed during the shear probes were completely recovered when the stresses moved inside the elastic region, as illustrated in Fig. 1, which shows data from a drained axial compression probe, at constant radial stress. The data scatter is around $\pm 0\cdot000\,03\%$ in terms of axial strains or $\pm 0\cdot05$ kPa in terms of stresses. In part, the data scatter results from the large quantity of data points logged, particularly

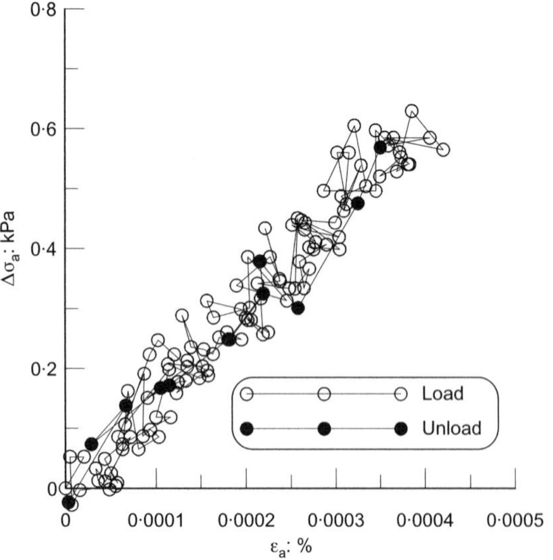

Fig. 1. Linear elastic response for an axial compression probe

during the loading phase. With data of this quality the elastic stiffness may be easily resolved to around ±3%. For probes reaching larger strains, the yield points for the elastic region could be identified from where the behaviour diverged from linearity. In each of the probes examining the elastic parameters the maximum stress changes did not exceed 2 kPa and usually the axial strains did not exceed 0·0025%, to avoid disturbance to the samples, as each sample was subjected to a number of probes. Because of the very small magnitude of the stresses and strains involved, the probe procedures were re-

quired to be particularly accurate to ensure good quality of the data and repeatability of the results.

In assessing the procedures used to achieve data of the quality shown in Fig. 1, several factors were considered, the most significant of which were that a condition of full drainage was maintained and that the sample response to a load could be measured while avoiding errors deriving from other external causes. The condition of full drainage of the London Clay required the use of very slow stress rates, which varied between 0·3 and 0·5 kPa/h depending on the lithology of the samples, so that a typical probe lasted about 4 or 5 h. The drainage was checked by means of a mid-height pore pressure probe, which registered no measurable pore pressure changes during the probes. The low stress

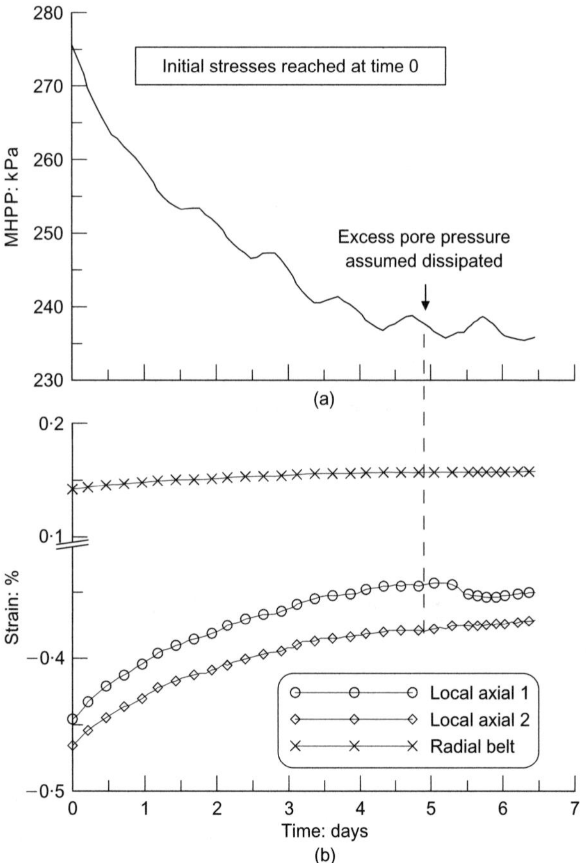

Fig. 2. (a) Pore pressure dissipation at mid-height (MHPP); (b) creep strain reduction with time

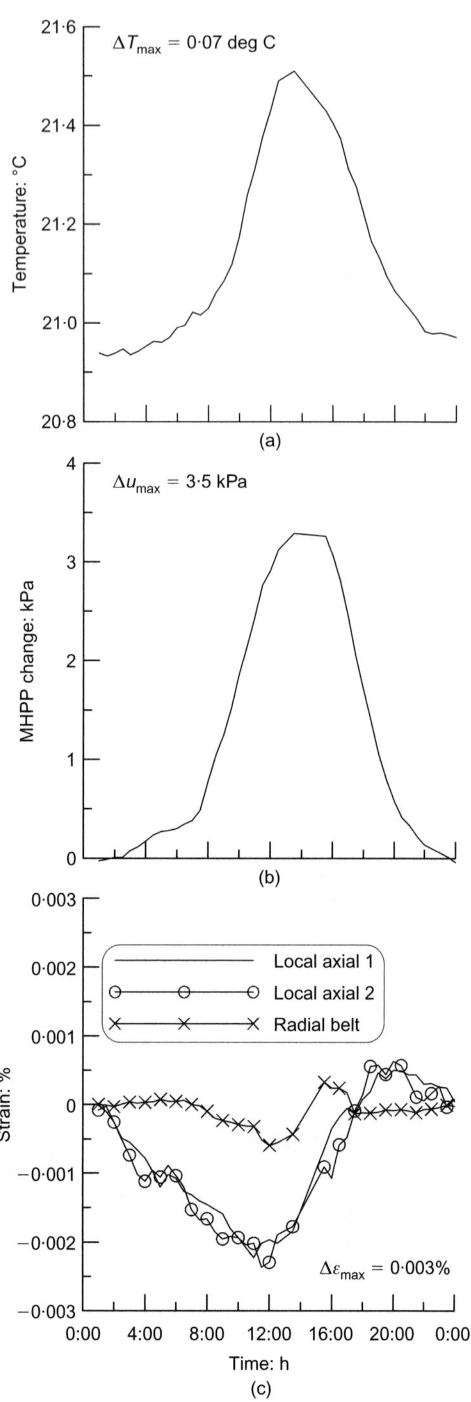

Fig. 3. Variation during a typical day of: (a) temperature; (b) mid-height pore pressure; (c) strains

rates had also to be compatible with a reasonable duration of the probes, because, owing to the small magnitude of the stress changes applied, secondary effects due to temperature or small amounts of drift in the transducers could affect the results significantly if they were too long. The low rates also meant that consideration had to be taken of the creep strains arising from the loading path the sample had experienced prior to the probes.

CREEP AND TEMPERATURE EFFECTS

The creep strains that arise from the approach stress path to the initial state for the probes will influence the measurements made during the probes (e.g. Jardine, 1992). Their effects were reduced until they were insignificant by holding the stresses constant for long periods before starting the probing stage. A typical criterion that is used is that the rate of creep should be less than a hundredth of the strain rate used during shearing (Jardine, 1992); however, because the strain rates were so small (typically between 0·0002 and 0·0008%/h), it was not possible to wait for a sufficiently long period of time to verify whether this criterion was fulfilled or not, and so the rest period simply lasted until the creep was not detectable. When the stress state of the sample had reached the required value, the change of strains and the pore pressure dissipation measured at the mid-height were monitored, as shown in Fig. 2. The magnitude of the strains can be seen to be very large compared with those applied during a probe. Initially, the strains were associated with the pore pressure dissipation, as an excess pore pressure of up to 5% of the current p' had been allowed during the consolidation stress path. The strain rates were further

allowed to reduce after the excess pore pressure had dissipated to insignificant values. Typically the total time between arriving at a stress state and starting the elastic probes was around a week, after which the strain changes were not measurable. As also seen in Fig. 2, the pore pressure usually showed a cyclic background noise superimposed on the overall reduction of pressure. These cycles are again large compared with the stress changes imposed during the probes. The regularity of this cyclic change and its continuance when the excess pore pressure had dissipated suggested that it was due to diurnal temperature changes.

Figure 3 shows the pore pressure and the strain behaviour during a typical day after the excess pore pressure had dissipated and the creep strains had reduced to negligible rates. The changes in temperature monitored by a thermometer installed inside the triaxial cell are also shown. The laboratory was temperature controlled, but a typical cyclic change of 0·7°C between night and day was still measured, which caused a cyclic variation of the mid-height pore water pressure of about 3·5 kPa and a variation of the strains of about 0·003%. This cyclic oscillation of the local transducers is again very significant for the probes to be undertaken, being about three times larger than the maximum strains to be measured during typical probes. The variation of the pore pressure and the variation of the LVDTs seemed almost in phase with the variation of temperature, although there might be some time lag in the response of the LVDTs.

Several checks were carried out to investigate whether these temperature effects correlated with the sample behaviour or with the sensitivity of the transducers to temperature. For the first check the drainage valve was closed and the diurnal variation of the base pore pressure was then found to

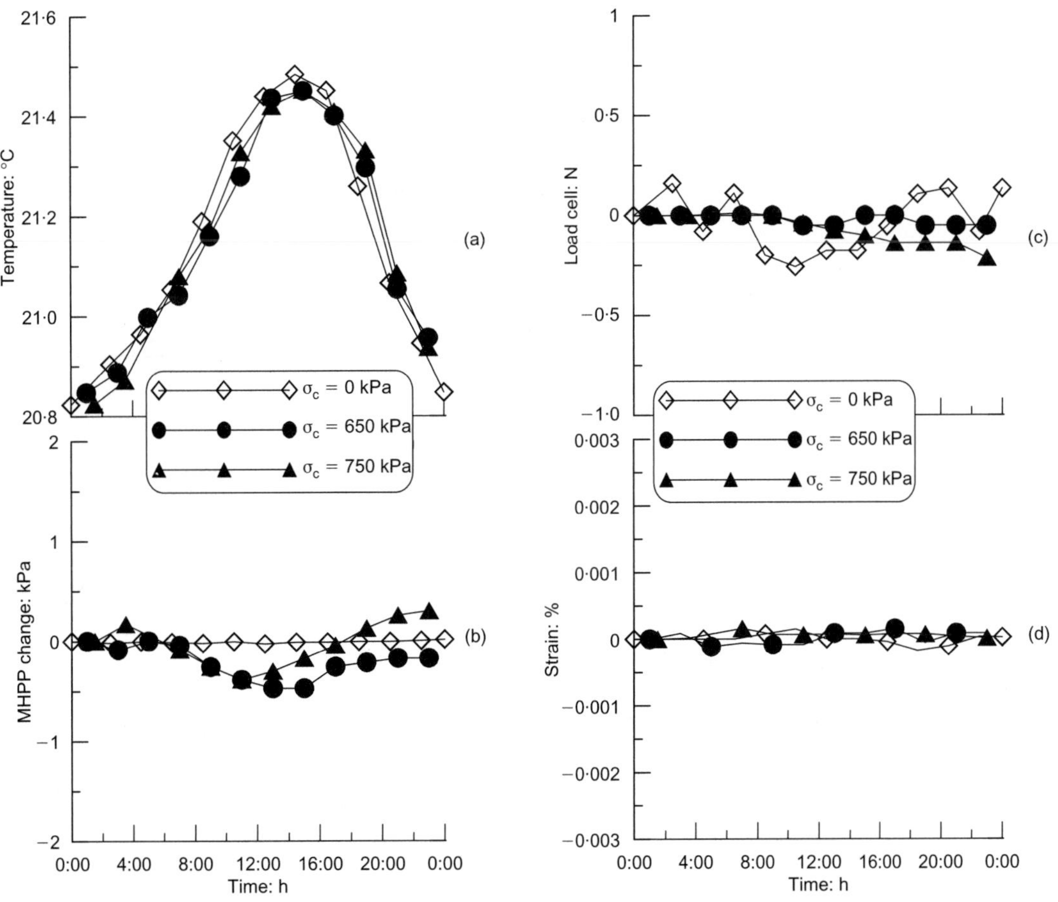

Fig. 4. Variation in a cell without sample of: (a) temperature; (b) mid-height pore pressure; (c) load cell; (d) strains at different pressures

be similar to that of the mid-height pore pressure probe, indicating that there was a real pore pressure variation in the sample and that the change was not simply a temperature sensitivity of the transducer. When the computer was controlling the cell pressure, no variations of the cell pressure readings would be expected, as any change would immediately be corrected by the computer via the cell pressure controller, but it was also found that if the computer control was switched off, so that the air pressure controller was not adjusted, the cell pressure transducer reading did not change, showing that the pore pressure variations were not due to changes in the total stress applied to the sample.

An analysis of the sensitivity of the transducers to temperature was conducted for about a week by using an empty triaxial cell pressurised at different cell pressures. In the cell, the LVDTs were held in clamps, with the armatures set at the zero volt position. The temperature inside the cell, the mid-height probe, the back-pressure transducer and the LVDTs were monitored for a few days under cell pressures of 0, 600 and 750 kPa. Data are shown in Fig. 4 for typical days. The temperature in the cell varied cyclically as during the test, but without a sample only the mid-height pore pressure transducer seemed to show a slight sensitivity to temperature. This agrees with the investigation conducted by De Campos (1984) who found that a temperature variation of about ±1 degC induced changes in the mid-height pore pressure readings of about ±0·5 kPa in this type of transducer, but negligible effects on the cell and base pore pressure transducers. The sensitivity of the mid-height probe to temperature is an effect that seems to increase with increasing pressure (Fig. 4(b)).

The LVDTs and load cell seemed, instead, completely insensitive to temperature at all pressures. The resolution of the LVDTs in this test was poorer than during the probing test as their voltages were not set to zero, which accounts for the larger step size than usual in the data. As the initial stress state of the samples for the probes was anisotropic, it was important to verify that the load cell reading was not sensitive to temperature, because if it had been, then any cyclic change of the load cell reading would have been compensated for by the control system adjusting the total axial stress applied to the sample, which might have caused the strains and pore pressure changes observed.

Campanella & Mitchell (1968) also observed that temperature changes induce changes in the reference zero of instruments. However, the maximum excursion of the mid-height probe found here, of only about 0·5 kPa, was much

lower than the 3·5 kPa measured with the transducer on the sample (Fig. 4(b)), and it was also in the opposite direction to the change seen with the sample present. In Fig. 5 the mid-height pore pressure changes are shown for both cases. This, together with the responses of the base pore pressure transducer and of the LVDTs, supported the idea that the sensitivity to temperature was associated with real changes of pore pressure inside the sample.

REDUCTION OF TEMPERATURE VARIATION

The only solution found to prevent these problems was to reduce the temperature variation inside the cell. The cell was therefore isolated by wrapping it in several layers of bubble

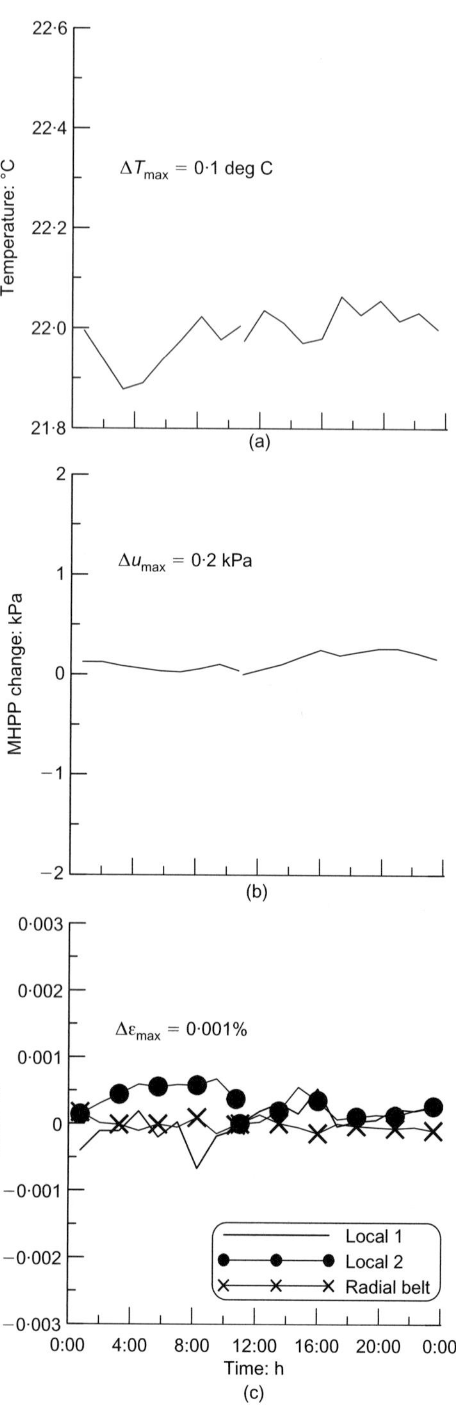

Fig. 6. Variation with time of (a) temperature; (b) mid-height pore pressure and (c) strains after the cell was wrapped in bubble paper and aluminium foil

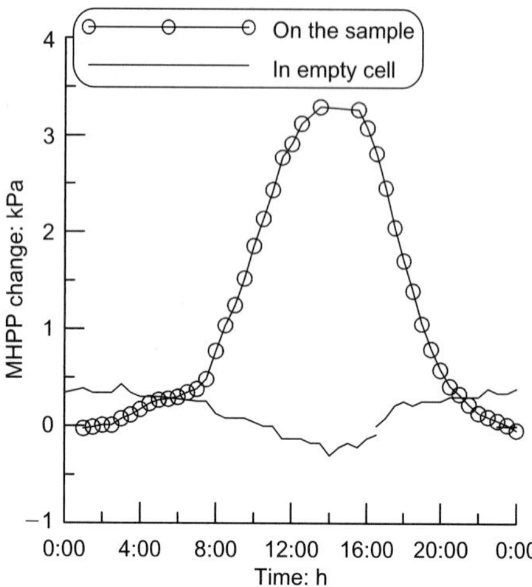

Fig. 5. Comparison between variation of mid-height pore pressure probe with temperature on the sample and in an empty cell

wrap and aluminium foil. Usually a minimum of three layers of bubble wrap was used, which covered the cell completely, including the top and the base, and reduced the effects of the laboratory temperature change. The bubble wrap was then covered by aluminium foil, with the shiny surface outwards, to reflect any incident light, even if the laboratory windows faced north. This reduced the temperature change inside the cell to less than 0·1 deg C, and therefore the pore pressure and the local strain oscillations reduced. The cell was wrapped in this way after the initial stresses were reached, so that a visual inspection of the samples could then be allowed during the approach stress paths, and for a few days before probing the cell temperature could be allowed to stabilise. In Fig. 6 the variation of the stresses and strains with time is presented for a typical day when the excess pore pressure had dissipated and the creep strains had reduced to negligible values. The cyclic variation of the transducers could not be completely eradicated, but the pore pressure change has reduced to less than 0·5 kPa, the strain excursions to around ±0·001%, and a period of time could be identified during the day when all the values were constant (e.g. between 7am and 12am in the example in Fig. 6). This period was sufficiently long to perform the probes in each test. Prior to the probing stage, the transducers were monitored for at least 24 h after the pore pressure had dissipated and the creep had stabilised, and the times of day when the strains were constant were identified and the probes were then conducted at these times.

CONCLUSIONS

The typical diurnal temperature variation of a typical temperature-controlled laboratory has been found to be insufficient for accurate small-strain probes to be conducted on low-permeability clays, for which drainage problems require slow loading rates. Cyclic variations of the pore pressure in the sample and its strains, induced by a diurnal temperature variation of 0·7 degC, were found to be of the same order of magnitude as the elastic yield strains of the clay. A simple means has been found to reduce these temperature effects dramatically, so that very accurate small-strain data can been obtained, without the need for a highly accurate and expensive temperature control system for the laboratory.

NOTATION

E_h Young's modulus in horizontal direction
E_v Young's modulus in vertical direction
F_a deviatoric force
G_{hv} shear modulus for horizontal propagation and vertical polarisation
G_{hh} shear modulus for horizontal propagation and horizontal polarisation
G_{eq} equivalent shear modulus measured from vertical stresses and vertical strains
J_{qp}, J_{pq} coupling moduli
K bulk modulus
p' mean normal effective stress $\{= (\sigma_a' + 2\sigma_r')/3\}$
q deviatoric stress $\{= \sigma_a' - \sigma_r'\}$
T temperature
u pore pressure
Δ change
ε_a axial strain
ν_{vh} Poisson's ratio for influence of vertical stress change on horizontal strain
ν_{hv} Poisson's ratio for influence of horizontal stress change on vertical strain
ν_{hh} Poisson's ratio for influence of horizontal stress change on horizontal strain
σ'_a axial effective stress
σ'_r radial effective stress

REFERENCES

Bishop, A. W. & Wesley, L. D. (1975). A hydraulic apparatus for controlled stress path testing. *Géotechnique* **25**, No. 4, 657–670.

Campanella, R. G. & Mitchell, J. K. (1968). Influence of temperature variations on soil behaviour. *J. Soil Mech. Found. Div. ASCE* **94**, No. SM3, 709–734.

Cuccovillo, T. & Coop, M. R. (1997). The measurements of local axial strains in triaxial tests using LVDTs. *Géotechnique* **47**, No. 1, 167–171.

De Campos, T. M. P. (1984). *Two low plasticity clays under cyclic and transient loading*. PhD thesis, University of London.

Hight, D. W. (1983). *Laboratory investigation on sea bed clays*. PhD thesis, University of London.

Gasparre, A. (2005). *Advanced laboratory characterization of London Clay*. PhD thesis, University of London.

Jardine, R. J. (1992). Some observations on the kinematic nature of soil stiffness. *Soils Found.* **32**, No. 2, 111–124.

Pennington, D. S., Nash, D. F. T. & Lings, M. L. (1997). Anisotropy of G_0 shear stiffness in Gault clay. *Géotechnique* **47**, No. 3, 391–398.

Puzrin, A. M. & Burland, J. B. (1998), Non-linear model of small-strain behaviour of soils. *Géotechnique* **20**, No. 4, 217–233.

Simpson, B., O'Riordan, N. J. & Croft, O. D. (1979). A computer model for the analysis of ground movements in London Clay. *Géotechnique* **29**, No. 2, 149–175.

Simpson, B., Calabresi, G., Sommer, H. & Wallays, M. (1981). Design parameters for stiff clays. *Proc.7th Eur. Conf. on Soil Mech. Geotech. Engng, Brighton* **5**, 91–125.

Tatsuoka, F., Santucci de Magistris, F., Hayano, K., Momoya, Y. & Koseki, J. (2000). Some new aspects of time effects on the stress–strain behaviour of stiff geomaterials. *Proc. 2nd Int. Conf. Hard Soils and Soft Rocks, Napoli*, **2**, 1285–1371.

Toll, D. G. (1993). *Durham University Control System Triax user manual*, internal report. University of Durham, internal report.